AF552624

INTRODUCTION TO CALCULUS FOR THE BIOLOGICAL AND HEALTH SCIENCES

Rodney D. Gentry
University of Guelph

ADDISON-WESLEY PUBLISHING COMPANY

Reading, Massachusetts
Menlo Park, California
London • Amsterdam
Don Mills, Ontario • Sydney

to
Patricia, Marjory, and Michael

Second printing, July 1978

 Printed in the United States of America. Published simultaneously in Canada. Library of Congress Catalog Card No. 77-79470.

ISBN 0-201-02477-2
BCDEFGHIJK-HA-8987654321

PREFACE

INTRODUCTION

This text is designed to provide the basic elements of calculus for students in the biological and health sciences. It was written in response to the increasing demand that students studying biological sciences or preparing for medical careers should have a firm understanding of, and functional ability in, basic calculus (and statistics). Previously, such students were either placed in the traditional "mainstream" calculus courses that were designed for mathematicians, engineers, and physical scientists, or they were given "short courses," which constituted diluted versions of the mainstream course (which often omitted important concepts). This approach has most often proved inadequate from the scientist's point of view, principally because the needs of biological science students and indeed their interests differ from those of the mathematics-physics-engineering students. With these problems in mind, this text was designed to provide an interesting, motivated introduction to calculus, which emphasizes not only the fundamental ideas of calculus, but also the techniques and concepts that are particularly important for biological applications.

All examples and exercises illustrating applications are biologically or medically oriented, to motivate the need for various calculus concepts and to make the text more relevant to basic biological-science students. Necessarily, most of the examples are simplified so that they are self-contained, requiring no previous biological knowledge, and can easily be translated into mathematical problems or illustrations. However, it should be emphasized that most of the examples or applied exercises are actual modeling problems that students will encounter in undergraduate courses and text books (ranging from soil science to physiology).

Keeping in mind the particular mathematical needs of biological and medical students, the text begins with a thorough review of elementary functions and graphing techniques. A section on difference equations is used to introduce the basic concepts of differential calculus. The derivation of the definite integral via discrete approximation is stressed and the traditional emphasis on integration techniques has been minimized. The chapter on differential equations stresses the qualitative as well as quantative aspects of solutions rather than the detailed techniques that are covered in second year O.D.E. courses. Similarly, the chapter on probability and statistics is designed to introduce the fundamental concepts and is not intended to provide all the details covered in a standard biostatistics course. Throughout the text, approximations are stressed and the use of calculators is encouraged.

THE TEXT

The choice of topics, emphasis, and ordering of material in the text is designed to provide a firm intuitive understanding of the basic calculus concepts that biologists or medical students are most likely to need and utilize. The first chapter contains a review of elementary functions and their graphs and algebra. Recognizing the important of log-log and semilog graphs in the biological sciences, these topics (which traditionally are ignored in calculus courses) are covered in a special section on graphing with different scales. Also included in the first chapter is a section that introduces some of the special functions (based upon the exponential function) that are consistently utilized in virtually every biological discipline.

Chapter 2 begins with an introduction to *sequences* and *series*. The material is designed to introduce the notation of sequences, the sigma notation for finite series, and the simplest finite sequences and series. (It is not designed to develop an understanding of infinite sequences or series and does not contain covergence tests.) It is intended to provide the functional awareness of the notation and concepts needed to develop the integral and difference equations so that the student is not stumbling over notation when trying to master these topics.

The last section of Chapter 2 introduces *difference equations*. While this topic is not a part of main stream calculus courses, it is probably the most important topic in the text. Difference equations are the basic concept utilized in over half the mathematical models in biological or medical sciences. The difference equation, through the difference quotient, provides the natural introduction to the derivative and the essential concepts of calculus. Furthermore, all computer models and all numerical computer routines are based upon difference concepts. In Section 11, the basic concepts of difference equations are introduced by examining the simplest linear equations. At this point, the principles of qualitative as well as quantative analysis are introduced.

Chapter 3 begins with an intuitive discussion of continuity. Then the difference quotient is examined and utilized to develop the *derivative* concept. The basic rules for differentiating, the Chain Rule, and higher derivatives are presented with emphasis on functional computational skills.

The use of calculus to solve extremal problems and in sketching graphs is presented in Chapter 4. Because of its wide usage in modeling, the *differential* is examined in Section 17, where its approximation of the difference is utilized in both theoretical and numerical problems. Differential notation is then used to illustrate implicit differentiation. In Section 18, Rolle's Theorem and the Mean Value Theorem are presented for mathematical completeness, since their utility is primarily in theoretical situations. Taylor's Theorem, on the other hand, is emphasized because it is constantly utilized in biological applications.

Chapter 5 examines the *integral*. Before examining the definite integral, the antiderivative, or indefinite integral, is presented as the reverse of the differentiation process. In essence this is simply practice in stating the derivative rules in a reverse form; however, the inverse of the Chain Rule is by no means a trivial concept.

Mathematicians as a whole have fallen into the pedagogical trap of presenting the definite integral as an area. While this approach is graphically and conceptually simple, it is misleading and frequently insufficient for students who will encounter models that utilize the integral to represent quantities other than areas. For this reason I have introduced the concept of integration by examining three problems whose solutions can be approximated by Riemann sums (and consequently expressed as an integral). Next, the definite integral is defined as a limit of such sums and, without dwelling on the limit aspects, the Fundamental Theorem of Calculus (FTC) is introduced so that specific examples can be considered. Then the illustration of the integral as an area is presented and the limit concept of its definition is explored. Finally, since in real applications very few integrals can be integrated using the FTC, the basics of numerical integration are discussed.

The integral definition of $\ln(x)$ is used in Chapter 6 to reexamine the calculus of the logarithm and exponential functions; this second look at these functions is justified by their extreme importance in describing biological models. A minimal number of integration techniques are presented in Section 24. In Section 25, indeterminant limits are examined and L'Hospital's Rule is given. While mathematicians tend to minimize this rule, it is frequently utilized in analyzing biological models. It is also necessary when examining singular integrals as discussed in the second part of the same section.

The goal in Chapter 7 is to provide a rapid introduction to both the quantitative and the qualitative aspects of ordinary differential equations. To accomplish this, the standard plethora of special equations and techniques traditionally covered in semester courses in O.D.E.'s have been omitted. Instead the solution of separable and constant coefficient equations are examined, and the basic qualitative aspects of solutions are presented. As most differential equations encountered in applications cannot be solved explicitly, Chapter 7 ends with a presentation of Euler's Method for solving initial-value problems numerically.

A brief introduction to probability and statistics is given in Chapter 8. The objectives of this chapter are to provide a quick introduction to the basic terminology and concepts of probability and statistics and to demonstrate the need and application of calculus when considering continuous probability or sample spaces. This chapter should be considered a preparation for a standard course in biostatistics. It should be noted that to keep the length and time requirements as minimal as possible the traditional "stumbling blocks" of permutations, combinations, and counting methods have been omitted. In my experience, these worthwhile topics are best taught after the basic probability concepts have been mastered, as their assimilation requires a considerable amount of time.

USAGE

The amount of mathematics included in biological-science programs is quite varied, ranging from only one course to two years of calculus, statistics, and computer programming. Consequently, this text was written so that it could accommodate many different courses depending on the requirements and available time. The choice of sections and the time alloted to each section will vary depending on the mathematical preparation of the students and their course of study. Some courses will need to start with precalculus material while others will be able to begin with Section 11 on difference equations. Introductory chapters on probability and statistics and on ordinary differential equations have been included to provide greater flexibility.

A minimal course for students with a good foundation in functions, relations, and sigma notation would include Sections 11 through 16 and 19 through 22, omitting certain subsections such as 11.4, 15.4, 21.3, and 22.3. This material could be covered in a quarter or semester course and, depending on the frequency of class meetings and student preparedness, could be supplemented by additional material selected from Sections 7, 23, 17, and 18.

For most courses, it would be best to include more precalculus material from Chapter 1; Sections 3 through 6 provide a review of the elementary functions and relations, while Section 7 will prove to be one of the most useful sections in the students' other science courses. Note that Section 8 should be approached on the basis of associating the particular functions with particular graph forms, just as the word "duck" is associated with a picture of a duck. Later, the properties presented are verified using calculus. As some science programs still have minimal mathematics prerequisites, many courses will want to start with Section 1 and proceed directly through the text, omitting or delaying certain more difficult subsections. This would require at least two quarter courses to complete most of the material in the first six chapters and portions of Chapter 7 or Chapter 8. A one-year course could cover all the material in the text.

It should be observed that the material on probability and statistics has been placed at the end of the text. However, this material can easily be integrated into the text at earlier points as indicated in the preface to the chapter.

ACKNOWLEDGEMENTS

In developing this text, the material (except for Chapter 8) was classroom tested at the University of Guelph over a three-year period in a course designed for biological and pre-veterinary students. The text was used in a multi-sections course with section sizes ranging from 80 to 150. During this period, numerous suggestions for improvement were gratefully received from the over 2500 students who used preliminary drafts and the 10 colleagues who have taught the course.

I would like to acknowledge these invaluable suggestions and to thank G. R. Chapman, J. Cunsolo, and J. Weiner for their contributions. I am indebted to Linda Kraemer, who typed the preliminary and final draft of the text with remarkable speed, accuracy, and cheerfulness.

I wish to thank several colleagues who have suggested improvements in the book: Jim Diederich (University of California, Davis), Jack L. Goldberg (University of Michigan), Samson M. Rosenzweig (Douglass College — Rutgers University), R. A. Ross (University of Toronto), Irwin Schochetman (Oakland University), Susann Shaw (San Francisco State University), James Turner (McGill University), and Stephen Willard (University of Alberta).

Guelph, Ontario
January 1978

R.D.G.

GENERAL SECTION DEPENDENCE

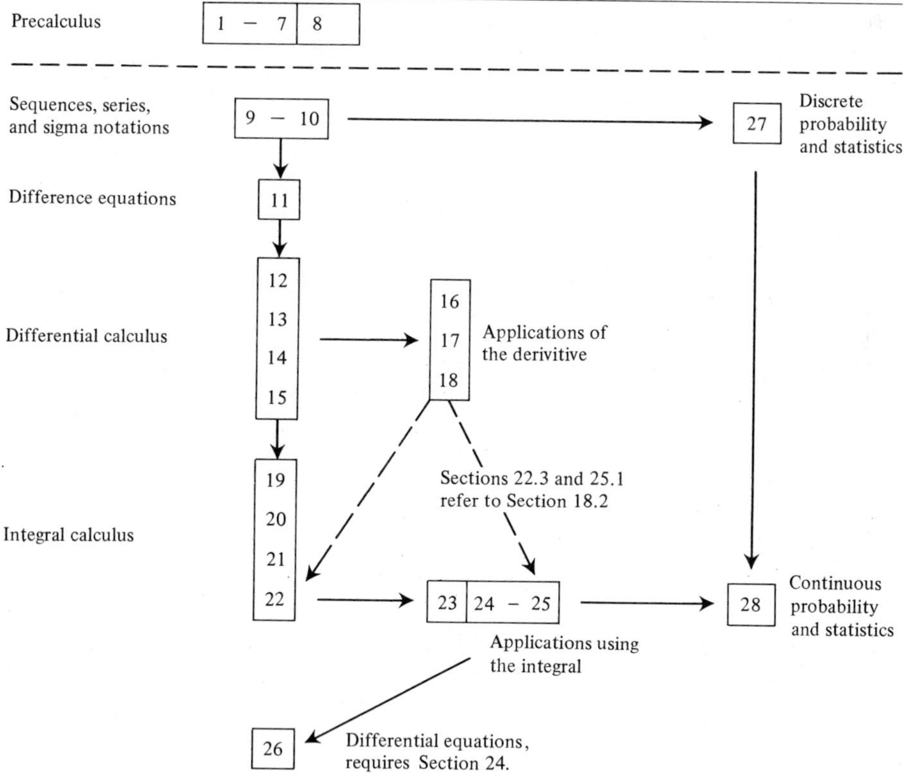

Arrows indicate dependence. Sections 16, 17, and 18 are independent of each other, as are Sections 23, 24, and 25.

CONTENTS

CONTENTS

CONTENTS

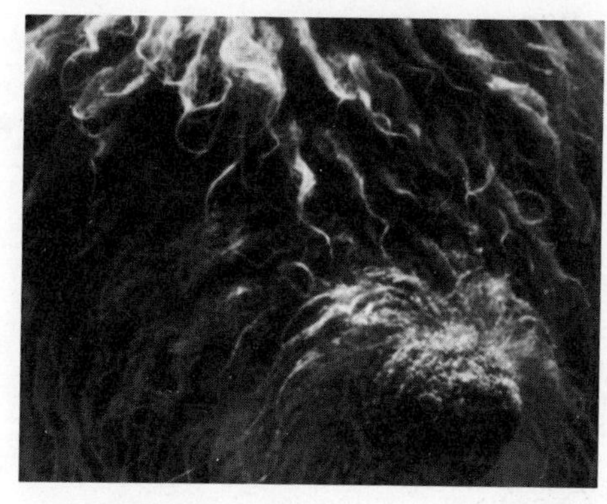

INTRODUCTION TO CALCULUS FOR THE BIOLOGICAL AND HEALTH SCIENCES

CHAPTER ONE

INTRODUCTION TO FUNCTIONS AND GRAPHING (PRECALCULUS)

Sections 1 and 2 should be review for most students. Section 3 may also be review but is very important, as the equation of a line is crucial for studying calculus, and the graphs of parabolas and cubic functions form the majority of the specific illustrations in the text. The composition of functions is generally one of the more difficult topics for students to grasp. Since understanding the composition of functions is critical for using the most important technique of Differential Calculus, the Chain Rule, the material of Section 4 is critical for the remainder of the text. Sections 5 and 6 introduce the functions e^x, ln (x) and the trigonometric functions. These functions are used throughout the text. Section 7 discusses graphing techniques including log-log and semilog graphs which are very important for biological and life science students. This is one section which will prove very useful in other undergraduate courses and is strongly recommended even though it is optional for the remainder of this text. Section 8 introduces special functions which are used in the biological sciences. This section is optional and the material cannot be covered in depth at this point. Later, using calculus, we will verify the assertions made in this section.

SECTION 1

SETS, RELATIONS, AND FUNCTIONS

INTRODUCTION

The principal object of this section is to introduce the basic concepts of functions. The word "function" is primarily used as a noun—for example, "The *function* of the eye is to see," or, in a more mathematical context, "The *function* of f is to square." Note that in each example there is an implied action; it is the action of a function which concerns us. Thus, we are inclined to use the verb form of the word function—for example, "The heart *functions* as a two-chamber pump," "The eye *functions* by transmitting neural impulses"; "f *functions* by multiplying its argument by itself." We shall confine ourselves to the investigation of the more specific aspects of a function, namely how a function may be defined mathematically. A function may be given a particular name or representation, but to actually describe a function we must describe the procedure by which it performs its activity, and the objects upon which it acts. To provide a vocabulary which will allow us to describe all functions, we must first introduce some fundamental concepts of set theory.

1.1 SETS

This section is intended as a quick review of basic set notation. Since elementary set theory has been so widely incorporated in precalculus mathematics, we introduce only those essential aspects of set theory which are necessary to develop the concept of a function.

Definition 1.1

i) A *set* is a well-defined collection of objects.
ii) An *element* of a set is an object which belongs to the collection comprising the set.

It is common practice to denote sets by capital letters, A, B, C, etc., and to denote elements of a set by small letters, a, b, c, x, y, etc. That an element x is a member of the set A is denoted by the expression

$$x \in A,$$

read "x is an element of the set A." When an element x does not belong to a set A, this is denoted by

$$x \notin A,$$

read "x is not an element of the set A."

We use two basic notations to describe sets. One is the roster notation which consists of a left brace followed by the elements of the set listed individually and separated by commas, followed by a right brace. Thus a set of common house pets would be denoted

$$\{\text{dog, cat, fish, bird}\}.$$

The second way to denote sets is to write a left brace followed by a representative symbol for elements of the set, a vertical bar, and a description specifying the exact values the representative element may assume, followed by a right brace. For example, the set of prime numbers may be expressed as

$$\{p \mid p \text{ is a prime number}\}.$$

This is read "the set of elements p such that p is a prime number."

Example 1.1

a) The following are sets:

$$\{\text{oats, wheat, sorghum, rye}\},$$

$$\{x \mid x \text{ is the name of a deciduous tree}\},$$

$$P = \{2, 3, 5, 7, 11, 13, 17, 19, \ldots\}, \qquad B = \{9, 3, 6, -7.1\},$$

$$R = \{x \mid x \text{ is a real number}\}.$$

b) The following are not sets as their descriptions employ ambiguous terms: $\{x \mid x \text{ is a big person}\}$; the collection of all sets; the collection of all intelligent animal species.

c) Given the set $A = \{1, 3, 7, 5\}$, the following statements are valid:

$$1 \in A,\ 4 \notin A,\ 6 \notin A,\ A \notin A,\ 5 \in A.$$ ◀

Definition 1.2

i) A set B *is a subset of the set* D if every element of B is also an element of D. The symbol $\subseteq$ is used to denote the relation of being a subset;

$$B \subseteq D$$

is read "B is a subset of D." If B is not a subset of D, this is denoted by $B \nsubseteq D$.

ii) The *empty set* ϕ is the set which contains no element. $\phi = \{\quad\}$. (ϕ is the Greek letter phi.)

iii) The *intersection of a set A and a set B* is the set of all elements which are elements of both set A and set B. The intersection symbol is $\cap$, and

$$A \cap B$$

is read "set A intersect set B." The intersection may be defined by the equation

$$A \cap B = \{x \mid x \in A \text{ and } x \in B\}.$$

iv) The *union of a set A and a set B* is the set of all elements which belong either to the set A or to the set B. The union symbol is $\cup$, and

$$A \cup B$$

is read "set A union set B." The union of two sets may be expressed by the equation

$$A \cup B = \{x | x \in A \text{ or } x \in B\}.$$

Note: In mathematics the word "or" is used inclusively. Thus "$x \in A$ or $x \in B$" includes the possibility that $x \in A$ and $x \in B$.

Example 1.2

a) Let $N = \{1, 2, 3, 4, \ldots\}$; that is, N is the set of all natural numbers. Let $E = \{2, 4, 6, 8, \ldots\}$, that is, E is the set of all even natural numbers. Let $C = \{x | x$ is a natural number between 1 and 10$\}$. Then,

$$\begin{aligned} C &= \{1, 2, 3, \ldots, 10\}, \\ E \cap C &= \{2, 4, 6, 8, 10\}, \\ N \cup C &= \{1, 2, 3, \ldots\} = N, \\ E \cup C &= \{1, 2, 3, 4, 5, 6, 7, 8, 9, 10, 12, 14, 16, \ldots\}. \end{aligned}$$

b) Let $A = \{\text{sheep, cat, dog, cow}\}$. Let $B = \{\text{sheep, cow}\}$, and $C = \{\text{cat}\}$. Then,

$$\begin{array}{llll} B \subseteq A, & C \subseteq A, & C \nsubseteq B, & \phi \subseteq A, \\ B \cap A = B, & B \cap C = \phi. & C \cap A = C, & \\ B \cup A = A, & B \cup C = \{\text{sheep, cow, cat}\}. & & \end{array}$$

◀

When we refer to inequalities it will be assumed that the elements are real numbers. Thus $\{x | x \geq 5\}$ is the set of all real numbers greater than or equal to 5. Similarly, the set $\{x | 3 \leq x < 5\}$ is the set of all real numbers greater than or equal to 3 but less than 5.

Definition 1.3. *Interval notation*

Let a and b denote real numbers. Then we define the following sets of real numbers as intervals in the manner indicated:

$\{a | a \leq x \leq b\} \equiv [a, b]$, "the closed interval a, b";

$\{x | a < x < b\} \equiv (a, b)$, "the open interval a, b";

$\{x | a \leq x < b\} \equiv [a, b)$, "the interval closed at a and open at b";

$\{x | a < x \leq b\} \equiv (a, b]$, "the interval open at a and closed at b."

The symbol "$+\infty$", read "plus infinity," and the symbol "$-\infty$", read "minus infinity," are *not* real numbers but are used to represent unbounded sets in the following infinite interval notation:

$$\{x \mid a \leq x\} \equiv [a, +\infty); \qquad \{x \mid a < x\} \equiv (a, +\infty);$$
$$\{x \mid x \leq a) \equiv (-\infty, a]; \qquad \{x \mid x < a) \equiv (-\infty, a).$$

Example 1.3 We illustrate interval notation and the terminology of Definition 1.3 by references to Fig. 1.1.

a) $[3, 5) = \{x \mid 3 \leqslant x < 5\}$

b) $(2, 4) = \{x \mid 2 < x < 4\}$

c) $[3, 5) \cap (2, 4) = [3, 4)$

d) $\{x \mid x^2 < 25\} = (-5, 5)$

e) $(1, 4) \cup [3, 5] = (1, 5]$

f) $(5, +\infty) \cap [-2, 6] = (5, 6]$

g) $(2, +\infty) \cap [4, +\infty) = [4, +\infty)$

h) $(-\infty, 3) \cap (0, +\infty) = (0, 3)$

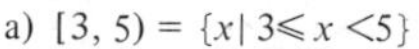

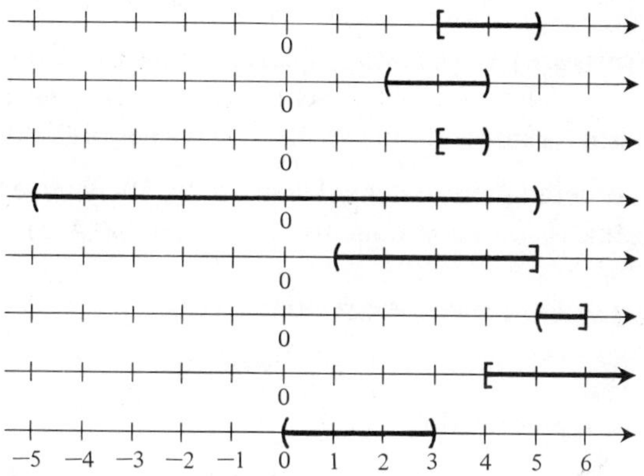

Figure 1.1

◀

1.2 RELATIONS

A relation is a rule which associates two or more objects in some specific way. We shall consider only binary relations—that is, relations which associate only two objects at one time. Typically a relation is specified by a statement of the form: *a* is related to *b* if ________. The way in which the blank is completed then specifies the particular relation.

Example 1.4 Consider the three relations:

a) *a* is related to *b* if *b* is a parent of *a*.
b) *a* is related to *b* if *b* has the same natural habitat as *a*.
c) *a* is related to *b* if *b* is twice *a*.

It is obvious that the three relations of this example apply to different types of objects *a* and *b*. Exactly what type of objects *a* and *b* may be is not clear in the example and thus relations (a), (b), and (c) are in fact not well defined. We must specify exactly what type of objects *a* and *b* represent to complete the definition of the relations. ◀

Definition 1.4 Let R denote a relation which relates objects of type a to objects of type b.

$$\text{Domain}(R) \equiv \{a \mid a \text{ is related by } R \text{ to some object } b\}.$$

$$\text{Range}(R) \equiv \{b \mid \text{for some } a \in \text{Domain}(R), a \text{ is related to } b \text{ by } R\}.$$

Relations are most easily described as sets of ordered pairs. The following definition defines an ordered pair using the same symbol used to denote an open interval. The usage that is being employed should always be clear from the context in which the symbol is used.

Definition 1.5 *An ordered pair* is a symbol of the form (a, b) read "the ordered pair a, b." The term a represents the *first component* of the ordered pair (a, b) and the term b represents the *second component* of the ordered pair (a, b).

Two ordered pairs are equal if and only if they have the same first components and the same second components. Thus $(3, 4) \neq (4, 3)$.

Using ordered pairs, the relation

$$R: a \text{ is related to } b \text{ if } \underline{\qquad\quad},$$

can be represented by the set

$$R = \{(a, b) \mid \underline{\qquad\quad}\},$$

where the ______ specifies "how" a is related to b. Relations may also be expressed as explicit sets of ordered pairs of objects.

a) The relation R_1 which relates a natural number n to its square, n^2, may be denoted by

$$R_1: n \text{ is related to } m \text{ if } m = n^2, n \text{ is a natural number},$$

or by

$$R_1 = \{(n, m) \mid m = n^2, n \text{ a natural number}\}.$$

b) A relation R_2 between the animal names of the set $A = \{\text{cow, sheep, cat, dog}\}$ is specified by

$$R_2 = \{(\text{cow, cow}), (\text{dog, dog}), (\text{cat, cat}), (\text{sheep, sheep}), (\text{dog, cat}), (\text{cat, dog}), (\text{sheep, cow}), (\text{cow, sheep})\}.$$

This relation relates two animal names in the set A if they are both barn animals or both companion animals.

c) Consider the relation $R_3 = \{(1, 2), (3, 2), (4, 3)\}$. We could write $(1, 2) \in R_3$ to express the fact that the relation R_3 relates 1 to 2.

d) The domain and range of the relations of parts (a), (b), and (c) are specified as follows:

Domain(R_1) = $\{x \mid x$ is a natural number$\}$.

Range(R_1) = $\{x \mid x$ is the square of a natural number$\} = \{1, 4, 9, 25, 36, \ldots\}$.

Domain(R_2) = R range (R_2) = set A given in (b).

Domain(R_3) = $\{1, 3, 4\}$.

Range(R_3) = $\{2, 3\}$. ◀

Relations are also specified by graphs that provide visual representations of the relations. To graph a relation, the domain of the relation is associated with a horizontal axis line and each element of the domain is associated with a particular point on this axis. Similarly, the range of the relation is associated with a vertical axis line and each element of the range is associated with a particular point on the vertical axis. An ordered pair of the relation is then associated with the point of intersection of two lines, a vertical line through the point associated with its first component and a horizontal line through the point associated with its second component. The graph of a relation consists of all the points associated with the ordered pairs of the relation.

Example 1.6

a) Graphs of the relations R_2 and R_3 of Example 1.5 are given in Fig. 1.2.

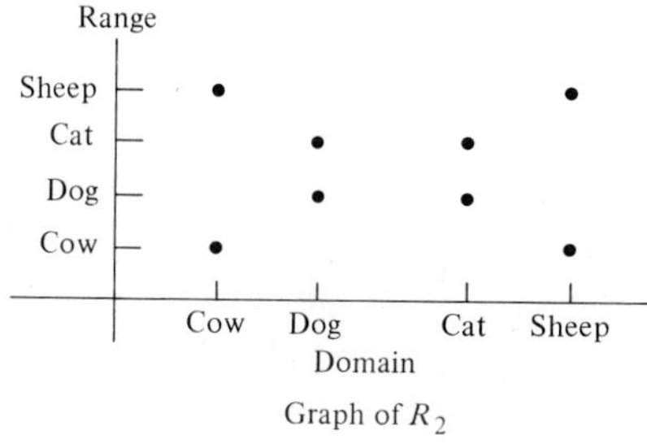

Graph of R_2

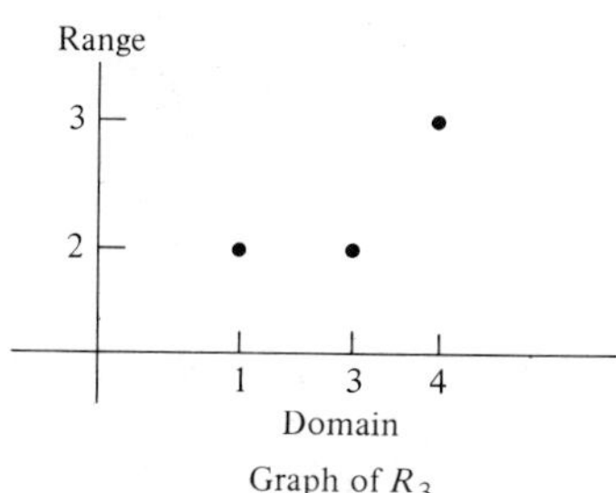

Graph of R_3

Figure 1.2

b) Consider the relation specified by the graph in Fig. 1.3, relating growing periods to plant height of potatoes. Due to the imprecision of the graphical information, it is impossible from such a graph to specify exactly the relation graphed. We can, however, estimate the relation by estimating the coordinates of the indicated points. The domain of this relation is the set $\{4, 6, 8, 10, 12, 14, 16\}$. An estimate of this relation would be the set of ordered pairs $\{(4, 0.10), (6, 0.42), (8, 0.70), (10, 0.80), (12, 0.79), (14, 0.79), (16, 0.78)\}$. The range of the estimated relation is $\{0.10, 0.42, 0.70, 0.78, 0.79, 0.80\}$.

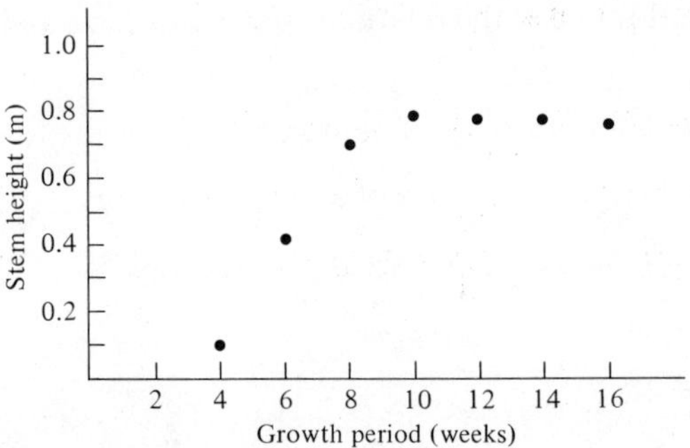

Figure 1.3

c) The relation describing the set of points on the circle of radius 5 in the $u - v$ coordinate system is $\{(u, v) | u^2 + v^2 = 25\}$. The graph of this relation is the circle in Fig. 1.4.

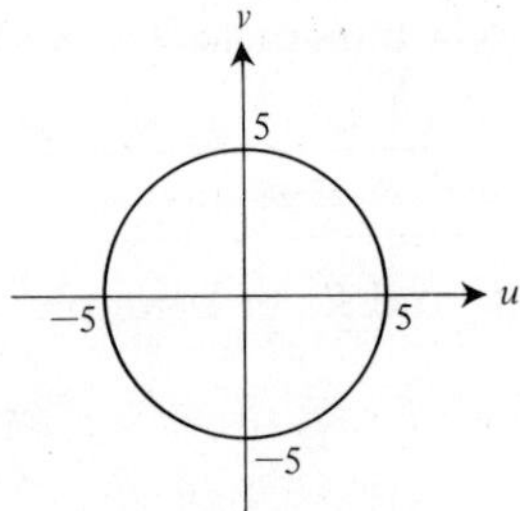

Figure 1.4

◀

Another common method of illustrating relations is to use tables. For example, consider Table 1.1. We can construct three different relations from this table, each having the common domain $\{a, b, c\}$. The relation Max which associates each harvest period with the maximum daily air temperature is given by

$$\text{Max} = \{(a, 18.8), (b, 18.1), (c, 20.1)\}.$$

Similarly, the relation Min which associated a harvest period with minimum daily air temperatures is given by

$$\text{Min} = \{(a, 13.1), (b, 12.0), (c, 10.7)\}.$$

And the relation TR which indicates an association between harvest periods and total rainfall is given by

$$TR = \{(a, 17.5), (b, 37.1), (c, 19.6)\}$$

Table 1.1

Harvest period	Average daily air temperature		Total rainfall (mm)
	Max °C	Min °C	
a	18.8	13.1	17.5
b	18.1	12.0	37.1
c	20.1	10.7	19.6

1.3 FUNCTIONS

A function is a special type of relation. A function is distinguished by the property that each element of the domain is related to only one element of the range. A more abstract mathematical definition of a function would read as follows.

Definition 1.6 A *function* is a set of ordered pairs, no two of which have the same first element.

An equivalent definition is: "A function f is a relation which associates with each element $x \in \text{Domain}(f)$ a unique element denoted $f(x)$ in the Range(f)."

The following relations are functions:

$\{(a, 1), (b, 2), (c, 3), (d, 1), (e, 2)\}$,

$\{(x, y) | x, y$ are birds and y is the mother of $x\}$,

$\{(t, T) | t$ is the time since an experiment began and T is the temperature of the laboratory at time $t\}$.

The following relations are not functions:

$\{(1, a), (2, b), (3, c), (1, d), (2, 3)\}$,

$\{(x, y) | x, y$ are dogs and y is a sibling of $x\}$,

$\{(u, v) | u, v$ are chemical elements with the same valence$\}$.

There is a simple test to determine when a graph represents a function.

Vertical-line test: If any vertical line meets the graph of a relation in more than one point then the relation is not a function. Otherwise, the graph represents a function.

Example 1.7

a) The graphs in Fig. 1.5 represent functions since each vertical line meets at most one point on the graph.
b) The graphs in Fig. 1.6 "fail" the vertical-line test and therefore do not represent functions.

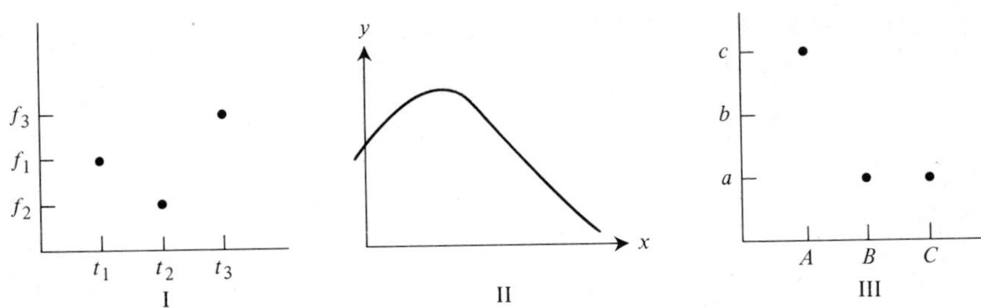

Figure 1.5

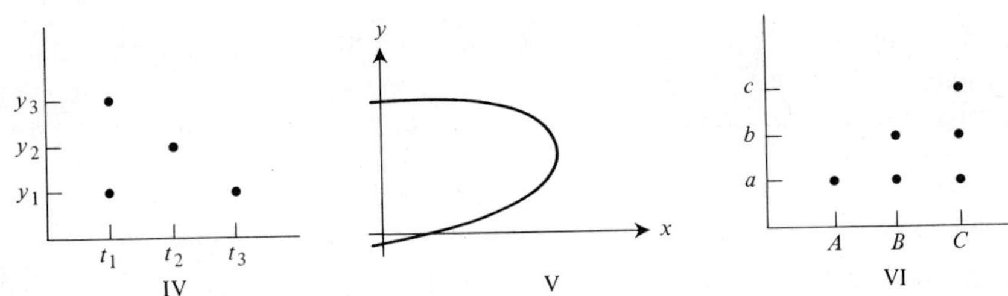

Figure 1.6

A function whose domain and range are sets of real numbers is called a *real function*. The most common method of denoting real functions is to simply indicate an equation which states the algebraic relationship between elements of the domain and range of the function. Thus the function

$$f = \{(x, y) \mid y = f(x)\}$$

is simply referred to by the equation

$$y = f(x),$$

read "y equals f of x." The two symbols f and $f(x)$ are both used to refer to the function. Technically $f(x)$ represents an element in the range of the function corresponding to an element x in the domain. In practice this technicality is frequently overlooked without causing confusion.

Example 1.8 Examples of real functions are given by the following equivalent notations.

a) The function $f = \{(x, y) \mid y = 2x\}$ is also specified by the equation $y = f(x)$, where $f(x) = 2x$.
b) The function $f(x) = x^2$ is equivalent to the function $f = \{(x, y) \mid y = x^2\}$ or simply $y = x^2$.

c) The function $f = \{(x, y) | y$ is 0 if x is an even natural number and y is 1 if x is an odd natural number$\}$ may also be specified by the equation

$$y = f(x) = \begin{cases} 0, & \text{if} \quad x \text{ is even,} \\ 1, & \text{if} \quad x \text{ is odd.} \end{cases}$$

d) Consider the function V which relates the volume of fluid flow through a vessel to the pressure p of the fluid. The basic relationship is given by an equation

$$V = C \cdot p,$$

where C is a constant dependent on the shape of the vessel. Using function notation this equation would be written as

$$V(p) = C \cdot p$$ ◀

Definition 1.7 *The graph* of a real function f is

$$\{(x, y) | x \in \text{Domain}(f) \text{ and } y = f(x)\}.$$

The graph of a function f is also referred to as "the graph of $f(x)$" or "the graph of the equation $y = f(x)$." A point (x_0, y_0) is on the graph of f if and only if $y_0 = f(x_0)$. For instance, the graph of $y = x^2$ is given in Fig. 1.7. The point (x_0, y_0) indicated on the graph represents a fixed but unspecified point on the graph. (x_0, y_0) could represent the point (0.9, 0.81) since $0.81 = (0.9)^2$. However, (x_0, y_0) could not be the point (1, 1.5) since $1.5 \neq (1)^2$.

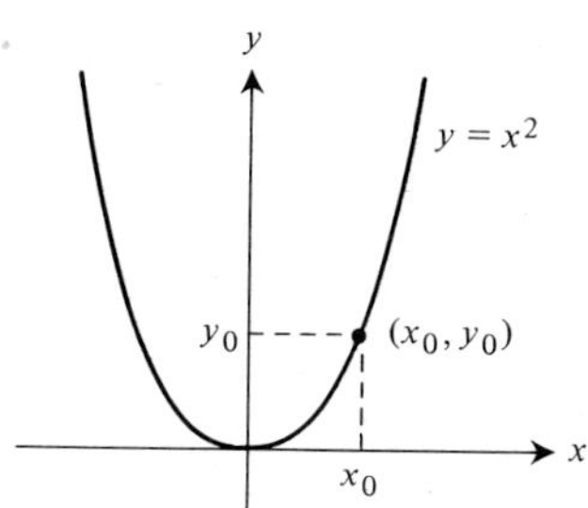

Figure 1.7

Example 1.9 *Antigen-antibody reactions*

An antibody is a protein which attaches itself to a blood-cell surface. Antibodies are "specific" to foreign substances called antigens. When an antigen is introduced to a cell, one of two results will occur.

1) If the cell has an antibody specific for the antigen these will combine and "kill" or immobilize the antigen which may be viewed as a parasite. This process usually results in destroying the cell to which the antibody is attached.
2) If the cell has no antibody specific to the antigen, the antigen will invade the cell as a parasite and the cell will remain intact as a host cell for the antigen.

To test for the presence of a specific antibody an investigator exposes a blood sample to a known antigen. She then estimates the percentage of blood cells which have been destroyed to determine the probable presence of a specific antibody. If P denotes the estimated percentage of cells destroyed, a simple classification would contain three categories

	Value of P		Conclusion
I	$P < 10\%$	$\leftrightarrow$	No antibody
II	$10\% \leq P \leq 90\%$	$\leftrightarrow$	Inconclusive experiment
III	$P > 90\%$	$\leftrightarrow$	Antibody definitely present

Thus each experiment results in estimating P and associating one of the three possible classifications I, II, or III. When studying genetic inheritance of antibodies vast numbers of experiments must be performed and the conclusions analyzed.

To allow computer analysis of the data, each of the categories is associated with a real number. This association defines a function X of the estimated percentage P. A typical choice is the function

$$X(P) = \begin{cases} 0, & \text{if} \quad P < 10\%, \\ 1, & \text{if} \quad 10\% \leq P \leq 90\%, \\ 2, & \text{if} \quad P > 90\%. \end{cases}$$

The domain of this function is the percentage interval [0, 100] and the range of the function X is $\{0, 1, 2\}$. The graph of the function X is indicated in Fig. 1.8.

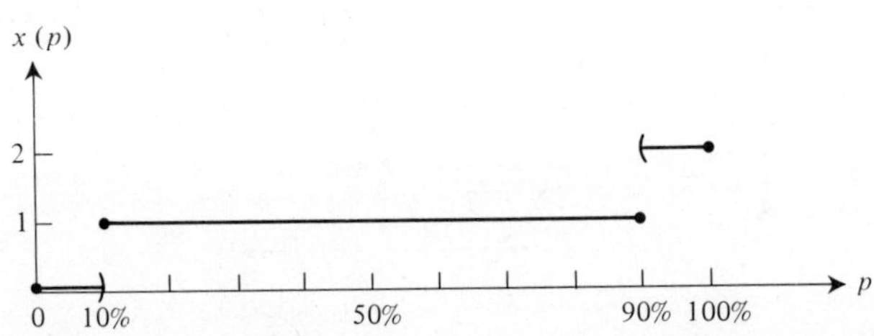

Figure 1.8

◀

We end this section with a remark about names of functions. Mathematicians traditionally make function names as short as possible. The most common mathematical names for functions are f, g, F, and G. Scientists, on the other hand, tend to use names which are as short as possible and convey some idea of what the function represents. Thus you will encounter function names such as pH, Vol, A_{tot}, V_{min}, $\mathring{Q}_{in}$, and L_A (for leaf area).

One of the goals of this text is to help the student become competent in calculus to the point that he or she can read and understand calculus concepts applied in biological disciplines. We have thus elected to introduce wherever possible appropriate notation in the form in which it would appear in a non-calculus text. Consequently, the student will need to develop the art of translating the notation used in specific problems to the notation used in general definitions and theorems.

EXERCISE SET 1

1.1 Express the following sets using braces.

a) The set of all ancestors of a particular individual, whose name is x.
b) The set consisting of the number of legs which a normal animal may have.
c) The set of temperatures at which pure H_2O is in a liquid form at sea level.
d) The set of even integers.
e) The set of numbers which are evenly divisible by 5.
f) The set of natural numbers which are multiples of 3.

1.2 Express the following sets using interval notation.

a) $\{x \in R \mid 0 \le x \le 8\}$ b) $\{x \in R \mid 0 < x \le 8\}$
c) $\{x \mid |x - 3| < 1\}$ d) $\{x \mid 3 \le x < 5\}$
e) $\{x \mid x > 3\}$ f) $\{x \mid x \ge 5\}$

1.3 Let $A = \{1, 2, 3, 4\}$; $B = \{0, 1, 2\}$; $C = \{3, 4, 5\}$. Determine the following sets:

a) $A \cap B$ b) $A \cup B$
c) $B \cap C$ d) $A \cup C$
e) $B \cap (A \cup C)$ f) $(B \cap A) \cup (B \cap C)$

1.4 Express the following relations in set notation.

a) The depth below sea level is related to the pressure of sea water at that depth.
b) The day of the year is related to the number of hours of daylight for that day at 45°N latitude.
c) Two animals are related if they are of the same phylum.
d) Two shrubs are related if they are of the family Philadelphus (commonly known as Mock-Orange).

1.5 Sketch a graph of the following relations. Indicate the domain and the range of the relation and whether it is a function.

a) An integer x between 1 and 10 is related to an integer y between 1 and 15 if x divides y evenly.
b) Two chemical elements are related if they are both inert gases.
c) Each integer is related to the number 1 if it is odd and otherwise to the number 0.
d) $\{(C, G), (G, C), (A, U), (U, A)\}$ (amino acids which form DNA.)
e) $\{(0, 0), (1, 2), (2, 1), (3, 0)\}$
f) Each number in the interval $[0, 10)$ is related to the nearest integer (round 0.5's up).

1.6 Indicate the domains and ranges of the relations graphed in Example 1.7, I, III, IV, VI.

1.7 Specify the domain and range of each function in Example 1.8.

1.8 On the same coordinate system graph the three relations derived in the text from Table 1.1. Denote the points of the relation Max by x, Min by 0, and TR by □.

1.9 Which of the following graphs are the graph of a function?

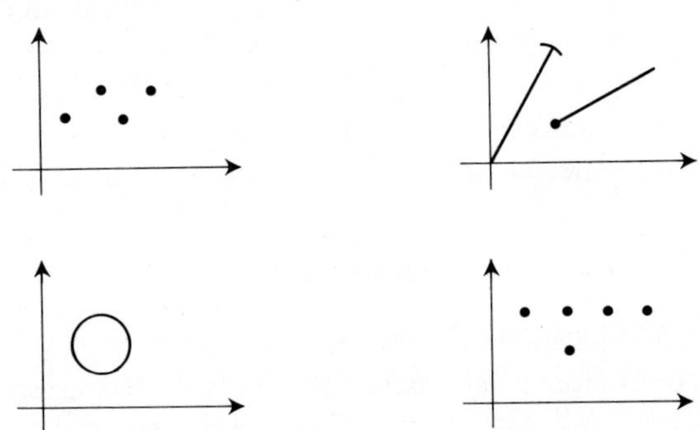

1.10 Determine the graph of the given function over the interval $[-5, 5]$.

a) $f(x) = \begin{cases} 1 & \text{if} \quad x \geq 0, \\ 0 & \text{if} \quad x < 0. \end{cases}$ b) $g(x) \begin{cases} 1 & \text{if} \quad x > 3, \\ 2 & \text{if} \quad 1 < x \leq 3, \\ -1 & \text{if} \quad x \leq 1. \end{cases}$

c) $V(x) =$ nearest integer to x.
d) $I(x) =$ greatest integer less than or equal to x.

1.11 Which of the following points lie on the graph of $y = 3x^2 - 1$?

a) $(1, 1)$ b) $(1, 2)$ c) $(2, 12)$ d) $(0, -1)$
e) $(-1, -4)$ f) $(-2, 11)$ g) $(\sqrt{2}, 5)$ h) $(-\sqrt{2}, -5)$

1.12 Determine the given sets and graph them on a number line.

a) $[3, 5) \cap [2, 4]$ b) $[3, 5) \cup [2, 4]$
c) $[-2, 5] \cap (5, 6)$ d) $[-2, 5] \cup (5, 6)$
e) $(-1, +\infty) \cap (2, 4)$ f) $(-1, +\infty) \cup (2, 4)$

1.13 State the domain and range of the function indicated by the given graphs. Describe the function by a set of ordered pairs.

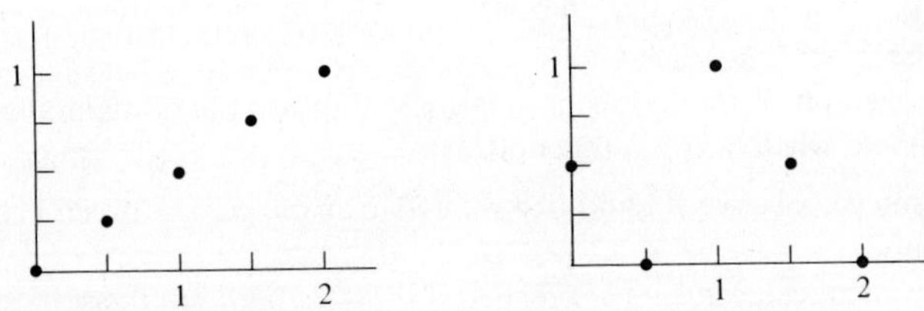

1.14 List all subsets of the set $\{a, b, c\}$.

1.15 How many subsets will a set of n elements have? To answer this, first count the subsets of the sets $\{\quad\}$, $\{1\}$, $\{1, 2\}$, $\{1, 2, 3\}$ and observe a pattern.

SECTION 2

ALGEBRA OF FUNCTIONS

2.1 THE ALGEBRA OF REAL FUNCTIONS

The operations of addition and multiplication of real numbers provide an "algebraic" structure for the set of real numbers. A similar algebraic structure for sets of real functions is provided by the operations of addition and multiplication of functions defined in the following way.

Definition 2.1 Let f and g be two real functions such that the set

$$D = \text{Domain}(f) \cap \text{Domain}(g) \neq \phi.$$

1. The sum of f and g is the function $f + g$ defined on the set D by the equation

$$(f + g)(x) = f(x) + g(x).$$

2. The product of f and g is the function $f \cdot g$ defined on the set D by the equation

$$(f \cdot g)(x) = f(x) \cdot g(x).$$

In most cases, the set D will be either a finite set or the union of a finite number of intervals. For any given set of numbers D, the set of all real functions defined on the set D will be denoted by $F(D)$, or simply F. The operations $+$ and $\cdot$ defined in Definition 2.1 satisfy properties analogous to those of addition and multiplication of real numbers. To avoid technicalities concerning the domains, we restrict our discussion to functions having a common domain D. The algebraic properties of the addition and multiplication of real functions are as follows.

P1 Closure: For any two functions $f, g \in F$, the sum $f + g \in F$ and the product $f \cdot g \in F$.

P2 Associative: If $f, g, h \in F$, then $(f + g) + h = f + (g + h)$ and $(f \cdot g) \cdot h = f \cdot (g \cdot h)$.

P3 Commutative: If $f, g \in F$, then $f + g = g + f$ and $f \cdot g = g \cdot f$.

P4 Identity: There exists an additive identity function $\mathbf{0} \in F$ with the property $\mathbf{0} + f = f + \mathbf{0} = f$ for every $f \in F$, and there exists a multiplicative identity $\mathbf{1} \in F$ with the property that $\mathbf{1} \cdot f = f \cdot \mathbf{1} = f$ for every $f \in F$.

P5 Inverse: For any $f \in F$, there exists an additive inverse, $-f$, such that $f + (-f) = (-f) + f = \mathbf{0}$, the additive identity, and for any $f \in F$, such that $f(x) \neq 0$ for all $x \in D$, there exists a multiplicative inverse, $1/f$, such that $f \cdot (1/f) = (1/f) \cdot f = \mathbf{1}$, the multiplicative identity.

The additive identity function **0** and the multiplicative identity function **1** of properties P4 and P5 are defined by

$$\mathbf{0}(x) = 0 \qquad \text{and} \qquad \mathbf{1}(x) = 1 \qquad \text{for all } x \in D.$$

Similarly, the additive inverse function $-f$ and the multiplicative inverse function $1/f$ are defined by

$$(-f)(x) = -f(x) \quad \text{for all } x \in D,$$

and

$$(1/f)(x) = 1/f(x) \qquad \text{for all } x \in D \qquad \text{such that } f(x) \neq 0.$$

When considering the arithmetic properties of real numbers all numbers have a multiplicative inverse except the number zero, which is the additive identity for real numbers. However, when considering functions we find that there exist functions which are not the additive identity function **0** and which do not have a multiplicative inverse defined on the same domain. For instance, the function

$$f(x) = \begin{cases} 3 & \text{if } 2 \leq x \leq 3, \\ 0 & \text{if } 3 < x \leq 4, \end{cases}$$

is defined on the interval [2, 4]. This function cannot have a multiplicative inverse function $1/f$ defined on the interval (3, 4] since $(1/f)(x) = 1/f(x)$ is not defined for $x \in (3, 4]$. Thus whenever we encounter a function f whose range contains the number zero, the function will not have a multiplicative inverse defined for values x such that $f(x) = 0$. Such functions do not have multiplicative inverses defined on their entire domain but do have multiplicative inverses defined on

$$\{x \in \text{Domain}(f) \mid f(x) \neq 0\}.$$

For the above function f the multiplicative inverse $(1/f)(x) = \frac{1}{3}$ is defined on the interval [2, 3).

Example 2.1 The addition and multiplication of functions may be illustrated by graphing the functions f and g and their sum and product on the same graph. In Fig. 2.1 the domain D is taken to be the interval [0, 2]. The functions f and g are given by the equations

$$f(x) = 2 \qquad \text{and} \qquad g(x) = 1 + x.$$

In Fig. 2.1(a) the sum $f + g$ and the product $f \cdot g$ are obtained by adding and multiplying, respectively, the ordinates of the graphs of f and g at each point x in the interval [0, 2]. In Fig. 2.1(b) the additive inverses of these functions are sketched and in Fig. 2.1(c) their multiplicative inverses are graphed.

Algebraically, the addition and multiplication of functions which are explicitly given by equations may be achieved by performing the corresponding operation on the right side of the equations. Thus for the functions f and g of this example we may compute $f + g$ and $f \cdot g$ as follows:

$$(f + g)(x) = f(x) + g(x) = 2 + (1 + x) = 3 + x$$

$$(f \cdot g)(x) = f(x) \cdot g(x) = 2 \cdot (1 + x) = 2 + 2x.$$

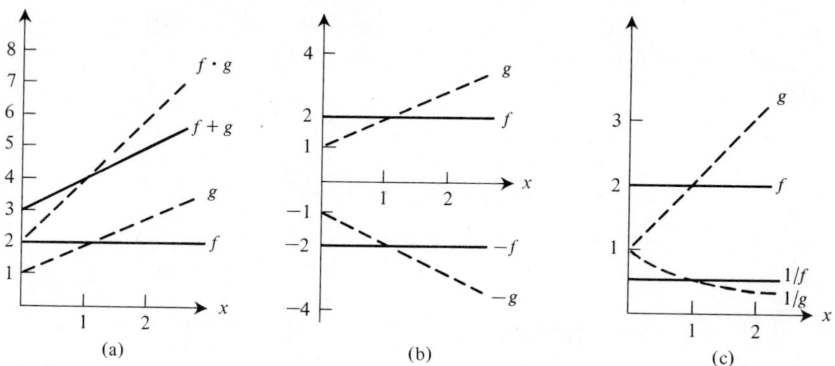

Figure 2.1

Similarly,

$$-f(x) = -2 \qquad \text{and} \qquad 1/f(x) = \tfrac{1}{2},$$
$$-g(x) = -1 - x \qquad \text{and} \qquad 1/g(x) = 1/(1 + x).$$

◀

Example 2.2 We compute the sum, product, and inverses (if they exist) for different pairs of functions.

a) Let f and g be the finite functions specified by the sets

$$f = \{(1, 3), (2, 5), (4, -2), (5, 3)\},$$
$$g = \{(-1, 2), (0, 2), (1, 0), (2, 2), (3, \tfrac{1}{2})\}.$$

The domain of $f + g$ and $f \cdot g$ is the set

$$D = \{1, 2, 4, 5\} \cap \{-1, 0, 1, 2, 3\} = \{1, 2\}.$$

The sum and product of f and g are therefore

$$f + g = \{(1, 3), (2, 7)\} \qquad \text{and} \qquad f \cdot g = \{(1, 0), (2, 10)\}.$$

The inverse functions have the same domain as the functions f and g, except $1/g$. Since $g(1) = 0$ the domain of $1/g$ is restricted to $\{-1, 0, 2, 3\}$.

$$-f = \{(1, -3), (2, -5), (4, 2), (5, -3)\}$$
$$1/f = \{(1, \tfrac{1}{3}), (2, \tfrac{1}{5}), (4, -\tfrac{1}{2}), (5, \tfrac{1}{3})\}$$
$$-g = \{(-1, -2), (0, -2), (1, 0), (2, -2), (3, -\tfrac{1}{2})\}$$
$$1/g = \{(-1, \tfrac{1}{2}), (0, \tfrac{1}{2}), (2, \tfrac{1}{2}), (3, 2)\}$$

b) Let $f(x) = x + 6$, $g(x) = x^2$. Since Domain(f) = Domain(g) = R, $f + g$ and $f \cdot g$ are defined for all real numbers. Algebraically $(f + g)(x) = x + 6 + x^2$, and $(f \cdot g)(x) = (x + 6)x^2 = x^3 + 6x^2$.

$$-f(x) = -x - 6 \quad \text{and} \quad 1/f(x) = \frac{1}{x + 6} \quad \text{for} \quad x \neq -6.$$

$$-g(x) = -x^2 \quad \text{and} \quad 1/g(x) = \frac{1}{x^2} \quad \text{for} \quad x \neq 0.$$ ◀

2.2 FUNCTIONS, VARIABLES, AND CONSTANTS

We have avoided talking about the terms "variable" and "constant" until now primarily because it is difficult to give a definition of these terms which will apply in all cases. This may be illustrated by considering the two gas laws known as Boyle's Law and Charles' Law. The symbols used in these "laws" are defined as follows:

V represents the volume of a gas:
T is the temperature of the gas;
P is the pressure of the gas;
R is a gas constant.

Both laws utilized the same equation:

$$V = RT/P.$$

Boyle's Law: If T is held constant, then the volume is "inversely proportional" to the pressure, i.e.,

$$V = \frac{\text{constant}}{P}.$$

Charles's Law: At a constant pressure, the volume of a given sample of gas is "directly proportional" to the temperature, i.e.,

$$V = \text{constant} \times T.$$

From both of the laws we obtain the volume V as a function. In Boyle's Law, T is treated as a constant and P is treated as a variable. Thus V is a function of P and, to emphasize this, the equation is sometimes written as

$$V(P) = RT_0/P,$$

where T_0 represents the constant temperature. In contrast, Charles' Law treats P as a constant and T as a variable. Thus from Charles' Law we obtain V as a function of the temperature T, and the equation describing this function becomes

$$V(T) = KT,$$

where K represents the constant R/P. These examples demonstrate that a term in an equation or expression may assume the role of a constant or that of a variable.

In fact, we could likewise treat either T or P as a function. By rearranging the equation in the form

$$P(T) = k_1 T,$$

where k_1 is held constant and represents the value of $1/RV$, we have expressed P as a function of the variable T.

A general guide as to the use of the words "variable" and "constant" is the following.

In the context of a given situation, a term is considered to be a *constant* if it is a real number or represents a real number and is not altered for any reason.

A term is called a *variable* if it is considered as the argument of a function and may assume any value in the domain of the function.

The most commonly used symbol for denoting a variable is the letter x. When the domain of a function corresponds to time intervals, the letter t is generally used to represent the variable. Other symbols frequently used to represent variables are the letters occurring at the end of the alphabet, s, u, v, w, y, and z. The letter y is often used in a special way. If a function f is specified by an equation $f(x) = x^2 - 3x + 2$, the equation is frequently expressed as $y = x^2 - 3x + 2$. In this context the variable y is dependent on the variable x and we thus refer to y as a *dependent variable* and to x as an *independent variable*. Thus the term "dependent variable" is a synonym for the term "function" in the sense that the dependent variable y in $y = x^2 - 3x + 2$ represents the function $f, f(x) = x^2 - 3x + 2$.

Most variables and constants which appear in scientific applications are given specific dimensions. When considering the sum or product of two functions, attention must be paid to the associated dimensions. If the dimensions are not compatible, the biological or physical interpretation attached to the sum (or product) must be questioned. For instance, functions used to describe spruce trees of age x might include: $h(x)$, the average height of a tree; $c(x)$, the average trunk circumference; and $v(x)$, the average volume of the trunk. The dimensions of h and c would be meters (m) while the dimension of v would be meters cubed (m^3). The sum $(h + c)(x)$ would be compatible and have the dimension meters. The product $(h \cdot c)(x)$ would have the dimension meters squared (m^2) and be proportional to the total bark area. The sum $(h + v)(x)$ would be mathematically acceptable but would involve the addition of terms with different dimensions. The quantities h and v are therefore incompatible for addition in a biological sense.

The equation

$$y = f(x)$$

implies that x is an independent variable and y is a dependent variable, y being dependent on x. Once the form of the function $f(x)$ is known, the independent variable x may be replaced by a constant such as $x = 2$ to evaluate $f(2)$. Frequently, we will have the need to replace x in the expression $f(x)$ by other variables or combination of variables resulting in expressions such as $f(z), f(2 + x), f(x + h)$, $f(\Delta x)$. We illustrate this for the function

$$y = f(x) = x^2 - x + 1.$$

To evaluate the y value corresponding to $x = 2$, we set

$$y = f(2) = 2^2 - 2 + 1 = 3.$$

If we set $x = z$ we obtain

$$f(z) = z^2 - z + 1.$$

If x is replaced by $2 + x$, the expression becomes

$$f(2 + x) = (2 + x)^2 - (2 + x) + 1,$$

which can be simplified algebraically to

$$f(2 + x) = (x^2 + 4x + 4) - (2 + x) + 1 = x^2 + 3x + 3.$$

If x is replaced by $x + h$, the expression becomes

$$f(x + h) = (x + h)^2 - (x + h) + 1 = x^2 + 2xh + h^2 - x - h + 1.$$

If Δx is considered as the argument of the function f, the expression becomes

$$f(\Delta x) = (\Delta x)^2 - \Delta x + 1.$$

Proportionality

It is extremely common to encounter proportionality arguments in the life sciences.

Definition 2.2 A quantity A is proportional to a quantity B, denoted

$$A \propto B,$$

if there exists a constant k such that

$$A = kB.$$

Example 2.3 Let $A(t)$ denote the distance a bird can fly in time t and $B(t)$ be the distance an insect can fly in time t. If $A(t) = 3t$ and $B(t) = 2t$, then A is proportional to B since $A = (\frac{3}{2})B$. But it would be inappropriate to imply that the distance flown by a bird depends on the distance flown by an insect. ◀

While mathematicians shy away from the symbol $\propto$ used to denote proportionality, scientists tend to employ it frequently, especially when the exact proportionality constant is not known. Thus the equation

$A \propto B$ is equivalent to $A = k \cdot B$, k a constant.

Proportionality arguments are frequently used to explain limitation of performance of various physiological features of animals.

Example 2.4 The following is an example of a proportionality argument in physiology. Each time a muscle is used it generates heat which must be dissipated.

Most body heat is lost through evaporation from the skin and lungs. The amount of muscle action, A, is proportional to the body mass, M. The body heat generated, H_g, is proportional to the muscle action A and hence to M. The body heat loss, H_l, is proportional to the surface area of the skin S. If an animal is not going to "burn up," the animal's heat loss must equal its heat gain. Hence, an animal's surface area S must be proportional to its total mass M.

Symbolically these statements can be represented as:

1. $\begin{matrix} A \propto M \\ H_g \propto A \end{matrix} \Rightarrow H_g \propto M,$

2. $\begin{matrix} H_l \propto S \\ H_g = H_l \end{matrix} \Rightarrow H_g \propto S,$

3. $\begin{matrix} H_g \propto S \\ H_g \propto M \end{matrix} \Rightarrow S \propto M.$

If an animal has the shape of a cube, of edge length L, then the animal's mass will be proportional to its volume, which is L^3. Its surface area will be proportional to L^2. Consequently to have $S \propto M$ would require $L^2 \propto L^3$ for a square-shaped animal. As L increases L^2 does not equal the same constant multiple of L^3, therefore L^2 is not proportional to L^3. Thus a square-shaped animal could not grow to be very large. In fact, to keep an animal in heat balance, i.e., to keep $S \propto M$, as the mass of the animal increases its surface area must increase at a faster rate. One way to accomplish this would be to grow cooling fins. What large animals have such an anatomy? ◀

EXERCISE SET 2

Exercises 2.1–2.4 will refer to the following pairs of functions:

a) $f = \{(1, 2), (2, 3), (3, -1), (5, 6)\}$.
   $g = \{(3, 1), (5, 0), (6, 1)\}$.

b) $f(x) = x - 1$ for $x \in [0, 5]$.

$$g(x) = \begin{cases} 3 & \text{if} \quad 0 \le x \le 5 \text{ and } x \text{ is an integer,} \\ 0 & \text{if} \quad 0 \le x \le 5 \text{ and } x \text{ is not an integer.} \end{cases}$$

c) The functions f and g specified in the following graph.

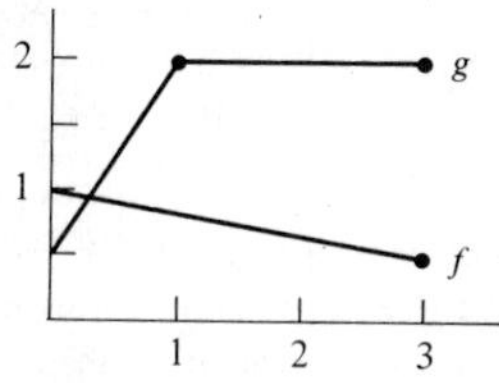

2.1 For each pair, state the domain of f, g, and $f + g$.

2.2 For each pair, indicate the functions $f + g$ and $f \cdot g$.

2.3 On one figure sketch the graph of the functions $f, g, f + g$ and $f \cdot g$ for the pair (a). Do the same for (b).

2.4 For each pair indicate the additive inverse of f and the multiplicative inverse of g.

2.5 Let $f(x) = x^2 - 2, g(x) = 3x + 1$, and $h(x) = \sqrt{x}$. Evaluate algebraically the following:

a) $(f + g)(x)$ b) $(f + h)(x)$
c) $(g \cdot h)(x)$ d) $(f \cdot g)(x)$
e) $(-f)(x)$ f) $(g + (-f))(x)$
g) $(1/h)(x)$ h) $((-g) \cdot (1/h))(x)$

2.6 In studying the water evaporation from lakes and reservoirs, the following Dalton Equation is often used:

$$E = f(\bar{u})(e_s - e_a).$$

Here E = reservoir evaporation (cm/day),
$\bar{u}$ = average wind speed (m/(sec)^2),
e_s = vapor pressure at the surface (millibars),
e_a = vapor pressure one meter above the surface (millibars).

a) Given that $f(\bar{u}) = 0.291\bar{u}/\sqrt{A}$, where

$$A = \text{surface area of the reservoir } (\text{m}^2),$$

describe which terms are constants, dependent variables, and independent variables in the Dalton Equation.

b) Rearrange the Dalton Equation using the function f of part (a) to give the average wind speed as a function of the evaporation.

2.7 A drug is administered to an animal daily in the animal's food. Let the amount of drug given on the nth day of treatment be $g(n) = 2/n$ units. The drug is either metabolized or excreted in the urine at the constant rate of $\frac{1}{2}$ unit/day. Thus the amount of drug in the animal on day n decreases by $E(n) = \frac{1}{2}$ units. The net change in the amount of drug present in the animal on the nth day of treatment is $d(n) = g(n) - E(n)$. The accumulation of the drug in the animal after n days is the sum, denoted by $A(n)$, of the daily change $d(n)$ over the days $1, 2, 3, \ldots, n$; i.e.,

$$A(n) = d(1) + d(2) + \cdots + d(n).$$

a) Evaluate the following:

$$E(2), g(2), d(2), E(3), g(3), d(3), A(1), A(3).$$

b) Give an equation expressing $d(n)$ as a function of n.

c) If the drug is administered for only two days when will the animal cease to contain the drug? Indicate the accumulation $A(n)$ for each day until the animal is drug free.

2.8 The size of a population is affected by two components, births and deaths. Assume that the number of births at generation n is given by $b(n) = A + k \cdot n$, where A and k are constants. Let the number of deaths at generation n be given by $d(n) = c \cdot n$, where c is a constant. The change in the population at generation n is the term $\Delta(n) = b(n) - d(n)$.

a) Write an explicit equation for the change $\Delta(n)$ in the population at generation n as a function of n.

b) Evaluate $b(n)$, $d(n)$, and $\Delta(n)$, when $A = 2$, $k = 2$, and $c = 1$ for $n = 1, 2, 3$.
c) The total population size at generation n is the initial population size plus the change in population at generation 1, 2, 3, . . . , up to n. At what generation will the population exceed 200 if it is initially 100 and the parameters are the same as in part (b)? What will the population be at this generation?

2.9 When is the surface area of a cube of edge length L larger than the volume of the cube?

2.10 Justify the conclusion of Example 2.4 algebraically by replacing each proportionality symbol by an equation. ($A \propto M$ means $A = k_1 M$, etc.)

2.11 An animal of length L will have a mass proportional to L^3. If the animal runs up a hill at a speed V it will have to expend energy, E, proportional to $V \cdot L^3$. However the energy available to the animal is proportional to L^2.

a) How is the speed V related to the animal's size?
b) Give an example which illustrates this conclusion.

2.12 If the amount of nutrient needed by a cell is proportional to its volume, why would this limit the size of a cell? (*Hint*: How would the cell obtain its nutrient?)

2.13 Let $f(x) = 3x + 2$ and $g(x) = x^2 + 2x - 4$. Evaluate the following.

a) $f(2)$ b) $f(-2)$ c) $g(2)$ d) $g(-2)$
e) $f(2 + x)$ f) $f(x - 2)$ g) $g(x + 2)$ h) $g(x - 2)$
i) $f(x + h)$ j) $g(x + h)$ k) $f(2x)$ l) $g(2x)$

2.14 a) What function $f(x)$ is equal to its additive inverse?
b) What function $f(x)$ is equal to its multiplicative inverse?
c) Does any real function f have the property that its additive inverse is the same as its multiplicative inverse? Explain why.

2.15 On what intervals will the following functions have multiplicative inverses?

a) $f(x) = 2x + 1$ b) $f(x) = \dfrac{1}{2x + 1}$ c) $f(x) = \dfrac{x + 1}{x - 2}$

d) $f(x) = \dfrac{x - 2}{x + 1}$ e) $f(x) = \dfrac{\sqrt{x}}{x - 2}$ f) $f(x) = \dfrac{x - 2}{\sqrt{x}}$

SECTION 3

POWER FUNCTIONS AND POLYNOMIALS

3.1 POWER FUNCTIONS $y = x^n$

In this section we examine the most basic and important function in mathematics, the power function. A power function is a function that has the form

$$f(x) = x^n, \tag{3.1}$$

where n is a natural number or zero. The power function in equation (3.1) is very frequently represented by the equation

$$y = x^n, \tag{3.2}$$

where the function value $f(x)$ is replaced by the dependent variable y.

To emphasize the dependence of the function f of equation (3.1) on the parameter n, throughout the rest of this section we will add a subscript n to the function name f. Thus we will use the notation

$$\begin{aligned} f_0(x) &= 1, \\ f_1(x) &= x^1, \\ f_2(x) &= x^2, \\ f_3(x) &= x^3, \\ &\vdots \\ f_n(x) &= x^n, \\ &\vdots \end{aligned} \tag{3.3}$$

To ensure that the functions (3.3) are well defined, i.e., that every person reading this will have exactly the same idea as to the meaning of the functions, the domain of each function f_n and the exact meaning of the symbol x^n must be specified. Our choice for the domain of these functions (and unless specified otherwise for all real functions considered in this text) will be the "natural domain."

Definition 3.1 The *natural domain* of a real function f is the largest subset D of the set of all real numbers such that $f(x)$ is defined for each element x of D.

Definition 3.2 Let x be a real number. The symbol x^n is defined for each natural number n as follows:

$$x^1 \equiv x,$$

and for each n greater than 1 define x^n by the relation

$$x^n \equiv x \cdot x^{n-1},$$

where it is assumed that $x^1, x^2, \ldots, x^{n-1}$ have been previously defined. If x is not zero, x^0 is defined by

$$x^0 \equiv 1.$$

The definition of x^0 excludes the case when $x = 0$ since mathematically 0^0 cannot be consistently defined. This is analogous to the situation of dividing by zero and the problems encountered trying to define the form 0/0. This creates a certain amount of inconvenience since the power functions $f_n(x) = x^n$ have domain R (the real numbers) for each $n = 1, 2, 3, \ldots$ and the domain of $y = x^0$ is $(-\infty, 0) \cup (0, +\infty)$. Thus, if we defined $f_0(x) = x^0$, we would have to recognize the difference in the domains each time we referred to these functions. To avoid this difficulty we have defined $f_0(x) = 1$ rather than $f_0(x) = x^0$.

To use a function it is necessary to be able to calculate the value of the function $f(x)$ for each x in its domain. (In the form $f(x)$, x is called the *argument* of the function f.) However, the ability to evaluate a function is generally not sufficient for the purpose of utilizing the function in applications. What is also needed is a knowledge of the properties that characterize the function near any fixed point in its domain and over its entire domain. Such behavior is most easily described using the tools and vocabulary of calculus. Without using calculus much information can be provided by sketching the graph of functions to present their behavior visually. Again, graphing is most easily accomplished with the aid of calculus, but for simple functions the basic techniques of "plotting points" and connecting the points is frequently sufficient. While this technique lacks finesse and can result in gross misrepresentations, when combined with basic concepts of inequalities it can lead to graphs that accurately reflect the "behavior" of the functions being graphed.

The Graph of $y = x^n$

The graph of $y = x^n$ for x in the interval $(0, +\infty)$ has the property that as the variable x increases so does the variable y. This behavior is called *monotone increasing*. (See Fig. 3.1.)

Definition 3.3 A function f is *monotone increasing* on an interval (a, b) if for x_1 and $x_2 \in (a, b)$,

$$x_1 < x_2 \qquad \text{implies that} \qquad f(x_1) < f(x_2).$$

A function f is *monotone decreasing* on an interval (a, b) if for x_1 and $x_2 \in (a, b)$,

$$x_1 < x_2 \qquad \text{implies that} \qquad f(x_1) > f(x_2).$$

Theorem 3.1 The function $f_n(x) = x^n$ is monotone increasing on $(0, +\infty)$ for each $n = 1, 2, 3, \ldots$

For example, $x_1 = 5 < 10 = x_2$ and hence $5^n < 10^n$ for $n = 1, 2, 3, \ldots$

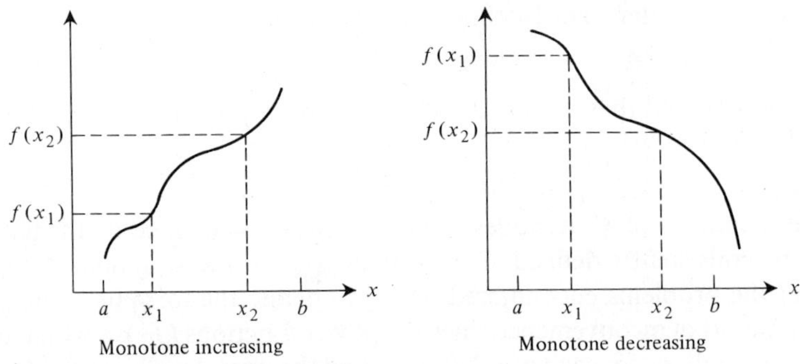

Figure 3.1

Theorem 3.2 Assume n and m are natural numbers with $n < m$. Then,

i) if $0 < x < 1, f_m(x) < f_n(x)$, that is, $x^m < x^n$;
ii) if $1 < x$, $f_n(x) < f_m(x)$, that is, $x^n < x^m$.

For example let $n = 2$ and $m = 3$. If $x = \frac{1}{2}$, then Theorem 3.2 (i) states that

$$f_3(\tfrac{1}{2}) < f_2(\tfrac{1}{2}), \quad \text{that is,} \quad (\tfrac{1}{2})^3 < (\tfrac{1}{2})^2.$$

If $x = 5$, then Theorem 3.2 (ii) states that

$$f_2(5) < f_3(5), \quad \text{that is,} \quad 5^2 < 5^3.$$

The proofs of these theorems rely on the properties of inequalities provided in Appendix 1. The graphical interpretation of these theorems is shown in Fig. 3.2.

1. As the value of x increases the value of y increases. Therefore the graph of $y = x^n$ never "turns down" as x increases from 0 to $+\infty$.
2. If $x > 1$ and $m > n$, the graph of $y = x^m$ will be "above" the graph of $y = x^n$.
3. If $0 < x < 1$ and $m > n$, the graph of $y = x^m$ will be "below" the graph of $y = x^n$.

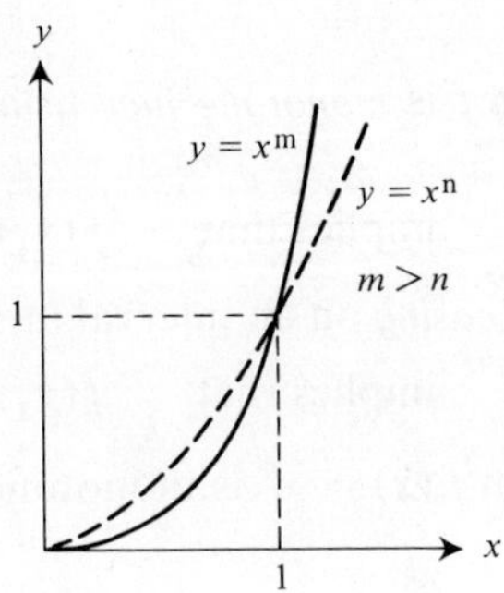

Figure 3.2

The graphs of $y = x^n$ for $n = 1, 2, 3, 4$ are illustrated in Fig. 3.3(a) and (b). Note that the graph of each function $y = x^n$ passes through the points (0, 0) and (1, 1) (except when $n = 0$).

The graph of the power functions $y = x^n$ for $x < 0$ are obtained from the graphs for $x > 0$ by utilizing the fact that the power functions are either "even" or "odd" functions.

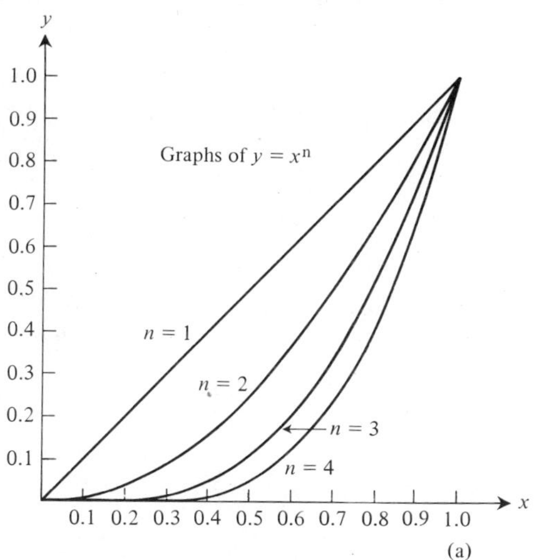

(a)

x \ n	1	2	3	4
0.1	0.1	0.01	0.001	0.001
0.2	0.2	0.04	0.008	0.0016
0.4	0.4	0.14	0.064	0.025
0.5	0.5	0.25	0.125	0.0625
0.8	0.8	0.64	0.512	0.4096
0.9	0.9	0.81	0.729	0.6581

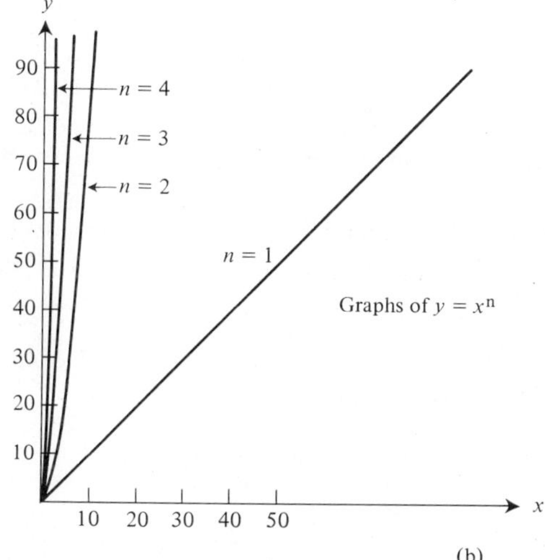

(b)

x \ n	1	2	3	4
2	2	4	8	16
4	4	16	64	256
5	5	25	125	625
6	6	36	216	1296
7	7	49	343	2401

Figure 3.3

Definition 3.4 A real function f is said to be

i) an *odd function* if $f(-x) = -f(x)$ for all $x \in R$;
ii) an *even function* if $f(-x) = f(x)$ for all $x \in R$.

By combining the associative and commutative properties, P2 and P3 of Section 2.1, and the fact that $-x = (-1) \cdot x$, the following theorem may be verified.

Theorem 3.3 The function $f_n(x) = x^n$ is an odd function if n is odd and an even function if n is even.

Even and odd functions may equivalently be defined by the symmetry which their graphs exhibit (see Fig. 3.4).

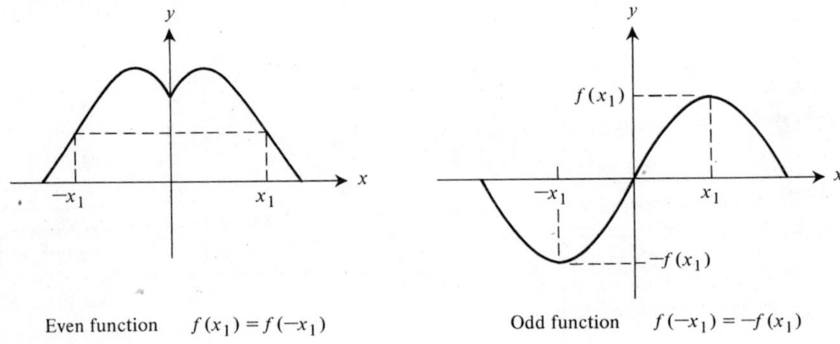

Even function $f(x_1) = f(-x_1)$ Odd function $f(-x_1) = -f(x_1)$

Figure 3.4

Definition 3.5

i) A real function f *is an even function* if its graph is symmetrical about the y-axis, i.e., a point (x_1, y_1) is on the graph of f if and only if the point $(-x_1, y_1)$ is on the graph of f.
ii) A real function f *is an odd function* if its graph is symmetrical about the origin—i.e., a point (x_1, y_1) is on the graph of f if and only if the point $(-x_1, -y_1)$ is on the graph of f.

Using Definition 3.5 and Theorem 3.3, the graph of $y = x^n$ for $x < 0$ is obtained from the graph of $y = x^n$ when $x > 0$ as follows. (See Fig. 3.5.)

If n is even, the graph is symmetric about the y-axis, so the graph is a mirror reflection about the y-axis.

If n is odd, the graph is symmetric about the origin, so the graph for $x < 0$ may be obtained by reflecting the graph for $x > 0$ about the y-axis and then about the x-axis.

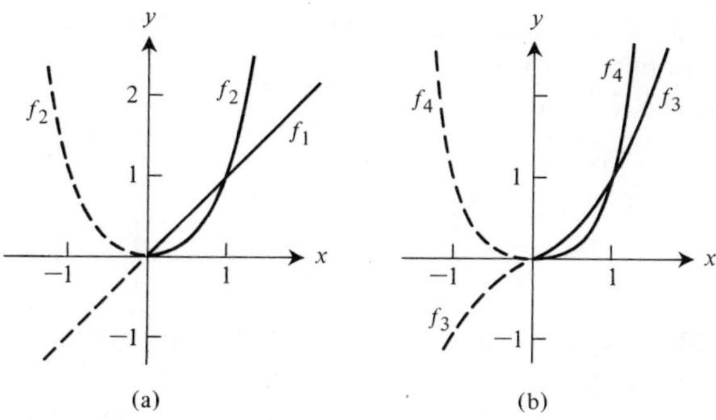

Figure 3.5

The basic features of the graph of power functions which will be repeatedly used are as follows.

1. For even values of n, the graph of $y = x^n$ is U-shaped and symmetric about $x = 0$.
2. For odd values of n, the graph of $y = x^n$ is the shape of a U which has been altered by rotating downward its left half; it is therefore symmetric about the origin and increasing on $(-\infty, +\infty)$.
3. As n increases, the graphs of $y = x^n$ become "flatter" on $(-1, 1)$ and "steeper" on $(-\infty, 1)$ and $(1, +\infty)$.

3.2 POLYNOMIAL FUNCTIONS

Polynomials are the most frequently encountered functions. The power functions are a special type of polynomial from which all other polynomials may be constructed. The main reason why polynomials play such an important part in mathematics is that every "relatively nice" function can be approximated by a polynomial function. In later sections the method of such approximation will be examined and the expression "relatively nice" will be explained. In this section we define the general polynomial and examine the properties of the simplest polynomials.

A polynomial is the sum of constant multiples of power functions. The largest exponent of a polynomial is called the degree of the polynomial. In most practical situations only simple polynomials of low degree are encountered. However, it is helpful to know the form of a general polynomial and the associated vocabulary. Thus we introduce the more general definition.

Definition 3.6 A *polynomial of degree n*, in the variable x, is an expression of the form

$$a_0 + a_1x + a_2x^2 + \cdots + a_{n-1}x^{n-1} + a_nx^n, \tag{3.4}$$

where the coefficients $a_0, a_1, a_2, \ldots, a_n$ are real numbers and $a_n \neq 0$.

A polynomial function is any function f whose value at a real number x_0 is given by a polynomial form of the type (3.4) evaluated at $x = x_0$. A polynomial function f of degree n is expressed either as

$$f(x) = a_0 + a_1x + a_2x^2 + \cdots + a_nx^n, \tag{3.5}$$

or as

$$y = a_0 + a_1x + a_2x^2 + \cdots + a_nx^n. \tag{3.6}$$

The function (3.5) is very general, and a direct attempt to analyze such a function or to sketch its graph would necessarily be complicated and abstract. Instead, we will examine polynomial functions of degree zero, one, two, and three and then observe only some general properties shared by polynomial functions of a higher degree. It is common practice to drop the term "function" and to refer to "the polynomial," or "polynomial equation" of the form (3.6). For convenience we shall adopt this policy when no confusion results.

Polynomials of Degree Zero

A polynomial equation of degree zero is an equation of the form

$$y = a_0, \tag{3.7}$$

where a_0 is a nonzero constant. The graph of equation (3.7) is a horizontal line, a_0 units from the x-axis. A function whose graph has this property is called a *constant function.* A constant function is uniquely determined by just one point on its graph, since it assumes the same y value, or functional value, at each argument x in its domain.

Polynomials of Degree One

A polynomial equation of degree one is of the form

$$y = a_0 + a_1x, \tag{3.8}$$

where a_0 and a_1 are real numbers with $a_1 \neq 0$. The graph of equation (3.8) is a straight line. When the constant a_1 is denoted by the "slope" m and the constant a_0 is denoted by the "y-intercept" b, the equation (3.8) is referred to as the

$$\boxed{\text{Slope-intercept equation of a line: } y = m \cdot x + b.} \tag{3.9}$$

To investigate the terms "slope" and "y-intercept," consider the graph of the line l given by (3.9):

$$l = \{(x, y) \mid y = mx + b\}.$$

Assume (x_1, y_1) and (x_2, y_2) are two distinct points on the line l as depicted in Fig. 3.6. Then the two equations

$$y_1 = m \cdot x_1 + b \qquad \text{and} \qquad y_2 = m \cdot x_2 + b$$

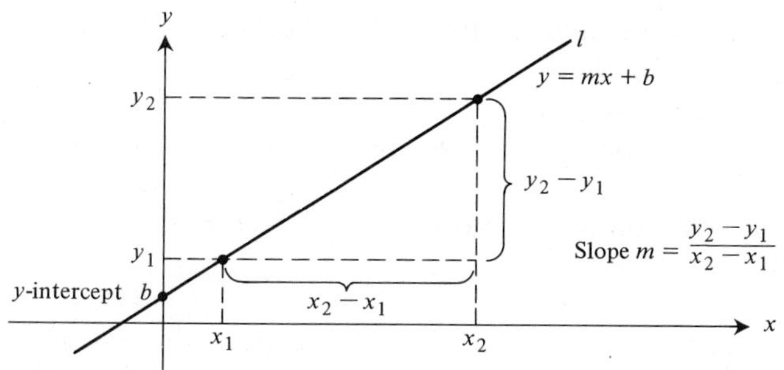

Figure 3.6

may be subtracted to yield the equation

$$y_2 - y_1 = (mx_2 + b) - (mx_1 + b) = m(x_2 - x_1).$$

Dividing by $x_2 - x_1$ (which does not equal zero since the points are distinct) results in the equation

$$\frac{y_2 - y_1}{x_2 - x_1} = m.$$

Since the right side of this equation is a constant, it does not depend on the choice of the points (x_1, y_1) and (x_2, y_2). This constant describes the *slope* or incline of the line l; it indicates the change in the y-value due to a unit change in the x-value. The number b is called the *y-intercept* since the line l crosses the y-axis at the point $(0, b)$.

Since the graph of a polynomial of degree one is a straight line, polynomial functions of degree one are called *linear functions*. Each linear function is uniquely determined by any two points on its graph. If (x_1, y_1) and (x_2, y_2) are two points on the graph of a line, then the slope m and y-intercept b are given by

$$m = \frac{y_2 - y_1}{x_2 - x_1} \quad \text{and} \quad b = y_1 - m_1 x_1.$$

The equation of the line is therefore given by

$$y = \left(\frac{y_2 - y_1}{x_2 - x_1}\right)x + \left[y_1 - \left(\frac{y_2 - y_1}{x_2 - x_1}\right)x_1\right].$$

Rearranging this equation we obtain the

Two-point equation of a line: $y - y_1 = \dfrac{y_2 - y_1}{x_2 - x_1}(x - x_1).$

(3.10)

When the slope of a line m and one point on the line (x_1, y_1) are known, the equation may be expressed in the

$$\boxed{\text{Point-slope equation of a line: } y = m(x - x_1) + y_1.} \tag{3.11}$$

The point-slope equation of a line will prove to be the most useful form in later sections of the text.

Example 3.1 The equation of a line may be determined from two bits of information: (a) two points, (b) a point and a slope, (c) a point and a y-intercept, or (d) the slope and a y-intercept.

a) The equation of the line through the points (1, 3) and (−1, 2) is given by formula (3.10) as

$$y - 3 = \frac{2-3}{(-1)-1}(x-1) \qquad \text{or} \qquad y - 3 = \tfrac{1}{2}(x-1).$$

b) The equation of the line with slope $m = \frac{1}{2}$ passing through the point $(\frac{1}{2}, -2)$ is given by formula (3.11) as

$$y = \tfrac{1}{2}(x - \tfrac{1}{2}) - 2.$$

c) The equation of the line passing through the point (3, −1) with y-intercept $b = 1$ is given by equation (3.10) using the point (0, 1) as a second point,

$$y + 1 = \frac{1-(-1)}{0-3}(x-3) \qquad \text{or} \qquad y + 1 = -(\tfrac{2}{3})(x-3)$$

d) The equation of the line with slope $m = 4$ and y-intercept $b = -2$ is given by equation (3.9) as $y = 4x - 2$.

These lines are sketched in Fig. 3.7.

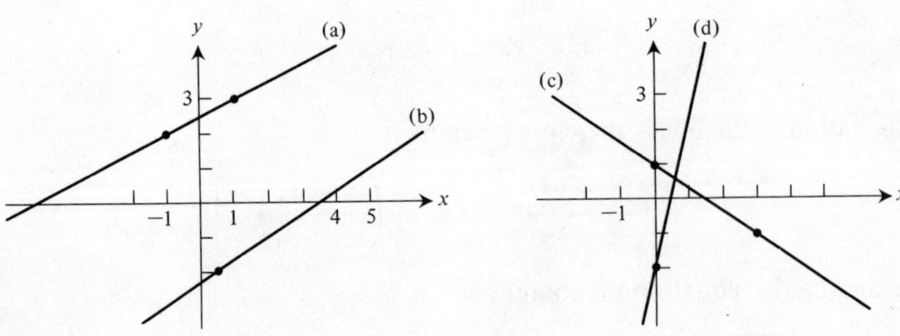

Figure 3.7

The importance and utility of linear graphs in describing biological and physical phenomena cannot be overstated. The utility of linear models—i.e., relationships between two components of an experiment or observation whose graphs are lines—lies in (i) the simplicity of the equation of a line and (ii) the ease with which the human eye can perceive a line which best fits a set of data points. The equation of a line is easily determined from its graph since any two points on its graph will uniquely determine the line and hence the equation of the line.

Example 3.2

a) The Fahrenheit and Celsius temperature scales are related by a linear function. To convert from Fahrenheit to Celsius, we express the temperature C° as a linear function of the temperature F°. We know two points on this line, the boiling temperatures of water (F° = 212 and C° = 100) and the freezing temperatures of water (F° = 32 and C° = 0). Thus the two-point equation (3.10) provides the equation of the line sketched in Fig. 3.8(a):

$$\text{C}^\circ - 0 = \frac{100 - 0}{212 - 32}(\text{F}^\circ - 32) \qquad \text{or} \qquad \text{C}^\circ = \tfrac{5}{9}(\text{F}^\circ - 32).$$

b) In chemistry it is often necessary to determine the x-intercept of a line. A frequently encountered linear relation is one with slope $m = K/V$ (K and V constants) and y-intercept $b = 1/V$. What is the x-intercept of this line? We can determine the x-intercept by substituting the point $(x_0, 0)$ into the slope intercept equation of the line: $y = (K/V)x + (1/V)$. Solving the equation $0 = (K/V)x_0 + (1/V)$ for x_0 gives the x-intercept $x_0 = -1/K$. This line is sketched in Fig. 3.8(b).

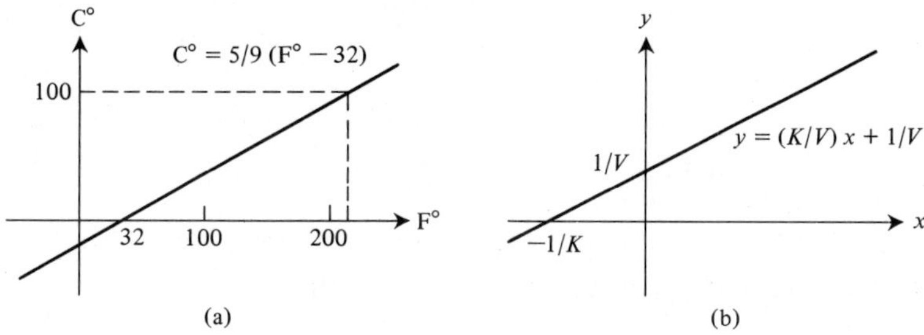

Figure 3.8

Polynomials of Degree Two

The general form of a second-degree polynomial is given by equation (3.6) with $n = 2$. Explicitly, the equation has the form

$$y = a_0 + a_1 x + a_2 x^2. \tag{3.12}$$

As remarked at the beginning of this section, polynomials are simply sums of multiples of power functions. For example, the equation (3.12) can be expressed using the power functions given in (3.3) as

$$y = a_0 f_0(x) + a_1 f_1(x) + a_2 f_2(x).$$

The power function $f_2(x) = x^2$ is in fact a special quadratic function having the form (3.12) with $a_0 = 0$, $a_1 = 0$, and $a_2 = 1$. One of the pleasant properties of quadratic functions is that the graph of any quadratic function has essentially the same basic form as the graph of the function $y = x^2$. To clarify what is meant by "essentially the same basic form," we rearrange the equation (3.12) by using a process called "completing the square." The end result is that the equation (3.12) may be expressed in the form

$$y = a(x - b)^2 + c, \tag{3.13}$$

where $a = a_2$, $b = -a_1/2a_2$, and $c = a_0 - (a_1{}^2/4a_2)$. For example,

$$y = 1 + 2x + 3x^2 \quad \text{is the same as} \quad y = 3(x + \tfrac{1}{3})^2 + \tfrac{2}{3};$$

and

$$y = 2 - x + 5x^2 \quad \text{is the same as} \quad y = 5(x - \tfrac{1}{10})^2 + \tfrac{39}{20}.$$

The graph of the equation (3.13) can be obtained from the graph of $y = x^2$ in three steps.

Step 1 Graph the function $y = ax^2$ by multiplying each y-coordinate on the graph of $y = x^2$ by a units. This has two effects. If $|a| < 1$, the graph becomes flatter, and if $|a| > 1$, it becomes steeper. If $a > 0$, it leaves the graph "opening up"; and, if $a < 0$, it inverts the graph so that it "opens down."

Step 2 Shift the graph of $y = ax^2$ horizontally b units. The graph should be shifted to the right if $b > 0$ and to the left if $b < 0$. This gives the graph of $y = a(x - b)^2$.

Step 3 Shift the graph obtained in step 2 vertically c units. The shift will be up if $c > 0$, down if $c < 0$. The result will be the graph of $y = a(x - b)^2 + c$.

Example 3.3 The three steps for graphing a quadratic are illustrated in Fig. 3.9 for the functions $y = \frac{1}{2}(x - 3)^2 + 1$ and $y = -2(x + 1)^2 - 2$.

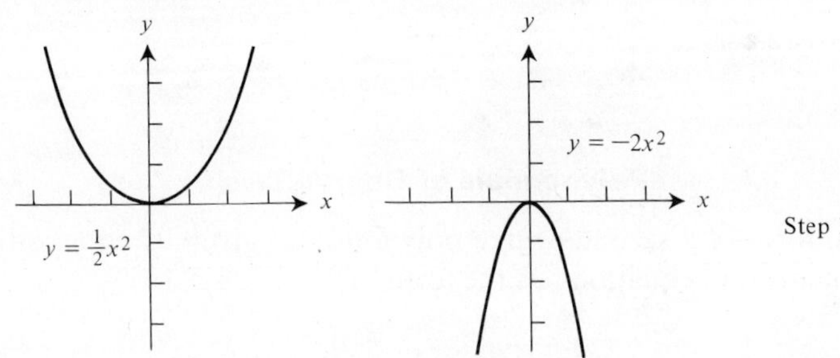

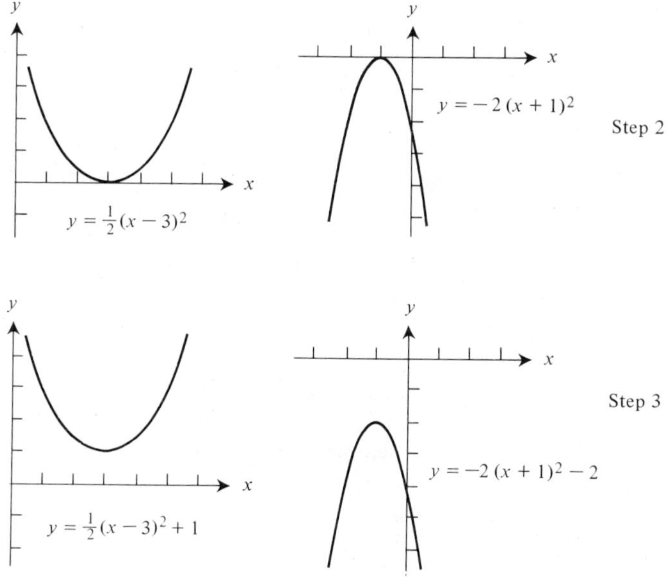

Figure 3.9

◀

The second-degree polynomial functions are referred to as quadratic or parabolic functions. The graph of a second-degree polynomial function is called a parabola. As the equation of a quadratic involves three parameters, three points on its graph are necessary to completely determine a quadratic. The general determination of the coefficients of a quadratic polynomial from points on its graph is complicated. However, if the graph of a parabola crosses the x-axis at points x_1 and x_2, then the equation of the parabola has the "factored form"

$$y = k(x - x_1)(x - x_2).$$

The constant k can be determined from any other point on the graph.

Example 3.4 The parabola passing through the points (2, 0), (3, 0), and (4, 1) is given by an equation of the form

$$y = k(x - 2)(x - 3).$$

The constant k is determined by substituting the point (4, 1) and solving:

$$1 = k(4 - 2)(4 - 3) = k \cdot 2$$

or

$$k = \tfrac{1}{2}.$$

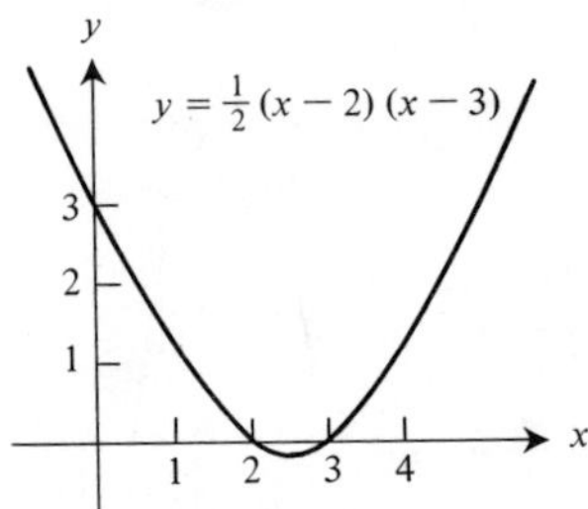

Figure 3.10

The graph of this parabola is given in Fig. 3.10. The equation of this graph is $y = \frac{1}{2}(x - 2)(x - 3)$. Put into the form (3.13) this equation is $y = \frac{1}{2}(x - \frac{5}{2})^2 - \frac{1}{8}$. Note this tells us that the graph is symmetric about $x_0 = \frac{5}{2}$, that it is opening up since the coefficient $\frac{1}{2}$ of the square term is positive, and that it has the value $y_0 = -\frac{1}{8}$ as its least y value at $x_0 = \frac{5}{2}$. ◀

Example 3.5 *Chemical kinetics*

Consider a undirectional bimolecular chemical reaction in which two compounds A and B combine to form a new compound X. Symbolically, this is expressed

$$A + B \rightarrow X.$$

This is the simplest type of second-order chemical reaction—i.e., a reaction in which the rate of formation of the compound X is given by a quadratic function. Using standard chemistry notation, let the initial concentration of the compounds A and B be denoted by $[A]_0$ and $[B]_0$, respectively, and let the concentration of X be denoted by $[X]$. The function R, which gives the rate of formation of X, has the form

$$R([X]) = k([A]_0 - [X])([B]_0 - [X]). \tag{3.14}$$

In equation (3.14) k is a proportionality constant and the terms $([A]_0 - [X])$ and $([B]_0 - [X])$ represent the concentration of the A and B compounds in the reaction mixture when the concentration of X has reached the level $[X]$. For instance, assume $k = 2$, $[A]_0 = 0.5$, $[B]_0 = 0.6$. Then equation (3.14) becomes

$$R([X]) = 2(0.5 - [X])(0.6 - [X]).$$

Expanding this equation, we obtain an equation of the general form (3.12),

$$R([X]) = 2[X]^2 - 2.2[X] + 0.6.$$

Completing the square, we can express this equation as

$$R([X]) = 2([X] - 0.55)^2 + (-0.005).$$

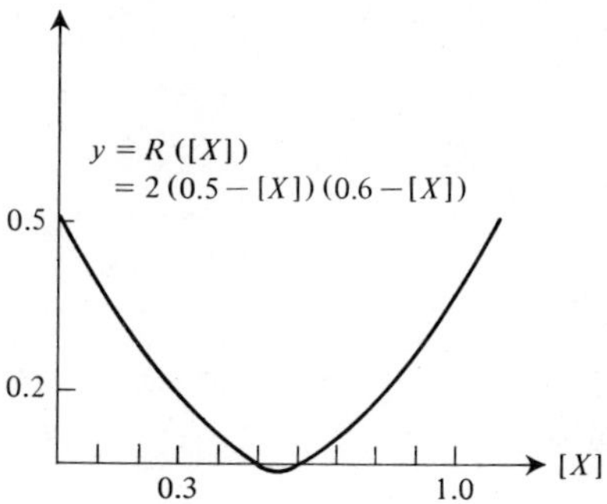

Figure 3.11

The graph of this equation is given in Fig. 3.11. Although the natural domain of the function R is the set of all real numbers, chemically, the concentration $[X]$ is limited to numbers between zero and one. ◀

Polynomials of Degree Three

A polynomial function of degree three is called a *cubic function* and its general form is given by the equation

$$y = a_0 + a_1x + a_2x^2 + a_3x^3. \tag{3.15}$$

As in the case of the quadratic polynomial, the term of highest degree, a_3x^3, dominates the function for large values of x. Thus the behavior and general characteristics of the graph of the equation (3.15) are similar to those of the function $f_3(x) = x^3$. However, unlike the quadratic function, the graph of the equation (3.15) cannot always be obtained through horizontal and vertical shifts of the graph of $y = x^3$. The reason for this is that the general cubic cannot be rearranged in the form of a cubic term plus a constant. The process of "completing the cube" results in a cubed term plus a linear term. The equation for completing a cube is

$$a_3x^3 + a_2x^2 + a_1x + a_0 = A(x - B)^3 + Cx + D. \tag{3.16}$$

where

$$A = a_3, \qquad B = \frac{-a_2}{3a_3},$$

$$C = a_1 - \frac{a_2{}^2}{3a_3}, \qquad D = a_0 - \frac{a_2{}^3}{27a_3{}^2}.$$

Thus the graph of a cubic polynomial is the sum of two graphs—one given by $y = A(x - B)^3$, which is obtained by shifting B units horizontally the graph of $y = Ax^3$, and one given by the line $y = Cx + D$. The net result is that the graph

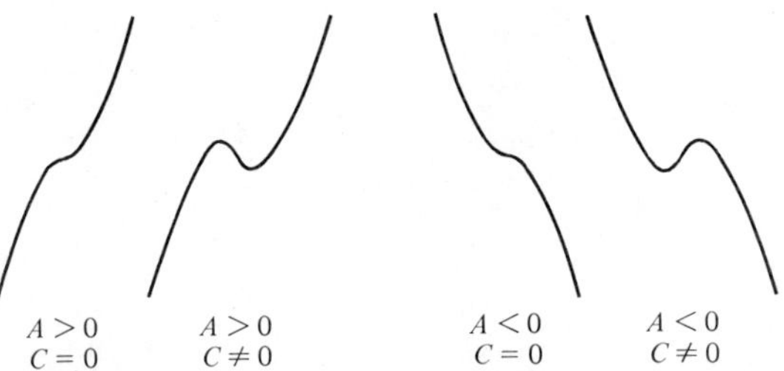

Figure 3.12

of the equation (3.15) will have one of the four forms indicated in Fig. 3.12. To sketch the graph of a particular cubic equation at this stage, you should plot several points, keeping in mind that the graph will have one of the four forms given in Fig. 3.12. Later, with the aid of calculus, we will be able to quickly determine the points where the graph of a cubic equation changes form.

Example 3.6 Two examples of cubic polynomials and their graphs are indicated together with representative points in Fig. 3.13.

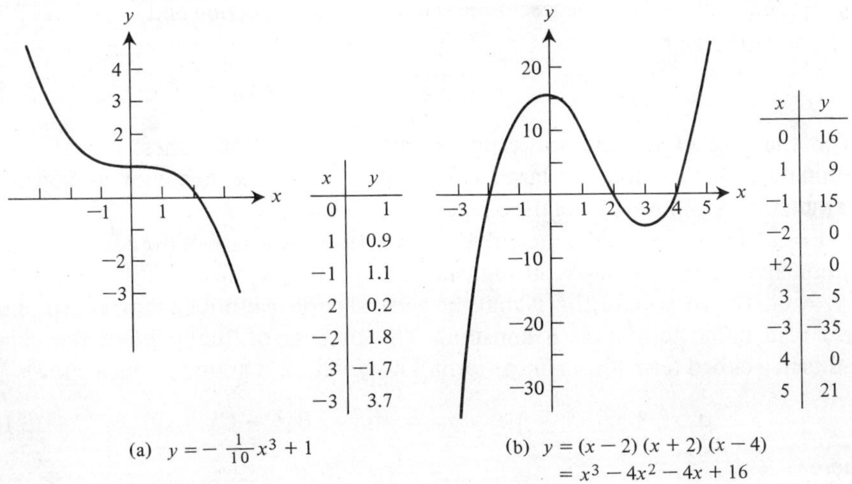

x	y
0	1
1	0.9
−1	1.1
2	0.2
−2	1.8
3	−1.7
−3	3.7

(a) $y = -\frac{1}{10}x^3 + 1$

x	y
0	16
1	9
−1	15
−2	0
+2	0
3	− 5
−3	−35
4	0
5	21

(b) $y = (x-2)(x+2)(x-4)$
$= x^3 - 4x^2 - 4x + 16$

Figure 3.13

◀

Example 3.7 Cubic equations frequently occur when relationships are established between linear measures and volumes or masses.

a) The simplest examples of such relationships are given by the equations expressing the volume of a cube as a function of an edge length, $V = s^3$, and the volume of a sphere as a function of its radius, $V = \frac{4}{3}\pi r^3$.

b) The total weight of a meal worm larva is related to the breadth of its head by a cubic relationship. As the larva grows its length and thickness remain proportional to its head breadth, b. Consequently, its weight, W, satisfies the relation $W = kb^3$ for a constant k.

c) In a study of the effects of injecting sex hormones into capons (young male chickens), the comb-growth of a capon was found to be a cubic function of the amount of hormone. The equation had the form $y = \alpha + \beta x^2 - \gamma x^3$, where α, β, and γ (read "alpha, beta, and gamma") represent particular positive constants, x the amount of injected hormones, and y the length of the comb.

d) The level of the hormone progesterone varies over the sexual cycle of an animal. Figure 3.14 indicates the level of progesterone over one cycle. As this curve decreases, increases, and then decreases again, the simplest function which would describe the progesterone level P as a function of the days d is a third-degree equation of the form

$$P = a_0 + a_1 d + a_2 d^2 + a_3 d^3.$$

This equation would be valid only for $0 \leq d \leq 30$.

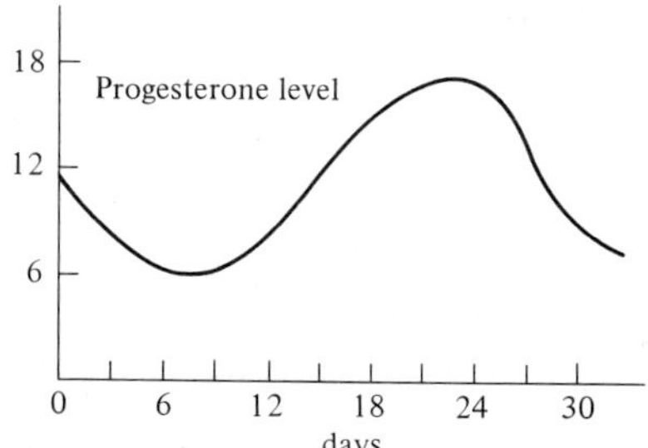

Figure 3.14

◀

Polynomials of Higher Degree

We have successively considered polynomial functions of degrees zero, one, two, and three. As the degree of the polynomial function increases, the complexity of the function also increases. A polynomial of degree n involves $n + 1$ constants $a_0, a_1, \ldots, a_n$ and its graph will be uniquely determined by $n + 1$ points. As the degree of a polynomial increases so does the degree of difficulty associated with graphing the polynomial. Even in the case $n = 3$ of a cubic polynomial, the graphing technique suggested above was not as explicit as that for quadratic functions. We would expect to encounter even greater difficulties when trying to graph a polynomial of degree four or greater.

A natural question to ask at this point is the following:

If we can graph polynomials of degrees 1, 2, and 3 and cannot easily graph polynomials of higher degrees, why should we not just use lines, parabolas, and cubics and ignore higher-degree curves?

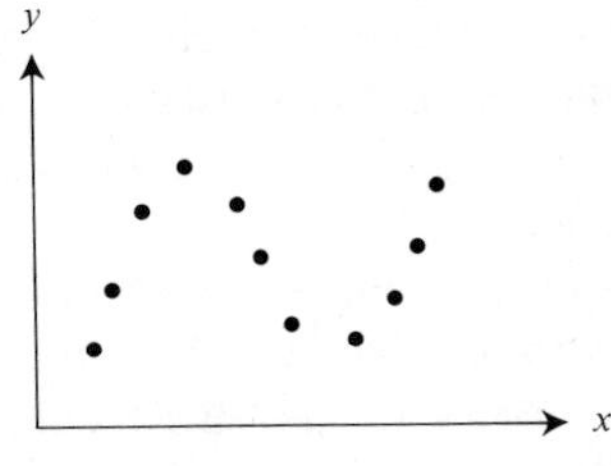

Figure 3.15

This question has two answers. First the choice of which function we use to describe a particular relation is not a free choice. If you wish to describe a relation that arises from an experiment, the function used must accurately reflect the data. For instance, the data given in Fig. 3.15 could not be adequately described by either a linear or a quadratic function. It would require a polynomial of degree three or higher. A cardinal rule is: *Always choose the simplest function that will adequately describe the relationship.* Therefore we would try to describe the function depicted in Fig. 3.15 by a cubic equation. If the data looked like that represented in Fig. 3.16, we would be forced to look for a polynomial function of degree four, such as $v = k(u - 1)^2(u - 5)^2$.

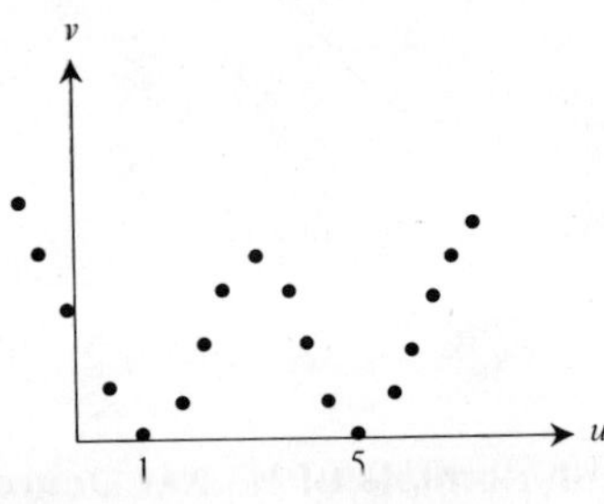

Figure 3.16

The second answer to the question is that whenever possible we do restrict ourselves to simple polynomials. In fact the study of calculus is frequently referred to as the study of straight lines. In many situations we describe a function by approximating it over small intervals by linear, quadratic, or cubic polynomials. See, for example, the relation described in Example 3.7(d).

Sometimes you may have to sketch the graph of a polynomial of a degree greater than three. Later, when we have developed the tools of calculus, this task can be made much easier. For the present we make a heuristic observation that may aid you if you need to graph higher-order equations. We observe that each time the degree of the equation is increased by one, the number of possible "turns" in the graph of the polynomial also increases by one. (See Fig. 3.17.)

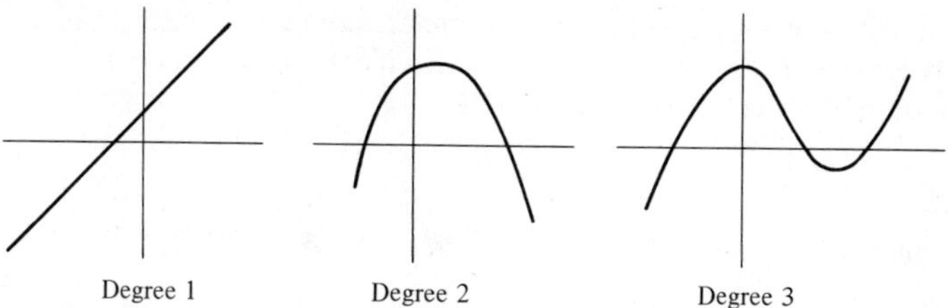

Figure 3.17

Thus for a fourth-degree polynomial, $y = a_0 + a_1x + \cdots + a_4x^4$, the possible forms of the graph for $a_4 > 0$ would be as shown in Fig. 3.18. These should be compared with the graph of the cubic polynomial shown in Fig. 3.12.

Figure 3.18

EXERCISE SET 3

3.1 Show that the function $f_2(x) = x^2$ is monotone increasing on the interval $(0, \infty)$.

3.2 Show that the function $f_2(x) = x^2$ is neither monotone increasing nor monotone decreasing on the interval $(-5, 5)$.

3.3 Show that $f_2(x) = x^2$ is an even function.

3.4 Show that $f_3(x) = x^3$ is an odd function.

3.5 Complete the graph of f for negative values of x given that (1) f is an even function and (2) f is an odd function.

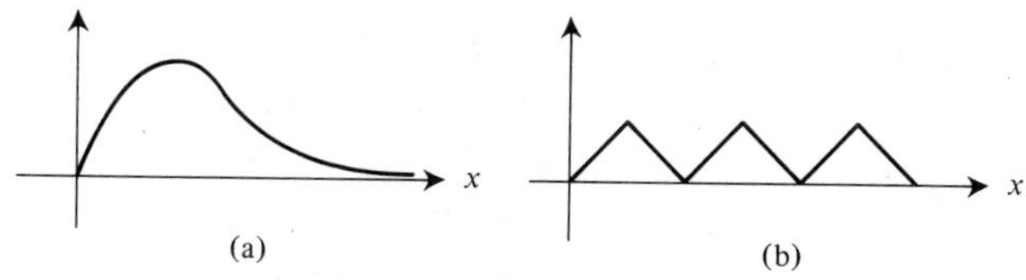

3.6 To examine how multiplication of a function by a constant either stretches or contracts the graph of a function, sketch on the same coordinate system the graphs of the functions f, $2 \cdot f$, $3 \cdot f$, $\frac{1}{2} \cdot f$, and $-2 \cdot f$, where the function f is given by the following.

a) $f(x) = x$ b) $f(x) = x^2$, c) $f(x) = x^3$

3.7 Determine the degree of the following polynomials.

a) $y = 2(x + 3)$ b) $y = (x + 3)(x - 2)$ c) $y = x - \frac{1}{2}(x + 2)$
d) $y = x^5 - 2x^2 + 5$ e) $y = 5x + \frac{1}{2}(1 - 10x)x$ f) $y = x^2 - 3x^3 + x$

3.8 Each of the following pairs of points determines a unique line which passes through them. (1) Determine the slope of the line; (2) determine the y-intercept of the line; (3) give the equation of the line and sketch its graph.

a) $(1, 0), (2, 5)$ b) $(3, -2), (-2, 3)$ c) $(5, 1), (6, 1)$
d) $(2, 2), (4, 1)$ e) $(2, 2), (4, 1)$ f) $(2.1, -2), (0.1, 3.4)$

3.9 Determine the equation of each of these lines.

a) The line passing through $(3, 3)$ with slope 6
b) The line with slope -3 and y-intercept 6
c) The line passing through $(3, 3)$ with slope $-\frac{1}{6}$
d) The line with slope 3 and y-intercept 6
e) The line with slope 5 passing through $(-2, 1.5)$
f) The line passing through $(-2.1, 3.4)$ with slope 0.7
g) The line with slope $\frac{3}{8}$ and x-intercept 4
h) The line with slope -0.05 and y-intercept 0.65

3.10 Sketch the lines of Exercise 3.9 on the same coordinate system.

3.11 If y is linearly related to x and z is linearly related to y, show that z is linearly related to x.

3.12 One characteristic of acute myeloblastic leukemia is the overproduction in the bone marrow of myeloblast cells, which eventually enter the blood supply. A model of the rate of production of myeloblast cells involves two parameters which are functions of the number N of cells already produced. Let these parameters be denoted by α and β. The relationship between α and β determines the growth rate.

If $\alpha > \beta$, there is overproduction and N increases.

If $\alpha < \beta$, there is underproduction and N decreases.

If $\alpha = \beta$, the corresponding population is the equilibrium population $\overline{N}$ at which the population of cells remains constant. Determine the equilibrium population $\overline{N}$, assuming that α and β are linear functions of N given by

$$\alpha(N) = -10^2N + 10^{14},$$

$$\beta(N) = -10N + 10^{13}.$$

3.13 If the slopes of the lines l_1, l_2, l_3, and l_4 are m_1, m_2, m_3, and m_4, respectively, how are the numbers m_1, m_2, m_3, and m_4 related?

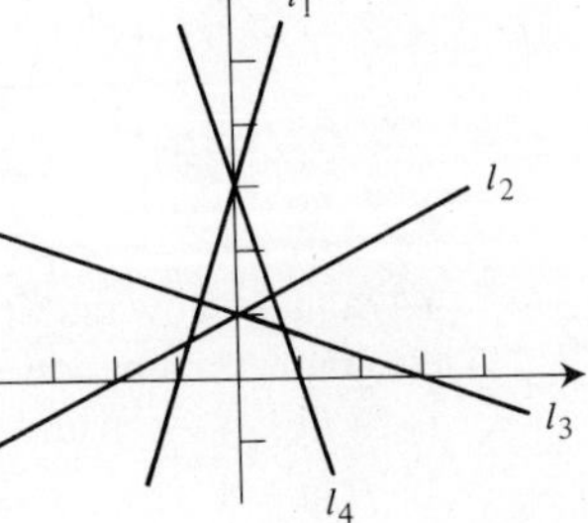

3.14 On the same axes sketch the graphs of the following.

a) $y = 4x^2$ b) $y = \frac{1}{4}x^2$
c) $y = \frac{1}{4}(x - 2)^2$ d) $y = \frac{1}{4}(x - 2)^2 + 1$

3.15 On the same axes sketch the graphs of the following.

a) $y = -x^2$ b) $y = -(x + 3)^2$
c) $y = -(x - 3)^2$ d) $y = -(x - 3)^2 + 2$

3.16 Express each of the following quadratics in the form of equation (3.13) and sketch its graph.

a) $y = x^2 - 6x + 10$ b) $y = x^2 + 2x + 1$
c) $y = (x - 3)(x - 1)$ d) $y = 3 - x^2$
e) $y = (2x + 1)(-x)/4$ f) $5y = x + 2 - x^2$
g) $x + y = x^2$ h) $y = (x - 2)(5 - x)$

3.17 For each of the following cubic equations do the following: (1) express the equation in the form on the right of equation (3.16), and (2) by plotting points, sketch its graph.

a) $y = x^3 - 1$ b) $y = (x - 1)^3$
c) $y = x^3 - 2x + 3$ d) $y = x^3 - x^2 + 2$
e) $y = -x^3 + x - x^2$ f) $y = -2x^3 + 6x^2 - 4$

3.18 Assume that the growth rate of a plant, $r(t)$, is a quadratic function of the number of days, t, since the seed was sown. If the plant is not growing at seeding and ceases to grow after 40 days, what is the equation for $r(t)$ if the plant's maximum growth rate is $r_{max} = 4$?

3.19 The following graph depicts three different sets of data. The data indicated by □'s and O's are scaled by the bottom and left scale, whereas the data given by × 's have the bottom and right scale. Assume that the data represent points on the graph of polynomial functions. Consider first the O's, then the × 's, and then the □'s.

a) Indicate the smallest possible degree of the polynomial which could represent the data.
b) Sketch a smooth curve which would approximate the graph of the polynomial represented by the data.
c) Using your sketched graphs and the data, give an equation which might reasonably describe the polynomials represented by the data.

Note: This is a subjective problem as you must *personally interpret* the coordinates of the data points, *sketch* the graphs, and determine the appropriate equations. Consequently, your results may differ slightly from another person's results.

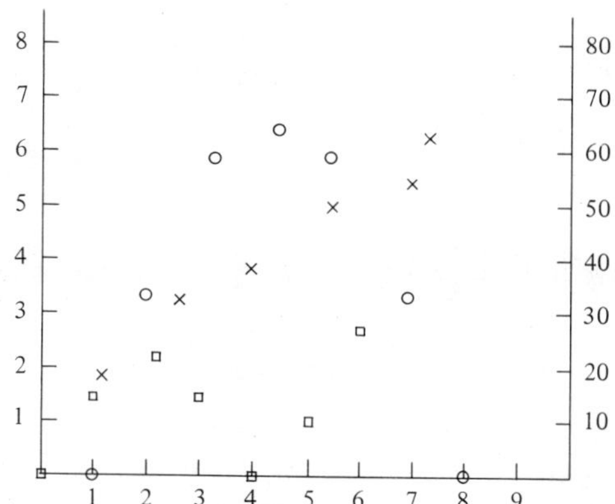

3.20 A population suffers from a contagious disease. Let the total size of the population be N and let the number of infected individuals on day n be $A(n)$.

a) If the number of noninfected individuals on day n is $S(n)$, how is $S(n)$ related to $A(n)$?
b) If there is a constant infection rate so that $A(n + 1) = A(n) + 5$, how many individuals will be infected after 12 days assuming that $A(0) = 2$?
c) Assuming that the number of new infected individuals is a constant proportion of the number of noninfected individuals, give an equation for the change in the number of infected individuals $\Delta(n) = A(n + 1) - A(n)$.

d) A more realistic model of the spread of an infectious disease would account for the number of individuals who recover from the disease. For a disease that does not produce immunity, the recovered individuals return to the pool of susceptible individuals, $S(n)$. Assume that the disease lasts for four days, the total population is $N = 100$, $A(0) = 2$, and the number of new infectives at day n is given by

$$\Delta(n) = (0.1) \cdot (S(n))$$

(rounded off to the nearest integer). Then the number of infectives is given by

$$A(n) = A(n-1) + \Delta(n-1) - \Delta(n-5).$$

Assuming no one was infected before day zero, make up a table indicating the following information for $n = 0, 1, 2, \ldots, 20$:

Day	Infectives	Susceptibles	New infectives
n	$A(n)$	$S(n)$	$\Delta(n)$

3.21 Assume that A and B combine to form X in a unidirectional biomolecular chemical reaction as described in Example 3.5. Assume that the kinetic constant k is 0.5 and the initial concentrations of A and B are $[A]_0 = 0.3$, $[B]_0 = 0.7$.

a) What is the equation for the rate of formation of X, $R([X])$?
b) When will the rate of formation of X be zero?
c) What is the maximum rate of formation of X and when does it occur?

SECTION 4

COMPOSITION OF FUNCTIONS

4.1 INTRODUCTION

One of the most basic techniques used in scientific investigation is to systematically analyze an event or process by separating the process into several subprocesses and analyzing each subprocess. The basic premise of this technique is that the overall process can be described by describing each subprocess and then describing how the subprocesses fit together. The simplest type of process described in this way is one where the component subprocesses combine in a sequential fashion to yield the overall process. In this case, the overall process is said to be a "composition" of its component processes. The mathematical description of a biological process is usually expressed by a function relating the initial and terminal constituents of the process. When a process is the "composition" of two or more subprocesses, the function modeling the process is obtained by sequentially combining the functions which describe the individual subprocesses. In this situation the function describing the overall process is a "composition" of the functions describing the subprocesses. The following examples illustrate the description of biological relationships as sequential compositions of intermediate relationships.

Example 4.1 The process of blood coagulation involves the sequential activation of many blood proteins called "clotting factors." A simplified model of this process is described by the following diagram.

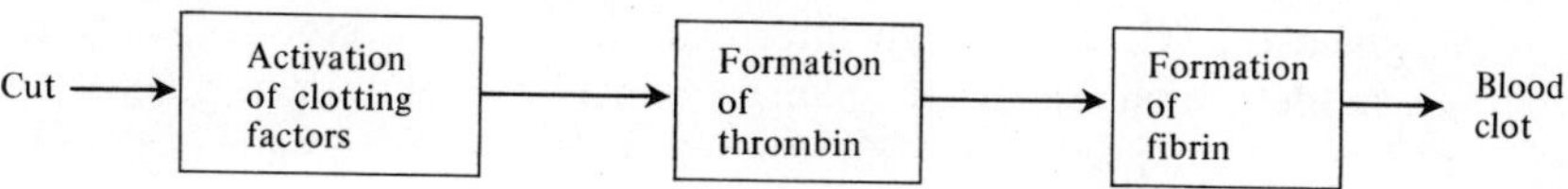

If you cut a finger, it initiates the activation of clotting factors, which leads to the formation of a chemical called thrombin. (The term thrombosis is used to describe blood clots in the circulatory system.) The generated thrombin then activates the formation of fibrin, a material which provides the structural stability of a blood clot. The amount of fibrin formed is a function of the amount of thrombin formed. Let F denote the amount of fibrin and let T denote the amount of thrombin. If we assume F is a power function of T, then

$$F = T^n, \qquad \text{for some positive integer } n.$$

Similarly, the amount of thrombin formed will be a function of the amount of the individual clotting factors which are activated. Classical hemophilia, or "bleeder's" disease, occurs when a person has an insufficient amount of one of the clotting factors. The most common form of hemophilia is due to a lack of the "Factor-VIII" clotting protein. If F_8 denotes the "activated" Factor-VIII level, the amount of thrombin generated will be proportional to a power function of F_8:

$$T = k \cdot F_8{}^m, \qquad m \text{ a positive integer and } k \text{ a constant.}$$

Consequently, the fibrin formed will also be a function of F_8.

$$F = (k \cdot F_8{}^m)^n \qquad \text{or} \qquad F = K \cdot F_8{}^{m \cdot n}, \qquad \text{where } K = k^n.$$

The graph of F as a function of F_8 is given in Fig. 4.1. Observe that very little fibrin is formed until the F_8 level exceeds the value 1. Thereafter, a very rapid increase in

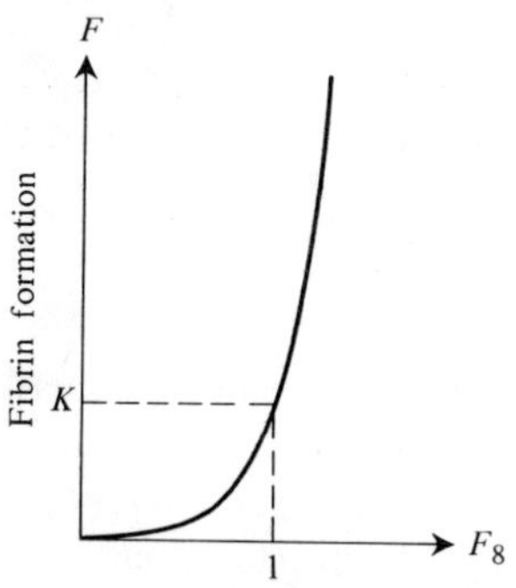

Figure 4.1

the fibrin levels occurs. This is necessary so that minor bruises will not lead to blood clots, but so that actual cuts will clot rapidly. ◀

Example 4.2 The hormone insulin affects most body tissue by stimulating the absorption of the sugar glucose into certain cells. The lack of sufficient insulin results in a diabetic state (*diabetes mellitus*). The basic laboratory test for diabetes is to check the glucose level of a patient's urine. The relation between insulin level and urine glucose level involves an intermediate step, the blood glucose level. Therefore a model which relates the changes in insulin level to urine glucose levels must involve two sequential steps—one step describes the blood glucose level as a function of insulin level; and a second step describes the urine glucose level as a function of blood glucose levels. ◀

Example 4.3 The sugar content of a sugar beet (*Beta vulgaris*) will vary according to the average temperature during its growing season. This variation is explained by the fact that the size of the beet root (dry-weight measurement) varies with the temperature, and that the sugar content is a function of the dry weight of the root. The dry weight must be used since the beet will vary considerably in water content after rain or irrigation. If w denotes dry weight, T denotes temperature, and S denotes sugar content, then w is a function of T and S is a function of w, hence S is a function of T. ◀

Example 4.4 The volume of red blood cells in an individual varies with the weight of the individual. A person's blood constitutes about 8 percent of his or her weight. The term "hematocrit" is used to describe the percentage content of red cells in a person's blood. Therefore, the volume occupied by an individual's red blood cells, R, may be determined from his or her weight, w, and hematocrit, h, as follows. Let V = volume of blood. Then,

$$V = 0.08w\left(\frac{\text{liter}}{\text{kilogram}}\right) \qquad \text{and} \qquad R = h \cdot V.$$

Therefore $R = h \cdot (0.08) \cdot w$. For instance, a person weighing 70 kilograms with a hematocrit $h = 45$ percent will have 2.52 liters of red cells in her or his circulation system since $R = (0.45)(0.08)(70) = 2.52$. ◀

Examples 4.1 and 4.4 illustrate the description of biological processes by functions and the sequential combination of the functions to arrive at an equation describing the complete process. These examples are straightforward in the sense that the composition of the equations describing the individual components amount to a simple substitution from one equation into the next equation. This is the fundamental concept of the composition of functions.

4.2 COMPOSITIONS OF FUNCTIONS: AN ALGEBRAIC APPROACH

In this section we examine the operation of composition, which, like the operations of addition and multiplication, combines two functions to form a third function. The addition and multiplication of functions are natural extensions of the operations of addition and multiplication of real numbers. The operation of composition of functions which we introduce in this section does not rely on the algebraic properties of the real numbers. Formally, given two functions f and g, we denote the composition of f with g using the symbol "$\circ$":

$$f \circ g \qquad \text{is read} \qquad \text{“}f \text{ composition } g.\text{”}$$

The function $f \circ g$ is evaluated by first evaluating $g(x)$ and then evaluating f at the number $g(x)$:

$$(f \circ g)(x) = f(g(x)). \tag{4.1}$$

Example 4.5 Consider the following illustrations of equation (4.1). In each case, the expression for $g(x)$ is substituted into the expression for $f(x)$ to obtain $(f \circ g)(x)$.

a) $f(x) = 2x + 1, g(x) = 3x;$
$(f \circ g)(x) = f(g(x)) = f(3x) = 2(3x) + 1 = 6x + 1.$

b) $f(x) = 3x, g(x) = 2x + 1;$
$(f \circ g)(x) = f(g(x)) = f(2x + 1) = 3(2x + 1) = 6x + 3.$

c) $f(x) = x - 3, g(x) = 1/x;\ \ (f \circ g)(x) = f(g(x)) = f(1/x) = 1/x - 3.$

d) $f(x) = 1/x, g(x) = x - 3;\ \ (f \circ g)(x) = f(g(x)) = f(x - 3) = 1/(x - 3).$

e) $f(x) = \sqrt{x}, g(x) = x - 4; (f \circ g)(x) = f(g(x)) = f(x - 4) = \sqrt{x - 4}.$

f) $f(x) = x - 4, g(x) = \sqrt{x};\ (f \circ g)(x) = f(g(x)) = f(\sqrt{x}) = \sqrt{x} - 4.$

To evaluate each of these composition functions at a specific x value, say $x_0 = 3$, we set $x = 3$ in the above equations. The results are:

a) $(f \circ g)(3) = 6 \cdot 3 + 1 = 19,$

b) $(f \circ g)(3) = 6 \cdot 3 + 3 = 21,$

c) $(f \circ g)(3) = \frac{1}{3} - 3 = -8/3,$

d) $(f \circ g)(x) = 1/(3 - 3) = 1/0$ (not defined!),

e) $(f \circ g)(3) = \sqrt{3 - 4} = \sqrt{-1}$ (not defined!),

f) $(f \circ g)(3) = \sqrt{3} - 4 \simeq -2.268.$

Examples (d) and (e) illustrate the fact that $(f \circ g)(x)$ may not be defined even though both $f(x)$ and $g(x)$ are defined. Also, observing that the functions f and g are reversed in examples (a) and (b), we see that the order in which two functions are "composed" can lead to different results. ◀

The observations made in Example 4.5 demonstrate that the domain of the composition function $f \circ g$ is not necessarily the same as the domain of either of the functions f or g.

Definition 4.1 Let f and g be two real functions. The *composition of f with g* is the function $f \circ g$ defined on the set

$$S = \{x \in \text{Domain}(g) | g(x) \in \text{Domain}(f)\}$$

by the equation

$$(f \circ g)(x) = f(g(x)) \qquad \textbf{Composition of } f \textbf{ with } g.$$

The composition $(f \circ g)(x_0)$ is evaluated sequentially by first evaluating $g(x_0)$ and then evaluating $f(x)$ with $x = g(x_0)$. Thus the Domain$(f \circ g) = S$ is a subset of Domain(g), and the Range$(f \circ g)$ is a subset of Range(f). Figure 4.2 depicts the composition $f \circ g$. The cross-marked regions denote the domain S of $f \circ g$, the image of the set S under g which is Domain$(f) \cap$ Range(g), and the Range$(f \circ g)$.

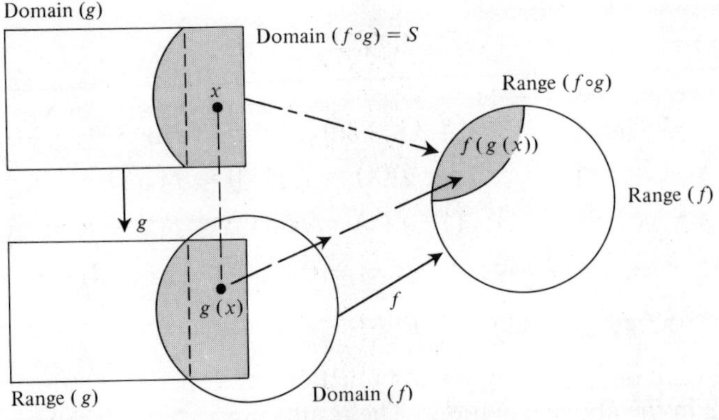

Figure 4.2

Example 4.6

a) Let $f = \{(1, 1), (2, 4), (3, 9)\}$ and $g = \{(1, 0), (2, 1), (3, 2)\}$. The domain of $f \circ g$ is $S = \{2, 3\}$. The composition function $f \circ g = \{(2, 1), (3, 4)\}$ and therefore the range of $f \circ g$ is the set $\{1, 4\}$.

b) Let $f(x) = \sqrt{x}$ and $g(x) = 3 - 2x$. Since the square root of a negative number is not a real number, Domain$(f) = \{x | x \geq 0\}$ and hence the domain of $f \circ g$ is the set

$$S = \{x | g(x) \geq 0\} = \{x | 3 - 2x \geq 0\} = \{x | x \leq \tfrac{3}{2}\}.$$

The function $(f \circ g)(x) = \sqrt{3 - 2x}$ and Range$(f \circ g) = [0, +\infty)$.

c) Let $f(x) = x/(x - 1)$ and $g(x) = x^2$. Division by zero is not permissible and thus the Domain$(f) = \{x \mid x \neq 1\}$. Consequently, the Domain$(f \circ g) = \{x \mid g(x) \neq 1\} = \{x \mid x^2 \neq 1\} = \{x \mid x \neq 1$ and $x \neq -1\}$. In this example, $(f \circ g)(x) = x^2/(x^2 - 1)$. ◀

Example 4.7 The amount of fish in a bay or lake is a function of the available free oxygen in the water. Let F denote the total biomass of fish in a given body of water. If O_2 denotes the available free oxygen level then

$$F(O_2) \equiv -A[O_2 - K]^2 + AK^2, \tag{4.2}$$

where K is the optimal level of free oxygen and A is a constant which depends on the size of the lake or bay. The oxygen in the water is generated by algae and plants such as seaweed. Thus the oxygen level is a function of the total biomass of algae, which we denote by x. The decay of algae consumes large amounts of oxygen and one model which reflects this is given by the following equation:

$$O_2(x) = -B(x - C)^2 + BC^2, \tag{4.3}$$

where C is the algae level which produces the maximum amount of oxygen and B is a constant which depends on the surface area of the lake or bay. The fish level F is thus a function of the algae level x. To obtain this function we take the composition $F \circ O_2$:

$$(F \circ O_2)(x) = F(O_2(x)) = -A[O_2(x) - K]^2 + AK^2,$$

or

$$F = -A[(-B(x - C)^2 + BC^2) - K]^2 + AK^2. \tag{4.4}$$

By squaring the terms, we see that F is a fourth degree polynomial function of x. It should also be noted that negative oxygen levels or biomass of fish are unreasonable biologically. Therefore the above equations are valid only if $F \geq 0$ and $O_2 \geq 0$:

$$\text{Domain}(F(O_2)) = \{O_2 \mid 0 \leq O_2 \leq 2K\},$$

and

$$\text{Domain}(O_2(x)) = \{x \mid 0 \leq x \leq 2c\}.$$

The domain of the composition $F \circ O_2$ is the set

$$S = \{x \mid 0 \leq x \leq 2c \text{ and } 0 \leq O_2(x) \leq 2K\}.$$

The set S will therefore depend on the parameters A, B, C, and K. ◀

The algebraic properties of the composition operation are similar to those of the addition and multiplication of functions with one major exception. The composition operation is not commutative. As a rule, $f \circ g$ does not equal $g \circ f$. For instance if $f(x) = x - 3$ and $g(x) = x^2$, then

$$(f \circ g)(x) = f(g(x)) = f(x^2) = x^2 - 3,$$

while

$$(g \circ f)(x) = g(f(x)) = g(x - 3) = (x - 3)^2.$$

The composition operation is associative and has an identity function

$$I(x) = x.$$

The identity function is characterized by the property that for any function f, for all $x \in \text{Domain}(f)$,

$$f \circ I(x) = f(x) \qquad \text{and} \qquad I \circ f(x) = f(x).$$

The algebraic property of having inverses corresponds to the important notion of reversibility. When two variables x and y are related by a function $y = f(x)$, it is assumed that y depends on x. Frequently in biological applications, it is desirable to reverse the relationship and express x as a function of y. This is easily accomplished when two variables are related by a linear relationship. For instance, assume the height, H, of a calf is related to age, A, during its first year by

$$H = (\tfrac{1}{12})A + 30.$$

To reverse this relationship, we solve for the age A:

$$A = 12H - 360.$$

However, when the relation between the variables is more complicated it may not be possible to reverse the relationship. For instance, the relationship (4.3) of Example 4.7 gives the dependence of oxygen levels on algae levels x. This relation is not reversible—that is, the algae level cannot be made a function of the oxygen level. The reason is that different algae levels correspond to the same oxygen levels.

In general, when we wish to reverse a functional relation between two variables, we must use the concept of an inverse function. The inverse of a function f with respect to the operation of composition is denoted by f^{-1}, read "f inverse," *not* "f to the minus one." The function f^{-1} can be used to reverse the relationship $y = f(x)$, i.e., to solve for x as a function of y. The formal definition of the inverse function rests on the property that f^{-1} essentially "undoes" the effect of the function f.

Definition 4.2 A function f is said to have an *inverse function* f^{-1} (with respect to composition) on a domain D if for all $x \in D$,

$$(f \circ f^{-1})(x) = x \qquad \text{and} \qquad (f^{-1} \circ f)(x) = x.$$

Thus,

$$\boxed{y = f(x) \qquad \text{is equivalent to} \qquad x = f^{-1}(y).} \tag{4.5}$$

Example 4.8 The equivalence (4.5) can be used to compute f^{-1} by solving $y = f(x)$ for x as a function of y, $f^{-1}(y)$. The symbol y is then replaced by x so that the argument of the function f^{-1} is x rather than y. We illustrate this technique and verify that the resulting function satisfies the conditions $(f \circ f^{-1})(x) = x$ and $(f^{-1} \circ f)(x) = x$.

a) If $f(x) = x + 6$, then solving $y = x + 6$ for x we have $x = y - 6$. Thus $f^{-1}(y) = y - 6$ and hence $f^{-1}(x) = x - 6$. To verify that this is the inverse of $f(x)$, we evaluate

$$(f \circ f^{-1})(x) = f(f^{-1}(x)) = f(x - 6) = (x - 6) + 6 = x,$$

and

$$(f^{-1} \circ f)(x) = f^{-1}(f(x)) = f^{-1}(x + 6) = (x + 6) - 6 = x.$$

b) If $g(x) = 2x$, then solving $y = 2x$ for x, we find $x = y/2$. Thus $g^{-1}(y) = y/2$ and $g^{-1}(x) = x/2$. To verify that this is indeed $g^{-1}(x)$, we evaluate

$$(g \circ g^{-1})(x) = g(g^{-1}(x)) = g(\tfrac{1}{2}x) = 2(\tfrac{1}{2}x) = x,$$

and

$$(g^{-1} \circ g)(x) = g^{-1}(g(x)) = g^{-1}(2x) = (\tfrac{1}{2})(2x) = x.$$

c) Let $h(x) = 2x + 6$; then solving $y = 2x + 6$ for x, we find $x = (y - 6)/2$. Thus $h^{-1}(y) = (y - 6)/2$ or $h^{-1}(x) = (x - 6)/2$. The verification that this gives h^{-1} is

$$(h \circ h^{-1})(x) = h(h^{-1}(x)) = h(\tfrac{1}{2}(x - 6)) = 2[\tfrac{1}{2}(x - 6)] + 6 = x,$$

and

$$(h^{-1} \circ h)(x) = h^{-1}(h(x)) = h^{-1}(2x + 6) = \tfrac{1}{2}((2x + 6) - 6) = x.$$

d) Let $f(x) = x^2 + 1$. Solving $y = x^2 + 1$ for x, we find $x = \pm\sqrt{y - 1}$. In order to make $f^{-1}(y)$ a function, we must pick only one of the two values $\sqrt{y - 1}$ or $-\sqrt{y - 1}$. Generally the positive value will be chosen. The function $f^{-1}(x) = \sqrt{x - 1}$ is only defined for $x \geq 1$. Therefore, on the domain $D = [1, +\infty)$, we find $f^{-1}(x) = \sqrt{x - 1}$, since for $x \in D$,

$$(f \circ f^{-1})(x) = f(f^{-1}(x)) = f(\sqrt{x - 1})$$
$$= (\sqrt{x - 1})^2 + 1 = x - 1 + 1 = x,$$

and

$$(f^{-1} \circ f)(x) = f^{-1}(f(x)) = f^{-1}(x^2 + 1) = \sqrt{(x^2 + 1) - 1} = \sqrt{x^2} = x.$$

◀

The function in part (c) of Example 4.8 is in fact the composition $h(x) = (f \circ g)(x)$, where $f(x)$ and $g(x)$ are the functions given in parts (a) and (b) of the same example. Furthermore, the function h^{-1} is also the composition of the two inverse functions g^{-1} and f^{-1}. But the order has been reversed, $h^{-1} = g^{-1} \circ f^{-1}$. That this is always the case when the individual inverses exist is the assertion of the next theorem.

Theorem 4.1 Assume $h = f \circ g$ and both f^{-1} and g^{-1} exist. Then h^{-1} also exists and

$$h^{-1} = (f \circ g)^{-1} = g^{-1} \circ f^{-1}.$$

Part of the hypothesis of Theorem 4.1 is the condition that f^{-1} and g^{-1} exist. Not all functions have an inverse with respect to composition. This may be seen by examining the following examples. Let the function $F = \{(1, 2), (3, 4), (2, 3)\}$. The function F has an inverse F^{-1} given by $F^{-1} = \{(2, 1), (4, 3), (3, 2)\}$. F and F^{-1} are depicted in Fig. 4.3.

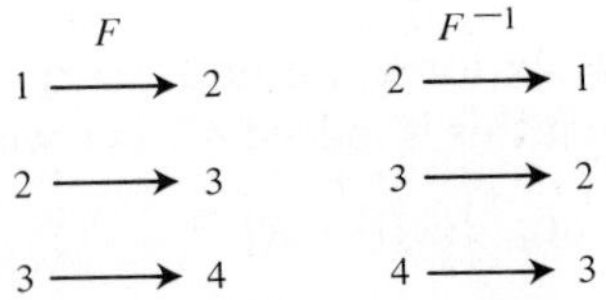

F: $1 \longrightarrow 2$, $2 \longrightarrow 3$, $3 \longrightarrow 4$

F^{-1}: $2 \longrightarrow 1$, $3 \longrightarrow 2$, $4 \longrightarrow 3$

Figure 4.3

However, consider the function $G = \{(1, 2), (2, 2), (3, 4)\}$. The function G does not have an inverse because it is not reversible. There is no way of deciding whether $G^{-1}(2)$ should be 1 or 2. $G^{-1}(2)$ cannot be both 1 and 2, for then G^{-1} would not be a function. This is illustrated in Fig. 4.4. The difficulty in constructing an inverse of the function G arises from the fact that the function G is not "one-to-one." G maps two numbers to one number; $G(1) = 2$ and $G(2) = 2$.

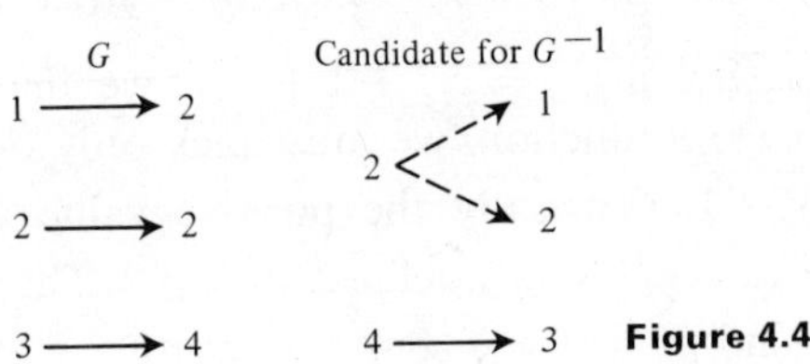

Figure 4.4

Definition 4.3 A function f is *one-to-one* if no element in the Range(f) is the image of two or more distinct elements in Domain(f), i.e., if $x_1, x_2 \in$ Domain(f) and $x_1 \neq x_2$, then $f(x_1) \neq f(x_2)$.

With this definition we can state a necessary and sufficient condition for a function f to have an inverse.

Theorem 4.2 A function f has an inverse if and only if it is one-to-one. If f^{-1} exists, then it may be expressed in the form

$$f^{-1} = \{(y, x) \mid y \in \text{Range}(f), f(x) = y\},$$

and the following statements are equivalent:

$$y = f(x) \qquad \text{and} \qquad f^{-1}(y) = x.$$

Theorem 4.2 is useful when we know that a function is one-to-one. The definition of monotone increasing and decreasing, Definition 3.3, immediately leads to the result that a function which is either monotone increasing or decreasing on an interval (a, b) is also one-to-one on the interval (a, b). Consequently, we have the following theorem.

Theorem 4.3 A function f has an inverse on an interval (a, b) if f is either monotone increasing or monotone decreasing on (a, b).

Example 4.9

a) Since the functions $y = x^n$ for $n = 1, 2, 3, \ldots$ are all monotone increasing on $[0, +\infty)$, they have inverses with respect to the composition on this interval.
b) Since the function $y = x^2$ is not one-to-one on its domain (for example, $f(2) = f(-2)$), it cannot have an inverse defined on all of R.
c) The power functions $y = x^n$ for n an odd natural number—i.e., $n = 1, 3, 5, 7, \ldots$ —are monotone increasing on $(-\infty, +\infty)$ and consequently have inverses on $(-\infty, +\infty)$.
d) The function

$$f(x) = \begin{cases} 2x + 1 & \text{if} \quad x < 0, \\ x^2 + 1 & \text{if} \quad x \geq 0, \end{cases}$$

is monotone increasing on $(-\infty, +\infty)$ and thus has an inverse f^{-1} defined on R. ◀

There is a simple procedure for determining geometrically from the graph of a function f whether f^{-1} exists and, if so, how to graph f^{-1}. First graph the function f on a square thin piece of paper. Then rotate the graph about the line $y = x$. This is accomplished physically by holding the paper by the corners in quadrants I and III and rotating the paper until you are looking at the back side of the page. Looking through the paper as it is held up to a strong light, you will observe that the x and y axes have changed position. The graph which you see through the paper, considering the horizontal axis (old y-axis) as the domain, will be the graph of f^{-1}. If this graph is not the graph of a function then the function f^{-1} does not exist.

A second geometric way to check whether a function has an inverse is to employ the horizontal-line test.

***Horizontal-line test*:** A function f has an inverse if and only if no horizontal line meets the graph of f more than once.

Example 4.10 In this example we indicate in Fig. 4.5 the graph of a function and how the graph would look after the rotating procedure previously discussed. Note that the horizontal and vertical axes have changed position.

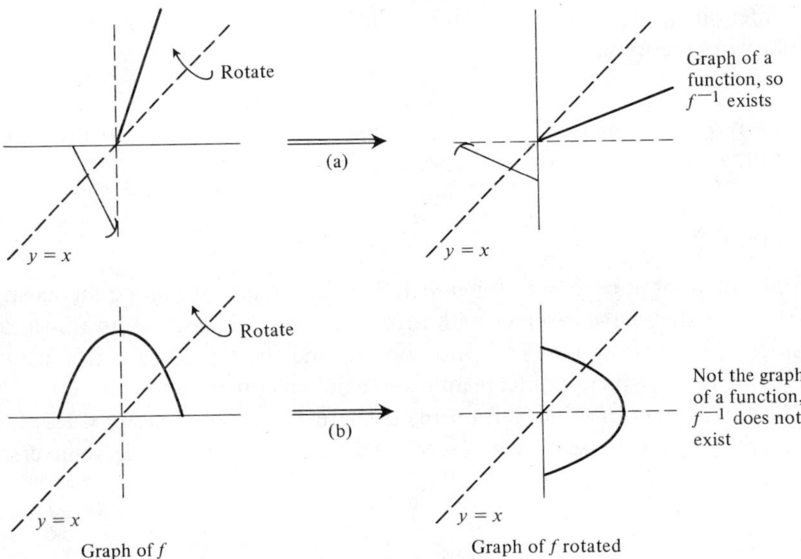

Figure 4.5

◀

4.3 ALLOMETRIC FUNCTIONS: $y = cx^k$

The analysis and description of biological processes is greatly facilitated if the variables representing the biological components of the process can be related by functional equations. One type of relationship which has a wide range of applications is the so called "allometric" relation. y is an allometric function of x if there exist constants c and k such that x and y satisfy the

$$\boxed{\textit{Allometric equation}: y = cx^k.} \tag{4.6}$$

The exponent k in this equation is not restricted to be an integer; indeed, it may take on any real value. After two examples illustrating the allometric equation, we will review the correct interpretation of x^k when k is a rational number.

Example 4.11 One frequent application of allometric functions is to indicate the relation between different dimensions or quantities of a plant which remains constant throughout the growth of the plant. An example is given by the following allometric relationships established for potatoes. If

W_F = fresh weight of a potato tuber,
N = number of cells per potato tuber,
S = starch content per cell,
P = protein content per cell,

then

$$N = (3.71 \times 10^6)W_F{}^{0.73},$$
$$S = (2.8 \times 10^{-8})W_F{}^{0.35},$$
$$P = (2.1 \times 10^{-9})W_F{}^{0.24}.$$

◀

Example 4.12 One measure of general physical fitness is given by the "basal metabolic rate" (BMR) which is defined as the rate of oxygen consumption during complete relaxation. For healthy animals, the BMR is related to the weight W of the animal by an allometric function

$$\text{BMR} = kW^r,$$

where k varies between species and r is between 0.67 and 0.75 depending on experimental conditions. ◀

The basic allometric function $y = x^k$ is a generalization of the power functions discussed in Section 2. Indeed, each power function $f_n(x) = x^n$ is an allometric function. While the definition of the function $y = x^k$ for an arbitrary real number k requires the introduction of limits, we can define $y = x^r$ where r is a rational number using the inverse and composition of the power functions $y = x^n, n = 1, 2, 3, \ldots$ We will define $y = x^r$ in three cases depending on the form of the rational number r:

Case 1 When r has the form $r = 1/n$.

Case 2 When r has the form $r = m/n$, where m and n are natural numbers with no common factors (i.e., m/n is in reduced form.)

Case 3 When r is a negative rational number $r = -m/n$.

Our discussion is motivated by the familiar "laws of exponents."

$$x^{a+b} = x^a \cdot x^b \qquad \text{and} \qquad (x^a)^b = x^{a \cdot b}$$

The basic technique is to define $y = x^{1/3}$ as the inverse with respect to composites of $y = x^3$. The term $y = x^{5/3}$ is defined as the composition of $y = x^5$ with $y = x^{1/3}$, giving $y = (x^{1/3})^5$; and $y = x^{-5/3}$ is defined as the reciprocal of $y = x^{5/3}$, $y = 1/x^{5/3}$. The general definition of x^r is as follows.

Case 1 If $r = 1/n$ and n is a natural number. The function $y = x^{1/n}$ is defined to be the inverse function (with respect to composition) of the power function $f_n(x) = x^n$. Thus, $x^{1/n} = f_n{}^{-1}(x)$. As noted in Example 4.9, the domain of the inverse function $f_n{}^{-1}(x)$ depends on the number n. Consequently,

$$\text{if } n \text{ is even,} \qquad \text{Domain}(x^{1/n}) = [0, +\infty),$$
$$\text{if } n \text{ is odd,} \qquad \text{Domain}(x^{1/n}) = (-\infty, +\infty).$$

The number $x^{1/n}$ is called the *nth root of* x. The restriction on the domain of $y = x^{1/n}$ extends to all natural numbers n the simple concept that the square root of a negative number does not exist (e.g., $(-4)^{1/2} \equiv \sqrt{-4}$ does not exist) but the cube root of a negative number does exist (e.g., $(-8)^{1/3} = -2$). By definition $(f^{-1} \circ f)(x) = x$ and $(f \circ f^{-1})(x) = x$. Therefore

$$(x^{1/n})^n = x \qquad \text{and} \qquad (x^n)^{1/n} = x.$$

Case 2 If $r = m/n$, where m and n are natural numbers with no common divisors. The function $y = x^{m/n}$ is defined as the composition of the two functions $f_m(x) = x^m$ and $f_{1/n}(x) = x^{1/n}$:

$$y = f_r(x) = x^{m/n} = (f_m \circ f_{1/n})(x) = (x^{1/n})^m.$$

Note that the domain of $y = x^{m/n}$ is therefore the domain of $y = x^{1/n}$. Hence $(-4)^{2/3}$ is defined but $(-4)^{3/2}$ is not defined.

Case 3 Negative exponent. Assume $r = -m/n$, where m and n are (positive) natural numbers with no common divisor. The function $y = x^r$ is defined to be the multiplicative inverse of the function $y = x^{m/n}$:

$$y = x^{-m/n} \equiv \frac{1}{x^{m/n}}.$$

The function $y = x^{-m/n}$ is not defined when $x = 0$. Therefore,

$$\text{if } n \text{ is even,} \qquad \text{Domain}(x^{-m/n}) = (0, +\infty);$$

$$\text{if } n \text{ is odd,} \qquad \text{Domain}(x^{-m/n}) = (-\infty, 0) \cup (0, +\infty).$$

Example 4.13 We evaluate $y = x^r$ for particular values of x and r.

a) To evaluate $y = 3^{5/2}$, we compute $3^{1/2}$ and raise this to the power 5.

$$3^{5/2} \simeq 15.59.$$

b) To evaluate $y = (-8)^{2/3}$, we first take the cube root of -8 and then square this number.

$$(-8)^{2/3} = (-2)^2 = 4.$$

c) The number $y = (-3)^{7/4}$ cannot be evaluated since $(-3)^{1/4}$ is not defined.
d) The number $y = 2^{-3}$ is $1/2^3 = \frac{1}{8}$.
e) The number $y = 4^{-3/2}$ is $1/(4^{3/2}) = \frac{1}{8}$.
f) The number $y = 5^{\sqrt{2}}$ cannot be evaluated using the method discussed above since $\sqrt{2}$ is not a rational number.
g) The number $y = 10^{0.32}$ can be evaluated since $0.32 = \frac{32}{100}$ or reduced to lowest terms $0.32 = \frac{8}{25}$. Thus, $y = (10^{1/25})^8 \simeq 2.09$. ◀

The number $(-4)^{2/3}$ is defined as $(-4)^{1/3}$ squared. It can be shown that this number, like $\sqrt{2}$, is an irrational number. Using logarithms or electronic calculators, we can obtain the approximate value of such numbers by giving the first part of their decimal expansion. For instance, we write $\sqrt{2} = 1.414213562$; and, similarly, the approximate value of $(-4)^{2/3}$ is 2.5198421—or, for many problems, just $(-4)^{2/3} = 2.52$.

To define x^α when α is an irrational number such as $\sqrt{2}$, we utilize the fact that α can be approximated by the first terms of its decimal expansion. For example, $\sqrt{2}$ is better and better approximated by the sequence of rational numbers 1, 1.4, 1.41, 1.414, 1.4142, 1.41421, ... For each of these rational numbers the value of x^r is defined. We therefore interpret $x^{\sqrt{2}}$ as the number which is better and better approximated by the sequence $x^1, x^{1.4}, x^{1.41}, x^{1.414}, x^{1.4142}, x^{1.41421}, \ldots$ $x^{\sqrt{2}}$ is defined only for $x \geq 0$. Similarly, for any irrational number α and all $x \geq 0$ the value of x^α can be defined.

EXERCISE SET 4

4.1 Compute $f \circ g$ and $g \circ f$ for the following pairs of functions:

a) $f = \{(1, 2), (2, 4), (3, 4), (4, -1)\}, g = \{(2, 3), (4, 3), (-1, 1)\}$.
b) $f(x) = x + 2, g(x) = x^2$.
c) $f(x) = x^2 + 2, g(x) = x - 3$.
d) $f(x) = 3x + 4, g(x) = (\frac{1}{3})x - 4$.
e) $f(x) = x^3 - 1, g(x) = x + 1$,
f) $f(x) = 3x^4, g(x) = 2x^3$.

4.2 Let $f(x) = x^3$ and $g(x) = 2x - 1$. Compute:

a) $(f \circ g)(x)$ and $(g \circ f)(x)$.
b) $f^{-1}(x)$ and $g^{-1}(x)$.
c) $(f \circ g)^{-1}(x)$ and $(g \circ f)^{-1}(x)$.

4.3 In reference to Example 4.2, assume the blood glucose level B_g is a linear function of the insulin level I with a negative slope. Assume the urine glucose level U_g is zero if the blood glucose level is less than a constant M and otherwise is proportional to the blood glucose level less the threshold level M.

a) Express B_g as a function of I and U_g as a function of B_g. Then express U_g as a function of I.
b) Given that the slope of the $B_g - I$ curve is -1 with intercept 6, the constant $M = 0.8$, and the $U_g - B_g$ proportionality constant is 2, simplify and graph the functions derived in part (a).

4.4 In reference to Example 4.1, sketch the graph of F versus F_8 if $n = 2$, $m = 3$, and $k = 10^{-2}$.

4.5 In reference to Example 4.3, assume the dry weight, W, is a quadratic function of temperature, t, with $W(0) = 0$, $W(50) = 0$, and $W(25) = 25$. Assume the sugar content S is proportional to W with proportionality constant $k = 0.35$.

a) Determine the function $W(t)$.
b) Express S as a function of t.
c) Graph $W(t)$ and $S(t)$ on the same graph.

4.6 Determine the domain $(f \circ g)$ and range $(f \circ g)$ for each pair of functions f and g given in Exercise 4.1.

4.7 Which of the following graphs represent functions which have inverses?

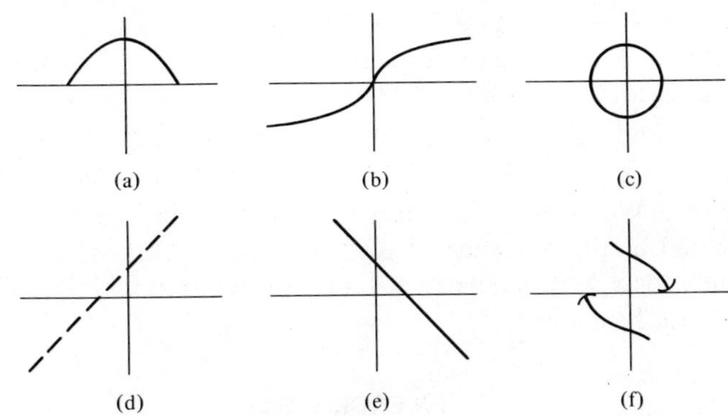

4.8 Determine the inverse of the given function f by solving the equations $y = f(x)$ for x and then interchanging the roles of x and y. State the domain of f and f^{-1}.

a) $f(x) = 2x$, b) $f(x) = 2x + 1$,
c) $f(x) = 2x^2, x \geq 0$, d) $f(x) = 2x^2 + 1, x \geq 0$
e) $f(x) = 2x^3$, f) $f(x) = 2x^3 + 1$,
g) $f(x) = 3x - 2$, h) $f(x) = 4x - 3$,
i) $f(x) = 1/x$, j) $f(x) = -x$.

4.9 Which of the following functions have inverses on the indicated set D? Sketch the graph of f and also f^{-1} when f^{-1} exists.

a) $f(x) = x^2, D = [0, 3)$ b) $f(x) = x^2, D = (-3, 0]$
c) $f(x) = x^3, D = [-3, 3]$ d) $f(x) = x^4, D = \{x | x^2 < 4\}$
e) $f(x) = x^4, D = (-3, 0)$ f) $f(x) = x^{10}, D = [-1, 1]$

4.10 The following functions may be expressed as a composition $(f \circ g)(x)$. Determine functions f and g which will yield the given function. (The solutions of this exercise are not unique.)

a) $y = x^2 + 2$ b) $y = (x + 3)^2 + 2$
c) $y = (x - 3)^5$ d) $y = -2x^2 + 2$
e) $y = x^2 + x^{1/2}$ f) $y = 5(x + 2)^5 - 4$

4.11 Evaluate the following numbers or indicate why you cannot evaluate them.

a) $8^{1/3}$ b) $16^{3/2}$ c) $-1^{3/2}$ d) $-8^{2/3}$
e) 3^{-2} f) $(-8)^{-1/3}$ g) $(-100)^{-3/2}$ h) $(-1)^{4/5}$

4.12 Decide which of the given pairs of numbers is the largest.

a) $-3, 2$ b) $2^{1/2}, 2^{1/3}$ c) $2^{1/2}, 3^{1/3}$
d) $0.5, (0.5)^2$ e) $(0.6)^{1/2}, 0.6$ f) $(1.6)^{1/2}, (1.6)^2$
g) $(-1)^2, (-1)^3$ h) $(-2)^2, (-2)^3$ i) $(-32)^{2/5}, (-32)^{3/5}$

4.13 Assume the number of lateral branches on a tree is $N(h) = 4h - 8$, where h is the height of the tree in feet. Assume the average number of leaves on a lateral branch is a function of the total number N of lateral branches given by $L = N^2 - N$. Express the average number of leaves on a tree of height h as a function of h.

4.14 In reference to Example 4.7, simplify and graph the equations (4.2) and (4.3) for the given parameter values.

a) $C = 10, B = 0.1, K = 20$, and $A = 0.04$.
b) $C = 10, B = 0.1, K = 10$, and $A = 0.2$.
c) $C = 10, B = 0.1, K = 5$, and $A = 0.25$.
d) $C = 10, B = 0.1, K = 4$, and $A = 0.5$.

4.15 Show that the composition of two allometric functions is an allometric function.

4.16 The DuBois formula is used in physiology to relate body weight W, height H, and surface area of the skin A. The relation is of the form.

$$A = cH^b \cdot W^a.$$

To facilitate the algebra assume $a = 0.25$, $b = 0.75$, and $c = 0.1$.

a) What is the surface area of a person of height 180 cm and weight 81 kg?
b) On the same graph sketch the allometric relationship between height and surface area for persons of fixed weight $W = 16$ and $W = 81$.
c) Express the height H as a function of surface area and weight.

SECTION 5

EXPONENTIAL AND LOGARITHMIC FUNCTIONS

5.1 EXPONENTIAL FUNCTIONS

In this section we discuss the "exponential functions" of the form $y = b^x$, where the independent variable x appears in the exponent.

Example 5.1 Consider a population of bacteria cells in a homogeneous environment containing all nutrients necessary for growth. If the population is synchronized so that all cells divide at the same time, the number of cells present after the nth division (i.e. the population of the nth generation) is denoted by $N(n)$ for $n = 0, 1, 2, \ldots$. As each cell of the nth generation divides to form two cells of the $(n + 1)$st generation, the equation

$$N(n + 1) = 2 \cdot N(n) \tag{5.1}$$

holds for $n = 0, 1, 2, \ldots$. Consequently, if N_0 denotes the initial population $N(0)$, then $N(1) = 2 \cdot N_0$, $N(2) = 2 \cdot N(1) = N_0 \cdot 2^2$, $N(3) = 2 \cdot N(2) = N_0 2^3$, and in general

$$N(n) = N_0 \cdot 2^n, \qquad n = 0, 1, 2, \ldots \tag{5.2}$$

To verify that the function given by equation (5.2) is a solution of equation (5.1) we observe that replacing n by $n + 1$ in equation (5.2) gives

$$N(n + 1) = N_0 2^{(n+1)} = N_0 2^n \cdot 2 = 2N(n).$$

If the length of each generation is one hour then the size of the population after t hours will be

$$N(t) = N_0 2^t, \qquad t = 1, 2, 3, \ldots$$

However, if the length of a generation is half an hour then the size of the bacteria population after t hours will be

$$N(t) = N_0 2^{2t}, \qquad t = 1, 2, 3, \ldots,$$

since the time t corresponds in this case to the $(2 \cdot t)$ generation. Generally, if the length of a generation is h hours, then the number of generations in one hour will be $k = 1/h$. Thus the size of the population after t hours will be the size after $k \cdot t$ generations or

$$N(t) = N_0 2^{kt}, \qquad t = 1, 2, 3, \ldots \tag{5.3}$$

The equation (5.3) has the form of a general exponential equation which is defined for all real values of t. ◀

Definition 5.1 For any *positive* number b the *exponential function* with *base* b is defined on $(-\infty, \infty)$ by the equation

$$y = b^t \qquad \text{or equivalently,} \qquad y = \exp_b(t),$$

read "y equals b to the power t" or "y equals the exponential, base b, of t." Here b is called the *base* of the exponential and t is called the *exponent*.

The most common type of exponential function is of the form

$$y = cb^{kt} \qquad \text{or} \qquad y = c \exp_b(kt),$$

where c and k denote constants. The exponential functions $y = b^t$ satisfy the following fundamental properties, which are derived from the basic laws of exponents and inequalities.

Properties of Exponentials

E-P1 Since b is positive, $b^t > 0$ for all t. Thus the domain of $y = b^t$ is $(-\infty, \infty)$ and the range of $y = b^t$ is $(0, +\infty)$.

E-P2 For any base $b > 0$, $b^0 = 1$ and $b^1 = b$.

E-P3 (Addition law for exponents.) For any two numbers t_1 and t_2 and all $b > 0$,

$$b^{(t_1 + t_2)} = b^{t_1} \cdot b^{t_2}.$$

E-P4 (Multiplication law for exponents.) For any two numbers t and k and all $b > 0$,

$$(b^t)^k = b^{(t \cdot k)}.$$

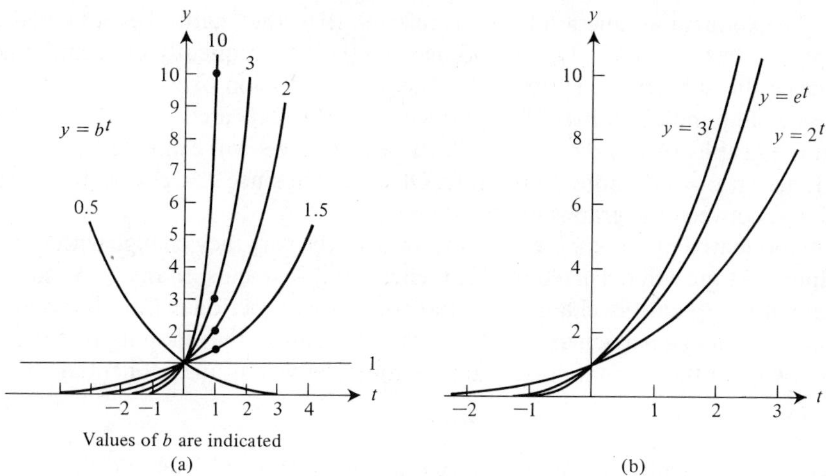

Figure 5.1

E-P5 If $b > 1$, then $y = b^t$ is a monotone increasing function on $(-\infty, +\infty)$. If $0 < b < 1$ then $y = b^t$ is a monotone decreasing function on $(-\infty, +\infty)$.

By evaluating b^t for integer values of t, we can plot points on the graph of $y = b^t$; then, connecting these points with a smooth curve, we can "sketch" the graph for all t values. The graphs of several exponential functions are indicated in Fig. 5.1(a). Observe that the graphs for $b > 1$ are increasing while for $0 < b < 1$ the graphs are decreasing. Also note that there is an "orderly" relation between the graphs for different base values b. For a number $t > 0$, say $t = 2$, as the base b increases the corresponding value $y = b^t$ increases. However when $t < 0$, say $t = -1$, as the base b increases the value $y = b^t$ decreases.

Base b	0.5	1	2	3	4	
$t = 2;\quad y = b^2$	0.25	1	4	9	16	Increasing values
$t = -1; y = b^{-1}$	2	1	$\frac{1}{2}$	$\frac{1}{3}$	$\frac{1}{4}$	Decreasing values

The exponential functions which are most frequently encountered are those with base 2, base 10, or base e. The number e is a very important irrational number which arises naturally in many physical and biological models. The number e cannot be defined without utilizing the concept of a limit. In Section 12 we define e as the limit of the term $(1 + \varepsilon)^{1/\varepsilon}$ as ε becomes small: and, in Section 22 we define e in terms of the area under a curve $y = 1/x$. For the present it is sufficient to think of e as a number approximately equal to 2.7182818275. For most calculations, $e \simeq 2.718$ is sufficient.

The exponential function base e is referred to as the "natural exponential" or simply the "exponential" function. Since $\exp_e(t)$ is so frequently utilized, for convenience the subscript e is omitted. Thus the expression $\exp(t)$ means e^t in the absence of an indicated base. The graph of $y = e^t$ is indicated in Fig. 5.1(b) along with the graphs of $y = 2^t$ and $y = 3^t$. In Section 8 we will examine the graph of the function $y = ce^{kt}$ more thoroughly. Observe that since $2 < e < 3$, the graph of $y = e^t$ is between the graphs of $y = 2^t$ and $y = 3^t$.

Exponential processes are characterized by the way they change with time. In Example 5.1 growth occurred in unit (generation) segments. Many biological and other natural quantities change instantaneously over a period of time. If the change in the amount of a particular quantity is proportional to the amount present, it can be shown that the amount present at any time t is given by an exponential function

$$A(t) = A_0 e^{kt}, \tag{5.4}$$

where A_0 represents the amount present at time $t = 0$ and k denotes the proportionality constant, or *rate constant.*

Example 5.2 *Radioactive decay*
Radioactive isotopes such as ^{14}C (carbon 14), ^{24}Na (sodium 24), ^{32}P (phosphorus 32), and ^{131}I (iodine 131) are widely used as tracers to study the movement or concentration of particular compounds both in vitro (i.e., in a laboratory) and in vivo (i.e., in their biological environment). The utility of these isotopes lies in the fact that they emit negatively charged particles and are thus easily monitored. However, these isotopes undergo transmutation in the process of emitting the particles and *decay* to nonradioactive forms. The amount of particles emitted is proportional to the amount of isotope present, which in turn indicates the amount of "labeled compound" present. To determine the amount of labeled compound present, it is necessary to account for the decay of the radioactive isotope. The basic assumption in radioactive assay is that the rate of decay is a constant $-k$ (the negative sign indicates that the amount present is decreasing). The number of radioactive atoms present at time t is given by

$$n(t) = n_0 e^{-kt}, \tag{5.5}$$

where n_0 is the initial number of radioactive atoms. The *half-life* of an isotope is the time, $t_{1/2}$, at which one-half of the initial number of atoms have decomposed. Thus the half-life $t_{1/2}$ is determined by the equation

$$n(t_{1/2}) = \tfrac{1}{2}n_0, \qquad \text{or equivalently,} \qquad e^{-kt_{1/2}} = \tfrac{1}{2}. \tag{5.6}$$

◀

Example 5.3 Colorimetry and spectrophotometry are commonly used techniques in chemistry and biochemistry. They utilize the laws of Lambert and Beer concerning the passage of light through solutions.

Lambert's Law relates the intensity I_0 of light incident to one surface of an absorbing substance, generally a solution, and the intensity I of the light transmitted through the solution. I is a function of the thickness l of the solution and is given by the equation

$$I = I_0 10^{-kl} \tag{5.7}$$

The constant k depends on the specific absorbing solution utilized and is called the *extinction coefficient*. By definition, k is the reciprocal of the thickness of the specific absorbing solution at which the intensity of the transmitted light is one-tenth of the intensity of the incident light.

Beer's Law concerns the relationship between the concentration, C(moles/liter), of a colored substance in a solution and the transmittance of monochromatic light through the solution. Again let I_0 denote the light intensity incident to a surface of the solution and let I denote the light intensity transmitted at a depth l. The relation between the depth and concentration and the transmitted light is given by

$$I = I_0 10^{-\varepsilon Cl} \tag{5.8}$$

The constant ε is the *molar extinction coefficient*; ε is a function of the solute being studied and the wavelength (i.e., color) of the transmitted light. In practice one of the constants C or ε is known and the other is obtained using equation (5.8) with experimentally determined values of I at different depths. ◀

Example 5.4 The extinction coefficient of reduced nicotinamide-adenine dinucleotide phosphate ($NADPH_2$) at a wavelength of 340 mμ is $\varepsilon = 6.22 \times 10^3$ cm^2. A 3-ml solution containing 0.2 μ mole $NADPH_2$ is placed in a cuvette of length 1.05 cm. (A cuvette is a receptacle used to hold a liquid sample in a spectrophotometer, which measures light intensities.) The percentage of light transmitted is $100 \times I/I_0$. Therefore the percentage of light transmitted through this solution at 340 mμ is

$$100 \times \frac{I(1.05)}{I_0} = 100 \times 10^{-(6.22 \times 10^3)\cdot C \cdot (1.05)}.$$

The concentration is converted to molarity as follows:

$$C = \frac{0.2\ \mu\text{M}}{3\text{ ml}} = \frac{0.2 \times 10^{-6}\text{ M}}{3 \times 10^{-3}\text{ l}} = 0.67 \times 10^{-4}\text{ M/l}.$$

Thus, using three-figure accuracy, we obtain the percentage

$$100 \cdot \frac{I(1.05)}{I_0} = \frac{100}{10^{-0.435}} = 36.7\%.$$ ◀

5.2 LOGARITHMS

In Example 5.2 the half-life of a radioactive substance was defined as the number $t_{1/2}$ representing the time necessary for one-half of a given amount of the substance to decay. The exact value of $t_{1/2}$ is thus determined by the equations in (5.6). But the equations in (5.6) are "implicit" equations for $t_{1/2}$. To determine the half life, we must solve the equation $\exp(-kt_{1/2}) = \frac{1}{2}$ explicitly for $t_{1/2}$. To solve this equation, we must introduce a new function which reverses the exponential process. According to property E-P5, for any base $b > 0$, but not equal to one, the exponential function $y = b^t$ is monotone on $(-\infty, +\infty)$. Hence, by Theorem 4.3 the exponential function $y = b^t$ has an inverse function (with respect to composition). The inverse of the exponential function is called the "logarithm" function.

Definition 5.2 For any positive real number $b \neq 1$, the *logarithm function with base b* is the inverse of the exponential function with base b. The logarithm function with base b is denoted by $\log_b(x)$, read "log base b of x," and defined by the relation

$$\boxed{y = \log_b(x) \qquad \text{if and only if} \qquad b^y = x.} \tag{5.9}$$

Before we discuss the algebraic properties and graphs of logarithm functions, let us first look at the use of logarithms to solve equations. To solve the equation

$$\exp_3(t) = 5 \qquad \text{or} \qquad 3^t = 5.$$

we must reverse the exponential function. By definition the function $\log_3$ is the inverse of the function $\exp_3$. Hence, $\log_3(\exp_3(t)) = t$. If we take the logarithm, base 3, of both sides of the above equation, we obtain

$$t = \log_3(\exp_3(t)) = \log_3(5).$$

Similarly, to find the half-life $t_{1/2}$ discussed in Example 5.2, we solve the equation

$$e^{-kt_{1/2}} = \tfrac{1}{2}$$

by first taking the logarithm, base e, of both sides of the equation and then dividing both sides by $-k$:

$$\log_e(e^{-kt_{1/2}}) = -kt_{1/2} = \log_e(\tfrac{1}{2}); \qquad t_{1/2} = -1/k \log_e(\tfrac{1}{2}).$$

The value of a logarithm, $\log_b(x)$, for particular values of b and x is usually not obvious. For this reason we provide (in Appendix II) tables giving values of $\log_b(x)$ for $b = e$ and $b = 10$. Occasionally, we can evaluate some logarithms without the aid of a table by using equation (5.9) and our knowledge of exponentials. For instance,

$$\begin{aligned}
\log_2(8) &= 3 && \text{since} && 2^3 = 8, \\
\log_{10}(0.01) &= -2 && \text{since} && 10^{-2} = 0.01 \\
\log_{\sqrt{2}}(16) &= 8 && \text{since} && (\sqrt{2})^8 = 16, \\
\log_7(1) &= 0 && \text{since} && 7^0 = 1, \\
\log_{1/2}(0.25) &= 2 && \text{since} && (\tfrac{1}{2})^2 = \tfrac{1}{4} = 0.25.
\end{aligned}$$

The logarithm functions, as a consequence of their definition in terms of exponential functions, satisfy properties corresponding to the properties E-P1 to E-P5 for exponential functions.

L-P1 Domain($\log_b$) = $(0, \infty)$, Range($\log_b$) = $(-\infty, \infty)$.

L-P2 $\log_b(1) = 0$ for all b, and $\log_b(b) = 1$.

L-P3 $\log_b(t_1 \cdot t_2) = \log_b(t_1) + \log_b(t_2)$.

L-P4 $\log_b(t^k) = k \cdot \log_b(t)$.

L-P5 If $b > 1$, $\log_b(t)$ is monotone increasing on $(0, \infty)$.
If $b < 1$, $\log_b(t)$ is monotone decreasing on $(0, \infty)$.

The properties of logarithms are illustrated by the following.

L-P1 $\log_3(5)$ is defined but $\log_3(-5)$ is not defined.

L-P2 $\log_6(1) = 0$ since $6^0 \equiv 1$, by definition of the zero exponent. $\log_6(6) = 1$ since $6^1 = 6$.

L-P3 $\log_2(8) = \log_2(4 \cdot 2) = \log_2(4) + \log_2(2) = 2 + 1 = 3$.

L-P4 $\log_2(8) = \log_2(2^3) = 3 \log_2(2) = 3 \cdot 1 = 3$.

L-P5 $\log_2$ is increasing while $\log_{1/2}$ is decreasing.

For instance, $4 < 8$ and

$$\log_2(4) = 2 < 3 = \log_2(8),$$

whereas

$$\log_{1/2}(4) = -2 > -3 = \log_{1/2}(8).$$

The most useful properties of logarithms for the purpose of arithmetic calculations are L-P3 and L-P4. By utilizing these properties we may reduce the operations of multiplication and exponentiation to the simpler operations of adding and multiplying, respectively. One particularly useful combination of these properties results in the formula,

$$\boxed{\log_b\left(\frac{A}{B}\right) = \log_b(A) - \log_b(B).}$$

This is accomplished by using the fact that when $k = -1$ in property L-P4,

$$\boxed{-\log_b(B) = \log_b(B^{-1}) = \log_b(1/B).}$$

Just as you are most likely to encounter only exponentials with bases 2, 10, and e, the logarithms most commonly utilized are of base 2, base 10, and base e. In fact, the occurrence of other bases is so rare that no confusion arises from the convention of writing

$$\boxed{\text{“the common logarithm,” } \log(x) \qquad \text{for } \log_{10}(x)}$$

and

$$\boxed{\text{“the natural logarithm,” } \ln(x) \qquad \text{for} \qquad \log_e(x).}$$

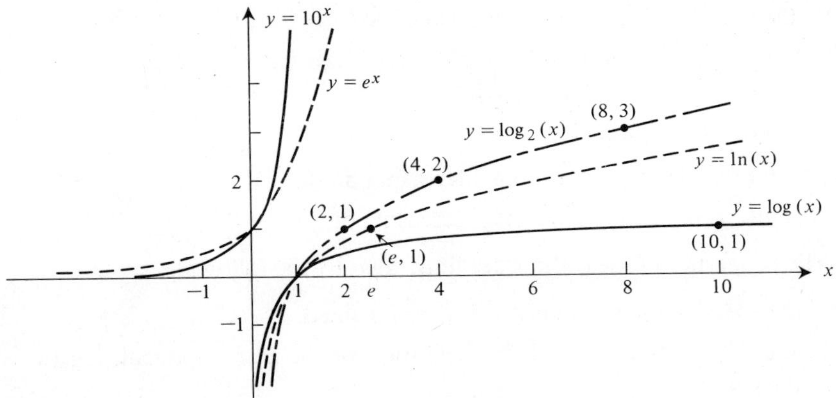

Figure 5.2

The graphs of $y = \log(x)$ and $y = \ln(x)$ are given in Fig. 5.2. These may be obtained geometrically from the graphs of $y = 10^x$ and $y = e^x$ by rotation about the line $y = x$ as discussed in Section 4, or analytically by plotting points using tabulated values of the logarithms.

Example 5.5 *Carbon-14 dating*
The radioactive isotope ^{14}C is present in all living matter. The main source of this isotope is carbon dioxide in the air, which contains ^{14}C isotopes as the result of cosmic radiation. In living material the level of ^{14}C is constant as the decay of the ^{14}C is counterbalanced by the assimilation of new ^{14}C. However, at death the assimilation stops and the ^{14}C present slowly disappears as it decays. The amount of ^{14}C in a sample t years after death is

$$A(t) = A_0 e^{-kt}, \qquad \textbf{Decay Equation}$$

where $-k$ is the decay rate and A_0 is the amount of ^{14}C present at the time of death, $t = 0$. The half-life of ^{14}C is $t_{1/2} = 5685$ years. The constant k is evaluated by solving the equation $A(t_{1/2}) = A_0/2$. Dividing the equation

$$A_0/2 = A_0 e^{-kt_{1/2}}$$

by A_0 gives

$$e^{-kt_{1/2}} = \tfrac{1}{2}.$$

Taking the natural logarithm of both sides gives

$$-kt_{1/2} = \ln(\tfrac{1}{2}) \qquad \text{or} \qquad k = -1/t_{1/2} \ln(\tfrac{1}{2}).$$

Since

$$-\ln(\tfrac{1}{2}) = \ln((\tfrac{1}{2})^{-1}) = \ln(2),$$

we find the constant

$$\boxed{k = \frac{\ln(2)}{t_{1/2}}.}$$

The constant $\ln(2) \simeq 0.6931$ and hence the decay rate for ^{14}C is $-k \simeq -0.0001219 = -1.219 \times 10^{-4}$. Using this value we can deduce that approximately 30 percent of the original ^{14}C will still be present in a sample 10,000 years after it ceases to live:

$$\frac{A(10{,}000)}{A_0} = e^{-k \cdot 10{,}000} \simeq 0.2954.$$

These values were obtained using a calculator. By use of Table A.3, the closest value we could obtain would be $A(10{,}000)/A_0 = \exp(-1.22) = 0.2952$. Similarly, we can determine how long it would take for 60 percent of the ^{14}C to decay by solving $A(t)/A_0 = 0.40$:

$$e^{-kt} = 0.40; \qquad -kt = \ln(0.4); \qquad t = -1/k \ln(0.4);$$

as $(1/0.4) = 2.5$, the value of t is thus

$$t = \frac{\ln(2.5)}{k} \simeq 7515 \text{ years.}$$ ◀

Example 5.6 In a study of relative growth of plant leaves, the leaf area L_A and the dry weight W of a leaf were found to increase at relatively constant rates; R_A for leaf area and R for dry weights. Thus the equations expressing L_A and W as functions of time t are

$$L_A(t) = L_A(0)e^{R_A t}$$

and

$$W(t) = W(0)e^{Rt}.$$

These relationships can also be described by linear equations if we solve explicitly for $R_A t$ and Rt:

$$R_A \cdot t = \ln\left[\frac{L_A(t)}{L_A(0)}\right] = \ln(L_A(t)) - \ln(L_A(0))$$

and

$$R \cdot t = \ln\left[\frac{W(t)}{W(0)}\right] = \ln(W(t)) - \ln(W(0)).$$

Upon rearrangement the first of these equations becomes

$$\ln(L_A(t)) = R_A t + \ln(L_A(0)),$$

which is a linear equation in t with slope R_A and intercept $\ln(L_A(0))$. By plotting data points corresponding to leaf area measurements for different times, a line can be fitted and the growth rate R_A determined from the slope of the line. The graph in

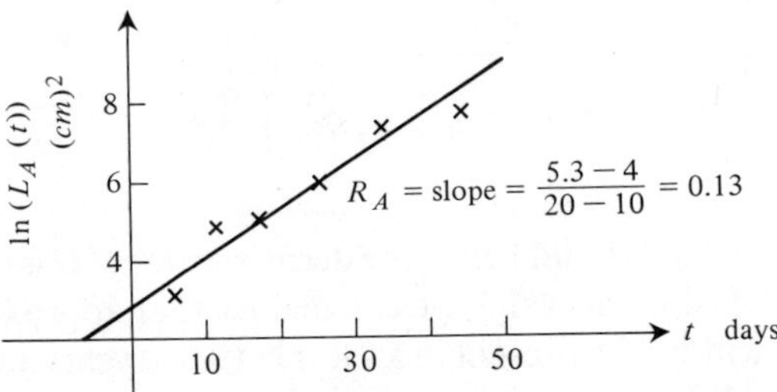

Figure 5.3

Fig. 5.3 illustrates this procedure for measurements on sunflowers. The value of $R_A = 0.13$ is obtained by observing that the fitted line appears to pass through the points (10, 4) and (20, 5.3). ◀

Example 5.7 In chemistry, acids and bases are distinguished on the basis of their hydrogen ion activity (H^+). In highly diluted material the hydrogen ion activity is equivalent to the hydrogen ion concentration, denoted by $[H^+]$. The number (H^+) is very frequently used to describe the acidity of a solution. For instance, the number (H^+) is about 10^{-5} for distilled water and 10^{-2} for stomach fluid. For the purpose of simplification and to provide an expanded scale of values, the term "pH" is introduced to denote the negative logarithm of (H^+),

$$\text{pH} \equiv -\log(H^+).$$

Thus the pH of distilled water is pH $= 5$ and for stomach fluid the pH $= 2$. The pH value of a solution can also be used to determine the "normality" of a solution. By definition, a solution is an "x-normal" solution if its pH $= -\log(x)$. For example a solution of hydrochloric acid which is completely dissociated will have one H^+ ion for each Cl^- ion. From the definition of pH, if the solution has pH $= 3$, the concentration of the acid will be the same as the concentration (H^+) of hydrogen ions. Thus the normality of this solution is

$$x = \exp_{10}(-\text{pH}) = 10^{-3} = 0.001.$$ ◀

Example 5.8 Beer's Law (Example 5.3) involves the molar extinction coefficient ε. The constant ε can be evaluated by measuring the transmitted light at a fixed wavelength and depth for two different concentrations of a solution. If I_1 and I_2 denote the transmitted light for concentrations C_1 and C_2, respectively, then

$$I_1 = I_0 10^{-\varepsilon C_1 l} \quad \text{and} \quad I_2 = I_0 10^{-\varepsilon C_2 l}.$$

Consequently,

$$\frac{I_1}{I_2} = 10^{-\varepsilon(C_1 - C_2)l}.$$

Solving this equation for ε, we find

$$\varepsilon = \frac{-1}{(C_1 - C_2)l}\log\left(\frac{I_1}{I_2}\right). \quad \blacktriangleleft$$

In practice the use of logarithms can be limited to a single base b. Scientists prefer to use $b = 10$ and work exclusively with $\log(x)$. One reason for this is the ease with which this function can be graphed, as we will see in Section 7. Mathematicians on the other hand prefer to use $b = e$ and work exclusively with $\ln(x)$. The reason for this is the extremely nice mathematical properties of the function $y = \ln(x)$, which will be explored in later sections. To illustrate how one base b is sufficient for all calculations, we solve the equation $3^t = 5$ for t using two different bases. If the base is $b = 10$, the value of t is obtained by solving as follows:

$$\log(3^t) = \log(5) \qquad \text{or} \qquad t\log(3) = \log(5),$$

hence,

$$t = \log(5)/\log(3).$$

If the base is $b = e$, the value of t is obtained by solving as follows:

$$\ln(3^t) = \ln(5) \qquad \text{or} \qquad t\ln(3) = \ln(5),$$

hence

$$t = \ln(5)/\ln(3).$$

The fact that these solutions are the same can also be verified from the general "change-of-base" formula relating logarithms of different bases:

$$\boxed{\log_a(x) = \frac{\log_b(x)}{\log_b(a)}.} \tag{5.10}$$

This formula is obtained as follows. If

$$y = \log_a(x),$$

then by definition,

$$a^y = x.$$

Taking logarithms, base b, we obtain

$$\log_b(a^y) = \log_b(x) \qquad \text{or} \qquad y\log_b(a) = \log_b(x).$$

Solving for y then gives the equation (5.10). The most common use of this change of base formula is to convert between $\ln(x)$ and $\log(x)$.

$$a = 10,\ b = e\colon \log(x) = \frac{\ln(x)}{\ln(10)} \simeq 0.4343\ln(x).$$

$$a = e,\quad b = 10\colon \ln(x) = \frac{\log(x)}{\log(e)} \simeq 2.3026\log(x).$$

EXERCISE SET 5

5.1 Determine which is the larger of the given pairs of numbers using general properties of exponentials and logarithms.

a) $2^5, 2^{\sqrt{5}}$ b) $2^a, \frac{1}{2}^a$
c) $e^x, \sqrt{10}^x$ d) $3^2, 2^3$
e) $3^5, 5^3$ f) $3, \log(3)$
g) $3, \ln(3)$ h) $a^x, \log_a(x), x > 0, a > 1$
i) $\log(5), \log(2\sqrt{5})$ j) $\log(536), 4$

5.2 Assume that a bacteria cell culture initially contained 3000 cells. Assume that the generation time for this culture is 20 minutes.

a) How large will the culture be in two days (assuming there are no deaths)?
b) If 20 percent of the cells die each generation, what will be the equation for the number of cells in the nth generation?
c) Under the assumption that 20 percent of the cells die each generation, what will be the size of the culture in two days? (*Hint*: To evaluate the number, use the fact that if $x = a^b$ then $x = \exp(b \ln(a))$.)

5.3 Sketch the graph of the given function.

a) $y = 2^x$ b) $y = 2^{x-2}$ c) $y = 2^t/4$
d) $y = 10^x$ e) $y = (0.1)^t$ f) $y = e^{-t}$

5.4 The half-life of ^{24}Na is approximately 15 hours. If the amount of ^{24}Na initially present is $n_0 = 400$, how much will be present after 10 hours; 15 hours; and 20 hours?

5.5 Assume a radioactive material is weighed at times $t = 2$ and $t = 5$ and the weights were 25 g and 10 g, respectively.

a) What is the decay rate of this substance?
b) What is the half-life $t_{1/2}$ of this substance?
c) How much of the material was present at time $t = 0$?
d) Graph the decay curve for this substance.

5.6 a) Compute the $t_{1/4}$ time of an exponential decay curve. (The $t_{1/4}$ time is the time in which one-fourth of the initial material will decay.)
b) Assuming that a substance has a $t_{1/4}$ time of 5 days, determine the initial amount of the substance given that 7 g remained after 12 days.

5.7 A body placed in water at 0°C will lose heat at a rate proportional to its temperature. The temperature t minutes after immersion will thus be given by $T(t) = T_0 e^{-kt}$, where the constant k reflects the surface area of the body and any insulation. If a human with body temperature 37°C fell into arctic waters at 0°C, she or he would lose consciousness if her or his body temperature reached 25°C. Assuming the decay rate is $-k = -0.02$, how long would a person survive in the water? If $-k = -0.2$, what would be the survival time?

5.8 To test the functioning of the thyroid gland, radioactive iodine ^{131}I is introduced into the body and its accumulation and decay in the gland are monitored. The half-life of ^{131}I is $t_{1/2} = 8$ days.

a) What fraction of a dose of ^{131}I will decay each day?
b) If the maximum permissible amount of ^{131}I is 5 units, how long after the administration of a dose of 4 units can a second 4-unit dose be safely administered?

c) If the maximum permissible amount of ^{131}I is 5 units, what is the maximum dose which can be administered every 24 hours when the initial dose is 5 units?

5.9 Graph the indicated function.

a) $y = \ln(x)$ b) $y = \log(x)$ c) $y = \log_2(x)$
d) $y = \log_3(x)$ e) $y = \log_2(2x)$ f) $y = \log(2x)$

5.10 Solve the given equation for x.

a) $35 = 10e^{-2x}$ b) $\log(2x) = 0.1$
c) $e^{3x} = 7$ d) $\ln(2x - 1) = 7$
e) $5^x = 10$ f) $e^x = 7e^{4x}$
g) $\log_3(5x) = 2$ h) $e^{3(x-2)} = \frac{1}{2}$
i) $\log(3) - \log(x) = 7$ j) $\ln(x + 1)^2 = 0$

5.11 Determine the molar extinction coefficient of a substance if a 0.8 concentration registered a value $I_1 = 530$ at a 3-cm depth and a 0.4 concentration registered a value $I_2 = 1060$ at the same depth and wavelength.

5.12 A cuvette of 2-cm depth and hence containing 2 ml of solution was used to measure the colored light absorbancy of a 0.18-mole solution of protein. At a given wavelength, μ, 16 percent of the light was transmitted. What is the *molar extinction coefficient* of the protein at wavelength μ?

5.13 Light transmitted through a solution registered 525 at a depth of 2 cm and 50 at a depth of 5 cm.

a) What is the extinction coefficient of the solution?
b) What was the incident intensity I_0?

5.14 What are the hydrogen ion concentrations of solutions having pH 6.5 and 9.0?

5.15 Find the pH of a 0.01-N solution of hydrochloric acid. Assume full dissociation.

5.16 In the equilibrium of an acid-base solution, the equation

$$\text{acid} \rightleftharpoons \text{base} + H^+$$

yields an algebraic equation

$$(A) = (1/K_a)(B)(H^+).$$

a) Using the term $pK_a \equiv -\log(K_a)$ write this equation in the form pH = ... (*Hint*: Take the logarithm of both sides.)
b) If an acid of concentration $(A) = 0.505$ and a base of concentration $(B) = 0.3$ are in equilibrium at a pH of 6.5, what is their equilibrium constant K_a?

5.17 Which solution has a higher concentration of H^+ ions, one at pH = 7.2 or one at pH = 6.5?

5.18 In reference to Example 5.2, assume the following data are presented for a sample of ^{32}P.

Time (days)	0	5	10	15	20
Counts/min	500	390	307	240	188

a) Give a guess as to the value of $t_{1/2}$.
b) Calculate a value of k from the data given.
c) Calculate the value of $t_{1/2}$ from the data given.
d) Use a log table to compute log(counts/min) of the given data.
e) Plot the log(counts/min) versus time(hours) on a graph.
f) The slope of the line which best fits the graph of (e) is

$$m = -\log(2)/t_{1/2}.$$

From the graph (e) compute $t_{1/2}$.

5.19 Use the $\log(x)$ tables in the appendix and the change of base formula for logarithms to evaluate the following numbers.

a) $\ln(3)$ b) $\log_2(3)$ c) $\log_3(3)$
d) $\ln(0.1)$ e) $\log_2(0.1)$ f) $\log_3(0.1)$
g) $\log_8(2)$ h) $\log_8(12)$ i) $\log_8(100)$

5.20 In some texts the term "antilog" is used as follows.

$$\text{The antilog }(y) = x \quad \text{if} \quad \log(x) = y.$$

(Thus the "antilog" is really the exponential base 10.) Compute the antilog of the following numbers: (a) $y = 2.8124$; (b) $y = -0.1119$; and (c) $y = -3.3580$.

5.21 Use logarithms to perform the indicated arithmetic operations.

a) 3048×0.2130 b) $(0.2130) \div (3048)$
c) $(0.2130) \div (0.0003281)$ d) $3048^{0.2130}$

5.22 The "loudness" of sound is measured on a scale of "bels" (after Alexander Graham Bell). The "loudness" of a sound L is related to the intensity of the sound I by the exponential relation

$$I = k10^L.$$

The "loudness" is a relative or subjective scale of measuring sound. To see this, let I_0 be the intensity of a sound just below the faintest sound which a person can hear (such as the sound of a pin dropped on grass). Define the "loudness" corresponding to I_0 to be $L_0 = 0$.

a) Show that the loudness L corresponding to any intensity I satisfies $L = \log(I/I_0)$ bels. A scale of decibels (db), in which the units are 1/10 as large as a bel, is usually used. Thus, measured in decibels, $L = 10 \log(I/I_0)$ db.
b) What is the "loudness" in decibels of a sound whose intensity is $3000 \times I_0$?
c) What is the increase in relative intensity between the sound of a watch tick at 20 db and the sound of normal conversation at 70 db?
d) What is the intensity of an automobile horn if it has a "loudness" of 100 db and $I_0 = 1$?

5.23 Earthquakes are measured on the Richter scale expressed in terms of a magnitude variable R. The intensity I of an earthquake vibration is an exponential function with base $b = 10$ of the Richter scale magnitude R.

a) Show that the magnitude R satisfies the equation $R = \log(I/I_0)$, where I_0 is the intensity of normal earth vibrations corresponding to $R_0 = 0$.
b) How much greater is the intensity of an earthquake measuring 7.2 on the Richter scale than one measuring 4.6?

SECTION 6

PERIODIC FUNCTIONS; THE TRIGONOMETRIC FUNCTIONS

6.1 PERIODIC FUNCTIONS

One of the most obvious and profound phenomena in the biological world is the cyclic repetition of particular events and processes. Examples of such repetition abound, from the annual seasons of the year to the constant replication of DNA in cells which then divide to form new cells. In this section we introduce basic terminology for describing analytically such repetition and then we define the trigonometric functions which are extensively used to describe cyclic processes.

The notion of repetition or cycling is associated with a linear variable, usually representing time. Thus, to represent a repetitive process analytically, we usually associate some measurable aspect of the process with the time variable t. The process is thus described by a function $f(t)$. The cyclic aspect of the process is observed analytically as the function repeats its values over intervals of time corresponding to the complete cycles of the process.

Example 6.1 Yeast cells, such as *Schizosaccharomyces pombe*, multiply by a process of mitosis in which each cell duplicates its DNA and then divides to form two sister cells. This is a cyclic process in which two stages can be measured: (1) the time of division of the cells and (2) the time it takes a cell to replicate its nucleus, called the nuclear division time. The cell grows linearly in length over the period from its birth to the time of nuclear division. It then ceases to grow during the time it takes to form new cell walls and divide. Thus, if the cycle of a cell is associated with an interval $[0, D]$, the graph of its length over one cycle would be as indicated in Fig. 6.1(a). If we follow one cell through repeated divisions, the graph of the cell's

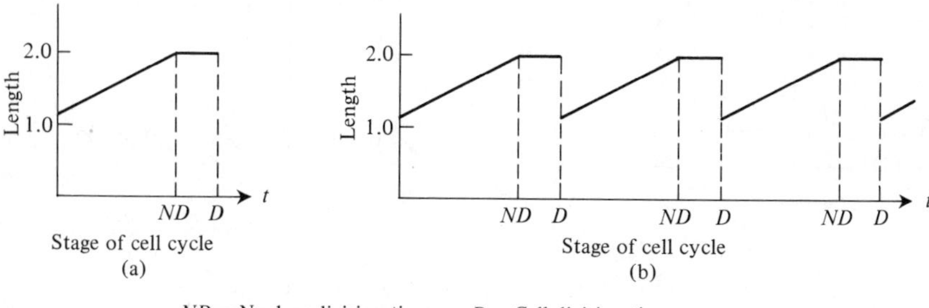

Figure 6.1

length would appear as in Fig. 6.1(b). If the cell cycle is of length 1.0 units of time and the nuclear division occurs at ND = 0.7 units of time, then the equation for the length of the cell as sketched in Fig. 6.1(a) is

$$f(t) = \begin{cases} (\frac{9}{7})t + 1.1 & \text{if} \quad 0 < t < 0.7, \\ 2.0 & \text{if} \quad 0.7 \le t < 1.0. \end{cases}$$

Note that we have avoided defining $f(t)$ at the time of cell division $t = 0$ and $t = 1.0$. The cell length varies from 1.1 to 2.0 in this model. ◀

Definition 6.1 A function f defined on a set of real numbers $S = [a, +\infty)$ is said to be *periodic on S with period T* if for each number $n = 1, 2, 3, \ldots$ and $t \ge a$,

$$f(t + nT) = f(t).$$

In particular a function f with period T satisfies $f(t + T) = f(t)$. Also, we note that a function f with period T is also a function with period $2T$ since $f(t + 2T) = f(t)$. Thus all multiples of the period of a function are also periods of the function.

Definition 6.2 The *basic period* (or simply *the period*) of a function f on $[a, +\infty)$ is the least positive number T, if it exists, such that $f(t + T) = f(t)$ for all $t \ge a$.

Example 6.2 The greatest integer function, denoted by $[\![x]\!]$, is defined to be the value of the greatest integer less than or equal to x. For example, $[\![3]\!] = 3$, since 3 is an integer, whereas $[\![2.9]\!] = 2$, since 2.9 is less than 3 but greater than 2. The greatest integer function is the same type of function as that commonly used to "round off" numbers to integers. However, the greatest integer function rounds all decimals down, as opposed to the common rounding practice of rounding up decimals ≥ 0.5 and rounding down decimals < 0.5. Thus $[\![3.71]\!] = 3$, $[\![82.4155]\!] = 82$, $[\![-1.2]\!] = -2$, $[\![-5.5]\!] = -6$, $[\![-3]\!] = -3$. The graph of $y = [\![x]\!]$ is given in Fig. 6.2(a). From the function $y = [\![x]\!]$, we may obtain a nice periodic function by setting $f(x) = x - [\![x]\!]$. The function $f(x)$ is a "sawtooth" function with period $T = 1$. We graph $f(x)$ in Fig. 6.2(b).

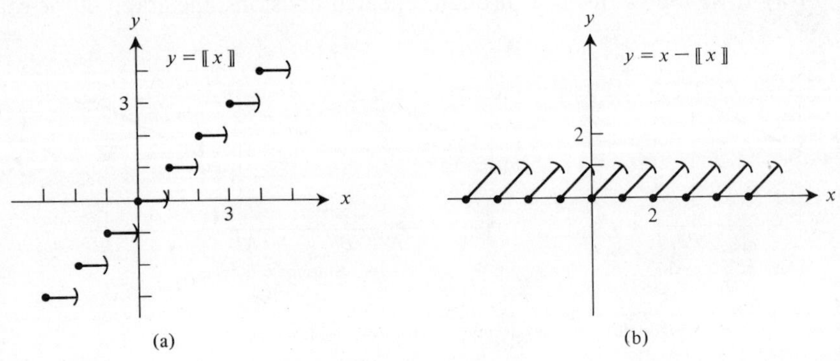

◀

Figure 6.2

6.2 THE FUNCTIONS SINE AND COSINE

The functions sine and cosine are periodic functions which are used to describe many natural periodic processes. To define these functions, we use a circle of radius one, which is the graph of the relation $x^2 + y^2 = 1$ in an x-y coordinate system. We associate with each real number t a point on the circle obtained by measuring $|t|$ units along the circle in a counterclockwise direction from the point (1, 0) if t is positive and in a clockwise direction from (1, 0) if t is negative. The resulting point on the circle has coordinates which we denote by $x(t)$ and $y(t)$. (See Fig. 6.3). We define the functions cosine and sine in terms of the coordinates $x(t)$ and $y(t)$ of the point on the circle in Fig. 6.3.

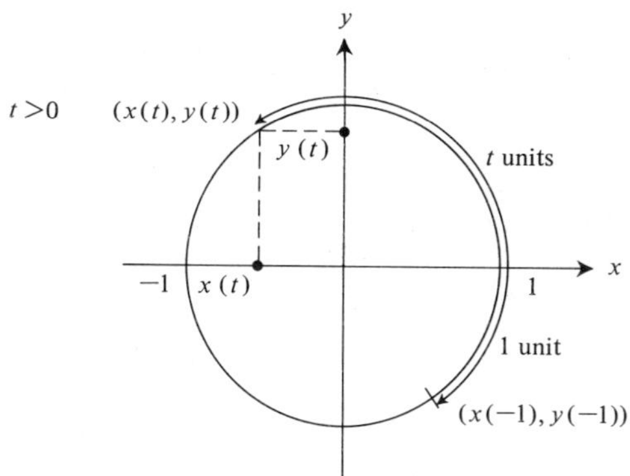

Figure 6.3

Definition 6.3

a) The function *cosine*, denoted by $\cos(t)$, is defined for all real t as the x-coordinate, $x(t)$, of the point on the circle $x^2 + y^2 = 1$, $|t|$ units along the circumference from the point (1, 0) measured counterclockwise if $t \geq 0$ and clockwise if $t < 0$.
b) The function *sine*, denoted by $\sin(t)$, is defined to be the y-coordinate $y(t)$ of the point indicated in part (a).
c) The argument t of the functions $\sin(t)$ and $\cos(t)$ is measured in terms of *radians* (since the unit of measure about the circle is the same as the radius of the circle).

The functions $\cos(t)$ and $\sin(t)$ may be graphed geometrically using their definition as depicted in Fig. 6.4. The functions $\cos(t)$ and $\sin(t)$ are both periodic since as t increases the point $(x(t), y(t))$ moves counterclockwise on the circle $x^2 + y^2 = 1$ and will thus duplicate its position each time a complete circumference of the circle has been measured. The circumference of a unit circle is 2π(two-pi) units. π is a particular irrational number approximated by $\pi \simeq 3.141593$. Since the period of the functions $\sin(t)$ and $\cos(t)$ is 2π, the number π is commonly used as a unit of measure on the t-axis when graphing these functions. Thus we indicated t values in Fig. 6.4 of 0, $\pm\pi/3$, $\pm\pi/2$, etc. (A table of values for the sine and cosine functions is given in Appendix II. Using the independent variable x rather than t, we have sketched the graph of $y = \sin(x)$ and $y = \cos(x)$ over an extended range of values in Fig. 6.9.)

This figure illustrates the relationship between the coordinates on the unit circle and the graphs of $\sin(t)$ and $\cos(t)$.

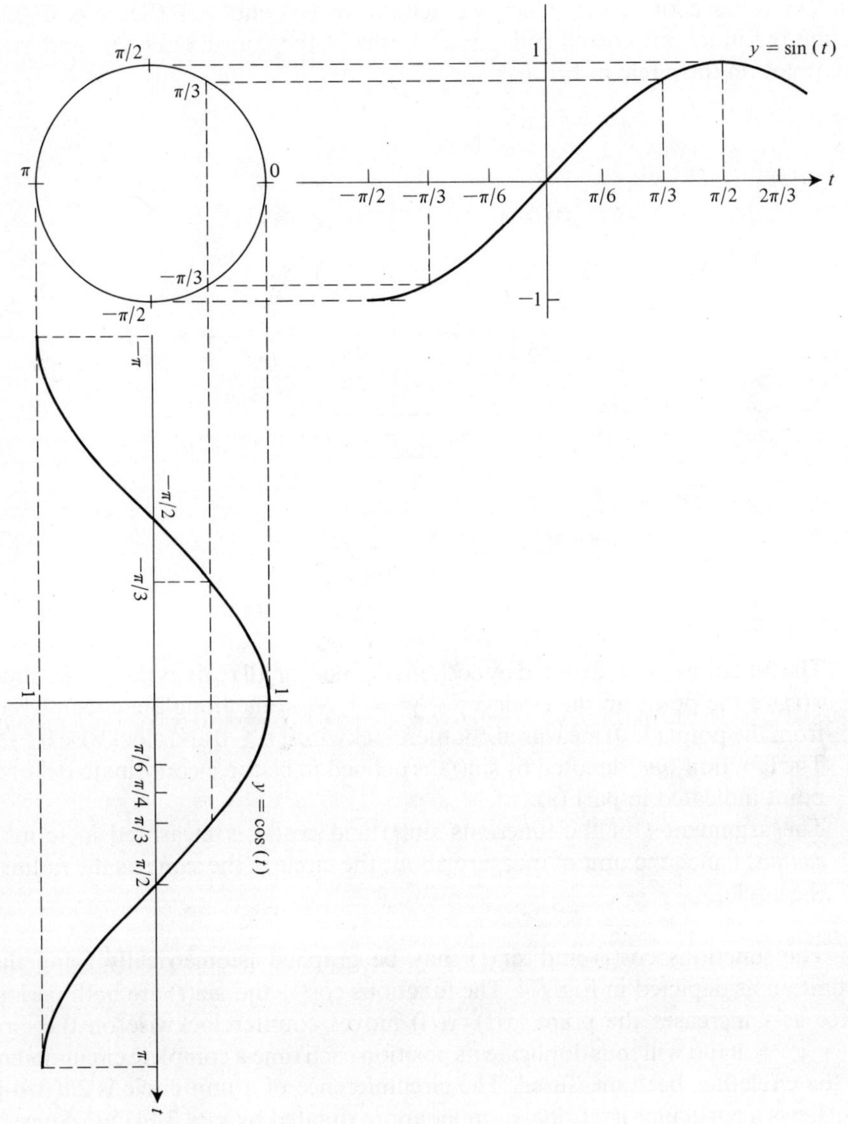

Figure 6.4

The points are projected horizontally to sketch $y = \sin(t)$ and vertically to sketch $y = \cos(t)$. Due to space limitation, only one-half of a cycle for $\sin(t)$ is sketched, while a full period of $\cos(t)$ is indicated.

The basic properties of the sine and cosine functions are as follows.

1. For all real x, $-1 \leq \sin(x) \leq 1$ and $-1 \leq \cos(x) \leq 1$.
2. Both $\sin(x)$ and $\cos(x)$ are periodic with period 2π: for each integer n and real number x, $\sin(x + 2n\pi) = \sin(x)$ and $\cos(x + 2n\pi) = \cos(x)$.
3. $y = \sin(x)$ is an odd function; for each number x, $\sin(-x) = -\sin(x)$.
   $y = \cos(x)$ is an even function; for each number x, $\cos(-x) = \cos(x)$.
4. $\text{Sin}(x) = 0$ for $x = 0, \pm\pi, \pm 2\pi, \ldots$
   $\text{Cos}(x) = 0$ for $x = \pm\pi/2, \pm 3\pi/2, \pm 5\pi/2, \ldots$
   $\text{Sin}(x) = 1$ for $x = -7\pi/2, -3\pi/2, \pi/2, 5\pi/2, \ldots$; i.e., $x = \pi/2 + n \cdot 2\pi$, n an integer.
   $\text{Sin}(x) = -1$ for $x = 3\pi/2 + n \cdot 2\pi$, n an integer.
   $\text{Cos}(x) = 1$ for $x = 0, \pm 2\pi, \pm 4\pi, \ldots$
   $\text{Cos}(x) = -1$ for $x = \pm\pi, \pm 3\pi, \pm 5\pi, \ldots$

Frequently the functions $\sin(x)$ and $\cos(x)$ are composed with the linear function $g(x) = ax + b$. This results in the functions

$$y = a\sin(x) + b \qquad \text{or} \qquad y = a\cos(x) + b \tag{6.1}$$

and

$$y = \sin(ax + b) \qquad \text{or} \qquad y = \cos(ax + b). \tag{6.2}$$

The graphs of the functions (6.1) and (6.2) are very similar to the graphs of $\sin(x)$ and $\cos(x)$.

The graph of $f(x) = a\sin(x) + b$ is obtained from the graph of $y = \sin(x)$ in two steps. First the y-values are multiplied by the number a to obtain the graph of $y = a\sin(x)$. Multiplying the basic sine function by the number a either stretches the y-values if $|a| > 1$, or compresses the y-values if $|a| < 1$, so that the amplitude of the graph is $|a|$; $-|a| \leq a\sin(x) \leq |a|$. If a is positive, the graph is left unchanged; but if a is negative, the graph is "flipped" or "rotated" about the x-axis. This first step is illustrated in Fig. 6.5. The second step is to shift the graph of $y = a\sin(x)$

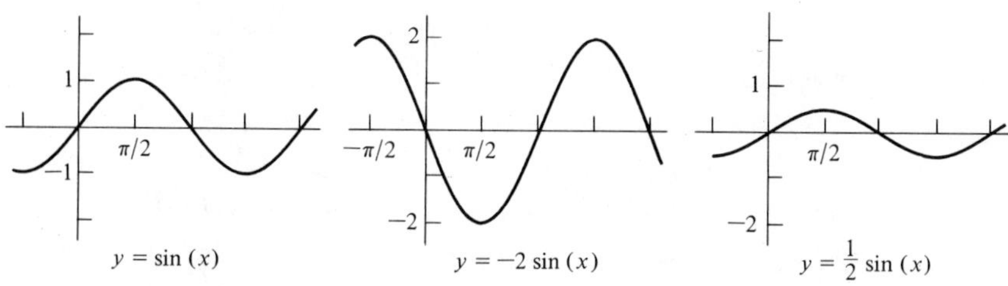

Figure 6.5

vertically b units. If $b > 0$, then the graph is shifted upward. If $b < 0$, the graph is shifted downward as illustrated in Fig. 6.6.

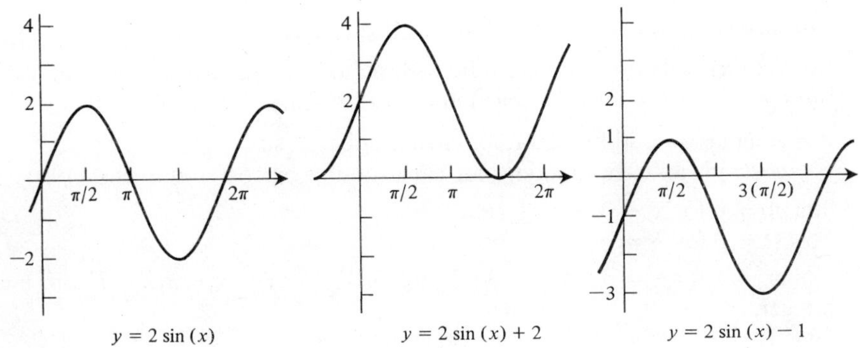

Figure 6.6

The visual difference between the graph of $y = \sin(x)$ and $y = \sin(ax + b)$ appears as a "stretching" or "compressing" of the x-axis combined with a horizontal shift. The first step in graphing $y = \sin(ax + b)$ is to graph $y = \sin(ax)$. Multiplying the x-values by the number a changes the period of the function $y = \sin(ax)$ to $T = 2\pi/|a|$ (this is verified below). Consequently if $|a| > 1$, the period is shortened and the graph is horizontally "compressed." If $|a| < 1$, then the period is increased and the graph is horizontally "stretched." If a is positive, the graph is not altered; but if a is negative, the graph is "flipped" or "rotated" about the y-axis as illustrated in Fig. 6.7.

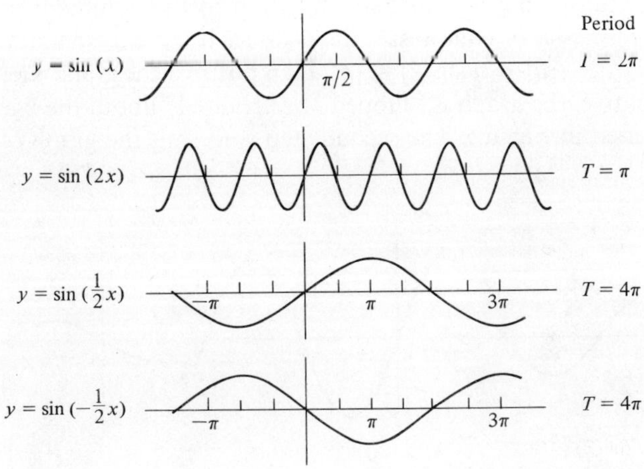

Figure 6.7

The graph of $y = \sin(ax + b)$ is then obtained from the graph of $y = \sin(ax)$ by horizontally shifting the graph $|b/a|$ units, to the left if b/a is positive and to the right if b/a is negative. In Fig. 6.8 we illustrate this shift.

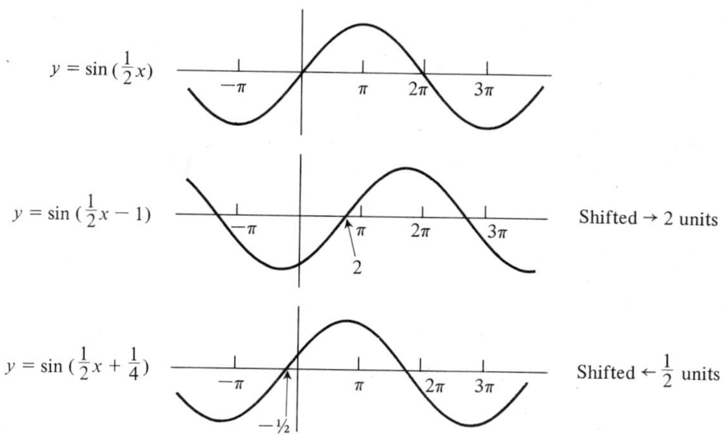

Figure 6.8

Algebraically, the period T of the function $g(x) = \sin(ax)$ is the smallest positive number T, such that $g(x) = g(T + x)$. That is,

$$\sin(ax) = \sin(a(T + x)) = \sin(aT + ax).$$

But the period of the function $y = \sin(x)$ is 2π, therefore

$$\sin(ax) = \sin(2n\pi + ax),$$

n an integer. Equating $\sin(aT + ax)$ and $\sin(2n\pi + ax)$ leads us to the identity $aT = 2\pi$ or $T = 2\pi/a$ if $a > 0$. If $a < 0$, since T must be positive, it follows that $T = 2\pi/-a$.

Comparing the graphs in Fig. 6.9(a) of $y = \sin(x)$ and $y = \cos(x)$, we see that the graphs are identical except for a horizontal shift. This reflects the algebraic relation between $\sin(x)$ and $\cos(x)$ expressed by the equations

$$\cos(x) = \sin(x + \pi/2) \qquad \text{and} \qquad \sin(x) = \cos(x - \pi/2). \tag{6.3}$$

The reader who has never encountered trigonometric functions may find the pace of this section too fast. We have assumed that the reader has experienced trigonometric functions and thus have only presented some highlights of these functions. The level of understanding which we will require in this text will be only a basic feeling for the form of the graphs of the functions in (6.1) and (6.2). If you have difficulty with these, we suggest that you work the exercises by imitating the examples sketched in Fig. 6.9.

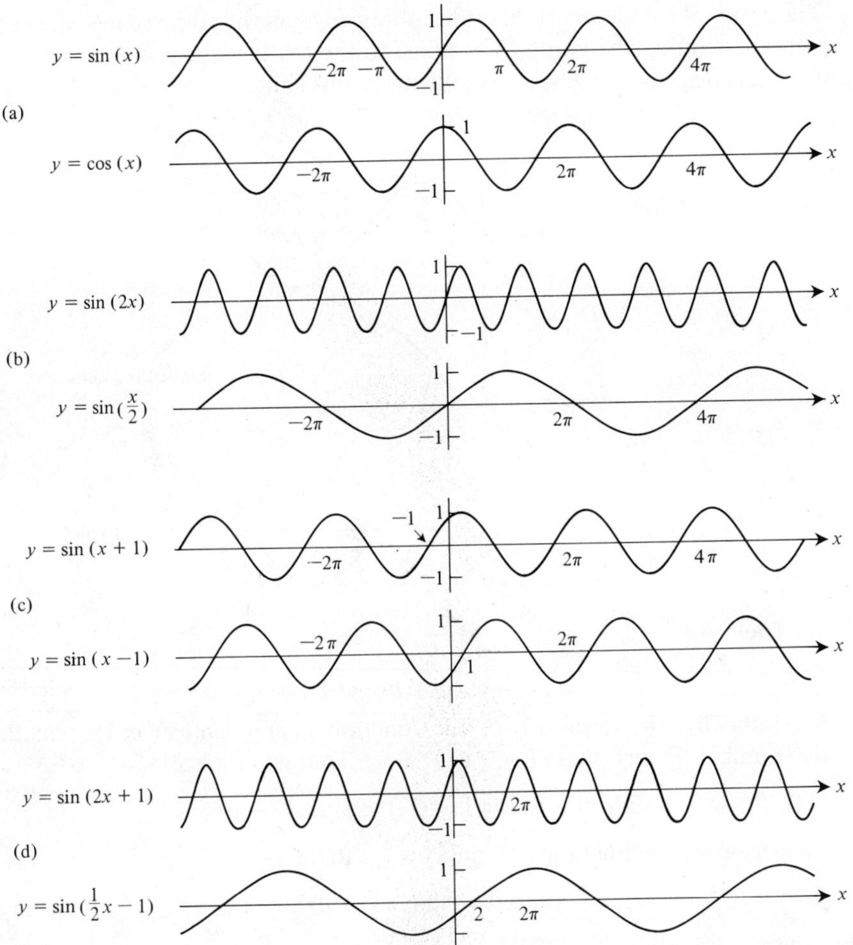

Figure 6.9

Example 6.3 Radiant energy is described by a sinusodial function which describes how the energy is propagated with time. It is thus described by the wave function

$$f(t) = A \sin\left(\frac{2\pi Ct}{W}\right) + \text{constant},$$

where W is the wavelength and A is the amplitude of the wave. The wavelength is divided by C, therefore the period T of the sine function is $y = \sin(2\pi Ct/W)$. The frequency f of a wave is the number of periods per unit time, usually denoted as cycles per second (cps).

$$f = C/W, \qquad C = \text{speed of light}, \qquad C \simeq 3 \times 10^{10}\ \text{cm/sec}.$$

The nature of radiant energy is determined by its wavelength as indicated in the following table. Note that low frequency corresponds to long wavelength.

Low frequency radio	10^4
	10^2
Ultra-high frequency radio	0
	10^{-2}
Radar microwaves	10^{-4}
Visible light	10^{-6}
	10^{-8}
X-rays	10^{-10}

◀

6.3 OTHER TRIGONOMETRIC FUNCTIONS

In this subsection we introduce four more trigonometric functions. These are each defined in terms of the functions sin(x) and cos(x). After defining these functions we indicate their graphs in Fig. 6.10. You should memorize the definition of each function and be able to associate each function with its graph. At this point we will not concentrate on the finer details concerning these functions and their graphs.

Definition 6.4

a) The *tangent function*, tan(x), is defined for x not an odd multiple of $\pi/2$ by

$$\tan(x) \equiv \frac{\sin(x)}{\cos(x)}.$$

b) The *cotangent function*, cot(x) (sometimes ctn(x)), is defined for x not a multiple of π by

$$\cot(x) \equiv \frac{\cos(x)}{\sin(x)}.$$

c) The *secant function*, sec(x), is defined for x not an odd multiple of $\pi/2$ by

$$\sec(x) \equiv \frac{1}{\cos(x)}.$$

d) The *cosecant function*, csc(x), is defined for x not a multiple of π by

$$\csc(x) \equiv \frac{1}{\sin(x)}.$$

Observe from Fig. 6.10 that tan(x) and cot(x) have period π while sec(x) and csc(x) have period 2π. From the definition of the functions sin(x) and cos(x) as coordinates on a unit circle we obtain the identity

$$\sin^2(x) + \cos^2(x) = 1. \tag{6.4}$$

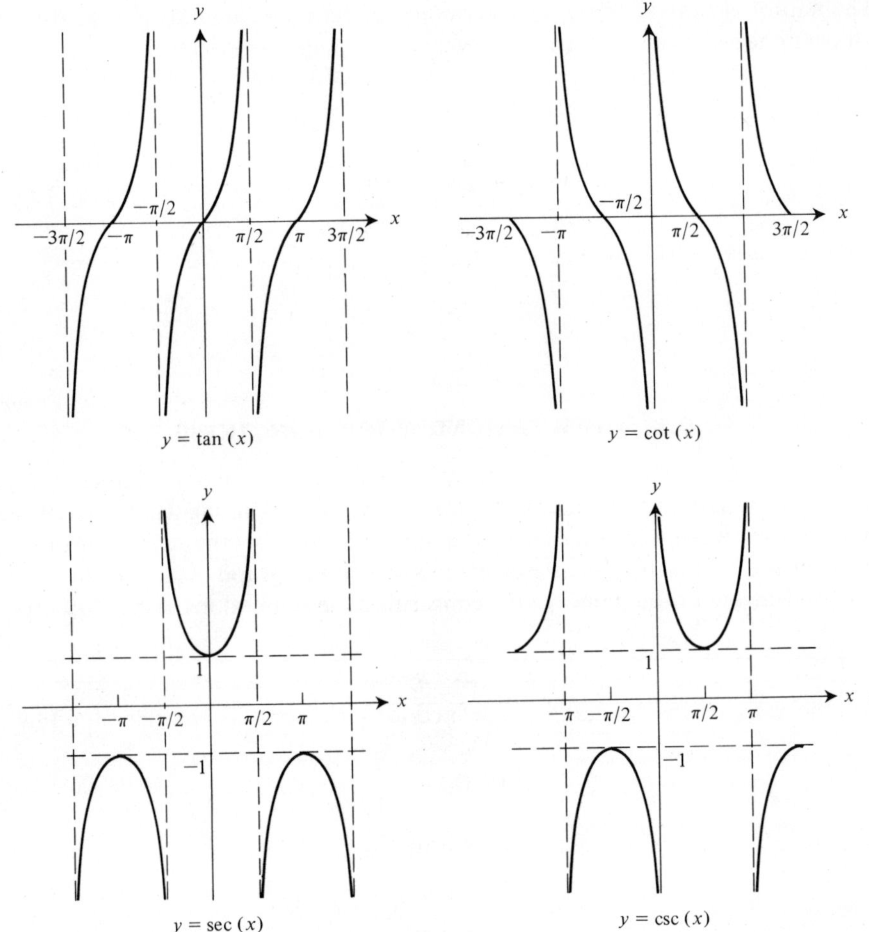

Figure 6.10

Dividing this equation by $\cos^2(x)$ and $\sin^2(x)$ results in the equations

$$\sec^2(x) = 1 + \tan^2(x) \tag{6.5}$$

and

$$\csc^2(x) = 1 + \cot^2(x). \tag{6.6}$$

Combining these equations with the shift equations (6.3), it is possible to obtain an extensive number of trigonometric identities. As we will not need these in later portions of this text, we do not present them here. The reader who encounters an identity or manipulation involving trigonometric functions is referred to any text on trigonometry or book of mathematical tables.

The independent variables x or t used in all of the above discussions are measured in terms of *radians*. The trigonometric functions can also be defined as functions of angles. This is usually accomplished with the aid of a right triangle as illustrated in Fig. 6.11. The unit of measurement of angles is *degree* (45 degrees = $\pi/4$ radians).

Pythagorean identities

$H^2 = A^2 + B^2$

$1 = \sin^2(\theta) + \cos^2(\theta)$

$\sec^2(\theta) = \tan^2(\theta) + 1$

$\csc^2(\theta) = 1 + \cot^2(\theta)$

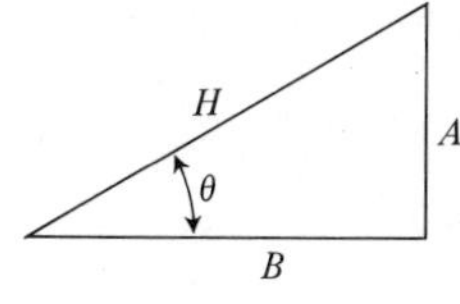

θ = angle measurement

Definition of trigonometric functions via the lengths of the sides of the depicted right triangle.

$\sin(\theta) = A/H$	$\csc(\theta) = H/A$
$\cos(\theta) = B/H$	$\sec(\theta) = H/B$
$\tan(\theta) = A/B$	$\cot(\theta) = B/A$

Figure 6.11

EXERCISE SET 6

6.1 List five naturally occurring periodic processes and indicate the basic period of each.

6.2 Sketch the graph of the function $y = f(x)$ if $f(x)$ is periodic with period $T = 2$ and over the interval $[0, 2)$ has the indicated form. Indicate the graph for $-2 \le x \le 6$.

a) $y = x$ b) $y = 2 - x$

c) $y = \begin{cases} 0 & \text{if} \quad 0 \le x < 1, \\ 1 & \text{if} \quad 1 \le x < 2. \end{cases}$ d) $y = \begin{cases} x & \text{if} \quad 0 \le x < 1, \\ 2 - x & \text{if} \quad 1 \le x < 2. \end{cases}$

e) $y = x(2 - x)$ f) $y = (x - 1)^2$

6.3 Sketch the graph of the following functions.

a) $y = 2[\![x]\!]$ b) $y = [\![2x]\!]$ c) $y = [\![x]\!] - x$
d) $y = (x - [\![x]\!])^2$ e) $y = [\![x + \frac{1}{2}]\!]$ f) $y = [\![x^2]\!]$

6.4 Determine the approximate value of the given numbers using the tables in Appendix II. The arguments are assumed to be in radians.

a) $\cos(0)$ b) $\cos(1)$ c) $\cos(2)$
d) $\sin(0)$ e) $\sin(1)$ f) $\sin(2)$
g) $\cos(\pi/4)$ h) $\sin(\pi/4)$ i) $\tan(\pi/4)$
j) $\sec(\pi/3)$ k) $\csc(\pi/3)$ l) $\cot(\pi/6)$.

6.5 Sketch the graph of the indicated functions.

a) $y = \cos(x)$ b) $y = 2\cos(x)$
c) $y = 2\cos(x) + 1$ d) $y = -2\cos(x) + 1$
e) $y = 2\cos(x) - 2$ f) $y = \frac{1}{2}\cos(x) - 2$
g) $y = 2\sin(x) + 1$ h) $y = \frac{1}{2}\sin(x) - 2$

6.6 Sketch the graph and determine the period T of the indicated function.

a) $y = \sin(x)$ b) $y = \sin(x + \pi)$
c) $y = \sin(x + \pi/2)$ d) $y = \cos(x + \pi)$
e) $y = \cos(2x)$ f) $y = \cos(\frac{1}{2}x)$
g) $y = \cos(x - \pi/4)$ h) $y = \cos(\frac{1}{2}x - \pi/4)$
i) $y = \sin(\pi x)$ j) $y = 3\cos(\pi/2 - x)$
k) $y = \sin(-x)$ l) $y = \cos(-x)$
m) $y = \sin(\pi x + \pi/2)$ n) $y = \sin(\pi x - \pi)$
o) $y = \cos(\pi x/3)$ p) $y = \cos(3\pi x)$

6.7 Let $f(x) = \cos(x)$, $g(x) = x^2 + 1$, $h(x) = \sin(x)$, and $k(x) = x - \pi$. Determine the following composition functions and evaluate each function at the indicated point x_0.

a) $(f \circ g)(x)$, $x_0 = 0$ b) $(g \circ f)(x)$, $x_0 = 0$
c) $(h \circ k)(x)$, $x_0 = \pi$ d) $(g \circ h)(x)$, $x_0 = 3\pi/2$
e) $(g \circ f \circ k)(x)$, $x_0 = \pi/2$ f) $(h \circ g)(x)$, $x_0 = 0$
g) $(g \circ f)(x)$, $x_0 = 0$ h) $(f \circ h)(x)$, $x_0 = 0$
i) $(h \circ f)(x)$, $x_0 = 3\pi/2$

6.8 Sketch the graph of the given functions and indicate the following: (i) the domain of the function; (ii) the period of the function.

a) $y = \tan(x)$ b) $y = \tan(2x)$ c) $y = \tan(\frac{1}{2}x)$
d) $y = \cot(\frac{1}{3}x)$ e) $y = 2\sec(x)$ f) $y = \csc(x - \pi/2)$

6.9 The volume of blood pumped from the heart is a periodic function of time. During the systolic phase of a heartbeat, the heart contracts and squeezes or pumps blood into the arterial system. During the diastolic phase of a heartbeat, the muscles of the heart relax, the heart fills with blood, and no blood is pumped. Assume the length of the diastolic phase is 2 units of time and the length of the systolic phase is 3 units of time.

a) What is the period of a heartbeat?
b) If the volume of blood pumped during the systolic phase has the form of a sine function, construct a function $V(t)$ which would describe the volume of blood pumped over one cycle of a heartbeat.
c) Graph the volume function, $V(t)$ of part (b), over several heartbeats.

6.10 Using the identities and a triangle as in Fig. 6.11 find the following:

a) $\cos(\theta)$ if $\sin(\theta) = x/2$ b) $\sin(\theta)$ if $\tan(\theta) = x/3$
c) $\tan(\theta)$ if $\sin(\theta) = x/4$ d) $\cos(\theta)$ if $\tan^2(\theta) = x^2/(x + 1)$

6.11 a) Sketch the graph of the function $f(t)$ in Example 6.3.
b) What is the frequency of a wave with wavelength $w = 0.0004$ cm?
c) Sketch the graph and give the equation of a wave function with frequency of 3 cps and intensity $f(t)$ varying between 4 and 8.

6.12 Blood pressure varies sinusodially between 80(mm Hg) and 120(mm Hg). Assume the period of the blood pressure is the same as that of a heartbeat but with a lag of one-half period. Give an equation for the blood pressure of a person with (a) 70 heartbeats/minute; (b) 80 heartbeats/minute; (c) 80 heart beats/minute but high blood pressure which is 20(mm Hg) above normal.

6.13 Set up and graph a sine function to describe the variation in average daily temperatures over several years. What is the period and amplitude of your function? (Use your local temperature.)

6.14 In reference to Example 6.3. The wavelength of visible light varies from 4000 Å to 7000 Å (Å is the symbol for the angstrom unit; one Å = 10^{-8} cm = 10^{-10} m.)

a) What is the range of the frequency of visible light?
b) Sketch the "wave" associated with a yellowish-orange light having wavelength 6000 Å if the amplitude of the wave is $\sqrt{2/L}$, where L is the basic period of the wave.

6.15 The vibration of a string may be described by sine functions. If the string is of fixed length L, such as a violin or guitar string, then the "fundamental" tone of the string is described by a sine function of period $2L$. The first overtone is described by a sine function of period $2L/2$. The second overtone is described by a sine function of period $2L/3$, and so forth.

a) Give the equation of the sine function associated with the fundamental tone and the first three overtones of a string of length $L = \pi$ and amplitude $A = 2$.
b) On the same graph sketch the functions of part (a) for $0 \leq x \leq L$.

SECTION 7

GRAPHS, SCALE, AND FORM

7.1 INTRODUCTION

The graph of a function f is defined to be the set of points (x, y) such that $f(x) = y$. (See Fig. 7.1(a).) Theoretically, the graph of a function and the equation of a function provide the same information. In practice this is not so. The material of this section is intended to investigate the amount of information provided by graphs and how the same function can be represented by apparently different yet equivalent graphs.

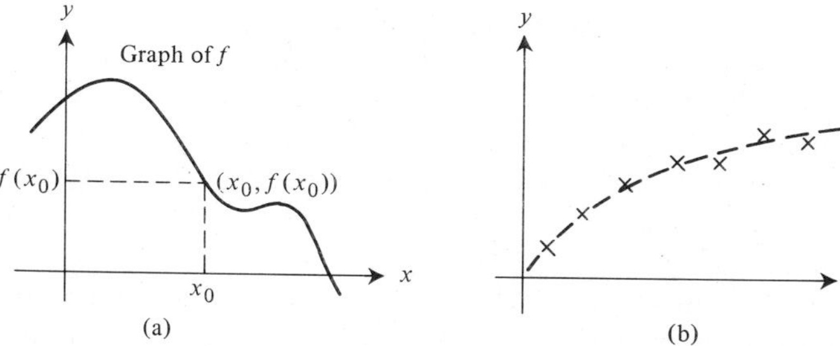

Figure 7.1

This section is not intended to provide special graphing techniques for general functions. This will be done after we have developed the tools of differential calculus. We do, however, examine the graphs of a few special types of functions—namely, allometric equations and exponential equations.

Equations of the form $y = f(x)$ provide a way of computing the value $f(x_0)$ of a function at any specific point x_0. With the aid of a computer, it is possible to evaluate a function at an extremely large number of points. However, the data generated are cumbersome and provide little help in understanding the behavior of a function over an interval of x values. The difficulty is one of assimilation; it is hard to comprehend a large volume of numerical data corresponding to the function's values.

A graph, on the other hand, provides an instant visual image of a function. Even when a graph is only partially indicated, say by points plotted from experimental data, the human mind can recognize and differentiate a considerable amount of information about the associated function. But the visual image of a graph is entirely dependent upon the choice of variables and scales represented by the coordinate axes. The visual image of the graph of a function or a set of data can vary widely, a fact which will prove useful but which can also lead to misinterpretation of graphs.

One of the more frequent mathematical problems which a scientist encounters is to determine the equation of a function from the graph of the function, or, more exactly, from a scattered collection of data points, which are assumed to be on or near the graph of the function. (See Fig. 7.1(b).) The problem of deriving the equation of a function from its graph is extremely difficult. We will consider a restricted version of this problem in which we know *a priori* the general form of the function. Hence our problem will be one of parameter fitting. For instance, we will know that the data satisfy an allometric equation of the form $y = cx^k$ and thus need to "fit" only the parameters c and k. The ease with which parameters can be determined from data points will depend upon the coordinate system used to represent the data as well as on the complexity of the function.

7.2 SCALING OF COORDINATE AXES

In mathematics courses it is the practice to represent graphs on standard x-y coordinate systems in which the scales used on both the axes are the same. There are good reasons for this practice, one of which is that, to preserve the common concepts of area, the axes must be on the same scale. In practice, however, the scientist seldom uses the same scale on both axes of a graph. Reasons for this include the fact that the variables represented on the axes are usually representing different biological or physical quantities. Frequently, the range of variables that are of experimental significance tend to vary markedly—for example, the geologist measures in units of kilometres while the biochemist measures in units of microlitres, 1 μl = 10^{-6} litres.

When sketching a graph, we try to satisfy three basic requirements:

1. That the graph clearly represent the data or function;
2. That the graph depict the data and the function it represents over the entire range of experimentally significant values; and,
3. That the graph be restricted to reasonable physical dimensions.

To satisfy these requirements, it is often necessary that we *rescale* the data or equivalently choose appropriate scales for the variables represented by the coordinate axes.

Rescaling is the process of introducing a new variable, say u, which is related to the original variable, say x, by a specified scaling function, $u = f(x)$. This results in a transformation $x \to u$ of the x-axis to the u-axis.

In most applications you will encounter only three basic scaling transformations:

Homogeneous scaling: $x \to u = k \cdot x$, k a constant,
Logarithmic scaling: $x \to u = \log(x)$,
Reciprocal scaling: $x \to u = 1/x$.

When using a uniformly scaled axis, the effect of each of these transformations is to alter the visual relationship between points on the x-axis.

Homogeneous scaling corresponds to the expansion or compression of the x-scale. For instance, the transformation $x \to u = 2x$ results in a stretching of the x-scale so that the visual distance between points is doubled. (See Fig. 7.2(a).)

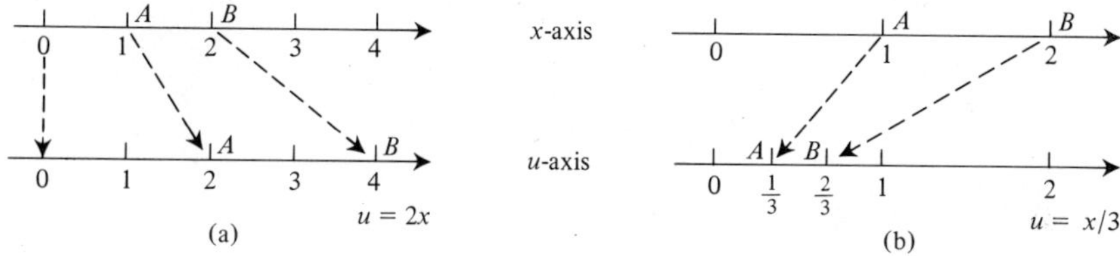

Figure 7.2

The transformation $x \to u = x/3$ results in a contraction of the x-scale so that the visual distance between points on the u-axis is reduced to one-third of the distance between the corresponding points on the x-axis. (See Fig. 7.2(b).) In general, the effect of a homogeneous scaling $x \to u = kx$ is to stretch the x-axis if $k > 1$, and to contract the x-axis if $0 < k < 1$. If k is negative, the transformation reverses the orientation of the axis in addition to either stretching or contracting the scale. (See Fig. 7.3.) Consequently, to represent small or close values, a constant $k > 1$ is chosen; and, to represent a large range of numbers, the scale is compressed by choosing $k < 1$.

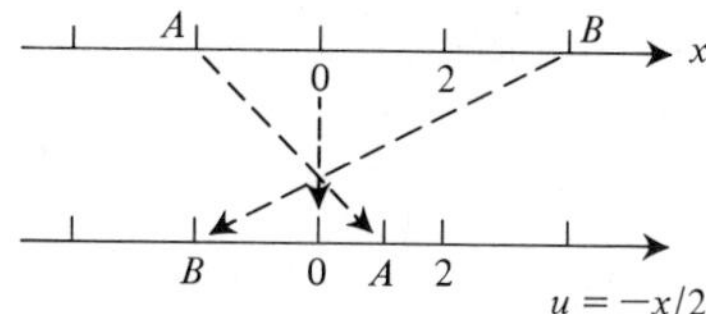

Figure 7.3

Example 7.1

a) An experiment consists of performing a test using various concentrations of a key chemical. The set of x-values corresponding to the test concentrations is $S = \{0.1, 0.25, 0.75, 0.9\}$. To represent this data set on a uniform scale (such as on ordinary graph paper), we could introduce a transformation $x \to u = 10x$. This transformation would result in a transformed set of values, $T = \{1, 2.5, 7.5, 9\}$.

b) To concisely represent the specific days associated with the migratory cycle of snow geese, we could introduce a compressed scale by dividing the day of the year by seven. The transformation $d \to w = d/7$ would transform a day scale into a week scale. ◀

Example 7.2 We sketch the graph of the exponential curve $y = 2^x$ for x in the interval $[0, 10]$. To accomplish this with the same scale on the x- and y-axis, we would require a large piece of paper because as x varies from 0 to 10, y would vary from 1 to 1024. To overcome this large range of y-values we plot the function on an x-v coordinate system, where the variable v is related to the y variable by the equation $v = 10^{-2}y$. The graph is sketched in Fig. 7.4. (Note that the curve sketched is $v = 10^{-2}2^x$.)

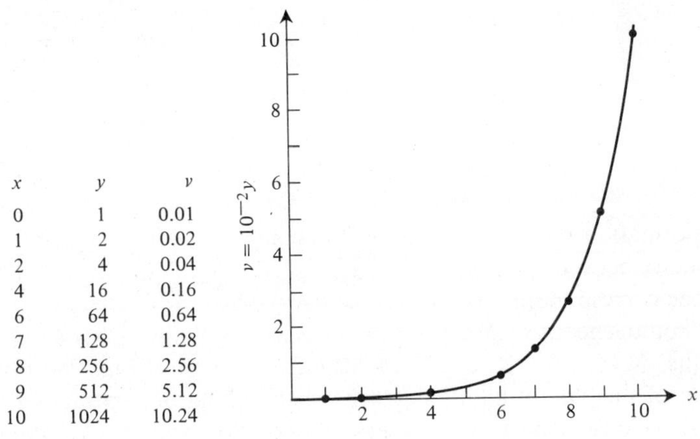

x	y	v
0	1	0.01
1	2	0.02
2	4	0.04
4	16	0.16
6	64	0.64
7	128	1.28
8	256	2.56
9	512	5.12
10	1024	10.24

Figure 7.4

◀

Logarithmic scaling is applicable only to positive data or x-values since Domain$(\log(x)) = (0, +\infty)$. The log transformation $x \to u = \log(x)$ is illustrated in Fig. 7.5, where we indicate the effect of the transformation on several x-values:

$$x = 0.1 \to u = \log(0.1) = -1;$$

$$x = 1 \to u = \log(1) = 0;$$

$$x = 5 \to u = \log(5) \simeq 0.7, \text{ etc.}$$

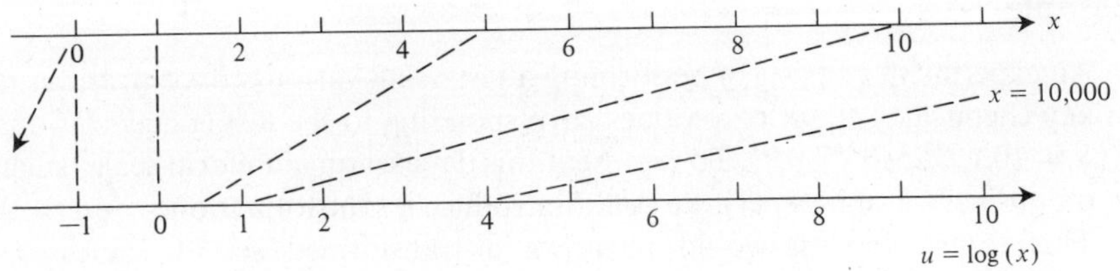

Figure 7.5

The four basic features of the log transformation are as follows.

1. It transforms the half-axis $(0, +\infty)$ into the entire real line $(-\infty, +\infty)$.
2. It transforms 1 to 0.
3. It compresses the x values in $(1, +\infty)$. (For instance, $x = 100 \to u = 2$ and $x = 10^5 \to u = 5$.)
4. It expands the x values in $(0, 1)$. (For example, $x = \frac{1}{10} \to u = -1$, and $x = 10^{-8} \to u = -8$.) ◀

Example 7.3 Table 7.1 indicates the half-life of several radioactive isotopes used in medicine. The symbol y indicates years and d indicates days. To represent these half-lives on a linear scale, it is necessary to make a transformation of the given $t_{1/2}$ values. The log transformation $t_{1/2} \to L = \log(t_{1/2})$ provides an ideal rescaling. We have indicated the L values corresponding to the $t_{1/2}$ (years) values and the representation of the isotopes on an L-axis. (The $t_{1/2}$ values given in days are first converted to years and then logarithmically transformed.) See also Fig. 7.6.

Table 7.1

Isotope	^{3}H	^{14}C	^{36}Cl	^{85}Kr	^{59}Fe	^{131}I
Half-life	$12.26y$	$5760y$	$3.1 \times 10^5 y$	$10.6y$	$45d$	$8d$
$L = \log(t_{1/2})$	1.09	3.76	5.49	1.03	-0.90	-1.65

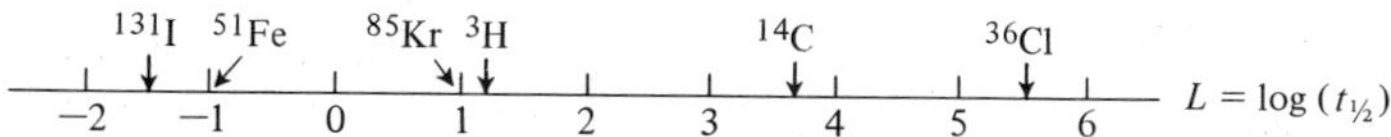

Figure 7.6

◀

The *reciprocal transformation* $x \to u = 1/x$ is defined for $x \neq 0$. This transformation, like the logarithmic transformation, stretches the x-values in the interval $(0, 1)$ and compresses the x-values in the interval $(1, +\infty)$. The basic properties of the reciprocal transformation are as follows.

1. It transforms the segments of the real line as follows:

$$x \in (0, 1) \to u \in (1, +\infty),$$
$$x \in (1, +\infty) \to u \in (0, 1),$$
$$x \in (-1, 0) \to u \in (-\infty, -1),$$
$$x \in (-\infty, -1) \to u \in (-1, 0).$$

2. The points $x = -1$ and $x = 1$ are left fixed by the transformation.
3. The orientation of positive (or negative) points is reversed. For instance, if $x_1 = 2$ and $x_2 = 3$, then $x_1 \to u_1 = \frac{1}{2}$ and $x_2 \to u_2 = \frac{1}{3}$; $x_1 < x_2$ but $u_1 > u_2$.

7.3 NONUNIFORM COORDINATE SYSTEMS

The first step in graphing any function or set of data is to establish an appropriate scale for the coordinate axes. If the same uniform scale is to be used on both axes, as was done in Fig. 7.4, it is usually necessary to introduce a transformation as illustrated in Example 7.2. Alternatively, the introduction of new scaled variables can be avoided by choosing different scales for each coordinate axis. For instance, the graph of $y = 2^x$ (given in Fig. 7.4) could be sketched as shown in Fig. 7.7. Observe that the unit value on the y-axis is 10^{-2} as large as the unit value on the x-axis.

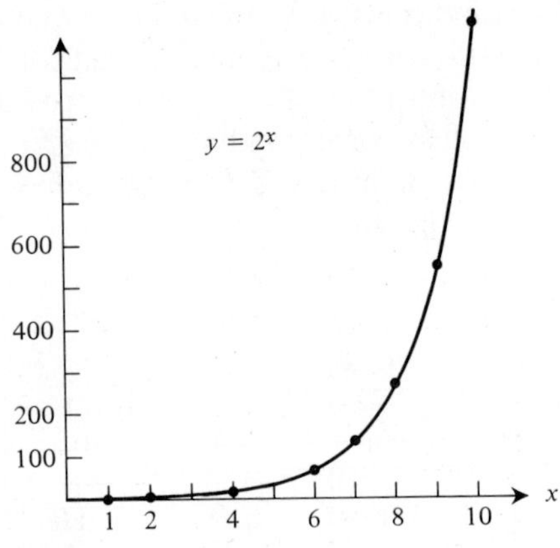

Figure 7.7

If you compare the curves in Figs. 7.4 and 7.7, you will see that they are identical. The only change is in the scale of the y-axis. The process of choosing a different scale for a coordinate axis is the *physical* analogy of the algebraic process of performing the transformation $y \to v = 10^{-2}y$. The term *physical* is used to emphasize the fact that it is the physical distance used to represent a basic unit on the scale which has been changed.

When both coordinate axes are rescaled, the shape of a graph can be drastically altered. For convenience and economic utilization of space, most graphs are scaled so that the physical dimensions of the axes are approximately equal. However, to emphasize a point or feature of a graph, the scales may be chosen to dictate the resulting visual image.

Example 7.4 The change in visual image of a graph due to the physical rescaling of the axis can be dramatic. In Fig. 7.8 the line $y = 5x + 2$ is sketched using two different scales—the one on the right minimizes the slope of the line, and the one on the left overaccentuates the slope of the line. When presenting an argument in behalf of a cause, the utilization of graphs with their instant visual image is a tried-and-tested technique. The principal reason why this is so and the reason why graphs frequently fall into disrespect is that they may be manipulated to "show" almost anything. The simplest and most efficient way to distort the visual impact of a graph is to physically change the scale as illustrated in Fig. 7.8.

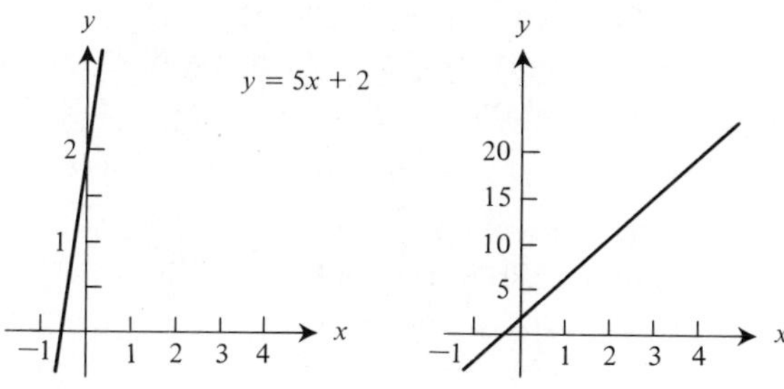

Figure 7.8

The Table 7.2 presents data indicating the average world yield of major grains (excluding rice) from 1960 to 1973. If you felt that the yield was increasing at a sufficient rate and wished to emphasize this, you might plot the data as shown in Figure 7.9(a). However, if you felt that the increase was not sufficient and wanted to present the data in a way which would emphasize this, you would probably use a graph similar to that in Fig. 7.9(b).

Table 7.2 Average world yield of major grains measured in quintals per hectare

Year	Yield	Year	Yield	Year	Yield
1960	13.9	1965	14.7	1970	17.3
1961	13.4	1966	16.2	1971	18.8
1962	14.3	1967	16.2	1972	18.5
1963	13.9	1968	16.7	1973	19.4
1964	14.5	1969	16.9		

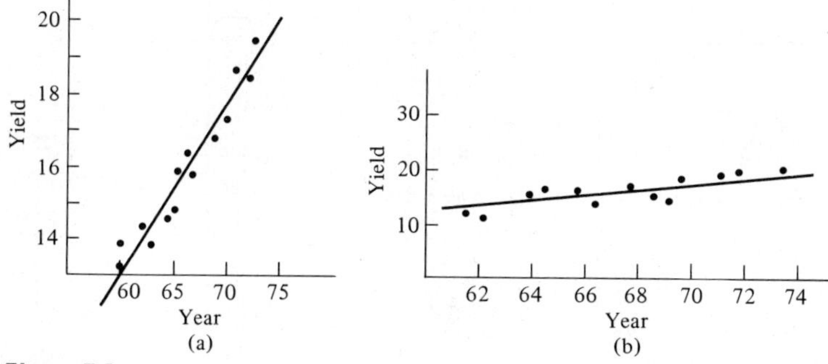

Figure 7.9

7.4 LOG–LOG GRAPHS

When both axes of a coordinate system are scaled by a logarithmic transformation, the resulting coordinate system is referred to as a *log–log* system. The representation of data or graphs of equations can be arrived at in two ways using log–log coordinates. One way is to algebraically transform x-y data into u-v data by using the two logarithmic transformations

$$x \to u = \log(x) \qquad \text{and} \qquad y \to v = \log(y).$$

The transformed points can then be plotted on a uniform u-v coordinate system. For instance, the x-y point (2, 1000) would be transformed into the u-v point $(\log(2), \log(1000)) = (0.3, 3)$ (approximately).

A second way to graph data using a log–log coordinate system, which is the physical analogy of the first method, is to physically rescale the coordinate axes to have distances measured in logarithmic units. On the x-axis a log scale is constructed by associating each number x_0 with a point on the x-axis that is physically $\log(x_0)$ units from the origin. Since the function $\log(x)$ is not linear, the physical distances between the x-values are not uniform. For instance, if the point corresponding to $x = 10$ is one unit on the x-axis (with a log scale!), then the point corresponding

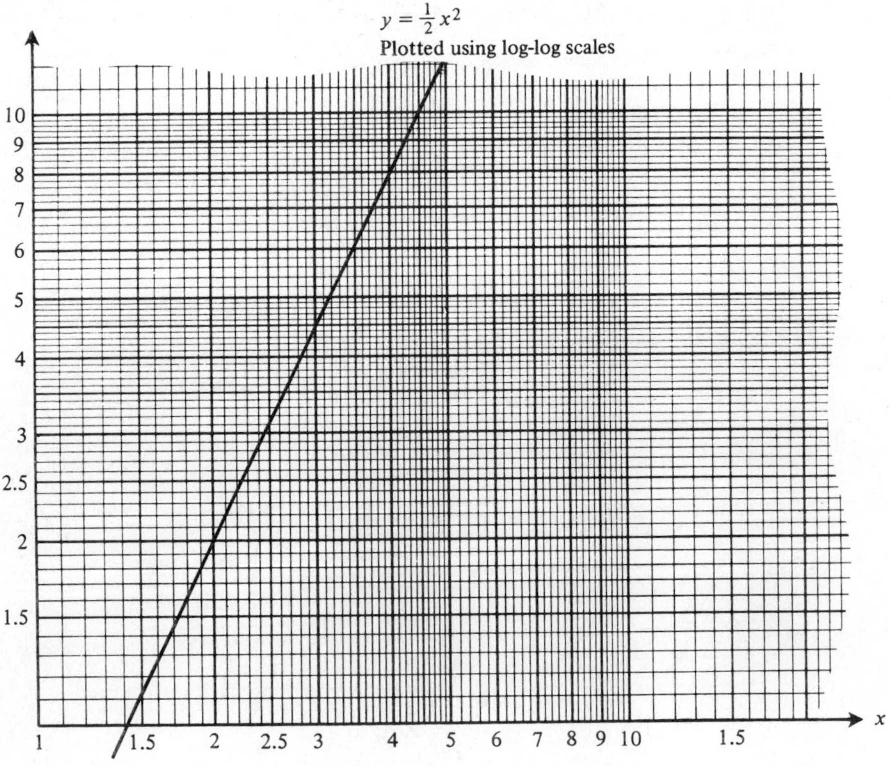

Figure 7.10

to $x = 100$ will be two units along the x-axis, and the point corresponding to $x = 200$ will be $\log(200) \simeq 2.3$ units along the x-axis (and similarly with the y-axis). Fortunately, it is not necessary to construct such log scales every time you wish to plot a log–log graph. Log–log graph paper, which has its units of measurement along each axis measured in logarithmic distances, is readily available. The markings on the axes correspond to the logarithmic distances from the origin.

The graphing of data on log–log scales is made very easy by the introduction of log–log graph paper. The lines on the paper are not evenly spaced as is the case with ordinary graph paper. Instead, they are distributed on a logarithmic scale. When you plot a point (x, y) on a log–log scale, the physical distances are not x and y units from the "origin" but are instead $\log(x)$ and $\log(y)$ units. Thus the point $(1, 1)$ is physically the point $(\log(1), \log(1)) = (0, 0)$ and corresponds to the "origin" on a log–log scale. Similarly, the point $(10, 100)$ is physically at a position $(\log(10), \log(100)) = (1, 2)$. As a consequence of the spatial distortion of the scales, to plot points on log–log graph paper, you do not need to evaluate the logarithm of each x or y value. Fig. 7.10 illustrates a log–log graph of the equation $y = \frac{1}{2}x^2$. We have plotted the points $(2, 2)$, $(3, \frac{9}{2})$, and $(4, 8)$. Observe that the graph of quadratic function $y = \frac{1}{2}x^2$ appears to be linear when graphed on log–log paper.

The following example illustrates the equivalence of the two methods presented for log–log graphing.

Example 7.5 We consider the following set of points in an x-y coordinate system:

$$s = \{(1, 2), (2, 10), (10, 5), (50, 10), (100, 5), (200, 2), (1000, 10)\}.$$

We plot these points using both of the methods discussed above.

a) We make the change of variable $x \to u = \log(x)$, $y \to v = \log(y)$. This is accomplished by looking up the logarithm of the numbers involved (rounding off) and expressing the set s as a new set $\hat{s}$ in the u-v coordinate system:

$$\hat{s} = \{(0, 0.3), (0.3, 1), (1, 0.7), (1.7, 1), (2, 0.7), (2.3, 0.3), (3, 1)\}.$$

We next graph the set $\hat{s}$ of the transformed values on a u-v coordinate system as shown in Fig. 7.11.

b) We plot the set s on log–log graph paper as indicated in Fig. 7.12.

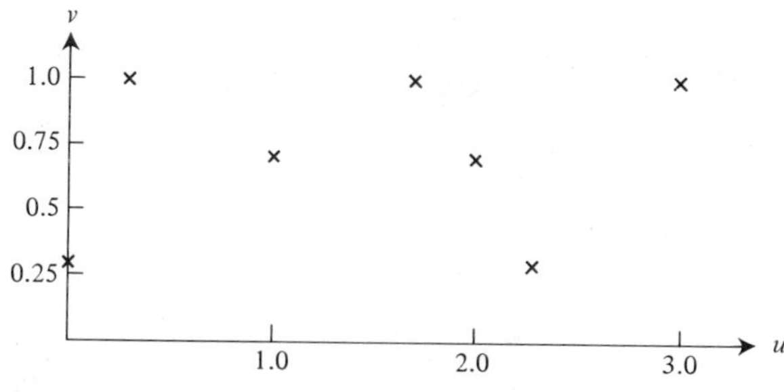

Figure 7.11

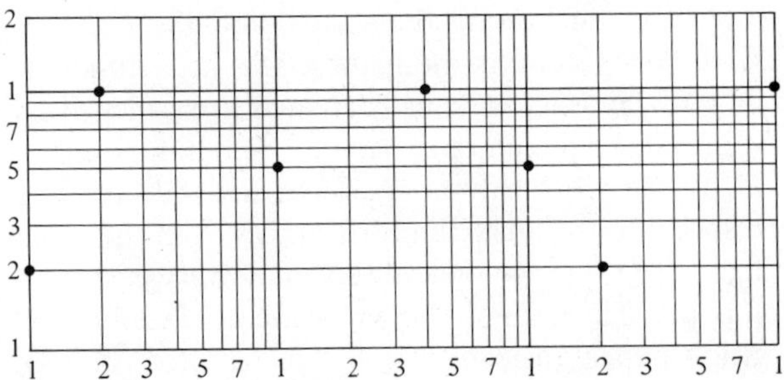

Figure 7.12

The graphs in Fig. 7.11 and Fig. 7.12 are identical except for the scales indicated. This is because the two methods of graphing are equivalent. In the first method (a), we had to look up the logarithms before we plotted the points; in the second method (b), the log values were not needed since the scales incorporated the process of the log transformation. Thus the convenience of using log–log graph paper lies in the fact that the conversion to a logarithmic scale has been done for us. ◀

The scales on the graph in Fig. 7.12 have the form given in Fig. 7.13. The only numbers on the scale are 1, 2, . . . , 9, which then repeat. This allows greater flexibility in graphing different numerical values. Whenever you use log–log paper, you must make a choice of both the basic scale unit and the "origin" of the graph—i.e., the point corresponding to (1, 1). Examples of different choices for the scale units would be as shown in Fig. 7.14.

Often only the scale values denoted by 1 (or 10) on commercial log scales are indicated on graphs. These points correspond to the different powers of 10. For example, a log scale used to indicate the wavelength of visible light (see Example 6.3) would be indicated as shown in Fig. 7.15. Point A on this scale would be 3×10^6; point B would be 2×10^7; and point C would be 7×10^7.

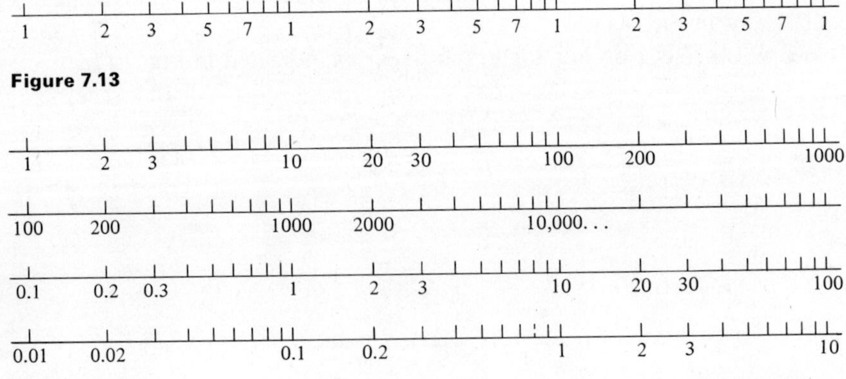

Figure 7.13

Figure 7.14

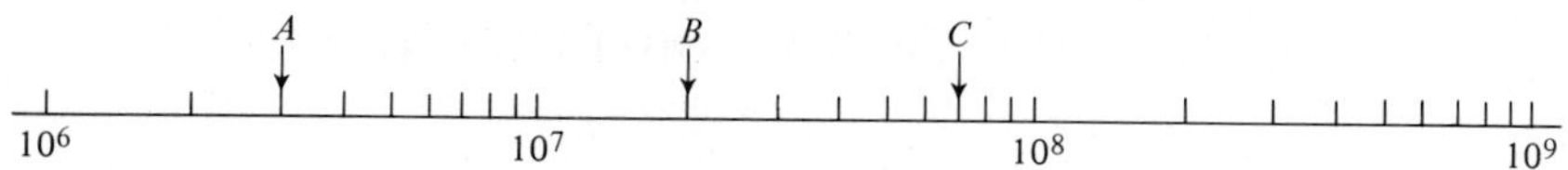

Figure 7.15

Linear Graphs on Log–Log Scales

The principal use of log–log scales is to graph allometric relations. The allometric equation

$$y = cx^k \tag{7.1}$$

is not easily graphed on a standard x-y coordinate system because (1) it is hard to evaluate x^k; (2) the graph quickly becomes too large (or too small); and, (3) it is difficult to sketch curved (i.e., nonlinear) graphs. A further drawback to graphing allometric relations on standard rectangular coordinates is that it is difficult to determine the parameters c and k of a particular relation from data points plotted on such graphs. To see why log–log coordinates are preferable for representing allometric relations, we rearrange the equation (7.1) by taking the logarithm of both sides to obtain the equation

$$\log(y) = k \log(x) + \log(c). \tag{7.2}$$

Now if we let $u = \log(x)$ and $v = \log(y)$, the equation (7.2) becomes the linear equation

$$v = ku + C, \tag{7.3}$$

where the v-intercept is $C = \log(c)$. The graph of the equation (7.3) on a u-v coordinate system, or equivalently on a $\log(x)$-$\log(y)$ coordinate system, is a straight line. If we plot experimental data expected to have the relationship (7.1), but use a log–log scale, then the parameter k will be the slope of the line which "best fits" the data, and the intercept of the line will be $\log(c)$.

Example 7.6

a) A graph of the equation $y = 3x^{1/2}$ on log–log graph paper is a line given by

$$\log(y) = \tfrac{1}{2} \log(x) + \log(3).$$

This is the equation of line (a) in Fig. 7.16.

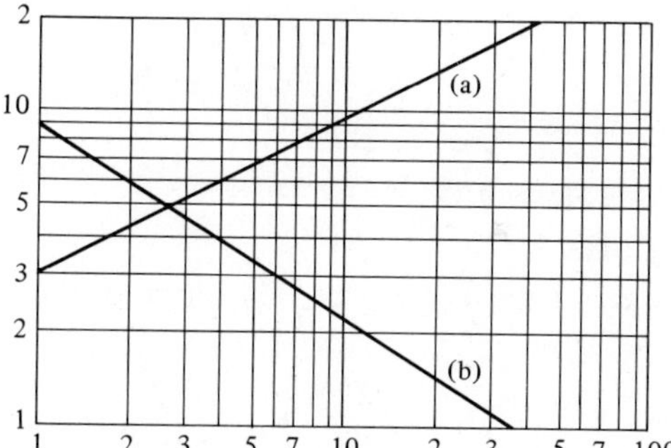

Figure 7.16

b) The line denoted (b) in Fig. 7.16 is the graph of an allometric relation $y = cx^k$. The value of $\log(c)$ is the $\log(y)$ intercept, thus $c = 9$. The value of k is the "physical slope" of the line, which is determined from the two points on the line

$$(\log (1), \log (9)) \qquad \text{and} \qquad (\log (6), \log (3)).$$

Therefore

$$k = \frac{\log(9) - \log(3)}{\log(1) - \log(6)} = \frac{\log(\frac{9}{3})}{\log(\frac{1}{6})} = \frac{\log(3)}{\log(\frac{1}{6})} = \frac{0.4771}{-0.7782} = -0.613.$$

Thus the line (b) represents the graph of

$$y = 9x^{-0.613}.$$ ◀

Example 7.7 In studying sense receptors (eyes, ears, nose, and tastebuds) in humans, it is necessary to establish a psycho-physical scale relating the perceived intensity, R, to the stimulus intensity S. The basic relation is thought to be an allometric function

$$R = cS^k,$$

where c and k are parameters. The parameter k is the most interesting, as it distinguishes between different sense modalities. For example, the value of k is approximately 0.33 for a visual brightness test and 0.60 for an odor test. Estimated values of k and c are usually obtained by plotting $\log(R)$ versus $\log(S)$ for a series of intensities S, and determining the slope and intercept of the line that "best fits" the data points. Figure 7.17 corresponds to a visual test as the value of $k = 0.33$ and

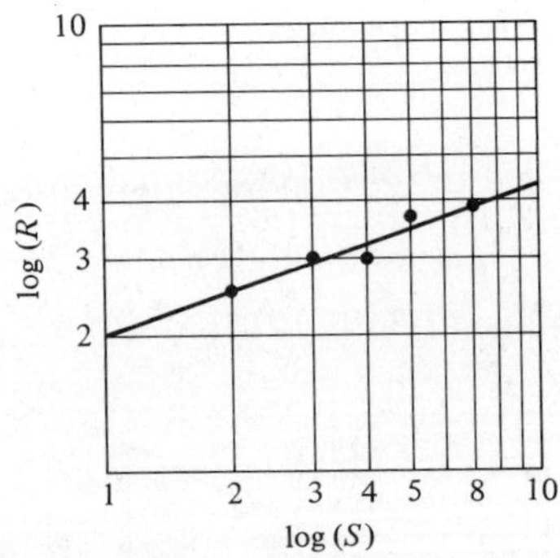

Figure 7.17

the value $c = 2$, since $c = \log(R)$ intercept. The slope of the line is given by the identity

$$k = \text{slope} = \frac{\log(R_1) - \log(R_2)}{\log(S_1) - \log(S_2)},$$

for any two points (S_1, R_1) and (S_2, R_2) on the line. Thus for $(S_1, R_1) = (8, 4)$ and $(S_2, R_2) = (1, 2)$, we find

$$k = \frac{\log(R_1/R_2)}{\log(S_1/S_2)} = \frac{\log(2)}{\log(8)} = \frac{\log(2)}{3\log(2)} = \frac{1}{3}. \blacktriangleleft$$

7.5 SEMILOG GRAPHS

In some cases it is not desirable to logarithmically rescale both the x and y variables. Scaling only one of the variables results in a semilog coordinate system having either log-uniform or uniform-log scales. Data may be graphed, using a semilog scale, by one of the two methods used to graph on the log–log scales—that is, either by first transforming the data and then plotting it on a uniform coordinate system, or by plotting the data directly on an x-y coordinate system, where one of the scales has been physically changed to a log scale. Semilog graph paper is readily available and has units on one axis measured in logarithmic distance and equally spaced units on the other axis (see Fig. 7.18).

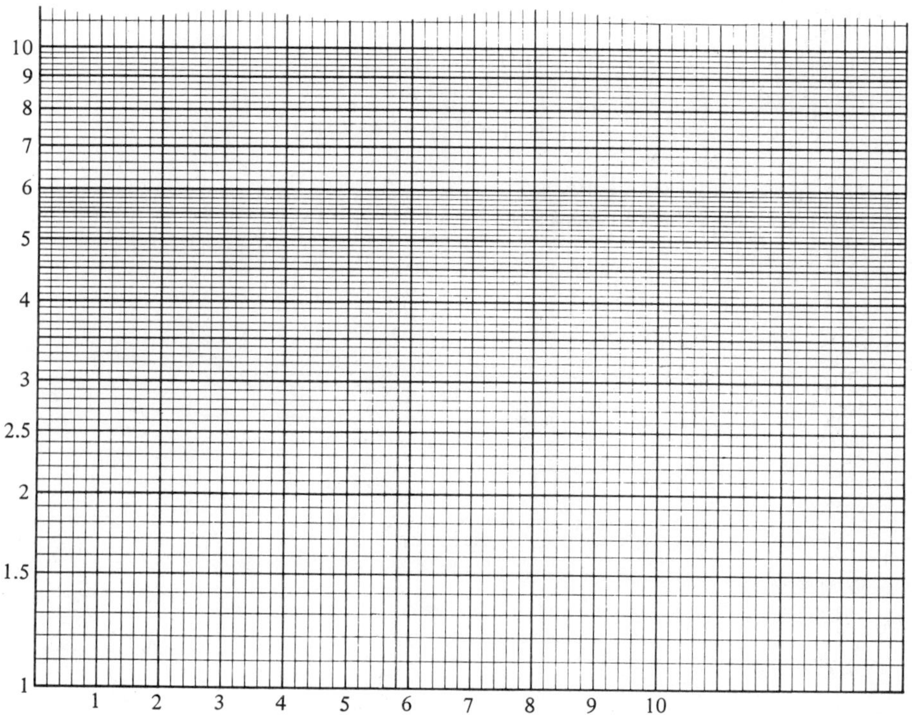

Figure 7.18

Linear Graphs on Semilog Scales

The main use of semilog scales is to linearize the graphs of exponential curves. Consider an exponential function of the form

$$y = c \cdot b^{kx}, \tag{7.4}$$

where b and c are positive constants and k is an arbitrary constant. If we take the logarithm of both sides of equation (7.4), we obtain the equation

$$\log(y) = k \cdot x \cdot \log(b) + \log(c). \tag{7.5}$$

When we make the log transformation $v = \log(y)$, the equation (7.5) becomes a linear equation

$$v = mx + C, \tag{7.6}$$

where the slope $m = k \cdot \log(b)$ and the v-intercept is $C = \log(c)$. In practice the base b is usually the number e or 10.

Example 7.8

a) The graph of the equation $y = 4 \cdot 10^{0.2x}$ is linear on a semilog coordinate system. The linear form of this equation is

$$\log(y) = 0.2 \log(10)x + \log(4),$$

or simply

$$\log(y) = 0.2x + \log(4).$$

This is the equation of line (a) in Fig. 7.19. Note that the log(y) intercept is log(4) and that the line passes through the point $(x, \log(y)) = (5, \log(40))$.

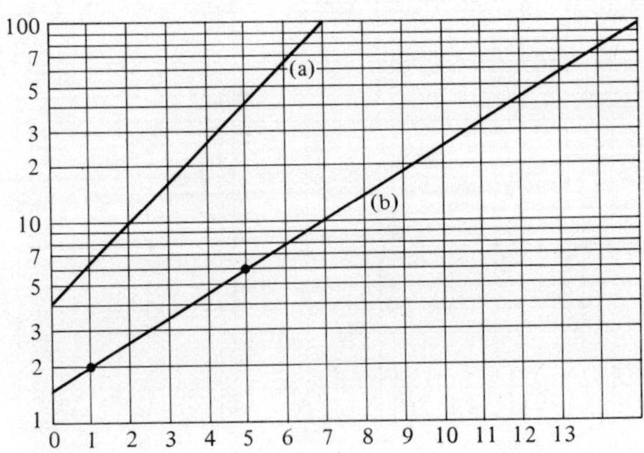

Figure 7.19

b) The line (b) in Fig. 7.19 is the graph of an exponential relationship: $y = cb^{kx}$. Using the linear form of this equation,

$$\log(y) = mx + \log(c),\ m = k \cdot \log(b),$$

we determine $\log(c) = \log(\frac{3}{2})$ and hence $c = \frac{3}{2}$. The slope m is the "physical slope" of the line,

$$m = \frac{\log(y_1) - \log(y_2)}{x_1 - x_2},$$

for any two points $(x_1, \log(y_1))$, $(x_2, \log(y_2))$ on the line. The line (b) passes through the two points indicated by dots. These are the points $(x_1, y_1) = (1, 2)$ and $(x_2, y_2) = (5, 6)$. Thus in this example we find that

$$m = \frac{\log(2) - \log(6)}{1 - 5} = \frac{\log(\frac{2}{6})}{-4} = \frac{\log(3)}{4} \simeq 0.1193.$$

Since $m = k \cdot \log(b)$, we must know one of the parameters k or b before we can determine the other. Assume $b = e$; $\log(b) = \log(e) \simeq 0.4343$. Then,

$$k = \frac{m}{\log(b)} \simeq \frac{0.1193}{0.4343} = 0.2747.$$

Thus the equation of the line (b) is (approximately)

$$y = \tfrac{3}{2}e^{0.2747x}.$$

Example 7.9 One type of population-growth model, which accounts for the natural variability in the birth rate, is described as a stochastic model. In a particular stochastic model of the growth of a population, it is assumed that the probability of an individual's giving birth to one offspring over a short interval of time is proportional to the length of the interval Δt. If the proportionality constant is r, then the average expected population size, denoted $\langle n(t)\rangle$, satisfies the exponential equation

$$\langle n(t)\rangle = N_0 e^{rt}.$$

To estimate the initial population N_0 and the probability constant r, experimental data are plotted on a $\log(\langle n(t)\rangle)$ versus t graph. An illustration is furnished by the Fig. 7.20. From the slope of the line of "best fit," and the $\log(\langle n(t)\rangle)$ intercept, the value of r is estimated to be 0.02 and the initial population N_0 is estimated to be 120. Data and calculations:

t	10	20	30	40
$\langle n(t)\rangle$	150	200	230	300

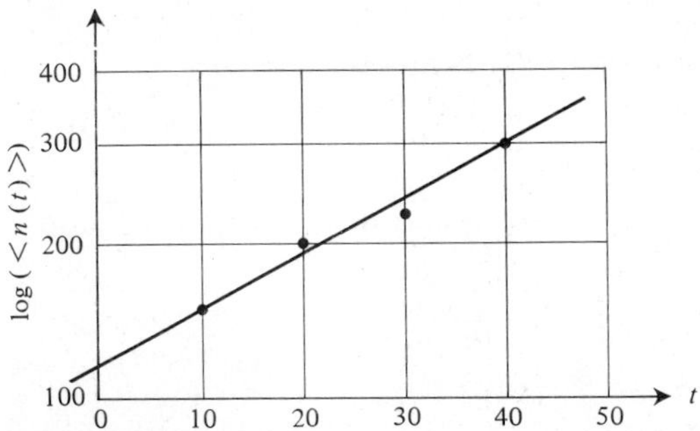

Figure 7.20

The slope of the line is

$$m \simeq \frac{\log(300) - \log(150)}{40 - 10} = \frac{\log(2)}{30}.$$

The $\log(\langle n(t)\rangle)$ intercept is estimated to be $\log(120)$. Thus $N_0 = 10^{\log(120)} = 120$. Finally,

$$r = \frac{\text{slope}}{\log(e)} = \frac{\log(2)}{30 \cdot \log(e)} \simeq \frac{0.3010}{30(0.4342)} \sim 0.02. \qquad \blacktriangleleft$$

7.6 RECIPROCAL SCALES

Another scaling transformation which compresses the x-values in $(1, +\infty)$ and expands the x-values in $(0, 1)$ is the reciprocal transformation

$$x \to u = 1/x.$$

This transformation maps the x interval $(1, +\infty)$ onto $(0, 1)$ and $(0, 1)$ onto $(1, +\infty)$. In the case of log–log or semilog coordinate systems, there were two equivalent ways of sketching graphs, either to perform the transformation of the x-y data points and plot the data directly on a u-v coordinate system or to plot the x-y data directly on an x-y coordinate system that has been physically rescaled (i.e., on log–log or semilog paper). Physically scaled graph paper is not available for reciprocal scaling. Hence the transformation of the data must be made before the data are plotted.

The reciprocal plotting of data is especially useful when plotting hyperbolas of a special form. The remainder of this section consists of a detailed analysis of how the graph of a particular hyperbola can be "linearized" by reciprocal rescaling.

The simple hyperbola sketched in Fig. 7.21(a) is the graph of $-xy = c, c > 0$. By shifting the graph b units to the left and a units upward, we obtain the graph of the hyperbola,

$$(a - y)(b + x) = c, \tag{7.7}$$

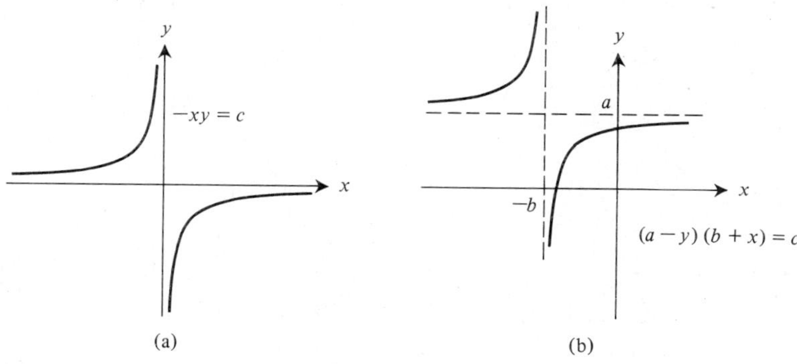

Figure 7.21

which is sketched in Fig. 7.21(b). The lines $x = -b$ and $y = a$ are *asymptotes* of this hyperbola.

We will consider only one "branch" of the hyperbola $(a - y)(b + x) = c$—that is, the "branch" sketched in Fig. 7.22(a) which has y-intercept $y_0 = a - c/b$

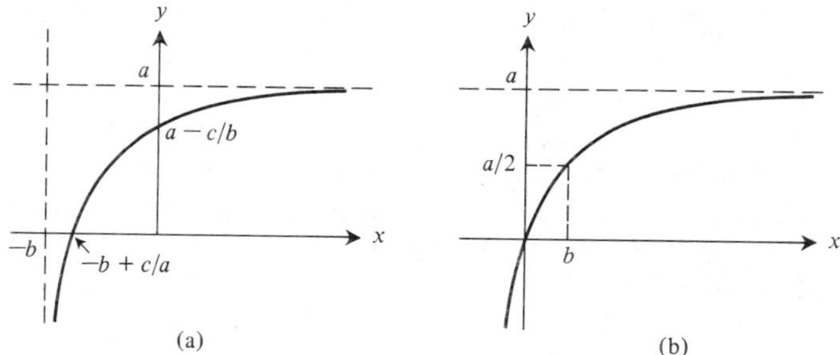

Figure 7.22

and x-intercept $x_0 = -b + c/a$. We further restrict the hyperbola so that the x-intercept is at the origin by requiring

$$c = b \cdot a.$$

In Fig. 7.22(b) we sketch the graph of the hyperbola

$$(a - y)(b + x) = b \cdot a. \tag{7.8}$$

This particular hyperbola is considered because it is very frequently used in mathematical models. Observe that y approaches zero as x approaches zero; but as x approaches $+\infty$, the y-value approaches the constant a.

Another interesting and useful feature of the curve (7.8) is that it passes through the point $(b, a/2)$. This fact can be used to estimate the parameters a and b from a curve sketched through data points that are assumed to have the hyperbolic

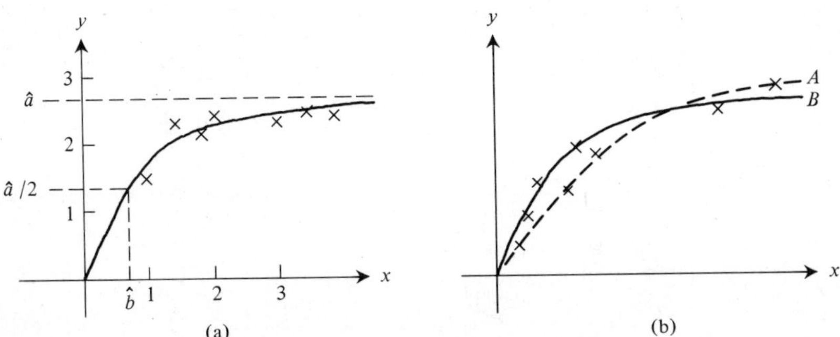

Figure 7.23

relations (7.8). For example, in Fig. 7.23(a) we have indicated experimental data values by ×'s and have made a free-hand sketch of a "hyperbolic-type" curve "through" these data points. Assuming the curve is of type (7.8), we estimate the value $\hat{a} = 2.5$ from the apparent asymptotic limit of the y-coordinate. The estimate $\hat{b} = 0.5$ is then obtained as the x-coordinate of the point on the curve having y-coordinate $\hat{a}/2$. You should immediately question the accuracy of these estimates. The above method involves two steps that are subject to errors. The first involves sketching a curve "through" nonlinear data values—it is unlikely that two individuals would sketch identical curves. A second place where error could arise is when estimating the limiting y-value, $\hat{a}$. In Fig. 7.23(b) we sketch two curves "through" the same data set—which curve is "correct" is not easily decided.

The difficulty encountered when we try to estimate the parameters of a hyperbolic curve from experimental data can be considerably reduced if we first linearize the equation of the hyperbola. The equation (7.8) can be linearized as follows.

Divide by $(b + x)$: $$a - y = \frac{ba}{b + x}$$

Solve for y: $$y = a - \frac{ba}{b + x}$$

Make a common denominator: $$y = \frac{(b + x)a - ba}{b + x} = \frac{ba + ax - ba}{b + x}$$

Cancel and invert: $$\frac{1}{y} = \frac{b + ax}{ax} = \frac{b}{ax} + \frac{x}{ax}$$

$$\frac{1}{y} = \left(\frac{b}{a}\right)\left(\frac{1}{x}\right) + \left(\frac{1}{a}\right)$$

This is now a linear equation between the reciprocals $1/y$ and $1/x$. If we make the reciprocal transformations

$$x \to U = \frac{1}{x}; \qquad y \to V = \frac{1}{y},$$

the equation of the hyperbola (7.8) becomes the linear equation

$$V = \left(\frac{b}{a}\right)U + \left(\frac{1}{a}\right). \tag{7.9}$$

The equation (7.9) is sketched in Fig. 7.24(a). Its V-intercept is $1/a$ and its U-intercept is $-1/b$. If data such as that used in Fig. 7.23(a) are first transformed by taking the reciprocals of each x- and y-coordinate, then plotted on a U-V coordinate system as indicated in Fig. 7.24(b) it will appear linear. In this case the values $\hat{a}$ and $\hat{b}$ are estimated from the U and V intercepts (or from the slope) of the "line of best fit." The advantage of this linearized method lies in our ability to fit straight lines through data better than we can fit curved lines through data.

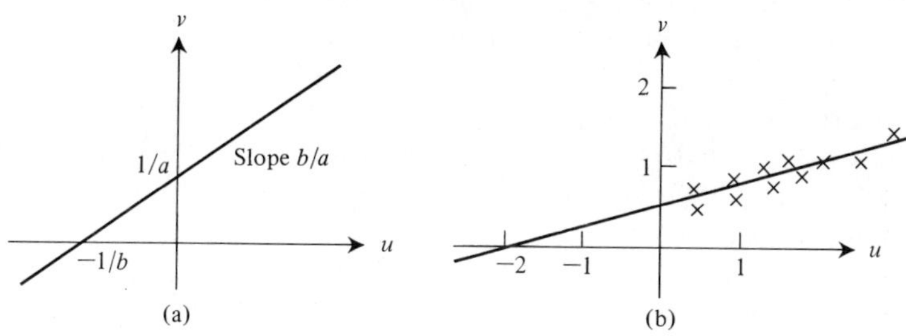

Figure 7.24

Example 7.10 In biochemistry, the study of enzyme kinetics extensively utilizes a hyperbolic relation of the form (7.8) with the variables v and s. The equation

$$(V - v)(K + s) = VK \tag{7.10}$$

is used to describe the kinetics of an enzyme reaction:

E	+	S	$\rightleftharpoons$	ES	$\longrightarrow$	E	+	P
Enzyme		Substrate		Complex		Enzyme		Product

Here v denotes the overall reaction rate and s denotes the substrate concentration. The constant V is the maximum velocity of the reaction and the constant K is called an equilibrium or dissociation constant, or the Michaelis constant, depending on experimental conditions. Equation (7.10) is called the Michaelis–Menten equation and is usually expressed in the form

$$v = \frac{V}{1 + \dfrac{K}{s}} \quad \text{or} \quad v = \frac{V \cdot s}{s + K}.$$

When the constants V and K are to be determined from experimental data, the Michaelis–Menten equation is linearized to have the form of equation (7.9).

$$\frac{1}{v} = \frac{K}{V}\left(\frac{1}{s}\right) + \frac{1}{V}.$$

The experimental data are then plotted on a $1/s$-$1/v$ coordinate system, and a straight line is fitted to the plotted points. The constants K and V are determined from the slope and axes intercepts. ◀

EXERCISE SET 7

7.1 Sketch the following functions on coordinate axes, which are homogeneously rescaled so that the graphs are approximately square in shape.

a) $y = 5x$, $x \in [0, 10]$, b) $y = 20x - 5$, $x \in [0, 10]$
c) $y = 3^x$, $x \in [0, 10]$, d) $y = \frac{1}{10}3^x$, $x \in [0, 10]$
e) $y = (\frac{1}{2})^x$, $x \in [0, 10]$, f) $y = e^{-x}$, $x \in [0, 100]$
g) $y = x^4$, $x \in [1, 5]$ h) $y = \log(x)$, $x \in [0.01, 0.1]$
i) $y = 5x^3$, $x \in [0, 10]$

7.2 Determine the parameters of the indicated equations which appear to "best fit" the data in the graphs.

a) $y = ax + b$

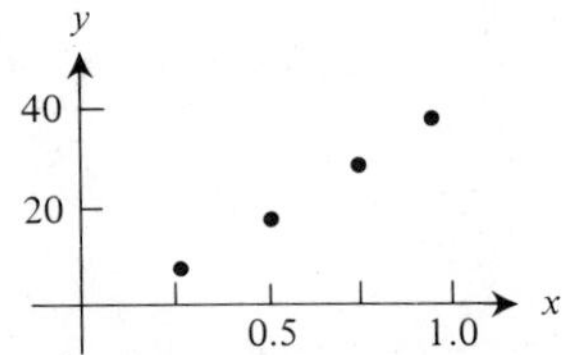

b) $y = mx + c$

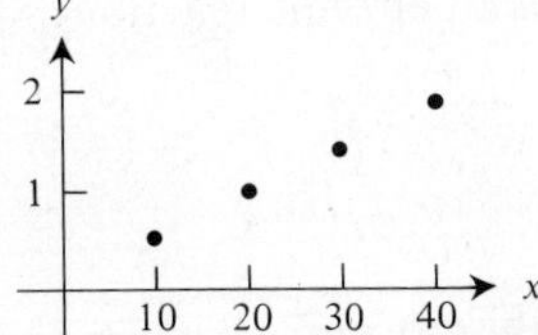

c) $y = x^a$
(a an integer)

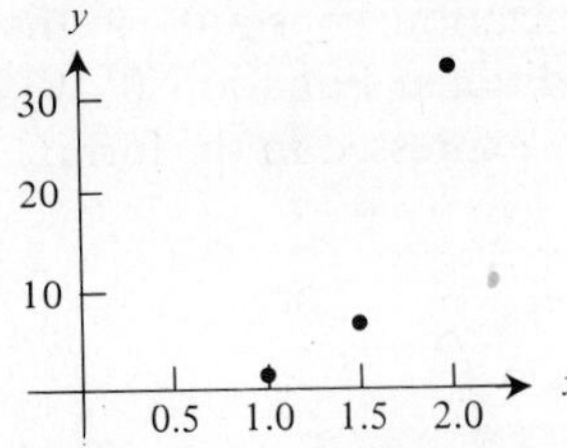

d) $y = a^x$
(a an integer)

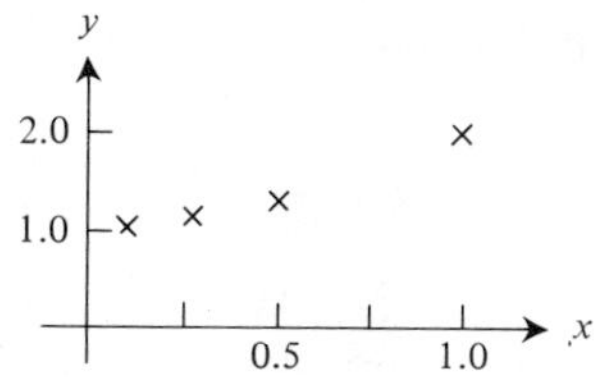

7.3 Sketch the following functions on graphs that are physically homogeneously scaled so that they are approximately square in shape.

a) $y = \sin(x), x \in [0, \pi]$
b) $y = \sin(x), x \in [0, 2\pi]$
c) $y = \sin(x), x \in [0, 4\pi]$
d) $y = \cos(2x), x \in [0, \pi]$
e) $y = \cos(2x), x \in [0, 2\pi]$
f) $y = \cos(2x), x \in [0, 3\pi]$

7.4 For the given set of data points S, do the following.

a) Plot the points on a graph with equal scales for both axes, x and y.
b) Rescale the points by dividing the x-values by the largest x-value and the y-values by the largest y-value; call the new data set $\hat{S}$.
c) Plot the point of $\hat{S}$ on the new u-v coordinate system.
d) Express as an equation the relation between the new u-v coordinates and the original x-y coordinates.

i) $S = \{(3, 5), (-2, 1), (4, 0), (6, 2), (0, -3)\}$
ii) $S = \{(0.05, 4), (0.01, 2), (0.025, 2), (0.03, 0)\}$
iii) $S = \{(0.2, 1.3), (0.15, 2.6), (0.05, 0.65), (0, 0)\}$

7.5 Plot the set of numbers $\{1, 3, 10, 28, 55, 80, 100\}$

a) on a uniformly scaled axis
b) on a logarithmically scaled axis,
c) on a reciprocal axis.

7.6 Repeat Exercise 7.5 for the set $\{0.056, 0.0001, 0.001, 0.02, 0.002, 0.005, 0.015\}$.

7.7 Give the coordinates of the points shown on the graph in Fig. 7.25.

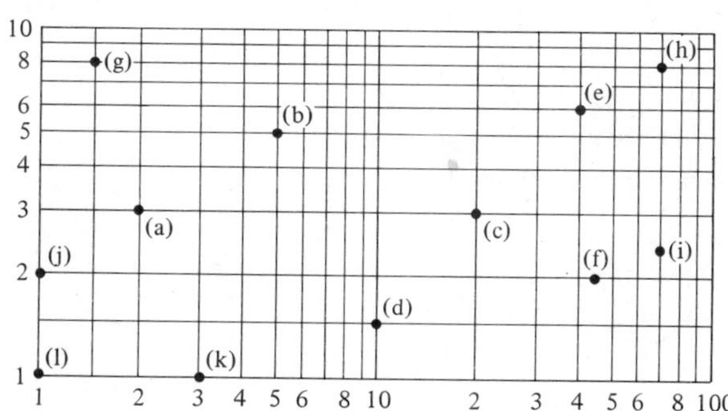

Figure 7.25

7.8 Plot the following points on a log–log graph (as in Fig. 7.25).

a) (2, 8)
b) (4, 16)
c) (28, 6)
d) (72, 25)
e) (55, 5)
f) (95, 42)
g) (6, 60)
h) (7.5, 40)
i) (1.8, 18)

7.9 Give the coordinates of the points indicated on the graph in Fig. 7.26.

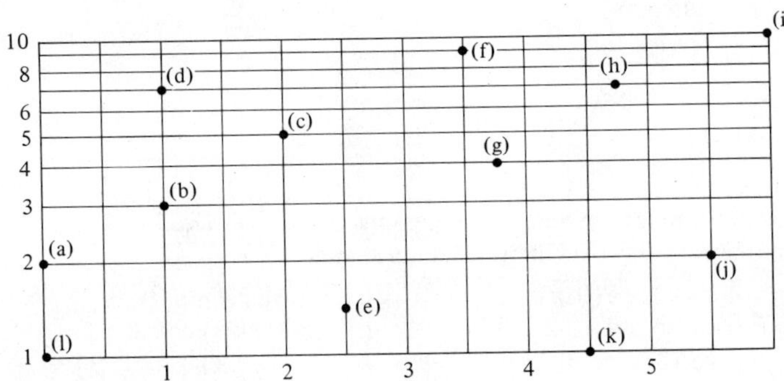

Figure 7.26

7.10 Plot the indicated points on a semilog graph (as in Fig. 7.26).

a) (3, 3) b) (3, 30) c) (3, 13)
d) (1, 6) e) (5, 55) f) (5.5, 27)
g) (5.6, 20) h) (4.7, 4.7) i) (2.3, 72)
j) (2.8, 1.2) k) (3.6, 2.5) l) (3.6, 25)

7.11 Determine a linear equation which is equivalent to the given equation and indicate on which type of physically scaled graph paper the graph of the original equation would be linear.

a) $y = 2^x$ b) $y = x^2$ c) $y = 3e^{2x}$
d) $y = (\frac{1}{2})^{-x}$ e) $y = 4x^5$ f) $x^2 \times y^2 = 7$
g) $xy = 4$ h) $y^3 = 8x^6$ i) $y = 2e^{x-2}$
j) $y = \sqrt{x}$ k) $y = \sqrt{x+2}$ l) $y = (x+1)/x$

7.12 Sketch the following functions in two ways: (1) on a standard x-y coordinate system; and, (2) on a coordinate system so that the graphs will be linear.

a) $y = \frac{3}{2}x^{1/2}$ b) $y = \frac{1}{10}x^{-2}$
c) $y = e^{1/2x}$ d) $y = e^{-1/2x}$
e) $y = 4x^{-2}$ f) $y = 3^{1/2x}$
g) $y = 2\exp_{10}(x)$ h) $y = 8 \cdot 2^{-x}$

7.13 Determine the equation of the lines on the graph in Fig. 7.27.

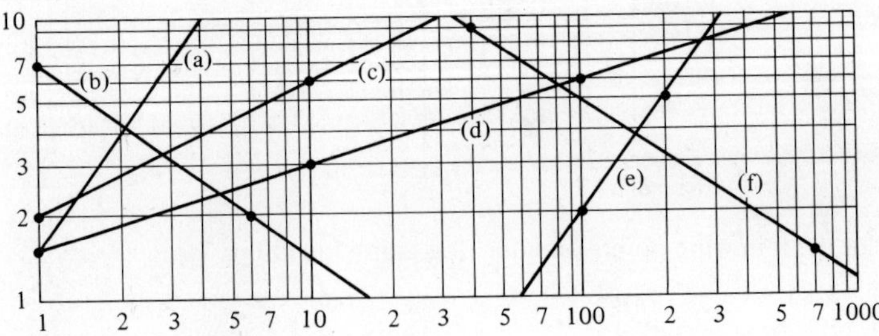

Figure 7.27

7.14 Determine the equation of the lines on the graph in Fig. 7.28, assuming the exponential base is $b = e$.

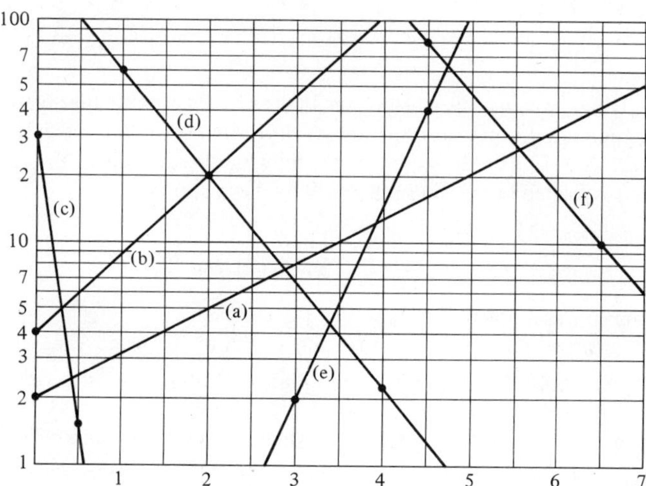

Figure 7.28

7.15 Solve the following equations for the indicated variable.

a) $y = c \log(kx)$; solve for k.
b) $y = \exp_{10}(kx + c)$; solve for x.
c) $\log(y) = a \ln(x)$; solve for y.
d) $\log(y) = a \ln(x^2)$; solve for x^a.
e) $\log(y) = 3x + 7$; solve for y.
f) $\ln(y) = 5 \ln(x) + 1$; solve for y.

7.16 Sketch the following data in two ways: (1) so that the line which "best fits" it will visually appear to have a steep slope; and, (2) so that the slope will visually appear to be small:

$$\{(0, 1), (1, 1.6), (2, 1.9), (3, 2.4), (4, 3.1)\}.$$

7.17 The following sketch indicates data of odor studies as in Example 7.7. Determine the relationship between R and S.

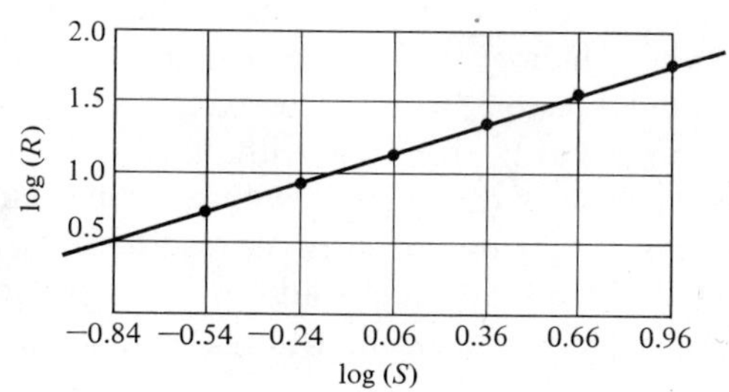

7.18 In reference to Example 7.9, the population size of a population satisfying the model was measured at $t = 2$ and $t = 5$. The values were $\langle n(2) \rangle = 609$ and $\langle n(5) \rangle = 4970$.

a) Estimate graphically the initial population size N_0 and the probability of birth constant r.

b) Using your estimated values, compute $\langle n(2) \rangle$ and $\langle n(5) \rangle$.

7.19 In reference to Example 7.10, plot the relation between v and s in two ways, on v-s and $1/v$-$1/s$ axes.

a) $K = 1, V = 1$
b) $K = 2.0 \times 10^{-3}, V = 1.75 \times 10^{-2}$
c) $K = 2.0 \times 10^{-3}, V = 1.75 \times 10^{2}$

7.20 In studying the metabolism of an animal, the rate of heat exchange between the animal's body and the air is an important factor. The insulation of an animal depends on its coat or fleece as well as on the temperature and wind speed. In sheep the thermal insulation factor I is related to the wind velocity V by the equation

$$I = \frac{1}{0.115 + 0.099V^5}.$$

a) Graph I as a function of V.
b) "Linearize" the above equation and sketch the corresponding linear graph.

7.21 a) Rearrange the Michaelis–Menten equation (7.10) so that it describes a linear relation on an s-s/v coordinate system.
b) Sketch the graph of the Michaelis–Menten equation on an s-s/v coordinate system for the values of the parameters K and V given in Exercise 7.19.

SECTION 8

SPECIAL FUNCTIONS FOR BIOLOGICAL APPLICATIONS

INTRODUCTION

In this section we will examine the basic properties of several functions that are widely applied in both biology and medicine. Each of these functions will involve the basic exponential function $y = e^t$ discussed in Section 5. We will present properties of the functions and their graphs without justification at this point. Later, after developing the tools of calculus, we will verify in examples and exercises the properties presented here. Consequently, in reading this section you should not be too concerned if you do not understand "why" a particular property is valid. Rather, you should accept the properties described in the same way that you accept the chemist's description of atomic structure or the cell biologist's description of nutrient transport through a cell wall.

What you should strive for in this section is (1) to develop the ability to plot points and sketch the graph of these particular functions with the use of tables or calculators, and (2) to learn to associate the basic graph forms with their associated functions and equations.

8.1 EXPONENTIAL GROWTH AND DECAY CURVES

I. The *exponential growth* equation has the form $y = ce^{kt}$, where c and k are positive constants. As we noted in Section 5, the constant k is called the *growth rate* and the constant c is called the *initial value*. The basic features of the graph of $y = ce^{kt}$ are as follows.

1. The graph has the y-intercept $y = c$.
2. The graph is monotone increasing.
3. As t approaches $+\infty$, y becomes unbounded.
4. As t approaches $-\infty$, y approaches zero.
5. If the constant k is increased, the graph becomes "steeper."
6. If the constant c is increased, the graph becomes "steeper" and is raised upward. (This is really a shift to the left.)
7. The y-coordinate is always positive (since $c > 0$).

In Fig. 8.1(a) we sketch the graph of the exponential growth curve for different values of c and fixed k. In Fig. 8.1(b) we sketch the graph of the exponential growth curve for fixed c and different growth rates.

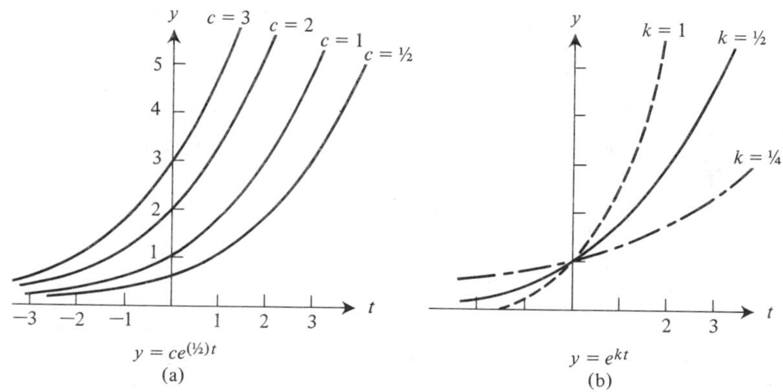

Figure 8.1

II. The *exponential decay equation* has the form $y = ce^{kt}$, where c is positive and k is negative. To emphasize that k is negative, we set $k = -\lambda$, where λ is a positive constant. The constant c is called the *initial value* and the factor λ (or $-\lambda$) is called the *decay* rate. The most frequent application of this equation is to the radioactive decay of material. The basic features of the graph of a decay curve $y = ce^{-\lambda t}$ are as follows.

1. The graph has the y-intercept, $y = c$.
2. The graph is monotone decreasing.

3. As t approaches $-\infty$, y becomes unbounded.
4. As t approaches $+\infty$, y approaches zero.
5. As λ increases, the graph "decreases faster."
6. As c increases, the graph "decreases faster" and is raised upward. (This is really a shift to the right.)
7. The y-coordinate is always positive (since $c > 0$).

In Fig. 8.2(a) we sketch the graph of decay curves with various initial values and fixed decay rates. In Fig. 8.2(b) we sketch the graph of decay curves with the same initial value and different decay rates. (Observe that these curves are the mirror image about the y-axis of the curves in Fig. 8.1.)

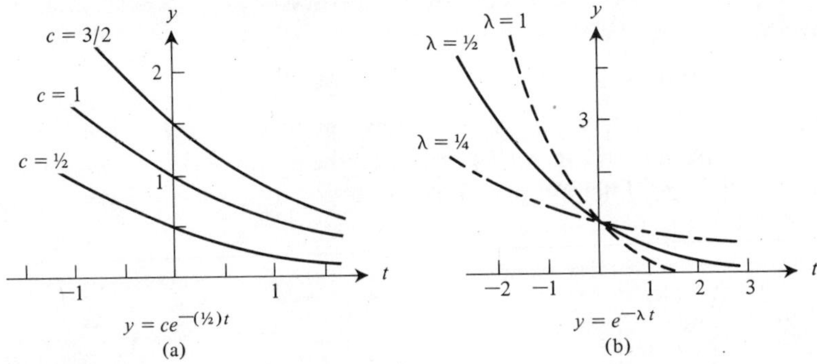

Figure 8.2

III. *Vertical shifts* in exponential growth and decay curves are the result of adding a constant to the exponential function; for instance,

$$y = A + ce^{kt} \qquad \text{or} \qquad Y = A + ce^{-\lambda t}.$$

The effect of adding a constant A to an exponential function is to shift the graph vertically. If $A > 0$, then the shift is upward as indicated in Fig. 8.3(a). If A is negative, the shift is downward. (See Exercise 8.2.) The only properties of the graph of an exponential curve altered by a vertical shift are (1) and (4). The y-intercept becomes $A + c$ and the asymptote is A instead of zero. You should verify these graphs by evaluating $y = A + e^{(1/2)t}$ at a number of t-values when A equals $0, \frac{1}{2}, \frac{3}{2}$, and 3. (Try $t = -2, -1, 0, 1, 2$.)

IV. *Horizontal shifts* in the graph of an exponential function are produced analytically by replacing the variable t with the term $(t - b)$. Thus the graph of the growth function $y = ce^{k(t-b)}$ has the identical form as the graph of $y = ce^{kt}$, except that the graph has been shifted horizontally b units. The direction of the shift will be to the right if $b > 0$ and to the left if $b < 0$. Consequently, in the property (1) the phrase "the y-intercept..." must be replaced with the phrase "the value of y at $x = b$ is c." In Fig. 8.3(b) we sketch several horizontal shifts of an exponential growth

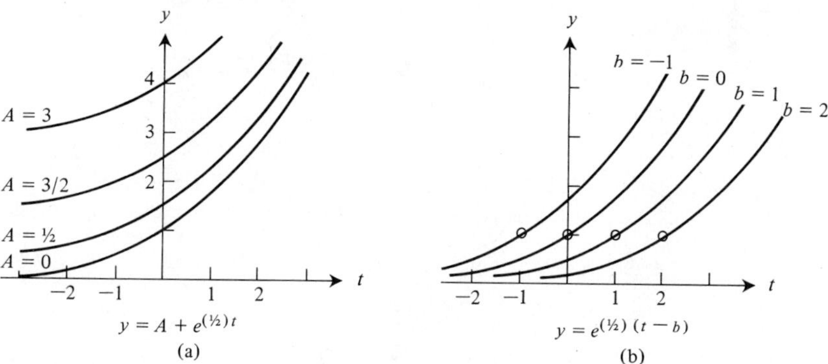

Figure 8.3

function $y = e^{(1/2)t}$. The point on each curve with $y = 1$ has been circled; observe that the corresponding x-coordinate in each case is the value of b.

A second way to produce a horizontal shift in the graph of an exponential function $y = ce^{kt}$ is to vary the constant c. The curves sketched in Fig. 8.2(a) all represent the same basic curve $y = e^{-(1/2)t}$ shifted vertically. To vary the initial constant c is equivalent to introducing a "delay term" b as discussed above. This equivalence is seen algebraically using the properties of exponentials as follows:

$$y = ce^{k(t-b)} \qquad \text{is equivalent to} \qquad y = ce^{kt-kb} = ce^{kt} \cdot e^{-kb}.$$

Thus,

$$y = c_1 e^{kt},$$

where

$$c_1 = c \cdot e^{-kb}.$$

Hence, to introduce a delay term by replacing t with $t - b$ is equivalent to multiplying the initial value c by the term e^{-kb}. Consequently, any function $y = ce^{kt}$ can be expressed as $y = e^{k(t-b)}$, where $b = -\ln(c)/k$.

8.2 SIGMOID OR "S-SHAPED" CURVES

The use of exponential curves to describe the growth of biological phenomenon is usually feasible for only a short time interval. Malthus, in 1798, observed that the exponential growth equation applied to world population would lead to an infinitely large population as time increased towards $+\infty$. In nature, organisms or populations do not grow unchecked forever. There are usually limiting growth factors, such as a limited food supply or a drop in reproduction rates when overcrowding occurs. What is observed in practice is that the number of a species or the size of an organ or plant will initially increase at an apparent exponential rate and will then slow down its growth rate as it approaches a maximum limiting size. Thus, over a long time period, the growth curve tends to have an S-shape or

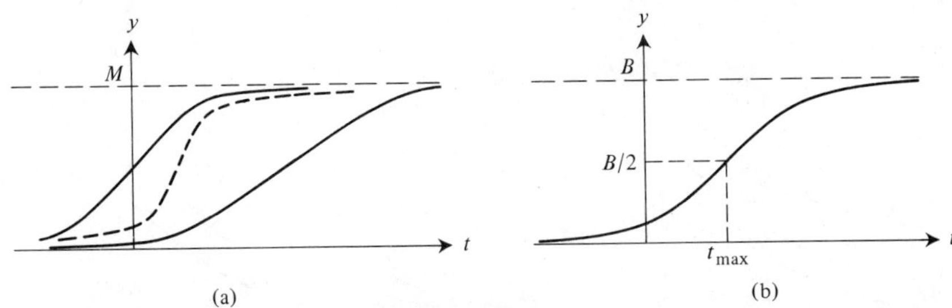

Figure 8.4

sigmoid shape as illustrated in Fig. 8.4(a). The graphs approach the limiting value M as t approaches $+\infty$. In this section we examine two different functions whose graphs have a sigmoid shape—the logistic equation and the Gompertz equation.

The Logistic Equation

The *logistic equation* occurs in studies of chemical kinetics, ecology, demography, and many other areas. P. F. Verhurst used the logistic equation in the 1840's to model world-population growth. In the 1920's it was used to describe various forms of biological growth by R. Pearl. The logistic equation can be manipulated into various forms. The following is a simple form of the equation:

$$\textit{Logistic equation}: y = f(t) = \frac{B}{1 + ce^{-\lambda B t}}. \tag{8.1}$$

The constants B, c, and λ are all assumed positive. The graph of equation (8.1) satisfies the following properties.

1. The graph is increasing for $t \in (-\infty, +\infty)$.
2. $0 < f(t) < B$ for all t.
3. The graph approaches the line $y = B$ asymptotically as t approaches $+\infty$.
4. The graph approaches the t-axes asymptotically as t approaches $-\infty$.
5. The graph has a y-intercept of $B/(1 + c)$.
6. The graph is sigmoidal in shape and symmetric about the point where its y value is one-half the maximum B, $y = B/2$.
7. At the point where $y = B/2$, the graph has its greatest rate of increase, and the corresponding t value is denoted by $t_{\max}$ or $t_{1/2}$.

The graph of the logistic equation (8.1) is sketched in Fig. 8.4(b). The $t_{\max}$ value, where the graph has its greatest rate of increase, can be obtained by solving the equation $f(t_{\max}) = B/2$,

$$\frac{B}{2} = \frac{B}{1 + ce^{-\lambda B t_{\max}}} \qquad \text{implies that} \qquad 1 + ce^{-\lambda B t_{\max}} = 2.$$

stalk. The actual equation will vary depending on the position of the leaf on the stock. The equation for the second leaf from the bottom is

$$W = \frac{24.6435}{1 + 2e^{-1.008t}}.$$

The graph of this function is given in Fig. 8.6. The points used to plot this graph are (0, 8.2), (0.5, 11.2), (1, 14.3), (2, 19.5), (3, 22.5), (4, 23.8), and (5, 24.3). The

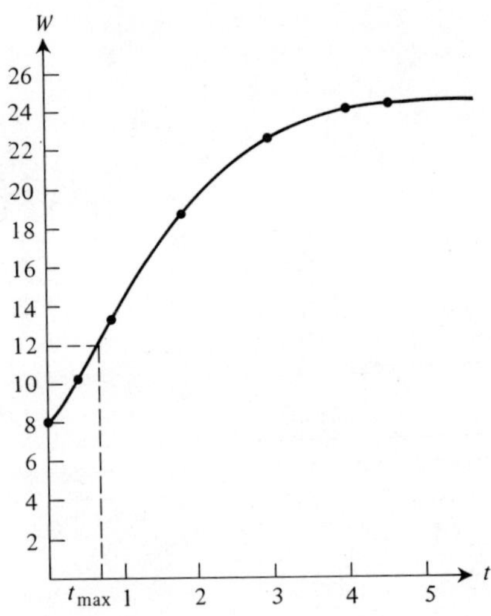

Figure 8.6

asymptotic limit is $W = 24.6$. The estimate $t_{\max} \simeq 0.7$ is obtained as the "eyeballed" t-coordinate corresponding to $W = 12.3$. Solving algebraically we find

$$t_{\max} = \ln(2)/1.008 \simeq 0.688.$$ ◀

The Gompertz Function

The *Gompertz function* was first introduced to model organisms whose growth curves are slightly different from those that the logistic equation would predict. Like the logistic equation, the basic Gompertz equation involves three positive parameters; c, k, and λ.

$$\text{\textit{Gompertz equation}: } y = f(t) = ce^{-ke^{-\lambda t}}. \tag{8.4}$$

The Gompertz function is thus an exponential function composed with another exponential function. If $g(t) = e^{-\lambda t}$ and $h(t) = ce^{-kt}$, then $f(t) = (h \circ g)(t)$. The properties of Gompertz curve are as follows.

1. The curve increases for all t.
2. $0 < f(t) < c$ for all t.
3. The graph approaches the line $y = c$ asymptotically as t approaches $+\infty$.
4. The graph approaches the t-axis asymptotically as t approaches $-\infty$.
5. The graph has the y-intercept $y_0 = ce^{-k}$.
6. The graph is sigmoid in shape.
7. The graph has its maximum rate of increase when $f(t) = c/e$.

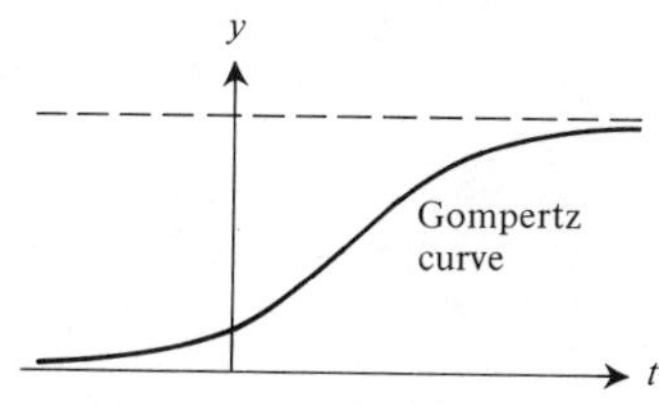

Figure 8.7

The graph of a Gompertz curve is given in Fig. 8.7. To evaluate the Gompertz function (8.4), two exponential evaluations are required. For instance, if $c = 2$, $k = 3$, and $\lambda = 4$, then the y-value corresponding to $t = 0.1$ would be obtained as follows:

$$e^{-\lambda t} = e^{-0.4} \simeq 0.67,$$

hence,

$$e^{-ke^{-\lambda t}} = e^{-3(0.67)} = e^{-2.01} \simeq 0.14.$$

Thus $y \simeq 2(0.14) = 0.28$.

8.3 PRODUCTS AND DIFFERENCES INVOLVING EXPONENTIAL DECAY FUNCTIONS

We consider two basic functions that are representative of a host of functions used in pharmacology, ecology, and chemistry. The graphs of these functions are characterized by initial growth, which changes into a decay that eventually approaches a limit value asymptotically.

We first consider the product of a linear term ct and an exponential $e^{-\lambda t}$. Let

$$\boxed{f(t) = cte^{-\lambda t}, \qquad \text{where} \qquad c > 0, \lambda > 0.} \tag{8.5}$$

The initial value of this function is $f(0) = 0$. For t near zero, $e^{-\lambda t}$ is near one, and equation (8.5) is initially dominated by the linear term ct. Thus, for small values of t, the graph resembles the graph of the line $y = ct$. However, as the value of t increases, the exponential term $e^{-\lambda t}$ becomes small and causes the graph of equation (8.5) to drop away from the line $y = ct$. Eventually, the effect of the decaying

8.2 Accurately sketch on the same graph the following functions.

a) $y = \frac{1}{4}e^t$ b) $y = \frac{1}{4}e^t + 1$ c) $y = \frac{1}{4}e^t - 1$
d) $y = \frac{1}{4}e^{(t-1)}$ e) $y = \frac{1}{4}e^{(t-1)} + 1$ f) $y = \frac{1}{4}e^{(t-1)} - 1$

8.3 Accurately sketch on the same graph the following functions.

a) $y = e^{-(1/4)t}$ b) $y = 1 + e^{-(1/4)t}$
c) $y = -e^{-(1/4)t}$ d) $y = 1 - e^{-(1/4)t}$

8.4 Express each of the equations in the form $y = e^{k(t-b)}$.

a) $y = 3e^{-2t}$ b) $y = 2e^{3t}$
c) $y = 3e^{5t}$ d) $y = \frac{1}{2}e^{2t}$
e) $y = 3e^{t-2}$ f) $y = 2e^{t-\ln(2)}$
g) $y = 5e^{2(t-\ln(3))}$ h) $y = 4e^{(1/2)(t+\ln(4))}$

8.5 Accurately sketch the following logistic curves. Indicate the values $t_{\max}$.

a) $y = \dfrac{10}{1 + 2e^{-t}}$ b) $y = \dfrac{8}{1 + 2e^{-(1/2)t}}$

c) $y = \dfrac{10}{1 + 4e^{-3t}}$ d) $y = \dfrac{8}{1 + 2e^{-t}}$

e) $y = 1 + \dfrac{10}{1 + 2e^{-3t}}$ f) $y = 2 + \dfrac{8}{1 + 3e^{-(1/2)t}}$

8.6 Sketch the graph of the given Gompertz curve.

a) $y = 10e^{-e^{-t}}$ b) $y = 20e^{-2e^{-t}}$ c) $y = 10e^{-e^{-2t}}$
d) $y = 20e^{-(1/2)e^{-t}}$ e) $g = 10e^{-e^{-(1/2)t}}$ f) $y = 20e^{-(1/2)e^{-(1/2)t}}$

8.7 On the same graph sketch both of the indicated functions.

a) $y = 10/(1 + 2e^{-0.1t})$ and $y = 10e^{-\ln(3)e^{-0.1t}}$
b) $y = 10/(1 + 4e^{-0.1t})$ and $y = 10e^{-0.1e^{-4t}}$

8.8 What is the y-coordinate of the point where the rate of increase of a Gompertz curve is maximum? Give this value for each equation in Exercise 8.6.

8.9 Accurately sketch the graph of the following functions (indicate the line $y = ct$ on the graph as discussed in Section 8.3).

a) $y = te^{-t}$ b) $y = te^{-2t}$ c) $y = te^{-(1/2)t}$
d) $y = 2e^{-t}$ e) $y = 2te^{-(1/2)t}$ f) $y = 3te^{-(1/4)t}$

8.10 The graph of the curve $y = cte^{-\lambda t}$ reaches its maximum at $t_{\max} = 1/\lambda$ and levels out at the value $t_i = 2/\lambda$. For each function in Exercise 8.9, compute the y-values corresponding to $t_{\max}$ and t_i and indicate these on the graph of Exercise 8.9.

8.11 Give the equation of the form (8.5) that has the indicated value $t_{\max}$ and $y_{\max}$.

a) $t_{\max} = 6$, $y_{\max} = 10$ b) $t_{\max} = 5$, $y_{\max} = 12$
c) $t_{\max} = 0.2$, $y_{\max} = 80$ d) $t_{\max} = 24$, $y_{\max} = 50$

8.12 a) Accurately sketch the graph of $y = c[1 - e^{-\lambda t}]$ for $\lambda = 0.1$, $\lambda = 1$, and $\lambda = 10$ if $c = 2$.

b) What value of t gives $y = c/2$ for the equation of part (a)? Indicate this t-value on the graphs of part (a).

8.13 Graph the following equations.

a) $y = 50(e^{-t} - e^{-2t})$ b) $y = 50(e^{-0.2t} - e^{-t})$
c) $y = 50(e^{-(1/2)t} - e^{-t})$ d) $y = 50(e^{-0.01t} - e^{-0.3t})$
e) $y = 50(e^{-(1/4)t} - e^{-t})$ f) $y = 50(e^{-\ln(2)t} - e^{-\ln(3)t})$

8.14 The t values where the graph of the equation (8.7) has its maximum and where it begins to level out are

$$t_{\max} = \frac{1}{\lambda_2 - \lambda_1} \ln\left(\frac{\lambda_2}{\lambda_1}\right) \quad \text{and} \quad t_i = \frac{2}{\lambda_2 - \lambda_1} \ln\left(\frac{\lambda_2}{\lambda_1}\right).$$

a) Compute the values of $t_{\max}$ and t_i for the functions of Exercise 8.13.
b) Evaluate $f(t)$ at the values $t_{\max}$ and t_i for each function f in Exercise 8.13.

8.15 a) Describe the graph of the equation (8.7) when $\lambda_1 > \lambda_2$.
b) Graph $y = 10(e^{-t} - e^{-0.1t})$,

8.16 In Example 8.1, equation (8.3):

a) Express the constant c of (8.3) in terms of $N(0)$.
b) At what time, t_m, is $N(t_m) = M/2$?
c) If $N(0) = 100$, $M = 500$, and $t_m = 10$, what is the constant λ equal to?
d) Sketch the growth curve $N(t)$ when $N(0) = 5$ and $M = 20$ for $\lambda = 2$ and for $\lambda = 4$.

8.17 What is the equation of a logistic curve which has a maximum increase at $t = 0$?

8.18 In Example 8.2 the equation for W is of the form $W = B/(1 + ce^{-\lambda Bt})$, where $c =$ the position of the leaf counted from the bottom, $\lambda B = (0.758 + 0.117c)^{-1}$, and $B = \exp(7.64 - 4.4\lambda B)$. Evaluate the coefficients and graph the equation for (a) the first and (b) the third leaf from the bottom.

8.19 Graph the following function which is used to describe the fraction of the substance creatinine that remains in intercellular fluid after an injection:

$$F = 0.5(e^{-0.18t} - e^{-1.65t}).$$

8.20 Assume that a drug is administered in a 50-mg dose and has an absorption rate of $K = 0.05$ and a clearence rate of $k = 2$. Assuming that the drug concentration is given by equation (8.9), graph the plasma concentration curve for this drug.

8.21 Under ideal conditions, the yeast *Saccharomyces cerevisiae* grows according to a logistic equation. One such equation is

$$N(t) = 13 \cdot (1 + e^{3.33 - 0.22t})^{-1}.$$

a) Express this equation in the form of equation (8.1).
b) What is the maximum size of a yeast culture growing according to this equation and when will the culture reach one-half of this size?
c) Graph this logistic equation.

8.22 The probability that a person will die in short time interval $[t, t + \Delta t]$ (for arbitrarily small values of Δt) is the *death density function*, which has the form

$$f(t) = \lambda^2 te^{-\lambda t}.$$

The parameter λ reflects the "hazards" that the person is expected to encounter and thus is a function of age, occupation, and health. Graph the function $f(t)$ when

a) $\lambda = 1$; b) $\lambda = 10$; c) $\lambda = 0.1$; d) $\lambda = 2$.

SECTION 9

SEQUENCE

9.1 SEQUENCES

The word "sequence" suggests a linearly ordered set of objects or events. For example, a laboratory manual prescribes a sequence of procedures to be followed; in studying cell biology, we learn that DNA specifies a sequence of nucleic acids which ensure identical replication of a molecule or cell; the coagulation of blood is the result of a specific sequence of enzymatic reactions. Each of these examples has in common the existence of a specific set of objects and an associated ordering of the objects. The ordering can be described by associating each object with a natural number to indicate its relative position in the context of the complete set of objects.

The mathematical definition of a sequence uses the concept of a function to transfer the natural ordering of the set of natural numbers

$$\mathbb{N} = \{1, 2, 3, 4, \ldots\}$$

to the objects which are "sequentially" ordered. The simplest mathematical definition of a sequence is a function f having Domain$(f) = \mathbb{N}$. This definition is deceptive in its simplicity and lack of descriptiveness. In practice there are two categories of sequences. One type of sequence, called a *finite sequence*, is associated with a process or event that terminates after a specific number of steps. A second type of sequence, called an *infinite sequence*, is associated with a process that does not terminate. To formalize the concepts of "finite" and "infinite" as applied to sequences, we associate with each natural number N a set consisting of the natural numbers less than or equal to N. We denote this set by the symbol $\underline{\mathbf{N}}$,

$$\underline{\mathbf{N}} = \{1, 2, 3, \ldots, N - 1, N\}.$$

A *finite sequence* is then a function whose domain is a set $\underline{\mathbf{N}}$, for some natural number N. An *infinite sequence* is a function whose domain is the set $\mathbb{N}$ of all natural numbers.

It is easier to give a precise definition of a sequence by first introducing special notation used to discuss sequences. The function notation $f(x)$ denotes the unique value the function f associates with an element x in its domain. This notation is very useful for general functions defined on intervals of real numbers. However, when working with functions defined on the set $\mathbb{N}$ of natural numbers, the notation $f(n)$ can be simplified. The simplification is accomplished by discarding the parentheses and writing the argument n as a subscript; for instance, we write f_n for $f(n)$, a_n for $a(n)$, g_n for $g(n)$, etc. Just as the standard function notation is $f(x)$, the most common sequence term is a_n. Using this notation, we can give a formal definition of a sequence.

Definition 9.1

a) A *sequence* $\{a_n\}$ is an ordered set of objects, the nth object being denoted by a_n, n a natural number.

b) If the index n assumes only the numbers $1, 2, \ldots, N$, the sequence $\{a_n\}$ is called a *finite sequence* of length N and is denoted by either

$$\{a_n\}_{n=1}^{N} \qquad \text{or} \qquad \{a_1, a_2, \ldots, a_N\},$$

read "the sequence a sub n, from n equals 1 to n equals capital N."

c) If the index n ranges over all natural numbers, 1, 2, 3, . . . , then the sequence $\{a_n\}$ is called an *infinite sequence* and is denoted by either

$$\{a_n\}_{n=1}^{\infty} \qquad \text{or} \qquad \{a_1, a_2, a_3, \ldots\},$$

read "the sequence a sub n, from n equals 1 to infinity."

Note: Technically, all sequences are associated with the natural counting numbers 1, 2, 3, . . . , and hence can be represented in a form $\{a_1, a_2, a_3, \ldots\}$. However, in some situations it is desirable to list a sequence using other subscripts. For instance, to refer to every third term of a sequence $\{a_n\}$, we would consider the sequence $\{a_3, a_6, a_9, \ldots\}$. In studying animal or cell populations, we encounter sequences where the subscript n is used to indicate the nth generation of the population. Generally, the first generation is not the initial population, but the population after one reproductive cycle; in this situation, the initial population is frequently referred to as the zeroth generation, and its size denoted by using the subscript zero. Thus a sequence of generation sizes may begin with a_0 instead of a_1, $\{a_0, a_1, a_2, \ldots\}$. In most situations, the initial term of a sequence $\{a_n\}$ will be a_1, unless otherwise noted.

In practice, the symbol used to denote a general term of a sequence is associated with the particular range of the sequence. Also, the symbol used as a subscript may vary; the common choices are the letters n, i, j, l. The following would be representative types of sequences:

$$\{\text{dog}_i\}_{i=1}^{50}, \qquad \{\text{Max}_k\}_{k=1}^{\infty}, \qquad \{x_j\}_{j=1}^{\infty}, \qquad \{A_n\}_{n=1}^{35}.$$

Example 9.1 In a study of the metabolic process of a seal, the seal's blubber was measured daily over a one-year period. An abstract way of referring to this collection of measurements would be to designate the measurement on the ith day by b_i. The sequence

$$\{b_i\}_{i=1}^{365}$$

would represent the set of all measurements for the one-year period. If we wished to observe weekly changes, we would consider the sequence of measurements

$$\{b_1, b_8, b_{15}, \ldots, b_{7\cdot n+1}, \ldots, b_{7\cdot 52+1} = b_{365}\}.$$

Example 9.5 *Proportional growth; constant rate of growth*
If a population grows in such a way that the population size of the $(n + 1)$st generation is a constant multiple of the population size of the nth generation, then the population is said to have *proportional growth* or *constant rate of growth.* These two terms may seem to be completely different, but as we shall see in a later section of the text they convey the same implications. The cell culture of Example 5.1 exhibited this type of growth. We can describe this form of growth by the equation

$$g_{n+1} = r \cdot g_n, \tag{9.3}$$

where g_n is the size of the nth generation and r is the constant rate of growth (or the *proportionality constant* of growth). Unless otherwise indicated, we will consider r to be a positive number. If we denote the "initial" population size by g_1, we can then employ the same procedure used in Example 9.4 to show that the size of the 6th generation is $g_6 = g_1 \cdot r^5$, and the size of the nth generation is

$$g_n = g_1 r^{(n-1)}. \tag{9.4}$$

As $y = r^{n-1}$ increases with n for $r > 1$ and decreases with n for $0 < r < 1$, we see that the population size that exhibits proportional growth either increases if $r > 1$ or decreases if $0 < r < 1$. ◀

Definition 9.3 A sequence $\{g_n\}$, where $g_n = gr^{(n-1)}$ for $n = 1, 2, 3, \ldots$, is called a *geometric sequence*. The constant r is the ratio g_{n+1}/g_n of two successive terms of the sequence, and the first term is the constant g, $g_1 = g$. The expanded form of the geometric sequence is

$$\{g, gr, gr^2, gr^3, \ldots, gr^{(n-1)}, \ldots\}.$$

Since two successive terms of a geometric sequence satisfy the equation (9.3), we can easily check any given sequence to determine whether it is in fact a geometric sequence. We need only to check the ratio g_{n+1}/g_n of successive terms. If these are *all* the same number r, then the sequence is a geometric sequence and the constant r is the common ratio. The constant g is then the first term of the sequence.

For example, the sequence $\{b_n\} = \{1, 2, 4, 6, 8, \ldots\}$ is not a geometric sequence since $b_3/b_2 = \frac{4}{2} = 2$ and $b_4/b_3 = \frac{6}{4} = 1.5$. The sequence $\{c_n\} = \{8, 4, 2, 1, \frac{1}{2}, \ldots\}$ is a geometric sequence with ratio $r = \frac{4}{8} = \frac{1}{2}$; the sequence is of the form $\{8(\frac{1}{2})^{n-1}\}$.

Example 9.6 *A model with delayed growth; Fibonacci sequences*
In this model, we consider the population size of a colony of rabbits, which are assumed to satify two basic assumptions concerning their rate of reproduction. First, we make the unrealistic assumption that each pair of productive rabbits produces one new pair every five weeks. Thus a generation in this model is five weeks. Second, we assume that each pair of new-born rabbits does not produce

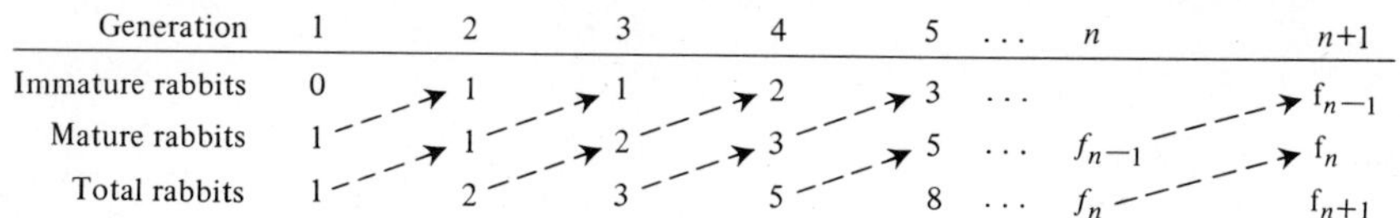

Figure 9.1

at the first five-week generation after their birth, but does produce at the second five-week generation after birth and at every subsequent five-week generation. Let f_n be the number of pairs of rabbits at generation n. The population at the $(n + 1)$st generation is then

$$f_{n+1} = f_n + b_n,$$

where b_n denotes the number of pairs born to the nth generation. Since only mature rabbits reproduce, not all of the f_n pairs give birth to new pairs. The mature rabbits in the nth generation are all rabbits that were present in the previous $(n - 1)$st generation—i.e., $b_n = f_{n-1}$. Thus a population whose first generation consists of f_1 mature rabbit pairs will result in a second generation of $f_2 = 2f_1$ rabbit pairs. (See Fig. 9.1.) And, for each generation, $n \geq 2$,

$$f_{n+1} = f_n + f_{n-1}. \tag{9.5}$$

Examples of the resulting sequence $\{f_n\}$ that satisfy equation (9.5) follow.

a) Let $f_1 = 1$, then $f_2 = 2$, $f_3 = 3$, $f_4 = 5, \ldots$, thus

$$\{f_n\} = \{1, 2, 3, 5, 8, 13, 21, 34, 55, 89, 144, \ldots\}.$$

b) Let $f_1 = 2$, then $f_2 = 4$, $f_3 = 6$, $f_4 = 10, \ldots$, thus

$$\{f_n\} = \{2, 4, 6, 8, 10, 16, 26, 42, 68, 110, \ldots\}.$$

c) Assume that a colony is started with three pairs of rabbits and, at the first generation, four more pairs of mature rabbits are added. We would then have $f_1 = 3$, $f_2 = 2 \cdot f_1 + 4 = 10$, $f_3 = 17$, and, for $n \geq 3$, $f_{n+1} = f_{n-1} + f_n$. Thus $f_4 = 27$, $f_5 = 44$, etc. The number of rabbit pairs would be given by the sequence $\{3, 10, 17, 27, 44, 71, 115, 186, 301, \ldots\}$. ◀

The sequence generated by equation (9.5) is called a *Fibonacci sequence* after the thirteenth century Italian mathematician Leonardo Fibonacci. As illustrated in part (c) of Example 9.6, the first two terms of a Fibonacci sequence can be arbitrary.

Definition 9.4 A sequence $\{f_n\}$ is called a Fibonacci sequence if f_1 and f_2 are specified numbers and, for $n \geq 2$, $f_{n+1} = f_n + f_{n-1}$.

We introduced the special sequences in this subsection via examples. The sequences may be defined independently of the examples. If you find it difficult

Hardy–Weinberg Laws describing its genetic proportionality. The population is inbred by repeatedly mating brothers and sisters of successive generations. To measure the effect of this inbreeding on the heterozygosity of the population, we define the *panmictic index*, P_t, as the ratio

$$P_t = \frac{H_t}{H_0}.$$

P_t is a relative measure of the heterozygosity of the tth generation. In genetics, it is shown that P_t is in fact the ratio of corresponding terms of a Fibonacci sequence and a geometric sequence. The Fibonacci sequence is $\{f_t\}$, given by $\{1, 2, 3, 5, 8, 13, \ldots\}$, and the geometric sequence is $\{g_t\}$, where $g_t = 2^{t-1}$. Hence

$$P_t = \frac{f_t}{g_t}, \tag{9.6}$$

and $P_1 = \frac{1}{1}$, $P_2 = \frac{2}{2}$, $P_3 = \frac{3}{4}$, $P_4 = \frac{5}{8}$, $P_5 = \frac{8}{16}$, $P_6 = \frac{13}{32}$, etc. The terms of the sequence $\{P_n\}$ become smaller as n increases. Why? In genetics, $I_t = 1 - P_t$ is called the *inbreeding coefficient* of the tth generation. ◀

A sequence, like other real functions, can be graphed to present a visual image of the sequence's behavior. This is often very helpful in illustrating trends in terms of infinite sequences.

Example 9.10 We illustrate the graph of several sequences as follows.

a) Let A be the finite sequence $\{1, 2, 3, 2, 1\}$. The graph of A is shown in Fig. 9.3.

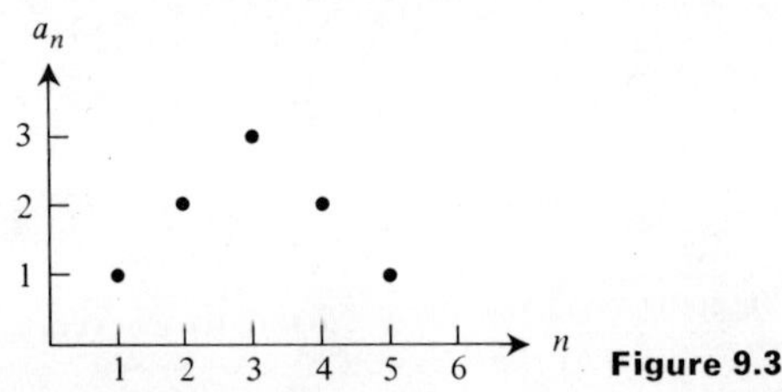

Figure 9.3

b) The initial portion of the graph of the geometric sequence $\{5(\frac{1}{2})^{(n-1)}\}$ is shown in Fig. 9.4.

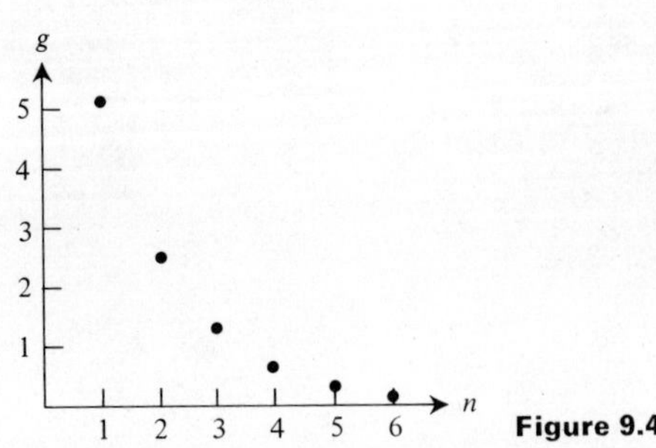

Figure 9.4

c) The sequence $\{a_n\}$, where $a_n = \sin(n\pi/2)$, has the graph illustrated in Fig. 9.5. $\{a_n\}$ is actually the sequence $\{1, 0, -1, 0, 1, 0, -1, \ldots\}$.

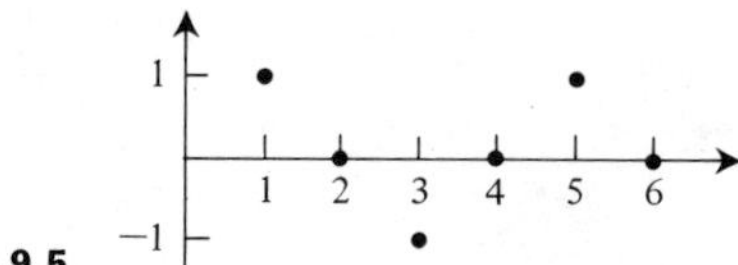

Figure 9.5

◀

9.4 THE LIMIT OF A SEQUENCE

Infinite sequences can be split into two classes. One class of infinite sequences consists of those sequences whose terms become very close to one number as the index n becomes larger; these sequences are said to be *convergent*. An example of a convergent sequence is the sequence $\{1/n\}$; as n increases $1/n$ becomes closer to zero. The other class of infinite sequences consists of sequences that are *not convergent*—that is, those sequences whose terms do not all become very close to one fixed number. An example of a sequence which does not converge is the alternating sequence $\{(-1)^n\}$. As n increases, the terms of this sequence alternate between $+1$ and -1 and hence do not all become close to a single number. When the terms of a nonconvergent sequence increase without bound—i.e., become larger than any finite number—then the sequence is said to be *divergent*. An example of a divergent sequence is the sequence $\{2^n\}$.

Example 9.11

a) The sequence

$$\{a_n\}, a_n = \frac{2n}{n+1}$$

is

$$\{1, \tfrac{4}{3}, \tfrac{6}{4}, \tfrac{8}{5}, \tfrac{10}{6}, \tfrac{12}{7}, \tfrac{14}{8}, \ldots\}.$$

As n becomes larger, the corresponding sequence term a_n increases; however, each term a_n is less than 2:

$$a_n = \frac{2n}{n+1} < \frac{2n+2}{n+1} = \frac{2(n+1)}{n+1} = 2.$$

The sequence $\{a_n\}$ in fact converges to 2 since the difference

$$2 - a_n = 2 - \frac{2n}{n+1} = \frac{(2n+2) - 2n}{n+1} = \frac{2}{n+1}$$

approaches zero as n approaches $+\infty$.

b) The sequence $\{b_n\}$, defined by $b_n = [1 + (-1)^n]/2$, does not converge as n approaches infinity since this sequence alternates between 0 and 1:

$$\{b_n\} = \{0, 1, 0, 1, 0, 1, \ldots\}.$$

9.7 This problem refers to the arithmetic sequence, $a_n = a + (n - 1)d$.

a) Can the term a_n ever be negative if $a > 0$? If so, when? Can you give an example?
b) If d is not an integer and $a = 0$, can a_n ever be an integer? If so, when? Can you give an example?
c) What is the equation of a function defined on $[0, \infty)$ that agrees with the arithmetic sequence at integer arguments?

9.8 This problem refers to the geometric sequence, $g_n = gr^{(n-1)}$.

a) Can the term g_n ever be negative? If so, when? Can you give an example?
b) If $g = 1$ and r is not an integer, can g_n ever be an integer? If so, when? Can you give an example?
c) Can you give two functions defined on $(0, +\infty)$ that agree with the geometric sequence at integer arguments?

9.9 Give the first six terms of the sequence $\{a_n\}_{n=1}^{\infty}$ specified by

a) $a_n = \sin(n)$ b) $a_n = \sin(n\pi)$
c) $a_n = \cos(n\pi)$ d) $a_n = \sec(n\pi)$
e) $a_n = \sin(\pi/2 \cos(n\pi))$ f) $a_n = \sin(n\pi/2)$
g) $a_n = \tan((2n + 1)\pi/4)$ h) $a_n = \cos(n\pi/2)\sin(n\pi/2)$

9.10 Many sequences are repetitive, corresponding to periodic functions. The *period* of a repetitive sequence is the length of the shortest cycle of numbers that repeat. Determine the period of the following sequences.

a) $\{3, 5, 6, 3, 5, 6, 3, \ldots\}$
b) The sequence of the unit digit of the positive even integers (6 is the unit digit of the number 546)
c) The sequence of digits in the decimal expansion of the function $\frac{1}{7}$
d) The sequence $\{\sin(n\pi/3)\}$
e) The sequence $\{\cos(n\pi)\}$

9.11 Let $A = \{1, 2, 3, 4\}$; $B = \{1, -1, 1, 0\}$; $C = \{-2, 3, -1, 2\}$. Evaluate the following.

a) $A + B$ b) $A - B$ c) $A + C$
d) $2A$ e) $2A - C$ f) $A \cdot B$
g) $(A \cdot B) + C$ h) $A \cdot (B + C)$ i) $3A - 2B + C^2$

9.12 Let A be an arithmetic sequence with $a = 95$ and $d = -5$. Let G be a geometric sequence with $g = 0.02$ and $r = 2$. Let F be a Fibonacci sequence with $f_1 = 3$ and $f_2 = 1$. Indicate the first five terms of the following sequences.

a) A b) G c) F
d) $A + G$ e) $A \cdot G$ f) $A - F$
g) $3G/F$ h) $G \cdot A$ i) $A + G + F$

9.13 Evaluate the following limits as $n \to \infty$.

a) $\lim_{n\to\infty} 3^n$ b) $\lim_{n\to\infty} (\frac{1}{2})^n$

c) $\lim_{n\to\infty} (1 + 0.1)^n$ d) $\lim_{n\to\infty} (1 - 0.1)^n$

e) $\lim_{n\to\infty} (-1 + 0.1)^n$ f) $\lim_{n\to\infty} (-1 - 0.1)^n$

g) $\lim_{n\to\infty} \sin(n\pi)$ h) $\lim_{n\to\infty} \sin(2n\pi)$

i) $\lim_{n\to\infty} \cos(n\pi)$ j) $\lim_{n\to\infty} \cos(2n\pi)$

9.14 What is the limit as $n \to \infty$ of (a) an arithemtic sequence and (b) a Fibonacci sequence?

9.15 The limit process commutes with the algebraic operations given in Definition 9.5. That is, if $\{a_n\} \to A$ and $\{b_n\} \to B$ as $n \to \infty$, then

(1) $\lim_{n\to\infty} \{a_n + b_n\} = A + B$

(2) $\lim_{n\to\infty} \{a_n \cdot b_n\} = A \cdot B$

(3) $\lim_{n\to\infty} \{k \cdot a_n\} = k \cdot A$

Using these equations, evaluate the limits of the following sequences.

a) $\{1 + (\frac{1}{2})^n\}$ b) $\{1 + (-\frac{1}{2})^n\}$
c) $\{\frac{3}{2^n}\}$ d) $\{(-0.1)^n \cdot (0.7)^n\}$
e) $\{4 \cdot (\frac{1}{3})^n - 1\}$ f) $\{\frac{1}{2} \sin((2n + 1)\pi)\}$

9.16 The roots, or zeros, of an equation of the form $f(x) - g(x) = 0$ are the x coordinates of the points where the graphs of $y = f(x)$ and $y = g(x)$ intersect. By sketching the graphs of the appropriate functions, decide whether the sequence of positive roots (in increasing order) of the given equation is finite or infinite.

a) $(x - 2)^2 - x = 0$ b) $\cos(x) - x = 0$
c) $\tan(x) - x = 0$ d) $e^x - \ln(x) = 0$
e) $\sin(x) - (x/18\pi) = 0$ f) $\sin(x) - 1/x = 0$

9.17 In Example 9.9, evaluate the first ten terms of the sequence $\{P_t\}$.

9.18 In Example 9.9, the ratio of successive relative heterozygosity indices, $R_t = P_t/P_{t-1}$, give a measure of the rates of change in heterozygosity for the population. Compute the first ten terms of the sequence, $\{R_t\}_{t=2}^{\infty}$. Do these numbers go to zero?

9.19 Compute the first ten terms of the sequence $\{\sin(n\pi/2)/n\}$. Is this sequence convergent?

9.20 The Watson–Crick model of the double helical structure of DNA describes the DNA molecule as two strands of nucleic acids arranged in a helical structure. Each strand contains a sequence for four bases: the pyrimidine bases, cytosine (C) and thymine (T); and the purine bases, adenine (A) and guanine (G). The sequence may be as long as one million bases. The linkage between the two strands is through the bases, and the classical model allows only four possible combinations. Using the above abbreviations, these are A-T, T-A, C-G, and G-C. Assume that one strand contained the sequence {A, T, C, T, C, A, T, G, A, G, G, C, T}.

a) If this sequence is expressed as $\{a_n\}$, what is the length of the sequence and what is a_7?
b) What would be the corresponding portion of the sequence of the other strand?

We first investigate the properties of a finite series, discuss several particular finite series, and then explore the concept of an infinite series in Section 10.3. It should be pointed out that the index symbol is often denoted by letters other than i. The most common choices for index symbols are n, j, k and l.

Example 10.1 We evaluate the following series by first expanding them and then adding their terms.

a) $\sum_{i=3}^{5} i^2 \qquad = 3^2 + 4^2 + 5^2 = 50$

b) $$\sum_{j=1}^{4} 2 \cdot 3^{(j-1)} = \underset{j=1}{2 \cdot 3^0} + \underset{j=2}{2 \cdot 3^1} + \underset{j=3}{2 \cdot 3^2} + \underset{j=4}{2 \cdot 3^3}$$
$$= 2(3^0 + 3^1 + 3^2 + 3^3) = 2 \cdot 40 = 80$$

c) $$\sum_{n=1}^{4} (2^n - n) = (2 - 1) + (2^2 - 2) + (2^3 - 3) + (2^4 - 4)$$
$$= (2 + 2^2 + 2^3 + 2^4) - (1 + 2 + 3 + 4)$$
$$= 30 - 10 = 20.$$

d) $$\sum_{i=1}^{4} 2 + (i - 1)3 = \underset{i=1}{2} + \underset{i=2}{(2 + 3)} + \underset{i=3}{(2 + 6)} + \underset{i=4}{(2 + 9)} = 26$$

e) The sum $1 + 3 + 5 + 7 + 9 = 25$ can be expressed as a series

$$\sum_{i=1}^{5} (1 + 2(i-1)).$$

The limit of the index was taken to be 5 as there were five terms in the sum. The form of the general term was recognized to be that of an arithmetic sequence since there is a constant difference between each term of the sum. Observe, however, that this sum could also be expressed as the series

$$\sum_{i=0}^{4} (1 + 2i).$$

Both series expressions are correct; which we would utilize depends on the circumstances of the problem in which we would need to express 25 as a series. ◀

Algebra of Finite Series

Since a finite series is defined in terms of the addition of real numbers, it is natural to expect that finite series may be added or multiplied by a constant. As each series is associated with a sequence, the algebraic properties of series are essentially the same as those defined for sequences in Definition 9.5.

Definition 10.3 Given two compatible series $\sum a_i$ and $\sum b_i$, having the same set of indices,

a) the *sum* $(\sum a_i) + (\sum b_i) \equiv \sum (a_i + b_i)$ (addition rule), and
b) for any constant k, $k(\sum a_i) \equiv \sum (ka_i)$ (distributive rule).

Note: The product of two series is not the series of the product. For instance, if $a_i = i$ and $b_i = 2^i$, then

$$\sum_{i=1}^{2} a_i = 1 + 2 = 3, \qquad \sum_{i=1}^{2} b_i = 2 + 4 = 6$$

and hence

$$\left(\sum_{i=1}^{2} a_i\right) \cdot \left(\sum_{i=1}^{2} b_i\right) = 3 \cdot 6 = 18.$$

But

$$\sum_{i=1}^{2} (a_i \cdot b_i) = 2 + 8 = 10.$$

The reason the sum of the product is not the product of the sums becomes clear if we multiply the series

$$\sum_{i=1}^{2} a_i \qquad \text{and} \qquad \sum_{i=1}^{2} b_i:$$

$$\sum_{i=1}^{2} a_1 \cdot \sum_{i=1}^{2} b_i = (a_1 + a_2)(b_1 + b_2) = (a_1 + a_2)b_1 + (a_1 + a_2)b_2$$

$$= a_1 b_1 + a_2 b_1 + a_1 b_2 + a_2 b_2 \neq a_1 b_1 + a_2 b_2.$$

Example 10.2 The addition and constant multiplication rules of Definition 10.3 are illustrated with the following series:

$$A = \sum_{i=1}^{4} i, \qquad B = \sum_{i=1}^{4} i^2, \qquad \text{and} \qquad C = \sum_{i=1}^{5} i - 2.$$

a) $$3A = 3 \cdot \sum_{i=1}^{4} i = \sum_{i=1}^{4} 3i$$

b) $$A + B = \sum_{i=1}^{4} i + \sum_{i=1}^{4} i^2 = \sum_{i=1}^{4} (i + i^2).$$

c) $$A - 2B = \sum_{i=1}^{4} i - 2 \sum_{i=1}^{4} i^2 = \sum_{i=1}^{4} (i - 2i^2)$$

d) $A + C$ cannot be expressed using the addition rule, as the index of C ranges from 1 to 5. But, since they represent finite series, A and C are numbers and hence $A + C$ is just another number. The problem here is one of expressing the numbers as similar series. This difficulty can be partially overcome as follows.

VI. *The geometric series* is the series $\sum g_n$, where the sequence $\{g_n\}$ is a geometric sequence; $g_n = gr^{n-1}$. The closed form of the geometric series is obtained algebraically as follows. First, let

$$S = \sum_{n=1}^{N} gr^{(n-1)}$$

be the sum of the geometric sequence. Then multiplying S by the constant r and utilizing the distributive rule, Definition 10.3(b), we have

$$r \cdot S = \sum_{n=1}^{N} gr^n.$$

Subtracting these two series, we obtain

$$(1 - r)S = S - rS = \sum_{n=1}^{N} [gr^{(n-1)} - gr^n].$$

Expanding this series, we have

$$(1 - r)S = \Big(\underset{n=1}{g - gr}\Big) + \Big(\underset{n=2}{gr - gr^2}\Big) + \Big(\underset{n=3}{gr^2 - gr^3}\Big) + \cdots + \Big(\underset{n=N}{gr^{N-1} - gr^N}\Big).$$

If we remove the parentheses, we see that the terms $-gr$ and $+gr$ cancel; similarly, $-gr^2$ and $+gr^2$ cancel. This is also the case for each term gr^n for $n = 1, 2, \ldots, N - 1$. After cancelling, all that remains is the first and last terms, g and $-gr^N$. We thus have $(1 - r)S = g - gr^N$. Replacing S by the series $\sum gr^{n-1}$ and dividing by $1 - r$ gives the *closed form of the geometric series*

$$\boxed{\sum_{n=1}^{N} gr^{(n-1)} = g\left[\frac{1 - r^N}{1 - r}\right], \qquad \text{if } r \neq 1.} \tag{10.9}$$

When $r = 1$, the geometric series is really a constant series whose sum is given by formula (10.4). Series (b) of Example 10.1 is a geometric series with $N = 4$, $r = 3$, and $g = 2$. By use of formula (10.9), the sum of this series is $2(1 - 3^4)/(1 - 3) = 2(1 - 81)/(-2) = 80$.

VII. *The collapsing series* is a particular type of series that is frequently encountered; it is distinguished by the fact that each term is in fact a difference between successive terms of a sequence. Thus a collapsing series has the form

$$\sum_{i=n}^{N} (a_i - a_{i-1}) \qquad \text{or} \qquad \sum_{i=n}^{N} (a_{i+1} - a_i).$$

The collapsing series is easy to evaluate for, as its name suggests, when expanded it collapses to the difference of just two terms. For example,

$$\sum_{i=n}^{N} (a_{i+1} - a_i) = (a_{n+1} - a_n) + (a_{n+2} - a_{n+1}) + (a_{n+3} - a_{n+2}) \\ + \cdots + (a_N - a_{N-1}) + (a_{N+1} - a_N).$$

Rearranging the terms, we can express the sum as

$$-a_n + (a_{n+1} - a_{n+1}) + (a_{n+2} - a_{n+2}) + \cdots + (a_N - a_N) + a_{N+1}.$$

Thus,

$$\sum_{i=n}^{N} (a_{i+1} - a_i) = a_{N+1} - a_n. \tag{10.10}$$

The use of a collapsing-series closed form will usually be restricted to more theoretical discussions, like the one above, in which we developed the closed form of the geometric series. In actual problems, such as

$$\sum_{n=1}^{N} [a_{n+1} - a_n],$$

where $a_n = 3n$, we would simply perform the subtraction and then sum the series:

$$\sum_{n=1}^{N} [3(n+1) - 3n] = \sum_{n=1}^{N} [(3n+3) - 3n] = \sum_{n=1}^{N} 3 = 3 \cdot N.$$

VIII. *An alternating series* is any series that can be expressed in the form

$$\sum_{i=n}^{N} a_i(-1)^i, \qquad \text{with all } a_i > 0.$$

The closed form of an alternating series depends on the sequence $\{a_i\}$. The simplest alternating series is the series $\sum(-1)^i$. By expanding such a series, grouping the terms in pairs of $+1$ and -1, the value of the series is seen to be a function of the index limits:

$$\sum_{i=1}^{N} (-1)^i = \begin{cases} -1 & \text{if } N \text{ is odd;} \\ 0 & \text{if } N \text{ is even.} \end{cases} \tag{10.11}$$

The validity of the above closed forms can be checked for particular values of the upper limit of summation N. Their verification for arbitrary natural numbers N can easily be proved using the principle of mathematical induction. We do not present these arguments here. The derivation of closed forms for series requires considerable algebraic skill and experience with finite series; there are no established methods which work for all series.

Example 10.3 The following series are evaluated using the closed forms given above and the addition and distributive rules of Definition 10.3.

a) Sum of a constant:

$$\sum_{n=1}^{7} c = 7 \cdot c;$$

$$\sum_{n=4}^{7} c = \sum_{n=1}^{7} c - \sum_{n=1}^{3} c = 7c - 3c = 4c.$$

f) Sum of geometric series: By use of equation (10.9),

$$\sum_{n=1}^{10} 6\left(\frac{1}{2}\right)^{n-1} = 6 + 3 + \frac{3}{2} + \cdots + \frac{6}{512} = 6\left[\frac{1 - (\frac{1}{2})^{10}}{1 - (\frac{1}{2})}\right]$$

$$= \frac{6 \cdot (1 - (\frac{1}{2})^{10})}{\frac{1}{2}} = 12(1 - (\tfrac{1}{2})^{10}) \simeq 11.99.$$

The geometric series

$$\sum_{n=0}^{7} 2 \cdot 3^n$$

can be evaluated by first shifting the index so that the exponent n becomes $k - 1$. To accomplish this, we set

$$n = k - 1 \qquad \text{or} \qquad k = n + 1.$$

Then as n ranges over 0, 1, 2, ..., 7, the index k will range over the values 1, 2, 3, ..., 8. The lower limit of summation becomes $k = 0 + 1 = 1$, while the upper limit of summation becomes $k = 7 + 1 = 8$. This gives the equivalence

$$\sum_{n=0}^{7} 2 \cdot 3^n = \sum_{k=1}^{8} 2 \cdot 3^{k-1}.$$

Using formula (10.9), we obtain the value of this geometric series as

$$2\left[\frac{1 - 3^8}{1 - 3}\right] = 2\left[\frac{1 - 6561}{-2}\right] = 6560.$$

The geometric series of the above form is frequently encountered. Using the shift technique employed above, we can derive the general closed form

$$\boxed{\sum_{n=0}^{N} ar^n = a\left[\frac{1 - r^{N+1}}{1 - r}\right],} \tag{10.12}$$

$$\sum_{j=0}^{9} 3 \cdot 2^j = 3\left(\frac{1 - 2^{10}}{1 - 2}\right) = 3\left(\frac{-1023}{-1}\right) = 3069.$$

g) Collapsing series: For any sequence $\{X_i\}$,

$$\sum_{i=1}^{10} (X_{i+1} - X_i) = (X_2 - X_1) + (X_3 - X_2) + \cdots + (X_{11} - X_{10})$$

$$= X_{11} - X_1;$$

$$\sum_{i=7}^{10} (X_{i+1} - X_i) = (X_8 - X_7) + (X_9 - X_8) + (X_{10} - X_9) + (X_{11} - X_{10})$$

$$= X_{11} - X_7.$$

◀

Example 10.4 Let $P(i)$ denote the size of the ith generation of a given population. If $\Delta P(i)$ is the change in population from the ith to the $(i+1)$st generation, then the total change in the population size from the first to the Nth generation is the sum of the changes $\Delta P(i)$ for $i = 1, 2, \ldots, N-1$. Expressed in series notation, this would read

$$\text{(Total change) } TC(N) = \sum_{i=1}^{N-1} \Delta P(i).$$

Since $\Delta P(i) = P(i+1) - P(i)$, the total change $TC(N)$ can also be represented as

$$TC(N) = \sum_{i=1}^{N-1} (P(i+1) - P(i)).$$

This is a collapsing series and thus the total change $TC(N) = P(N) - P(1)$.

a) Suppose the population growth is specified by defining its change $\Delta P(n)$ at generation n to be

$$\Delta P(n) = \log(n).$$

Then the population size at generation $N+1$ is

$$\begin{aligned} P(N+1) &= P(1) + \sum_{i=1}^{N} \Delta P(i) \\ &= P(1) + \sum_{i=1}^{N} \log(i) \\ &= P(1) + \log(N!). \end{aligned}$$

The symbol "$N!$" is read "N factorial" and is defined as the product of the first N natural numbers,

$$N! = 1 \cdot 2 \cdot 3 \cdot \cdots \cdot (N-1) \cdot N, \qquad \text{for} \qquad N = 1, 2, 3, \ldots$$

$$0! = 1 \text{ by definition.}$$

b) If the population size at generation n is $P(n) = P_0 r^n$, the population is growing geometrically. The change in a geometric population at generations n is

$$\begin{aligned} \Delta P(n) = P(n+1) - P(n) &= P_0 r^{n+1} - P_0 r^n \\ &= P_0(r^{n+1} - r^n) = [P_0(r-1)]r^n. \end{aligned}$$

Thus the sequence of changes, $\{\Delta P(n)\}$, is also a geometric sequence. ◀

10.3 INFINITE SERIES

A finite series

$$\sum_{n=1}^{N} a_n$$

is a number obtained by adding together the terms $a_1, a_2, \ldots, a_N$. An infinite series,

$$\sum_{n=1}^{\infty} a_n, \qquad (10.2)$$

is not a number. As stated in Definition 10.1, an infinite series is a symbol having the form (10.2). In some instances, the infinite series (10.2) will have a unique associated number, in which case the series will be said to be *convergent* and to converge to the associated number. If there is not a unique number to which a series converges, it is said to be *divergent*. If the concept of a convergent series is to generalize the concept of a finite series, then the number associated with a convergent series should generalize the concept of adding all the terms of the series.

Example 10.5 Consider the infinite geometric series

$$\sum_{n=1}^{\infty} (\tfrac{1}{2})^{n-1} = 1 + \tfrac{1}{2} + \tfrac{1}{4} + \tfrac{1}{8} + \cdots \tag{10.13}$$

Let S_N denote the sum of the first N terms of this infinite series. Then,

$$S_1 = 1; \qquad S_2 = 1 + \tfrac{1}{2} = \tfrac{3}{2};$$

$$S_3 = 1 + \tfrac{1}{2} + \tfrac{1}{4} = \tfrac{7}{4}; \qquad S_4 = 1 + \tfrac{1}{2} + \tfrac{1}{4} + \tfrac{1}{8} = \tfrac{15}{8}.$$

In general, using formula (10.9),

$$S_N = \sum_{n=1}^{N} \left(\frac{1}{2}\right)^{n-1} = \frac{1 - (\frac{1}{2})^N}{1 - \frac{1}{2}} = 2 - \left(\frac{1}{2}\right)^{N-1}.$$

As the upper limit of summation N becomes larger and larger, the sum S_N becomes closer and closer to the number 2, since the terms $(\frac{1}{2})^{N-1}$ become negligible; $(\frac{1}{2})^{100} \simeq 7.8 \times 10^{-31}$. We are thus led to associate the series (10.13) with the number 2, and say that the

$$\sum_{n=1}^{\infty} \left(\frac{1}{2}\right)^{n-1} \quad \text{converges to 2.}$$

◀

Example 10.6 Consider the infinite series

$$\sum_{n=1}^{\infty} \frac{1}{\sqrt{n}} = 1 + \frac{1}{\sqrt{2}} + \frac{1}{\sqrt{3}} + \cdots$$

The sum of the first N terms of this series is

$$S_N = \sum_{n=1}^{N} \frac{1}{\sqrt{n}} = 1 + \frac{1}{\sqrt{2}} + \frac{1}{\sqrt{3}} + \cdots + \frac{1}{\sqrt{N}}. \tag{10.14}$$

Since

$$1 < \sqrt{2} < \sqrt{3} < \cdots < \sqrt{N},$$

we conclude that

$$\frac{1}{1} > \frac{1}{\sqrt{N}}; \qquad \frac{1}{\sqrt{2}} > \frac{1}{\sqrt{N}}; \qquad \frac{1}{\sqrt{3}} > \frac{1}{\sqrt{N}}, \text{ etc.}$$

Consequently, replacing each term $1/\sqrt{n}$ in the equation (10.14) by $1/\sqrt{N}$, we obtain the inequality

$$S_N > \underbrace{\frac{1}{\sqrt{N}} + \frac{1}{\sqrt{N}} + \frac{1}{\sqrt{N}} + \cdots + \frac{1}{\sqrt{N}}}_{N \text{ times}} = N\left(\frac{1}{\sqrt{N}}\right) = \sqrt{N}.$$

For example, if $N = 100$, we have

$$S_{100} = 1 + \frac{1}{\sqrt{2}} + \frac{1}{\sqrt{3}} + \cdots + \frac{1}{\sqrt{99}} + \frac{1}{\sqrt{100}}$$

$$> \frac{1}{\sqrt{100}} + \frac{1}{\sqrt{100}} + \frac{1}{\sqrt{100}} + \cdots + \frac{1}{\sqrt{100}} + \frac{1}{\sqrt{100}} = \sqrt{100} = 10.$$

As the index N becomes larger and larger, the value of S_N must also become larger, since it is always greater than $\sqrt{N}$. Consequently, we do not associate a finite value with the infinite series $\sum 1/\sqrt{n}$. Instead we say that the series is *not convergent* and, in fact, it *diverges* to $+\infty$. ◀

The above examples illustrate how the convergence of an infinite series is analyzed. Basically, a sequence $\{S_N\}$ of partial sums of the series is generated and the series is said to "converge," "not converge," or "diverge" accordingly as the sequence $\{S_N\}$ converges, does not converge, or diverges.

Definition 10.4

a) For each natural number n, the *nth partial sum* of the series

$$\sum_{i=1}^{\infty} a_i$$

is the finite series

$$S_n = \sum_{i=1}^{n} a_i.$$

b) A series

$$\sum_{i=1}^{\infty} a_i$$

is said to be *convergent*, and to converge to the number A if its sequence of partial sums $\{S_n\}$ converges to the number A.

c) A series

$$\sum_{i=1}^{\infty} a_i$$

is said to be *divergent* if its sequence of partial sums $\{S_n\}$ does not converge.

Example 10.7

a) The series

$$\sum_{i=1}^{\infty} 1$$

is divergent. For each $n = 1, 2, 3, \ldots$, the nth partial sum of this series is

$$S_n = \sum_{i=1}^{n} 1 = n.$$

Consequently, the sequence $\{S_n\} = \{n\}$ diverges to $+\infty$.

b) The series

$$\sum_{i=1}^{\infty} (-1)^i$$

is also divergent. The partial sums of this series are equal to either -1 or 0.

$$S_1 = -1, \qquad S_2 = -1 + 1 = 0, \qquad S_3 = -1 + 1 - 1 = -1, \text{ etc.}$$

Hence $\{S_n\}$ does not converge.

c) The series

$$\sum_{i=1}^{\infty} (1/i^2)$$

converges to the number $\pi^2/6$. This fact was known in 1736, but even the simplest proofs are very difficult. We compute a few terms in the sequence of partial sums, $\{S_n\}$, to illustrate this convergence.

$$S_1 = \tfrac{1}{1} = 1$$

$$S_2 = (\tfrac{1}{1}) + (\tfrac{1}{4}) = 1.25$$

$$S_3 = (\tfrac{1}{1}) + (\tfrac{1}{4}) + (\tfrac{1}{9}) = 1.25 + 0.11111 = 1.36111$$

$$S_4 = (\tfrac{1}{1}) + (\tfrac{1}{4}) + (\tfrac{1}{9}) + (\tfrac{1}{16}) = 1.36111 + 0.625 = 1.42361$$

$$S_5 = S_4 + (\tfrac{1}{25}) = 1.42361 + 0.04 = 1.46361$$

$$S_6 = S_5 + (\tfrac{1}{36}) = 1.46361 + 0.02777 = 1.49138$$

$$S_7 = S_6 + (\tfrac{1}{49}) = 1.49138 + 0.02040 = 1.51179$$

By repeating this process, we obtain the value $S_{50} = 1.62513$. The value of $\pi^2/6$ is approximately 1.64493; thus the difference between S_{50} and $\pi^2/6$ is 0.01980,

$$\sum_{i=1}^{\infty} \left(\frac{1}{i^2}\right) - \sum_{i=1}^{50} \left(\frac{1}{i^2}\right) \simeq 0.01980. \qquad ◀$$

The series in Example 10.7(c) converges to the number $\pi^2/6$. The *rate of convergence* is very slow for hand calculations, and it would require a genius to recognize the limit by simply calculating partial sums. However, using a computer, it would be a simple exercise to evaluate the terms S_n until they did not change in the 5th decimal place. The value S_n at this stage would then be an approximation of the limit of the series.

To obtain the exact limit of an infinite series which converges, it is usually necessary to represent the partial sum

$$S_N = \sum_{i=1}^{N} a_i$$

in a closed form—that is, to represent S_N as a function of N, such as $S_N = N^2$ or $S_N = 1 - (1/N)$. This is possible only for very special series and requires considerable experience in manipulating series and sequences.

EXERCISE SET 10

10.1 Evaluate the following series by expanding and then adding.

a) $\sum_{i=1}^{6}(i + 1)$ b) $\sum_{i=3}^{10}(i^2 + 2)$

c) $\sum_{k=1}^{10}(2k^2 - k)$ d) $\sum_{k=3}^{8}(k^3 + (-1)^k)$

e) $\sum_{i=-2}^{4} i^2$ f) $\sum_{j=2}^{5}\log(j)$

g) $\sum_{k=1}^{4} t_k$ h) $\sum_{i=1}^{5}(t_i - t_{i-1})$

10.2 Write the following series as expanded sums.

a) $\sum_{i=1}^{20}(1 + i)^{-1}$ b) $\sum_{k=3}^{7}(k + 1)^2$

c) $\sum_{j=1}^{N} x_j^* \Delta(j)$ d) $\sum_{l=-1}^{5} 3^l$

e) $\sum_{i=1}^{8}(i + 1) - (i - 1)$ f) $\sum_{j=1}^{N} j \cdot \Delta x$

10.3 Express the following expanded sums using sigma notation.

a) $1 + 4 + 9 + 16 + 25$
b) $\frac{1}{2} + \frac{1}{4} + \frac{1}{8} + \frac{1}{16} + \cdots + \frac{1}{1024}$
c) $5 + 7 + 9 + 11 + 13$
d) $a_1 + b_1 + 2a_2 + b_2 + 3a_3 + b_3 + \cdots$
e) $(a + 3) + (2a + 4) + (3a + 5) + (4a + 6)$
f) $3 + 1 + 3 + 1 + 3 + 1 + 3 + 1$
g) $(5 - 4) + (6 - 5) + (7 - 6) + (8 - 7)$

10.4 Using the formulas of Section 10.2, evaluate the following series.

a) $\sum_{n=1}^{30} 7$ b) $\sum_{n=1}^{30} n$ c) $\sum_{n=1}^{30} n^2$

d) $\sum_{n=1}^{30} n^3$ e) $\sum_{n=1}^{30} 2 + (n-1)3$ f) $\sum_{n=1}^{30} 4n$

g) $\sum_{n=1}^{30} 4n^2 - n$ h) $\sum_{n=1}^{30} (-1)^n$ i) $\sum_{n=1}^{30} (n+1)^2$

10.5 Evaluate the following series.

a) $\sum_{n=1}^{10} 3 \cdot 2^{(n-1)}$ b) $\sum_{n=1}^{10} 3(\frac{1}{2})^{(n-1)}$ c) $\sum_{n=1}^{10} 3(-\frac{1}{2})^{(n-1)}$

d) $\sum_{n=1}^{10} 2^n$ e) $\sum_{n=0}^{10} a^n$ f) $\sum_{n=0}^{10} b^{n-1}$

10.6 Evaluate the following series for the indicated parameters.

a) $\sum_{i=1}^{N} ar^{i-1}$; $a = 3$, $r = 2$, $N = 7$.

b) $\sum_{i=1}^{N} ar^{i-1}$; $a = 3$, $r = 10^{-1}$, $N = 7$.

c) $\sum_{i=1}^{N} ar^{i-1}$; $a = -2$, $r = 6/10$, $N = 7$.

d) $\sum_{j=3}^{N} aj^2$; $a = 2$, $N = 5$.

e) $\sum_{k=2}^{N} (k+2)^2$; $N = 7$.

f) $\sum_{j=1}^{N} (aj + b) \cdot \Delta x$; $N = 10$, $a = 3$, $b = 2$, $\Delta x = 10^{-5}$.

10.7 A series can be considered as a function of its upper limit of summation. Let

$$S(n) = \sum_{i=1}^{n} a_i.$$

a) Compute $S(10)$ if $a_i = 1$.
b) Compute $S(n)$ if $a_i = i + 3$.
c) If $a_i = 2^i$, compute $S(10)$ and $S(100)$.
d) If $a_i = (i+1)h$, where h is a constant, compute $S(n)$.
e) If $h = 1/n$ in part (d), what value does $S(n)$ have?
f) Show that $S(n)$ is a monotone-increasing function of N if $a_i > 0$ for all i.

10.8 Which of the special series considered in Section 10.2 are convergent if $N \to \infty$? Indicate what restrictions must be placed on the parameters for the series to converge.

10.9 Evaluate the 10th partial sum $S(10)$ of the following series. Indicate whether the series appears to converge and if so what its limit might be.

a) $\sum_{n=1}^{\infty} n^2$ b) $\sum_{n=1}^{\infty} (\frac{1}{2})^n$ c) $\sum_{n=1}^{\infty} \frac{1}{n}$

d) $\sum_{n=1}^{\infty} \sin(n\pi/2)$ e) $\sum_{n=1}^{\infty} (-2)^{-n}$ f) $\sum_{n=1}^{\infty} \frac{\sin(n\pi/2)}{n}$

10.10 A simple way to verify equation (10.5) is to construct a rectangular region of dimensions N by $N + 1$ and to systematically show that the series is equal to one-half the area of the rectangle. Draw a sketch to illustrate this when $N = 4$.

10.11 A folk tale dating from early Egyptian times recounts the story of a slave who saved the Pharaoh's life. The grateful Pharaoh offered the slave a reward of his choosing. The slave seemed reasonable when he asked for one grain of wheat on the first day of the month, two grains of wheat on the second day, four grains of wheat on the third day, and so on, doubling the number of grains each day until the end of one month. The Pharaoh was overjoyed by the slave's meager request and granted it immediately. Why was he a foolish Pharaoh? If the month had 31 days, how many grains of wheat would the slave have accumulated by the end of the month? Convert this amount to an approximate amount as powers of 10.

10.12 A veterinarian prescribes a drug-dosage scheme for an animal in which the animal is given 10 grams of the drug on the first day and the dosage is halved on each successive day for one week. What is the total amount of drug which should be prescribed to complete this dosage scheme?

10.13 Using the rabbit population of Example 9.6, express the total rabbit production as a collapsing series.

10.14 If a population increases in size by an amount $\Delta P(t) = t^2 - 2$ at generation t and has an initial size of 102, what will its size be after 10 generations? After N generations?

10.15 Give a closed form of the series

$$\sum_{i=n}^{N} (a_i - a_{i-1}).$$

Use $n = 1$ for a special case.

10.16 A drug is administered in 5-mg doses each day. The patient metabolizes 4.9 mg of the drug on the first day, so that 0.1 mg remains in circulation when the second dose is administered. Assume that, on successive days, the patient metabolizes one-half of the amount of the drug that was metabolized on the preceding day.

a) How much of the drug will be in the patient's blood after four days?
b) If the drug is given for only two days, when will the patient's blood be completely free of the drug?

10.17 Referring to Exercise 5.8, what is the maximum permissible dose if all doses are assumed to be the same, including the initial dose? (*Note*: The initial dose is not assumed to be five units.)

SECTION 11

DIFFERENCE EQUATIONS

11.1 DIFFERENCE EQUATIONS

In this section, we introduce the basic concepts of difference equations. Our principal examples, used to motivate and illustrate difference equations, will be discrete models of populations. However, population models are not the only applications of difference equations. Difference equations are the essential basic mathematical components of all "finite" or discrete mathematical models. As we shall see in the following sections, the fundamental concept of a difference equation, namely the change in a quantity between "generations," translates naturally into the "derivative," which is the fundamental concept of differential calculus. The connection between difference equations and calculus is very important. The entire theory of calculus can be presented without using difference equations; yet, when one attempts to apply calculus to real problems that require real solutions, frequently, the only way to obtain a solution is to utilize a computer. This requires translating the calculus concepts into discrete step-by-step procedures that are described by difference equations. We shall not explore this connection between difference equations and calculus, but simply mention it at the outset so that you do not develop the misconception that difference equations are used just to describe populations.

A discrete population model describes the size of a population X at each generation. The complete description of the population's size is given by a sequence $\{X_n\}$, where X_n denotes the size of the population at generation n for $n = 1, 2, 3, \ldots$ X_0 is used to denote the initial size of the population from which the first generation develops. The objective of a population model is either to predict future population sizes or to estimate present population sizes, such as those of fish or seals, which are difficult to observe. Thus, a population model seldom specifies population size $\{X_n\}$. Instead, some other feature of the population is observed, and the model attempts to obtain the sizes $\{X_n\}$ by analyzing this aspect of the population.

The one aspect of a population that is most directly related to its size is how it reproduces and dies. Most population models describe the birth and death processes of a population over single generations. Therefore, population models begin with the mode of change in the population from one generation to the next. The change from the nth generation to the $(n + 1)$st generation is denoted by the difference

$$\Delta X_n \equiv X_{n+1} - X_n \qquad \textbf{Forward Difference.} \tag{11.1}$$

Equations describing this change are called *difference equations.*

Example 11.1 A new park is designated to have a buffalo herd. Due to normal mortality, the herd size is expected to decrease by 5 percent each year. If the initial herd size is 85 buffalo and each year 38 additional buffalo are added to the herd, what will be its eventual size? To determine the eventual size, we construct a model of the population as follows.

First, let X_n denote the herd size at generation n. $X_0 = 85$ denotes the initial herd size. Based upon the above information, the change ΔX_n in the herd size will be the same at each generation X_n:

$$\Delta X_n = \underset{\text{5 percent deaths}}{\underset{\uparrow}{-0.05X_n}} + \underset{\text{New stock}}{\underset{\uparrow}{38.}} \tag{11.2}$$

Using equation (11.1), we can rearrange this equation into the form

$$X_{n+1} = 0.95X_n + 38.$$

From this formula we can sequentially evaluate the herd size at generations $n = 1, 2, 3, \ldots$.

$$X_1 = 0.95X_0 + 38 = 118.75$$

$$X_2 = 0.95X_1 + 38 \simeq 150.81$$

$$X_3 = 0.95X_2 + 38 \simeq 181.27$$

$$X_4 = 0.95X_3 + 38 \ldots \text{etc.}$$

We do not round off to integer values since the death rate is assumed to be an "average" value.

This procedure is not very helpful if we want to know the herd size X_{50} of the 50th generation. Instead, let us develop a formula for X_n as a function of n. To do this, we again begin to evaluate the terms $X_1, X_2, X_3, \ldots,$ but this time we do not perform the arithmetic. Instead, we will look for simplifications and a pattern that will indicate how to evaluate X_n for arbitrary n.

$$\begin{aligned}
X_1 &= 0.95X_0 + 38 \\
X_2 &= 0.95X_1 + 38 \\
&= (0.95)[0.95X_0 + 38] + 38 \\
&= (0.95)^2X_0 + (0.95)38 + 38 \\
X_3 &= 0.95X_2 + 38 \\
&= (0.95)[(0.95)^2X_0 + (0.95)38 + 38] + 38 \\
&= (0.95)^3X_0 + (0.95)^2 38 + (0.95)38 + 38
\end{aligned}$$

A pattern begins to emerge in the formula for X_3. It suggests that the formula for X_4 is

$$X_4 = (0.95)^4X_0 + (0.95)^3 38 + (0.95)^2 38 + (0.95)38 + 38.$$

By substituting X_3 into $X_4 = (0.95)X_3 + 38$, we find that the above formula is correct. To generalize this formula to an arbitrary value of n, we first express it in a simple form by using summation notation,

$$X_4 = (0.95)^4 X_0 + \sum_{i=1}^{4} 38(0.95)^{i-1}.$$

From this equation, we obtain the formula for X_n by replacing the number 4 with n:

$$X_n = (0.95)^n X_0 + \sum_{i=1}^{n} 38(0.95)^{i-1}.$$

To actually evaluate a term X_n, we can use equation (10.9) to evaluate the geometric series

$$\sum_{i=1}^{n} 38(0.95)^{i-1} = 38\left[\frac{1-(0.95)^n}{1-(0.95)}\right].$$

Substituting this expression into the above equation, we obtain

$$X_n = (0.95)^n X_0 + 38\left[\frac{1-(0.95)^n}{1-0.95}\right].$$

Using a calculator to evaluate the term $(0.95)^n$, this formula gives $X_{50} \simeq 708$.

How do we answer the question, "What will be the eventual size of the herd?" We simply take the limit of X_n as $n \to +\infty$. Using formula (9.7) with $a = 0.95$, we see that

$$\lim_{n\to\infty} (0.95)^n = 0.$$

The limit of the herd size X_n is thus

$$\begin{aligned}\lim_{n\to\infty} X_n &= X_0 \cdot \lim_{n\to\infty} (0.95)^n + 38\left[\frac{1-\lim_{n\to\infty}(0.95)^n}{1-0.95}\right]\\ &= 38\left[\frac{1}{1-0.95}\right] = 760.\end{aligned}$$

This limiting value is called the *steady-state* value of the population and is denoted by $\bar{X}$.

If instead we had asked the question "At what size will the buffalo herd remain constant?" we could have answered the question simply by using equation (11.2). If the herd size is constant, the change ΔX_n will be zero. Thus, equating the right side of equation (11.2) to zero, we obtain the *equilibrium* size X_E:

$$\begin{aligned}0 &= -0.05X_E + 38;\\ X_E &= \frac{38}{0.05} = 760.\end{aligned}$$

In more complicated models, the equilibrium value X_E and the limiting steady-state value $\bar{X}$ need not be equal. ◀

Example 11.2 We list several common equations for growth mechanisms, where X_n indicates the population size at generation n and $\Delta X_n = X_{n+1} - X_n$ is the change in the population size from the nth to the $(n + 1)$st generation.

a) Constant growth: $\Delta X_n = c$.
b) Proportional growth: $\Delta X_n = cX_n$.
c) Geometric growth: $\Delta X_n = \Delta_0 r^n$, Δ_0 is a constant.
d) Logarithmic growth: $\Delta X_n = \log(n)$.
e) Delay growth: $\Delta X_n = f(X_{n-t})$, where f is a function to be specified and t is the lag-time or delay-time.
f) *Logistic* growth: $\Delta X_n = cX_n(X_E - X_n)$, where X_E is an equilibrium population.
g) Stochastic or probabilistic growth: $\Delta X_n = pX_n$, where p is the probability of each individual of the population "giving birth" to a new individual. ◀

Each of the equations of Example 11.2 is a difference equation. The operator Δ used in these equations is called a forward difference operator since it reflects a forward change in the population size, X_n.

Definition 11.1 The *forward difference operator*, Δ, associates with each sequence $\{X_n\}$ a sequence $\{\Delta X_n\}$, where

$$\Delta X_n = X_{n+1} - X_n.$$

Definition 11.2 A *difference equation* associated with a sequence $\{X_n\}$ is an equation that involves two or more arbitrary terms of the sequence.

The following are examples of difference equations:

$$X_n = X_{n-1}; \qquad X_{k+2} = X_k; \qquad X_n + n^2 X_{n+1} = X_{n-3}.$$

The following equations are not difference equations;

$$X_n = 3 \text{ (only one term)}, \qquad \text{and} \qquad X_5 = X_6 - X_2 \text{ (no arbitrary terms)}.$$

While a difference equation may be stated using arbitrary index symbols (usually n, $n + 1$, etc.), the equation is not altered by changing the indices as long as they maintain the same relationship. For example, each of the following represents the same difference equation:

$$\Delta X_n = 2X_n; \qquad X_{n+1} = 3X_n; \qquad X_n = 3X_{n-1}; \qquad X_k = 3X_{k-1}.$$

It is the relationship between the terms that characterizes the difference equation. To classify difference equations having the same general form, we associate with each equation a number called its *order*. The order of a difference equation is similar to the degree of a polynomial.

Definition 11.3 The *order* of a difference equation is the difference, $N_2 - N_1$, between the largest index of the equation, N_2, and the smallest index of the equation, N_1.

Example 11.3 We indicate examples of difference equations and their order denoted by the letter O.

a) $X_{n+3} = X_n + 1$; $\quad O = (n + 3) - n = 3$
b) $X_{n+1} - X_n = 1 + r^n$; $\quad O = (n + 1) - n = 1$
c) $X_{n+1} - X_{n-1} = 2X_n \cdot h$; $\quad O = (n + 1) - (n - 1) = 2$
d) $X_{n+1} = X_n + hX_{n-1} + h^2 X_{n-2}$; $\quad O = (n + 1) - (n - 2) = 3$
e) $X_n = X_{n+4} - 3n^2 X_{n+1}$; $\quad O = (n + 4) - (n) = 4$
f) $\Delta X_n = kX_n - X_{n-3}$; $\quad O = (n + 1) - (n - 3) = 4$
g) $X_{n+1} = X_n^2 - 2(X_{n-2})^2$; $\quad O = (n + 1) - (n - 2) = 3$
h) $y_{n+2} = y_n + y_{n+1}$; $\quad O = (n + 2) - (n) = 2$
i) $X_n = 3n + 7$; $\quad O = n - n = 0$* ◀

The difference equation (f) of Example 11.3 is expressed using the forward difference operator Δ. Equation (f) could be equivalently expressed as

$$X_{n+1} - X_n = kX_n - X_{n-3} \quad \text{or} \quad X_{n+1} = (k + 1)X_n - X_{n-3}.$$

Similarly, any difference equation can be written in an equivalent form that employs the difference operator Δ. To demonstrate how this may be accomplished, we consider the third order difference equation

$$y_{n+2} - y_{n-1} = 2y_n + n^2. \tag{11.3}$$

To write equation (11.3) in the form

$$\Delta y_k = \underline{\qquad\qquad},$$

we first solve the equation (11.3) explicitly for the term having the highest subscript, y_{n+2}, obtaining the equation

$$y_{n+2} = 2y_n + y_{n-1} + n^2. \tag{11.4}$$

Next, we change the subscript in equation (11.4) so that the highest subscript has the form $k + 1$. To achieve this, we introduce a new index variable, k, related to the old index variable, n, in such a way that the term y_{n+2} becomes y_{k+1}. That is, we set $n + 2 = k + 1$ and hence have $n = k - 1$. Substituting this representation of n into the equation (11.4) yields the equation

$$y_{k+1} = 2y_{k-1} + y_{k-2} + (k - 1)^2.$$

Subtracting y_k from both sides of this equation, we obtain an equation of the desired form:

$$\Delta y_k = -y_k + 2y_{k-1} + y_{k-2} + (k - 1)^2.$$

---

* This is not a difference equation, since it involves only one sequence term.

11.2 FIRST-ORDER DIFFERENCE EQUATIONS

The simplest type of difference equation is a first-order difference equation. The general form of a first-order difference equation is

$$X_{n+1} = F(n, X_n), \tag{11.5}$$

where $F(n, X_n)$ is a function involving n, X_n, and constants.

Example 11.4 Examples of first-order difference equations are given by the following.

a) $X_{n+1} = (X_n + 1)^n$
b) $X_{n+1} = X_n + \sqrt{X_n}$
c) $X_{n+1} = aX_n + b$
d) $X_{n+1} = aX_n^2 + bX_n + C$
e) $X_{n+1} = X_n^{-1}$
f) $\Delta X_n = 2X_n^{-3}$
g) $\Delta X_n = 8X_{n+1} - 2X_n$ ◀

Definition 11.4 A solution of the difference equation (11.5) is a sequence $\{X_n\}$ such that, for each choice of $n = 1, 2, 3, \ldots$, the equation (11.5) is a true identity when X_n and X_{n+1} are the nth and $(n + 1)$st terms of the sequence $\{X_n\}$.

The sequence $\{X_n\}$, where $X_n = 3 \cdot 2^{-n}$, is a solution of the difference equation $X_{n+1} = \frac{1}{2}X_n$. This may be verified by showing that X_{n+1} equals $\frac{1}{2}X_n$:

$$X_{n+1} = 3 \cdot 2^{-(n+1)} = 3 \cdot 2^{-n-1} = (3 \cdot 2^{-n})\tfrac{1}{2} = \tfrac{1}{2}X_n.$$

The sequence $\{X_n\}$, where $X_n = 3 \cdot 4^{-n}$, is not a solution of the difference equation $X_{n+1} = \frac{1}{2}X_n$. This may be demonstrated by showing that X_{n+1} and $\frac{1}{2}X_n$ are not equal for this sequence:

$$X_{n+1} = 3 \cdot 4^{-(n+1)} = 3 \cdot 4^{-n-1} = (3 \cdot 4^{-n})(\tfrac{1}{4}) \neq \tfrac{1}{2}X_n.$$

Example 11.5 The solution of a difference equation is not unique. Consider the following examples:

a) The difference equation $X_{n+1} = X_n + 2$ has the solution $\{3, 5, 7, 9, \ldots\}$ and the solution $\{-4, -2, 0, 2, 4, \ldots\}$.
b) The difference equation $X_{n+1} \cdot X_n = 1$ has the solution $\{X_n\}$, where $X_n = 1$ for all n, and the solution $\{X_n\}$, where

$$X_n = \begin{cases} \frac{1}{2} & \text{if } n \text{ is odd,} \\ 2 & \text{if } n \text{ is even.} \end{cases}$$

c) The difference equation $\Delta X_n = 0$ has as a solution the constant sequence $\{X_n\}$, where $X_n = k$, a constant. Thus each choice of k results in a different solution. ◀

It is not possible to give a general solution or method of solution for all first-order difference equations. However, it is known that each equation of the form (11.5) will have either no solution or a general solution that involves one arbitrary parameter. Each particular solution will then have the form of the general solution with a specific value assigned to the parameter. In this subsection, we examine the general solution of *linear equations* with constant coefficients.

Definition 11.5

a) A *first-order linear constant-coefficient difference equation* has the form

$$X_{n+1} = aX_n + b, \tag{11.6}$$

where a and b are constants.

b) The linear difference equation (11.6) is *homogeneous* if $b \equiv 0$.

We consider first the homogeneous equation

$$Y_{n+1} = aY_n \quad \textbf{Homogeneous first-order equation.} \tag{11.7}$$

The general solution of equation (11.7) can be derived by iteratively expressing each term in the solution sequence, $\{Y_n\}$, as a multiple of the term preceding it. If we assume that $Y_1 = aY_0$, where Y_0 is a constant parameter, then by equation (11.7),

$$\begin{aligned} Y_2 &= aY_1 &&= a(aY_0) &&= a^2 Y_0; \\ Y_3 &= aY_2 &&= a(a^2 Y_0) &&= a^3 Y_0; \\ &\vdots \\ Y_n &= aY_{n-1} &&= a(a^{n-1} Y_0) &&= a^n Y_0. \end{aligned}$$

Consequently, we find:

The general solution of $Y_{n+1} = aY_n$ is: $Y_n = Y_0 \cdot a^n$.

A particular solution is obtained when the *initial value*, Y_0, is replaced by a specific number. The proportional growth model in Example 11.2(b) is defined by an equation of the form (11.7). This is the same model given in Example 5.1 when we discussed a cell population.

Example 11.6 The simplest model described by a liner homogeneous difference equation is one in which the birth and death rates of a population are assumed to be directly proportional to the size of the population. If B denotes the constant birth rate and D denotes the constant death rate, the change in the population is given by

$$\Delta X_n = \underset{\text{Births}}{BX_n} - \underset{\text{Deaths}}{DX_n}.$$

Thus the population is described by the equation

$$X_{n+1} = [1 + (B - D)]X_n.$$

The solution of this equation has the form

$$X_n = X_0[1 + (B - D)]^n.$$

If the birth rate exceeds the death rate, then $B - D$ is positive and $a = 1 + (B - D)$ is greater than 1. Consequently, the population will increase in size, growing geometrically. ◀

Next, consider the nonhomogeneous first-order difference equation,

$X_{n+1} = aX_n + b$	**Nonhomogeneous first-order difference equation.**	(11.8)

The general solution of equation (11.8) is a sequence, $\{X_n\}$, which depends on an initial value X_0. To determine an explicit formula for the nth term of the solution sequence, we first examine the form of the terms X_2, X_3, and X_4. We assume that the term X_1 may be represented in terms of an initial value X_0, as

$$X_1 = aX_0 + b.$$

Then, using the defining relation (11.8), we compute

$$X_2 = a(X_1) + b = a(aX_0 + b) + b;$$
$$X_2 = a^2X_0 + ab + b.$$

Likewise,

$$X_3 = a(X_2) + b = a(a^2X_0 + ab + b) + b;$$
$$X_3 = a^3X_0 + a^2b + ab + b.$$

Continuing in this fashion, the term X_4 is determined as follows:

$$X_4 = a(X_3) + b = a(a^3X_0 + a^2b + ab + b) + b;$$
$$X_4 = a^4X_0 + a^3b + a^2b + ab + b.$$

To continue this approach in evaluating X_{100} would be very simple; however, it would consume a lot of paper and time to write out the necessary sums. An alternative is to introduce summation notation to express the sums in a compact form. To introduce an appropriate form, we look at the terms already computed. Note that each term, X_2, X_3, and X_4, is the sum of two components. The first component has the form a^nX_0. The second component is a sum of terms, each of which has the form $b \cdot a^{i-1}$. Thus each term X_n, $n = 2, 3, 4$, is the sum of an exponential term and a geometric series:

$$X_2 = X_0a^2 + \sum_{i=1}^{2} ba^{i-1};$$

$$X_3 = X_0a^3 + \sum_{i=1}^{3} ba^{i-1};$$

$$X_4 = X_0a^4 + \sum_{i=1}^{4} ba^{i-1}.$$

The general term X_n should have a form that reduces to the above equations, when $n = 2, 3$, or 4. Such a form is obtained by replacing the number four in the last equation with the letter n:

$$X_n = X_0 a^n + \sum_{i=1}^{n} b a^{i-1}. \tag{11.9}$$

Using the closed form of the geometric series, equations (10.9) and (10.4), the equation (11.9) may be written in the equivalent form:

The general solution of $X_{n+1} = aX_n + b$ is:

$$X_n = \begin{cases} X_0 a^n + b\left[\dfrac{1 - a^n}{1 - a}\right] & \text{if } a \neq 1; \\ X_0 + n \cdot b & \text{if } a = 1. \end{cases} \tag{11.10}$$

To solve a particular problem, we need only to determine parameters a and b and to write down the corresponding solution. The constant X_0 is then determined by specifying the value of X_n for any particular index n.

Example 11.7

a) A population is regulated so that it increases by 120 units each year. If the 1970 population was 2000, what will be the population size in the year 1990? *Solution*: The population size x_n satisfies $\Delta x_n = 120$. Thus, $x_{n+1} = x_n + 120$. The solution of this equation is given by (11.10) with $a = 1$ and $b = 120$. Hence, $x_n = x_0 + 120n$. The initial population is $x_0 = 2000$ in 1970 and the population in 1990 is then

$$x_{20} = 2000 + 120(20) = 4400.$$

b) A trust fund is established to support student scholarships at the rate of \$100,000 per year. If the fund initially contained \$900,000, which is invested at a 10 percent rate of interest compounded annually, when will the fund "go broke?" *Solution*: The monies in the fund n years after it is established, denoted by f_n, satisfies the first order difference equation

$$f_{n+1} = (1 + 0.1)f_n - 10^5.$$

The -10^5 is the monies spent each year and the $(1 + 0.1)f_n$ is the amount available from the previous year plus the interest. The solution of this difference equation is given by (11.10), with $a = 1.1$, $b = -10^5$, and $f_0 = 9 \times 10^5$ (here we associate f_n with the term X_n in (11.10)),

$$f_n = 9 \cdot 10^5 (1.1)^n - 10^5\left(\frac{1 - (1.1)^n}{1 - 1.1}\right).$$

This equation reduces to

$$f_n = 10^6 - 10^5 \cdot (1.1)^n = 10^5(10 - (1.1)^n).$$

Hence, f_n will be negative when

$$(1.1)^n > 10.$$

To determine the least positive integer n for which this inequality is valid, we take the logarithm of both sides of the inequality

$$n \log (1.1) > \log(10) = 1.$$

Since $\log(1.1) \simeq 0.04139$, this requires $n > 1/0.04139$. As n is an integer, it follows that $n \geq 25$. Hence, the endowment fund will be depleted in the 25th year. ◀

Example 11.8 In this example, we consider an isolated population that has no predators and neither gains members through immigration nor loses members through emigration. Such a population is sometimes called an *island population.*

a) Under ideal conditions such a population would grow proportionally, thus satisfying the equation $X_{n+1} = (1 + c)X_n$. This would lead to a population size $X_n = X_0(1 + c)^n$, which would increase without bound if $c > 0$.
b) However, it would be more realistic to assume that the island provides a constant but limited food supply. Assume that the food supply is measured by the maximum number of individuals it can sustain. Let this number be A. The growth of the population is then governed by the food supply. If the population, X_n, exceeds the value A, it will deplete the food supply; there will be a shortage of food and the population will decrease in size. If X_n is less than A then the population will increase in size. The amount of increase or decrease in the population should be a function of how close the population size, X_n, is to the value A. One model which reflects these assumptions is obtained by assuming the change in the population, ΔX_n, is proportional to $A - X_n$:

$$\Delta X_n = k(A - X_n). \tag{11.11}$$

For instance, if $k = 3/2$ and $A = 10$, then equation (11.11) is

$$\Delta X_n = \tfrac{3}{2}(10 - X_n).$$

Expressing this as a linear difference equation in the form of equation (11.8), we have

$$X_{n+1} = (-\tfrac{1}{2})X_n + 15.$$

Thus, $a = -\frac{1}{2}$ and $b = 15$. The solution is given by formula (11.10)

$$X_n = X_0(-\tfrac{1}{2})^n + 15\left[\frac{1 - (-1/2)^n}{1 - (-1/2)}\right].$$

This equation simplifies to

$$X_n = (-\tfrac{1}{2})^n(X_0 - 10) + 10. \tag{11.12}$$

The term X_0 gives the initial population size. We observe that the population size fluctuates about the value $\bar{X} = 10$, since the term $(-\frac{1}{2})^n(X_0 - 10)$ alternates between positive and negative values. ◀

In the event $a \neq 1$, the solution of $X_{n+1} = aX_n + b$ is given by

$$X_n = X_0 a^n + b\left(\frac{1 - a^n}{1 - a}\right).$$

We note that this equation expresses X_n as the sum of two components. The first component, $X_0 a^n$, contains an arbitrary parameter X_0, called the *initial value*. This component is the general solution of the associated homogeneous equation

$$X_{n+1} = aX_n.$$

The second component does not contain the parameter X_0 and is a particular solution of $X_{n+1} = aX_n + b$. To verify this, we set $X_n = b(1 - a^n)/(1 - a)$, then

$$X_{n+1} = b\left[\frac{1 - a^{n+1}}{1 - a}\right] = a \cdot b\left[\frac{1 - a^n}{1 - a}\right] + b = aX_n + b.$$

Consequently, the general solution X_n has the form

$$X_n = \begin{pmatrix}\text{General solution of} \\ X_{n+1} = aX_n\end{pmatrix} + \begin{pmatrix}\text{Particular solution of} \\ X_{n+1} = aX_n + b\end{pmatrix}. \qquad (11.13)$$

In the situation where the equation $X_{n+1} = aX_n + b$ is used to describe a population, the coefficient a is associated with the rates of birth and death of the population and the term b is associated with external interactions affecting the population size. A simple type of association would be to let

a = (rate of birth) − (rate of death);

b = (immigration) − (emigration);

or $\quad b$ = new stock added each year;

or $\quad b$ = amount harvested each generation.

In this setting, the population at generation n can be viewed as being composed of two components, one component reflecting the *natural* progress of the population in the absence of any outside influence (the term $X_0 a^n$) and the second component reflecting the outside influence (the term $b(1 - a^n)/(1 - a)$). These are the same two components indicated in equation (11.13).

Example 11.9 A bacteria population doubles its size every generation. However, the sampling procedure used to monitor the culture requires a withdrawal of 500 cells at each generation. If the culture initially contains 10^4 cells, the equation describing its size at each generation is obtained as follows.

Let X_n equal the number of cells at generation n. The model of the culture is described by letting

$$X_0 = 10^4 \qquad \text{and} \qquad X_{n+1} = 2(X_n - 500).$$

This equation is simplified to the form

$$X_{n+1} = 2X_n - 10^3.$$

Setting $a = 2$ and $b = -10^3$ in equation (11.10) with X_0 as above, the solution is

$$X_n = 10^4 \cdot 2^n - 10^3\left[\frac{1 - 2^n}{1 - 2}\right] = 2^n(10^4 - 10^3) + 10^3.$$ ◀

11.3 THE STEADY STATE AND EQUILIBRIUM SOLUTIONS OF DIFFERENCE EQUATIONS

When a difference equation is used to model the dynamics of a biological phenomenon—for instance, a cancer tumor or an endangered species, there are two types of information that can be derived from the model. One type of information is called *quantitative information*, and is provided by an explicit solution of the difference equation. When the parameters are known, the solution enables a scientist to state a number that describes the phenomenon at a particular stage. For instance, the scientist might say that, according to a particular model, in 30 days a tumor will have a size $X_{30} = 6.2$ g. Note, however, that such a numerical quantity can only be provided when the value of all parameters in the model are known.

A second type of information, called *qualitative information*, provides knowledge about the behavior of the phenomenon in more general terms. For instance, a physician would be interested in whether a tumor is going to persist or regress. Since a model always involves errors, the physician does not worry about exact quantitative values; that a tumor is 5.1 g or 4.8 g is not as important as whether the tumor will remain at about 5 g, will grow, or will diminish. Such alternatives represent qualitative information, which a model may provide. Often qualitative information about the solution of a difference equation can be provided without knowing the value of all parameters in the model.

One qualitative aspect of a solution sequence $\{X_n\}$ of a difference equation is the *steady-state* of the solution. The steady state is the limiting value of the phenomenon described by the difference equation. The steady state, denoted by $\overline{X}$, is defined by (provided the limit exists)

$$\overline{X} = \lim_{n \to \infty} X_n \qquad \textbf{Steady state.}$$

If the limit does not exist, the sequence $\{X_n\}$ does not have a steady state. You will observe that the steady state is simply another name for the *limit of a sequence*.

The steady state of the solution of the linear first-order difference equation (11.8) is obtained from the solution (11.10). The dependence of the steady state on the coefficient a is most easily seen by rearranging equation (11.10), when $a \neq 1$, in the alternative form

$$X_n = a^n\left[X_0 - \frac{b}{1-a}\right] + \frac{b}{1-a}. \tag{11.14}$$

In this form it is clear that the steady state $\bar{X}$ will depend on the limit of a^n, as $n \to \infty$. From equation (9.7), we find that this limit is $\pm\infty$, 0, or does not exist. Thus, if $X_0 - (b/(1-a)) \neq 0$, the steady state $\bar{X}$ is $+\infty$, $-\infty$, $b/(1-a)$ or does not exist, depending on the value of a. If $X_0 - (b/(1-a)) = 0$, equation (11.14) reduced to $X_n = b/(1-a)$ and the steady state $\bar{X}$ will equal $b(1-a)$, again with $a \neq 1$. When $a = 1$, the solution is $X_n = X_0 + nb$ and hence the steady state depends on the value of b. In this case, $\bar{X}$ will be $-\infty$, X_0, or $+\infty$, depending on whether b is negative, zero, or positive. We combine this analysis of the steady state of the first-order difference equation in the following theorem.

Theorem 11.1 The steady state $\bar{X}$ of the difference equation

$$X_{n+1} = aX_n + b, \qquad a \text{ and } b \text{ constants,}$$

is given by the following.

i) If $a = 1$,

$$\bar{X} = \lim_{n\to\infty} X_0 + nb = \begin{cases} +\infty & \text{if } b > 0; \\ X_0 & \text{if } b = 0; \\ -\infty & \text{if } b < 0. \end{cases}$$

ii) If $a \neq 1$,

$$\bar{X} = \lim_{n\to\infty} a^n\left[X_0 - \frac{b}{1-a}\right] + \frac{b}{1-a}$$

$$= \begin{cases} +\infty & \text{if } a > 1 \quad \text{and} \quad X_0 - \dfrac{b}{1-a} > 0; \\ -\infty & \text{if } a > 1 \quad \text{and} \quad X_0 - \dfrac{b}{1-a} < 0; \\ \dfrac{b}{1-a} & \text{if } |a| < 1 \quad \text{or} \quad X_0 - \dfrac{b}{1-a} = 0; \\ \text{does not exist} & \text{if } a \leq -1 \quad \text{and} \quad X_0 - \dfrac{b}{1-a} \neq 0. \end{cases}$$

Example 11.10 The following illustrate the use of Theorem 11.1 to evaluate steady-state values $\bar{X}$.

a) $x_{n+1} = 3x_n - 2,\ x_0 = 1:\ \bar{X} = \lim_{n\to\infty} x_n = 1$

b) $x_{n+1} = 5x_n + 2,\ x_0 = 1:\ \bar{X} = \lim_{n\to\infty} x_n = \infty$

c) $x_{n+1} = \frac{1}{2}x_n - 3,\ x_0 = 2:\ \bar{X} = \lim_{n\to\infty} x_n = -6$

d) $x_{n+1} = \frac{1}{2}x_n - 3,\ x_0 = 12:\ \bar{X} = \lim_{n\to\infty} x_n = -6$

e) $x_{n+1} = \frac{1}{4}x_n + 1,\ x_0 = 2:\ \bar{X} = \lim_{n\to\infty} x_n = \frac{4}{3}$

f) $x_{n+1} = x_n - 2,\ x_0 = 8:\ \bar{X} = \lim_{n\to\infty} x_n = \infty$

g) $x_{n+1} = 2x_n - 10,\ x_0 = 2:\ \bar{X} = \lim_{n\to\infty} x_n = -\infty$ ◀

Example 11.11 To visualize a population that has a negative growth rate, let X_n denote the difference between a population's size and that of a desired population size M. Then (within reason) negative values of X_n do not indicate a negative population, but simply a population that is below the desired mean population M. In this setting, consider the behavior of a population satisfying

$$X_{n+1} = -0.9X_n + 3.$$

In the absence of the additional three units, this population would oscillate about the number M, since the terms X_n would alternate between negative and positive values approaching zero. However, with the three additional units, the population is described by the solution

$$X_n = (-0.9)^n(X_0 - \tfrac{30}{19}) + \tfrac{30}{19}.$$

Its steady state is $\bar{X} = 30/19$. Thus, this population will exceed the desired population M by $\frac{30}{19}$ units irrespective of the initial population size. ◀

An *equilibrium* or *stationary value* of a difference equation is a value X_E such that once the solution X_n equals X_E it does not change. Thus, X_E is a stationary value of a difference equation with solution $\{X_n\}$, if $X_n = X_E$ implies that $X_i = X_E$ for all $i > n$. The stationary values X_E are obtained by setting all sequence terms equal to X_E in the difference equation and solving the resulting equation.

Example 11.12 We determine the stationary, or equilibrium, values X_E for the following equations.

a) $X_{n+1} = 3X_n + 2: X_E = 3X_E + 2 \Rightarrow X_E = -1$
b) $X_{n+1} = 1/X_n: X_E = 1/X_E \Rightarrow X_E = 1$ or $X_E = -1$
c) $X_{n+2} - X_{n+1} - 2X_n = 6: X_E - X_E - 2X_E = 6 \Rightarrow X_E = -3$
d) $X_{n+1} = aX_n + b: X_E = aX_E + b \Rightarrow X_E = b/1 - a$ ◀

We have extensively utilized population dynamics models to illustrate difference equations and their solutions. Such models are usually not too difficult to visualize since we are constantly exposed to examples of changing populations. However, we do not want to imply that these are the only or the most important application of difference equations. Difference equations are frequently used in many other areas such as medicine, pharmacology, psychology, economics, etc. However, probably the most important application of difference equations is in the numerical solution of problems using a computer or electronic calculator. While the mathematics of these applications is beyond the scope of this text, we illustrate one form of application in the following example.

Example 11.13 How does the $\sqrt{x}$ button on an electronic calculator work? It works by finding the stationary value of a difference equation. The equation is

$$X_{n+1} = \frac{1}{2}\left(X_n + \frac{x}{X_n}\right),$$

where x is the number whose square root we wish to compute. How does a computer or calculator find a stationary value? Not by solving for the stationary values of the equation, since this would require a knowledge of $\sqrt{x}$. Instead, it sequentially computes the terms $X_2, X_3, X_4, \ldots$ After each term is computed, the machine checks to see if the new term is the same as the last term. If so, the calculations stop and the last term is displayed as the answer. If not, the process continues. The value of X_1 must be provided and, in this case, is taken to be the number x. To see how this works, we compute $\sqrt{6}$ as follows:

$$X_1 = 6$$

$$X_2 = \frac{1}{2}\left(X_1 + \frac{6}{X_1}\right) = 3.5$$

$$X_3 = \frac{1}{2}\left(X_2 + \frac{6}{X_2}\right) = 2.607142857$$

$$X_4 = \frac{1}{2}\left(X_3 + \frac{6}{X_3}\right) = 2.454256360$$

$$X_5 = \frac{1}{2}\left(X_4 + \frac{6}{X_4}\right) = 2.449489743$$

$$X_6 = \frac{1}{2}\left(X_5 + \frac{6}{X_5}\right) = 2.449489743$$

Thus, since $X_6 = X_5$, we stop and record to 8 decimal places,

$$\sqrt{6} = 2.44948974\underline{3}.$$

The $\underline{3}$ at the end cannot be guaranteed to be correct and some calculators may not indicate this 9th decimal place. ◀

11.4 SECOND-ORDER LINEAR DIFFERENCE EQUATIONS*

The most general second-order linear difference equation can be expressed in the form

$$X_{n+2} + \alpha X_{n+1} + \beta X_n = \gamma \qquad (11.15)$$

Nonhomogeneous second-order difference equation.

We will assume that the coefficients α(alpha) and β(beta) and the parameter γ(gamma) are constants throughout this subsection. The equation (11.15) is nonhomogeneous if $\gamma \neq 0$. For ease of reference, we will refer to the homogeneous form of equation (11.15) by

$$X_{n+2} + \alpha X_{n+1} + \beta X_n = 0 \qquad (11.16)$$

Homogeneous second-order difference equation.

The solution of equation (11.16) will be a generalization of the solution of the first-order homogeneous equation (11.7). Since the solution of equation (11.7) is a geometric sequence, we assume that a solution of equation (11.16) is also a geometric sequence and then determine the value of the geometric term. Assume that

$$X_n = C\lambda^n, \quad \lambda \text{ a constant to be determined.}$$

Then $X_{n+1} = C\lambda^{n+1}$ and $X_{n+2} = C\lambda^{n+2}$. Consequently, for the sequence $\{X_n\}$ to satisfy equation (11.16), λ and C must be chosen to satisfy the equation

$$C\lambda^{n+2} + \alpha C\lambda^{n+1} + \beta C\lambda^n = 0.$$

If we factor out the term $C\lambda^n$, this equation is equivalent to either $C\lambda^n = 0$ or $\lambda^2 + \alpha\lambda + \beta = 0$. $C\lambda^n = 0$ means $X_n = 0$ for all n and hence is not considered. Since α and β are constants, the quadratic auxiliary equation

$$\lambda^2 + \alpha\lambda + \beta = 0$$

has exactly two solutions that are given by the quadratic formula

$$\lambda_1 = (\tfrac{1}{2})(-\alpha + \sqrt{\alpha^2 - 4\beta}) \qquad \text{and} \qquad \lambda_2 = (\tfrac{1}{2})(-\alpha - \sqrt{\alpha^2 - 4\beta}).$$

I. If $\alpha^2 - 4\beta > 0$, then λ_1 and λ_2 are both real and distinct numbers (distinct roots).

II. If $\alpha^2 - 4\beta = 0$, then λ_1 and λ_2 are both equal to $-\alpha/2$ (repeated roots).

III. If $\alpha^2 - 4\beta < 0$, then λ_1 and λ_2 are complex numbers involving the term i, which has the property that $i^2 = -1$ (complex roots).

---

* This material is optional and is not utilized elsewhere in the text except when compared to second-order differential equations in Section 26. It is used in many applications, particularly in ecological and population models that separate populations into age groups.

Since the equation (11.16) is a second-order equation, its most general solution will contain two arbitrary parameters. We shall denote these by C_1 and C_2. The form of the general solution of equation (11.16) will depend on which one of the three alternatives (I, II, or III) occurs.

Form 1 If $\alpha^2 - 4\beta > 0$, then the most general solution of (11.16) is the sequence $\{X_n\}$, where for each $n = 1, 2, 3, \ldots,$

$$X_n = C_1\lambda_1{}^n + C_2\lambda_2{}^n;$$
$$\lambda_1 = \tfrac{1}{2}(-\alpha + \sqrt{\alpha^2 - 4\beta});$$
$$\lambda_2 = \tfrac{1}{2}(-\alpha - \sqrt{\alpha^2 - 4\beta}). \tag{11.17}$$

Form 2 If $\alpha^2 - 4\beta = 0$, then the most general solution of (11.16) is the sequence $\{X_n\}$, where for each $n = 1, 2, 3, \ldots,$

$$X_n = (C_1 + C_2 n)\lambda^2, \qquad \text{and} \qquad \lambda = \frac{-\alpha}{2}. \tag{11.18}$$

Form 3 If $\alpha^2 - 4\beta < 0$, then the roots λ_1 and λ_2 (as in Form I) are both complex numbers. If we do not object to complex numbers, then the Form I is mathematically acceptable. However, if we are modeling a population, we do not expect the population size to involve nonreal numbers. Using complex number algebra, the complex number i can be eliminated from the solution with the result that the solution $\{X_n\}$ is given by

$$X_n = r^n[C_1 \cos(n\theta) + C_2 \sin(n\theta)], \tag{11.19}$$

where $r = \sqrt{\beta}$ and $\cos(\theta) = -\alpha/2\sqrt{\beta}$.

Example 11.14 We illustrate the above forms of the general solution of a homogeneous second-order linear difference equation.

a) $X_{n+2} + 7X_{n+1} + 12X_n = 0$; $\alpha = 7$, $\beta = 12$, and $\alpha^2 - 4\beta = 1 > 0$. The auxiliary equation $\lambda^2 + 7\lambda + 12 = 0$ has real roots, $\lambda_1 = -3$ and $\lambda_2 = -4$. Hence, the general solution is

$$X_n = C_1(-3)^n + C_2(-4)^n.$$

b) $X_{n+2} - 6X_{n+1} + 9X_n = 0$; $\alpha = -6$, $\beta = 9$, and $\alpha^2 - 4\beta = 0$. The auxilary equation $\lambda^2 - 6\lambda + 9 = 0$ has real repeated roots, $\lambda_1 = \lambda_2 = 3$. Hence, the general solution is

$$X_n = (C_1 + C_2 n)3^n.$$

c) $X_{n+2} - 2X_{n+1} + 4X_n = 0$; $\alpha = -2$, $\beta = 4$, and $\alpha^2 - 4\beta = -4 < 0$. The auxiliary equation $\lambda^2 - 2 + 4 = 0$ has complex roots, $\lambda_1 = 1 + 2i$ and $\lambda_2 = 1 - 2i$. These are not directly useable, but the equation (11.19) gives the general solution

$$X_n = 2^n\left(C_1 \cos\left(\frac{n\pi}{3}\right) + C_2\left(\frac{n\pi}{3}\right)\right)$$

with $r = \sqrt{\beta} = 2$ and $\theta = \pi/3$, since $\cos(\theta) = 1/2$. ◀

For first-order equations, we let the initial constant, X_0, play the role of the parameter. For second-order equations, we will assume that the initial constant, X_0, represents the initial population size and is the first term of the solution sequence—i.e., we assume that the equation (11.16) and *all* of its solutions hold for $n \geq 0$.

Each form of the general solution of equation (11.16) contains two arbitrary parameters. The constant C_1 and C_2 of equations (11.17), (11.18), and (11.19) can be determined by specifying any two terms of the sequence $\{X_n\}$. This will lead to two equations in C_1 and C_2, which can be solved for specific values of C_1 and C_2.

Example 11.15 For each solution in Example 11.14, we determine the constants C_1 and C_2 such that $X_0 = 1$ and $X_1 = 2$.

a) The general solution is $X_n = C_1(-3)^n + C_2(-4)^n$. Thus $X_0 = C_1 + C_2$ and $X_1 = -3C_1 - 4C_2$. Hence we set

$$C_1 + C_2 = 1 \qquad \text{and} \qquad -3C_1 - 4C_2 = 2.$$

Solving these equations simultaneously, we find $C_1 = 6$ and $C_2 = -5$. Thus the particular solution is $X_n = 6(-3)^n - 5(-4)^n$.

b) The general solution is $X_n = (C_1 + C_2 n)3^n$. Hence $X_0 = C_1$ and $X_1 = 3(C_1 + C_2)$. Setting

$$C_1 = 1 \qquad \text{and} \qquad 3(C_1 + C_2) = 2,$$

we find $C_1 = 1$ and $C_2 = -1/3$. The particular solution is therefore $X_n = (1 - (n/3))3^n$.

c) The general solution is $X_n = 2^n[C_1 \cos(n\pi/3) + C_2(n\pi/3)]$. Thus $X_0 = C_1$ and $X_1 = C_1 + \sqrt{3}C_2$, since $\cos(\pi/3) = \frac{1}{2}$ and $\sin(\pi/3) = \sqrt{3}/2$. Setting

$$C_1 = 1 \qquad \text{and} \qquad C_1 + \sqrt{3}C_2 = 2,$$

we find $C_2 = 1/\sqrt{3}$. The particular solution is therefore

$$X_n = 2^n[\cos(n\pi/3) + (1/\sqrt{3})\sin(n\pi/3)].$$ ◀

Next we solve the nonhomogeneous second-order difference equation, (11.15). The basic theorem for such equations states that the most general solution of equation (11.15) is of the form $\{X_n\}$, where

$$X_n = {}_hX_n + {}_pX_n,$$

and $\{{}_hX_n\}$ is the general solution of the associated homogeneous equation (11.16) and $\{{}_pX_n\}$ is *any* particular solution of equation (11.15). Thus, if we can solve the associated homogeneous equation in general and determine just one solution of the nonhomogeneous equation, we can then find all solutions of the nonhomogeneous equation. We thus concentrate on finding a particular solution of equation

(11.15). If such a solution is given by a constant k—i.e., ${}_pX_n = k$ for all n—then k must satisfy

$$k + \alpha k + \beta k = \gamma \qquad \text{or} \qquad k = \frac{\gamma}{1 + \alpha + \beta}.$$

Consequently, if $1 + \alpha + \beta \neq 0$, a particular solution of equation (11.15) is given by

$${}_pX_n = \frac{\gamma}{1 + \alpha + \beta}.$$

If $1 + \alpha + \beta = 0$, we try to find a solution of the form ${}_pX_n = k \cdot n$. This leads to the equation

$$k(n + 2) + \alpha k(n + 1) + \beta k \cdot n = \gamma.$$

Gathering terms, we find $k \cdot n(1 + \alpha + \beta) + 2k + \alpha k = \gamma$. Since $(1 + \alpha + \beta) = 0$, k must satisfy $k = \gamma/(2 + \alpha)$. Consequently, if $1 + \alpha + \beta = 0$ and $2 + \alpha \neq 0$, a particular solution of equation (11.13) is

$${}_pX_n = \frac{\gamma \cdot n}{2 + \alpha}.$$

If, however, $1 + \alpha + \beta = 0$ and $2 + \alpha = 0$, then neither of the above equations provides a particular solution. In this case, $\alpha = -2$ and $\beta = 1$. There is only one particular solution sequence in this situation, which is given by

$${}_pX_n = \frac{\gamma \cdot n^2}{2}.$$

To verify this, you need only substitute this sequence into the equation

$$X_{n+2} - 2X_{n+1} + X_n = \gamma.$$

The above arguments may be summarized as follows. The general solution of the nonhomogeneous equation

$$X_{n+2} + \alpha X_{n+1} + \beta X_n = \gamma$$

has the form

$$X_n = {}_hX_n + {}_pX_n, \tag{11.20}$$

where $\{{}_pX_n\}$ is given by the following:

$$\text{if} \quad 1 + \alpha + \beta \neq 0, \qquad {}_pX_n = \frac{\gamma}{1 + \alpha + \beta}, \; n = 0, 1, \ldots;$$

$$\text{if} \quad 1 + \alpha + \beta = 0, 2 + \alpha \neq 0, \qquad {}_pX_n = \frac{\gamma \cdot n}{2 + \alpha}, \; n = 0, 1, \ldots;$$

$$\text{if} \quad \alpha = -2, \beta = 1, \qquad {}_pX_n = \frac{\gamma \cdot n^2}{2}, \; n = 0, 1, 2, \ldots;$$

and $\{{}_hX_n\}$ satisfies the homogeneous equation $X_{n+2} + \alpha X_{n+1} + \beta X_n = 0$.

Example 11.16 To describe the genetic effects of inbreeding, the geneticist uses a parameter α_n. This parameter is the probability that two genes of an individual are identical by descent (i.e., they may both be traced back to a common ancestor). The subscript n refers to the nth generation and α_n denotes this probability for an individual selected at random from the nth generation. In genetics it is shown that the sequence $\{\alpha_n\}$ satisfies the difference equation

$$(1 - \alpha_n) = \left(1 - \frac{1}{N_e}\right)(1 - \alpha_{n-1}) + \frac{1}{2N_e}(1 - \alpha_{n-2}). \tag{11.21}$$

The term N_e is the effective population size that is assumed to be constant over all generations. The equation (11.21) has the form of equation (11.16) upon rearrangement:

$$\alpha_{n+2} + \left(\frac{1}{N_e} - 1\right)\alpha_{n+1} - \frac{1}{2N_e}\alpha_n = \frac{1}{2N_e},$$

where

$$X_n = \alpha_n; \quad \alpha = \frac{1}{N_e} - 1; \quad \beta = \frac{-1}{2N_e}; \quad \text{and} \quad \gamma = \frac{1}{2N_e}.$$

This equation has the particular solution ${}_p\alpha_n = 1$, and the solution of the associated homogeneous equation is

$${}_h\alpha_n = C_1\lambda_1{}^n + C_2\lambda_2{}^n,$$

where λ_1 and λ_2 are determined by

$$\lambda = \frac{1}{2}\left[\left(1 - \frac{1}{N_e}\right) \pm \sqrt{1 + \left(\frac{1}{N_e{}^2}\right)}\right].$$

The general solution α_n is therefore given by

$$\alpha_n = C_1\lambda_1{}^n + C_2\lambda_2{}^n + 1.$$

If $N_e = 2$, then

$$\alpha_n = C_1\left(\frac{1 + \sqrt{5}}{4}\right)^n + C_2\left(\frac{1 - \sqrt{5}}{4}\right)^n + 1.$$ ◀

EXERCISE SET 11

11.1 State the order of each difference equation and whether or not it is linear.

a) $\Delta X_n = 3$ b) $\Delta X_n = X_{n-2}$ c) $\Delta X_n = 2X_n X_{n+2}$
d) $\Delta X_{n+1} = X_n$ e) $X_{n+2} - X_n = 0$ f) $X_{n+1} = X_n(X_n - X_{n-2})$
g) $X_{n+1} - X_n = -X_n$ h) $X_{n+1} = \sin(X_n)$

11.2 Compute X_5 when $X_0 = 3$ for the following equations.

a) $X_{n+1} = 2X_n - 1$ b) $X_{n+1} = \frac{1}{2}X_k + 3$ c) $X_{n+1} = X_n^2 - X_n$
d) $X_{n+1} = \sin(\pi X_0/2)$ e) $\Delta X_n = X_n$ f) $\Delta X_n = -\frac{1}{2}X_n$
g) $\Delta X_n = X_n(10 - X_n)$ h) $\Delta X_n = -0.1$

11.3 Compute X_5 when $X_0 = 3$ and $X_1 = 2$ for the following equations.

a) $X_{n+1} = X_n + X_{n-1}$ b) $X_{n+2} - 2X_n = 0$ c) $X_{n+2} = X_{n+1} - X_n$
d) $\Delta X_{n+1} = \Delta X_n$ e) $\Delta X_n = X_{n-1} + 5$ f) $\Delta X_n = X_{n+2} - X_n$

11.4 Give the general solutions of the following equations.

a) $X_{n+1} = 3X_n - 2$ b) $X_{n+1} = \frac{1}{2}X_n - 2$ c) $X_{n+1} = 0.5X_n + 1$
d) $X_{n+1} = \frac{1}{2}X_n$ e) $\Delta X_n = 5$ f) $\Delta X_n = -1.5X_n + 3$
g) $\Delta X_n = -0.5X_n + 3$ h) $\Delta X_n = \log(n)$ *Hint*: Substitute $X_n = \log(y_n)$.

11.5 For each equation in Exercise 11.4 give the value of X_{10} when $X_0 = 5$.

11.6 Write each difference equation in Example 11.2 as an explicit equation for X_{n+1}.

11.7 Write each equation of Example 11.3, except (f) and (i), in a different form by first changing the subscript so that the largest subscript is of the form $k + 1$. Then write each equation using the forward difference operator Δ in a form similar to the equations of Example 11.2.

11.8 Determine the general solution of the growth equations (a) through (d) of Example 11.2.

11.9 A population changes according to the difference equation $x_{n+1} = kx_n$.

a) If $k = \frac{3}{2}$ and $x_{10} = 1000$, what is the initial value x_0 associated with this growth?
b) If a population grows from $x_1 = 10$ to $x_5 = 10^4$, what is the growth factor k?
c) If k is greater than one, at what generation will the population size be twice the size of the first generation, x_1?
d) If $0 < k < 1$, at what generation will the population size be one-half the size of the first generation?

11.10 Find the solution of the following difference equations.

a) $y_{n+1} = 3y_n - 10$; $y_0 = 10$ b) $x_{n+1} = 2x_n + 10$; $x_1 = 100$
c) $\Delta x_n = x_n - 30$; $x_0 = 100$ d) $f_{n+1} = 3f_n - 10$; $f_0 = 6$
e) $s_{n+1} = h/2(s_n + s_{n+1}) + s_n$; $s_0 = 1$ f) $t_{n+1} = \frac{1}{4}t_n - 1$; $t_0 = 100$

11.11 a) Express the difference equation

$$\frac{\Delta y_n}{h} = ay_n + b$$

as an explicit equation for y_{n+1}.

b) What is the solution of this difference equation?

11.12 Derive and solve an equation to describe the sum of money in an account drawing 5 percent interest.

11.13 Derive and solve an equation to describe an epidemic in which no one recovers from the disease and in which each infective has a 20 percent chance of infecting an additional person each day.

11.14 An insect colony increases in size daily. The change in population size is made up of two components, birth and death. The birth component is proportional to the population size and the death component is a constant amount each day.

a) If the number of births is $2 \cdot x_n$ and the number of deaths is -10, what is the smallest initial population x_0 at which the colony will not become extinct? What will its steady-state value be?
b) If the initial population is $x_0 = 100$, the number of births is $(0.1)x_n$, and the number of deaths is -5, when will the population reach a size 10^{50}?

11.15 Determine the steady state of the solution of the following difference equations.

a) $X_{n+1} = X_n - 3;\ X_0 = 10$ b) $X_{n+1} = \frac{1}{2}X_n - 1;\ X_0 = 10$
c) $X_{n+1} = \frac{1}{2}X_n - 10;\ X_0 = 0$ d) $X_{n+1} = 3X_n + 1;\ X_0 = 0$
e) $X_{n+1} = 2X_n - 1;\ X_0 = 1$ f) $\Delta X_n = 2X_n + 1;\ X_0 = 1$
g) $\Delta X_n = \frac{1}{2}X_n + 1;\ X_0 = 1$ h) $\Delta X_n = \frac{1}{2}X_n + 1;\ X_0 = 2.1$

11.16 Which solutions of Exercise 11.4 will approach a steady state when $X_0 = 5$?

11.17 Describe the bacteria population X_n of Example 11.9 if the population only increases by 50 percent each generation.

11.18 How many buffalo must be added each year to the herd described in Example 11.1 if the desired eventual size of the herd is 1000?

11.19 Use the difference equation derived in Example 11.13 to compute the square root of the following numbers to three decimal places.

a) 5 b) 10 c) 9 d) 12 e) −4 Try it any way.

11.20 The following formula may be used to compute the cube root of a number x:

$$X_{n+1} = \frac{2}{3}X_n + \frac{1}{3}\left(\frac{x}{X_n^2}\right).$$

Use this formula to compute the cube root of the following numbers. Let $X_1 = x$.

a) 8 b) 4 c) 10 d) 25 e) −4

11.21 The *logistic difference equation* is defined as $\Delta X_n = kX_n(E - X_n)$.

a) What is the order of the logistic equation?
b) Is the logistic equation linear?
c) Compute the terms $X_1, X_2, \ldots, X_5$ when $k = 0.1$, $E = 10$, and $X_0 = 1$.
d) Do the terms X_n in part (c) appear to approach a limit? If so what is it?
e) Repeat parts (c) and (d) with $k = \frac{1}{2}$, $E = 10$, and $X_0 = 20$.

11.22 a) Give the general solution of the equation (11.11) of Example 11.8.
b) How does the constant k in equation (11.11) effect the steady-state population of this model? Can you explain this in nonmathematical terms?
c) Why would an equation of the form $\Delta X_n = k(A - X_n)^2$ not be suitable for modeling the population described? Explain your answer biologically and mathematically.
d) Would the equation $\Delta(X_n) = k(A - X_n)^3$ be reasonable to describe this model? You cannot solve this equation but you can compute several terms of the sequence $\{X_n\}$ using it. Assume $X_0 = 8$, $k = \frac{3}{2}$, and $A = 10$. What would X_4 be? Now assume that $X_0 = 9\frac{1}{2}$, $k = \frac{3}{2}$, and $A = 10$. What would X_4 be?

The remaining exercises are on second-order difference equations as discussed in Section 11.4.

11.23 Find the general solution of the following difference equations.

a) $X_{n+2} + 6X_{n-1} - 3X_n = 0$ b) $2X_n - 3X_{n-2} = 0$
c) $X_{n+1} - 5X_n + X_{n-1} = 0$ d) $\Delta X_n - \Delta X_{n-1} = 0$
e) $4X_{n+2} - 3X_{n+1} - X_n = 0$ f) $X_{n+2} - 3X_{n+1} - X_n = 0$
g) $-X_{n-1} + X_{n-2} + X_{n-3} = 0$ h) $X_{n+1} = X_{n-1} + 3X_n$

11.24 Find the most general solution of the following equations.

a) $X_{n+2} + X_{n+1} - 6X_n = 4$ b) $Y_{n+2} + 3Y_{n+1} - 4Y_n = 10$
c) $Z_{n+2} - 2Z_{n+1} + Z_n = 4$ d) $X_{n+2} + X_{n+1} - 2X_n = 2$
e) $Y_{n+2} - Y_{n+1} - 4Y_n = 3$ f) $Z_{n+2} + 3Z_{n+1} + 9Z_n = 2$

11.25 Find the particular solution of the corresponding equation in Exercise 11.24 that satisfies the given initial conditions.

a) $X_0 = 2, X_1 = 1$ b) $Y_0 = 2, Y_1 = 1$ c) $Z_0 = 1, Z_1 = 0$
d) $X_0 = 2, X_1 = 2$ e) $Y_0 = 1, Y_1 = 2$ f) $Z_0 = 2, Z_1 = 0$

11.26 If a population grows in a medium so that its change at any generation is three times its change at the preceding generation plus five units, find (a) the general equation describing the population at generation n and (b) the size of the population of the nth generation when $X_0 = 1$ and $X_2 = 12$.

11.27 Find a second-order difference equation that is satisfied by the sequence $\{X^n\}$.

a) $X_n = 2 \cdot 3^n + (-1)^n$ b) $X_n = (5 - 2n)(\frac{1}{2})^n$
c) $X_n = 2^n(\cos(n\pi/2) + \sin(n\pi/2))$ d) $X_n = 2 \cdot 3^n + (-1)^n + 2$
e) $X_n = (5 - 2n)(\frac{1}{2})^n + 2$ f) $X_n = 2^n(\cos(n\pi/2) + \sin(n\pi/2)) + 2$

11.28 In Example 11.16

a) Compute the constants C_1 and C_2 if $N_e = 2$, $\alpha_0 = 0$, and $\alpha_1 = 0$.
b) Compute the terms $1 - \alpha_n$ for $n = 0, 1, 2, \ldots, 5$ under the assumptions of (*a*).
c) Compare your results in part (b) with the sequence $\{P_n\}$ in Example 9.9.

11.29 a) State and solve the second-order difference equation satisfied by a Fibonacci sequence, $\{f_n\}$.
b) Give the constants C_1 and C_2 corresponding to the sequence with $f_1 = 1$ and $f_2 = 2$.

CHAPTER THREE

DIFFERENTIAL CALCULUS

SECTION 12
LIMITS AND CONTINUITY

SECTION 13
THE DIFFERENCE QUOTIENT

SECTION 14
THE DERIVATIVE

SECTION 15
THE CHAIN RULE AND HIGHER DERIVATIVES

Section 12 introduces the intuitive concepts of the limit of a function. The basic properties of continuous functions are introduced in connection with the graph of a function. The difference quotient is introduced in Section 13. The basic calculational aspects and interpretations of difference quotients are presented as a prelude to introducing the derivative. In Section 14, the derivative is introduced along with its interpretations as the limit of a difference quotient. The basic properties of the derivative are intuitively introduced and the formulas for the derivatives of the elementary functions are quickly and nonrigorously "developed." (In Section 23, we develop formulas for the derivative of e^x and $\ln(x)$ in a more rigorous manner.) These derivatives are introduced at this point because we feel it is important for students to learn the techniques of calculus by using the functions that they will encounter in actual applications. In Section 15, the Chain Rule is introduced and generalized derivative formulas are developed. Higher derivatives, related rates, and the derivative of an inverse (optional topic) are also presented.

SECTION 12

LIMITS AND CONTINUITY

12.1 INTRODUCTION TO LIMITS

There are two basic types of mathematical models used to describe or "model" biological phenomena—discrete models and continuous models. Discrete models are used to describe *discrete* phenomena, which may be characterized by a sequence, $\{X_n\}$, giving the value of a quantity X associated with a corresponding sequence of time—for instance, the size of a population at successive generations or the annual production of grain. The principal mathematical concept used to describe and analyze discrete models is the difference equation that we explored in Section 11. Continuous phenomena are characterized by the fact that they do not involve sudden changes. Continuous models are described by *continuous* functions, $f(x)$, defined for all values of x in some interval. The principal mathematical concept used to describe and analyze continuous models is that of the limit of a function $f(x)$, as x approaches a specific number. Calculus may be described as the utilization of the limit concept to study and describe continuous phenomena. In this section, we introduce the notions of a *limit* and a *continuous* function.

An Intuitive Definition of a Limit. Given a function, $f(x)$, and a fixed point, $x = b$, we intuitively say that the limit of $f(x)$, as x approaches b, is equal to a number A if the value $f(x)$ is very close to the number A for all values of x close to b.

To illustrate this concept, consider the two functions $f(x) = 2x + 1$ and $g(x) = [\![x]\!]$. (The greatest integer function $[\![x]\!]$ is defined in Section 6; $[\![x]\!]$ is the greatest integer value less than or equal to x. Thus $[\![x]\!]$ rounds off numbers by rounding all decimals down: $[\![3.7]\!] = 3$.) Let $b = 1$, and consider the values $f(x)$ and $g(x)$ for different x values near $b = 1$. We list several such values in Table 12.1. From Table 12.1, we observe that as the distance between x and $b = 1$ becomes smaller, the value of $f(x)$ becomes closer to $A = 3$. However, as the distance between x and $b = 1$ becomes smaller, the value of $g(x)$ remains 1 if $x > 1$, or 0 if $x < 1$. The values of $g(x)$ do not all lie very close to one number A.

Table 12.1

x	0.5	0.9	0.99	0.999	1.1	1.01	1.001
$f(x)$	2	2.8	2.98	2.998	3.2	3.02	3.002
$g(x)$	0	0	0	0	1	1	1

Intuitively, we would say that $f(x)$ has the limit 3 as x approaches 1 and $g(x)$ does not have a limit as x approaches 1. The notation

$$\lim_{x \to b} f(x)$$

denotes the *limit of* $f(x)$ *as* x *approaches* b.

With this notation, we would express these limits as

$$\lim_{x \to 1} f(x) = 3,$$

and

$$\lim_{x \to 1} g(x) \text{ does not exist.}$$

The function $g(x) = [\![x]\!]$ does not have a limit at $x = 1$ because the graph of $y = g(x)$ "jumps" from $y = 0$ to $y = 1$ at $x = 1$. (See Fig. 12.1.)

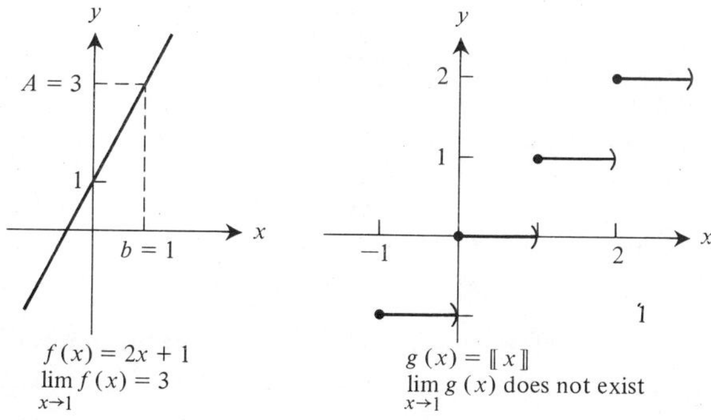

Figure 12.1

A function f may have a limit at $x = b$ even though $f(b)$ is not defined. Consider the function

$$f(x) = \frac{x^2 + x - 6}{x - 2}.$$

Since division by zero is not defined, the Domain(f) = $(-\infty, 2) \cup (2, +\infty)$. Hence f is not defined at the point $b = 2$. Computing the value of $f(x)$ for numbers x near 2, we obtain

$$f(2.1) = 5.1; \quad f(1.9) = 4.9; \quad f(2.01) = 5.01; \quad f(1.99) = 4.99.$$

Since for $x \neq 2$,

$$f(x) = \frac{(x + 3)(x - 2)}{(x - 2)} = x + 3,$$

we can see that the graph of $f(x)$ is the line $y = x + 3$, except for the point with $x = 2$. Because $f(2)$ is not defined, the graph has a "hole" at $x = 2$. (See Fig. 12.2.)

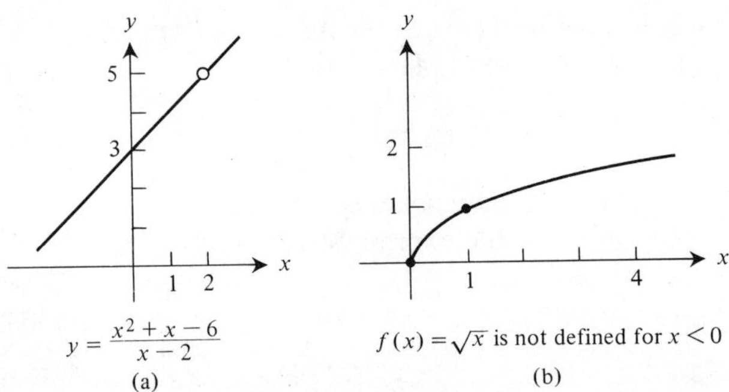

Figure 12.2

As the values of x becomes closer and closer to the point $b = 2$, the corresponding values $f(x)$ become closer and closer to $A = 5$. Therefore, we say that the limit as x approaches 2 of $f(x)$ is 5:

$$\lim_{x \to 2} f(x) = \lim_{x \to 2} \frac{x^2 + x - 6}{x - 2} = 5.$$

Note that we did not simply evaluate $f(x)$ at $x = 2$. If we had attempted to do this, we would have

$$f(2) = \frac{2^2 + 2 - 6}{2 - 2} = \frac{0}{0},$$

which is not defined!

What is the limit of the function $f(x) = x^{1/2}$ as x approaches 0? If we proceed as we did in the above examples, we first evaluate $f(x)$ for numbers x near 0; $f(0.1) = \sqrt{0.1} \simeq 0.316$; $f(0.01) = 0.1$; $f(0.001) \simeq 0.0316$. These values of $f(x)$ seem to be getting closer to zero as x approaches zero. However, in the above examples we considered values of x both above and below the limit point b. For the function $g(x) = [\![x]\!]$, the values of $g(x)$ were markedly different for x values above and below $b = 1$. This led us to conclude that the function g does not have a limit at $x = 1$. Therefore, in the present example, we must consider values of x less than zero. But $f(-0.1) = \sqrt{-0.1}$ does not exist. Since the domain of f is $[0, \infty)$, f is not defined for any $x < 0$. We conclude that $\lim_{x \to 0} f(x)$ does not exist.

You might argue that the function clearly does approach zero as x approaches 0 through positive values. To state this observation using *limit vocabulary*, we introduce the notion of a *one-sided limit*. In this case, we write

$$\lim_{x \to 0^+} f(x) = 0,$$

to denote that as x approaches 0 from above, $f(x)$ approaches 0.

Returning to the function $g(x) = [\![x]\!]$ considered above, we can describe the limiting behavior of $g(x)$ as x approaches 1 from the right or above by

$$\lim_{x \to 1^+} [\![x]\!] = 1.$$

To describe the limiting value of $g(x)$ as x approaches 1 from the left or below, we write

$$\lim_{x \to 1^-} [\![x]\!] = 0.$$

Example 12.1 The tensile strength of a muscle fiber is defined as the maximum force that the muscle fiber can sustain without "breaking." One method of measuring this force is to affix a fiber to a stand by one end and stretch the fiber by pulling on the other end. If x denotes the distance the fiber has been stretched, then the force acting on the fiber can be measured at each value of x. (See Fig. 12.3.) Denote this

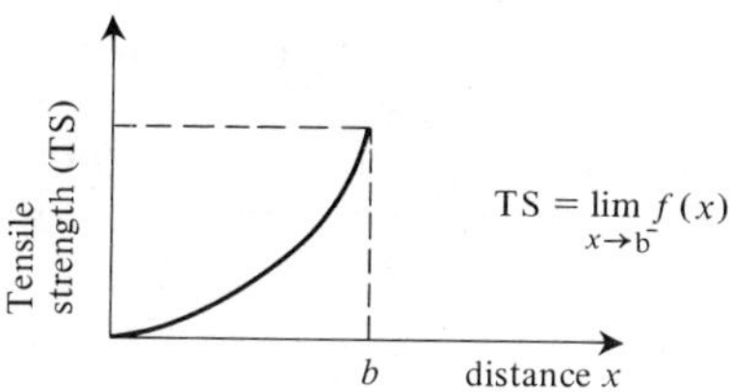

Figure 12.3

force by $f(x)$. At some distance $x = b$, the fiber will break. At the exact point at which the fiber breaks the force $f(b)$ cannot be measured. Hence the tensile strength (TS) is defined as the one sided limit,

$$TS = \lim_{x \to b^-} f(x).$$ ◀

An intuitive definition of a continuous function is a function that has a "smooth" graph containing no "jumps" or "holes."

The function $f(x) = 2x + 1$ is continuous at $x = 1$ since $\lim_{x \to 1} f(x) = 3$ and $f(1) = 3$. The function $g(x) = [\![x]\!]$ is not continuous at $x = 1$ because $\lim_{x \to 1} g(x)$ does not exist. The graph of $g(x)$, as sketched in Fig. 12.1, has a jump at $x = 1$. The function

$$f(x) = \frac{x^2 + x - 6}{x - 2}$$

has a limit at $x = 2$, but it is not continuous at $x = 2$ because $f(2)$ does not exist. In this case, the graph has a "hole" at $x = 2$.

12.2 THE DEFINITION AND PROPERTIES OF LIMITS

In this subsection, we first give a more complete, but nonrigorous, definition of the limit concept. We then state the basic algebraic properties of limits. Intuitively, the definition of the limit of $f(x)$ as x approaches b is a number A if we can guarantee that all values $f(x)$ will be close to the number A when x is close to the number b.

This requires $f(x)$ to be defined for all x in an interval containing the number b but $f(b)$ need not exist. The difficulty in defining the limit concept in a nonrigorous manner is to express the concept of being "close." We do this as follows.

The distance between two numbers is the absolute value of their difference.

$$\text{The distance between } f(x) \text{ and } A = |f(x) - A|.$$

To say that $f(x)$ is close to A, we want to be able to specify any tolerance limit and ensure that the distance between $f(x)$ and A will be smaller than this tolerance limit for all x sufficiently close to b. How close a number x is to the number b is measured by the distance between x and b.

$$\text{The distance between } x \text{ and } b = |x - b|.$$

The distances $|f(x) - A|$ and $|x - b|$ are illustrated in Fig. 12.4(a). In Fig. 12.4(b), a tolerance interval about the number A is indicated for a given tolerance of three units. The interval is $(A - 3, A + 3)$. As indicated in the graph, $f(x_1)$ is within three units of the number A, and $f(x_2)$ is not less than three units from A.

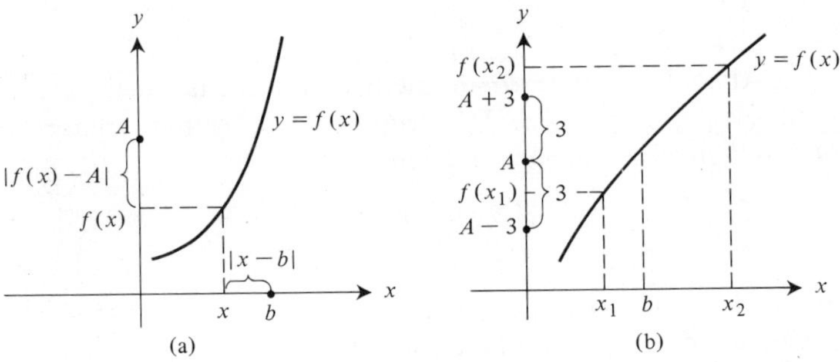

Figure 12.4

When the limiting value of $f(x)$ is considered as x approaches the limit point b from only one side, the resulting limit is called a *one-sided limit.*

If only x values greater than b are considered, then the limit as x approaches b is denoted by $x \to b^+$, read "x approaches b from above" or "x approaches b from the right."

If only x values less than b are considered, then the limit as x approaches b is denoted by $x \to b^-$, read "x approaches b from the left" or "x approaches b from below."

The following are intended to provide more concrete intuitive definitions of the limit concept. We shall not use these definitions to *prove* particular limits exist, but we shall use them as intuitive guides to help determine particular limits. (See Fig. 12.5).

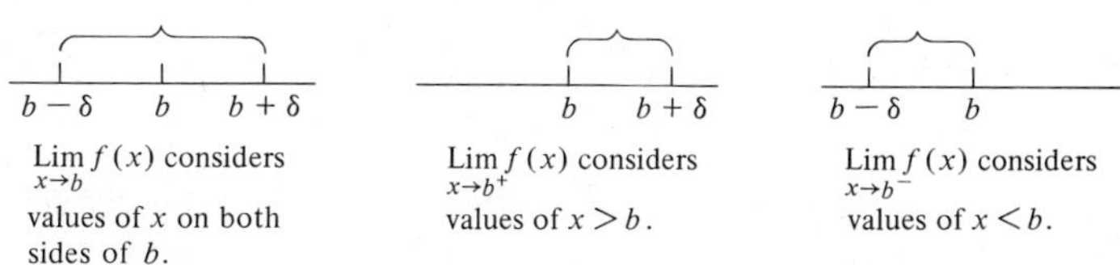

Figure 12.5

Definition 12.1 *The limit as x approaches b of $f(x)$ is the number A* (denoted $\lim_{x \to b} f(x) = A$), if for any given tolerance limit the distance between $f(x)$ and A is less than the tolerance limit for all x, except $x = b$, whose distance from b is less than an appropriate corresponding distance.

Definition 12.2 *The limit as x approaches b from the right of $f(x)$ is the number A* (denoted $\lim_{x \to b^+} f(x) = A$), if for any given tolerance limit the distance between $f(x)$ and A is less than the tolerance limit for all x greater than b, whose distance from b is less than an appropriate corresponding distance.

Definition 12.3 *The limit as x approaches b from the left of $f(x)$ is the number A* (denoted $\lim_{x \to b^-} f(x) = A$), if for any given tolerance limit the distance between $f(x)$ and A is less than the tolerance limit for all x less than b, whose distance from b is less than an appropriate corresponding distance.

Example 12.2

a) Consider $\lim_{x \to 3} 2x$. Here, $f(x) = 2x$ and $b = 3$. The limit should intuitively be seen to be $A = 6$. We illustrate the *tolerance* concept of the above definitions by choosing a tolerance of $\varepsilon = 0.01$. The requirement that $f(x)$ is less than 0.01 units from $A = 6$ is thus

$$|f(x) - 6| < 0.01.$$

To determine how close x must be to $b = 3$ for this inequality to hold, we first expand the inequality,

$$-0.01 < f(x) - A < 0.01 \qquad \text{or} \qquad -0.01 < 2x - 6 < 0.01.$$

We next reduce the term $f(x) - A$ to x, just as we would solve $2x - 6 = y$ for x. Adding six to each term, we have

$$6 - 0.01 < 2x < 6 + 0.01.$$

Dividing by two, we obtain the inequality

$$3 - 0.005 < x < 3 + 0.005.$$

Subtracting $b = 3$ from each term gives us an inequality for $x - b$,

$$-0.005 < x - 3 < 0.005.$$

The last inequality states that

$$|x - 3| < 0.005.$$

Thus if the distance between x and 3 is less than 0.005, the distance between $f(x)$ and 6 will be less than 0.01.

b) Consider the function defined by

$$f(x) = \begin{cases} 2x & \text{if} \quad x \leq 3; \\ 4 - x & \text{if} \quad x > 3. \end{cases}$$

To consider the limit of $f(x)$ at $b = 3$, we must use one-sided limits since f is defined differently for $x > 3$ than for $x < 3$. The limit from the left is intuitively

$$\lim_{x \to 3^-} f(x) = \lim_{x \to 3^-} 2x = 6.$$

Note that $\lim_{x \to 3^-}$ is not the same as $\lim_{x \to -3}$. Similarly, the limit from the right is

$$\lim_{x \to 3^+} f(x) = \lim_{x \to 3^+} 4 - x = 1.$$

Comparing these two limits, we see that

$$\lim_{x \to 3^-} f(x) \neq \lim_{x \to 3^+} f(x).$$

Because these limits are not equal, we may conclude that $\lim_{x \to 3} f(x)$ does not exist since $f(x)$ cannot be arbitrarily close to both 6 and 1 as x approaches 3. ◀

Theorem 12.1 The $\lim_{x \to b} f(x)$ exists and equals A if and only if (i) both $\lim_{x \to b^+} f(x)$ and $\lim_{x \to b^-} f(x)$ exist and (ii) $\lim_{x \to b^-} f(x) = \lim_{x \to b^+} f(x) = A$.

Example 12.3 Consider the following function:

$$f(x) = \begin{cases} x^2 & \text{if} \quad x < 0; \\ 2x & \text{if} \quad 0 \leq x < 1; \\ x - 1 & \text{if} \quad x \geq 1. \end{cases}$$

The graph of the function f is shown in Fig. 12.6. The interesting x-values, as far as limits are concerned, are $x = 0$ and $x = 1$. We examine the limits of the

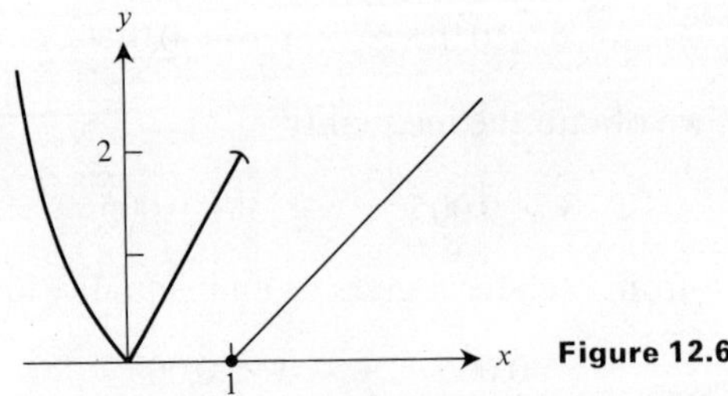

Figure 12.6

function at 0 and 1 by considering both one-sided limits. Intuitively, at $x = 0$,

$$\lim_{x \to 0^-} f(x) = \lim_{x \to 0^-} x^2 = 0, \qquad \text{and} \qquad \lim_{x \to 0^+} f(x) = \lim_{x \to 0^+} 2x = 0.$$

Since these limits are equal, the limit as x approaches 0 of $f(x)$ exists:

$$\lim_{x \to 0} f(x) = 0.$$

At $x = 1$,

$$\lim_{x \to 1^-} f(x) = \lim_{x \to 1^-} 2x = 2, \qquad \text{and} \qquad \lim_{x \to 1^+} f(x) = \lim_{x \to 1^+} x - 1 = 0.$$

Since these one-sided limits are not equal, we concluded that

$$\lim_{x \to 1} f(x) \text{ does not exist.}$$

◀

Example 12.4 Many chemical substances can assume more than one physical state. For instance, water can be a liquid, a solid, or a gas. At a constant temperature, the state of water depends on the volume, V, and the pressure, P. Figure 12.7

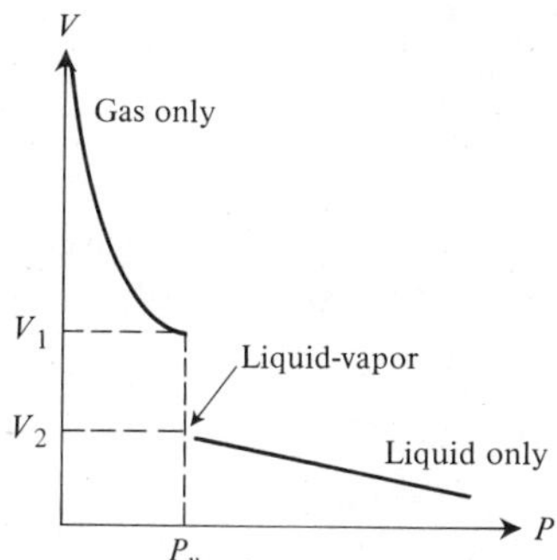

Figure 12.7

illustrates the relationship between pressure, volume, and liquid-gas state. For P less than the vapour pressure P_v, the water is in a gaseous form and satisfies Boyle's Law, $V \propto (1/P)$. For P greater than P_v, the water is in a liquid form and the volume is linearly related to the pressure. Considering V as a function of P, as indicated in Fig. 12.7, we have

$$\lim_{P \to P_v^+} V(P) = V_2, \qquad \text{and} \qquad \lim_{P \to P_v^-} V(P) = V_1.$$

Since $V_2 \neq V_1$, we concluded that $\lim_{P \to P_v} V(P)$ does not exist. ◀

To evaluate a limit of a complicated function $F(x)$, $\lim_{x \to b} F(x)$, we will use the following theorems. These theorems state that the limit process commutes with the basic arithmetic operations. That is, we can perform the arithmetic operations or evaluate the limits in any order with the same result.

Theorem 12.2 If $\lim_{x\to b} f(x) = A$ and $\lim_{x\to b} g(x) = B$, then

$$\lim_{x\to b} [f(x) + g(x)] = \lim_{x\to b} f(x) + \lim_{x\to b} g(x) = A + B.$$

The limit of the sum is the sum of the limits.

Theorem 12.3 If $\lim_{x\to b} f(x) = A$ and k is a constant, then

$$\lim_{x\to b} [k \cdot f(x)] = k \cdot A.$$

Theorem 12.4 If $\lim_{x\to b} f(x) = A$ and $\lim_{x\to b} g(x) = B$, then

$$\lim_{x\to b} [f(x) \cdot g(x)] = \left[\lim_{x\to b} f(x)\right] \times \left[\lim_{x\to b} g(x)\right] = A \cdot B.$$

The limit of the product is the product of the limits.

Theorem 12.5 If $\lim_{x\to b} f(x) = A$, $\lim_{x\to b} g(x) = B$, and $B \neq 0$, then

$$\lim_{x\to b} \left[\frac{f(x)}{g(x)}\right] = \frac{\lim_{x\to b} f(x)}{\lim_{x\to b} g(x)} = \frac{A}{B}.$$

The limit of the quotient is the quotient of the limits.

Since $\lim_{x\to b}(x) = b$, we can use the above theorems to compute the limit as $x \to b$ of any polynomial function. For instance, if $f(x) = x^2 - 3x$, then

$$\begin{aligned}
\lim_{x\to b} f(x) &= \lim_{x\to b} (x^2 - 3x) \\
&= \lim_{x\to b} (x^2) + \lim_{x\to b} (-3x) \\
&= \left[\lim_{x\to b} (x) \cdot \lim_{x\to b} (x)\right] + (-3) \lim_{x\to b} (x) \\
&= [b \cdot b] - 3 \cdot b = b^2 - 3b = f(b).
\end{aligned}$$

We state the generalization of this process as a theorem.

Theorem 12.6 If $f(x)$ is a polynomial function, then for any number b,

$$\lim_{x\to b} f(x) = f(b).$$

For instance,

$$\lim_{x\to -2} x^5 - 3x + 7 = (-2)^5 - 3(-2) + 7 = -19.$$

A rational function is a function $R(x)$, which is the ratio of two polynomials $R(x) = P(x)/Q(x)$. If $Q(b) \neq 0$, then the last two theorems verify the following theorem.

Theorem 12.7 If $R(x) = P(x)/Q(x)$, where P and Q are polynomials and $Q(b) \neq 0$, then

$$\lim_{x \to b} R(x) = R(b) = P(b)/Q(b).$$

Each of the theorems of this section apply verbatim when the $\lim_{x \to b}$ is replaced by either $\lim_{x \to b^+}$ or $\lim_{x \to b^-}$.

Example 12.5 The following limits are evaluated using the above theorems.

a) $\lim\limits_{x \to 5} x^2 + 3x - 9 = 5^2 + 3 \cdot 5 - 9 = 31$

b) $\lim\limits_{x \to 2^-} x^4 - x^3 = 2^4 - 2^3 = 8$

c) $\lim\limits_{x \to -2} x^4 - x^3 = (-2)^4 - (-2)^3 = 24$

d) $\lim\limits_{x \to 3} \left[\dfrac{x^2 - x}{4 + x^3}\right] = \dfrac{3^2 - 3}{4 + 3^3} = \dfrac{6}{31}$

e) $\lim\limits_{x \to 4} \dfrac{x^2 + 6 - 5x}{x - 4}$ cannot be evaluated using Theorem 12.5 since the limit of the denominator is 0 as $x \to 4$. ◀

12.3 CONTINUITY

The conclusion of Theorem 12.7 states what we might intuitively expect, that the limit of the value $f(x)$ as x approaches b is $f(b)$. In the theorem, however, the function f is assumed to be a polynomial function. This conclusion is not valid for every function f, since $f(b)$ may not exist or $\lim_{x \to b} f(x)$ may not exist. Those functions for which it does hold are *nice* functions. As we shall see, they have many useful properties and include most elementary functions. Rather than use the term "nice" to describe these functions, we use the term "continuous," since their graph will not have "jumps" or "holes."

Definition 12.4 *A function f is continuous at the number b* if

$$\lim_{x \to b} f(x) = f(b).$$

The definition of continuity requires the following three conditions to hold:

1. $f(b)$ must exist. 2. $\lim\limits_{x \to b} f(x)$ must exist 3. $\lim\limits_{x \to b} f(x) = f(b)$.	$\Longleftrightarrow$	f is continuous at $x = b$.

Since the limit of a function at a point b may exist without the function being defined at the limit point b, there are three ways in which a function can fail to be continuous at a point.

A function f is not continuous at a point b if any of the following hold.

1. $f(b)$ is not defined.
2. $f(b)$ exists; $\lim_{x \to b} f(x)$ exists but $\lim_{x \to b} f(x) \neq f(b)$.
3. $f(b)$ exists; $\lim_{x \to b} f(x)$ does not exist.

A point b at which f is not continuous is called a *point of discontinuity* of the function f.

Example 12.6 The following functions fail to be continuous at the points indicated.

a) Let

$$f(x) = \frac{x^2 - 2x - 3}{x - 3}$$

and $b = 3$; $f(3)$ is not defined so f cannot be continuous at $x = 3$. However, $\lim_{x \to 3} f(x)$ does exist:

$$\lim_{x \to 3}\left(\frac{x^2 - 2x - 3}{x - 3}\right) = \lim_{x \to 3} \frac{(x + 1)(x - 3)}{(x - 3)} = \lim_{x \to 3}(x + 1) = 4.$$

b) Let $f(x) = [\![\sin(x)]\!]$ and $b = \pi/2$. The graph of $f(x)$ is given in Fig. 12.8. An equivalent way of describing the function f is as follows:

$$f(x) = \begin{cases} 1 & \text{if} \quad \sin(x) = 1; \\ 0 & \text{if} \quad 0 \leq \sin(x) < 1; \\ -1 & \text{if} \quad \sin(x) < 0. \end{cases}$$

As $f(\pi/2) = 1$, but $\lim_{x \to \pi/2} f(x) = 0$, both $f(b)$ and $\lim_{x \to b} f(x)$ exist, but are not equal. Consequently, f is not continuous at $x = \pi/2$.

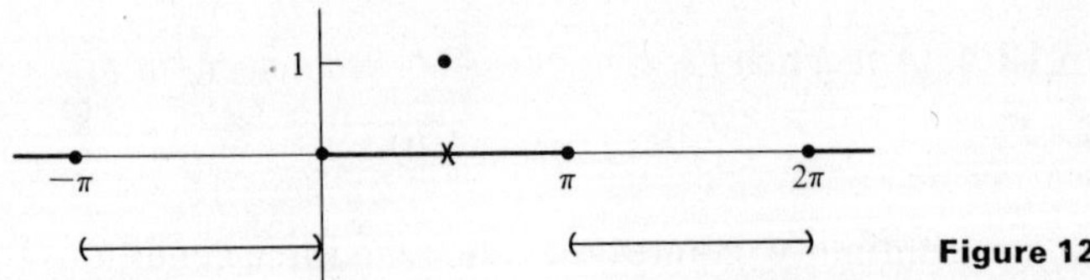

Figure 12.8

c) Let $f(x) = |x|/x$ if $x \neq 0$ and $f(0) = 0$. The function f is also given by the following:

$$f(x) = \begin{cases} 1 & \text{if} \quad x > 0; \\ 0 & \text{if} \quad x = 0; \\ -1 & \text{if} \quad x < 0. \end{cases}$$

The only x value where the function f is not constant is at $x = 0$. The $\lim_{x \to 0} f(x)$ does not exist, since for $x > 0$, $f(x)$ is 1, and for $x < 0$, $f(x)$ is -1. Therefore, near $x = 0$, the values of $f(x)$ are not all near the same number. The function f is not continuous at $x = 0$. ◀

The concept of continuity is a *local* concept. Whether or not a function is continuous at a point $x = b$ depends only on the value of $f(b)$ and the values $f(x)$ for all x close to b.

To describe the set of values where a function f is continuous, we give the following definition.

Definition 12.5 *A function f is continuous on the open interval,* (α, β), if f is continuous at each point b in the interval.

Frequently, it is necessary to discuss a function on a closed interval, $[\alpha, \beta]$. The definition of continuity utilizes the ordinary limit of a function as $x \to b$. We cannot use the ordinary definition of a limit when discussing the endpoints, α and β, since the function is not defined or considered in a full neighborhood of these endpoints. To accommodate this difficulty, we make the following convention.

Definition 12.6 *A function f is continuous on a closed interval,* $[\alpha, \beta]$, if
i) f is continuous at each point of the open interval, (α, β), and
ii) $\lim_{x \to \alpha^+} f(x) = f(\alpha)$ and $\lim_{x \to \beta^-} f(x) = f(\beta)$.

Example 12.7

a) The function f, defined by $f(x) = x^{-1}$ for $x \neq 0$ and $f(0) = 0$, is continuous on the open interval $(0, 1)$, but not on the closed interval $[0, 1]$, since $\lim_{x \to 0^+} f(x) = +\infty$ and not $f(0)$.
b) The function defined by

$$f(x) = \begin{cases} |x| & \text{if} \quad |x| \leq 3; \\ 2 & \text{if} \quad |x| > 3; \end{cases}$$

is not continuous at $x = -3$ or $x = 3$ (see Fig. 12.9); f is continuous on $[-3, 3]$, since $\lim_{x \to 3^-} f(x) = f(3)$ and $\lim_{x \to -3^+} f(x) = f(-3)$; $f(x)$ is not continuous on $[3, 10]$, since $\lim_{x \to 3^+} f(x) = 2 \neq f(3)$.

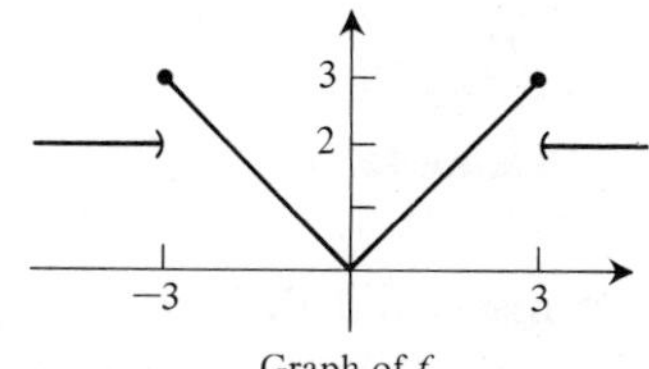

Figure 12.9 Graph of f

◀

Example 12.8 The amount of air a person breathes depends on the oxygen requirements of the body. During exercise, muscle action requires a large amount of oxygen and consequently results in an increased breathing rate. The amount of air a person breathes is measured by *pulmonary ventilation* (denoted PV), which is defined as the product of the lung volume (tidal volume) and the number of breaths per minute. Figure 12.10 indicates an experimentally determined curve that

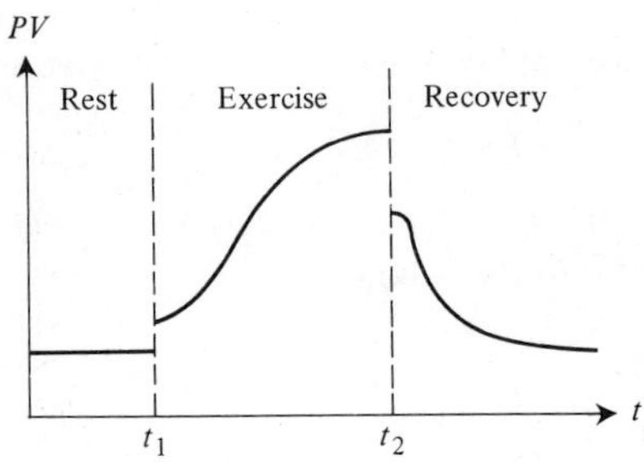

Figure 12.10

represents the pulmonary ventilation, PV, as a function of time over an interval in which a person experiences vigorous exercise. The function $PV(t)$ is not continuous at the times t_1 and t_2, which correspond to the initiation and termination of the exercise. The discontinuity at t_2 is usually twice as great as the discontinuity at t_1. The curve in Fig. 12.10 is the type of curve usually presented in physiology texts.

The discontinuity arises because experimental methods require the measurement of the PV volume over relatively long intervals. If, however, the measurements of air volume breathed were made "continuously," the graph would be continuous as depicted in Fig. 12.11.

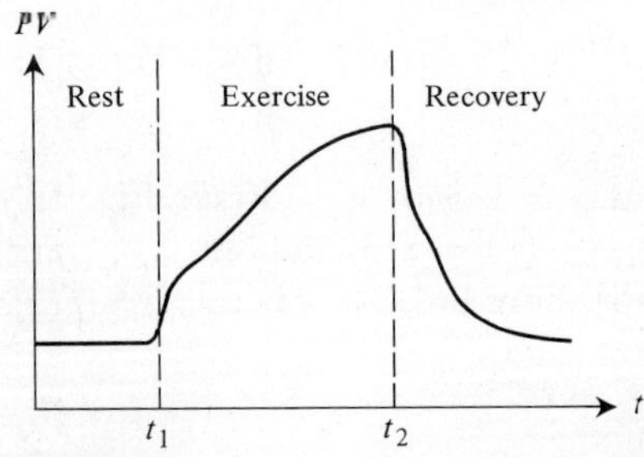

Figure 12.11 ◀

Discontinuities correspond to "jumps" or "holes" in the graph of the function. Most biological processes do not involve sudden changes; consequently, the functions used to describe biological processes will usually be continuous functions.

Theorem 12.6 tells us that polynomials are continuous everywhere. Similarly Theorem 12.7 states that rational functions are continuous everywhere that they are defined. (If their denominators are zero they are not defined.) So far, the only examples we have given of discontinuous functions have been either rational functions or functions defined by more than one equation. None of the more common functions have been used to illustrate discontinuities. The reason for this is that they are all continuous.

Theorem 12.8 The following functions are continuous on the indicated intervals.
a) $a^x:(-\infty, +\infty)$, any $a > 0$
b) $\log_a(x)$: $(0, \infty)$, $a > 1$
c) $\cos(x)$ and $\sin(x)$: $(-\infty, +\infty)$
d) $\cot(x)$ and $\csc(x)$: $x \neq n\pi$, n an integer
e) $\tan(x)$ and $\sec(x)$: $x \neq (n + \frac{1}{2})\pi$, n an integer
f) $|x|$: $(-\infty, +\infty)$

The concept of continuity rests on the limit concept. Therefore, each of the algebraic properties of limits has a corresponding property for continuous functions.

Theorem 12.9 If $f(x)$ and $g(x)$ are continuous at $x = b$, then
a) $f(x) + g(x)$ is continuous at $x = b$;
b) $k \cdot f(x)$ is continuous at $x = b$, for k constant;
c) $f(x) \cdot g(x)$ is continuous at $x = b$; and
d) $\dfrac{f(x)}{g(x)}$ is continuous at $x = b$ if $g(b) \neq 0$.

Example 12.9 Using the algebraic properties of Theorem 12.9 and the continuity of the elementary functions, we evaluate the following limits.

a) $\lim\limits_{x\to 3} 2^x - x = 2^3 - 3 = 5$

b) $\lim\limits_{x\to 1}[\cos(x) - \ln(x)] = [\cos(1) - \ln(1)] = \cos(1) \simeq 0.54$

c) $\lim\limits_{x\to -2} \dfrac{2x-1}{e^x} = \dfrac{2(-2)-1}{e^{-2}} = -5e^2 \simeq -36.9$

d) $\lim\limits_{x\to 2^-} \dfrac{2x-1}{e^x} = \dfrac{2(2)-1}{e^2} = 3e^{-2} \simeq 0.4$

e) $\lim\limits_{t\to 5}[2t \cdot \log(t)] = [2(5) \cdot \log(5)] = 10 \cdot \log(5) \simeq 7.0$

f) $\lim\limits_{x\to \pi/2} x\cot(x) = (\pi/2)\cot(\pi/2) = 0$ ◀

The operation of composition is not mentioned in the above list of algebraic properties of limits. The reason for this is that the limit process and composition of functions do not commute in all cases. For $\lim(f \circ g)$ to equal $f \circ \lim g$, it is necessary that the second function of the composition, f, be continuous.

Theorem 12.10

a) If $\lim_{x \to b} g(x) = B$ and f is continuous at $x = B$, then

$$\lim_{x \to b} (f \circ g)(x) = f(\lim_{x \to b} g(x)) = f(B).$$

b) If f and g are continuous for all x, then $f \circ g$ is continuous for all x.

Example 12.10 The following functions are compositions of two or more elementary functions. We evaluate their limits, using Theorem 12.10, as follows.

a) $\displaystyle \lim_{x \to 3} e^{2x} = \exp\left\{\lim_{x \to 3}(2x)\right\} = e^6 \simeq 403.4.$

b) $\displaystyle \lim_{x \to \pi} \cos(x/3) = \cos\left(\lim_{x \to \pi}(x/3)\right) = \cos(\pi/3) = 0.5.$

c) $\displaystyle \lim_{x \to 2} e^{x^2 - \ln(x)} = \exp\left\{\lim_{x \to 2}(x^2 - \ln(x))\right\} = e^{4 - \ln(2)} = \tfrac{1}{2}e^4 \simeq 27.3.$

d) $\displaystyle \lim_{x \to 3^-} \cos(\sqrt{3 - x}) = \cos\left(\lim_{x \to 3^-} \sqrt{3 - x}\right) = \cos(0) = 1.$

Note that the one-sided limit is necessary in this example since $\sqrt{3 - x}$ is not defined for $x > 3$. ◀

Example 12.11 The growth functions discussed in Section 8 are all continuous functions since they are either the sum, product, or composition of continuous functions.

a) The exponential growth function, $y = ce^{kt}$, is continuous since e^{kt} is the composition of the continuous functions $f(t) = e^t$ and $g(t) = kt$.
b) Similarly, the Gompertz function

$$y = ce^{-ke^{-\lambda t}}$$

is the composition of two exponential functions and is thus continuous. $y = (f \circ g)(t)$, where

$$g(t) = e^{-\lambda t} \qquad \text{and} \qquad f(t) = ce^{-kt}.$$

c) The function $y(t) = 2^{[\![t/2]\!]}$ is sketched in Fig. 12.12. The greatest integer function $[\![t/2]\!]$ is constant, except when $t/2$ is an integer. Consequently, $y(t)$ is continuous on each half-open interval $[\alpha, \beta)$, where $\alpha = 2n$ and $\beta = 2(n + 1)$ for $n = 0, 1, 2, \ldots$

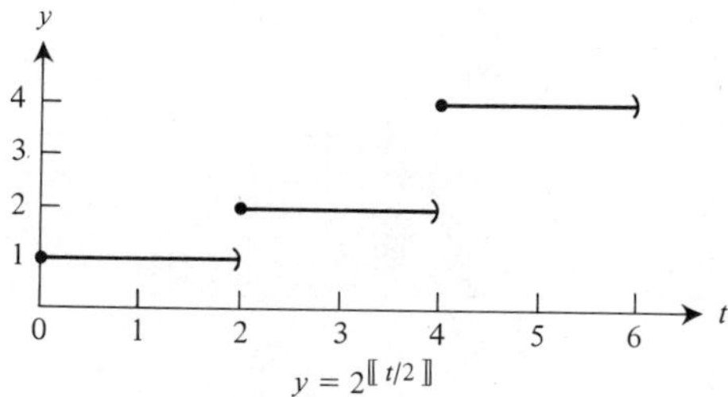

Figure 12.12

◀

12.4 THE GEOMETRICAL INTERPRETATION OF LIMITS AND CONTINUITY

The geometrical picture that corresponds to Definition 13.1 of a limit is illustrated in Fig. 12.13. In this figure, we indicate an interval about the point A, $(A - \varepsilon, A + \varepsilon)$, corresponding to a tolerance limit ε. Associated with each interval is a horizontal band. Each point within this band has a y-coordinate that satisfies the inequality

$$|y - A| < \varepsilon.$$

In each part of Fig. 12.13, two vertical bands are depicted. These correspond to the *punctured* neighborhood about $x = b$, consisting of the two open intervals $(b - \delta, b)$ and $(b, b + \delta)$. Each point within these vertical bands has an x-coordinate which satisfies the inequality.

$$0 < |x - b| < \delta.$$

Geometrical definition of a limit: The limit of $f(x)$ as $x \to b$ is the number A if for each choice of the horizontal band centered about $y = A$ (a choice of a tolerance $\varepsilon > 0$); there is a choice of vertical bands, corresponding to a punctured neighbourhood of b (a choice of a maximum distance δ between x and b) such that the graph of the function being within the vertical bands also lies within the horizontal bands.

In Fig. 12.13(a), we illustrate a horizontal band together with a suitable choice for the vertical bands. Observe that the graph of $f(x)$ for $b - \delta < x < b + \delta$ is completely contained in the rectangles determined by the intersection of the horizontal and vertical bands.

In Fig. 12.13(b), the illustrated choice of vertical bands does not work. Observe that for the indicated number x_1, the corresponding point on the graph $(x_1, f(x_1))$ is within the vertical bands, but not within the horizontal band. This is not to say that $\lim_{x \to b} f(x)$ does not equal the value A. If we make the vertical bands much narrower by choosing a smaller bound for the distance between x and b, we can satisfy the geometrical condition of the above definition.

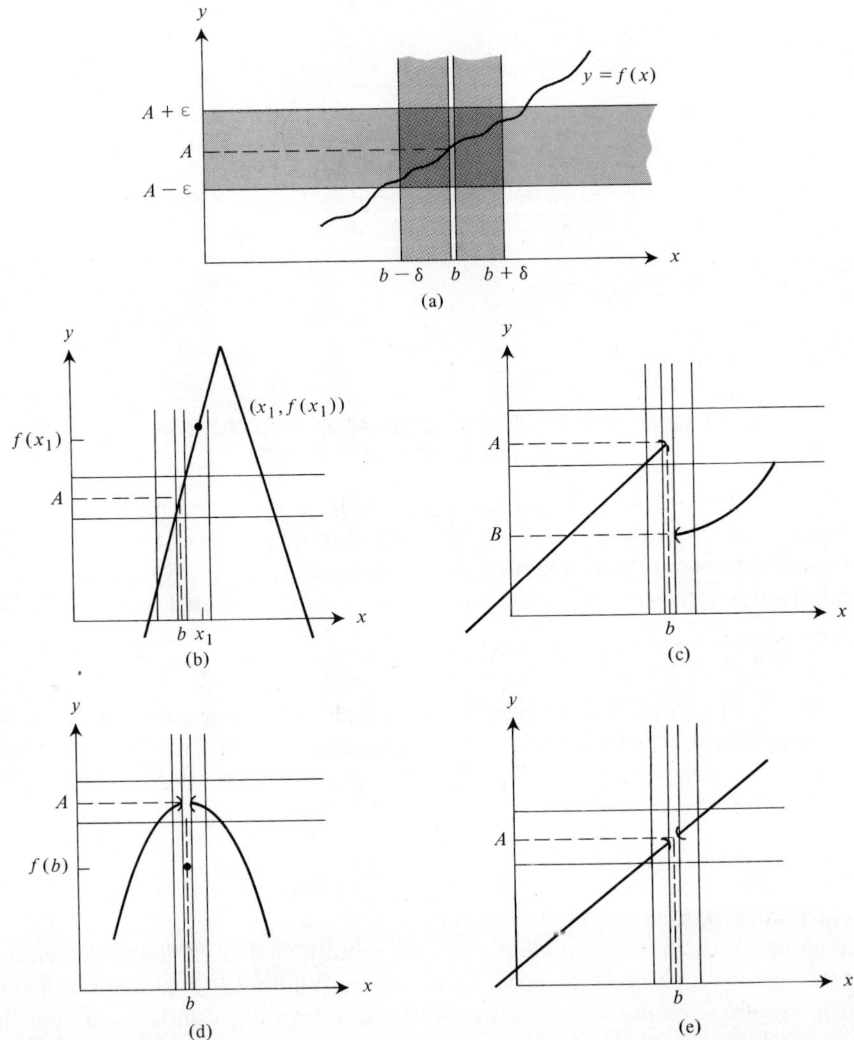

Figure 12.13

The function f sketched in Fig. 12.13(c) does not have a limit at the indicated number b. The graph of $y = f(x)$ appears to approach A as $x \to b^-$ (x approaches b from the left) and to approach B as $x \to b^+$ (x approaches b from the right). Consequently, for any small tolerance limit ε, no matter how small we make the bound δ for the distance between x and b, the geometric picture will be as illustrated. As the point $(x, f(x))$ is not in the horizontal band for every value of x in the interval $(b, b + \delta)$, we cannot satisfy the limit requirement by choosing smaller vertical bands.

The functions sketched in Fig. 12.13(d) and (e) provide examples of where the $\lim_{x\to b} f(x)$ exists, but the function $f(x)$ is not continuous at $x = b$. In Fig. 12.13 (d), the function is defined at $x = b$, while the graph in Fig. 12.13(e) is that of a function not defined at $x = b$. The graphs in Figs. 12.13(d) and (e) cannot be sketched without lifting the pen from the paper.

As a rule,

> A function is continuous on an interval if you can sketch its graph over the interval without lifting your pen from the paper.

Observe that in Fig. 12.13 each function sketched has both one-sided limits ($\lim_{x\to b^+} f(x)$ and $\lim_{x\to b^-} f(x)$) at the indicated point, b. The only function that does not have a limit at $x = b$ is the function in Fig, 12.13(c). For this function f, from the graph, we obtain the one-sided limits

$$\lim_{x\to b^-} f(x) = A \qquad \text{and} \qquad \lim_{x\to b^+} f(x) = B.$$

12.5 CONTINUOUS FUNCTIONS ON A CLOSED INTERVAL $[a, b]$

We finish this section with two basic theorems that concern continuous functions on closed intervals. We mentioned in the previous section that a function is continuous if you can sketch its graph without raising your pen from the paper. As a consequence of this, we have the following two simple observations.

1. In tracing the graph of a continuous function over an interval, $[a, b]$, your pen must pass through the *highest* and *lowest* points on the graph.
2. There is a point on the graph of a continuous function that corresponds to every y value between $f(a)$ and $f(b)$, the y values at the endpoints a and b.

The first observation is stated mathematically as the following *Extreme Value Theorem.*

Theorem 12.11 If f is continuous on $[a, b]$, then there are two points, x_M and x_m, in $[a, b]$ such that for all $x \in [a, b]$,

$$f(x_m) \le f(x) \le f(x_M).$$

x_M is called a *maximum point* of f, and x_m is called a *minimum point* of f. $M = f(x_M)$ is the *maximum value* and $m = f(x_m)$ is the *minimum value* of f over $[a, b]$.

The *Extreme Value Theorem* is illustrated in Fig. 12.14. Note that in this illustration there are two maximum points, x_{M1} and x_{M2} which correspond to the maximum value M.

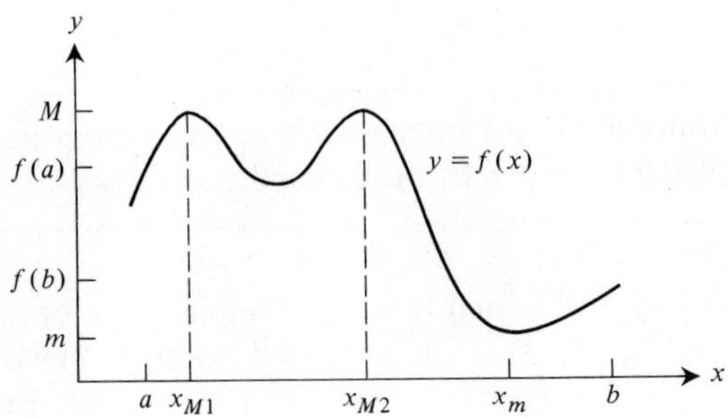

Figure 12.14

The second of the above observations is formally stated as a theorem, called the *Intermediate Value Theorem* (IVT).

Theorem 12.12 If f is continuous on $[a, b]$, and y_0 is a number such that $f(a) < y_0 < f(b)$ (or $f(b) \leq y_0 < f(a)$), then there is a point $x_0 \in (a, b)$ such that $f(x_0) = y_0$.

In Fig. 12.15, we indicate two numbers y_0 and y_1 on the y-axis. Both numbers are between $f(a)$ and $f(b)$. There is one x value, x_1, which corresponds to y_1; $y_1 = f(x_1)$. There are two x values, x_2 and x_3, which correspond to y_0; $f(x_2) = y_0$ and $f(x_3) = y_0$.

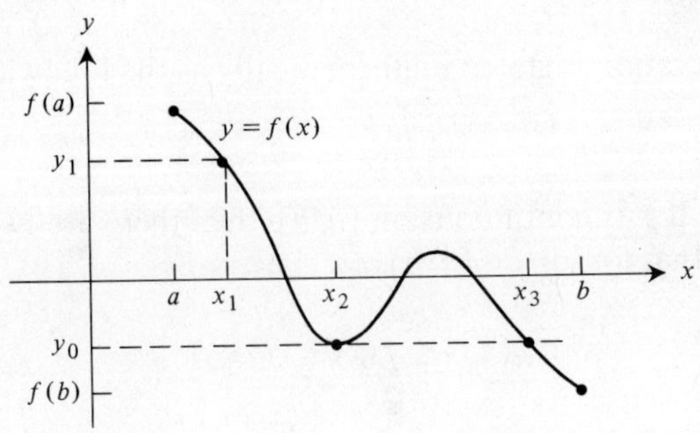

Figure 12.15

The Theorem 12.12 can be used to guarantee that solutions of equations exist and, in fact, are contained in particular intervals. Suppose that we wish to solve $f(x_0) = 0$ for a complicated function $y = f(x)$. If we can find two numbers, a and b, such that $f(a) > 0$ and $f(b) < 0$ (or $f(a) < 0$ and $f(b) > 0$), then the number $y_0 = 0$ is between $f(a)$ and $f(b)$. Consequently, by the above theorem, we know that there is at least one x value in the interval (a, b) such that $f(x) = y_0 = 0$.

Example 12.12 Does the equation $\cos(\pi x) = -(x + (\frac{1}{3}))$ have a solution? The answer is yes. To see why, we observe that a solution of this equation is a number x_0 such that $f(x_0) = 0$, where $f(x)$ is the function obtained when we rearrange the equation so that it has the form $f(x) = 0$. In this case,

$$f(x) = \cos(\pi x) + x + (\tfrac{1}{3}).$$

Since f is the sum of continuous functions, it is continuous for all real values of x. As $f(1) = \frac{1}{3}$ and $f(-1) = -\frac{5}{3}$, $f(-1) < 0 < f(1)$, we may employ Theorem 12.12 with $y_0 = 0$ to conclude that there is at least one value $x_0 \in (-1, 1)$ such that $f(x_0) = 0$. ◀

Example 12.13 The rate of photosynthesis in a cucumber leaf varies as the leaf grows. Assuming that the photosynthesis rate P is a continuous function of time, we know that it will attain a maximum value, P_{max}, at some time, t_{max}, in the growth period of the plant. When the rate of photosynthesis reaches its maximum value, it decreases; this induces a slowing in the growth rate of the leaf. Experimental data for cucumber leaves indicate that the photosynthesis rate has the form of a double-exponential decay function, and the area of the leaf $A(t)$ is found to be given by a logistic function. (See Section 8, equations (8.1) and (8.7).) Thus the graph of these functions would appear as in Fig. 12.16, with the maximum of the $P(t)$ curve occurring at the same time as the maximum rate of increase in the $A(t)$ curve.

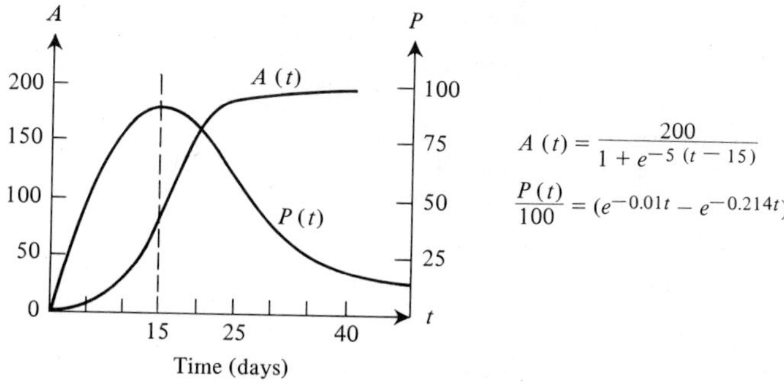

$$A(t) = \frac{200}{1 + e^{-5(t-15)}}$$

$$\frac{P(t)}{100} = (e^{-0.01t} - e^{-0.214t})$$

Figure 12.16

◀

EXERCISE SET 12

12.1 Intuitively and using the theorems of this section, determine the following limits, if they exist.

a) $\lim_{x \to 2} (x - 1)(x + 3)$ b) $\lim_{t \to -3} t^2 - 2t + 1$

c) $\lim_{z \to 0} (z + 1)/(z - 1)$ d) $\lim_{x \to 0} (1 + x)/x$

e) $\lim_{t \to 0} (t^2 + 2t)/t$ f) $\lim_{z \to 2} (z^2 - 6z + 8)/(z - 2)$

g) $\lim_{h \to 0} h - 2^h$ h) $\lim_{h \to 0} e^h - e^{-h}$

i) $\lim_{x \to 0} [\![x]\!]$ j) $\lim_{x \to 0} [\![x + \frac{1}{2}]\!]$

12.2 Use the method illustrated in Example 12.2(a) to find a number δ such that $|f(x) - A|$ is less than $\varepsilon = 0.1$ when $|x - b| < \delta$.

a) $f(x) = 3x;\ A = 6, b = 2$ b) $f(x) = 3x + 1;\ A = 7, b = 2$

c) $f(x) = x^2;\ A = 4, b = 2$ d) $f(x) = \sqrt{x};\quad A = 2, b = 4$

e) $f(x) = e^x;\ A = 1, b = 0$

12.3 Compute the one-sided limits $\lim_{x \to x_0^+} f(x)$ and $\lim_{x \to x_0^-} f(x)$ and sketch the graphs of the functions near the point x_0.

a) $f(x) = |x - 1|,\quad x_0 = 1$

b) $f(x) = 1/x,\quad x_0 = 0$

c) $f(x) = \begin{cases} x & \text{if} \quad x < 0; \\ 2x & \text{if} \quad x > 0, \quad x_0 = 0 \end{cases}$

d) $f(x) = x/|x|,\quad x_0 = 0$

e) $f(x) = \begin{cases} |x| & \text{if} \quad |x| < 1; \\ 1 + x^2 & \text{if} \quad x > 1, \qquad x_0 = -1, 0 \text{ and } 1; \\ 1 - x^2 & \text{if} \quad x < -1. \end{cases}$

12.4 Which of the functions in Exercise 12.3 are continuous at the specified value x_0?

12.5 Which of the following would be modeled by a continuous function?

a) The height of a plant as a function of time
b) The velocity of a rock dropped from a cliff
c) The number of deer in a specific herd
d) The amount of rain measured at a weather station
e) The volume of a bacteria cell
f) The heartbeat of a human

12.6 Sketch each of the given functions and determine where they are not continuous.

a) $f(x) = 3x^2 - 7$ b) $f(x) = |x - 2|$

c) $g(x) = \begin{cases} x + 1 & \text{if} \quad |x| < 1; \\ 2 & \text{if} \quad |x| \geq 1 \end{cases}$ d) $g(x) = \begin{cases} x & \text{if} \quad x \text{ is an integer}; \\ 0 & \text{if} \quad x \text{ is not an integer} \end{cases}$

e) $m(h) = \dfrac{h^2 - 6h + 5}{h - 1}$

f) $V(t) = -16t^2$

g) $f(t) = \dfrac{2t}{|t|}$

h) $g(t) = \dfrac{t^2}{|t|}$

i) $g(x) = |2x|$

j) $g(x) = x - |x|$

k) $f(x) = \begin{cases} |x| & \text{if} \quad x < 0; \\ x^2 & \text{if} \quad x \geq 0 \end{cases}$

l) $f(x) = \begin{cases} |x| & \text{if} \quad x < 0; \\ x^2 + 1 & \text{if} \quad x \geq 0 \end{cases}$

12.7 Evaluate the indicated limits.

a) $\lim\limits_{x \to \pi/2} x \cos(x)$

b) $\lim\limits_{t \to -1} \sin(t)$

c) $\lim\limits_{x \to 0^+} \tan(x)$

d) $\lim\limits_{x \to 0^+} \cot(x)$

e) $\lim\limits_{x \to \pi/2^-} \sec(x)$

f) $\lim\limits_{x \to \pi/2^-} \csc(x)$

g) $\lim\limits_{x \to 0} x \sin(x)$

h) $\lim\limits_{x \to 0} \sin(x) \cdot \csc(x)$

i) $\lim\limits_{x \to 0} \sin(\cos(x))$

j) $\lim\limits_{x \to 0} \dfrac{\sin(x)}{x}$

12.8 Which of the following graphs do not represent a continuous function?

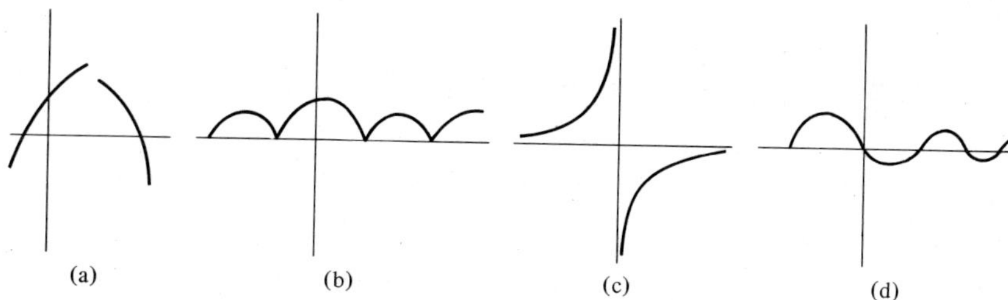

12.9 Sketch a graph of a function satisfying the indicated conditions and give the equation of the function you sketched.

a) $\lim\limits_{x \to 2^+} f(x) = 3$; $\lim\limits_{x \to 2^-} f(x) = 1$

b) $\lim\limits_{x \to 2^+} f(x) = 3$ and f is continuous at 2.

c) $\lim\limits_{x \to 1} f(x) = 4$, but f is not continuous at 1.

d) f is periodic with period $T = 2$ and is continuous with $f(0) = 1$

12.10 A drug is administered every four hours in doses of 2 mg. The drug is neutralized at an exponential rate with rate constant $k = -0.5$. Not all of the drug is neutralized before the next dose is administered. Denote by $A(t)$, the amount of drug in a patient's system at time t (hours).

a) Compute $\lim_{t\to 4^-} A(t)$. This is the drug left over from the first dose, which must be included in the computation of $A(t)$ for $t \in [4, 8)$.
b) Give a formula and sketch the graph of $A(t)$ over the successive intervals $[0, 4)$, $[4, 8)$, $[8, 12)$.
c) Determine the residual amount of drug in the patient's system at the end of the nth dosage period. (Do this for $n = 1, 2, 3$, and 4 first.)

12.11 a) Sketch several points on the graph of the following function

$$f(x) = \begin{cases} x & \text{if} \quad x \text{ is rational;} \\ 0 & \text{if} \quad x \text{ is irrational.} \end{cases}$$

b) Explain why f is not continuous at any point $b \neq 0$ and draw a sketch to illustrate your argument.
c) Explain why f is continuous at $x = 0$.

12.12 In the text, we defined what it means to say "f is continuous on an interval of the form $[\alpha, \beta]$ or (α, β)." Give a definition that would apply to the intervals $[\alpha, \beta)$ and $(\alpha, \beta]$.

12.13 Which of the following equations has a solution on the interval $[-1, 2]$? Indicate why.

a) $e^x - 10x = 0$
b) $xe^x = 2$
c) $\cos(x^2) + x = 0$
d) $e^x = \ln(x + 2)$
e) $x^5 - 3x + 1 = 0$
f) $x^2 - 2 = 0$

12.14 Compute the following limits, if they exist.

a) $\lim\limits_{h\to 0}(3h^2 - 2h + 7)(8h - 9h^2 + 2)$
b) $\lim\limits_{x\to 2}(x^2 - 2x - 6) + (x - 3)/(x + 2)$
c) $\lim\limits_{t\to 1} \dfrac{t^2 - 2t + t^5}{t^2 - 8t - 9}$
d) $\lim\limits_{h\to 0} \dfrac{h + 2h^4}{h(h + 2)}$
e) $\lim\limits_{x\to 3^-} \dfrac{x^2 - 2x + 6}{x - 3}$
f) $\lim\limits_{t\to 0^+} 5\log(t + 1) - t^2/(t + 1)$

12.15 Often a function f is not continuous at a point b because $f(b)$ is not defined, but $\lim_{x\to b} f(b)$ does exist. In this case, it is possible to alter the definition of the function f so as to make f continuous at the point b. We simply define $f(b) \equiv \lim_{x\to b} f(x)$. Redefine the following functions to make them continuous at the indicated point b or indicate why this is not possible. Graph the functions near the point b.

a) $f(x) = \dfrac{x^2 - 9}{x - 3}, b = 3.$
b) $f(x) = \dfrac{x - 4}{-\sqrt{x} + 2}, b = 4.$
c) $f(x) = [\![\sin(x)]\!], b = \pi/2.$
d) $f(x) = (x^2 - x - 6)/(x^2 - 5x + 6), b = 2.$
e) $f(x) = 1/x, b = 0.$
f) $f(x) = \dfrac{\sin(x)}{x}, b = 0.$
g) $f(x) = \sin(1/x), b = 0.$
h) $f(x) = x\sin(1/x), b = 0.$

SECTION 13

THE DIFFERENCE QUOTIENT

13.1 INTRODUCTION TO DIFFERENCE QUOTIENTS

One of the essential features of all living organisms is the ability to grow and reproduce. Consequently, mathematics in the life sciences is most frequently used to describe the growth or changes of an organism or, in general, of any biological phenomenon. The process of describing mathematically a particular biological phenomenon is known as *mathematical modeling*. There are four basic steps in the modeling process: (1) identifying the biological components using mathematical variables, (2) establishing relationships between the variables, (3) analyzing these relationships, and (4) interpreting the biological significance of the mathematical analysis.

The first step in any modeling process is to associate *independent variables* with those aspects of the phenomenon being described that do not appear to depend on other components; time is the most common independent variable. Other quantities, such as physical dimensions or the amount of nutrients available to an organism, may be independent variables. In general, any components that can be controlled in an experimental situation will be independent variables. After this is done, all other measurable aspects of the phenomenon must be represented by *dependent variables*. A dependent variable is really a function of one or more independent variables.

The second step in the modeling process is describing the functional relationship between the dependent variables and the independent variables. In some situations, these relationships are simple or well-established, and we can express them explicitly. Some examples are given by the volume of a spherically shaped cell,

$$V(r) = (\tfrac{4}{3})\pi r^3,$$

where r is the radius of the cell, or the average volume of blood plasma in an adult human,

$$V(w) = 41w(\text{ml}),$$

where w is the weight of the individual in kilograms. However, in most modeling situations, the exact functional relationship between a dependent variable, say y, and an independent variable, say t, is not known *a priori* and must be derived. Indeed the whole point of a mathematical model is to determine the mathematical relationships that are not known at the outset. How this may be accomplished will vary with the complexity of the model.

In general, the approach used in all mathematical models is to describe an unknown functional relationship, $y = f(t)$, by an alternative equation that indicates how the quantity y changes due to changes in the independent variable t.

The analysis of the mathematical model, which constitutes the third step in modeling, consists of solving the equation describing the change in y to obtain the relationship $y = f(t)$. The final step then is to interpret the algebraic equation $y = f(t)$ in terms of the biological phenomena being studied.

Equations describing how different changes are related play a central role in the process of mathematical modeling. The purpose of this discussion is to identify this important role as a prelude to examining such equations in detail; it is not intended to develop techniques of mathematical modeling per se.

The change in a variable y due to a change in the variable t from a value t_0 to a value $t_0 + h$ is the difference:

$$y(t_0 + h) - y(t_0).$$

Here $y(t)$ denotes the value y associated with the number t. To describe the relative change in y due to a change in t, the difference $y(t_0 + h) - y(t_0)$ is divided by thc difference in t values, $(t_0 + h) - t_0$. This gives a difference quotient:

$$\frac{y(t_0 + h) - y(t_0)}{(t_0 + h) - t_0}.$$

In this section, we explore the mathematics of difference quotients; in Section 13.3, we explore their use to describe changes in discrete phenomena. In Section 14, difference quotients will form the foundation of the concept of a *derivative*, which describes continuous or instantaneous changes.

The Greek capital letter Δ (delta) is traditionally used to denote *differences*; Δt denotes a *difference* in the variable t; and, Δf denotes a *difference* in the values of the function f.

If the variable t is changed by an increment h from a point $t = a$ to $t = a + h$, the corresponding difference is

$$\boxed{\Delta t = (a + h) - a = h.}$$

The corresponding difference in the value of the function $f(t)$ is

$$\boxed{\Delta f = f(a + h) - f(a).}$$

Definition 13.1 For a given function, f, reference number $t = a$ and real number h, *the difference quotient of f at a with increment h* is the number

$$\boxed{\frac{\Delta f(a)}{\Delta t} = \frac{f(a + h) - f(a)}{h} \qquad \textbf{Difference quotient.}} \tag{13.1}$$

The variable t is frequently replaced by the variable x. The increment h is often denoted by Δt, or Δx, depending on the variable. Thus alternative forms of equation (13.1) would include the following, where $|_{x=a}$ and $|_{t=a}$ are used to indicate the reference number a:

$$\left.\frac{\Delta f(x)}{\Delta x}\right|_{x=a} = \frac{f(a+h) - f(a)}{h};$$

$$\frac{\Delta f(a)}{\Delta x} = \frac{f(a+\Delta x) - f(a)}{\Delta x};$$

$$\left.\frac{\Delta f}{\Delta t}\right|_{t=a} = \frac{f(a+\Delta t) - f(a)}{\Delta t}.$$

When the number a is not specifically known, the variable t (or x) is usually employed to represent difference quotients in the form

$$\frac{\Delta f(t)}{\Delta t} = \frac{f(t+\Delta t) - f(t)}{\Delta t}; \quad \text{or} \quad \frac{\Delta f}{\Delta x} = \frac{f(x+\Delta x) - f(x)}{\Delta x}.$$

The difference quotient is a real number. We can evaluate the difference quotient if we are given (1) the function, (2) the number at which the quotient is evaluated, and (3) the increment that appears as the denominator of the difference quotient. For some problems, it is possible to use an arbitrary symbol to represent the evaluation point or increment. When this is done, the resulting number will be expressed in terms of the arbitrary symbol.

Example 13.1 The difference quotient is computed for the function, point, and increment indicated.

a) $f(x) = x^2 - x + 2; x = 3; h = 0.1;$

$$\frac{\Delta f(3)}{\Delta x} = \frac{f(3+0.1) - f(3)}{0.1} = 5.1, \text{ since } f(3.1) = 8.51 \text{ and } f(3) = 8.$$

b) $g(t) = t - \sqrt{t}; t = 3; h = -1;$

$$\frac{\Delta g(3)}{\Delta t} = \frac{g(3-1) - g(3)}{-1} = g(3) - g(2) \simeq 0.68216,$$

since $g(3) \simeq 1.26794$ and $g(2) \simeq 0.58578$.

c) $f(t) = 2t - 3; t = t_0, h = 0.2;$

$$\begin{aligned}\frac{\Delta f(t_0)}{\Delta t} &= \frac{f(t_0+0.2) - f(t_0)}{0.2}\\ &= \frac{[2(t_0+0.2) - 3] - [2t_0 - 3]}{0.2}\\ &= \frac{[2t_0 + 0.4 - 3] - [2t_0 - 3]}{0.2} = \frac{0.4}{0.2} = 2.\end{aligned}$$

d) $f(x) = x^2 + 1; x = x_0; \Delta x = 0.2;$

$$\left.\frac{\Delta f}{\Delta x}\right|_{x_0} = \frac{f(x_0 + 0.2) - f(x_0)}{0.2}$$

$$= \frac{[(x_0 + 0.2)^2 + 1] - [x_0{}^2 + 1]}{0.2}$$

$$= \frac{[x_0{}^2 + 0.4x_0 + 0.04 + 1] - [x_0{}^2 + 1]}{0.2}$$

$$= \frac{0.4x_0 + 0.04}{0.2} = 2x_0 + 0.2.$$

e) $g(x) = x^2 + 1; x = x_0; \Delta x = h;$

$$\left.\frac{\Delta g}{\Delta h}\right|_{x_0} = \frac{g(x_0 + h) - g(x_0)}{h}$$

$$= \frac{[(x_0 + h)^2 + 1] - [x_0{}^2 + 1]}{h}$$

$$= \frac{[x_0{}^2 + 2hx_0 + h^2 + 1] - [x_0{}^2 + 1]}{h}$$

$$= \frac{2hx_0 + h^2}{h} = \frac{(2x_0 + h)h}{h} = 2x_0 + h.$$ ◀

Example 13.2 To illustrate the dependence of the difference quotient on each of the components, a function, an x-value, and an increment, consider the following examples in which each is varied individually.

a) $f(x) = x^2; a = 2; h = 0.1; \dfrac{\Delta f(a)}{\Delta x} = 4.1$

b) $f(x) = x^2, a = 2; h = 0.2; \dfrac{\Delta f(a)}{\Delta x} = 4.2$

c) $f(x) = x^2, a = 3; h = 0.1; \dfrac{\Delta f(a)}{\Delta x} = 6.1$

d) $f(x) = x^2, a = 3; h = 0.2; \dfrac{\Delta f(a)}{\Delta x} = 6.2$

e) $f(x) = x^3, a = 2; h = 0.1; \dfrac{\Delta f(a)}{\Delta x} = 12.6$

f) $f(x) = x^3, a = 2; h = 0.2; \dfrac{\Delta f(a)}{\Delta x} = 13.24$

g) $f(x) = x^3$, $a = 3$; $h = 0.1$; $\dfrac{\Delta f(a)}{\Delta x} = 27.91$

h) $f(x) = x^3$, $a = 3$; $h = 0.2$; $\dfrac{\Delta f(x)}{\Delta x} = 28.84$ ◀

13.2 INTERPRETATIONS OF THE DIFFERENCE QUOTIENT

There are two common ways to interpret the difference quotient $\Delta f/\Delta t$. One way, which is very useful when the function f describes a biological or physical quantity is as an *average rate of change* of the quantity. The second way, the more common approach in calculus classes, is to interpret the difference quotient geometrically using the graph of the associated function. Each interpretation has its own advantages over the other. The graphing approach is easily depicted by a diagram (hence its popularity), while the approach that we consider first lends itself to a greater variety of applications.

I. *The difference quotient as an average rate of change*

To say that the difference quotient (13.1) is the "average rate of change of the function f over the interval $[a, a + h]$," assuming that $h > 0$, seems reasonable since the word "rate" indicates a ratio, proportion, or quotient, and the equation (13.1) is a quotient of "changes." The term "average" is used to indicate that the difference quotient is representative of the rate of change of f at all values of $t \in [a, a + h]$ rather than particular rates of change. (Similarly, the average of the numbers 1, 2, 4, and 5 is the number 3, which represents the four numbers collectively but does not equal any one of the numbers.)

The rate of change of a function is analogous to the velocity of a moving object. If a "jogger" runs at a velocity or rate of 9.6 kilometers per hour, then, over a time interval of one-half hour, she or he would have jogged 4.8 km; over an interval of 1.2 hours, she or he would have jogged 11.52 km; and, over an interval of h hours, she or he would have jogged $h \times (9.6)$ km. Similarly, if a function $f(t)$ is known to change at an average rate of R units over the interval $[a, a + h]$, then the change in the value of $f(t)$ from $t = a$ to $t = a + h$ would be the length of the interval times the rate R:

$$f(a + h) - f(a) = h \times R$$

Change = (time interval) × (average rate of change).

Thus we find the *average rate of change*

$$\boxed{R = \frac{f(a + h) - f(a)}{h} \equiv \frac{\Delta f(a)}{\Delta t}.}$$

Example 13.3

a) Assume a quantity is described as a function of time by $y = f(t)$. If the average rate of change of the quantity over the interval $[a, a + h]$ is the same constant value for every reference point $t = a$ and increment h, then $f(t)$ is a linear function. To see this, let M be the constant value of the difference quotient of the function f. Let $a = 0$, and consider the difference quotient

$$\frac{\Delta f(0)}{\Delta t} = \frac{f(0 + h) - f(0)}{h} = M.$$

Solving for $f(h) = f(0 + h)$ gives us

$$f(h) = M \cdot h + f(0).$$

Since this equation is assumed to hold for all increments, we can replace the symbol h by t and $f(t)$ by y to obtain

$$y = M \cdot t + f(0),$$

which is the standard equation of a line.

b) The amount of a drug circulating in an animal's system was found to decrease at an average rate proportional to the time between measurement. Let $C(t)$ denote the drug concentration at time t. Then the difference quotient for the function $C(t)$, representing the rate at which the drug is "cleared from circulation," is given by

$$\frac{\Delta C(t_0)}{\Delta t} = -k\,\Delta t,$$

for some positive constant k. With $\Delta C(t_0) = C(t_0 + \Delta t) - C(t_0)$, this equation is equivalent to

$$C(t_0 + \Delta t) = C(t_0) - k(\Delta t)^2.$$

Setting $t_0 = 0$ and $\Delta t = T$, we obtain the concentration of the drug at any time T as

$$C(T) = C(0) - kT^2.$$ ◀

Example 13.4 One of the basic laws governing the diffusion of materials was given by A. Fick in 1855. Fick's Law states that the rate of diffusion of a substance past a point will be proportional to the concentration gradient of the substance at the point. Assume that a substance diffuses through a uniform tube of cross-sectional area A, or equivalently across a membrane with surface area A. To describe the diffusion of the substance past a point x, let $Q(t)$ denote the amount of the substance at time t between x and $x + \Delta x$ for a small value of Δx. (See Fig. 13.1.) The average rate of diffusion past the point x over an interval $[t, t + \Delta t]$ is the difference quotient:

$$\textit{Average flux}: \frac{\Delta Q}{\Delta t} = \frac{Q(t + \Delta t) - Q(t)}{\Delta t}.$$

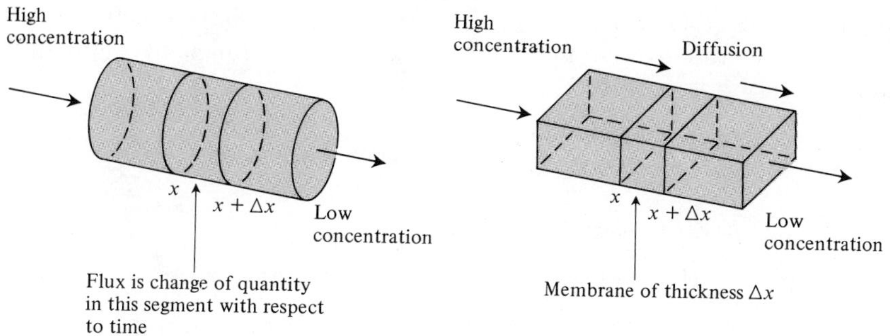

Figure 13.1

Let $C(x)$ denote the concentration of the substance at position x. The concentration gradient of the substance is the average rate of change in the concentration:

$$\textit{Concentration gradient}: \frac{\Delta C(x)}{\Delta x} = \frac{C(x + \Delta x) - C(x)}{\Delta x}.$$

Fick's First Law (in one dimension) states:

$$\frac{\Delta Q(t)}{\Delta t} = D \cdot A \frac{\Delta C(x)}{\Delta x},$$

where D is the *diffusion constant* and varies depending on the substance, solute, and experimental condition. ◀

Example 13.5 In biological systems, the rate of change of a quantity is sometimes available even when the function describing the quantity is not available. In this situation, values of the associated function can be approximated using the rate of change. To illustrate this, consider the volume of water uptake by a plant. The actual volume of water entering or leaving a leaf or stem area is very difficult to measure directly. The change in the water present—i.e., the flow rate—can be measured over short time intervals with instruments that correlate the flow rate with changes in electrical potentials.

Assume that the average flow rate is specified over subintervals of the time interval [0, 2] as indicated in Table 13.1. Let $V(t)$ denote the volume of H_2O at time t. Assuming that the volume $V(0)$ at time $t = 0$ is known to be 10, we can

Table 13.1

Interval	[0, 0.5]	[0.5, 0.75]	[0.75, 1.5]	[1.5, 2]
Average flow rate	0.5	2	−1	0

approximate the volume $V(2)$ at $t = 2$ as follows. The volume function $V(t)$ has an average flow rate over an interval $[a, b]$ given by the difference quotient with increment $\Delta t = b - a$. Since $V(a + \Delta t) = V(a + b - a) = V(b)$,

$$\frac{\Delta V(a)}{\Delta t} = \frac{V(b) - V(a)}{b - a} = \text{average flow rate over } [a, b].$$

Thus, solving for $V(b)$, we find that

$$V(b) = V(a) + (b - a) \times \text{(average flow rate)}.$$

Assuming that $V(0) = 10$, by successively choosing the interval $[a, b]$ as the intervals in the table, we can compute the volume at the endpoints of the subintervals in a sequential manner to obtain $V(2)$.

$$V(b) = V(a) + (b - a)\left(\frac{\Delta V}{\Delta t}\right)$$

$$\begin{aligned}
a = 0, \quad b = 0.5&: \quad V(0.5) = 10 + (0.5)(0.5) = 10.25 \\
a = 0.5, \quad b = 0.75&: \quad V(0.75) = 10.25 + (0.25)(2) = 10.75 \\
a = 0.75, \quad b = 1.5&: \quad V(1.5) = 10.75 + (0.75)(-1) = 10.0 \\
a = 1.5, \quad b = 2&: \quad V(2) = 10.0 + (0.5)(0) = 10.0
\end{aligned}$$

The calculated values of $V(t)$ are plotted in Fig. 13.2. Note that we have connected the computed points with straight lines; there is no justification for this. The function $V(t)$ would be linear over the interior of the intervals, (0, 5), (0.5, 0.75), (0.75, 1.5), and (1.5, 2), only if the flow rate remained constant over these subintervals.

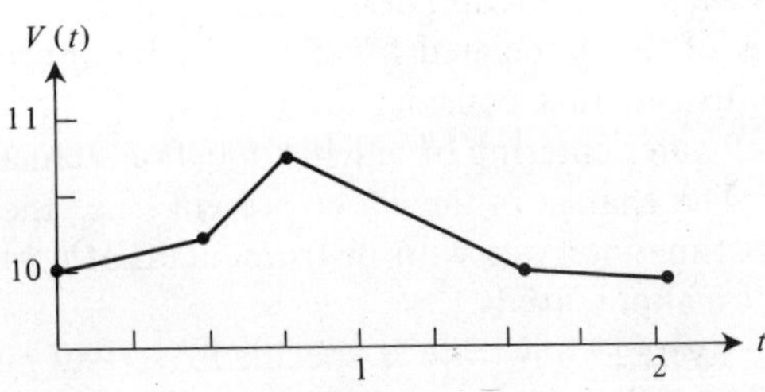

Figure 13.2

II. *The difference quotient as the slope of a secant line*

The geometrical interpretation of the difference quotient associates the difference quotient with the slope of a secant line. A secant line is a line that passes through two points on the graph of a function. A given function f, an x-value, x_0, and an increment, Δx, determine a unique secant line (see Fig. 13.3) passing through the points $(x_0, f(x_0))$ and $(x_0 + \Delta x, f(x_0 + \Delta x))$ on the graph of $y = f(x)$. The

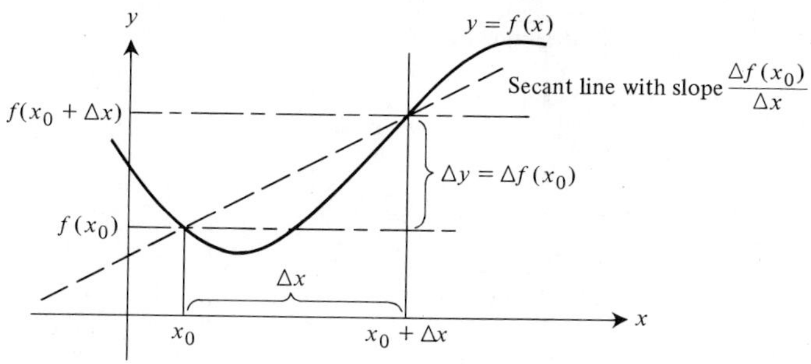

Figure 13.3

slope of this secant line is given by the change in the y-coordinates, Δy, divided by the change in the x-coordinates, Δx. Using the two points $(x_0, f(x_0))$ and $(x_0 + \Delta x, f(x_0 + \Delta x))$, we compute the slope m of the secant line:

$$m = \frac{\Delta y}{\Delta x} = \frac{f(x_0 + \Delta x) - f(x_0)}{\Delta x} = \frac{\Delta f(x_0)}{\Delta x}.$$

Thus each choice of the increment Δx leads to a secant line through the point $(x_0, f(x_0))$. *The point slope equation of the secant line* is given by the equation

$$\boxed{y = \left[\frac{\Delta f(x_0)}{\Delta x}\right] \cdot (x - x_0) + f(x_0).} \tag{13.2}$$

The slope of the secant line in Fig. 13.3 is positive. However, if we were to choose a much smaller increment Δx, then the secant line through the point $(x_0, f(x_0))$ would have a negative slope for the function illustrated.

Example 13.6 A boat sets out from a point A on one side of a river and "heads straight for a point B" on the opposite side of the river. However, the current in the river is stronger in the center than it is close to the shore. As a result, the boat does not travel directly across the river, but along a U-shaped path. Assume the path traveled by the boat is described by the graph of a function, $f(x)$. We assume that the river is l units wide and the variable x measures the distance from the starting point at the river bank. The function f must satisfy $f(0) = f(l)$, since A and B are opposite each other on the river bank. (The model works equally well when this is not the case. See Fig. 13.4.) The direction in which the boat is steered at any point along its path is along the secant line between the point $(x_0, f(x_0))$, where the boat is, and the point $B = (l, f(l))$, where the boat is headed. This direction is the slope of the secant line given by the difference quotient

$$m = \frac{f(l) - f(x_0)}{l - x_0}.$$

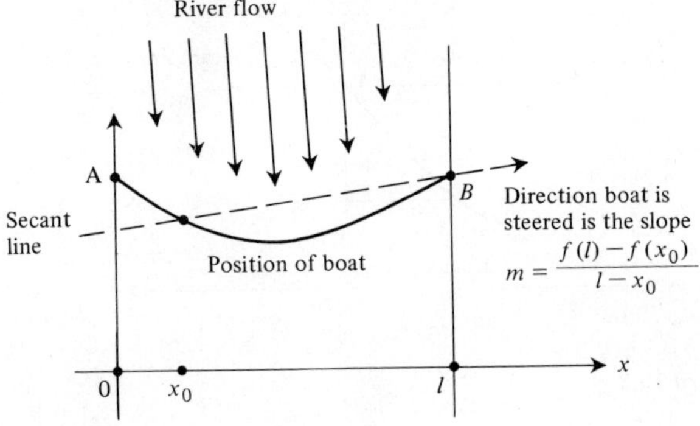

Figure 13.4

Example 13.7

a) Let $f(x) = x^2$ and $x_0 = 1$. The secant lines through $(x_0, f(x_0)) = (1, 1)$, corresponding to the increments $\Delta x = -1, -\frac{1}{2}, \frac{1}{2}$, and 1, are obtained from the equation (13.2):

$$y = \frac{\Delta f(1)}{\Delta x}(x - 1) + 1.$$

Table 13.2 gives the equations of these lines which are in Fig. 13.5.

Table 13.2

Line	Δx	$\frac{\Delta f(1)}{\Delta x}$	Equation of the line	
1)	-1.0	1	$y = (x - 1) + 1,$	$y = x$
2)	-0.5	1.5	$y = 1.5(x - 1) + 1,$	$y = 1.5x - 0.5$
3)	0.5	2.5	$y = (2.5)(x - 1) + 1,$	$y = 2.5x - 1.5$
4)	1.0	3	$y = 3(x - 1) + 1,$	$y = 3x - 2$

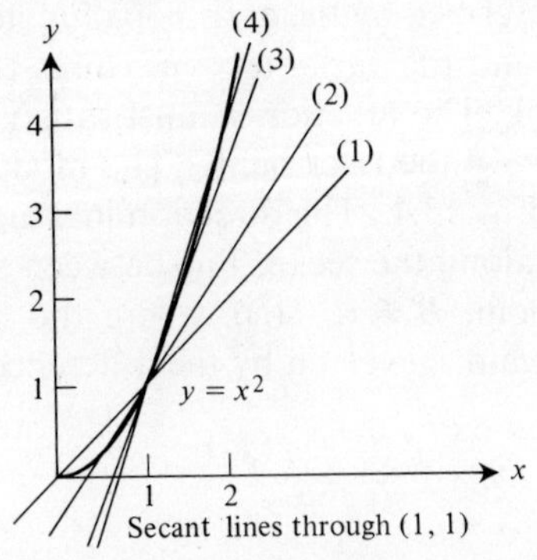

Figure 13.5

b) Let $f(x) = \ln(x)$ and $x_0 = 1$. Then $(x_0, f(x_0)) = (1, 0)$ and

$$(x_0 + \Delta x, f(x_0 + \Delta x)) = (1 + \Delta x, \ln(1 + \Delta x)).$$

The difference quotient

$$\frac{\Delta f(1)}{\Delta x} = \frac{\ln(1 + \Delta x) - \ln(1)}{\Delta x} = \frac{\ln(1 + \Delta x)}{\Delta x},$$

since $\ln(1) = 0$. Therefore, the secant line through $(1, 0)$, determined by an increment Δx is

$$y = \left[\frac{\ln(1 + \Delta x)}{\Delta x}\right](x - 1).$$

Table 13.3

Δx	5	1	0.5	0.10	-0.5
$\frac{\Delta(f(x_0))}{\Delta x}$	0.4	0.7	0.8	0.95	1.4

The secant lines corresponding to the list of increments Δx in Table 13.3 are sketched in Fig. 13.6.

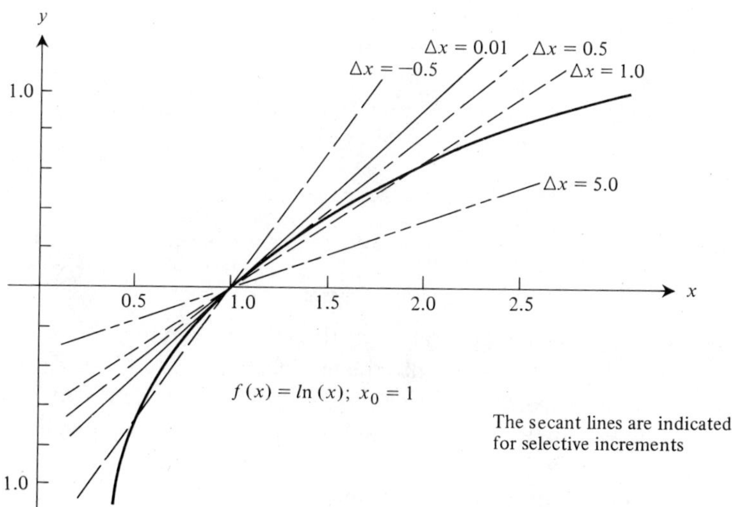

Figure 13.6

◀

13.3 DISCRETE AND CONTINUOUS MODELS

In this subsection, we examine the role of the difference quotient in discrete population models and how such models may be used to describe continuously changing populations.

By the term "discrete model" we mean a model that specifies the state of the population at a specific sequence of times, rather than for all values of time. The sequence of times at which a population, X, is specified will generally correspond to natural biological events, such as the start of each breeding season; but it may be a simple sequence corresponding to regular observations such as daily samples.

In the following, we will refer to the discrete model of a population, X, by a sequence, $\{X_n\}$, and we will refer to X_n as the size of the population at generation n. In Section 11, this type of model was examined by simply using the length of a generation as a basic unit of measurement; this avoided any reference to time other than as the nth generation.

A more realistic type of model may be obtained by representing the length of each generation by a number h measured in terms of some standard unit of measure. Typical standard units of measure are seconds, hours, years, etc. The independent variable t is used in the following to represent time. Starting a model at time $t = 0$, we find that the successive generations then occur at times

$$t_1 = h, t_2 = 2h, t_3 = 3h, \ldots, t_n = n \cdot h, \ldots$$

When the size of the population, X, at time $t \geq 0$ is given by a function $X(t)$, the corresponding sequence giving the population size for successive generations is

$$\{X(h), X(2h), X(3h), \ldots\}.$$

This is the sequence $\{X_n\}$, where $X_n = X(n \cdot h)$. The change in the population size from generation n to generation $n + 1$ is the

$$\boxed{\frac{\text{change}}{\text{generation}} = \Delta X_n = X_{n+1} - X_n.}$$

This reflects the change per generation at the nth generation.

Introducing the time variable into the model via the generation time h changes the frame of reference from generations to units of time. Consequently, if the scale of a model is determined by a specific unit of time, then the change in population considered as a change per unit of time is the change per generation multiplied by the number of generations per unit time:

$$\frac{\text{change}}{\text{unit }(t)} = \frac{\text{change}}{\text{generation}} \cdot \frac{\text{generation}}{\text{unit }(t)}.$$

When the length of each generation is constant, say of length h, then the number of generations in one unit of time is $1/h$:

$$\frac{\text{generation}}{\text{unit }(t)} = \frac{1}{h}.$$

In this case the *change per unit time at the nth generation* (or equivalently at time $t_n = n \cdot h$) is given by the difference quotient

$$\frac{\Delta X_n}{\Delta t_n} = \frac{X_{n+1} - X_n}{t_{n+1} - t_n} = \frac{X_{n+1} - X_n}{h}. \tag{13.3}$$

Note that

$$t_{n+1} - t_n = (n + 1) \cdot h - n \cdot h = h.$$

The difference quotient (13.3) may be used in each of the growth models discussed in Section 11 in place of the difference, ΔX_n. This leads to new growth models that depend on the generation time, h.

Example 13.8

a) Consider the *generation-time-dependent form* of the *constant-growth model* in which the average rate of change per unit time is a constant, c. Then,

$$\frac{\Delta X_n}{\Delta t_n} = c \qquad \text{or} \qquad X_{n+1} = X_n + c \cdot h.$$

This is a first order difference equation. The sequence, $\{X_n\}$, may be generated by assuming that

$$X_1 = X_0 + ch.$$

Then,

$$X_2 = X_1 + ch = X_0 + ch + ch = X_0 + 2ch.$$

The nth generation size, X_n, is readily found to be

$$X_n = X_0 + n(ch).$$

b) Consider the *generation-time-dependent form* of the *proportional-growth model* in which the average rate of change per unit time corresponding to the nth generation is proportional to X_n. The model is specified by the equations

$$\frac{\Delta X_n}{\Delta t_n} = cX_n \qquad \text{or} \qquad X_{n+1} = (1 + ch)X_n,$$

which are difference equations describing a geometrically growing population. If

$$X_1 = X_0(1 + ch), \qquad \text{then} \qquad X_2 = (1 + ch)X_1 = X_0(1 + ch)^2;$$

and, for $n = 1, 2, 3, \ldots$, we have

$$X_n = X_0(1 + ch)^n. \tag{13.4}$$

◀

The sequence $\{X_n\}$ specifies the population's size at the sequence of generation times,

$$t_1 = h, \qquad t_2 = 2h, \qquad t_3 = 3h, \qquad \ldots,$$

which depend on the generation length h. This type of model is very useful when studying such populations as migrating birds or caribou, which clearly exhibit discrete generations, or even synchronized cell cultures, which exhibit uniform reproduction. But this type of model is not satisfactory when studying such continuously changing populations as bacteria cultures, which are not synchronized. One way to approach the description or modeling of such populations is to assume that they change discretely, say with a generation h, and then to let the length of the generation approach zero. We illustrate this approach using the proportional growth model of Example 13.8(b).

Exponential Growth

Assume a population, X, grows in such a manner that at any time, t, it is increasing at an *instantaneous rate* that is proportional to its size $X(t)$:

$$\text{Rate of change} = c \cdot X(t),\ c \text{ a constant.}$$

The population can be modeled by first assuming that it is a discrete population having generation length h. This leads to the population model of Example 13.8(b) in which the average rate of change over each generation is proportional to the population's size. The discrete model of the population is given by equation (13.4), which describes the size of the nth generation of the population:

$$X_n = X_0(1 + ch)^n, \tag{13.4}$$

where

$$X_0 = \text{initial population;}$$
$$c = \text{proportional growth rate constant;}$$
$$h = \text{generation length.}$$

To obtain a continuous-growth model with *instant growth*, we consider the effect of allowing the generation length h to approach zero in the discrete model. To describe the corresponding population with instant growth, we cannot simply allow h to approach zero in equation (13.4). This would result in a population that is constant:

$$\lim_{h \to 0} X_n = \lim_{h \to 0} X_0(1 + ch)^n = X_0.$$

The problem here is that X_n is the population size at time $t_n = n \cdot h$. Letting $h \to 0$, we simply get the population size at $t = 0$. To circumvent this difficulty, we must reformulate the solution of the discrete model to specify the population at a specific time, T. This is accomplished by assuming that T is a multiple of the generation length h—i.e., that there is a positive integer N such that

$$T = N \cdot h.$$

Then the population size at time T is that of the Nth generation:

$$X(T) = X_N = X_0(1 + ch)^N.$$

As $N = T/h$ this may be expressed as

$$X(T) = X_0[(1 + ch)^{(1/h)}]^T. \tag{13.5}$$

What is the population size at time T if the generation length h becomes smaller? The answer may be easier to visualize if we introduce a new parameter $\varepsilon = ch$ (ε is the Greek letter epsilon). Writing $1/h = c/ch = (1/\varepsilon)c$, we find that the equation for $X(T)$ may be written in the form

$$X(T) = X_0[(1 + \varepsilon)^{(1/\varepsilon)}]^{cT}.$$

Since ε is a constant multiple of h, as h approaches zero, $h \to 0$, ε also approaches zero, $\varepsilon \to 0$. Thus the population at time T, corresponding to a populatión whose generation time $h \to 0$, is given by

$$X(T) = \lim_{\varepsilon \to 0} X_0[(1 + \varepsilon)^{(1/\varepsilon)}]^{cT}.$$

This limit depends on the value of $(1 + \varepsilon)^{(1/\varepsilon)}$, as $\varepsilon \to 0$. This is one of the fundamental limits in mathematics and can be shown to be equal to the number e, the same number that is the base of the *natural* logarithm;

$$\lim_{\varepsilon \to 0} (1 + \varepsilon)^{(1/\varepsilon)} = e. \tag{13.6}$$

Consequently, the population $X(t)$ is given by

$$X(T) = X_0 e^{cT}. \tag{13.7}$$

Thus we have used a finite difference model to arrive at a description of a population that reproduces or grows instantaneously (having a generation time approaching zero). Our solution corresponds to the exponential-growth function discussed in Section 5.

Example 13.9 To illustrate the formula (13.5), consider a cell population that divides at each generation and has an initial size of 10^4 cells. Using equation (13.5) with $c = 2$, the size of the population at $T = 3$ for various generations lengths h is indicated in Table 13.4, where the relative values $x(3)/10^4$ have been rounded to the nearest integer.

Table 13.4

Generation length h:	1	10^{-1}	10^{-2}	10^{-3}	10^{-4}	10^{-5}
Relative population $X(3)/10^4$:	27	237	380	401	403	403

The corresponding continuous growth model for a population that reproduces by fission, such as cell division, is given by equation (13.7) with $c = 2$,

$$X(t) = X_0 e^{2t}.$$

As $e^6 \simeq 403.4$, the continuous-growth model gives the value of $X(3)/10^4$ as 403 (rounded off). Thus the discrete model with generation length $h = 10^{-3}$ would give an error of two for this value, while values of $h \leq 10^{-4}$ would result in an error less than one. ◀

In the above discussion we approached the problem of describing instantaneous growth by first constructing a model with generations of length h. This gave a difference equation, which was then solved to give the population at the nth generation as a function involving the length h. We then expressed $X(T)$ as X_N and took $\lim_{h \to 0} X_N$. This limit gave us the continuous-model description of $X(T)$. This process is indicated in Fig. 13.7, where the blank in the difference equation is completed according to the assumed or experimentally determined rate of change of the population.

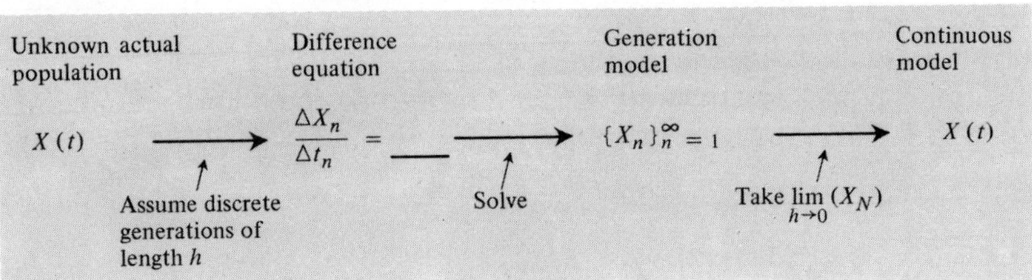

Figure 13.7

Can this process be accomplished in any other way? The answer is yes. The last two steps, *solve* and *take the limit as* $h \to 0$, can be interchanged. This would result in first taking the $\lim_{h \to 0}$ of the difference quotient $\Delta X/\Delta t$, and then solving the resulting equation,

$$\lim_{h \to 0} \frac{\Delta X}{\Delta t} = \_\_\_\_\_.$$

The limit of the difference quotient is the most important concept in calculus and is the principal topic of the next Section 14.

SUMMARY

1. *Notation*: Δt denotes a *change* in the variable t. If the change is due to an *increment* h, then

$$\Delta t = (t + h) - t = h.$$

The change in a function $f(t)$, due to an increment h, is the *difference*

$$\Delta f(t) = f(t+h) - f(t).$$

2. *Difference quotient*: The difference quotient of $f(t)$ at $t = a$ with increment h is

$$\frac{\Delta f(a)}{\Delta t} = \frac{f(a+h) - f(a)}{h}.$$

3. *Interpretation of the difference quotient*:
   a) If $f(t)$ specifies a measurable quantity then the difference quotient $\Delta f(a)/\Delta t$ specifies an *average rate of change* over the interval $[a, a+h]$.

Examples

$f(t)$ = population $\longrightarrow$ $\dfrac{\Delta f(a)}{\Delta t}$ = average growth rate

$f(t)$ = distance $\longrightarrow$ $\dfrac{\Delta f(a)}{\Delta t}$ = average velocity

$C(t)$ = drug concentration $\longrightarrow$ $\dfrac{\Delta C(a)}{\Delta t}$ = average rate of drug clearance

$C(x)$ = concentration in solution $\longrightarrow$ $\dfrac{\Delta C(a)}{\Delta t}$ = average concentration gradient

$Q(t)$ = amount of solute in a membrane $\longrightarrow$ $\dfrac{\Delta Q(a)}{\Delta t}$ = average diffusion rate

$V(t)$ = volume of fluid $\longrightarrow$ $\dfrac{\Delta V(a)}{\Delta t}$ = average flow rate

If $f(x)$ is any function, then the difference quotient $\Delta f(x_0)/\Delta x$ is the slope of the secant line through the points

$$(x_0, f(x_0)) \quad \text{and} \quad (x_0 + \Delta x, f(x_0 + \Delta x)).$$

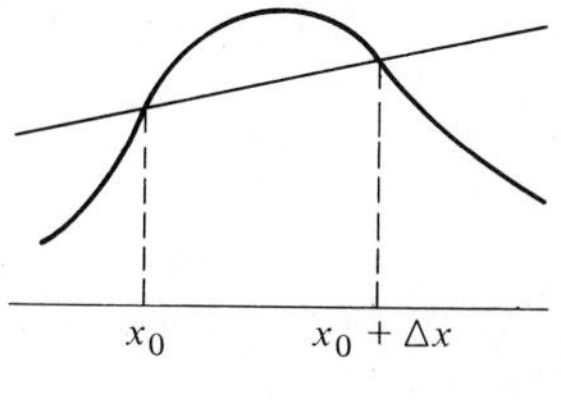

Secant line

$$y = \left(\frac{\Delta f(x_0)}{\Delta x}\right)(x - x_0) + f(x_0)$$

x_0 $\quad$ $x_0 + \Delta x$

4. *Discrete population models*:
   a) Assume that the population reproduces with a generation length h. The time of each generation will be

$$t_0 = 0 \text{ (initial population)}, t_1 = h, t_2 = 2h, \ldots, t_n = nh \text{ (}n\text{th generation)}.$$

   b) Denote the population size of the nth generation by X_n. Then the change in the population between the nth and $(n + 1)$st generations is

$$\Delta X_n = X_{n+1} - X_n.$$

   c) The change in the population per unit of time at the nth generation is the difference quotient:

$$\frac{\Delta X_n}{\Delta t_n} = \frac{X_{n+1} - X_n}{h}.$$

   d) The discrete proportional growth model assumes that

$$\frac{\Delta X_n}{\Delta t_n} = cX_n.$$

   The solution of the model is

$$X_n = X_0(1 + ch)^n;$$

   where X_0 = initial population, c = proportionality constant, and n = generation
   e) The continuous proportional growth model is obtained from the discrete model by taking the limit $h \to 0$. The result is *exponential growth equation*:

$$X(T) = X_0 e^{cT}.$$

EXERCISE SET 13

13.1 Compute the difference quotient for the function, reference point, and increment indicated.

a) $f(x) = 2x; x = 3, h = 0.1$
b) $f(x) = \log(x); x = 2, h = 0.1$
c) $g(x) = e^x; x = 0, h = 0.2$
d) $g(x) = (x + 2)/x; x = 1, \Delta x = 0.5$
e) $f(x) = (x + 2)/(x - 2); x = 1.9, \Delta x = 0.05$
f) $f(x) = (x + 2)/(x - 2); x = 1.9, \Delta x = 0.15$

13.2 Compute the difference quotients indicated, simplifying as much as possible.

a) $\dfrac{\Delta f(x_0)}{\Delta x}; f(x) = 2x - 8$ b) $\left.\dfrac{\Delta f(t)}{\Delta t}\right|_{t=t_0}; f(t) = t^2 + 3$

c) $\dfrac{\Delta g(t)}{\Delta t}; \quad g(t) = 1/t$ d) $\dfrac{\Delta f(x_0)}{\Delta x}; \Delta x = h, f(x) = x^3 + 3$

e) $\frac{\Delta g(z_0)}{\Delta z}$; $g(z) = z^{(1/2)}$ f) $\frac{\Delta f(t)}{\Delta t}$; $f(t) = 2^t$

g) $\frac{\Delta f(x)}{\Delta x}$; $f(x) = e^x$ h) $\frac{\Delta g(x)}{\Delta x}$; $g(x) = e^{-x}$

13.3 Determine the average rate of change of the given function over the interval [1, 3].

a) $f(t) = t^2$ b) $f(t) = t^{-1}$ c) $f(t) = \sin(\pi t)$ d) $f(t) = \log_3(t)$
e) $f(t) = 2^t$ f) $f(t) = t(3 - t)$ g) $g(t) = e^t$ h) $g(t) = e^{2t}$

13.4 A jogger runs at 1 km per hour for 20 minutes and then runs for another 10 minutes at 7 km per hour. How far has she or he run?

13.5 A corn plant, which was fed a growth hormone, was measured at the start of an experiment and found to be 0.8 m in height. Over the next week of the experiment, the plant grew at an average rate of 0.15 cm per day; over the second week of the experiment, it grew at a rate of 0.05 m per day; and, during the third week of the experiment it grew at 0.02 m per day.

a) What was the height of the corn plant at the end of the three weeks?
b) What was its average rate of growth over the three-week period. Determine the answer using two different methods.

13.6 A drug is administered to a patient. The average rate of change of the drug concentration in the patient's blood is monitored over 10 minute intervals for two hours. Sketch a broken line graph to indicate a possible function describing the drug concentration if the concentration at $t = 0$ is zero and the average rates of change over the intervals [0, 10], [10, 20], etc. are 0.2, 1.5, 2.0, 1.8, 1.8, 1.6, 1.4, 0.8, 0.2, 0.0, -1.0, -3.5, respectively.

13.7 Give the equation of the secant line through the point (0, 3) on the graph of $y = x^4 + 3$ as determined by the following increments.

a) $\Delta x = 1$ b) $\Delta x = 0.1$ c) $\Delta x = 0.01$

13.8 For each function in Exercise 13.3 give the equation of the secant line, which is determined by the point $t_0 = 1$ with the following increments.

a) $\Delta t = 1$ b) $\Delta t = -1/2$ c) $\Delta t = 1/2$

13.9 Compute the equation of the secant line through the point $(x_0, f(x_0))$, which is determined by the given increment, Δx.

a) $f(x) = \sin(x)$; $x_0 = 0$, $\Delta x = \pi/2$
b) $f(x) = \tan(x)$; $x_0 = 0$, $\Delta x = \pi/4$
c) $f(x) = \cos(\pi x)$; $x_0 = 0$, $\Delta x = -1/2$
d) $f(x) = -\sin(\pi x)$; $x_0 = 1$, $\Delta x = 1/3$
e) $f(x) = 2\sec(x)$; $x_0 = \pi/4$, $\Delta x = -\pi/4$
f) $f(x) = \sin(x) + \cos(x)$; $x_0 = \pi/4$, $\Delta x = -\pi/4$

13.10 Graph the secant lines in Exercise 13.8.

13.11 a) Write the equation in Example 13.4 (called Fick's Law) when the fluid is in a tube of radius 2 and the diffusion coefficient is 0.7.

b) If the concentration C in the example is constant, show that the diffusion is zero.
c) If the concentration C is linear show that the quantity function Q is also linear.

13.12 Graph the secant lines in Exercise 13.9.

13.13 This problem considers a fox chasing a hare. Initially, the fox is at the point (0, 10) and the hare is at the point (10, 0) of a coordinate system. Assume that the hare runs 1 unit/min in a positive direction along the x-axis. Assume that the fox runs 2 units/min in a straight line, changing direction each minute. Assume that the fox always runs in the direction of the point where the hare was at the start of each minute of time.

a) What will be the positions of the fox and hare after one minute? Illustrate these on a graph.
b) What will be the positions of the fox and hare after three minutes? Illustrate these on a graph.
c) Repeat parts (a) and (b) assuming that the fox is "smart" and runs in the direction of the point where the hare will be at the end of each minute of time.

13.14 Assume that the average rate of change in the concentration of a drug in a patient's blood over the nth hour interval $I_n = [n, n + 1]$ is $r_n = 10^{-2}(5 - n)$g/l. Assume that the concentration at $t = 1$ is $C(1) = 50$ mg/l.

a) Plot the concentration until the drug level in the patient's blood disappears.
b) What will be the maximum recorded drug concentration?

13.15 Compute the average rate of change of the logistics function $f(t) = 3/(1 + 2e^{-5t})$ over the interval $[0, t_{2/3}]$ when $t_{2/3}$ is the time at which the graph reaches $\frac{2}{3}$ of its limiting value, ($\lim_{t\to\infty} f(t) = 3$).

13.16 A goat bounds up a hill described by the function $h(x) = 10 - (x - 5)^2$. If the goat leaps one vertical unit in each bound, how many bounds will it take the goat to reach the top of the hill? What will be the longest and shortest horizontal distances covered in one jump.

13.17 Use the equation (13.5) to evaluate $X(1)$ using a model with $X_0 = 1$ and $c = 1$ for generation lengths $h = 1, h = \frac{1}{2}, h = \frac{1}{5}, h = \frac{1}{10}$, and $h = \frac{1}{100}$.

13.18 Repeat Exercise 13.17 with $X_0 = 10$ and $c = 3$.

13.19 a) Derive the equation for $X(T)$ similar to equation (13.5) when the population has constant growth as illustrated in Example 13.8, part (a).
b) Using part (a), what would be the population at $T = 3$, corresponding to instantaneous growth and $C = 5$, if $X(0) = 35$?

13.20 a) Show that the difference quotient of the function $f(t) = \ln(t)$ has the form

$$\frac{\Delta \ln(a)}{\Delta t} = \ln[(1 + (h/a)^{(1/h)}].$$

b) Show that

$$\lim_{h\to 0} \frac{\Delta \ln(a)}{\Delta t} = \frac{1}{a}$$

by first setting $\varepsilon = h/a$ and using the limit (13.6).
c) Using the parts (a) and (b), what is the equation describing a population whose instantaneous growth rate is the reciprocal of the population's size?

SECTION 14

THE DERIVATIVE

14.1 INSTANTANEOUS RATE OF CHANGE AND THE TANGENT LINE

In Section 13, the numerical value of a difference quotient was associated with two concepts: the average rate of change of a function over an interval, and the slope of a secant line to the graph of the function. The *average rate of change concept* is useful when modeling populations with distinct generations. However, to model a population that grows continuously, it is necessary to consider the effect of letting the generation length, h, approach zero.

A similar situation arises with the geometrical interpretation of the difference quotient. If we attempt to estimate the actual direction of change of a graph, $y = f(x)$, at a fixed point, $(x_0, f(x_0))$, we may approximate the graph of $y = f(x)$ by a secant line passing through $(x_0, f(x_0))$. An intuitive concept of the actual direction of the graph at $(x_0, f(x_0))$ is the direction in which a vehicle would travel if it were speeding along the graph in the positive direction of the x-axis and, at the point $(x_0, f(x_0))$, lost the ability to steer or turn. The resulting direction of the vehicle would not be that of the secant line associated with an increment Δx. The slope of the line traveled would, however, be better and better approximated by the slope $\Delta f(x_0)/\Delta x$ of the secant line as the increment Δx approaches zero. This is illustrated in Fig. 14.1, where the line l_1 associated with an increment Δx_1 is a poor approximation of the function, $y = f(x)$. The line l_2 associated with the smaller increment Δx_2 is a much better approximation of the graph of $y = f(x)$ for x near x_0. If we considered smaller and smaller increments Δx, the corresponding secant lines would approach a single line, called the *tangent line*. The slope of the

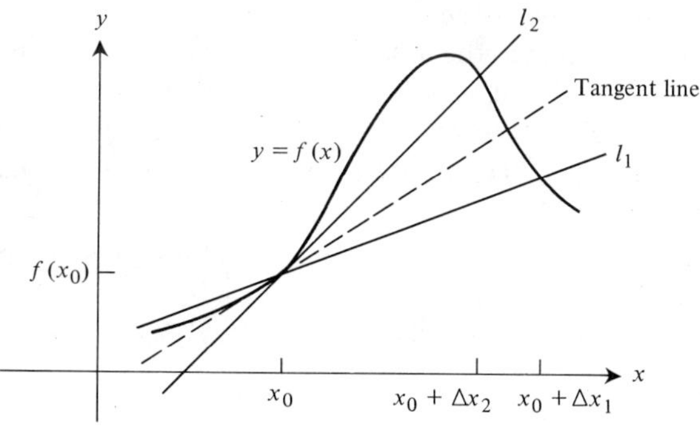

Figure 14.1

tangent line would then be taken as the limit, as $\Delta x \to 0$, of the slopes of the associated secant lines.

The above models or approximations, which utilize the difference quotient $\Delta f(x)/\Delta x$, lead naturally to examining the limit of the difference quotient $\Delta f/\Delta x$ as Δx approaches zero:

$$\lim_{\Delta x \to 0} \frac{\Delta f(x_0)}{\Delta x} = \lim_{h \to 0} \frac{f(x_0 + h) - f(x_0)}{h}. \tag{14.1}$$

When the limit (14.1) exists, it is a real number that is determined by the function f and the fixed number x_0. In reference to the two models discussed, this limiting value is given two interpretations.

Definition 14.1 If $f(x)$ is a function defined in a neighborhood of the number $x = x_0$, then the *instantaneous rate of change of f at x_0* is

$$\lim_{h \to 0} \frac{f(x_0 + h) - f(x_0)}{h}, \qquad \textbf{Instantaneous rate of change}$$

if the limit exists.

Definition 14.2 If $f(x)$ is a function defined in a neighborhood of the number $x = x_0$, then the *tangent line to the graph* of $y = f(x)$ at point $(x_0, f(x_0))$ is the line

$$y = m(x - x_0) + f(x_0), \qquad \textbf{Tangent line}$$

with slope

$$m = \lim_{h \to 0} \frac{f(x_0 + h) - f(x_0)}{h}, \qquad \textbf{Slope of tangent line}$$

if it exists. If the limit is $\pm\infty$ and f is continuous at x_0, then the graph will have a vertical tangent, $x = x_0$, at $(x_0, f(x_0))$. If the limit does not exist, then the graph does not have a tangent line at the point $(x_0, f(x_0))$.

In Section 13, we described the difference quotient as an average rate of change over an interval. Letting $h \to 0$, we know that the length of the interval over which the average is computed goes to zero. In the limit, the quotient represents the rate of change at the point x_0. Hence the term "instantaneous rate of change" is introduced.

The difference quotient is the slope of the secant line through $(x_0, f(x_0))$, corresponding to the increment $\Delta x = h$. Thus, letting $h \to 0$ in Definition 14.2 corresponds to defining the tangent line at $(x_0, f(x_0))$ as the limit of the secant lines.

Example 14.1

a) A tumor is estimated to have a total mass of $6 \times 10^{-3} \cdot t^2$ gm, t days after its discovery. How fast is the tumor growing on the eighth day? *Solution*: The instantaneous rate of growth (IRG) on the eighth day is given by Definition 14.1, with $f(t) = 6 \cdot 10^{-3}t^2$.

$$\begin{aligned}
\text{IRG} &= \lim_{h\to 0} \frac{f(8+h) - f(8)}{h} \\
&= \lim_{h\to 0} \frac{6\cdot 10^{-3}(8+h)^2 - 6\cdot 10^{-3}(8)^2}{h} \\
&= \lim_{h\to 0} \frac{6\cdot 10^{-3}[8^2 + 16h + h^2 - 8^2]}{h} \\
&= \lim_{h\to 0} \frac{6\cdot 10^{-3}[16+h]h}{h} \\
&= \lim_{h\to 0} 6\cdot 10^{-3}[16+h] \\
&= 9.6\cdot 10^{-2} \text{ gm/day.}
\end{aligned}$$

b) What is the equation of the tangent line to the graph of $y = (x-2)(x-4)$ at the point $(3, -1)$? *Solution*: the tangent line is the line $y = m(x-3) - 1$, where the slope m is defined by Definition 14.2, with $x_0 = 3$ as

$$\begin{aligned}
m &= \lim_{h\to 0} \frac{y(3+h) - y(3)}{h} \\
&= \lim_{h\to 0} \frac{(3+h-2)(3+h-4) - (3-2)(3-4)}{h} \\
&= \lim_{h\to 0} \frac{(h+1)(h-1)+1}{h} \\
&= \lim_{h\to 0} \frac{h^2 - 1 + 1}{h} = \lim_{h\to 0} h = 0.
\end{aligned}$$

Therefore the tangent to the graph at the point $(3, -1)$ is a horizontal line, $y = -1$.

If we repeat the above calculations with $x_0 = 2$, we obtain the slope of the tangent line to the graph of f at the point $(2, 0)$:

$$\begin{aligned}
\text{Slope } m &= \lim_{h\to 0} \frac{y(2+h) - y(2)}{h} \\
&= \lim_{h\to 0} \frac{(2+h-2)(2+h-4) - (2-2)(2-4)}{h} \\
&= \lim_{h\to 0} \frac{h(h-2)}{h} \\
&= \lim_{h\to 0} h - 2 = -2.
\end{aligned}$$

The tangent line is $y = -2(x-2)$. These lines are sketched in Fig. 14.2.

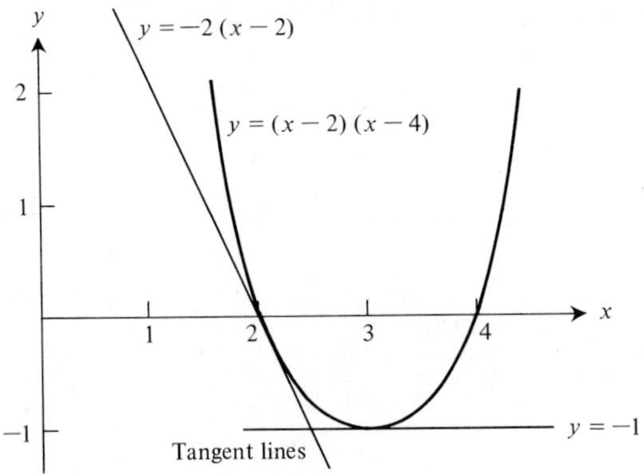

Figure 14.2

◀

To compute the limits in Example 14.1 we could not simply apply the quotient rule of limits, Theorem 12.6. This is because the denominator h approaches the limit zero. Consequently, to evaluate the limits, we employed the same technique:

1. First we evaluated the numerator, $f(x_0 + h) - f(x_0)$.
2. Then we manipulated the numerator algebraically until we had factored a term h.
3. We then had a quotient of the form:

$$\frac{h(\text{terms involving } x \text{ and } h)}{h}.$$

Consequently, we cancelled the h's and evaluated the limit of the terms left in the numerator.

In evaluating such limits, step (2) can present difficulties. For simple functions, we will be able to perform the necessary manipulations. For most functions, the algebraic manipulations will be beyond our capabilities. Indeed, if $f(x) = e^x$ or $f(x)$ is a trigonometric function, it is impossible to "factor out an h and cancel." To evaluate the limit of a difference quotient involving such functions requires careful consideration using basic properties of the function rather than complicated algebra. To evaluate the limit for more complicated functions, either we algebraically reduce them to forms that only involve the elementary functions or we simply rely on theorems that state the necessary results without proof.

Example 14.2 In Example 13.6, a boat crossing a river with a swift current was always assumed to head directly for a fixed point, B, on the opposite shore. The boat, therefore, "heads" along the secant line, passing through the point B and its present position. But, due to the current, the direction in which it actually

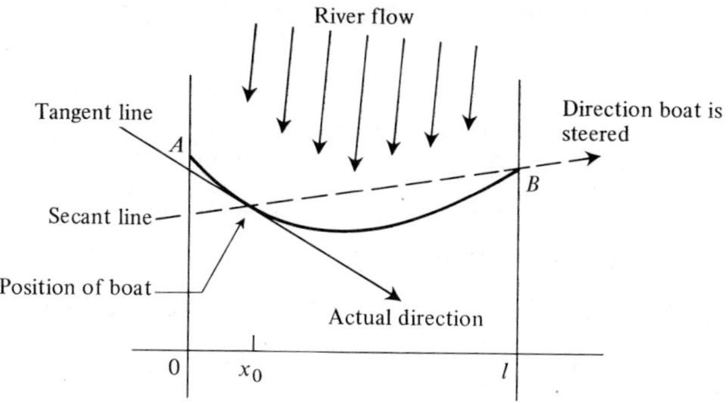

Figure 14.3

travels is the direction of the tangent line to its path. Let its path be given by $y = f(x)$. When the boat is x_0 metres from the shore, the actual direction the boat is traveling is given by the slope of the tangent line through $(x_0, f(x_0))$:

$$\lim_{h \to 0} \frac{f(x_0 + h) - f(x_0)}{h}.$$

The tangent line and actual direction of the boat are indicated in Fig. 14.3.

◀

Example 14.3 Assume that an object moves along a tract and the distance of the object from some reference point at time t is given by a function $f(t)$. Then the *velocity* of the object at time t_0 is defined to be the instantaneous rate of change of f at t_0:

$$\text{Velocity} = V(t) = \lim_{h \to 0} \frac{f(t + h) - f(t)}{h}.$$

The velocity may be positive or negative, depending on the orientation of the reference point. The *speed* at which the object is traveling is the absolute value of the velocity:

$$\text{Speed} = S(t) = |V(t)|.$$

The *acceleration* of the object is defined to be the instantaneous rate of change of the velocity:

$$\text{Acceleration} = A(t) = \lim_{h \to 0} \frac{V(t + h) - V(t)}{h}.$$

◀

14.2 THE DERIVATIVE

The limit (14.1) is a mathematical concept. For any function f and point x_0, we can evaluate the limit (14.1) if it exists. To give a mathematical name to this limit, independent of any models, we introduce the term "derivative."

Definition 14.3 If the function f is defined in a neighborhood of $x = x_0$, *the derivative of f at* x_0 is the limit,

$$\boxed{\frac{d}{dx}f(x_0) \equiv \lim_{h\to 0} \frac{f(x_0 + h) - f(x_0)}{h}, \text{ if it exists.}}$$

If the limit exists, f is said to be *differentiable* at x_0.

If the derivative of f at x_0 exists, it is denoted by the following equivalent symbols:

$f'(x_0)$, read "f prime of x_0";

$\dfrac{d}{dx}f(x_0)$, read "the derivative of f with respect to the variable x at the point x_0";

$\left.\dfrac{df}{dx}\right|_{x=x_0}$, which is read the same as $\dfrac{d}{dx}f(x_0)$;

$D_x f(x_0)$, which is read the same as $\dfrac{d}{dx}f(x_0)$.

When the variable is a time variable, t, another alternative form for the derivative is the symbol

$$\mathring{f}(t_0) = \frac{d}{dt}f(t_0),$$

where the dot or small circle over the function indicates the derivative of f with respect to time.

Example 14.4 The derivative of the function $f(x) = x^2 - x$ is computed at the values $x = 0$, $x = 3$, and $x = x_0$.

a) $$f'(0) = \lim_{h\to 0} \frac{f(0 + h) - f(0)}{h} = \lim_{h\to 0} \frac{h^2 - h}{h}$$

$$= \lim_{h\to 0} (h - 1) = -1$$

b) $$\frac{df(3)}{dx} = \lim_{h\to 0} \frac{f(3 + h) - f(3)}{h}$$

$$= \lim_{h\to 0} \frac{[(3 + h)^2 - (3 + h)] - [9 - 3]}{h}$$

$$= \lim_{h\to 0} \frac{9 + 6h + h^2 - 3 - h - 6}{h}$$

$$= \lim_{h\to 0} \frac{5h + h^2}{h} = \lim_{h\to 0} (5 + h) = 5$$

c) $D_x(x_0) = \lim_{h \to 0} \dfrac{f(x_0 + h) - f(x_0)}{h}$

$$= \lim_{h \to 0} \frac{[(x_0 + h)^2 - (x_0 + h)] - [x_0{}^2 - x_0]}{h}$$

$$= \lim_{h \to 0} \frac{(x_0{}^2 + 2x_0 h + h^2) - (x_0 + h) - (x_0{}^2 - x_0)}{h}$$

$$= \lim_{h \to 0} \frac{(2x_0 - 1)h + h^2}{h}$$

$$= \lim_{h \to 0}[(2x_0 - 1) + h] = 2x_0 - 1$$

◀

In Example 14.4, part (c), we computed the derivative $D_x(x_0)$ at an unspecified value x_0. The resulting equation is $f'(x_0) = 2x_0 - 1$. The subscript zero is only added to the x symbol to emphasize that the value of x is held constant throughout the limit evaluation process. If the actual value of x_0 is not crucial, we can simply forget about the subscript and, by doing so, we obtain the formula $f'(x) = 2x - 1$. This defines a new function, called *the derivative of f*. Having a formula for the derivative of a function, it is unnecessary for us to repeatedly evaluate limits to determine the derivative of the function at specific points. Had we first obtained the formula, $f'(x) = 2x - 1$, in part (c) of Example 14.4, the limit calculation of parts (a) and (b) would not have been necessary. We would simply evaluate $f'(x)$ at $x = 0$ and $x = 3$ to obtain

$$f'(0) = 2 \cdot 0 - 1 = -1 \qquad \text{and} \qquad f'(3) = 2 \cdot 3 - 1 = 5.$$

The process of taking a derivative of a function can be visualized as performing an operation on the function. Recall that the operations of multiplication, addition, and composition are called *binary operations*, since they required two functions to form a third function. An operation that transforms a single given function into another function is called a *transformation*. The capital D notation for denoting the derivative is convenient for describing the derivative as an "operator." Symbolically, we can represent the differentiation operation by

$$D_x(f) \longrightarrow f'(x).$$

The differential operator, D_x, transforms a function $f(x)$, which is *differentiable* into a new function $f'(x)$, called its *derivative*.

Definition 14.4 A function $F(x)$ is *differentiable* on an open interval (a, b) if at each point $x_0 \in (a, b)$, the derivative of F at x_0, $F'(x_0)$ exists.

Definition 14.5 Given a function F that is differentiable on an interval (a, b), the function f is called the *derivative of F* on (a, b), if for each $x \in (a, b)$:

$$D_x F(x) = f(x).$$

The function $f(x) = x^2 - x$ considered in Example 14.4 is differentiable on every interval (a, b), since its derivative, $f'(x) = 2x - 1$, is defined for all real numbers, x.

The derivative of a constant function is zero. This should be obvious when we visualize the derivative as the slope of the tangent line. If $f(x) = k$, a constant, then the graph of f is the horizontal line $y = k$. The tangent line is simply the same as the graph of f, a horizontal line with slope zero. Algebraically, we may compute the derivative as follows. If $f(x) = k$ for all x, then $f(x + h) = k$ also, and

$$f'(x) = \lim_{h\to 0} \frac{f(x + h) - f(x)}{h} = \lim_{h\to 0} \frac{k - k}{h}$$

$$= \lim_{h\to 0} \frac{0}{h} = \lim_{h\to 0} 0 = 0.$$

Thus,

$$\boxed{\frac{dk}{dx} = 0, \text{ for any constant } k.}$$

The polynomial functions are the simplest elementary functions to graph. Consequently, the first functions studied in algebra and calculus are usually the polynomials. To compute the derivative of a polynomial function, we utilize the fact that polynomials are linear combinations of power functions. Thus we first need to develop a derivative formula for the power function $f(x) = x^r$. We will derive the derivative when $r = 1$, 2, and 3 and, from these examples, we will extrapolate to a general formula without proof.

$\underline{r = 1}$ The derivative of $f(x) = x$:

$$\boxed{\frac{dx}{dx} = 1}$$

$$f'(x) = \lim_{h\to 0} \frac{f(x + h) - f(x)}{h}$$

$$= \lim_{h\to 0} \frac{(x + h) - x}{h}$$

$$= \lim_{h\to 0} \frac{h}{h} = \lim_{h\to 0} 1 = 1.$$

$\underline{r = 2}$ The derivative of $f(x) = x^2$:

$$\boxed{\frac{dx^2}{dx} = 2x}$$

$$f'(x) = \lim_{h\to 0} \frac{f(x + h) - f(x)}{h}$$

$$= \lim_{h\to 0} \frac{(x + h)^2 - x^2}{h} = \lim_{h\to 0} \frac{x^2 + 2xh + h^2 - x^2}{h}$$

$$= \lim_{h\to 0} \frac{2xh + h^2}{h} = \lim_{h\to 0} \frac{(2x + h)h}{h}$$

$$= \lim_{h\to 0} (2x + h) = 2x.$$

$\underline{r = 3}$ The derivative of $f(x) = x^3$:

$$\boxed{\frac{dx^3}{dx} = 3x^2}$$

$$\begin{aligned} f'(x) &= \lim_{h\to 0} \frac{f(x+h) - f(x)}{h} \\ &= \lim_{h\to 0} \frac{(x+h)^3 - x^3}{h} \\ &= \lim_{h\to 0} \frac{x^3 + 3x^2h + 3xh^2 + h^3 - x^3}{h} \\ &= \lim_{h\to 0} 3x^2 + 3xh + h^2 = 3x^2. \end{aligned}$$

The pattern exhibited by the derivatives of $y = x$, $y = x^2$, and $y = x^3$ suggests the following *power formula.*

Theorem 14.1 The derivative of the power function $f(x) = x^r$, r a real number, is given by

$$\boxed{D_x(x^r) = rx^{r-1} \qquad \textbf{Power rule}} \tag{14.2}$$

for all x, such that x^{r-1} is defined.

Example 14.5 The following illustrate the power rule (14.2) for computing derivatives.

If $f(x) = x^5$, $f'(x) = 5x^4$.

If $f(x) = x^{(3/2)}, f'(x) = (\frac{3}{2})x^{(3/2)-1} = (\frac{3}{2})x^{(1/2)}$.

If $f(x) = x^{-2}$, $f'(x) = -2x^{-3}$.

If $f(x) = x^{(2/3)}$, $f'(x) = (\frac{2}{3})x^{((2/3)-1)} = (\frac{2}{3})x^{-(1/3)}$.

If $f(x) = x^3$, $f'(5) = 3(5)^2 = 75$.

If $g(x) = x^{(5/2)}$, $g'(4) = \frac{5}{2}(4)^{(3/2)} = \frac{5}{2}(8) = 20$.

If $f(x) = 1$, $D_x f(x) = D_x(x^0) = 0 \cdot x^{0-1} = 0$. ◀

To evaluate the derivative of a polynomial function, such as $f(x) = 3x^2 - x^5 + 2$, we would like to be able to use the power rule to compute the derivatives of x^2 and x^5 and then combine these to give the derivative of $f(x)$. To do this, we must know how the operation of taking a derivative is related to the operations of addition, multiplication, and composition. The derivative is defined through a limiting process. Consequently, it is not surprising to find that the algebraic properties of the derivative are similar to those of the limit process.

Consider the derivative of the sum of two functions: $D_x[f(x) + g(x)]$. Using Definition 14.3, we find that

$$D_x[f(x) + g(x)] = \lim_{h \to 0}\left[\frac{(f(x+h) + g(x+h)) - (f(x) + g(x))}{h}\right]$$

$$= \lim_{h \to 0}\left[\frac{f(x+h) - f(x) + g(x+h) - g(x)}{h}\right]$$

$$= \lim_{h \to 0}\left[\frac{f(x+h) - f(x)}{h} + \frac{g(x+h) - g(x)}{h}\right]$$

$$(\textit{By Theorem 12.2}) = \lim_{h \to 0}\left[\frac{f(x+h) - f(x)}{h}\right] + \lim_{h \to 0}\left[\frac{g(x+h) - g(x)}{h}\right]$$

$$(\textit{By definition}) \qquad = D_x f(x) + D_x g(x).$$

We state this result as a theorem.

Theorem 14.2 If f and g are differentiable, then so is their sum, $f + g$, and

$$D_x[f(x) + g(x)] = D_x f(x) + D_x g(x) \qquad \textbf{Sum rule.}$$

The derivative of the sum is the sum of the derivatives.

An argument similar to the one given above can be presented to justify commuting the derivative process with *scalar* multiplication (a scalar is a constant).

Theorem 14.3 If c is a constant and f is differentiable, then

$$D_x[c \cdot f(x)] = c \cdot D_x f(x) \qquad \textbf{Constant rule.}$$

The derivative of a constant times a function is the constant times the derivative of the function.

That the derivative of a constant is zero may also be deduced by combining the constant rule and the power rule with $r = 0$, $x_0 = 1$:

$$\frac{dc}{dx} = c\frac{d1}{dx} = c\frac{dx^0}{dx} = c \cdot 0x^{-1} = 0.$$

Thus, for any constant c,

$$\frac{dc}{dx} = 0 \qquad \textbf{The derivative of a constant is zero.}$$

Example 14.6 We illustrate the use of the *constant product* and *sum* rules to simplify the evaluation of derivatives.

a) *Constant product rule*:

If $f(x) = 6x^2$, then $f'(x) = 6D_x(x^2) = 6 \cdot 2x = 12x$.

If $f(x) = -3x^{-1}$, then $D_x f(x) = -3D_x(x^{-1}) = 3x^{-2}$.

If $f(x) = \frac{1}{2}x^2$, then $f'(5) = \frac{1}{2} \cdot 2(5) = 5$.

b) *Sum rule*:

$$D_x(2x + 5x^{-3}) = D_x(2x) + D_x(5x^{-3}) = 2 - 15x^{-4}.$$

If $f(x) = 3x^2 - 2x + 1$, then

$$f'(x) = D_x(3x^2) + D_x(-2x) + D_x(1) = 6x - 2.$$

$$D_x[10x^5 - 3x^\pi] = 10D_x(x^5) - 3D_x(x^\pi) = 50x^4 - 3\pi x^{\pi-1}.$$ ◀

In the Example 14.6, we have employed a common and very convenient short-cut in notation. If a function is given by an equation, such as $f(x) = x^2 + 2x - 3$, the derivative of the function f is the function $f'(x) = 2x + 2$. However, to convey the same information we may supress the function notation, $f(x)$ and $f'(x)$, and simply write

$$D_x(x^2 + 2x - 3) = 2x + 2.$$

Or, equivalently,

$$\frac{d}{dx}(x^2 + 2x - 3) = 2x + 2.$$

Frequently we encounter functions of the form

$$y = (\text{some algebraic expression in } x),$$

where the variable y is treated as a function of x, $y = y(x)$, and the dependence on x is usually not explicitly noted in the form $y(x)$. In this case we denote the derivative of the function y with respect to x by

$$y', \quad \frac{dy}{dx}, \quad \text{or} \quad D_x(y).$$

For example, if $y = 5x^{-3} + 2x^2$, then y' (read "y prime") is given by $y' = -15x^{-4} + 4x$.

Using the sum rule, Theorem 14.2, and the constant multiple rule, Theorem 14.3, we can take the derivative of any polynomial by taking the derivatives of each term of the polynomial. For instance, if $f(x) = 3x^2 - x^5 + 2$, then

$$\begin{aligned} f'(x) &= 3D_x(x^2) - D_x(x^5) + D_x(2) \\ &= 3 \cdot 2x - 5x^4 + 0. \end{aligned}$$

Theorem 14.4 The derivative of the polynomial function is obtained by differentiating each term individually.

If $f(x) = a_0 + a_1x + a_2x^2 + \cdots + a_nx^n$,
then $f'(x) = a_1 + 2a_2x + 3a_3x^2 + \cdots + na_nx^{n-1}$ **Polynomial rule.**

The verification of the *sum rule* and the *constant multiplication rule* are straightforward. This is because of the algebraic facts:

$$\frac{A+B}{h} = \frac{A}{h} + \frac{B}{h} \quad \text{and} \quad \frac{C \cdot A}{h} = C \cdot \left(\frac{A}{h}\right).$$

When we examine the derivative of a product of two functions, we experience a problem because

$$\frac{A \cdot B}{h} \neq \frac{A}{h}\frac{B}{h}.$$

As a result, we must use a standard mathematical trick to obtain a *product rule* for derivatives. By definition,

$$D_x[f(x) \cdot g(x)] = \lim_{h \to 0} \frac{f(x+h) \cdot g(x+h) - f(x) \cdot g(x)}{h}.$$

We introduce into the numerator the number zero, expressed as

$$0 = -f(x) \cdot g(x+h) + f(x) \cdot g(x+h).$$

This results in the limit

$$\lim_{h \to 0} \frac{f(x+h) \cdot g(x+h) \overbrace{- f(x) \cdot g(x+h) + f(x) \cdot g(x+h)}^{\text{this term is zero}} - f(x) \cdot g(x)}{h}.$$

Factoring, we obtain two terms

$$\lim_{h \to 0} \left\{ \frac{[f(x+h) - f(x)]g(x+h)}{h} + \frac{f(x)[g(x+h) - g(x)]}{h} \right\}.$$

Expressing this as the sum of two limits (Limit Theorem 12.2), we get

$$\lim_{h \to 0} \left\{ \frac{[f(x+h) - f(x)]g(x+h)}{h} \right\} + \lim_{h \to 0} \left\{ \frac{f(x)[g(x+h) - g(x)]}{h} \right\}.$$

Then, expressing the limit of each product as a product of limits (Limit Theorem 12.4), this becomes

$$\left[\lim_{h \to 0} \frac{f(x+h) - f(x)}{h}\right]\left[\lim_{h \to 0} g(x+h)\right] + \left[\lim_{h \to 0} f(x)\right]\left[\lim_{h \to 0} \frac{g(x+h) - g(x)}{h}\right].$$

If g is continuous, then $\lim_{h\to 0} g(x+h) = g(x)$. Assuming that f and g are differentiable, in the limit the above expression becomes

$$[f(x)g(x)]' = f'(x)\cdot g(x) + f(x)\cdot g'(x).$$

We state this result as a theorem.

Theorem 14.5 If f and g are differentiable functions, then so is their product $f \cdot g$ and

$$D_x[f(x)\cdot g(x)] = f(x)\cdot D_x g(x) + g(x)\cdot D_x f(x) \qquad \textbf{Product rule.}$$

The derivative of f times g is f times the derivative of g, plus g times the derivative of f.

Example 14.7 The following derivatives are evaluated using the *Product rule*:

a) $$\begin{aligned} D_x[(3x+1)(2x)] &= (3x+1)D_x(2x) + (2x)D_x(3x+1) \\ &= (3x+1)\cdot 2 + (2x)\cdot 3 \\ &= 12x + 2. \end{aligned}$$

b) $$\begin{aligned} D_x[x^{(3/4)}(2x+5x^{-3})] &= x^{(3/4)}D_x(2x+5x^{-3}) + (2x+5x^{-3})D_x(x^{(3/4)}) \\ &= x^{(3/4)}(2-15x^{-4}) + (2x+5x^{-3})(\tfrac{3}{4}x^{-1/4}) \end{aligned}$$

c) $$\begin{aligned} D_x[(x^2-2)x^{-3}] &= (x^2-2)D_x(x^{-3}) + x^{-3}D_x(x^2-2) \\ &= (x^2-2)(-3x^{-4}) + x^{-3}(2x) \\ &= 6x^{-4} - x^{-2}. \end{aligned}$$

d) $$\begin{aligned} D_x[(x^5-3x^2+2)(x^4+2x)] &= (x^5-3x^2+2)D_x(x^4+2x) \\ &\quad + (x^4+2x)D_x(x^5-3x^2+2) \\ &= (x^5-3x^2+2)(4x^3+2) \\ &\quad + (x^4+2x)(5x^4-6x). \end{aligned}$$

e) For simple polynomial functions, the product rule can be avoided by multiplying the terms and then differentiating:

$$\begin{aligned} D_x[(3x+1)2x] &= D_x(6x^2+2x) = 12x+2; \\ D_x[x^{(3/4)}(2x+5x^{-3})] &= D_x(2x^{(7/4)} + 5x^{-(9/4)}) \\ &= 2\cdot\tfrac{7}{4}\cdot x^{(3/4)} + 5(-\tfrac{9}{4})x^{-(13/14)} \\ &= \tfrac{7}{2}x^{(3/4)} - \tfrac{45}{4}x^{-(13/14)}. \end{aligned}$$

◀

In the heuristic argument used to justify the form of the product rule, we required $g(x)$ to be continuous. The reason that this assumption does not appear in the hypothesis of the product rule, Theorem 14.5, is because all differentiable functions are also continuous functions.

Theorem 14.6 If f is differentiable at a point x_0, then f is continuous at x_0.

That the converse of Theorem 14.6 is not true (i.e., that not all continuous functions are differentiable), is illustrated by the function $f(x) = |x|$. The absolute value function is continuous at $x = 0$, since $f(0) = 0$,

$$\lim_{x \to 0^+} f(x) = \lim_{x \to 0^+} |x| = \lim_{x \to 0^+} x = 0$$

and

$$\lim_{x \to 0^-} f(x) = \lim_{x \to 0^-} |x| = \lim_{x \to 0^-} -x = 0.$$

The absolute value function is not differentiable at $x = 0$. The reason that it is not differentiable is because the difference quotient, $\Delta f/\Delta x$, does not have a limit at $x = 0$:

$$\lim_{h \to 0^+} \frac{f(0 + h) - f(0)}{h} = \lim_{h \to 0^+} \frac{|h| - |0|}{h} = \lim_{h \to 0^-} \frac{h}{h} = 1,$$

whereas

$$\lim_{h \to 0^-} \frac{f(0 + h) - f(0)}{h} = \lim_{h \to 0^-} \frac{|h| - |0|}{h} = \lim_{h \to 0^-} \frac{-h}{h} = -1.$$

Since the left and right limits do not agree, the $\lim_{h \to 0} \Delta f(0)/\Delta x$ does not exist. The geometrical reason why the limit does not exist is because the graph of the absolute value function has a *corner* at $x = 0$. Geometrically, it is impossible to define a tangent line to the graph at $x = 0$. As illustrated in Fig. 14.4, the slope of the graph for positive x is $+1$, while the slope for negative x is -1.

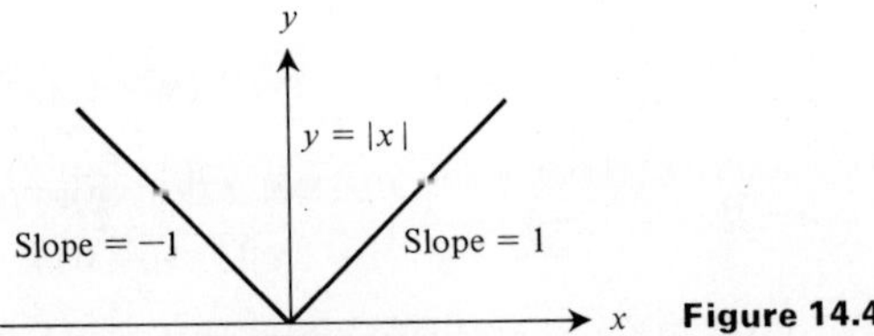

Figure 14.4

To compute the derivative of a rational function—i.e., the ratio of two polynomials, we need to evaluate the derivative of a quotient. The next theorem is known as the *quotient rule*. In Section 15, we will show that the quotient rule is a special case of the product rule.

Theorem 14.7 If f and g are differentiable and $g(x) \neq 0$, then

$$\frac{d}{dx}\left(\frac{f(x)}{g(x)}\right) = \frac{g(x)\dfrac{d}{dx}f(x) - f(x)\dfrac{d}{dx}g(x)}{[g(x)]^2},$$

or simply

$$\left[\frac{f(x)}{g(x)}\right]' = \frac{g(x)f'(x) - f(x)g'(x)}{[g(x)]^2} \qquad \textbf{Quotient rule.}$$

Example 14.8

a) To evaluate $D_x[3x/(x^2 + 1)]$ using the quotient rule, Theorem 14.7, set

$$f(x) = 3x \qquad \text{and} \qquad g(x) = x^2 + 1.$$

Then $f'(x) = 3$ and $g'(x) = 2x$. The formula

$$D_x[f(x)/g(x)] = \frac{g(x)f'(x) - f(x)g'(x)}{[g(x)]^2}$$

becomes

$$D_x[3x/(x^2 + 1)] = \frac{(x^2 + 1)(3) - (3x)(2x)}{(x^2 + 1)^2} = \frac{-3x^2 + 3}{(x^2 + 1)^2}.$$

b) The derivative of

$$y = \frac{3 + 2x}{1 - 5x}$$

is obtained from the quotient with $f(x) = 3 + 2x$, $f'(x) = 2$ and $g(x) = 1 - 5x$, $g'(x) = -5$:

$$y' = \frac{(1 - 5x)(2) - (3 + 2x)(-5)}{(1 - 5x)^2} = \frac{17}{(1 - 5x)^2}.$$

c) To evaluate

$$\frac{d}{dt}\frac{3t^2 + 1}{t - 2t^5},$$

set

$$f(t) = 3t^2 + 1 \qquad \text{and} \qquad g(t) = t - 2t^5.$$

Then $f'(t) = 6t$ and $g'(t) = 1 - 10t^4$ and the quotient rule gives us

$$\frac{d}{dt}\frac{3t^2 + 1}{t - 2t^5} = \frac{(t - 2t^5)(6t) - (3t^2 + 1)(1 - 10t^4)}{(t - 2t^5)^2}$$

$$= \frac{18t^6 + 10t^4 + 3t^2 - 3}{(t - 2t^5)^2}.$$

◀

Example 14.9 The number of forest fires in a particular region can be expressed as a function of the number of days x since the last measurable rainfall. However, the fires fall into two categories: (1) those caused by nature (i.e., lightning) and (2) those attributable to man. Denote the number of fires due to natural causes by $N(x)$ and the number caused by man by $M(x)$. The total number of fires is therefore

$F(x) = N(x) + M(x)$. The proportion of fires that are caused by man is thus $R(x) = M(x)/F(x)$. The rate of change in the relative number of man-related fires is (by the quotient rule)

$$R'(x) = D_x\left[\frac{M(x)}{F(x)}\right] = \frac{F(x)D_xM(x) - M(x)D_xF(x)}{[F(x)]^2}.$$

If we substitute $M'(x) + N'(x)$ for $F'(x)$, this equation becomes

$$R'(x) = \frac{[M(x) + N(x)]M'(x) - M(x)[M'(x) + N'(x)]}{[M(x) + N(x)]^2}.$$

Or, upon simplifying, we find that

$$R'(x) = \frac{N(x)M'(x) - M(x)N'(x)}{[M(x) + N(x)]^2}.$$

If $N(x) = (0.1)(x - 1)$ and $M(x) = (0.4)x^2$, then $F(x) = 0.4x^2 + 0.1x - 0.1$. The rate of change in the proportion of man related fires is thus

$$R'(x) = \frac{[(0.1)(x - 1)\cdot 0.8x] - [0.4x^2 \cdot 0.1]}{[F(x)]^2}$$

$$= \frac{(0.04)[x^2 - 2x]}{[0.4x^2 + 0.1x - 0.1]^2}.$$

◀

Example 14.10 In cold regions, weather forecasts frequently include a *wind-chill factor*. This is a term used to convey additional cooling due to a more rapid heat exchange when the wind is blowing. The *wind-chill index*, Q_H, is defined by

$$Q_H = (10\sqrt{u} + 10.45 - u)(33 - T),$$

where u is the average wind speed in metres per second, and T is the temperature in degrees Celsius. The number 33 is the temperature of normal human skin. For instance, a temperature of $T = -18°\text{C}$ combined with a wind speed of 22.3 m/sec (about 50 mph) would have a wind-chill index of

$$Q_H = (10\sqrt{22.3} + 10.45 - 22.3)(33 - (-18) \simeq 1804.$$

The *wind-chill corrected temperature* that is usually given in a weather forecast is defined to be the temperature, T_c, which would give the same wind-chill index with a wind speed of 2.2 m/sec (5 mph). (The wind speed of 2.2 m/sec is arbitrary and corresponds to a normal, calm day.) Thus for the above conditions we solve

$$Q_H = (10\sqrt{u} + 10.45 - u)(33 - T_c) = 1804$$

for T_c with $u = 2.2$.

$$1804 = (10\sqrt{2.2} + 10.45 - 2.2)(33 - T_c),$$

which leads to the corrected temperature $T_c \simeq -45°C$. To determine how the wind-chill index Q_H changes with respect to changes in wind velocity u, at a fixed temperature T, we compute

$$\frac{dQ_H}{du} = D_u[(10\sqrt{u} + 10.45 - u)(33 - T)]$$

$$= (5u^{-(1/2)} - 1)(33 - T).$$

The term $(33 - T)$ is treated as a constant! Consequently the wind-chill index increases at a rate that depends on the reciprocal of $\sqrt{u}$. ◀

14.3 THE DERIVATIVE OF THE TRIGONOMETRIC FUNCTIONS

In this subsection, we present (without proof) the derivatives of the trigonometric functions. First, we investigate one particular derivative, $D_x \sin(x)$, at the point $x_0 = 0$. This derivative will hinge on the limit as $h \to 0$ of $\sin(h)/h$. This limit forms the basis upon which the derivatives of all the trigonometric functions can be evaluated. Since it is frequently encountered in other problems, its evaluation is of independent interest.

To evaluate the limit,

$$\lim_{h \to 0} \frac{\sin(h)}{h},$$

we cannot simply "factor an h out of the numerator $\sin(h)$ and then cancel." Instead, we use a geometrical argument. By definition, $\sin(h)$ is the y-coordinate on the circle $x^2 + y^2 = 1$, which is h units along the arc of the circle from (0, 1). We assume that $h > 0$ to make the argument simple. In Fig. 14.5, we have sketched a portion of the circle $x^2 + y^2 = 1$, and superimposed the triangles OAB and OAC.

The triangles OAB and OAC are similar triangles since they are both right triangles and have the common angle at the origin. Consequently, the ratios of the

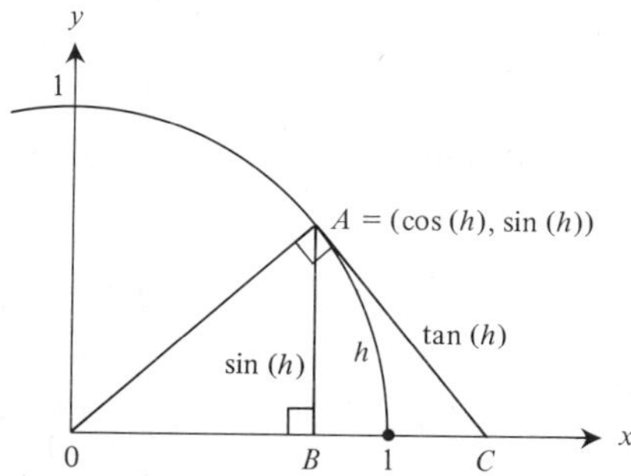

Figure 14.5

lengths of their sides have the same values:

$$\frac{BA}{OB} = \frac{CA}{AO}.$$

The length of OA is 1, since it is a radius of the unit circle. The length $BA = \sin(h)$, and $OB = \cos(h)$. Hence the length

$$CA = \tan(h).$$

From the diagram, we see geometrically that h is the length of the arc from (0, 1) to $A = (\cos(h), \sin(h))$. Consequently, the number h is between the lengths of the line segments BA and CA. Algebraically, this is stated by the inequality

$$\sin(h) < h < \tan(h).$$

Dividing by $\sin(h)$, we have

$$1 < \frac{h}{\sin(h)} < \frac{1}{\cos(h)} \qquad \text{or} \qquad \cos(h) < \frac{\sin(h)}{h} < 1.$$

As $\cos(x)$ is a continuous function, we know that $\lim_{h\to 0} \cos(h) = \cos(0) = 1$. Hence $\sin(h)/h$ is between 1 and a number approaching 1. Taking the limit as $h \to 0$, we obtain

$$\lim_{h\to 0} \frac{\sin(h)}{h} = 1. \tag{14.3}$$

We can compute the derivative of $\sin(x)$ at $x = 0$ using equation (14.3). From the Definition 14.3 of the derivative, we have

$$\begin{aligned} \frac{d}{dx}\sin(0) &= \lim_{h\to 0} \frac{\sin(h) - \sin(0)}{h} \\ &= \lim_{h\to 0} \frac{\sin(h)}{h} = 1. \end{aligned}$$

If we try to compute $D_x \sin(x)$ at an arbitrary value x, we are confronted with the limit

$$\frac{d}{dx}\sin(x) = \lim_{h\to 0} \frac{\sin(x + h) - \sin(x)}{h}$$

We cannot factor out an h term from the numerator of this limit. However, if trigonometric relations beyond those required in this text are used, the difference quotient can be rearranged so that its limit can be evaluated. The resulting limit will have the form

$$\frac{d}{dx}\sin(x) = \lim_{h\to 0}[\cos(x + \tfrac{1}{2}h)] \cdot \lim_{h\to 0}\left[\frac{\sin(\frac{1}{2}h)}{\frac{1}{2}h}\right].$$

Letting $h_0 = \frac{1}{2}h$, we find that this is equivalent to

$$\frac{d}{dx}\sin(x) = \lim_{h_0\to 0}[\cos(x + h_0)] \cdot \lim_{h_0\to 0} \frac{\sin(h_0)}{h_0}.$$

The first limit is $\cos(x)$, since the cosine function is continuous. By (14.3), the second limit is 1. Hence,

$$D_x \sin(x) = \cos(x).$$

The above arguments are not mathematically rigorous. They are only intended to provide the heuristic introduction to the derivatives of trigonometric functions so that the following theorem might not appear completely arbitrary.

Theorem 14.8 The trigonometric functions are differentiable at each point x for which they are defined. Their derivatives are given as follows.

$$\begin{aligned} D_x \sin(x) &= \cos(x), \\ D_x \cos(x) &= -\sin(x), \\ D_x \tan(x) &= \sec^2(x), \\ D_x \cot(x) &= -\csc^2(x), \\ D_x \sec(x) &= \sec(x)\tan(x), \\ D_x \csc(x) &= -\csc(x)\cot(x). \end{aligned}$$

Observe that the derivatives of $\cos(x)$, $\cot(x)$, and $\csc(x)$ have a negative sign as a part of their formula. The omission of this negative sign is one of the most common errors in computing derivatives. Here, the rule is simple: if the function begins with *co*, there is a negative sign in its derivative; otherwise, there is not a negative sign.

Example 14.11 We use the formula of Theorem 14.8 to evaluate the following derivatives.

a) If $f(x) = \sin(x) + \cos(x)$, then, by the sum rule,

$$f'(x) = \frac{d}{dx}\sin(x) + \frac{d}{dx}\cos(x) = \cos(x) - \sin(x).$$

b) If $f(x) = 3\tan(x) + \cot(x)$, then, by the sum rule,

$$D_x f(x) = 3D_x \tan(x) + D_x \cot(x) = 3\sec^2(x) - \csc^2(x).$$

c) If $f(x) = x \cdot \sin(x)$, then $f'(x) = x \cdot D_x \sin(x) + \sin(x)D_x(x)$:

$$f'(x) = x \cdot \cos(x) + \sin(x).$$

d) If $f(x) = \cos(x)/(x^2 + 1)$, then, by the quotient rule,

$$\begin{aligned} f'(x) &= \frac{(x^2+1) \cdot D_x \cos(x) - \cos(x) \cdot D_x(x^2+1)}{(x^2+1)^2} \\ &= \frac{-\sin(x) \cdot (x^2+1) - \cos(x) \cdot 2x}{(x^2+1)^2}. \end{aligned}$$

e) If $g(x) = \sin(x)/\cos(x)$ then, by the quotient rule,

$$g'(x) = \frac{\cos(x) \cdot D_x \sin(x) - \sin(x) \cdot D_x \cos(x)}{\cos^2(x)}$$

$$= \frac{\cos(x) \cdot \cos(x) - \sin(x) \cdot (-\sin(x))}{\cos^2(x)}$$

$$= \frac{\cos^2(x) + \sin^2(x)}{\cos^2(x)} = \frac{1}{\cos^2(x)} = \sec^2(x).$$

But $g(x) = \tan(x)$, hence we have developed the formula $D_x \tan(x) = \sec^2(x)$ from the formulas for $D_x \cos(x)$ and $D_x \sin(x)$.

f) $D_x 3[\csc(x)] = 3D_x \csc(x) = -3 \csc(x) \cot(x)$. ◀

In the above formulas, we used x to denote the independent variable. When dealing with trigonometric functions, the most common symbol for the radian measure is θ (theta). Consequently, in applications, the equations are more likely to look like

$$D_\theta \sin(\theta) = \cos(\theta) \qquad \text{or} \qquad F(\theta) = \tfrac{1}{2}\tan(\theta),\ F'(\theta) = \tfrac{1}{2}\sec^2(\theta).$$

Example 14.12 Galileo described a simple pendulum's motion in terms of the deviations of the pendulum arm from the vertical position. (See Fig. 14.6.) Denote the angle determined by the pendulum's position and the vertical position by θ radians. Gravity exerts a force acting on the mass of the pendulum in the direction of the tangent to the path of the pendulum. This force is

$$F = -M \cdot g \sin(\theta),$$

where g denotes a universal gravity constant. Due to this force, the angular acceleration of the pendulum is found to be

$$A = -g/l \sin(\theta),$$

where the l is the length of the pendulum. The instantaneous rate of change in the angular acceleration of the pendulum is, consequently,

$$\frac{d}{d\theta} A(\theta) = -g/l \cos(\theta).$$

These equations treat the movement of the pendulum as a function of the angular displacement θ. It is more practical to consider the angular displacement θ as a function of time, $\theta(t)$, and to describe the movements of the pendulum as a function of time. This would introduce the composition of two functions $\theta(t)$ and $A(\theta)$. To describe the change in A as a function of time, we would have to compute

$$D_t A(\theta(t)).$$

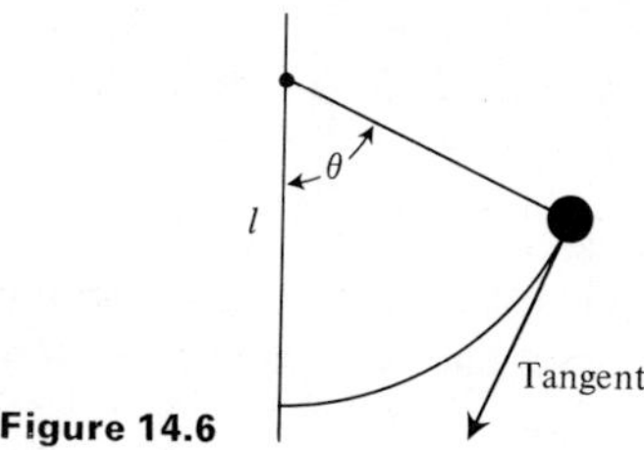

Figure 14.6

If $\theta(t) = t^2$, this would be the derivative

$$D_t[-g/l \sin(t^2)].$$

This derivative cannot be obtained using the formula of Theorem 14.8, since the variable of integration, t, is different from the argument t^2 of the sine function. In Section 15, we introduce a technique for computing such derivatives. ◀

14.4 THE DERIVATIVE OF e^x AND $\ln(x)$

In Section 13.3, we examined the description of a population, X, whose instantaneous rate of growth at any time t is proportional to the size of the population, $X(t)$. Using the derivative $X'(t)$ to represent the instantaneous rate of growth, we describe the model by the equation

$$X'(t) = cX(t). \tag{14.4}$$

Using a discrete model of the population X, we obtained a description of the population in the form (equation (13.4))

$$(\text{Size at generation } n)\ X_n = X_0(1 + ch)^n.$$

However, by expressing X_n as $X(T)$, where $T = nh$, the model became (equation (13.5))

$$X(T) = X_0[(1 + ch)^{(1/h)}]^T.$$

Setting $\varepsilon = ch$ and hence $1/h = (1/\varepsilon)c$, we find that this becomes

$$X(T) = X_0[(1 + \varepsilon)^{(1/\varepsilon)}]^{cT}.$$

As h approaches zero, ε must also approach zero. Thus the instantaneous-growth model is given by

$$X(T) = X_0\left(\lim_{\varepsilon \to 0}(1 + \varepsilon)^{(1/\varepsilon)}\right)^{cT}.$$

Accepting the fact that

$$\lim_{\varepsilon \to 0}(1 + \varepsilon)^{(1/\varepsilon)} = e, \qquad \text{the base of the natural logarithm,} \tag{14.5}$$

we find the function X to be (equation (13.7))

$$X(T) = X_0 e^{cT}. \tag{14.6}$$

Review Section 5 if you are not familiar with or have forgotten the properties of e^x and $\ln(x)$.

The conclusion to be drawn from the above discussion is that the derivative of the exponential function, $X_0 e^{cT}$, satisfies the equation (14.4). With $T = t$, this means that

$$D_t(X_0 e^{ct}) = cX_0 e^{ct}.$$

Using the constant rule, we can factor the constant X_0 past the derivative process;

$$X_0 D_t e^{ct} = cX_0 e^{ct},$$

and, upon canceling the X_0 terms, we obtain

$$D_t e^{ct} = ce^{ct}. \tag{14.7}$$

When $c = 1$, equation (14.6) gives the derivative of the exponential function, $y = e^t$. The above argument was not completely rigorous; it required the use of the unverified limit (14.5) and, in fact, worked from the derivative back to the function. In Section 23, the derivative of $y = e^t$ will be verified directly; but, for the present, we will use the above discussion as a heuristic justification of the following theorem.

Theorem 14.9 The derivative of the exponential function, $y = e^x$ is $y' = e^x$.

$$\frac{d}{dx} e^x = e^x \qquad \textbf{Derivative of } e^x. \tag{14.8}$$

Example 14.13 The following derivatives are evaluated by using equations (14.7) and (14.8).

a) $D_x e^{3x} = 3e^{3x} \, (c = 3)$

b) $D_x e^{-x} = (-1)e^{-x} \, (c = -1)$

c) $D_x[x^2 e^x] = x^2 D_x e^x + e^x D_x x^2$

$$= x^2 \cdot e^x + e^x \cdot 2x = (x^2 + 2x)e^x$$

d) $D_x[\cos(x)e^{2x}] = \cos(x)D_x e^{2x} + e^{2x} D_x \cos(x)$

$$= \cos(x) \cdot 2e^{2x} + e^{2x}(-\sin(x))$$

$$= [2\cos(x) - \sin(x)]e^{2x}$$

e) $D_x[e^{-2x} - e^{-10x}] = D_x e^{-2x} - D_x e^{-10x}$

$$= -2e^{-2x} - (-10e^{-10x})$$

$$= -2e^{-2x} + 10e^{-10x}$$

◀

The derivative of the natural logarithm function, $f(x) = \ln(x)$, can be evaluated using the definition

$$\frac{d}{dx}\ln(x) = \lim_{h\to 0}\frac{\ln(x+h) - \ln(x)}{h},$$

if we accept the limit (14.5). The method of evaluating this derivative is to rewrite the difference quotient as follows:

$$\begin{aligned}
\frac{\Delta \ln(x)}{\Delta x} &= \frac{\ln(x+h) - \ln(x)}{h} \\
&= \frac{\ln[(x+h)/x]}{h} && \left(\text{Since } \ln(A) - \ln(B) = \ln\left(\frac{A}{B}\right)\right) \\
&= \frac{1}{h}\ln\left(1 + \frac{h}{x}\right) && \text{(Algebra)} \\
&= \ln\left[\left(1 + \frac{h}{x}\right)^{1/h}\right] && (\text{Since } a\ln(b) = \ln(b^a)) \\
&= \ln[(1+\varepsilon)^{(1/\varepsilon)\cdot(1/x)}] \ . && \left(\text{Setting } \varepsilon = \frac{h}{x}; \quad h = \varepsilon x; \therefore \frac{1}{h} = \frac{1}{\varepsilon}\cdot\frac{1}{x}\right). \\
&= \frac{1}{x}\ln[(1+\varepsilon)^{1/\varepsilon}].
\end{aligned}$$

Hence, as $h \to 0$, the variable $\varepsilon \to 0$, and

$$\frac{d}{dx}\ln(x) = \lim_{\varepsilon\to 0}\frac{1}{x}\ln[(1+\varepsilon)^{(1/\varepsilon)}].$$

Since $\ln(x)$ is a continuous function, we can evaluate the limit as

$$\frac{d}{dx}\ln(x) = \frac{1}{x}\ln\left[\lim_{\varepsilon\to 0}(1+\varepsilon)^{1/\varepsilon}\right].$$

Using the limit (14.5), we find that this becomes

$$\frac{d}{dx}\ln(x) = \frac{1}{x}\ln(e) = \frac{1}{x},$$

since, by definition, $\ln(e) = 1$.

Again the above argument required the use of the limit (14.5), which has not been substantiated. The same result, that $D_x \ln(x) = 1/x$, will be verified in Section 23, where we will present a definition of the natural logarithm function and the number e without using the limit (14.5). The verification of the limit (14.5) will be given in Section 25, but it will depend on knowing in advance the derivative of $\ln(x)$. For the present, we simply state the derivative formula for $\ln(x)$.

Theorem 14.10 The derivative of the natural logarithm function $\ln(x)$ is

$$\frac{d}{dx}\ln(x) = \frac{1}{x} \qquad \textbf{Derivative of } \ln(x). \tag{14.9}$$

Note: x must be positive.

Example 14.14 The derivative formula (14.9), combined with the sum and constant rules, enables us to evaluate what appears to be more complicated derivatives.

a) $D_x \ln(x^3)$ can be evaluated using the property that $\ln(x^r) = r\ln(x)$. We have

$$D_x \ln(x^3) = D_x[3\ln(x)] = 3D_x \ln(x) = 3\cdot\frac{1}{x}.$$

b) $D_x \ln(\pi x)$ is evaluated by using the property that $\ln(A\cdot B) = \ln(A) + \ln(B)$;

$$\begin{aligned} D_x \ln(\pi x) &= D_x[\ln(\pi) + \ln(x)] \\ &= D_x \ln(\pi) + D_x \ln(x). \end{aligned}$$

As $\ln(\pi)$ is a constant (approximately 1.15), $D_x \ln(\pi) = 0$; and, by (14.9), $D_x \ln(x) = 1/x$. Thus

$$D_x \ln(\pi x) = 0 + \frac{1}{x} = \frac{1}{x}.$$

c)
$$\begin{aligned} D_x(x^3 \ln(x)) &= x^3 D_x \ln(x) + \ln(x) D_x x^3 \\ &= x^3\cdot\frac{1}{x} + \ln(x)\cdot 3x^2 \\ &= x^2[1 + 3\ln(x)]. \end{aligned}$$

◀

SUMMARY

In this Section, we have introduced a considerable number of rules and formulas for computing derivatives. We list these again using more abbreviated notation.

1. *Definitions*: The three main definitions of this section each refer to the same limit. We list them together to emphasize that they result in different names for the same concept.

If

$$L = \lim_{h\to 0}\frac{f(x_0 + h) - f(x_0)}{h}$$

exists, then the following applies.

L is the *instantaneous rate of change* of f at x_0, shortened to *the rate of change of f at x_0* (Definition 14.1).

L is the *slope of the tangent line* to the graph of f at $(x_0, f(x_0))$ (Definition 14.2).

L is the *derivative of f* at x_0 (Definition 14.3).

2. *Notations*: The following symbols are used to denote the derivative of a function:

$$f'(x_0), \qquad \frac{d}{dx}f(x), \qquad D_x f(x), \qquad \frac{d}{dt}f(t) \equiv \mathring{f}(t),$$

or simply

$$f', \qquad y', \qquad \frac{df}{dx}, \qquad D_x f, \qquad \text{or} \qquad \mathring{y}.$$

3. *Derivative rules*: In the following rules, it is assumed that the functions are differentiable and c is a constant.

Sum rule: $(f + g)' = f' + g'$

Constant rule: $(cf)' = cf'$

Product rule: $(f \cdot g)' = f \cdot g' + g \cdot f'$

Quotient rule: $(f/g)' = (g \cdot f' - f \cdot g')/g^2$

4. *Derivatives of particular functions*

Power function: $\dfrac{d}{dx}x^r = r \cdot x^{r-1}$

In particular, for any constant c,

$$\frac{d}{dx}c = 0, \qquad \frac{d}{cx}cx = c, \qquad \frac{d}{dx}cx^2 = 2cx, \qquad \frac{d}{dx}cx^3 = 3cx^2.$$

Polynomial function: $\dfrac{d}{dx}\sum_{i=0}^{n} a_i x^i = \sum_{i=1}^{n} i \cdot a_i x^{i-1}$.

Trigonometric functions: $D_x \sin(x) = \cos(x)$

$D_x \cos(x) = -\sin(x)$

$D_x \tan(x) = \sec^2(x)$

$D_x \cot(x) = -\csc^2(x)$

$D_x \sec(x) = \sec(x) \cdot \tan(x)$

$D_x \csc(x) = -\csc(x) \cdot \cot(x)$

Exponential function: $D_x e^x = e^x$

Natural logarithm function: $D_x \ln(x) = 1/x$

A WORD OF ADVICE

If you are experiencing calculus for the first time, stop at this point and commit the above rules to memory. It will be impossible for you to continue through this text successfully without complete mastery of these formulas.

View calculus as another language and the above formulas as a new vocabulary to be memorized. Once memorized, these formulas, like the vocabulary of a second language, will gain meaning through the context in which they are experienced.

EXERCISE SET 14

14.1 Determine the tangent line to the graph of the function at the point indicated.

a) $f(x) = 3x - 1; x_0 = 1$
b) $g(x) = 3x^2 - 1; x_0 = 0$
c) $f(t) = t^2 - 3t + 2; t_0 = 2$
d) $g(t) = t^2 - 4t + 4; t_0 = 2$
e) $V(t) = 1 + 2t + 3t^2; t_0 = 0$
f) $R(t) = \pi \cdot t^2; t_0 = 2$

14.2 For each of the following functions: (1) determine where the function is zero, and (2) compute the instantaneous rate of change of the function at its zero values.

a) $f(x) = x^2 - 4$
b) $f(t) = 3t + 2$
c) $M(t) = t^3 - 8$
d) $g(t) = t^2 - t - 6$
e) $g(x) = 2x - x^2$
f) $f(x) = (x - 2)^2$

14.3 Using Definition 14.3, compute the derivative of $f(x) = 2x + 3$ at $x = 1$, $x = 2$, and $x = 0$.

14.4 Using Definition 14.2, compute the slope of the tangent line to the graph of $f(x) = x^2 + x$ at $(x_0, f(x_0))$ when $x_0 = 0$, 1, and 2.

14.5 Using Definition 14.1, compute the instantaneous rate of change of the function $f(x) = \sqrt{x}$ at the points $x_0 = 1$, $x_0 = 2$, and $x_0 = 3$.

14.6 A population grows in such a way that its mass at time t is $M(t) = 3t(10 - t)$.

a) What is the instantaneous rate of change of the population's mass at $t = 4$?
b) When will the instantaneous rate of change of the population's mass be 15?
c) When will the population cease to grow?
d) What is the maximum size of the population?

14.7 Sketch the graph of $y = x^3 - 3x + 1$ by plotting the points corresponding to $x = -2$, $-1, 0, 1$ and then sketch in the tangent lines to the graph at these points. As part of your solution give the equation of each tangent line.

14.8 Find y' for the given function y.

a) $y = 3x^2 - 2x + 1$
b) $y = 5x^4 - 2x^5$
c) $y = 3x^3 - 7x^{10}$
d) $y = \frac{1}{2}x^3 - 2x^2 + 5$
e) $y = x^{(3/2)} + 1$
f) $y = x^{(1/2)} + 1$
g) $y = x^{(1/4)} + x$
h) $y = x^{(1/2)} + x$
i) $y = 4x^{(3/4)} - x^2$
j) $y = x^{1.1} - 3x^2$
k) $y = x^{-(2/3)} + x^{(2/3)}$
l) $y = x^{(5/2)} - x^{(2/5)}$.

14.9 Let $f(x) = x^2 + 1$, $g(x) = 3x - 2$, and $h(x) = x^{-5}$. Evaluate the following derivatives.

a) $\frac{d}{dx}(f(x) \cdot g(x))$ b) $\frac{d}{dx}(g(x) \cdot f(x))$ c) $\frac{d}{dx}(h(x) \cdot g(x))$

d) $\frac{d}{dx}(f(x) \cdot g(x) \cdot h(x))$ e) $\frac{d}{dx}\left(\frac{1}{f(x)}\right)$ f) $\frac{d}{dx}\left(\frac{1}{g(x)}\right)$

g) $\frac{d}{dx}\left(\frac{1}{h(x)}\right)$ h) $\frac{d}{dx}\left(\frac{f(x)}{g(x)}\right)$ i) $\frac{d}{dx}\left(\frac{g(x)}{f(x)}\right)$.

j) $\frac{d}{dx}\left(\frac{h(x)}{f(x)}\right)$ k) $\frac{d}{dx}\left(\frac{f(x) + g(x)}{h(x)}\right)$ l) $\frac{d}{dx}\left(\frac{g(x) \cdot h(x)}{f(x)}\right)$

14.10 Assume that the derivatives of the function f, g, h are given by $f'(x) = x^2 - 2$, $g'(x) = 2x - 1$, $h'(x) = e^x$. Compute the following.

a) $D_x(f + g)(x)$ b) $D_x(3f(x) - h(x))$
c) $D_x(h(x) - g(x))$ d) $D_x(h(x) + f(x) + g(x))$
e) $D_x(f(x) - 2h(x))$ f) $D_x(h(x) - f(x) - g(x))$

14.11 Evaluate the following derivatives.

a) $D_x[\cos(x) + \tan(x)]$ b) $D_\theta[2\sin(\theta) - \theta^2]$ c) $\frac{d}{d\theta}[\sin(\theta)\cos(\theta)]$

d) $D_\theta\left[\csc(\theta) - \frac{1}{\sin(\theta)}\right]$ e) $\frac{d}{dx}\left[\frac{3\sin(x) + 1}{\cos(x)}\right]$ f) $\frac{d}{dt}\left[\frac{\sin(t)}{t}\right]$

g) $D_t[t^2 \sin(t)]$ h) $\frac{d}{dx}[\cos(x) + \sin(x)]^2$

14.12 Use the formula (14.7) to evaluate the following.

a) $D_x e^{3x}$ b) $D_x e^{-x}$ c) $D_x e^{-3x}$
d) $D_x[e^{4x} + e^x]$ e) $D_x e^{4x+1}$ f) $D_x e^{-4x+1}$
g) $D_x(xe^x)$ h) $D_x[\sin(x)e^x]$ i) $D_x[e^x \ln(x)]$

14.13 Use the formula (14.9) and the properties of logarithms to evaluate the following.

a) $D_x[3\ln(x)]$ b) $D_x[\ln(x) + 3]$ c) $D_x[\ln(x^2)]$
d) $D_x \ln(3x)$ e) $D_x \ln(4x)$ f) $D_x \ln(2/x)$
g) $D_x x\ln(x)$ h) $D_x[\sin(x)\ln(x)]$ i) $D_x[x^2 \ln(x)]$

14.14 Evaluate the derivative y' in two ways: (1) by using the product rule and (2) by multiplying out the expression and then differentiating.

a) $y = (3x + 2)(x^2 - 1)$ b) $f(x) = x^{(1/2)} \cdot (x + x^2)$
c) $y = (t^2 - 3)(t^2 + 3)$ d) $f(t) = \cos(t) \cdot \tan(t)$
e) $y = (t + 1)^2$ f) $g(t) = t^{(3/2)}(t^{(1/2)} + t^{(5/2)})$.

14.15 A particle moves along a path in an oscillatory fashion where its distance from a fixed point is given by (a) $f(t) = \sin(t)$ and (b) $f(t) = t^2 \sin(t)$. Compute its velocity and acceleration at times $t_0 = 0$, $t_0 = \pi/2$, $t_0 = \pi$.

14.16 To test for diabetes, a patient is subjected to a large quantity of sugar. The amount of glucose in the patient's urine is then measured over an interval $[0, T]$. If the amount of glucose is given by $g(t) = 10 - (0.6)t^2$, where t denotes hours, at what rate is the patient metabolizing the sugar 2 hours after the test begins?

14.17 The mass of a cell is related to its diameter, d, by the equation $M = \rho\pi(d/2)^2$. The constant ρ (rho) represents the specific density of the cell and is usually close to 1. When the diameter $d = 0.3$ cm, what is the rate of change in the cell mass as d increases?

14.18 The Michaelis–Menten equation describing the velocity of an enzyme reaction, v, as a function of the substrate concentration, s, is

$$v = \frac{V}{1 + K/s}.$$

V is the maximum velocity and K is a constant. Compute the rate of change of the velocity v as a function of s.

14.19 In Fig. 14.7, we indicate the percent of depletion of potassium ions from the soil as a function of the density of barley roots. To estimate the rate of change in the depletion at specific root densities, $r_0 = 1$, $r_0 = 2$, and $r_0 = \frac{1}{2}$ for each point r_0:

1. Sketch a tangent line to the graph through $(r_0, f(r_0))$.
2. Estimate two points on the tangent line and from these compute the slope of the tangent line.
3. Take these slopes as estimated rates of change.

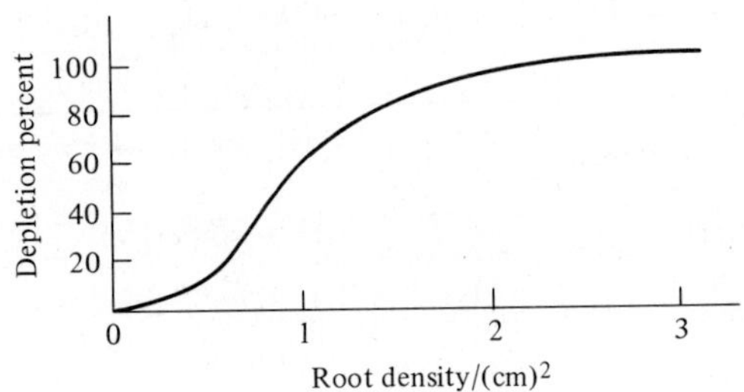

Figure 14.7

14.20 In describing "laminar" fluid flow in a tube such as a blood vessel, the velocity, V, of the fluid is a function of the distance, r, of the fluid from the center of the tube. Poiseuille's formula describes this flow; it involves the following parameters:

R = radius of the tube
l = length of the tube
P = pressure differential between the ends of the tube
η = a viscosity constant

The equation is

$$V = \frac{P}{4\eta l}(R^2 - r^2).$$

a) How would you describe the function $V(r)$ in terms of elementary functions?
b) Compute the rate of change of V as a function of r.
c) Compute the rate of change of V as a function of the tube length l.
d) Sketch the graph of V as a function of r assuming that, $l = 7$, $P = 14$, $\eta = 0.9$, and $R = 1.5$.

14.21 The amount of leaf area that a tree limb can support is a function of the angle, θ, that the limb makes with the trunk of the tree. Assume that this function has the form

$$L(\theta) = k \cdot \theta \cos(\theta), \qquad k > 0 \qquad 0 \leq \theta \leq \pi/2.$$

a) Compute $L'(\theta)$.
b) Evaluate $L'(\theta)$ for multiples of $\pi/12$.
c) Estimate the size of θ that will yield a maximum value of L.

14.22 A yeast cell is assumed to be cylindrical, with radius r and length l.

a) Derive an equation for the volume of the cell.
b) What is the rate of volume change, if the length is constant and the radius varies?
c) If the length is always $\sqrt{3}$ times larger than the radius, how does the volume change as a function of the radius?

14.23 The linear distance a frog can leap is a function of the angle of inclination at which it leaves the ground. Assume that the function is $l(\theta) = \theta(10 - \sec(\theta))$.

a) Compute the rate of change of $l(\theta)$.
b) Evaluate $l(\theta)$ and $l'(\theta)$ at $\theta = 0$, $\pi/3$, and $\pi/4$.

14.24 With reference to Example 14.10, do the following.

a) Determine a function $T_c(u)$ which describes the "wind-chill-corrected" temperature as a function of the wind velocity, u. (Assume T is fixed.)
b) Compute the rate of change in the corrected temperature, T_c, due to a change in wind speed, u.
c) What is the corrected temperature, T_c, corresponding to $-10°$C, with a wind speed of $u = 9$ m/sec?
d) At what rate is the corrected temperature, T_c, dropping under the conditions of part (c) if wind velocity begins to increase?

14.25 One-sided derivatives can be defined as follows.

$$D_x^{+} f(x_0) = \lim_{h \to 0^+} \frac{f(x_0 + h) - f(x_0)}{h}$$

$$D_x^{-} f(x_0) = \lim_{h \to 0^-} \frac{f(x_0 + h) - f(x_0)}{h}$$

Compute $D_x^{+} f(x_0)$ and $D_x^{-} f(x_0)$ at the indicated points and graph the functions.

a) $f(x) = 2x + 1$, $x_0 = 1$ b) $f(x) = [\![x]\!]$, $x_0 = 3$

c) $f(x) = |x|$, $x_0 = 0$ d) $f(x) = x^2$, $x_0 = 0$

e) $f(x) = \begin{cases} x^2 + 1 & \text{if} \quad x < 1 \\ -2x + 4 & \text{if} \quad x \geq 1 \end{cases} \qquad x_0 = 1$

f) $f(x) = \begin{cases} 2x & \text{if} \quad x < 0 \\ \frac{1}{2}x & \text{if} \quad x \geq 0 \end{cases} \qquad x_0 = 0$

g) $f(x) = [\![x]\!]/x$, $x_0 = 1$.

14.26 Give a limit proof of Theorem 14.3.

14.27 The size of a human eye pupil is related to the amount of light incident to the retina of the eye. The equation describing this relationship is

$$A = \frac{40I^{-0.4} + 23.7}{I^{-0.4} + 3.95},$$

where

A = area of pupil of the eye, and
I = quantity of visible radiant energy per unit of time incident on the retina of the eye.

Compute the rate of change in pupil area corresponding to a change in light intensity.

14.28 Evaluate the difference quotient corresponding to $h = 0.1$, $h = 0.01$, and $h = 0.001$ to obtain approximations of $f'(x_0)$; determine the error of each approximation.

a) $f(x) = x^2$, $x_0 = 2$;　b) $f(x) = \sin(x)$, $x_0 = \pi/3$;
c) $f(x) = 10^x$, $x_0 = 0$;　d) $f(x) = e^x$; $x_0 = 0$;
e) $f(x) = \ln(x)$, $x_0 = 1$.

SECTION 15

THE CHAIN RULE AND HIGHER DERIVATIVES

15.1 THE DERIVATIVE OF A COMPOSITION

In this section, we develop a formula for computing the derivative of the composition of two functions. This formula is probably the most frequently used "tool" of differential calculus. It is called the *Chain Rule*. Before we give the Chain Rule, we first give several examples that illustrate the use of the composition function in formulating mathematical models.

Introductory Examples Involving the Derivative of a Composition

Example 15.1 *A parasite model*
Parasites are animals or organisms that live on or in another organism, called a *host*. Parasites can either be helpful or harmful to their host. (Ruminant animals such as sheep are dependent on parasites to complete their digestive process.) Parasites are frequently employed to biologically control pests. One such parasite destroys the eggs of a spider. If the number of spiders in an area is H and the relative number of parasites is P, then the number H is a function of P;

$$H(P) = M(1 - 2P^3).$$

M is the maximum host population. However this parasite can only reproduce when the temperature is between 24 and 30°C. Consequently, the relative number of parasites is a function of the temperature, t. Assume that

$$P(t) = (t - 24)(30 - t)/9.$$

Then, although the spider population is not sensitive to the temperature t, its population size H is affected by the temperature. This can be described by the following composition equation:

$$\hat{H}(t) = H \circ P(t) = H(P(t)) \qquad \text{for} \qquad t \in [24, 30].$$

If the temperature is 28°C, is the spider population increasing or decreasing, and at what rate? To answer this, we need to evaluate $D_t\hat{H}(28)$, the derivative $D_t(H \circ P)(t)$ of the composition at $t = 28$. ◀

Example 15.2 *Nerve impulses*
A nerve impulse is translated into muscular movement. In a simplified model of nerve function, a nerve generates an electrical impulse in response to a stimulus. The transmission of an electrical impulse, v (in millivolts $= 10^{-3}$ volts), liberates acetylcholine ions at nerve junctions. We denote the relative number of acetylcholine ions by $\boldsymbol{a}$. The transmission of the nerve impulse to the muscle is then a function of $\boldsymbol{a}$. It is an increasing function, $R(\boldsymbol{a})$, which becomes zero once $\boldsymbol{a}$ reaches a critical level. Assume that

$$\boldsymbol{a}(v) = v^n, \qquad \text{where } n > 1 \qquad \text{and} \qquad R(\boldsymbol{a}) = \begin{cases} \dfrac{3\boldsymbol{a}}{\boldsymbol{a} + 2} & \text{if} \quad 0 \le \boldsymbol{a} \le 1; \\ 0 & \text{if} \quad \boldsymbol{a} > 1. \end{cases}$$

The muscle reaction is thus a function of the electrical nerve impulse v:

$$\hat{R}(v) = (R \circ \boldsymbol{a})(v) = R(\boldsymbol{a}(v)).$$ ◀

Example 15.3 *Gazelle population*
The size of a gazelle herd is a function of the edible grasses within its grazing territory. The amount of grass is estimated by sampling techniques to be x tons. The size of the gazelle herd is assumed to be

$$g(x) = \begin{cases} 0 & \text{if} \quad x < m; \\ (x - m)^2 - (x - m) + 2 & \text{if} \quad x \ge m, \end{cases}$$

where m is the minimum amount of grass necessary to sustain a pair of gazelles. But the amount x of grass is a function of the total rainfall, r, over the grazing region. Assume that

$$x(r) = 40r - r^2.$$

Then the size of the gazelle herd is a function of the rainfall;

$$\hat{g}(r) = (g \circ x)(r) = g(x(r)).$$ ◀

Example 15.4 *Gaseous pressure of oxygen in body tissue*
The tissues of your body contain oxygen in a gaseous form. The pressure, P, of the oxygen in the tissues is a function of the percentage saturation, S, of oxygen in the red blood cells, which carry the oxygen from the lungs to the body tissue. The saturation level of the arterial blood system is sensitive to radical changes in the oxygen level of the air inhaled into the lungs. Oxygen levels of the air decrease with altitude. Consequently, as you climb a mountain or ride in an unpressurized airplane, the change in altitude, $\boldsymbol{a}$, results in a change in the oxygen saturation level, S. The graphs in Fig. 15.1 illustrate these relationships. As a consequence, the gaseous pressure of oxygen in body tissue is a function of altitude. This can be denoted by a composition function $\hat{P} = (P \circ S)(\boldsymbol{a})$.

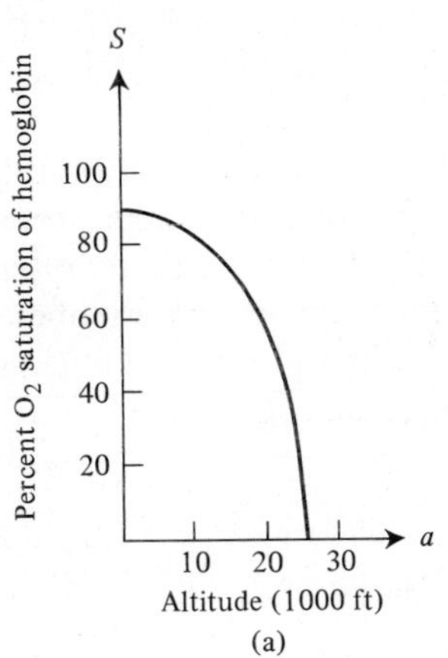

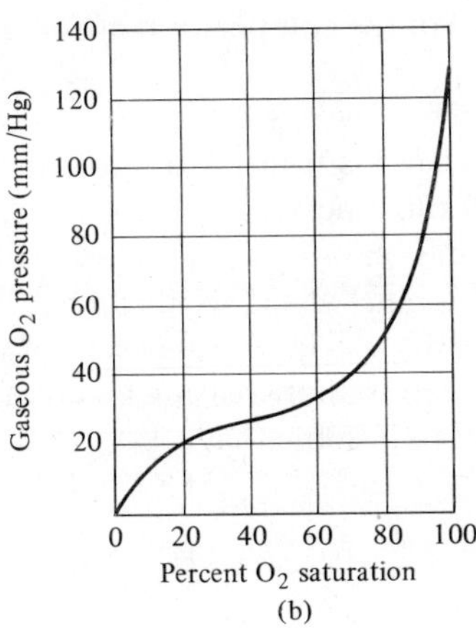

Figure 15.1

◀

Example 15.5 *A two-compartment model*
This example considers a situation which occurs frequently in many different biological areas. A process is visualized as consisting of two sequential actions. Each action is assumed to be the function of a biological compartment. We can regulate or determine the quantity x that enters the first compartment. This quantity is changed into a quantity u, which then enters the second compartment. The second compartment transforms quantity u into a quantity y.

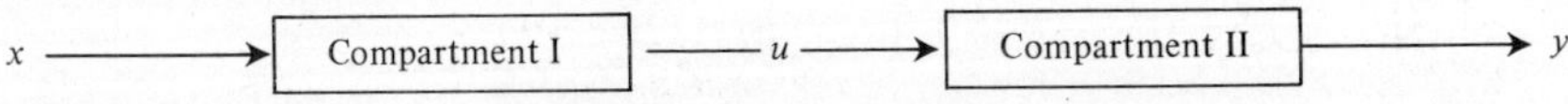

Mathematically, we represent the effect of each compartment by a function:

$$\text{Compartment I: } u = g(x),$$
$$\text{Compartment II: } y = f(u).$$

The effect of the two-compartment system is thus a composition of functions,

$$y = (f \circ g)(x).$$

Compartment models are used extensively in medicine, pharmacology, developmental biology, and agricultural modeling, where it is impossible biologically to separate the functions of individual components of a process. ◀

The Derivative of a Composition

In practice the actual functions f and g in Example 15.5 are not known. The theory of compartmental models considers the determination of the functions, f and g, and their rates of change, f' and g', when the only measurable quantity is the instantaneous rate of change $D_x(f \circ g)(x)$ of the composition. In this section we examine how $D_x(f \circ g)(x)$ is related to the derivatives $D_x f$ and $D_x g$.

Assume that f and g are differentiable functions. By definition, the derivative of the composition $f \circ g$ is

$$\lim_{h \to 0} \frac{(f \circ g)(x + h) - (f \circ g)(x)}{h} = \lim_{h \to 0} \frac{f(g(x + h)) - f(g(x))}{h}. \tag{15.1}$$

If we multiply and divide by the expression $g(x + h) - g(x)$, we obtain the equivalent limit,

$$\lim_{h \to 0} \left[\frac{f(g(x + h)) - f(g(x))}{g(x + h) - g(x)} \cdot \frac{g(x + h) - g(x)}{h} \right]. \tag{15.2}$$

Equation (15.2) appears to be the limit of a product of two difference quotients, one of the form $\Delta g/\Delta x$ and the other resembling $\Delta f/\Delta g$. As we are not accustomed to using the symbol g as an independent variable, at this point we introduce a new variable u, where $u = u(x) = g(x)$. The term $\Delta u = g(x + h) - g(x)$, and hence

$$u + \Delta u = g(x) + (g(x + h) - g(x)) = g(x + h).$$

With this notation, equation (15.2) has the following form:

$$\lim_{h \to 0} \left[\frac{f(u + \Delta u) - f(u)}{\Delta u} \cdot \frac{u(x + h) - u(x)}{h} \right].$$

Expressing this limit as the product of the limits of the two terms, we have

$$\lim_{h \to 0} \left[\frac{f(u + \Delta u) - f(u)}{\Delta u} \right] \cdot \lim_{h \to 0} \left[\frac{u(x + h) - u(x)}{h} \right].$$

The second limit is $u'(x)$, or $g'(x)$, since $u = g(x)$. The first limit can be evaluated by observing that when $u = g(x)$ is a continuous function,

$$\lim_{h \to 0} u(x + h) = u(x)$$

and hence

$$\lim_{h \to 0} \Delta u = \lim_{h \to 0} [u(x + h) - u(x)] = 0.$$

Consequently,

$$\lim_{h\to 0}\frac{f(u+\Delta u)-f(u)}{\Delta u}=\lim_{\Delta u\to 0}\frac{f(u+\Delta u)-f(u)}{\Delta u}=f'(u).$$

Combining the above equations, we find

$$D_x(f\circ g)(x)=f'(u)\cdot u'(x),\qquad \text{where } u=g(x). \tag{15.3}$$

If we substitute $u = g(x)$, we obtain

$$D_x(f\circ g)(x)=D_x[f(g(x))]=f'(g(x))\cdot g'(x). \tag{15.4}$$

This nonrigorous development establishes the basic fomula, (15.3) or (15.4), known as the *Chain Rule*.

Theorem 15.1 *The Chain Rule*
If f and g are differentiable functions, then so is $f\circ g$. The derivative of $(f\circ g)(x)$ with respect to the variable x is

$$D_x(f\circ g)(x)=f'(g(x))\cdot g'(x)\qquad \textbf{Chain Rule.}$$

Example 15.6 We use the Chain Rule to compute $D_x(f\circ g)(x)$ and $D_x(g\circ f)(x)$ for the indicated functions.

a) Let $f(x) = 3x + 1$ and $g(x) = x^2$; then $f'(x) = 3$ and $g'(x) = 2x$. Using the Chain Rule, we evaluate the derivatives:

$$D_x(f\circ g)(x)=f'(g(x))\cdot g'(x)=3\cdot 2x=6x.$$
$$D_x(g\circ f)(x)=g'(f(x))\cdot f'(x)=2(f(x))\cdot 3=6(3x+1).$$

b) Let $f(x) = x^3 - 4x + 2$ and $g(x) = x^{(1/2)}$; then $f'(x) = 3x^2 - 4$ and $g'(x) = (\frac{1}{2})x^{-(1/2)}$. Consequently,

$$\begin{aligned}D_x(f\circ g)(x)&=f'(g(x))\cdot g'(x)=[3(g(x))^2-4]\cdot(\tfrac{1}{2})x^{-(1/2)}\\&=[3x-4]\cdot(\tfrac{1}{2})x^{-(1/2)};\end{aligned}$$

$$\begin{aligned}D_x(g\circ f)(x)&=g'(f(x))\cdot f'(x)=\tfrac{1}{2}(f(x))^{-(1/2)}\cdot(3x^2-4)\\&=\frac{3x^2-4}{2\sqrt{x^3-4x+2}}.\end{aligned}$$

c) Let $f(x) = x^2 - 1$ and $g(x) = \cos(x)$; then $f'(x) = 2x$ and $g'(x) = -\sin(x)$. Consequently,

$$D_x(f\circ g)(x)=f'(g(x))\cdot g'(x)=(2\cos(x))(-\sin(x))=-2\cos(x)\sin(x).$$
$$D_x(g\circ f)(x)=g'(f(x))\cdot f'(x)=-\sin(x^2-1)\cdot 2x.$$

◀

A WORD OF ADVICE

The Chain Rule will be used in over 75 percent of the problems that require differentiation. You must develop a complete mastery of the Chain Rule to progress further in the study of calculus. Memorize it now!

One of the most difficult problems encountered when using the Chain Rule is to determine the functions f and g in a problem where they are not specified. The basic procedure is simple. You numerically evaluate any function in a logical manner. Each step in evaluating a complex function corresponds to a composition of functions. Given a function such as $y = (x^2 + 1)^{(3/2)}$, the method of expressing y as a composition is to ask yourself how you would evaluate y if $x = 3$. First you would evaluate $3^2 + 1 = 10$; hence let $g(x) = x^2 + 1$. Then you would evaluate $10^{(3/2)}$; hence let $f(u) = u^{(3/2)}$. Combining these, we write $y = (f \circ g)(x)$, where $g(x) = x^2 + 1$ and $f(x) = x^{(3/2)}$.

Practice is the only sure way to gain familiarity with a technique. The exercises contain many problems set in a form that makes the application of the Chain Rule apparent; work as many exercises as you can and then try making up your own problems.

Example 15.7 We compute y' using the Chain Rule for the following functions.

a) $y = (2x + 1)^3 - (2x + 1)^{(1/2)}$. Let $u = g(x) = 2x + 1$ and $f(u) = u^3 - u^{(1/2)}$. Then $y = (f \circ g)(x)$ and $dy/dx = (df/du \cdot du/dx)$. Therefore

$$y' = [3u^2 - \tfrac{1}{2}u^{-(1/2)}] \cdot (2) = [3(2x + 1)^2 - \tfrac{1}{2}(2x + 1)^{-(1/2)}]2.$$

b) $y = (x^3 - x + 5)^{10} + 3(x^3 - x + 5)$. Let

$$u = g(x) = x^3 - x + 5 \qquad \text{and} \qquad f(x) = x^{10} + 3x.$$

Then $y = (f \circ g)(x)$ and $D_x y = D_u f(u) \cdot D_x g(x)$. Therefore

$$y' = (10u^9 + 3)(3x^2 - 1) = (10(x^3 - x + 5)^9 + 3)(3x^2 - 1).$$

c) $y = \cos(x^2 - 3) + [1 + \sin(x)]^3$. We evaluate y' by computing the derivative of each term separately. The first term is $y_1 = \cos(x^2 - 3)$. To evaluate y'_1, we let

$$g_1(x) = x^2 - 3 \qquad \text{and} \qquad f_1(x) = \cos(x).$$

Then

$$y'_1 = f'_1(g_1(x)) \cdot g'_1(x) = -\sin(x^2 - 3) \cdot 2x.$$

The second term is $y_2 = [1 + \sin(x)]^3$. To evaluate y'_2, we let

$$g_2(x) = 1 + \sin(x) \qquad \text{and} \qquad f_2(x) = x^3.$$

Then

$$y_2' = f_2'(g_2(x)) \cdot g_2'(x) = 3[1 + \sin(x)]^2 \cos(x).$$

Consequently,

$$y' = -2x \sin(x^2 - 3) + 3 \cos(x)[1 + \sin(x)]^2.$$

d) $y = e^{3x^2-2}$. Let

$$u = g(x) = 3x^2 - 2 \qquad \text{and} \qquad f(x) = e^x.$$

Then, $y = (f \circ g)(x)$ and $D_x y = D_u f(u) \cdot D_x u(x)$:

$$y' = e^u \cdot 6x = 6xe^{3x^2-2}.$$

e) $y = \ln(\cos(x))$. Let

$$u = g(x) = \cos(x) \qquad \text{and} \qquad f(x) = \ln(x).$$

Then $y = (f \circ g)(x)$ and

$$\frac{dy}{dx} = \frac{df(u)}{du} \cdot \frac{du}{dx} = \frac{1}{u} \cdot -\sin(x) = \frac{-\sin(x)}{\cos(x)} = -\tan(x). \quad ◀$$

In the above examples, we encountered terms having the form $(g(x))^r$. If we apply the Chain Rule to compute y' for the function $y = (g(x))^r$, we obtain the formula known as the *generalized power formula* for derivatives.

Theorem 15.2 If g is differentiable and r is any real number then

$$\frac{d}{dx}(g(x))^r = r \cdot (g(x))^{r-1} \cdot g'(x) \qquad \textbf{Generalized Power Rule.}$$

A special case of the generalized power rule occurs when $r = -1$. The resulting formula is

$$\frac{d}{dx}(g(x))^{-1} = -1(g(x))^{-2}g'(x)$$

or

$$\frac{d}{dx}\left(\frac{1}{g(x)}\right) = \frac{-g'(x)}{(g(x))^2}. \tag{15.5}$$

The formula (15.5) may be combined with the *product rule* to derive the *quotient rule* for differentiation since

$$D_x\left[\frac{f(x)}{g(x)}\right] = D_x[f(x) \cdot (g(x))^{-1}].$$

We leave the verification as an exercise.

Each of the basic derivative formulas presented in Section 14 has a corresponding *generalized form* that is obtained by applying the Chain Rule. While the

Chain Rule can be used to evaluate particular derivatives, as illustrated in Example 15.7, the following *generalized rules* are easy to remember and can be used without constructing an explicit composition $y = (f \circ g)(x)$.

Generalized Trigonometric Derivatives

$$D_x \sin(u(x)) = \cos(u(x)) \cdot u'(x).$$

$$D_x \cos(u(x)) = -\sin(u(x)) \cdot u'(x).$$

$$D_x \tan(u(x)) = \sec^2(u(x)) \cdot u'(x).$$

$$D_x \cot(u(x)) = -\csc^2(u(x)) \cdot u'(x).$$

$$D_x \sec(u(x)) = \sec(u(x)) \cdot \tan(u(x)) \cdot u'(x).$$

$$D_x \csc(u(x)) = -\csc(u(x)) \cdot \cot(u(x)) \cdot u'(x).$$

Generalized Exponential Derivative

$$D_x e^{u(x)} = u'(x)e^{u(x)}$$

Generalized Derivative of ln(*x*)

$$D_x \ln(u(x)) = \frac{u'(x)}{u(x)}.$$

Example 15.8 The following derivatives are evaluated using the above generalized formulas.

a) $D_x \cos(x^2 - 1) = -\sin(x^2 - 1) \cdot 2x.$

b) $D_x \sin(3x - x^2) = \cos(3x - x^2) \cdot (3 - 2x).$

c) $D_x \tan(x^2 + 5x - 2) = \sec^2(x^2 + 5x - 2) \cdot (2x + 5).$

d) $$D_x \cot\left(\frac{x}{x+1}\right) = -\csc^2\left(\frac{x}{x+1}\right) \cdot D_x\left(\frac{x}{x+1}\right)$$
$$= -\csc^2\left(\frac{x}{x+1}\right) \cdot \frac{1}{(x+1)^2}.$$

e) $D_x e^{x^2-2x+3} = e^{x^2-2x+3} \cdot (2x - 2).$

f) $D_x \ln(x^2 - 2x + 3) = \dfrac{(2x - 2)}{(x^2 - 2x + 3)}.$

g) $D_x \ln(x + \sin(x)) = \dfrac{1 + \cos(x)}{x + \sin(x)}.$ ◀

15.2 RELATED RATES

The derivative of a function gives the instantaneous rate of change of the function. This is the rate of change due to a change in the variable of differentiation. The Chain Rule states that the rate of change of a composition is the product of two rates of change. In Section 15.1, we considered several examples that illustrate the way one function or variable can be related to another variable through an intermediary term. These relations were described using the composition operation. In the composition, $y = (f \circ g)(x)$, the change in the variable y is related to the change in the initial variable x through a product of changes. The rates of change are thus "related."

The term "related rates" is used to describe rate problems that involve an intermediary term and hence a composition of functions. All related rate problems are similar in nature. They are typically "word problems," that present in the text of a sentence or paragraph the following essential information:

I. An initial variable, x.
II. A second variable u, related to x by a function: $u = g(x)$. (15.6)
III. A third variable y, related to u by a function: $y = f(u)$.

Related-rates problems generally have the same form; they ask

"How is a change in y related to a change in x?"

or

"What is the change in y as x changes?"

or

"When y is changing at rate r, and u is u_0, how is u changing?"

The solutions of related-rates problems are thus always obtained from the same equation, expressing the change in y caused by a change in x:

$$\frac{dy}{dx} = \frac{d(f \circ g)(x)}{dx} = f'(g(x)) \cdot g'(x).$$

What then is difficult about related-rates problems? Certainly not the calculus, since it only involves the simple differentiation of a composition, $f \circ g$. The difficult part of related-rates problems is determining the variables and equations as specified in I, II, and III of (15.6). Frequently this is a reading problem. However, it is very helpful to know what you are "expected to read" before you read. In each problem you are looking for the functions as indicated in (15.6). For most problems, the variables will have different names. (Read each example in Section 15.1 and see if you can pick out the variables and functions before they are stated.) Many problems will contain references to formulas of geometry and measurement. These are not calculus topics and you are frequently expected to know them from previous experience. If you do not know the required formulas, look for them in a handbook of mathematical tables and formulas, or in a scientific reference book.

Example 15.9

a) A stone is tossed into a pool of still water. The result is seen as a ring of concentric ripples, expanding from the point of impact. If the crest of the first ripple, t seconds after impact, is $(1.3) \cdot t$ metres from the center, at what rate is the area inside the outer ring increasing after 2 seconds?

Solution: The radius of the circle is r, which is related to the time variable t by $r = g(t) = 1.3t$. The area, A, is related to the radius variable by $A = f(r) = \pi r^2$. Hence, as a function of time, $A = (f \circ g)(t) = \pi(1.3t)^2 = 1.69\pi t^2$. The rate of change of A as a function of time is $\mathring{A}(t) = 2(1.69)\pi t$. Therefore, at $t = 2$, $\mathring{A}(2) = 6.76\pi$.

b) A spherical balloon is being inflated. If the rate of change in the volume is 0.3π when the radius is 5, what is the rate of change in the radius at this point?

Solution: The volume $V = 4\pi r^3/3$. The radius is an unknown function of time, $r = r(t)$. By the Chain Rule,

$$\frac{dV}{dt} = \frac{dV}{dr} \cdot \frac{dr}{dt} = V'(r(t)) \cdot r'(t).$$

Since

$$\frac{dV}{dr} = 4\pi r^2,$$

when $r = 5$,

$$\frac{dV}{dr} = 100\pi.$$

By assumption, when $r = 5$,

$$\frac{dV}{dt} = 0.3\pi.$$

Consequently, when $r = 5$,

$$r' = \frac{dV/dt}{dV/dr} = \frac{0.3\pi}{100\pi} = 3 \times 10^{-4}.$$

The exact form of the function $r(t)$ is still unknown; we did not need to know its form to determine its derivative. ◀

Example 15.10 One of the contributing factors to the acumulation of lipid (fat) deposits on the inside of blood vessels is the fact that the flow of blood near the vessels walls is much slower than the flow of blood in the center of the vessel. Due to the slower flow rate near the walls, the lipid molecules have a greater chance of becoming attached to the vessel walls and eventually leading to a "heart attack."

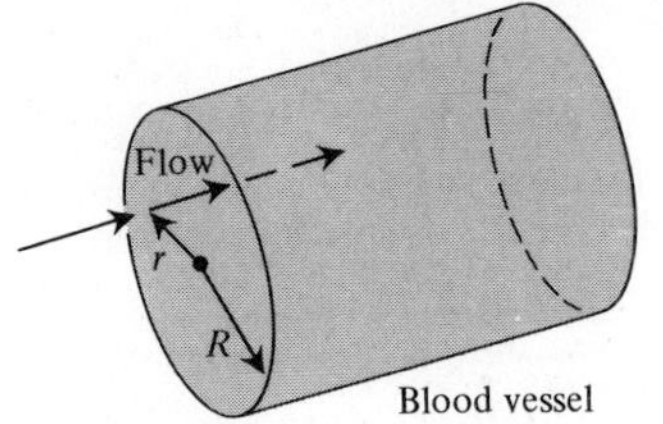

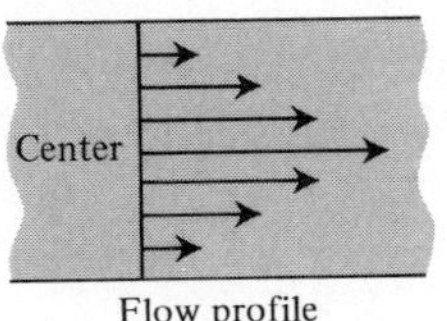

Figure 15.2

(See Fig. 15.2.) The function that describes the velocity of blood flow as a function of the distance, r, from the center of a vessel is known as Poiseuille's Law:

$$V = \frac{\rho}{4\lambda\eta}(R^2 - r^2),$$

where

V = velocity,
R = radius of the vessel,
r = distance from the center,
ρ, λ, η are physical constants corresponding to pressure, length, and viscosity.

The outer radius of a blood vessel can be changed by administering drugs, which either constrict the vessel (R decreases) or dilate the vessel (R increases). Aspirin dilates blood vessels. Assume that an individual "took two aspirins on a doctor's orders" and, as a result, the radius R of her or his blood arteries increased in size at a rate of

$$\frac{dR}{dt} = 2.10^{-4} \text{ cm/min}.$$

At what rate would the velocity of the blood flow be changing? The answer is obtained by computing the derivative of V using the Chain Rule:

$$\frac{dV}{dt} = \frac{dV}{dR} \cdot \frac{dR}{dt}.$$

Since

$$\frac{dV}{dR} = 2\left(\frac{\rho}{4\lambda\eta}\right) \cdot R,$$

the rate of change in V is

$$\frac{dV}{dt} = 2\left(\frac{\rho}{4\lambda\eta}\right)R \cdot 2 \cdot 10^{-4} \text{ cm/min}.$$

If $R = 0.02$ cm and $(\rho/4\lambda\eta) = 1$, then

$$\frac{dV}{dt} = 8 \cdot 10^{-6} \text{ cm/min.}$$

Note that the rate of change of the velocity does not depend on the distance, r, from the center of the vessel. ◀

15.3 HIGHER DERIVATIVES

The velocity of a particle moving along a path is defined to be $V(t) = D_t f(t)$, where $f(t)$ denotes the particle's distance from a fixed point at time t. The acceleration of the particle is then defined as the rate of change of the velocity: $A(t) = D_t V(t)$. Consequently, we can compute the acceleration from the position function $f(t)$: $A(t) = D_t(D_t f(t))$. The acceleration is the "second derivative" of the position function.

Definition 15.1 The *second derivative of a function* f, at a point x_0, is defined as the limit

$$\lim_{h \to 0} \frac{f'(x_0 + h) - f'(x_0)}{h}, \text{ if it exists.}$$

The second derivative of a function f is denoted by the following equivalent symbols:

$f''(x_0)$: f double prime of x_0.

$\dfrac{d^2 f(x_0)}{dx^2}$: the second derivative of f at x_0 or d by dx squared of f at x_0.

$D_x{}^2 f(x_0)$: the second derivative of f at x_0.

$f^{(2)}(x_0)$: the second derivative of f at x_0.

Or, when the variable is a time variable t,

$\ddot{f}(t)$: the second time derivative of f.

A function f is said to be "twice differentiable" or to have a "second derivative" if $f''(x)$ exists for all x in an open interval. Note that a necessary condition for the limit of Definition 15.1 to hold is that $f'(x)$ exists for all x in a neighborhood of x_0. A function is sometimes called *smooth* if its second derivative exists. An example of a function that is differentiable, but for which the second derivative does not exist, is given by $f(x) = x^{(4/3)}$ at $x_0 = 0$: $f'(x) = (\frac{4}{3})x^{(1/3)}$ and hence $f'(0) = 0$. By the power rule, $f''(x) = D_x(f'(x)) = D_x(\frac{4}{3}x^{(1/3)}) = \frac{4}{9}x^{-(2/3)}$. However, at $x_0 = 0$, $f''(x)$ does not exist, since this would require that $f''(0) = \frac{4}{9}(0^{-(2/3)})$, and $0^{-(2/3)}$ does not exist.

Example 15.11 We compute the second derivative, applying the standard rules for differentiation to the first derivative.

a) $f(x) = x^3 - 2x + 3; f'(x) = 3x^2 - 2; f''(x) = 6x$

b) $g(x) = \cos(x); g'(x) = -\sin(x); g''(x) = -\cos(x)$

c) $f(x) = 6x + 4; D_x(f(x)) = 6; D_x^2(f(x)) = 0$

d) $g(t) = t^2 - t^5; \dot{g}(t) = 2t - 5t^4; \ddot{g}(t) = 2 - 20t^3$

e) $f(t) = t^{-2}; f'(t) = -2t^{-3}; f''(t) = 6t^{-4}$

f) $F(x) = (x^2 - 2)^3; F'(x) = 3(x^2 - 2)^2 \cdot 2x; F''(x) = 6(x^2 - 2)^2 + 24x^2(x^2 - 2)$

g) $f(x) = e^{2x}; f'(x) = 2e^{2x}; f''(x) = 4e^{2x}$

h) $f(x) = \ln(x); f'(x) = 1/x; f''(x) = -1/x^2$ ◀

Having taken the derivative twice to obtain the second derivative, we can repeat the process. The third derivative is defined to be the derivative of the second derivative. And successively, we can compute the fourth, fifth, and higher order derivatives. Rather than denote the fifth derivative by $f'''''(x)$, we use the notation $f^{(5)}(x)$.

Definition 15.2 If *the nth derivative of* f, $f^{(n)}$, is defined on a neighborhood of a point x_0, then the $(n + 1)$st derivative of f at x_0 is

$$f^{(n+1)}(x_0) = \lim_{h \to 0} \frac{f^{(n)}(x_0 + h) - f^{(n)}(x_0)}{h},$$

if the limit exists. The following symbols are used to denote the nth derivative of f at x_0.

$$\frac{d^n f(x_0)}{dx^n}; \quad D_x^n f(x_0); \quad f^{(n)}(x_0).$$

Example 15.12 We compute the fifth derivative of the given functions.

a) $f(x) = x^5 - x^7; \quad f'(x) = 5x^4 - 7x^6;$

$f^{(2)}(x) = 20x^3 - 42x^5; \quad f^{(3)}(x) = 60x^2 - 210x^4;$

$f^{(4)}(x) = 120x - 840x^3; \quad f^{(5)}(x) = 120 - 2520x^2.$

b) $g(x) = x^2 - 3x + x^3; \quad D_x g(x) = 2x - 3 + 3x^2;$

$D_x^2 g(x) = 2 + 6x; \quad D_x^3 g(x) = 6;$

$D_x^4 g(x) = 0; \quad D_x^5 g(x) = 0.$

c) $f(t) = \cos(2t)$; $\quad \dfrac{df(t)}{dt} = -2\sin(2t)$;

$$\frac{d^2f(t)}{dt^2} = -4\cos(2t); \quad \frac{d^3f(t)}{dt^3} = 8\sin(2t);$$

$$\frac{d^4f(t)}{dt^4} = 16\cos(2t); \quad \frac{d^5f(t)}{dt^5} = -32\sin(2t).$$ ◀

Observe that, each time we compute the derivative of a polynomial, the degree of the polynomial is decreased by one. Consequently, in part (b) of the above example, since $g(t)$ was of degree 3 to begin with, the derivative of g became zero before we reached $D_x^{\,5}g(x)$. Also observe how the successive derivatives of the sine and cosine functions in part (c) were again cosine and sine functions. As a result of this, we can express identities involving the derivatives of the function $f(x) = \cos(2x)$. The equation $D_x^{\,2}f(x) + 4f(x) = 0$ is called a *differential equation.* If we let $y = f(x)$, this equation looks like $y'' + 4y = 0$.

Example 15.13 A laboratory technique called *gel electrophoresis* is used to separate proteins in a solution. The technique consists of placing the solution at the top of a glass column containing an agarose gelatin-type substance and passing an electrical current through the column from top to bottom. The electrical current creates a flow of the proteins from the top of the column to the bottom. The rate of flow depends on (1) the constant electrical voltage V, (2) the molecular size M of the protein, and (3) the total negative charge C of the protein. The technique can separate protein of different molecular weight or charge because smaller or more highly charged molecules will flow faster down the column than larger or less highly charged proteins, respectively. In Fig. 15.3, a column is illustrated at the termination of an experiment. The three "bands," A, B, and C, represent different

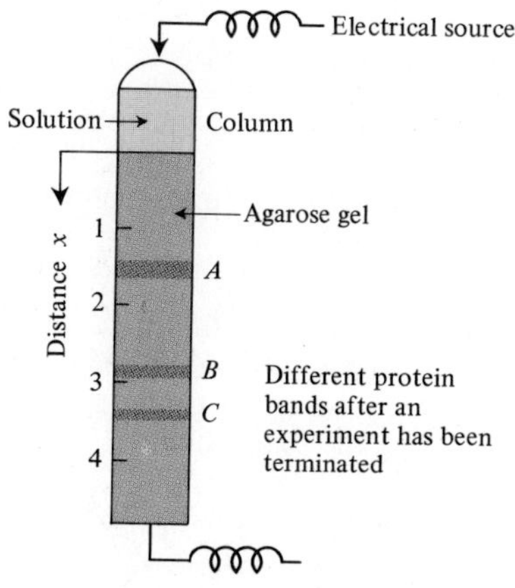

Figure 15.3

proteins contained in the original solution. To decide which bands correspond to which proteins assumed to be in the original solution is a standard biochemical problem. In some instances, the distance X that a protein will travel can be expressed as a function of time and the experimental conditions. A simple function, which would account for the basic experimental observations, is

$$X(t) = k_1 CVt + \left(\frac{g}{M} - k_2\right)t^2,$$

where g, k_1, and k_2 are constant, which depends on the characteristics of the agarose gel. (Note that M must be greater than g/k_2.) The velocity with which a particular protein band travels down the column would be

$$\text{(velocity) } \mathring{X}(t) = k_1 CV + 2\left(\frac{g}{M} - k_2\right)t.$$

The acceleration of a band would be the constant

$$\text{(acceleration) } \mathring{\mathring{X}}(t) = 2\left(\frac{g}{M} - k_2\right).$$

◀

15.4 THE DERIVATIVE OF THE INVERSE OF A FUNCTION

The inverse of a function $f(x)$ is the function $f^{-1}(x)$ defined by the identities

$$(f \circ f^{-1})(x) = x \qquad \text{and} \qquad (f^{-1} \circ f)(x) = x.$$

From these identities, we can derive a relationship between the derivatives of the functions f and f^{-1}. We set

$$y = (f \circ f^{-1})(x), \qquad \text{then} \qquad y = x.$$

The derivative of y, with respect to x, is simply

$$y' = 1.$$

However, using the Chain Rule, we can compute y' in a more difficult manner. We let $u = f^{-1}(x)$. Then $y = f(u)$ and

$$\frac{dy}{dx} = \frac{df(u)}{du} \cdot \frac{du}{dx}.$$

But since $dy/dx = 1$ this leads to the equation

$$\frac{du}{dx} = \frac{1}{\dfrac{df(u)}{du}}.$$

Replacing u with $f^{-1}(x)$ and $df(u)/du$ by $f'(u)$ or $f'(f^{-1}(x))$, we obtain the equation

$$\boxed{D_x f^{-1}(x) = \frac{1}{f'(f^{-1}(x))}.} \tag{15.7}$$

As an example of equation (15.7), let $f(x) = x^2$; then $f'(x) = 2x$. Hence, $f'(f^{-1}(x)) = 2f^{-1}(x)$ and

$$\frac{df^{-1}(x)}{dx} = \frac{1}{2(f^{-1}(x))}.$$

Since the inverse of $f(x) = x^2$ is $f^{-1}(x) = x^{(1/2)}$, this gives

$$\frac{d(x^{(1/2)})}{dx} = \frac{1}{2(x^{(1/2)})},$$

in agreement with the power rule. If $g(x) = x^3$, then $g'(x) = 3x^2$ and, by (15.7),

$$\frac{dg^{-1}(x)}{dx} = \frac{1}{3(g^{-1}(x))^2}.$$

As $g^{-1}(x) = x^{(1/3)}$, this agrees with the power rule

$$\frac{dx^{(1/3)}}{dx} = \frac{1}{3(x^{(1/3)})^2} = (\tfrac{1}{3})x^{-(2/3)}.$$

The formula (15.7) will be most useful in the situation where the derivative of f is known and the derivative of f^{-1} is not directly discernible.

Example 15.14 Consider the function

$$f(x) = \tan(x).$$

From its graph, Fig. 15.4(a), we see that the function $\tan(x)$ is monotone increasing on the interval $(-\pi/2, \pi/2)$. Hence, by the Theorem 4.3, there exists an inverse function of $f(x) = \tan(x)$. We do not have an expression for this inverse function in terms of other elementary functions. Therefore, we simply give it a new name, "the inverse tangent function." We denote this function by

$$\tan^{-1}(x) \qquad \text{or} \qquad \arctan(x).$$

The graph of $y = \tan^{-1}(x)$ is obtained from the graph of $y = \tan(x)$ by rotating about the line $y = x$ and then interchanging the labels on the coordinates axes. There should be no confusion between $\tan^{-1}(x)$ and $1/\tan(x)$, since $1/\tan(x)$ would be denoted as $\cot(x)$. Without knowing more about this inverse function, we can compute its derivative: If we set

$$y = \tan^{-1}(x);$$

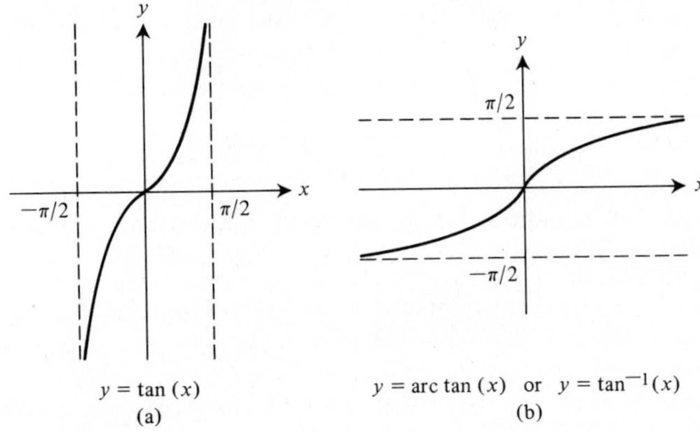

Figure 15.4

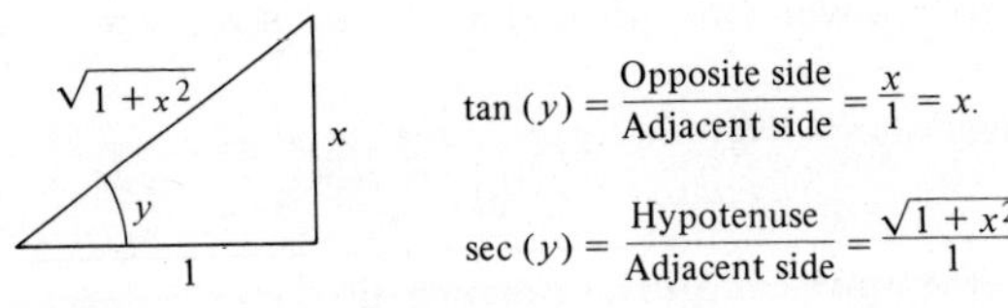

Figure 15.5

then, since $D_x(\tan(x)) = \sec^2(x)$, we have from equation (15.7) with $f(x) = \tan(x)$ and $f^{-1}(x) = y$,

$$D_x(y) = \frac{1}{\sec^2(y)}. \tag{15.8}$$

To evaluate the right side of this equation, we use the relations

$$y = \tan^{-1}(x) \Leftrightarrow \tan(y) = x.$$

(See Theorem 4.2.) We then use the relationship between the trigonometric function and the sides of a triangle as illustrated in Fig. 15.5 to establish that if

$$\tan(y) = x \qquad \text{then} \qquad \sec(y) = \sqrt{1 + x^2}.$$

Consequently, equation (15.8) can be expressed as

$$D_x \tan^{-1}(x) = \frac{1}{1 + x^2}. \tag{15.9}$$

◀

Example 15.15 The exponential function e^x and natural logarithm function $\ln(x)$ are defined as the inverse of each other. If

$$f(x) = e^x \qquad \text{then by definition} \qquad f^{-1}(x) = \ln(x).$$

If we apply the formula (15.7) for the derivative of f^{-1} to this choice of $f(x)$, we obtain

$$\begin{aligned} D_x \ln(x) = D_x f^{-1}(x) &= \frac{1}{f'(f^{-1}(x))} \\ &= \frac{1}{e^{(\ln(x))}} \qquad \text{(Since } f'(x) = e^x.\text{)} \\ &= \frac{1}{x} \qquad \text{(Since } e^{\ln(x)} = x.\text{)} \end{aligned}$$

Thus the derivative $D_x \ln(x)$ can be obtained from the identity $D_x e^x = e^x$. ◀

SUMMARY

1. *Basic formulas:*

$$\frac{d}{dx}(f \circ g)(x) = f'(g(x)) \cdot g'(x) \qquad \textbf{Chain rule.}$$

$$D_x[u(x)]^r = r[u(x)]^{r-1} \cdot u'(x) \qquad \textbf{Generalized power rule.}$$

The generalized trigonometric formulas appear exactly like the standard trigonometric derivative formulas, except that they are multiplied by the derivative $u'(x)$. For instance, $D_x \tan(u(x)) = \sec^2(u(x)) \cdot u'(x)$.

$$D_x e^{u(x)} = e^{u(x)} \cdot u'(x) \qquad \textbf{Generalized exponential rule.}$$

$$D_x \ln(u(x)) = \frac{1}{u(x)} u'(x) = \frac{u'(x)}{u(x)} \qquad \textbf{Generalized natural logarithm derivative.}$$

2. *Related rate problems*: These are composition of functions problems of the type $u = g(x)$, $y = f(u)$; hence $y = (f \circ g)(x)$. Use the equation

$$\frac{dy}{dx} = \frac{dy}{du} \cdot \frac{du}{dx}.$$

Most solutions require $\dfrac{dy}{du} = \dfrac{(dy/dx)}{(du/dx)}$.

3. *Higher derivatives*:

$$\frac{df'(x)}{dx} = f''(x) \qquad \textbf{Second derivative.}$$

Notation:

$$f''(x) = D_x{}^2 f(x) = \frac{d^2 f(x)}{dx^2}$$

The *n*th derivative

$$\frac{d^n f(x)}{dx^n} = f^{(n)}(x)$$

is obtained by taking the derivative of f, n-times. For example, if $f(x) = x^5$, then $f'(x) = 5x^4$, $f''(x) = 20x^3$, $f^{(3)}(x) = 60x^2$, $f^{(4)}(x) = 120x$, $f^{(5)}(x) = 120$, $f^{(6)} = 0$, and $f^{(n)}(x) = 0$ for $n = 7, 8, \ldots$

EXERCISE SET 15

15.1 Compute the indicated derivatives.

a) $\dfrac{df(u)}{du}$, if $f(u) = 2u^2 + 3u - 2$

b) $D_z g(z)$, if $g(z) = (z + 3)^2 - 1$

c) $D_v f(v)$, if $f(v) = (2v + 1)^5(v^2 - 2)$

d) $\dfrac{dg(t)}{dt}$, if $g(t) = (t^2 - 2t)^2 + (6t + 1)$

e) $D_u y$, if $y(u) = u^{(1/2)} + u^2$

f) $\dfrac{dy(u)}{du}$, if $y(x) = 3x^2 - x^{-1}$

g) $\dfrac{dg(u)}{du}$, if $g(t) = (t^2 + 5t - 1)^7$

h) $\dfrac{dg(v)}{dv}$, if $g(v) = v\cos(v)$

i) $\dfrac{df(u)}{du}$, if $f(u) = e^u$

j) $\dfrac{dg(u)}{du}$, if $g(u) = \ln(u)$

15.2 Compute the compositions, $f \circ g$ and $g \circ f$, for the indicated functions.

a) $f(x) = x^2 + 2$; $g(x) = 2x$
b) $f(x) = 3x - 7$; $g(x) = -x + 2$
c) $f(x) = x + \sqrt{x}$; $g(x) = (x + 2)^2$
d) $f(x) = e^x$; $g(x) = x^2 + 1$
e) $f(x) = \ln(x + 1)$; $g(x) = (x + 1)(x - 1)$
f) $f(x) = x^{-1}$; $g(x) = e^{x^2}$
g) $f(x) = (3x - 2)^3$; $g(x) = x^{(1/2)} - x^{-(1/2)}$
h) $f(x) = \sin(x)$; $g(x) = 1 - x^2$

15.3 Express the following functions as compositions, $y = f \circ g$. Indicate $f(x)$, $g(x)$ and verify that your choice is valid. (There are no unique answers.)

a) $y = (2x + 1)^2 + (2x + 1)^{(1/2)}$
b) $y = (2x + 1)^2 + x + \frac{1}{2}$
c) $y = (x^2 + 1)e^{x^2+1}$
d) $y = e^{3x-2} + 3x - 2$
e) $y = (x + 1)(x - 1)/(1 + x^2)$
f) $y = x^2 + 6x + 90 - 2\ln(x + 3)$
g) $y = e^x + e^{e^x+x}$

15.4 Determine dy/dx for the given function $y(x)$.

a) $y = (3x + 1)^2$
b) $y = (2x^2 + 2)^2$
c) $y = (x^5 + 1)^3$
d) $y = (x^4 + x^3 + 1)^3$
e) $y = (x^2 + 3x - 1)^{(1/2)}$
f) $y = (2x + 1)^{-(1/2)}$
g) $y = (x + 3)^6$
h) $y = (x + 3)^6(2x - 1)^2$
i) $y = (2x + 1)^4 + (x^2 - 1)^3$
j) $y = (2x + 1)^2/3x^2$
k) $y = (x^3 - 2x - 8)/(5x + 1)^3$
l) $y = (x^3 - 2x - 8)(5x + 1)^{-3}$
m) $y = (x^5 - 1)^2(x^2 - 1)^5$
n) $y = 1/(x + 1)(x - 1)$
o) $y = 1/(x + 1)^2(x - 1)$
p) $y = (x^2 + 2)/(x + 1)^2(2x - 1)$

15.5 Evaluate the following derivatives.

a) $D_x \cos^2(x)$
b) $D_x \cos(x^2)$
c) $D_x \sin(1 + x^2)$
d) $D_x(\sin(x) + 1)^2$
e) $D_x \tan(1 - 2x)$
f) $D_x(\cot(x) - x)^{(1/2)}$
g) $D_t \sin(\sin(t))$
h) $D_t(\sin^2(t) - 3\sin(t) + 7)$

15.6 Evaluate the following derivatives.

a) $D_x e^{2x}$
b) $D_x \ln(2x)$
c) $D_x e^{2x^2}$
d) $D_x \ln(2x^2)$
e) $D_x e^{2x-x^3}$

f) $D_x \ln(2x - x^3)$
g) $D_x e^{(x-1)^2}$
h) $D_x \ln[(x - 1)^2]$
i) $D_x e^{\sin(x)}$
j) $D_x \ln(\sin(x))$
k) $D_x[xe^{x^2}]$
l) $D_x[x \ln(x^2)]$
m) $D_x[\sin(x)e^{-x}]$
n) $D_x[\sin(x)\ln(2x)]$
o) $D_x \sin(e^x)$
p) $D_x \sin(\ln(x))$
q) $D_x(e^{e^x})$
r) $D_x \ln(\ln(x))$
s) $D_x e^{2e^{3x}}$
t) $D_x \ln[\sin(x)\cos(x)]$

15.7 Give the equation of the tangent line to the graph of the function at the point $(x_0, f(x_0))$.

a) $f(x) = (1 - x^2)^{(1/2)}; x_0 = \frac{1}{2}$
b) $f(x) = \sin(x^2); x_0 = \pi/4$
c) $f(x) = (x - 2)^2 + (1/(x - 2)); x_0 = 3$
d) $f(x) = \sqrt{1 + \sec^2(x)}; x_0 = 0$
e) $f(x) = e^{2x}; x_0 = 0$
f) $f(x) = \ln(x); x_0 = 1$

15.8 Evaluate $D_x(f \circ g)(x)$ for the functions of Exercise 15.2.

15.9 Compute the derivative y' for the functions y of Exercise 15.3.

15.10 Give the answer to the question posed in Example 15.1.

15.11 Referring to Example 15.2, answer the following questions.

a) What is the rate of change in muscle reaction, $\hat{R}$ due to a change in stimuli, v?
b) If $n = 3$, what is the rate of change of $\hat{R}$ when $\boldsymbol{a} = 0.4$?

15.12 Referring to Example 15.3, give the formula for the gazelle population as a function of rainfall.

a) If $m = 10$ in the model, what is the minimum rainfall necessary to keep the herd from disappearing? What is the maximum rainfall in this case?
b) If $m = 10$, what is the rate of change of the herd size due to a change in rainfall.

15.13 Referring to Example 15.4, estimate the following.

a) The oxygen pressure in the tissue if you climb to 15,000 ft.
b) The slopes of the tangent lines corresponding to $a = 15{,}000$ ft in graphs (a) and (b) of Fig. 15.1.
c) Using the estimates in part (b), estimate the rate of change in your oxygen pressure as you climb past the 15,000-ft level on a mountain.

15.14 Use the formula (15.5) and the product rule to obtain the quotient rule for differentiation.

15.15 Compute the third derivative of the given functions.

a) $f(x) = x^3 + x^2 + x + 1$
b) $f(x) = x^2 - x^{-2}$
c) $f(x) = x^{(1/2)}$
d) $f(x) = (x^2 + 1)^4$
e) $f(x) = \sqrt{1 - x^2}$
f) $f(x) = x(1 + x)^{-1}$
g) $f(x) = \sin(x)$
h) $f(x) = \sin(3x)$
i) $f(x) = \tan(x)$
j) $f(x) = \cos(x^2)$
k) $f(x) = e^{x^2}$
l) $f(x) = \ln(x^2)$

15.16 Determine the velocity and acceleration of a particle which oscillates about the origin on the x-axis and whose distance from the origin at time t is (a) $X(t) = t\sin(t)$, (b) $X(t) = t\sin(1/t)$.

15.17 Referring to Example 4.7, give the rate of change in the fish biomass, F, due to a change in the algae levels of the bay, using the Chain Rule.

15.18 Show that, if y is an allometric function of u and u is an allometric function of x, the rate of change of y as a function of x is also represented by an allometric equation.

15.19 An ice cube melts uniformly so that its shape is always cubic. If the rate of melting is -36 cc/min when its edge dimension is 100 cm, at what rate is the side length decreasing?

15.20 If a fish tank having base dimensions 28 by 1.2 metres is being filled at the rate of 45 l/hr, at what rate is the water level rising?

15.21 A fish is assumed to grow so that its length is always twice its height and 10 times its width. At what rate is the fish's mass increasing if its width is increasing at 2 cm/week?

15.22 A conical filter funnel is filled at a rate of 500 ml/min. The liquid drains out of the funnel at a rate of 300 ml/min. When the height of the liquid in the funnel is 10 cm, at what rate is the liquid rising in the funnel? Assume that at its neck, the angle of the funnel is 30° from the vertical.

15.23 The bottom portion of an Ehrlenmeyer flask has an inverted conical shape. Assume that the radius at a height h is $r = 2.5(30 - h)$cm. What is the rate at which the flask is being filled if the height, h, is increasing at the rate 4 cm/min when $h = 10$ cm?

15.24 A pulley is hung in a loft 10 metres high. A rope 25 metres long is used to hoist hay bales to an upper level. This is accomplished by pulling the free end of the rope as a person walks away from the lifting point.

a) How is the distance of the person pulling the rope from the lifting point related to the height of the bale being lifted? Assume that the end of the rope is held 1 metre off the ground.
b) If the person walks away at a rate of 20 metres/min, how fast is the bale being lifted?

15.25 Since seals are air breathing mammals, when they dive for food they must efficiently utilize their oxygen supply. One way in which this is done is by minimizing their heat loss to the cold water by restricting their peripheral blood flow. A simple model describing this phenomenon is to assume that the blood flow, f_B, is a linear function of the depth, x, of the seal.

a) Give an equation for $f_B(x)$ if $f_B(0) = 50$ and $f_B(10) = 25$.
b) Under the assumptions of part (a), at what rate is a seal descending on a dive when its blood flow is decreasing at a rate of -30?
c) Assuming that the function $f_B(x)$ is a more complicated but unknown function, if it is found that at some time, t_0, $\mathring{f}_B(t_0) = -70$ and $\mathring{x}(t_0) = 10$, what is the rate of change in f_B with respect to depth at this time, t_0?

15.26 A model for "molecular weight gel filtration" is similar to that discussed in Example 15.13. However, in "molecular weight gel filtration" no electrical current is used, and the function specifying the total distance traveled in a fixed time by a band of molecules having molecular weight, W, is a linear function of the negative log of W:

$$d = -a \log(W) + b.$$

a) What is the rate of change in the distance, d, due to a change in molecular weight, W?
b) If Ribonuclease A with Mol weight 13,700 traveled $d = 0.7$ units and Ovalbumin with Mol weight 45,000 traveled $d = 0.45$ units, how far will Chymotrypsinogen A with Mol weight 75,000 travel?

15.27 Each of the trigonometric functions has an inverse if it is considered on a restricted domain. Using the graph of the trigonometric functions in Fig. 6.10, we see that the inverses can be defined as follows:

$$\sin(x),\ x \in [-\pi/2, \pi/2] \text{ has an inverse, } \sin^{-1}(x) \equiv \arcsin(x).$$

$$\cos(x),\ x \in [0, \pi] \text{ has an inverse, } \cos^{-1}(x) \equiv \arccos(x).$$

$$\tan(x),\ x \in (-\pi/2, \pi/2) \text{ has an inverse, } \tan^{-1}(x) \equiv \arctan(x).$$

$$\cot(x),\ x \in (0, \pi) \text{ has an inverse, } \cot^{-1}(x) \equiv \text{arccot}(x).$$

$$\sec(x),\ x \in [0, \pi/2) \cup (\pi/2, \pi] \text{ has an inverse, } \sec^{-1}(x) \equiv \text{arcsec}(x).$$

$$\csc(x),\ x \in [-\pi/2, 0) \cup (0, \pi/2] \text{ has an inverse } \csc^{-1}(x) \equiv \text{arccsc}(x).$$

a) By rotating the graphs in Fig. 6.10, sketch the graph of the inverse trigonometric functions.
b) Using a diagram similar to that in Fig. 15.5, determine

i) $\cos(y)$ if $\tan(y) = x$.
ii) $\sin(y)$ if $\cos(y) = x$.
iii) $\sec(y)$ if $\tan(y) = x/2$.
iv) $\tan(y)$ if $\sin(y) = 4/x$.
v) $\cot(y)$ if $\sin(y) = x$.
vi) $\csc(y)$ if $\sec(y) = x$.

c) Using the method illustrated in Example 15.14, verify each of the following derivative formulas.

$$D_x \sin^{-1}(x) = \frac{1}{\sqrt{1 - x^2}} \qquad D_x \cos^{-1}(x) = \frac{-1}{\sqrt{1 - x^2}}$$

$$D_x \tan^{-1}(x) = \frac{1}{1 + x^2} \qquad D_x \cot^{-1}(x) = \frac{-1}{1 + x^2}$$

$$D_x \sec^{-1}(x) = \frac{1}{x\sqrt{x^2 - 1}} \qquad D_x \csc^{-1}(x) = \frac{-1}{x\sqrt{x^2 - 1}}$$

d) Using the above formulas and the Chain Rule, evaluate the following derivatives.

i) $D_x \sin^{-1}(3x)$ ii) $D_x \cos^{-1}(x^2)$ iii) $D_x \tan^{-1}(1 - 2x)$

iv) $D_x \cot^{-1}(3x^2)$ v) $D_x \sec^{-1}(4x - 1)$ vi) $D_x \csc^{-1}(\sqrt{x})$

CHAPTER FOUR

APPLICATIONS USING THE DERIVATIVE

SECTION 16
MAXIMUM AND MINIMUM PROBLEMS AND CURVE SKETCHING

SECTION 17
DIFFERENTIALS; IMPLICIT DIFFERENTIATION

SECTION 18
THE MEAN VALUE THEOREM; TAYLOR'S THEOREM AND TAYLOR'S POLYNOMIAL APPROXIMATIONS

The most common application of the derivative is to determine the extreme values of a function. In Section 16, maximum and minimum problems are examined in conjunction with graphing techniques that utilize information provided by the first and second derivative. In Section 17, the differential is introduced. This is an important topic as the differential approximation is useful both theoretically and in practical applications. The calculus of differentials also leads to the method of implicit differentiation, which is essential when examining relationships that are not in the standard explicit functional form. The first subsection of Section 18 examines two mathematically important theorems, Rolle's Theorem and the Mean Value Theorem; while of theoretical importance, these theorems are not encountered in most applications. The second subsection considers Taylor's Theorem and Taylor's Polynomial Approximations of functions. This topic is very important and is utilized in an extremely large number of applications that range from simplifying biological models to numerical approximations. This chapter may be deferred until after Chapter 5 and is optional for most of the material in Chapter 6.

SECTION 16

MAXIMUM AND MINIMUM PROBLEMS AND CURVE SKETCHING

16.1 USING THE FIRST DERIVATIVE TO DESCRIBE A FUNCTION

Functions are described in two ways: algebraically, a function is described quantitatively by an equation $y = f(x)$, which tells us how to evaluate the function for any number x; a second way a function is described is to indicate qualitative aspects of the function or, equivalently, of its graph. The type of qualitative features that are usually used to describe functions are indicated by terms such as "increasing," "decreasing," "maximum," "minimum," or "curved upward." Most people have an intuitive notion of what these terms mean; indeed we utilized this type of vocabulary to describe functions in the preceding portions of the text (especially, in Section 8 on special functions.) In this subsection, we investigate the mathematical meaning of such terms and how they are related to the derivative of a function. In particular, we present a very useful method, called the *First Derivative Test*, for identifying maximum and minimum values of a function.

The principal concepts upon which all of the arguments of this section rest are the following.

Concept 1 The derivative of a function $f(x)$ evaluated at a point x_0 gives the slope of the tangent line to the graph of f at $(x_0, f(x_0))$.

Concept 2 The graph of a differentiable function is "locally linear" in the sense that the graph of $y = f(x)$ for x near the number x_0 is essentially the same as the graph of the tangent line to the graph of f at $(x_0, f(x_0))$.

Concept 3 The graph of the line $y = ax + b$ is increasing if $a > 0$, decreasing if $a < 0$, and horizontal if $a = 0$.

The terms "increasing" and "decreasing" are used to describe how a function changes over an interval. Assume that a function $f(x)$ is defined on an interval (a, b). If for any two points, x_1 and x_2, in the interval (a, b)

$$x_1 < x_2 \text{ implies } f(x_1) < f(x_2),$$

then $f(x)$ is *increasing* on (a, b); and if

$$x_1 < x_2 \text{ implies } f(x_1) > f(x_2),$$

then $f(x)$ is *decreasing* on (a, b).

These definitions refer to the behavior of a function over an interval. The derivative describes the behavior of a function *at a point*. It seems reasonable to expect that the terms "increasing" and "decreasing" can be defined at a point x_0. The key to providing a definition is the association between the graph of $y = f(x)$ and the tangent line at $(x_0, f(x_0))$. We assume that the function $f(x)$ is differentiable on an interval (a, b) containing the point x_0. The tangent line to the graph of f at $(x_0, f(x_0))$ is given by

$$y = f'(x_0)(x - x_0) + f(x_0).$$

If $f'(x_0)$ is positive, the tangent line is increasing and, if $f'(x_0)$ is negative, the tangent line is decreasing. Using Concept 2, we are led to make the following definition.

Definition 16.1 A differentiable function f is said to be

i) *increasing at* x_0 if $f'(x_0) > 0$,
ii) *decreasing at* x_0 if $f'(x_0) < 0$,
iii) *stationary at* x_0 if $f'(x_0) = 0$. (See Fig. 16.1.)

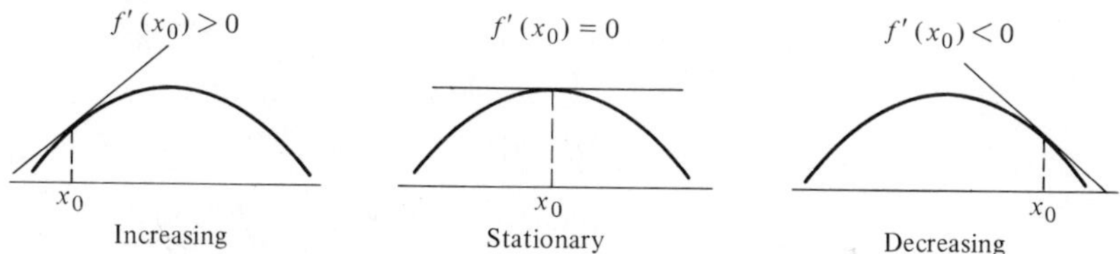

Figure 16.1

The next theorem states that the terms increasing and decreasing at a point, as defined in Definition 16.1, correspond to the interval concepts of increasing and decreasing.

Theorem 16.1

i) If $f'(x) > 0$ for all $x \in (a, b)$, then the function f is increasing on (a, b).
ii) If $f'(x) < 0$ for all $x \in (a, b)$, then the function f is decreasing on (a, b).
iii) If $f'(x) = 0$ for all $x \in (a, b)$, then $f(x)$ is a constant value for all $x \in (a, b)$.

Example 16.1

a) The function $y = a(x - b)^2 + c$ is a quadratic whose graph is "u-shaped" if $a > 0$. To verify this, we observe that $y' = 2a(x - b)$.
   i) If $x < b$, then $y' < 0$, and the graph is decreasing.
   ii) If $x > b$, then $y' > 0$, and the graph is increasing. The point, $x_0 = b$, is a stationary point as $y'(b) = 0$. (See Fig. 16.2.)

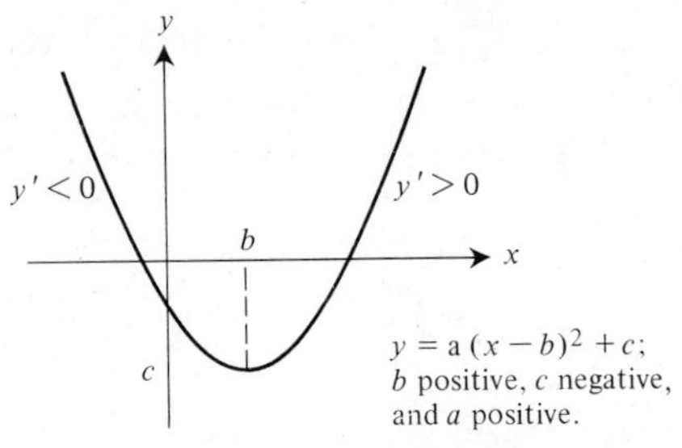

Figure 16.2

b) Let

$$f(x) = \tfrac{1}{3}x^3 - \tfrac{3}{2}x^2 + 2x.$$

To decide where $f(x)$ is increasing or decreasing, we first determine the stationary points, where $f'(x) = 0$. Since

$$f'(x) = x^2 - 3x + 2 = (x - 1)(x - 2),$$

$f'(x) = 0$ at two points,

$$x_1 = 1 \qquad \text{and} \qquad x_2 = 2.$$

To evaluate the sign of $f'(x)$ for various values of x, we use the factored form of $f'(x)$ and the sign of each factor. *Remember*:

$$\text{(positive)} \times \text{(positive)} = \text{positive}$$

$$\text{(negative)} \times \text{(negative)} = \text{positive}$$

$$\text{(positive)} \times \text{(negative)} = \text{negative}$$

Using the fact that

$x - 1$ is positive if $x > 1$ and negative if $x < 1$,

together with the fact that

$x - 2$ is positive if $x > 2$ and negative if $x < 2$,

we may determine the sign of $f'(x)$ as illustrated in Fig. 16.3

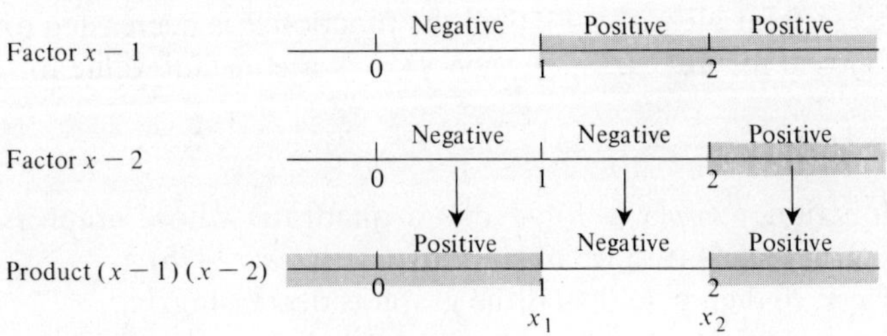

Figure 16.3

Thus

$$f'(x) = (x - 1)(x - 2) \text{ is positive if } x < 1 \text{ or } x > 2,$$

and

$$f'(x) = (x - 1)(x - 2) \text{ is negative if } 1 < x < 2.$$

Consequently, the function

f is increasing on $(-\infty, 1) \cup (2, +\infty)$, since $f'(x) > 0$;

f is decreasing on $(1, 2)$, since $f'(x) < 0$.

The function f is the cubic sketched in the Fig. 16.4. The graph is stationary at the points $(x_1, f(x_1)) = (1, \frac{5}{6})$ and $(x_2, f(x_2)) = (2, \frac{2}{3})$.

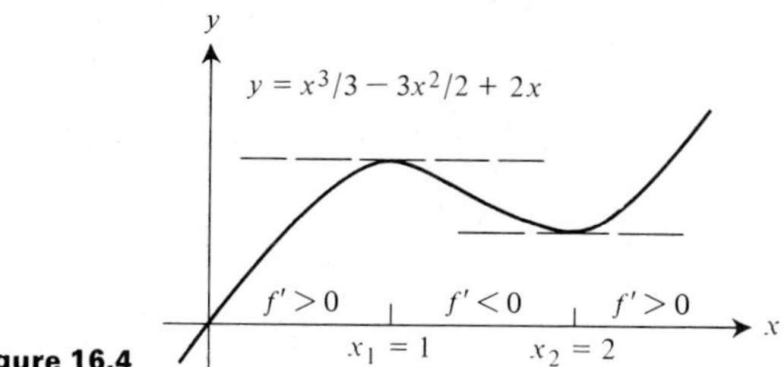

Figure 16.4

c) The double exponential decay function

$$f(x) = e^{-t} - e^{-2t}$$

is typical of a type of function used to describe drug and hormone concentrations after intravenous injections. The graph of this function is indicated in Fig. 16.5. Observe that it is increasing for $0 < t < t_{\max}$ and decreasing for $t > t_{\max}$; the point $t_{\max}$ is a stationary point on the graph. To verify this, we first compute the derivative of $f(x)$,

$$f'(x) = -e^{-t} + 2e^{-2t}.$$

The point $t_{\max}$ is obtained by solving the equation $f'(t) = 0$. This is accomplished by first factoring $f'(t)$:

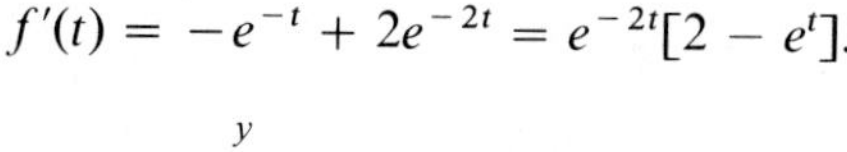

$$f'(t) = -e^{-t} + 2e^{-2t} = e^{-2t}[2 - e^{t}].$$

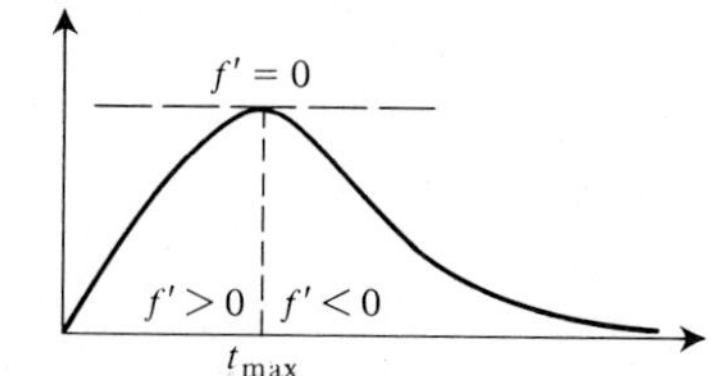

Figure 16.5

Since e^{-2t} can never equal zero, the only way $f'(t)$ can equal zero is for the second factor, $2 - e^t$, to be zero. Thus we solve

$$2 - e^t = 0 \qquad \text{or} \qquad e^t = 2,$$

by taking the natural log of both sides of the equation. This gives the value

$$t_{\max} = \ln(2).$$

Since e^t is an increasing function,

$$\text{if } t < \ln(2), \qquad \text{then } e^t < e^{\ln(2)} = 2 \qquad \text{and} \qquad 0 < 2 - e^t.$$

Hence, using the factored form of $f'(t)$, we find for $0 < t < \ln(2)$,

$$f'(t) = e^{-2t}[2 - e^t] > e^{2t}[0] = 0;$$

that is,

$$f'(t) > 0 \text{ on } (0, t_{\max}).$$

Similarly,

$$\text{if } t > \ln(2), \qquad \text{then } e^t > 2 \qquad \text{and} \qquad 2 - e^t < 0;$$

hence

$$f'(t) = e^{-2t}[2 - e^t] < e^{-2t}[0] = 0;$$

that is,

$$f'(t) < 0 \qquad \text{on} \qquad (t_{\max}, +\infty).$$

d) The logistic equation

$$f(t) = \frac{B}{1 + ce^{-\lambda Bt}}$$

is used to describe sigmoidal phenomena in virtually every scientific discipline. Examples of its applications range from observation of the pupil dilation of pup seals to the growth pattern of cereal crops. (See Fig. 16.6.) One of the

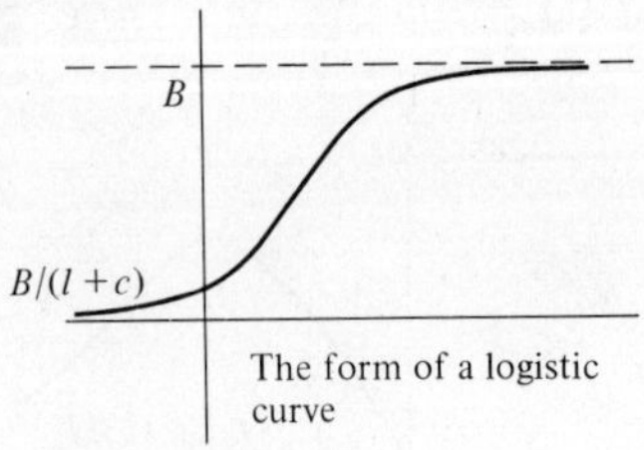

Figure 16.6

important features of a logistic curve is that its graph is increasing for all values of t. This can be verified by evaluating the derivative $f'(t)$, remembering that B, c, and λ are positive constants,

$$f'(t) = \frac{c \cdot \lambda \cdot B^2 \cdot e^{-\lambda Bt}}{[1 + ce^{-\lambda Bt}]^2}.$$

Since the denominator is positive, the exponential $e^{-\lambda Bt}$ is positive, and the constants, λ and c, are positive, the derivative $f'(t)$ is positive for all values of t. ◀

Example 16.2 The amount of water in a lake fluctuates over an annual period due to seasonal changes in rainfall. Assume that the volume of the lake over a one-year period is given by

$$V(t) = V_0 - (A/\pi)\cos(\pi t) - Bt,$$

where V_0 denotes the average natural lake volume, $-(A/\pi)\cos(\pi t)$ denotes the fluctuation in volume due to rainfall, and $-Bt$ denotes a constant release of lake water through a hydroelectric plant.

A recreation authority, which oversees the public use of the lake, is interested in determining when the lake level will be raising and lowering. The lake level is proportional to the lake volume; hence, when the volume decreases the lake level is lowering. The change in V is

$$\mathring{V}(t) = A\sin(\pi t) - B.$$

Thus $\mathring{V}(t) < 0$ when $A\sin(\pi t) < B$ or $\sin(\pi t) < B/A$. As A represents the maximum inflow, it is reasonable to expect $A > B$. (Otherwise, the lake continually drains and will empty.) Over a one-year period, the lake level drops during the periods $(0, t_1)$ and $(t_2, 1)$, where the points t_1 and t_2 are the times when $\sin(\pi t) = B/A$. These values may be obtained graphically as the ordinates of the points, where the curves, $y = \sin(\pi t)$ and $y = B/A$, intersect. This is depicted in Figure 16.7

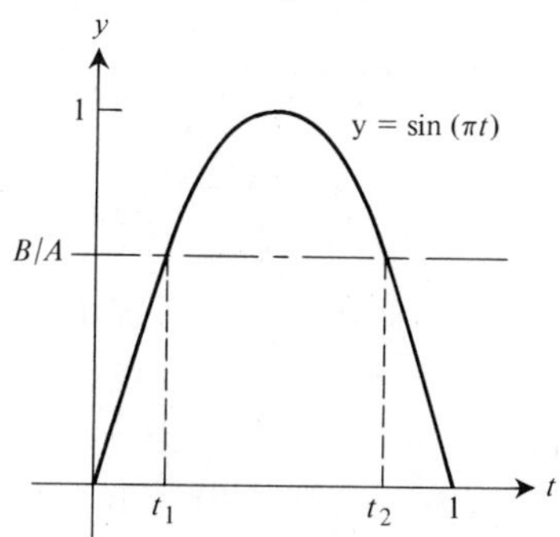

Figure 16.7

These values may also be obtained from a table of the sine function by looking for the value B/A in the body of the table and choosing πt_1, πt_2 as the corresponding "t-values." Another alternative is to use the inverse sine function, arcsin or $\sin^{-1}$, and then let $\pi t_1 = \sin^{-1}(B/A)$ and $t_2 = 1 - t_1$. (See Exercise 15.27.) ◀

When the graph of a function changes from increasing to decreasing or conversely, the graph has an "extremum." By this we mean that the function assumes either a "maximum" or "minimum" value at the point of change. Such values are more cautiously referred to as "local" extrema, as a function may assume greater or lesser values elsewhere. To be more precise, we give the following definition of the terms "extremum" and "maximum" or "minimum."

Definition 16.2 Let f be defined on (a, b) and $x_0 \in (a, b)$.

i) *The function f has a (local) maximum at* $x = x_0$ if $f(x) \leq f(x_0)$ for all x in some open interval containing x_0.
ii) *The function f has a (local) minimum at* $x = x_0$ if $f(x) \geq f(x_0)$ for all x in some open interval containing x_0.
iii) *The function f has an extremum at* $x = x_0$ if f has either a (local) maximum or a (local) minimum at $x = x_0$.

The function specified by the graph in Fig. 16.8(a) has four extrema in the interval (a, b). The x-values denoted by x_1 and x_3 correspond to (local) maximum values of f. The points denoted by x_2 and x_4 correspond to (local) minimum values of f.

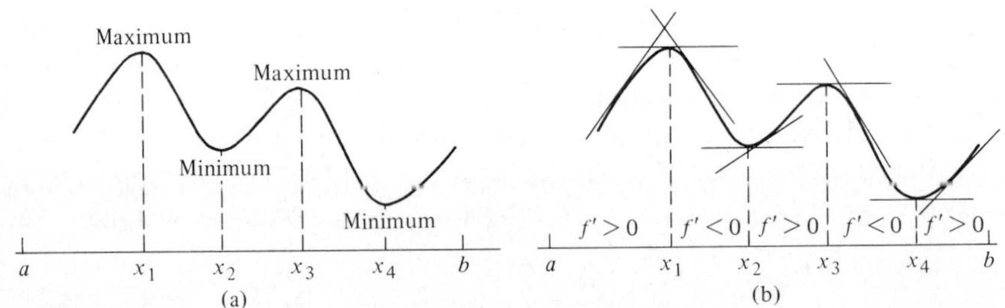

Figure 16.8

In Fig. 16.8(b), we have sketched several tangent lines to the same graph. The four points, x_1, x_2, x_3, and x_4, together with the endpoints, a and b, partition the interval, $[a, b]$, into five subintervals. Note that the sign of the derivative, f', remains constant over each subinterval as indicated. Also note that the tangent lines at the points x_1, x_2, x_3, and x_4, are all horizontal. This reflects the fact that the derivative f is zero at these points. To see why this is true, observe that the point x_1 was chosen as the point where the graph of f changed from increasing to decreasing. For $x < x_1$, $f'(x)$ is positive and for $x > x_1$, $f'(x)$ is negative. If f' is continuous, then as x increases from a value less than x_1 to a value greater than x_1, the values of $f'(x)$ must change at x_1 from positive to negative. Therefore $f'(x_1)$ must be zero. The same argument holds for the other extrema.

Theorem 16.2 If f is differentiable on (a, b) and has an extremum at $x = x_0$, then $f'(x_0) = 0$.

The converse of this theorem is not valid; it is not true that $f'(x_0) = 0$ implies that the graph of f has a maximum or a minimum at x_0. The function $f(x) = x^3$ satisfies $f'(0) = 0$, but the point $x_0 = 0$ corresponds to neither a maximum nor a minimum value of f. Since $f'(x) = 3x^2 > 0$ for all $x \neq 0$, the graph of f is increasing both to the left and to the right of x_0.

In general, to decide whether or not a function has an extremum at a point x_0, we use the following "first derivative test." This test applies to functions that are assumed to have derivatives for all x near a *critical point* x_0.

Definition 16.3 Assume that f is continuous on the interval (a, b). A point x_0 in (a, b) is a *critical point of f* if either
i) $f'(x_0) = 0$ (a stationary point) or
ii) $f'(x_0)$ does not exist.

An example of a critical point that corresponds to condition (ii), $f'(x_0)$ does not exist, is the point $x_0 = 3$ for the function $f(x) = (x - 3)^{(1/3)}$;

$$f'(x) = \frac{1}{3(x - 3)^{(2/3)}},$$

hence $f'(3)$ does not exist.

The First Derivative Test

1. Let x_0 be a critical point of $f(x)$ such that $f'(x)$ exists for all $x \neq x_0$ in an open interval containing x_0.
2. When restricting x to values near x_0,

   A) if $f'(x) < 0$ for $x < x_0$ and $f'(x) > 0$ for $x > x_0$, then f has a (local) minimum at x_0;

   B) if $f'(x) > 0$ for $x < x_0$ and $f'(x) < 0$ for $x > x_0$, then f has a (local) maximum at x_0;

   C) if the sign of $f'(x)$ is the same for $x > x_0$ and for $x < x_0$, then f has neither a maximum nor a minimum at x_0.

Example 16.3 The sketches in the Fig. 16.9 illustrate the three possibilities, (A), (B), and (C), of the *First Derivative Test*. The critical point, $x_0 = 0$, corresponds to a stationary value in (a), (c), and (e) and to an undefined derivative in graphs (b), (d), and (f).

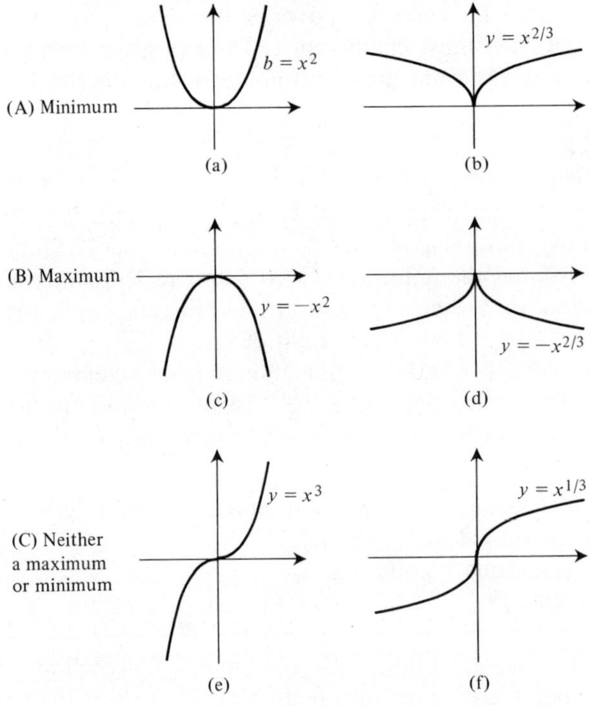

Figure 16.9

Example 16.4 For each of the following functions, we identify all critical points and use the First Derivative Test to determine whether or not they correspond to extrema of the function $f(x)$.

a) Let $f(x) = x^2 - 3x + 10$. Then

$$f'(x) = 2x - 3$$

is defined for all x, and the only critical point is the root, $x_1 = \frac{3}{2}$, of the equation $f'(x) = 0$, $2x - 3 = 0$.

$$\text{For } x < \tfrac{3}{2},\ 2x < 3, \text{ and } 2x - 3 < 0; \text{ hence } f'(x) < 0.$$

$$\text{For } x > \tfrac{3}{2},\ 2x > 3, \text{ and } 2x - 3 > 0; \text{ hence } f'(x) > 0.$$

Consequently, condition A of the First Derivative Test is satisfied and the function f has a maximum at $x_0 = \frac{3}{2}$, $f(x_0) = 7.75$. (See Fig. 16.11(a).)

b) Let $f(x) = xe^{-(x/4)}$. The derivative of f is computed using the product rule:

$$f'(x) = e^{-(x/4)} - (\tfrac{1}{4})xe^{-(x/4)} = e^{-(x/4)}[1 - (x/4)];$$

$f'(x)$ is defined for all x. Setting $f'(x) = 0$ is equivalent to

$$1 - (x/4) = 0 \qquad \text{or} \qquad x_0 = 4.$$

Using the factored form of $f'(x)$, as $e^{-(x/4)}$ is positive for all x, we see that $f'(x)$ has the same sign as the term $1 - (x/4)$.

$$\text{For } x < 4,\ (x/4) < 1, \text{ and } 1 - (x/4) > 0; \text{ hence } f'(x) > 0.$$

$$\text{For } x > 4,\ (x/4) \geq 1, \text{ and } 1 - (x/4) < 0; \text{ hence } f'(x) < 0.$$

Consequently, condition B of the First Derivative Test is satisfied and the function f has a maximum at $x_0 = 4$, $f(x_0) = 4e^{-1} \simeq 1.5$.

c) Let $f(x) = 2x^3 - 3x^2 - 6x + 1$. Then

$$f'(x) = 6x^2 - 6x - 6;$$

$f'(x)$ is defined for all x. The critical points corresponding to $f'(x) = 0$ are the two roots, x_1 and x_2, of the quadratic equation

$$x^2 - x - 1 = 0.$$

Using the quadratic formula, we find that these roots are

$$x_1 = \tfrac{1}{2}(1 + \sqrt{5}) \simeq 1.6 \qquad \text{and} \qquad x_2 = \tfrac{1}{2}(1 - \sqrt{5}) \simeq -0.6.$$

Using the fact that the derivative $f'(x)$ factors into the form

$$f'(x) = 6(x - x_1)(x - x_2),$$

we see that the sign of $f'(x)$ depends on the sign of the two factors $x - x_1$ and $x - x_2$. Using the diagram in Figure. 16.10, we come to the conclusion that

$$f'(x) > \text{ on } (-\infty, x_2) \cup (x_1, +\infty)\ (f \text{ is increasing});$$

$$f'(x) < 0 \text{ on } (x_2, x_1) \qquad\qquad (f \text{ is decreasing}).$$

Consequently, the First Derivative Test tells us that the function, f, has a maximum at x_2, $f(x_2) \simeq 3.1$ and a minimum at x_1, $f(x_1) \simeq -8.1$.

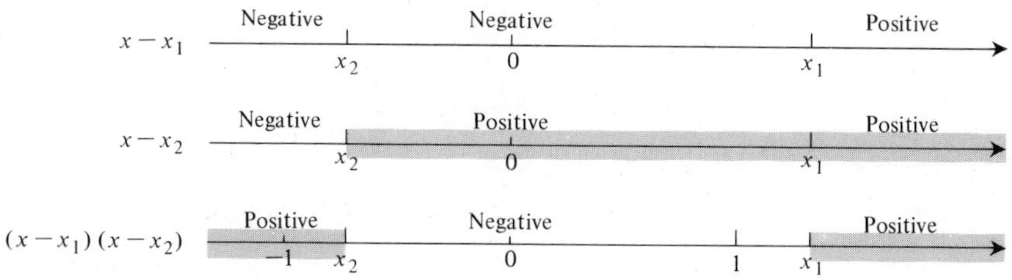

Figure 16.10

d) A more difficult example is given by the function

$$f(x) = x^{1/3}(x - 4)^{2/3}.$$

The derivative of f is easily evaluated using the product rule; however, to investigate its sign, we need to factor it as illustrated:

$$f'(x) = x^{1/3} \cdot (\tfrac{2}{3})(x - 4)^{-1/3} + (x - 4)^{2/3}(\tfrac{1}{3})x^{-2/3}.$$

Since $x^{1/3} = x \cdot x^{-2/3}$ and $(x - 4)^{2/3} = (x - 4)(x - 4)^{-1/3}$,

$$f'(x) = x^{-2/3} \cdot (x - 4)^{-1/3}[(\tfrac{2}{3})x + (\tfrac{1}{3})(x - 4)].$$

Adding the terms inside the square brackets gives us

$$f'(x) = x^{-2/3}(x - 4)^{-1/3}(x - (\tfrac{4}{3})),$$

or

$$f'(x) = \frac{x - (\frac{4}{3})}{x^{(2/3)}(x - 4)^{1/3}}.$$

From this form of $f'(x)$, we see that the function f has three critical points.

The point $x_0 = \frac{4}{3}$	corresponds to	$f'(x) = 0$.
The point $z_1 = 0$	corresponds to	$f'(x)$ does not exist.
The point $z_2 = 4$	corresponds to	$f'(x)$ does not exist.

Rather than examining the sign of the derivative $f'(x)$ for arbitrary x, we utilize the following observation:

> The sign of $f'(x)$ can only change at a critical point.

Therefore, to evaluate the sign of $f'(x)$ for all x between two critical points, we need to check only the sign at one value between the critical points. In

$f'(-1) > 0$ | $f'(1) > 0$ | $f'(2) < 0$ | $f'(5) > 0$

-1, 0 (z_1), 1, (x_0), 2, 3, 4 (z_2), 5

this example, the critical points create four intervals; in each interval, we pick any number x^* and evaluate $f'(x^*)$ to determine the sign of $f'(x)$ on the interval.

Interval	x^*	$f'(x^*)$		
$(-\infty, 0)$	$x^* = -1$	$f'(-1) = \dfrac{-1 - (\frac{4}{3})}{(-1)^{2/3}(-1 - 4)^{1/3}}$	$= \dfrac{-\frac{7}{3}}{(-5)^{1/3}}$	> 0
$(0, \frac{4}{3})$	$x^* = 1$	$f'(1) = \dfrac{1 - (\frac{4}{3})}{(1)^{2/3}(1 - 4)^{1/3}}$	$= \dfrac{-\frac{2}{3}}{(-3)^{1/3}}$	> 0
$(\frac{4}{3}, 4)$	$x^* = 2$	$f'(2) = \dfrac{2 - (\frac{4}{3})}{(2)^{2/3}(2 - 4)^{1/3}}$	$= \dfrac{(\frac{1}{3})}{2^{2/3}(-2)^{1/3}}$	< 0
$(4, +\infty)$	$x^* = 5$	$f'(5) = \dfrac{5 - (\frac{4}{3})}{(5)^{2/3}(5 - 4)^{1/3}}$	$= \dfrac{\frac{11}{3}}{5^{2/3}}$	> 0

Using this information, we conclude that

$z_1 = 0$ corresponds to neither a maximum nor a minimum of f (Case C of the First Derivative Test), since f' does not change sign at 0; $f(0) = 0$.

$x_0 = \frac{4}{3}$ corresponds to a maximum of f (Case B of the First Derivative Test) since f' changes from positive to negative at $\frac{4}{3}$; $f(\frac{4}{3}) \simeq 3.0$.

$z_2 = 4$ corresponds to a minimum of f (Case A of the First Derivative Test) since f' changes from negative to positive at 4; $f(4) = 0$.

(See Fig. 16.11.)

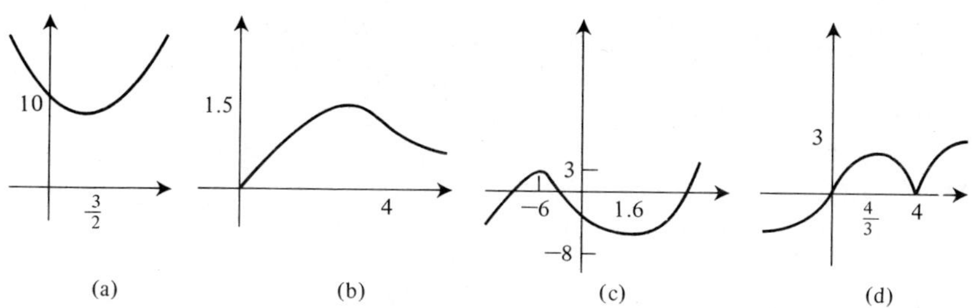

Figure 16.11

◀

Example 16.5

a) A population that varies seasonally is described by the equation $P(t) = P_0 e^{\cos(\pi t)}$, $t \geq 0$. The extremes of the population size occur at the time values t for which $P'(t) = 0$;

$$P'(t) = -\pi \sin(\pi t) P_0 e^{\cos(\pi t)}.$$

$P'(t) = 0$ for the sequence of values $t_n = n$, $n = 0, 1, 2, \ldots$ (the zeros of $\sin(\pi t)$). From a physical point of view, each zero of $P'(t)$ corresponds to either a maximum or a minimum population size. If n is odd, the population at time $t = n$ is at its minimum size, $P_0 e^{-1}$. If n is even, the population at time $t = n$ is at its maximum size, $P_0 e$. This may be verified by using the First Derivative Test since the sign of $P'(t)$ is the same sign as the function $-\sin(\pi t)$.

b) The population size in part (a) oscillates about a middle value, $P = P_0(e + e^{-1})/2$. We can model a population, which actually grows in an oscillatory manner, by introducing a multiple of the exponential term. The function, $P_1(t) = P_0 \cdot te^{\cos(\pi t)}$ is such a function. The extrema of the population P_1 occur when the derivative $P_1'(t) = 0$. Since

$$P_1'(t) = P_0 e^{\cos(\pi - t)}(1 - t\pi \sin \pi t),$$

the zeros of P_1' are the solutions of the equation

$$\sin(\pi t) = 1/\pi t.$$

This equation has a sequence of positive solutions $\{t_1\}_{i=1}^{\infty}$. These are the ordinates of the points where the curves $y = \sin(\pi t)$ and $y = 1/\pi t$ intersect. The first root is $t_1 \simeq 1.11416$, and it can be shown that for large n, $t_n \simeq n\pi + ((-1)^n/n)$. ◀

Optimality principles and problems abound in both theoretical and experimental biology. For instance, in physiology, the optimal distance between branching in blood-vessels is determined to minimize turbulence and maximize the blood flow. In ecology, interspecies relations develop to maximize the utilization of available resources. In cell biology, the shape of a cell is determined to maximize its ability to receive nutrient and to minimize the amount of energy needed for cell function. In experiments, the biochemist tries to maximize the yield of an enzyme being purified; the soil scientist attempts to determine optimal soil conditions for specific crops; and the pharmacologist tries to determine optimal dosages for specific drugs.

The mathematical formulation of all optimization problems is the same. The quantity that is to be maximized or minimized is expressed as a function of another quantity which is taken as an independent variable. This gives an equation $y = f(x)$. The mathematical problem is to determine the extreme values of y for *feasible* values of x. By this, we mean that the set of x values is to be limited or *constrained* to some set, S. In practice, the set S corresponds to values of x that are biologically possible. For instance, the optimal wing span of a bird might be determined by maximizing the lift capacity as a function of the wing span, x; in this case, x must be a positive value, and the set S would be the interval $(0, +\infty)$. The constraint will frequently be expressed as an inequality. For instance, the constraint, $x^2 \geq 4$, is equivalent to requiring that $x \in S = (-\infty, -2] \cup [2, +\infty)$.

A function $f(x)$ may have several local maximum and minimum values on a given set S. The largest maximum value is called a *global maximum* and the smallest minimum value is called a *global minimum*. These terms are defined in reference to a specific set S of x values as follows. (See Fig. 16.12.)

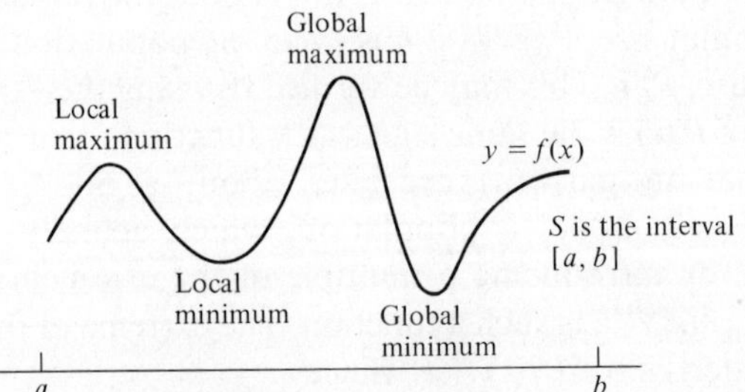

Figure 16.12

Definition 16.4 A function f is said to have

i) a (*global*) *maximum*, M, *on the set* S if, for some $x_M \in S$, $f(x_M) = M$ and, for all $x \in S$, $f(x) \leq M$;

ii) a *(global) minimum, m, on the set S* if, for some $x_m \in S$, $f(x_m) = m$ and, for all $x \in S$, $f(x) \geq m$.

iii) The points x_M and x_m of (i) and (ii) are called *(global) maximum* and *(global) minimum points* of f on S; and the numbers M and m are called the *(global) maximum* value and the *(global) minimum* value of f on S.

The problem of determining the global maximum or minimum of a function $f(x)$ on a set S does not always have a solution.

For instance, the function $f(x) = x$ does not have a maximum value on the set $S = (0, 1)$. The values of $f(x)$ are all less than one and, for any point $x_0 \in (0, 1)$, we can find another point x_1 such that $x_0 < x_1 < 1$ and hence $f(x_0) < f(x_1)$.

Another example is given by the function $f(x)$ defined on the set $S = [0, 2]$ by $f(x) = x - [\![x]\!]$. (This function is graphed in Fig. 6.2.) Again, all values of $f(x)$ are less than one, and there is no value x_0 such that $f(x) \leq f(x_0)$ for all $x \in [0, 2]$.

These two examples involve situations in which either the constraint set S is not a closed interval or the function $f(x)$ is not continuous. When S is a closed interval $[a, b]$ and $f(x)$ is continuous on $[a, b]$ then the Extreme Value Theorem 12.11 assures us that f will have both a global maximum and a global minimum on the set S. By the First Derivative Test and Theorem 16.2, we know that these extrema will either occur at the zeros of the derivative f' or where the derivative does not exist. Consequently, the global extrema can be easily obtained using the following procedure.

Method for Determining the Global Extrema of a Continuous Function *f* on an Interval [*a*, *b*]

Step 1 Compute $f'(x)$ and simplify if possible.

Step 2 Determine the critical points of f on $[a, b]$. These consist of the following:

i) the endpoints, $x = a$ and $x = b$;

ii) the points in $[a, b]$ where $f'(x) = 0$, denoted by $x_1, x_2, \ldots$;

iii) the points in $[a, b]$ where $f'(x)$ does not exist, denoted by $z_1, z_2, \ldots$

Step 3 Evaluate $f(x)$ for all of the critical points $\{a, b, x_1, x_2, \ldots, z_1, z_2, \ldots\}$.

Note: if $f'(x)$ exists for all x, there will not be points $z_1, z_2, \ldots$ Also, if $f'(x) = 0$ has no solutions, there will not be points $x_1, x_2, \ldots$

Step 4 The global maximum is the value

$$M = \text{largest value } f(x) \text{ evaluated in Step 3.}$$

The global minimum

$$m = \text{smallest value } f(x) \text{ evaluated in Step 3.}$$

Example 16.6

a) We determine the extrema of the function $f(x) = x^{1/3}(x - 2)$ on the interval $[-8, 8]$.

Step 1 $f'(x) = (\frac{1}{3})x^{-2/3}(x - 2) + x^{1/3}$. Factoring out the term $x^{-2/3}$, this becomes $f'(x) = x^{-2/3}[(\frac{1}{3})(x - 2) + x]$. Hence

$$f'(x) = \frac{4x - 2}{3x^{2/3}}.$$

Step 2 The critical points are as follows:

i) $a = -8, b = 8$;

ii) since $f'(x) = 0$ for $x = \frac{1}{2}$, $x_1 = \frac{1}{2}$;

iii) as $f'(0)$ is not defined ($0^{-2/3}$ is not defined), $z_1 = 0$.

Step 3

$$f(-8) = (-8)^{1/3}(-8 - 2) = (-2)(-10) = 20.$$

$$f(8) = (8)^{1/3}(8 - 2) = (2)(6) = 12.$$

$$f(\tfrac{1}{2}) = (\tfrac{1}{2})^{1/3}(\tfrac{1}{2} - 2) \simeq -1.19.$$

$$f(0) = 0.$$

Step 4 Maximum $M = 20$ occurs at the left endpoint, $a = -8$. Minimum $m = -(\frac{3}{2})(\frac{1}{2})^{1/3} \simeq -1.19$ occurs at the point $x_1 = \frac{1}{2}$.

b) We determine the extrema of $f(x) = (x - 2)^2(x - 4)^2$ over the interval $[1, 6]$.

Step 1

$$\begin{aligned} f'(x) &= 2(x - 2)(x - 4)^2 + 2(x - 2)^2(x - 4) \\ &= 2(x - 2)(x - 4)[(x - 4) + (x - 2)] \\ &= 2(x - 2)(x - 4)(2x - 6). \end{aligned}$$

Step 2 The critical points are as follows:

i) $a = 1, b = 6$;

ii) $f'(x) = 0$ for $x_1 = 2$, $x_2 = 4$ and $x_3 = 3$;

iii) $f'(x)$ is defined for all x so there are no z values.

Step 3

$$f(1) = (-1)^2(-3)^2 = 9.$$

$$f(6) = (6 - 2)^2(6 - 4)^2 = 64.$$

$$f(2) = (2 - 2)^2(2 - 4)^2 = 0.$$

$$f(4) = (4 - 2)^2(4 - 4)^2 = 0.$$

$$f(3) = (3 - 2)^2(3 - 4)^2 = 1.$$

Step 4 Maximum $M = 64$. Minimum $m = 0$. The maximum occurs at $b = 6$. The minimum occurs at two points, $x_1 = 2$ and $x_2 = 4$.

c) We determine the extrema of $g(t) = -2t^3 + 15t^2 - 36t$ over $[1, 3.5]$.

Step 1 $g'(t) = -6t^2 + 30t - 36 = -6(t^2 - 5t + 6) = -6(t - 2)(t - 3)$.

Step 2 The critical points are as follows:
i) $a = 1, n = 3.5$;
ii) $g'(t) = 0$ for $t_1 = 2$ and $t_2 = 3$;
iii) $g(t)$ exists for all t; no z values.

Step 3 $g(1) = -23; g(3.5) = -28; g(2) = -28; g(3) = -27$.

Step 4 Maximum $M = -23$. Minimum $m = -28$. The maximum occurs at $t = 0$; the minimum occurs at $t = 2$ and $t = 3.5$. ◀

The above procedure has two main limitations. First, it applies only to closed intervals $[a, b]$ and is only practical when there are few critical points. A second limitation is that Step 3 of the method requires the evaluation of $f(x)$ at the critical points. If the function f is complicated or the critical points are not rational numbers, the evaluations can be difficult. But, more importantly, in the theoretical development of a model, the critical points may depend on unknown or unspecified parameters. When this happens, the exact numerical values of f at these critical points cannot be obtained. For the mathematical analysis of a model, it may be essential to determine which critical points, if any, correspond to extrema. To handle situations in which these limitations present difficulties, we need a stronger result than that of Theorem 16.2. In the next subsection, we develop a test to decide if a critical point corresponds to a maximum or a minimum value of the function by looking at the second derivative.

16.2 USING THE SECOND DERIVATIVE TO DESCRIBE A FUNCTION

What distinguishes a maximum point from a minimum point on a graph? From the first derivative, they cannot be distinguished since both correspond to a vanishing derivative. Geometrically, the difference is obvious and is easily described. At a minimum point, a graph is U-shaped and at a maximum point the graph resembles an inverted U. How can we mathematically describe the concept of being U-shaped or the shape of an inverted U? To describe these concepts we introduce a new term, "concave." A graph will be called *concave upward* if it is U-shaped and *concave downward* if it has the shape of an inverted U. The definition of concave upward and concave downward is based on the relationship between the graph of $y = f(x)$, and the tangent line to the graph at a point $(x_0, f(x_0))$. In Fig. 16.13, we illustrate the fact that a U-shaped curve, concave upward, is above each tangent line, while a curve with an inverted-U shape, concave downward, is always below each tangent line. The algebraic description of these properties is given in the following definition, using the equation

$$y = f'(x_0)(x - x_0) + f(x_0),$$

of the tangent line at $(x_0, f(x_0))$.

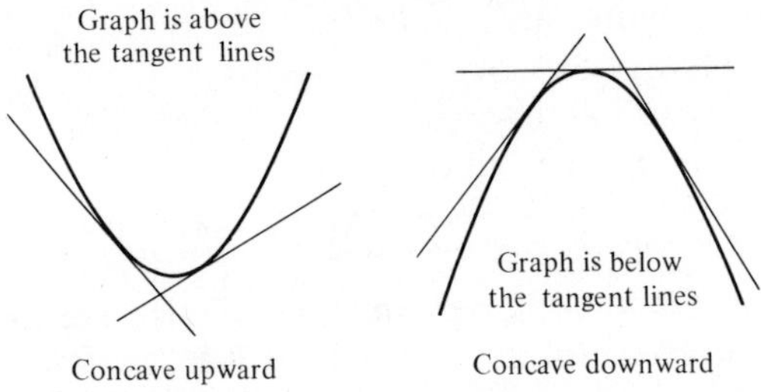

Figure 16.13

Definition 16.5 Let f be differentiable at $x = x_0$.

i) *The graph of f is concave downward at $x = x_0$* if for all x in some open interval containing x_0,
$$f(x) \leq f'(x_0)(x - x_0) + f(x_0)$$
(i.e., the graph of f is below the tangent line).

ii) *The graph of f is concave upward at $x = x_0$* if for all x in some open interval containing x_0,
$$f(x) \geq f'(x_0)(x - x_0) + f(x_0)$$
(i.e., the graph of f is above the tangent line).

iii) *A function f is concave downward (or upward) on an interval* (a, b) if f is concave downward (or upward) at each point in the interval.

Sometimes the term "concave" is used to mean *concave downward* and the term "convex" is introduced to mean *concave upward*. Definitions of these terms may be given in which the function f is not assumed to be differentiable and thus may not have a tangent line at $(x_0, f(x_0))$.

Since a graph will be concave downward at maximum points and concave upward at minimum points, we may use the concavity of a graph to decide whether a critical point corresponds to a maximum or minimum. To do this, we must be able to determine whether a graph is concave upward or concave downward. The key to determining the concavity of a graph is the following observation.

As a point moves along a curve, $y = f(x)$, past a maximum point, $(x_0, f(x_0))$, the slope of the tangent line to the curve at the moving point must decrease from a positive value to a negative value. The slope of the tangent line is the derivative $f'(x)$. Thus $f'(x)$ must be decreasing on the interval where the graph of f is concave downward. By Theorem 16.1, when the derivative $f'(x)$ is decreasing, its derivative $D_x(f'(x))$ must be negative. Hence, if the graph of f is concave downward, the second derivative $f''(x) = D_x(f'(x))$ must be negative. A similar argument shows that when the graph of f is concave upward (near a minimum), then the second derivative $f''(x)$ must be positive. These conclusions are valid even when there are no extrema.

Theorem 16.3 Assume that f has a second derivative on an interval (a, b) and $x_0 \in (a, b)$.

1. If $f''(x_0) > 0$, then the graph of f is concave upward at $x = x_0$
2. If $f''(x_0) < 0$, then the graph of f is concave downward at $x = x_0$.

Example 16.7

a) The graph of the parabola $y = 3x^2 + 2x - 1$ is concave upward for all x. To verify this we compute the derivatives, $y' = 6x + 2$ and $y'' = 6$. Therefore, $y'' > 0$ for all x. (See Fig. 16.14(a).)

b) To examine the concavity of the graph of $y = x^3 - 4x^2 + 4x + 3$, we compute the derivatives $y' = 3x^2 - 8x + 4$ and $y'' = 6x - 8$. Since $y'' > 0$ if $x > \frac{4}{3}$, and $y'' < 0$ if $x < \frac{4}{3}$, we conclude that the graph is concave upward on the interval $(\frac{4}{3}, +\infty)$ and concave downward on the interval $(-\infty, \frac{4}{3})$. Theorem 16.3 provides no information about the concavity of the graph when $x = \frac{4}{3}$. (See Fig. 16.14(b).)

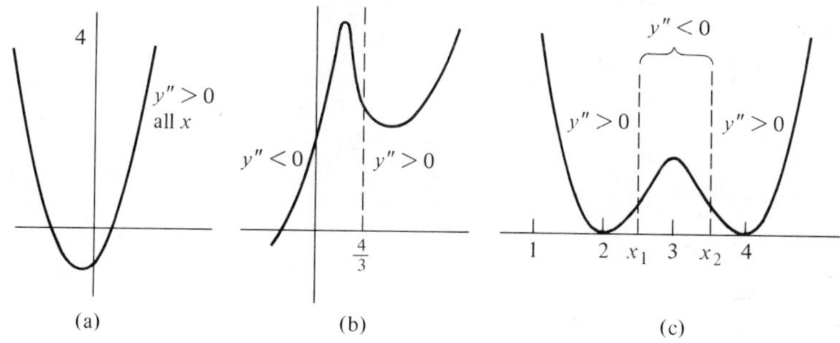

Figure 16.14

c) To examine the concavity of the graph of $y = (x - 2)^2(x - 4)^2$, we must compute $y''(x)$. As in Example 16.6(b), $y' = 2(x - 2)(x - 4)(2x - 6)$. Hence, applying the product rule twice, we find

$$\begin{aligned} y'' &= 2[(x - 4)(2x - 6) + (x - 2)(2x - 6) + 2(x - 2)(x - 4)] \\ &= 4[(x^2 - 7x + 12) + (x^2 - 5x + 6) + (x^2 - 6x + 8)] \\ &= 4(3x^2 - 18x + 26). \end{aligned}$$

The second derivative is a quadratic function of x with a positive x^2 coefficient. Therefore, $y''(x)$ will be positive on two intervals $(-\infty, x_1)$ and $(x_2, +\infty)$ and negative on the interval (x_1, x_2), where $x_1 < x_2$ are the roots of $y''(x) = 0$. (See Fig. 16.15.) Using the quadratic formula, we find that the roots of $y''(x) = 0$ are $x_1 = 3 - (1/\sqrt{3})$ and $x_2 = 3 + (1/\sqrt{3})$. Therefore the graph of $y = (x - 2)^2(x - 4)^2$ is concave downward on the interval $(3 - (1/\sqrt{3}),$

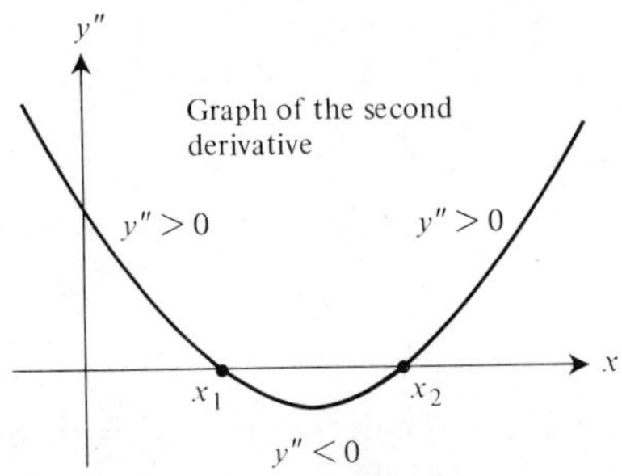

Figure 16.15

$3 + (1/\sqrt{3})$) and concave upward on the intervals $(-\infty, 3 - (1/\sqrt{3}))$ and $(3 + (1/\sqrt{3}), +\infty)$. (See Fig. 16.14(c).) ◀

Combining Theorems 16.2 and 16.3, we obtain the following result known as the *Second Derivative Test*. (See Fig. 16.16(a).)

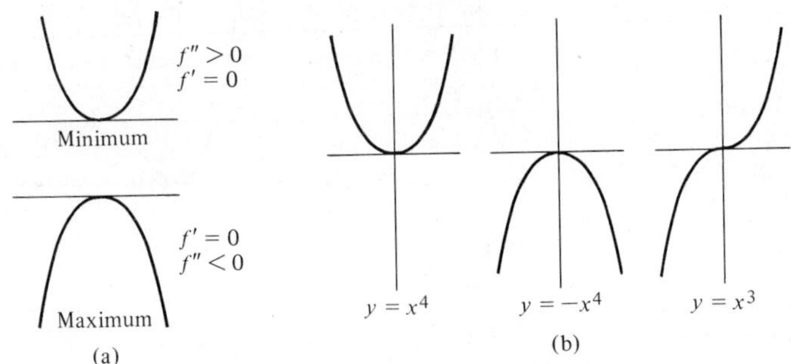

Figure 16.16

Theorem 16.4 Assume that the function f is twice differentiable on an interval (a, b)—that is, $f'(x)$ and $f''(x)$ exist. If $x_0 \in (a, b)$ and $f'(x_0) = 0$, then

i) if $f''(x_0) < 0$, f has a local maximum at x_0:
ii) if $f''(x_0) > 0$, f has a local minimum at x_0:
iii) if $f''(x_0) = 0$, no conclusion may be made—that is, f may have a maximum, a minimum, or no extremum at x_0.

The necessary ambiguity of conclusion (iii) of Theorem 16.4 is demonstrated by considering the three functions, $f_1(x) = x^4$, $f_2(x) = -x^4$, and $f_3(x) = x^3$. At the critical point, $x_0 = 0$, the first and second derivatives of each of these functions are zero. However, as illustrated in Fig. 16.16(b), f_1 has a minimum, f_2 has a maximum, and f_3 has neither a maximum nor a minimum at $x_0 = 0$.

Example 16.8 In Example 8.3, the concentration of a drug circulating in the blood t hours after administration of a single dose was described by the equation

$$C(t) = 122(e^{-0.18t} - e^{-t}).$$

To verify that this function has only one maximum value for $t > 0$, we apply the Second Derivative Test. First, we compute

$$C'(t) = 122(-0.18e^{-0.18t} + e^{-t}).$$

The solution of $C'(t) = 0$ is obtained by factoring $C'(t)$:

$$C'(t) = 122e^{-0.18t}[-0.18 + e^{-0.82t}].$$

From this equation, we see that $C'(t) = 0$ implies that

$$-0.18 + e^{-0.82t} = 0.$$

Solving this equation, we find that the only critical point is

$$t_0 = -\ln(0.18)/0.82.$$

Next, we compute the second derivative,

$$C''(t) = 122[(0.18)^2 e^{-0.18t} - e^{-t}],$$

or

$$C''(t) = 122e^{-0.18t}[(0.18)^2 - e^{-0.82t}].$$

To evaluate $C''(t_0)$, we use the identity

$$e^{-0.82t_0} = e^{-(0.82)(-\ln(0.18)/0.82)}$$
$$= e^{\ln(0.18)} = 0.18.$$

This leads to the equation

$$C''(t_0) = 122e^{-0.18t_0}[(0.18)^2 - (0.18)].$$

Since $0.18 < 1$, $(0.18)^2 < 0.18$, and hence $C''(t_0) < 0$. By Theorem 16.4 part (i), the function $C(t)$ must have a maximum value at t_0. As this is the only critical value, the maximum must also be a global maximum, ($\lim_{t \to +\infty} c(t) = 0$). ◀

Example 16.9 The following function indicates the relationship between the percentage of red-blood cells, called the *hematocrit level H*, and the oxygen, O_2, supplied by an individual's blood. A more extensive discussion is included in Example 23.4.

$$O_2 = mH\mathring{Q}_{pl}e^{-K_1 H}.$$

Treating K_1, m, and $\mathring{Q}_{pl}$ as constants, the derivative $D_H(O_2)$ is computed using the product rule; upon simplification this becomes

$$\frac{dO_2}{dH} = [1 - K_1 H]m\mathring{Q}_{pl}e^{-K_1 H}.$$

From this equation, we find that the only critical value of H is $H_0 = 1/K_1$. The second derivative

$$\frac{d^2O_2}{dH^2} = -K_1 m\mathring{Q}_{pl}e^{-K_1H} - K_1[1 - K_1H]m\mathring{Q}_{pl}e^{-K_1H}$$

$$= -K_1 m\mathring{Q}_{pl}e^{-K_1H} - K_1\left(\frac{dO_2}{dH}\right).$$

Consequently, the second derivative evaluated at H_0 is given by

$$\frac{d^2O_2}{dH^2}(H_0) = -K_1 m\mathring{Q}_{pl}e^{-K_1H_0} - K_1\left(\frac{d}{dH}O_2(H_0)\right)$$

$$= -K_1 m\mathring{Q}_{pl}e^{-1}.$$

Since $K_1 > 0$, $m > 0$, $\mathring{Q}_{pl} > 0$, and $e^{-1} > 0$, we find

$$\frac{d^2}{dH^2}O_2(H_0) < 0.$$

Thus H_0 must be the optimal hematocrit level to provide a maximum supply of oxygen. This is illustrated in Fig. 16.17.

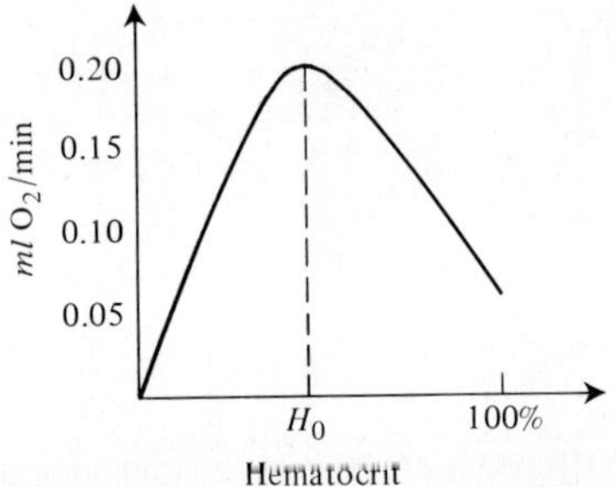

Figure 16.17

16.3 CURVE SKETCHING USING CALCULUS

In this section, we indicate how the First and Second Derivative Tests may be used to sketch the graph of a function. We first introduce one additional concept to describe the points where the second derivative is zero.

In Fig. 16.18, we illustrate the same function depicted in Fig. 16.8(a). The three x-values indicated by σ_1, σ_2, and σ_3 partition the interval $[a, b]$ into four subintervals, (a, σ_1), (σ_1, σ_2), (σ_2, σ_3), and (σ_3, b). Over each subinterval, the sign of the second derivative remains constant. Over the intervals, (a, σ_1) and (σ_2, σ_3), the second derivative is negative, $f''(x) < 0$, and hence the graph is concave downwards on these intervals. On the intervals (σ_1, σ_2) and (σ_3, b), the second derivative is positive, $f''(x) > 0$, and the graph is concave upward. (We are assuming that the second derivative exists for the function f depicted in the graph.)

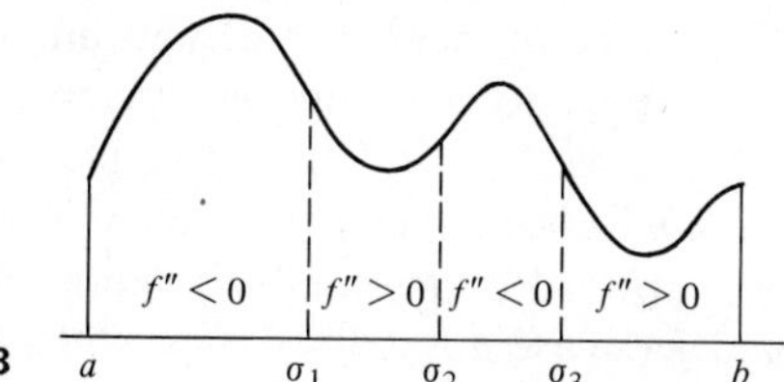

Figure 16.18

The points where the function f has extrema are zeros of $f'(x)$, since they are the points where the derivative changes sign. Similarly, the points where the second derivative changes sign are zeros of $f''(x)$, provided f'' is continuous. Thus the points σ_1, σ_2, and σ_3, corresponding to where the graph of f changes its concavity (from upwards to downwards or conversely) are zeros of f''—that is, $f''(\sigma_1) = 0$, $f''(\sigma_2) = 0$, and $f''(\sigma_3) = 0$.

Again, as in the case of extreme points, you might be tempted to conjecture that the concavity of a function changes whenever the second derivative has a zero. That this is not true may be demonstrated by considering the function $f(x) = x^4$. The graph of $y = x^4$ is concave upward for all x. For nonzero x, we may use Theorem 16.3, since $f''(x) = 12x^2$ is always positive if $x \neq 0$. If $x = 0$, $f''(0) = 0$, and Theorem 16.3 is not applicable. However, the graph of f lies above the tangent line, $y = 0$, for all $x \neq 0$. Using Definition 16.5, we conclude that the graph is concave upward at $x = 0$. Consequently, $f''(0) = 0$, but the concavity of the graph does not change at $x = 0$.

The points at which the graph of a function changes its concavity are useful when sketching the graph. These are the points where you stop your pen so that you can change the flow of the graph. They are referred to as *inflection points*. (See Fig. 16.19.)

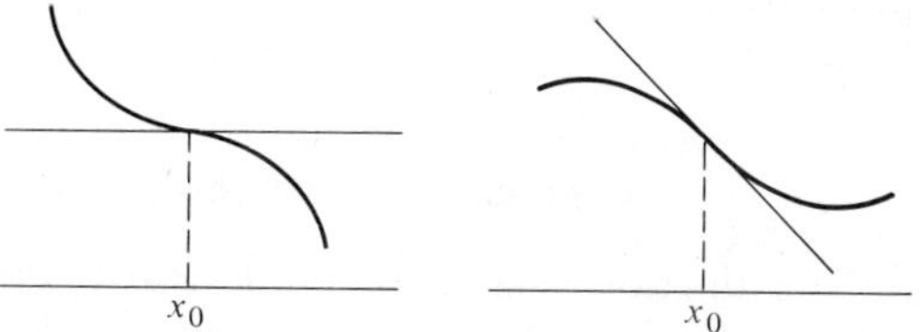

Figure 16.19 Horizontal inflection point Nonhorizontal inflection point

Definition 16.6 *The graph of a function f has an inflection at the point* $(x_0, f(x_0))$ if there exists a $\delta > 0$ such that the graph of f is concave upwards (or downwards) in the interval $(x_0 - \delta, x_0)$ and the graph of f is concave downwards (or upwards), respectively, in the interval $(x_0, x_0 + \delta)$. That is, f has an inflection at the *inflection point* x_0 if the graph of f changes its concavity at $x = x_0$.

Theorem 16.5 If f has a second derivative on an interval (a, b) containing the point x_0, and f has an inflection at x_0, then $f''(x_0) = 0$.

Theorem 16.5 is more restrictive than you might expect at first glance. For instance, it does not apply to the function $f(x) = x^{1/3}$. As $f'(x) = (\frac{1}{3})x^{-2/3}$ and $f''(x) = (-\frac{2}{9})x^{-5/3}$, we see that $f''(x)$ is positive for $x \in (-\infty, 0)$ and $f''(x)$ is negative for $x \in (0, \infty)$. Hence the graph of $y = x^{1/3}$ changes its concavity at $x = 0$ from being concave upward on $(-\infty, 0)$ to being concave downward on $(0, \infty)$. However, $f''(x)$ and $f'(x)$ are not defined at $x = 0$.

Example 16.10

a) The function, $y = \sin(x)$, has an inflection point at each multiple of π; $y'(x) = \cos(x)$ and $y''(x) = -\sin(x)$. The function $y''(x)$ changes sign at each multiple of π, and hence these points are inflection points.

b) A quadratic function has no inflection points. If y is a quadratic, then y has the form $y = a_0 + a_1x + a_2x^2$, where $a_2 \neq 0$. Then $y'(x) = a_1 + 2a_2x$ and $y''(x) = 2a_2$. Hence y'' is of constant sign, and the graph of the quadratic is concave upward if $a_2 > 0$ and concave downward if $a_2 < 0$.

c) The cubic $y = (x - 2)^3$ has an inflection point at $x = 2$. Since $y''(x) = 6(x - 2)$, $y''(2) = 0$, and $y''(x) > 0$, if $x > 2$, and $y''(x) > 0$, if $x < 2$. Thus the graph of y is concave downward if $x < 2$ and concave upward if $x > 2$. ◀

The conclusions of Theorems 16.1 to 16.5 indicate relationships between the sign of the first and second derivative of a function f and the shape of the graph of f. The diagram given in Fig. 16.20 illustrates the possible shapes of the graph of a function f corresponding to the nine possible combinations of the signs of f' and f''. To simplify the diagram, we have assumed that $f(x)$ is positive and $f''(x)$ is continuous on an interval containing x_0.

The following general procedure for sketching the graph of a function over a given interval $[a, b]$ reflects the results of the theorems of this section.

Method of Curve Sketching

Assumptions: The function f has a second derivative defined on the open interval (a, b) and f is continuous on the closed interval $[a, b]$.

Step 1 Compute $f'(x)$ and determine all roots of $f'(x) = 0$ in the interval $[a, b]$. Denote these by $x_1, x_2, \ldots, x_{n_1}$. (Usually n_1 will be small.) Plot the points $(x_i, f(x_i))$ for $i = 1, 2, \ldots, n_1$, and indicate the sign of f' in between the points, x_i.

Step 2 Compute $f''(x)$ and determine all roots of $f''(x) = 0$ in the interval $[a, b]$. Denote these by $\sigma_1, \sigma_2, \ldots, \sigma_{n_2}$. Plot the points $(\sigma_i, f(\sigma_i))$ for $i = 1, 2, \ldots, n_2$. Compute $f'(\sigma_i)$ and sketch (in pencil) a portion of the tangent line,

$$y = f'(\sigma_i)(x - \sigma_i) + f(\sigma_i) \qquad \text{for } i = 1, 2, \ldots, n_2.$$

Indicate the sign of f'' in between the points σ_i.

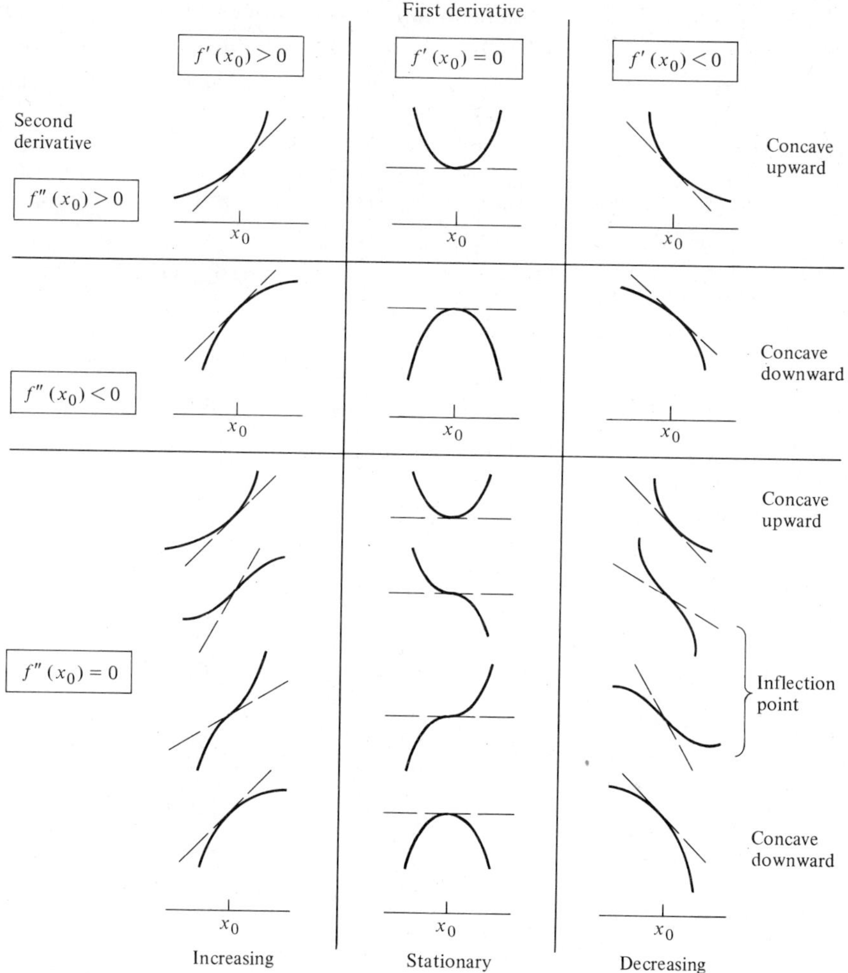

Figure 16.20

Step 3 Evaluate $f''(x_i)$ for $i = 1, 2, \ldots, n_1$.

a) If $f''(x_i)$ is positive, sketch (in pencil) a concave upward curve at $(x_i, f(x_i))$.

b) If $f''(x_i)$ is negative, sketch (in pencil) a concave downward curve at $(x_i, f(x_i))$.

c) If $f''(x_i)$ is zero, evaluate $f''(x)$ for two x values close to x_i, one greater than x_i and one less than x_i. If these values are both positive or both negative, f has either a minimum or maximum at x_i (you cannot tell which). If they differ in sign, x_i is an inflection point. Sketch (in pencil) the appropriate curve. (See Fig. 16.20.)

Step 4 Plot the points $(a, f(a))$ and $(b, f(b))$. If the form of the graph is dubious with the above information, plot additional points. For each additional point, x_0, plot $(x_0, f(x_0))$ and sketch the tangent line $y = f'(x_0)(x - x_0) + f(x_0)$. Also evaluate $f''(x_0)$ and sketch a curve similar to the corresponding curve in Fig. 16.20.

Step 5 Connect the plotted points using the pencil sketches as an aid. Then erase the pencil sketches.

Example 16.11 We illustrate the above method by sketching the graph of $f(x) = x^3 - 9x^2 + 24x$ over the interval $[a, b] = [0, 5]$.

Step 1 $f'(x) = 3x^2 - 18x + 24$. To solve $f'(x) = 0$, we factor $f'(x)$ and obtain

$$3(x - 4)(x - 2) = 0.$$

The roots are $x_1 = 4$ and $x_2 = 2$. The plotted values are all indicated in Fig. 16.21; $f(2) = 20$ and $f(4) = 16$; $f'(x) < 0$ on $(2, 4)$; otherwise it is positive.

Step 2 $f''(x) = 6x - 18 = 6(x - 3)$. Therefore $f''(x) = 0$ for $x = \sigma_1 = 3$; $f(3) = 18$ and $f'(3) = -3$, so the tangent line is

$$y = -3(x - 3) + 18.$$

As $f''(x) < 0$ for $x < 3$ and $f''(x) > 0$ for $x > 3$, $\sigma_1 = 3$ is an inflection point of the graph.

Step 3 $f''(4) = 6 > 0$, so the graph has a local minimum point at $(4, 16)$; $f''(2) = -6 < 0$, so the graph has a local maximum point at $(2, 20)$.

Step 4 $f(0) = 0$ and $f(5) = 20$.

Step 5 From the points plotted above, we fill in the graph as indicated in Fig. 16.21.

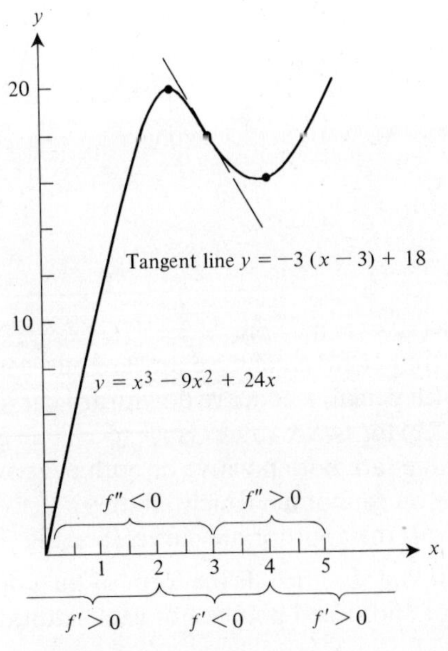

Figure 16.21 ◀

Example 16.12 We sketch the graph of $f(x) = xe^{-x}$ over $[0, 10]$ using the above method.

Step 1 $f'(x) = -xe^{-x} + e^{-x} = (1 - x)e^{-x}$. Therefore, $f'(x) = 0$ has one root, $x_1 = 1$; $f(1) = e^{-1} \simeq 0.368$; $f'(x) > 0$ if $0 < x < 1$ and $f'(x) < 0$ if $x > 1$, since the sign of f' is the same as the sign of $(1 - x)$.

Step 2 $f''(x) = -(1 - x)e^{-x} - e^{-x} = (x - 2)e^{-x}$. Therefore, $f''(x) = 0$ at $x = \sigma_1 = 2$; $f(2) = 2e^{-2} \simeq 0.271$. As $f'(2) = -e^{-2} \simeq -0.135$, the tangent line at $(2, f(2))$ is

$$y = -e^{-2}(x - 2) + 2e^{-2} = -(x + 4)e^{-2}.$$

As $f''(x) < 0$ for $x < 2$ and $f''(x) > 0$ for $x > 2$, $\sigma_1 = 2$ is an inflection point.

Step 3 $f''(1) = -e^{-1} < 0$; hence the graph has a maximum value at $(1, e^{-1})$.

Step 4 $f(0) = 0$ and $f(10) = 10e^{-10} \simeq 0.0004$. As the form of the graph is not especially clear from the points plotted, we plot more points. (This is frequently necessary when the number of zeros of f' and f'' is small.)

Step 5 The points plotted are indicated on the graph in Fig. 16.22, where the function is sketched.

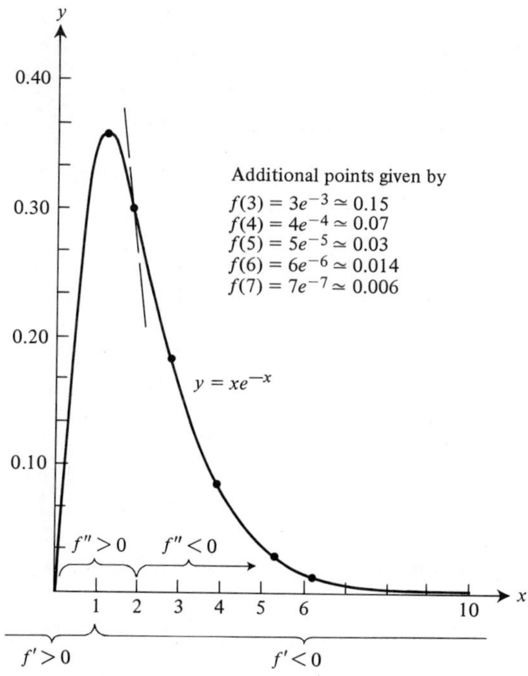

Figure 16.22

◀

Example 16.13 We sketch the graph of the function,

$$g(t) = t^3 + t^2 - t + 1.$$

Step 1 $g'(t) = 3t^2 + 2t - 1 = (3t - 1)(t + 1)$. Thus

$$g'(t) = 0 \qquad \text{for } t_1 = \tfrac{1}{3} \qquad \text{and} \qquad t_2 = -1.$$

$$g(\tfrac{1}{3}) = (\tfrac{1}{27}) + (\tfrac{1}{9}) - (\tfrac{1}{3}) + 1 = \tfrac{22}{27} \simeq 0.8.$$

$$g(-1) = -1 + 1 + 1 + 1 = 2.$$

$$g'(t) < 0 \qquad \text{if } -1 < t < \tfrac{1}{3}, \qquad \text{otherwise } g'(t) > 0.$$

Step 2 $g''(t) = 6t + 2$. Therefore $g''(t) = 0$ when $t = \sigma_1 = -\frac{1}{3}$.

$$g(-\tfrac{1}{3}) = (-\tfrac{1}{27}) + (\tfrac{1}{9}) + (\tfrac{1}{3}) + 1 = \tfrac{38}{27} \simeq 1.4.$$

Since $g''(t) < 0$ for $t < -\frac{1}{3}$ and $g''(t) > 0$ for $t > -\frac{1}{3}$, we conclude that the graph has an inflection point at $t = -\frac{1}{3}$. The tangent line at the point $(-\frac{1}{3}, \frac{38}{27})$ is

$$y = (-\tfrac{4}{3})(t + (\tfrac{1}{3})) + (\tfrac{38}{27})$$

since $g'(-\frac{1}{3}) = (\frac{1}{3}) - (\frac{2}{3}) - 1 = -\frac{4}{3}$.

Step 3 $g''(\frac{1}{3}) = 4$, so the graph has a local minimum at $t_1 = \frac{1}{3}$; $g''(-1) = -4$, so the graph has a local maximum at $t_2 = -1$.

Step 4 In this example, we did not specify an interval $[a, b]$. Instead, we simply note that as $t \to +\infty$, $g(t) \to +\infty$, and as $t \to -\infty$, $g(t) \to -\infty$.

Step 5 The points determined above are sketched in Fig. 16.23.

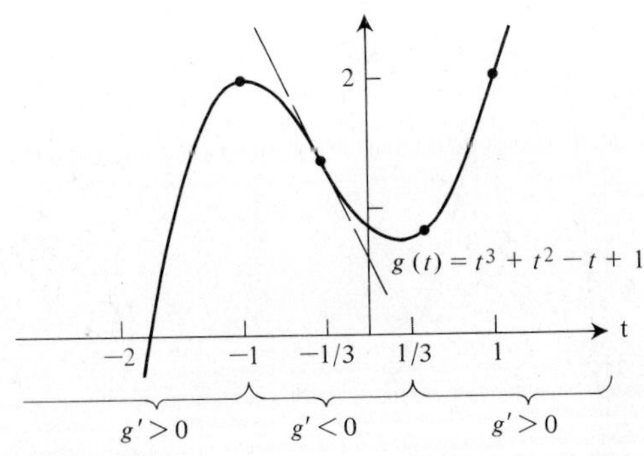

Figure 16.23

◀

SUMMARY

1. *Definitions*:
   a) f is *increasing on* (a, b) if $x_1 < x_2 \Rightarrow f(x_1) < f(x_2)$.
   b) f is *decreasing on* (a, b) if $x_1 < x_2 \Rightarrow f(x_1) > f(x_2)$.

c) f is *increasing at* x_0 if $f'(x_0) > 0$.
d) f is *decreasing at* x_0 if $f(x_0) < 0$.
e) f is *stationary at* x_0 if $f'(x_0) = 0$.
f) f has a (*local*) *maximum* at x_0 if, for all x near x_0, $f(x) \leq f(x_0)$.
g) f has a (*local*) *minimum* at x_0 if, for all x near x_0, $f(x) \geq f(x_0)$.
h) x_0 is a *critical point of* f if $f'(x_0) = 0$ or $f'(x_0)$ does not exist.
i) f has a (*global*) *maximum*, M, on a set S if there is a point x_M in S such that $M = f(x_M)$, and $f(x) \leq M$ for all $x \in S$.
j) f has a (*global*) *minimum*, m, on a set S if there is a point x_m in S such that $m = f(x_m)$ and $f(x) \geq m$ for all $x \in S$.
k) The graph of f is *concave downward at* x_0 if the graph of f is below the tangent line at $(x_0, f(x_0))$.
l) The graph of f is *concave upward at* x_0 if the graph of f is above the tangent line at $(x_0, f(x_0))$.
m) The graph of f has an *inflection* at $(x_0, f(x_0))$ if the graph changes its concavity at x_0.

2. *Theorems*:

16.1 If $f'(x) > 0$ on (a, b), f is increasing on (a, b).
If $f'(x) < 0$ on (a, b), f is decreasing on (a, b).
If $f'(x) = 0$ on (a, b), f is constant on (a, b).

16.2 If f has an extremum at x_0, then $f'(x_0) = 0$ (provided $f'(x_0)$ exists!).

16.3 If $f''(x_0) > 0$, the graph of f is concave upward at x_0.
If $f''(x_0) < 0$, the graph of f is concave downward at x_0.

16.4 If $f'(x_0) = 0$ and
i) $f''(x_0) < 0$, f has a local maximum at x_0;
ii) $f''(x_0) > 0$, f has a local minimum at x_0;
iii) $f''(x_0) = 0$ (no conclusion may be made).

16.5 If f has an inflection at $(x_0, f(x_0))$, then $f''(x_0) = 0$ (provided $f''(x_0)$ exists!).

3. *Techniques*:

a) *First Derivative Test*: Check the sign of $f'(x)$ for x on both sides of a critical point x_0.
i) $f(x_0)$ is a maximum if f' changes from positive to negative at x_0.
ii) $f(x_0)$ is a minimum if f' changes from negative to positive at x_0.
iii) $f(x_0)$ is neither a maximum nor a minimum if f' does not change sign at x_0.

b) *Finding maximum and minimum on an interval* $[a, b]$:
i) First determine all critical points of f on $[a, b]$.
ii) Evaluate $f(x)$ at all critical points plus the endpoints, $x = a$ and $x = b$.
iii) The maximum M = largest evaluated value of $f(x)$.
iv) The minimum m = smallest evaluated value of $f(x)$.

c) *Second Derivative Test*:
   i) Determine all critical points of f.
   ii) Evaluate $f''(x_0)$ for each critical point x_0.
   iii) Use theorem (d) in the Summary to determine if f has a maximum or minimum at x_0.

d) *Curve sketching*:
   i) First determine the critical points of f, and indicate where the graph is increasing and decreasing.
   ii) Determine the possible inflection points where $f'' = 0$; determine if the critical points are maximum or minimum points; indicate the concavity of the function.
   iii) Combine the above information with the plotting of the above points to sketch the graph. Use Fig. 16.20 as a guide.

EXERCISE SET 16

16.1 Determine if the given function is increasing or decreasing at $x = 2$.

a) $f(x) = x^2 - 2x + 3$
b) $f(x) = 3e^{x^2}$
c) $f(x) = (x^2 - 2)\ln(x)$
d) $f(x) = \dfrac{x+1}{x-1}$
e) $f(x) = 1 + \dfrac{1}{x}$
f) $f(x) = x\left(1 + \dfrac{1}{x}\right)$
g) $f(x) = x^4 - 3x^3$
h) $f(x) = x(1 - \ln(x))$
i) $f(x) = \cos(x)$
j) $f(x) = \tan(x)$
k) $f(x) = \sec(x)$
l) $f(x) = x^2e^{-x}$
m) $f(x) = x^2e^{\cos(x)}$
n) $f(x) = \ln(\cos^2(x))$
o) $f(x) = e^{-x^2/2}$
p) $f(x) = x^2/(x - 1)$

16.2 Determine the intervals on which the given function is increasing.

a) $y = \sin(t - \pi)$
b) $y = x^2e^{-x}$
c) $y = 4x^3 - 3x^2 - 2x + 6$
d) $y = (x - 2)^2(x + 2)^2$
e) $y = 3(e^{-2t} - e^{-t})$
f) $y = \ln(x)/x$

16.3 List all critical points for the function $f(x)$ and use the First Derivative Test to determine if they correspond to extrema.

a) $f(x) = x^2 - 1$
b) $f(x) = x^2 - x$
c) $f(x) = x^3 - x$
d) $f(x) = x^3 - 12x + 6$
e) $f(x) = -6x^3 - 15x^2 - 36x + 3$
f) $f(x) = xe^{-2x}$
g) $f(x) = e^{-2x} - e^{-4x}$
h) $f(x) = e^{-5x} - e^{-x}$
i) $f(x) = \cos(x)$
j) $f(x) = x^2e^{-x}$
k) $f(x) = (x + 2)^2(x - 2)^2$
l) $f(x) = e^{-x^2}$
m) $f(x) = e^{-(x-2)^2}$
n) $f(x) = 1/(1 + x^2)$
o) $f(x) = x^{(2/5)}$

16.4 Determine all extrema of the functions indicated over the given interval. Also indicate the critical points.

a) $f(x) = 2x - 3$, $[0, 5]$.
b) $f(x) = (2x - 1)^2$, $[0, 5]$.
c) $f(x) = (x - 2)^3$, $[0, 5]$.
d) $y = 2x^3 - 9x^2 + 12x$, $[0, 2.5]$.
e) $f(x) = x^2(x - 2)^2$, $[-1, 1]$.
f) $f(x) = x^2(x - 2)^2$, $[1, 3]$.
g) $y = 1 - x^{(2/3)}$, $[-1, 2]$.
h) $g(x) = \sqrt{x^2 + 2}$, $[-1, 2]$.
i) $g(x) = e^{-x^2}$, $[-1, 1]$.
j) $g(x) = 5 - x^2$, $[-2, 2]$.
k) $g(x) = \ln(5 - x^2)$, $[-2, 2]$.

16.5 Determine the extrema of the function subject to the indicated constraint.

a) $f(x) = x(x + 2)(x - 3)$, $|x| \leq 4$.
b) $f(\theta) = 1 - 2\sin(\theta)$, $|\tan(\theta)| \leq 1$.
c) $g(t) = t\sin(t)$, $|t| \leq \pi/2$.
d) $h(s) = 2s^2 - s^4$, $|s - 1| \leq 1$.
e) $g(t) = t(\frac{1}{2})^t$, $t^2 \leq 2$.
f) $T(x) = \ln(1 + x^2)$, $|x| \leq 1$.
g) $f(\theta) = \sin(\theta)\cos(\theta)$, $|\theta| \leq \pi/2$.

16.6 The yield of a crop is a function of the time between planting and harvesting. If the yield is given by

$$y(t) = \begin{cases} (t - 30)(50 - t) & \text{if} \quad 30 \leq t \leq 50; \\ 0 & \text{if} \quad t < 30 \quad \text{or} \quad t > 50, \end{cases}$$

what is the maximum yield and on what harvest day t will it be reached?

16.7 A farmer has 10 kilometres of fence and wishes to fence a rectangular field with one fence dividing the field in the middle. What is the largest field he or she can fence? Give the largest area and the dimensions of the field.

16.8 A cell culture has a rate of growth $r(t) = 36t - t^2$. What is its maximum rate of growth? When is the maximum size of the culture reached?

16.9 Assume that an animal has the shape of a rectangular solid with width equal to one-half of its length and height equal to one-half of its width.

a) Give an equation representing the animal's volume minus its surface area as a function of its width.
b) What are the dimensions of the animal for which its volume minus its surface area is the smallest possible value?

16.10 Determine the minimum and maximum areas of a right triangle having a constant perimeter equal to 10.

16.11 A peach orchard is planted on a 10-acre site. The mature production of each tree is 300 lb when fewer than 65 trees are planted per acre. However, when the number of trees per acre exceeds 64, overcrowding occurs and the yield per tree is $(300 - (n - 64))$ lb, where n is the number of trees per acre. What is the number of trees that should be planted on the site to maximize the harvest yield?

16.12 Determine the extrema of the function $f(x) = x^n e^{-x}$ for n an integer and $x \geq 0$.

16.13 When an animal runs, its center of gravity moves up and down in a horizontal plane. These movements correspond to two phases of motion—that is, a stepping phase, where the animal's legs propel it forward, and a floating phase, where it is being carried by its momentum, which brings its legs forward to begin another stepping phase. The distances over which these phases occur are denoted by s and F as indicated in the graph. The ratio $J = F/s$ is a measure of how smooth or jerky the animal's motion appears. It can be shown that J is related to the power, P, necessary for the animal to run at a speed, V.

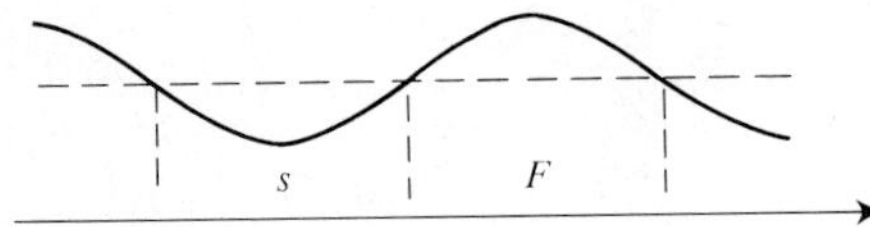

This relationship is given by

$$P = AJ\frac{L^4}{V} + B\frac{V^3L^2}{1+J},$$

where L is the length of the animal and A and B are constants.

a) Show that the optimal gait as described by J, which minimizes the required power, P, is an increasing function of the velocity, V, and a decreasing function of the animal's length, L.

b) How would the optimal value of J be related when a horse walks, trots, canters, and gallops?

c) How would the optimal values of J for a mouse, a dog, a cow, and an elephant be related?

16.14 Determine the critical points of the function, f. Use the Second Derivative Test to determine whether they correspond to local maximum or local minimum points. Determine the inflection points of the graph of f.

a) $f(x) = (x-2)^2$
b) $f(x) = 1 - x^2$
c) $f(x) = (x-2)(2-x)$
d) $f(x) = (x-2)(x-3)(x-4) + 6$
e) $f(x) = x^2e^{-x}$
f) $f(x) = x - x^2$
g) $f(x) = e^{-4x} - e^{-3x}$
h) $f(x) = e^{-x^2}$
i) $f(x) = xe^{-x^2}$
j) $f(x) = x^4 - x$

16.15 Sketch the graph of the function $f(x) = \frac{1}{30}(x^3 - 12x^2 + 45x + 3)$ over the interval $[0, 6]$.

16.16 Sketch the graph of $y = x^2e^{-x}$ over $[-1, 5]$.

16.17 Sketch the graph of the function $f(t) = 1 + 2e^{-2t} - e^{-t}$ over the interval $[0, \ln(5)]$. *Hint*: To find the roots of $f(t) = 0$, use the property $(e^A + ce^B) = e^A(1 + ce^{B-A})$.

16.18 Sketch the graph of the logistic equation $f(t) = 3/(1 + 2e^{-t})$ over the interval $[0, 5]$.

16.19 Sketch the graph of $f(t) = t + 2/t$.

16.20 Sketch the graph of $f(t) = 1/(1 + t^2)$.

16.21 Sketch the graph of $f(t) = e^{-t^2}$.

16.22 Sketch the graph of $f(t) = 10e^{-0.01e^{-t}}$.

16.23 For the function specified by Fig. 16.24, estimate the following:

a) where f changes sign, zeros of f;
b) where f' changes sign, zeros of f';
c) where f'' changes sign, zeros of f'';
d) which points are maximum points;
e) which points are minimum points;
f) which points are points of inflection.

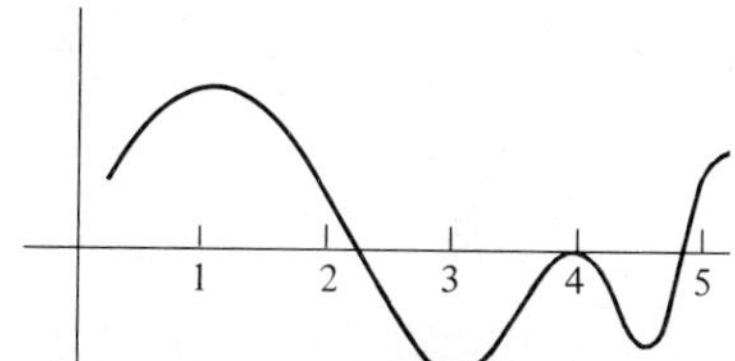

Figure 16.24

16.24 Sketch the graph of $y = 2e^{-x^2}$ over $[-5, 5]$.

16.25 Sketch the graph of a function f over the interval $[2, 6]$, satisfying the following conditions simultaneously:

a) $f(2) = 3, f(6) = 1$
b) $f'(3) = 0, f''(3) < 0$
c) $f'(4) = 0, f''(4) = 0, f(4) = 2$
d) $f'(5) = 0, f''(5) = 3, f(5) = -1$

16.26 Sketch the graph of the logistic curve and verify the properties stated in Section 8.

$$y = \frac{B}{1 + ce^{-\lambda Bt}}, \quad B > 0, c > 0, \lambda > 0.$$

16.27 a) Sketch the graph of the Gompertz curve, $y = e^{-ke^{-\lambda x}}$, $k > 0, \lambda > 0$. Indicate where it is concave upward and concave downward.

b) Verify that its maximum rate of increase occurs at its inflection point.

16.28 Sketch the graph of the functions examined in Example 16.6.

16.29 a) Sketch the graph of the double exponential decay curve

$$y = c(e^{-\lambda_1 t} - e^{-\lambda_2 t}),$$

where $c > 0, \lambda_1 > 0, \lambda_2 > 0$, and $\lambda_2 > \lambda_1$.

b) Verify the value of t_{max} given in Exercise 8.14.

16.30 Sketch the graph and determine the local extrema and inflection points of the following.

a) $y = x^{(1/3)}$
b) $y = x^{(2/3)}$
c) $y = (x - 2)^{(3/5)}$
d) $y = (x + 2)^{(4/5)}$
e) $y = x^{(1/3)}(x + 2)^{(1/3)}$
f) $y = x^{(1/3)}(x - 2)^{(3/5)}$
g) $y = x^{(2/3)}(x - 2)^{(3/5)}$
h) $y = x^{(4/3)}(x - 2)^{(1/5)}$

16.31 Sketch the graph and indicate the extrema and inflection points of the following curves over the interval $[-\pi/2, \pi/2]$.

a) $y = x + \sin(x)$
b) $y = e^{\sin(x)}$
c) $y = \sin(x)\cos(x)$
d) $y = \sin^2(x)$
e) $y = \sin(x) - \cos(x)$
f) $y = \sin^2(x) - \cos(x)$
g) $y = \cos(x) + \sin(x)$
h) $y = \sin(x)/x$

16.32 Using the methods of this section, sketch the graphs of the functions given in the following exercises of Section 8 ((a) and (b) functions only).

a) Exercise 8.6
b) Exercise 8.7
c) Exercise 8.9
d) Exercise 8.13
e) Exercise 8.5

SECTION 17

DIFFERENTIALS; IMPLICIT DIFFERENTIATION

17.1 THE DIFFERENTIAL

The terms "difference" and "derivative" have been used to describe the change in the value of a function and the instantaneous rate of change of a function. In this subsection, we introduce an additional term, "differential," to describe a function's behavior.

The term "differential" is applied to both independent variables, x, t, θ, etc., and to functions or dependent variables, $f(x)$, y, etc. The differential of an independent variable is a number that represents a change or increment in the independent variable. In the preceding sections of the text we used the symbols h and Δx to denote an increment or change in the independent variable x. In this section, we call such changes *differentials* and represent them by the symbol "dx."

Definition 17.1 The *differential of an independent variable* is a number that represents an increment of change in the variable.

The differential of an independent variable x is denoted by the symbol dx, and dx is also an independent variable. The differential of the variable t is denoted dt. The differential of the variable θ is denoted $d\theta$, etc.

The differential of a function $f(x)$ is defined so that it provides an approximation of the difference Δf. For a fixed point x_0 and differential dx, the change in the value of $f(x)$ from $x = x_0$ to $x = x_0 + dx$ is given by the difference

$$\Delta f(x_0) = f(x_0 + dx) - f(x_0). \tag{17.1}$$

Note that the value of $\Delta f(x_0)$ depends on the number dx. The difference quotient of f at x_0, due to the differential dx, is

$$\frac{\Delta f(x_0)}{\Delta x} = \frac{f(x_0 + dx) - f(x_0)}{dx},$$

where

$$\Delta x = (x_0 + dx) - x_0 = dx.$$

The derivative $f'(x_0)$ is defined as the limit of the difference quotient as the differential dx approaches zero.

$$f'(x_0) = \lim_{dx \to 0} \frac{f(x_0 + dx) - f(x_0)}{dx}. \tag{17.2}$$

If the value of dx is small, the derivative is approximated by the difference quotient,

$$f'(x_0) \simeq \frac{\Delta f(x_0)}{dx}.$$

This is the same as saying that the difference $\Delta f(x_0)$ is approximated by $f'(x_0)dx$:

$$\Delta f(x_0) \simeq f'(x_0)\,dx. \tag{17.3}$$

The differential of the function f is defined to be the approximate value of the difference Δf given by $f' \cdot dx$.

Definition 17.2 The *differential df* of a differentiable function f, determined by a point, x_0, and a differential, dx, is the number $f'(x_0) \cdot dx$. The differential, df, is a function of the two independent variables, x_0 and dx. To emphasize this dependence, we write

$$df = df(x_0, dx) = f'(x_0)\,dx \qquad \textbf{Differential of } f. \tag{17.4}$$

The values of the differential, df, and the difference, Δf, both depend on the reference point x_0 and the fundamental increment denoted by the differential, dx. We illustrate this dependence in the following example.

Example 17.1 The difference, Δf, and differential, df, are evaluated using the function

$$f(x) = x^2$$

and the indicated values of x_0 and dx. As $f'(x) = 2x$, the formula (17.4) for the differential in this case is

$$df = 2x_0\,dx.$$

We also indicate the error in the approximation $\Delta f \simeq df$ which is given by

$$E = |\Delta f - df|. \tag{17.5}$$

a) Let $x_0 = 2$ and $dx = 3$. Then the difference is

$$\Delta f(x_0) = f(2 + 3) - f(2) = 25 - 4 = 21.$$

The differential is

$$df = 2 \cdot 2 \cdot 3 = 12.$$

Consequently,

$$E = |21 - 12| = 9.$$

b) Let $x_0 = 2$ and $dx = 0.1$. Then the difference is

$$\Delta f = f(2 \cdot 1) - f(2) = 4.41 - 4 = 0.41.$$

The differential is

$$df = 2 \cdot 2 \cdot 0.1 = 0.4.$$

The resulting error is

$$E = |0.41 - 0.4| = 0.01.$$

c) Let $x_0 = 2$ and $dx = -0.22$. The difference is

$$\Delta f = f(1.78) - f(2) = 3.1684 - 4 = -0.8316.$$

The differential is

$$df = 2 \cdot 2(-0.22) = -0.88.$$

The error term is thus

$$E = |-0.8316 - (-0.88)| = 0.0484.$$

d) Let $x_0 = 6$ and $dx = -0.22$. Then the difference is

$$\Delta f = f(5.78) - f(6) = 33.4084 - 36 = -2.5916.$$

The differential in this case is

$$df = 2 \cdot 6(-0.22) = -2.64.$$

The error term is then

$$E = |-2.5916 - (-2.64)| = 0.0484.$$ ◀

Example 17.2 The differential, df, and difference, Δf, are computed for the given function with $x_0 = 2$ and $dx = 0.1$. The error E is then evaluated.

a) Let $f(x) = \ln(x)$. Then $f'(x) = 1/x$. The difference is

$$\Delta f = \ln(2.1) - \ln(2) = 0.7419 - 0.6931 = 0.0488.$$

The differential is $df = (1/x_0)\,dx$:

$$df = (\tfrac{1}{2})0.1 = 0.05.$$

The error term is

$$E = |0.0488 - 0.05| = 0.0012.$$

b) Let $f(x) = e^x$. Then $f'(x) = e^x$. The difference is

$$\Delta f = e^{2.1} - e^2 = 8.1661 - 7.3891 = 0.7770.$$

The differential is $df = e^{x_0}\,dx$:

$$df = e^2(0.1) = 0.7389.$$

The error term is

$$E = |0.7770 - 0.7389| = 0.0381.$$

c) Let $f(x) = \cos(x)$. Then $f'(x) = -\sin(x)$. The difference is

$$\Delta f = \cos(2.1) - \cos(2) = (-0.5048) - (-0.4161) = -0.0887.$$

The differential is $df = -\sin(x_0)dx$:

$$df = -\sin(2)(0.1) = (-0.9092)(0.1) = -0.0909.$$

The error term is

$$E = |-0.0887 - (-0.0909)| = 0.0022.$$ ◀

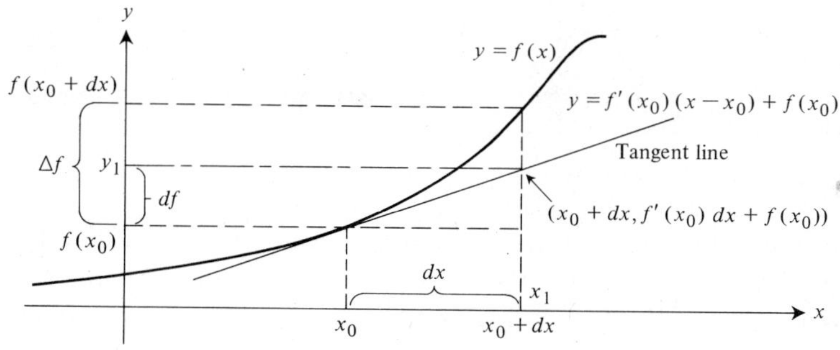

Figure 17.1

The relationship between the difference, Δf, and the differential, df, is illustrated by the graph in Fig. 17.1. We have sketched a curve, $y = f(x)$, and the tangent line to this curve at $(x_0, f(x_0))$. The equation of the tangent line is

$$y = f'(x_0)(x - x_0) + f(x_0),$$

and hence the y-coordinate of the point on the tangent line with x-coordinate, $x_1 = x_0 + dx$, is

$$y_1 = f'(x_0)dx + f(x_0).$$

The change in the y-coordinate on the tangent line from $x = x_0$ to $x = x_0 + dx$ is the differential of f:

$$\text{Change in } y \text{ values on tangent line: } y_1 - f(x_0) = [f'(x_0)dx + f(x_0)] - f(x_0) = f'(x_0)dx = df.$$

The change in the y-coordinate on the graph of $y = f(x)$ from $x = x_0$ to $x = x_0 + dx$ is the difference, Δf:

$$\text{Change in } y \text{ values on the graph } y = f(x)\text{: } \Delta f(x_0) = f(x_0 + dx) - f(x_0).$$

The difference is the change of the function values and the differential is the change of the tangent line values; thus the differential approximates the difference in the sense that the tangent line approximates the graph of the function.

If the differential dx is small, the differential df provides a reasonable approximation of the difference Δf. Algebraically, the approximate $\Delta f \simeq df$ allows us to approximate the value $f(x_0 + dx)$ of the function f at $x_0 + dx$. The approximation equation

$$\Delta f = f(x_0 + dx) - f(x_0) \simeq f'(x_0)dx = df$$

may be solved for $f(x_0 + dx)$ to obtain

$$f(x_0 + dx) \simeq f(x_0) + f'(x_0)dx \qquad \textbf{Differential approximation formula.} \tag{17.6}$$

Equation (17.6) approximates the y-coordinate on the graph of f at $x_0 + dx$ by the y-coordinate on the tangent line through $(x_0, f(x_0))$.

One application of the differential approximation formula is to approximate numbers such as $\sqrt{50}$ or $7^{2/3}$. Each of these numbers requires an "evaluation," which is not easily performed. We do not recognize the square root of 50; however, there is a number close to 50 whose square root we can easily compute. The number is $x_0 = 49$, and evaluating $\sqrt{49} = 7$ is not difficult. The differential approximation formula (17.6) can be used to approximate $\sqrt{50}$ using the $\sqrt{49}$. We illustrate this application and simultaneously present the general method to be used in this type of application.

General method	Specific method to evaluate $\sqrt{50}$
1. First represent the number y to be evaluated as a function, $y = f(x)$; indicate the value of x and the form of $f(x)$.	$\sqrt{50} = f(x)$, where $x = 50$ and $f(x) = \sqrt{x}$.
2. Next determine a point x_0 near x such that both $f(x_0)$ and $f'(x_0)$ are easily evaluated.	If $x_0 = 49$, $f(x_0) = \sqrt{49} = 7$. Since $f'(x) = \dfrac{1}{2\sqrt{x}}$, $f'(x_0) = \dfrac{1}{2\sqrt{49}} = \dfrac{1}{14}$.
3. Set $dx = x - x_0$. (The order is important!) Use the differential approximation (17.6) to estimate $f(x) = f(x_0 + dx)$: $f(x) \simeq f(x_0) + f'(x_0)dx$ or $f(x) \simeq f(x_0) + f'(x_0)(x - x_0)$.	$dx = 50 - 49 = 1$. $f(50) = f(49 + 1) \simeq f(49) + f'(49)(1)$: $\sqrt{50} \simeq \sqrt{49} + \dfrac{1}{2\sqrt{49}}(1)$; $\sqrt{50} \simeq 7 + \dfrac{1}{14} \simeq 7.0714$. Since the true value of $\sqrt{50}$ to four places is $\sqrt{50} = 7.0710$, we find the above approximation is accurate to three decimal places.

Example 17.3 The number $7^{2/3}$ can be approximated as follows:

$$7^{2/3} = f(x), \qquad \text{where } x = 7 \qquad \text{and } f(x) = x^{2/3}.$$

Since the cube root of 8 is 2,

$$f(8) = 8^{2/3} = (8^{1/3})^2 = 2^2 = 4.$$

The derivative $f'(x) = (2/3)x^{-1/3}$ is also easily evaluated at 8,

$$f'(8) = (\tfrac{2}{3})(8)^{-1/3} = (\tfrac{2}{3})(2)^{-1} = \tfrac{1}{3}.$$

Thus we set

$$x_0 = 8 \qquad \text{and} \qquad dx = x - x_0 = 7 - 8 = -1.$$

The approximation formula (14.7) gives

$$7^{2/3} = f(x) \simeq f(x_0) + f'(x_0)dx = 4 + (\tfrac{1}{3})(-1)$$

or

$$7^{2/3} \simeq 3.666\ldots$$

◀

Using Differential Approximations to Simplify Functions

Another application of the differential approximation formula, which is more practical than evaluating numbers that are easily obtained using a hand calculator, is to simplify complex equations. When a biological phenomenon is mathematically modeled, the equations describing the model are frequently too complex to be *solved* or analyzed in the form in which they first appear. One method that is often used to "reduce" or simplify equations is to replace a function by its differential approximation. For instance, in the evaluation of $\sqrt{50}$, we used the approximation

$$\sqrt{x_0 + dx} \simeq \sqrt{x_0} + \frac{1}{2\sqrt{x_0}}\,dx.$$

Setting $x_0 = 1$ and $dx = x$, we obtain the formula

$$\sqrt{1 + x} \simeq 1 + \frac{x}{2}.$$

The error in this approximation will depend on the value of $x(=dx)$; the smaller x is the better the approximation will be. Consequently, if a theoretical equation involves the expression $\sqrt{1 + x}$ and x is known to be a small quantity, then the square root $\sqrt{1 + x}$ could be replaced by $1 + x/2$. As an example, suppose x is known to be small; then the function

$$f(x) = x + \sqrt{1 + x}$$

could be replaced by

$$F(x) = x + 1 + \frac{x}{2} = 1 + \frac{3x}{2}.$$

Example 17.4 The following approximation formulas are obtained from equation (17.6) by choosing the value x_0 as indicated. The value x corresponding to dx is assumed to be small, usually less than 0.1; otherwise the errors involved in these approximations may be relatively large.

a) $\sqrt{1+x} \simeq 1 + \frac{x}{2}; x_0 = 1$ b) $e^x \simeq 1 + x; x_0 = 0$

c) $(1+x)^n \simeq 1 + nx; x_0 = 1$ d) $\ln(1+x) \simeq x; x_0 = 1$

e) $\sin(x) \simeq x; x_0 = 0$ f) $\tan(x) \simeq x; x_0 = 0$

g) $\frac{x}{a-x} \simeq \frac{x}{a}; x_0 = 0$ h) $1 + ax + bx^2 \simeq 1 + ax; x_0 = 0$

We illustrate the derivation of formulas (c) and (f) and leave the others to be done as exercises.

c) Let $g(x_0) = x_0{}^n$. Then $g'(x_0) = nx_0{}^{n-1}$ and the differential approximation formula for $g(x_0 + dx)$ is

$$(x_0 + dx)^n \simeq x_0{}^n + n \cdot x_0{}^{n-1} \cdot dx.$$

Setting $x_0 = 1$ and $dx = x$, we obtain

$$(1 + x)^n \simeq 1 + nx.$$

f) Let $g(x_0) = \tan(x_0)$. Then $g'(x_0) = \sec^2(x_0)$ and the differential approximation formula for $g(x_0 + dx)$ is

$$\tan(x_0 + dx) \simeq \tan(x_0) + \sec^2(x_0)dx.$$

Setting $x_0 = 0$ and $dx = x$, we obtain

$$\tan(x) \simeq 0 + 1 \cdot x = x.$$ ◀

Example 17.5 In Example 14.8, we discussed the motion of a pendulum in terms of its angle $\theta(t)$ from the vertical at time t. The angular velocity and the angular acceleration of the pendulum are the first and second derivatives

$$V(t) = \theta'(t), \qquad A(t) = \theta''(t).$$

The force F acting on the pendulum due to gravity is a function of $\sin(\theta(t))$. The equation that describes the relation between the force, velocity, and acceleration is

$$ml^2\theta''(t) + G\theta'(t) + mgl\sin(\theta(t)) = 0,$$

where m, l, g, and G are constants. This is a differential equation whose solution cannot be expressed in terms of elementary functions. However, if we approximate $\sin(\theta(t))$ by $\theta(t)$, as in (e) of Example 17.4, we obtain the equation

$$ml^2\theta''(t) + G\theta'(t) + mgl\theta(t) = 0. \tag{17.7}$$

This equation can be solved by an elementary method, which will be discussed in Section 26. Consequently, the use of a differential approximation has reduced a problem that cannot be solved explicitly to one whose solution is easily determined. ◀

Example 17.6 In genetics, when two allelic genes, A and a, have probability of survival P and $P(1 - k)$, respectively, the constant k is called the *selective disadvantage* of the genotype aa. In estimating the value of k, the ratio of the number of AA and Aa genotypes to the number of aa genotypes in the nth generation of a population is denoted by U_n. The values of U_n are related to the constant k by the difference equation

$$\Delta U_n = \frac{U_n k}{U_n + k + 1}.$$

Solving for k, we find

$$k = (U_n + 1)\frac{\Delta U_n}{U_n - \Delta U_n}.$$

This formula is simplified by using the differential approximation (g) of Example 17.4 with $x = \Delta U_n$ and $a = U_n$:

$$\frac{\Delta U_n}{U_n - \Delta U_n} \simeq \frac{\Delta U_n}{U_n}.$$

Substituting this into the equation for k, we obtain the approximation

$$k \simeq \left(1 + \frac{1}{U_n}\right)\Delta U_n.$$

This is the equation for k found in genetics texts. ◀

Example 17.7 We present two standard applications of differential approximations to simplify equations in *chemical kinetics.*

a) In Example 3.5, we discussed the simple unidirectional biomolecular chemical reaction: $A + B \to X$. The rate equation specifying the formation of the product X was given by (3.14). If the concentration of the product is $X(t)$, then the rate is the derivative $\mathring{X}(t)$ and the time-dependent equation corresponding to (3.14) is the equation

$$\begin{aligned}\mathring{X}(t) &= k(A_0 - X(t))(B_0 - X(t)).\\ &= k[A_0 B_0 - (A_0 + B_0)X(t) + X^2(t)].\end{aligned} \tag{17.8}$$

The chemical concentrations have been denoted without the bracket notation; $[A]_0 = A_0$, etc. Equation (17.8) is a nonlinear differential equation, as it expresses the derivative, $\mathring{X}$, as a quadratic function of X. When t is small, since $X(0) = 0$, the value of $X(t)$ will be small and the equation (17.8) may be "linearized." This is accomplished by replacing the quadratic term by its differential approximation as indicated in Example 17.4(h).

The resulting differential equation is used to describe "first-order kinetics,"

$$\mathring{X}(t) = kA_0B_0 - k(A_0 + B_0)X(t). \tag{17.9}$$

A solution of equation (17.9) is

$$X(t) = e^{-k(A_0+B_0)t} + \left[\frac{A_0B_0}{A_0 + B_0}\right]. \tag{17.10}$$

This may be verified by differentiating (17.10) and substituting into (17.9).

b) The chemical kinetics of a reaction in which an enzyme, E, combines with a substrate, S, to form an enzyme-substrate complex, C, which then converts to a product, P,

$$E + S \rightleftarrows C \longrightarrow P,$$

can be described by a simultaneous system of three differential equations. These equations describe the change in the concentrations of the substrate and complex, and the rate of formation of the product. The product is the most easily observed aspect of the reaction. The initial rate of formation of the product is called the *reaction velocity* and is denoted by v. Assuming that the reaction is proceeding with the amount of complex C remaining constant, the initial velocity can be expressed as

$$v = \frac{V_{\max}}{2E_T}\left\{A - \sqrt{A^2 - 4E_TS_T}\right\}, \tag{17.11}$$

where

$$A = K_M + S_T + E_T.$$

E_T and S_T denote the total concentration of enzyme and substrate in the reaction solution; K_M is the *Michaelis constant*, which measures the ratio of the rate of dissociation to the rate of formation of the complex C. $V_{\max}$ is another constant which gives the maximum value of the velocity v. The value of v given in (23.11) may be simplified by using the differential approximation

$$\sqrt{A^2 + dx} \simeq \sqrt{A^2} + \frac{1}{2\sqrt{A^2}}\,dx = A + \frac{dx}{2A}.$$

With $dx = -4E_TS_T$ (which is assumed small), this approximation applied to (17.11) yields, upon simplifying,

$$v = \frac{V_{\max}S_T}{K_m + S_T + E_T}. \tag{17.12}$$

This is the standard equation found in biochemistry texts. If the amount of enzyme E_T is small compared to the amount of substrate S_T, then this equation can be further approximated by

$$v = \frac{V_{\max}S_T}{K_m + S_T}. \tag{17.13}$$

This is the Michaelis–Menten equation discussed in Section 7, Example 17.10. When this curve is plotted on reciprocal axes, the resulting linear graph is called a *Lineweaver–Burk curve.* From this curve, the values of V_{max} and K_M can be obtained from the intercepts. (See Fig. 17.2.)

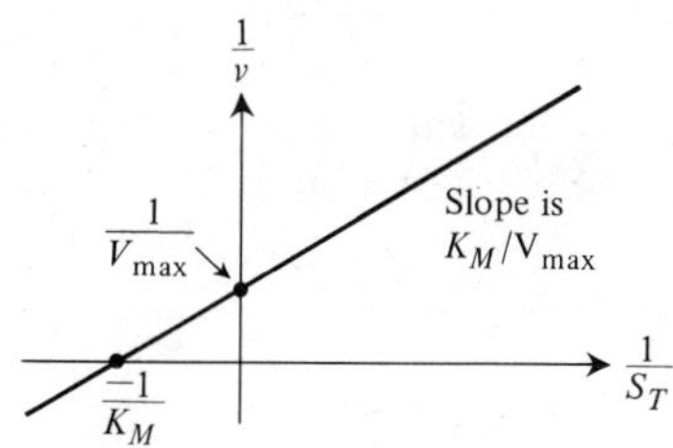

Figure 17.2

◀

The differential approximation (17.4) can also be used to estimate the error in a calculated value $f(x)$ due to an error in determining the value of the variable x. If a quantity whose true value is x is measured with an error dx, the measured value is $x + dx$. This error in measuring x will result in a subsequent error in evaluating $f(x)$ using $x + dx$ rather than x. Using the measurement $x + dx$, the calculated value is $f(x + dx)$ and the error E is the difference Δf. The error can therefore be approximated by the differential df:

$$E = |f(x + dx) - f(x)| \simeq |f'(x)dx| = |df|.$$

Example 17.8 If the radius of a sphere is used to compute its volume, $V = (\frac{4}{3})\pi r^3$, then an error, dr, in measuring r will result in an error $E = \Delta V$ in the volume. This error is approximated by

$$E \simeq dV = V'(r)dr = 4\pi r^2 dr.$$

Consequently, if a sphere that should have a radius $r = 5$m were manufactured with an imprecision of 2 percent or

$$dr = 5 \times 0.02 = 0.1 \text{ m},$$

then the volume of the sphere would be altered from its designated volume

$$V = (\tfrac{4}{3})\pi \cdot 5^3 \simeq 523.6 \text{ m}^3$$

by the approximate error

$$dV = 4 \cdot 5^2 \cdot (0.1) \simeq 31.4 \text{ m}^3.$$

This is an error of approximately 6 percent. ◀

Newton's Method

One further application of the differential approximation (17.6) is known as *Newton's Method* for finding "zeros" of a function. The problem is to find a zero of a given function f. By this, we mean a value z, such that

$$f(z) = 0.$$

We begin by approximating the value z by a number x_0 (usually by guessing). The differential approximation (17.6) with $dx = z - x_0$ becomes

$$f(z) \simeq f(x_0) + f'(x_0)(z - x_0).$$

Since $f(z) = 0$, by assumption, we obtain

$$0 \simeq f(x_0) + f'(x_0)(z - x_0).$$

Solving for z, we find

$$z \simeq x_0 - \frac{f(x_0)}{f'(x_0)}.$$

This equation provides an approximation of the unknown value z in terms of the value x_0. We set x_1 equal to this approximation:

$$x_1 = x_0 - \frac{f(x_0)}{f'(x_0)},$$

and evaluate $f(x_1)$. If $f(x_1) = 0$, we have found a root; $z = x_1$. If $f(x_1) \neq 0$, we can repeat the above process using x_1 in place of x_0 as the estimate of z. The resulting approximation is

$$z \simeq x_1 - \frac{f(x_1)}{f'(x_1)}.$$

We set

$$x_2 = x_1 - \frac{f(x_1)}{f'(x_1)}.$$

If x_2 does not result in a zero of f, we may continue the process. This leads to a sequence $\{x_n\}$ generated by the first-order difference equation,

$$x_{n+1} = x_n - \frac{f(x_n)}{f'(x_n)}. \qquad \textbf{Basic equation of Newton's method.} \qquad (17.14)$$

It can be shown that for functions, f, with f'/f'' bounded, the sequence $\{x_n\}$ will converge to a zero, z, of f. Graphically, this process is illustrated in Fig. 17.3. Newton's Method consists of approximating the zero z of the function f by the zero x_{n+1} of the tangent line to the graph of f at the point $(x_n, f(x_n))$.

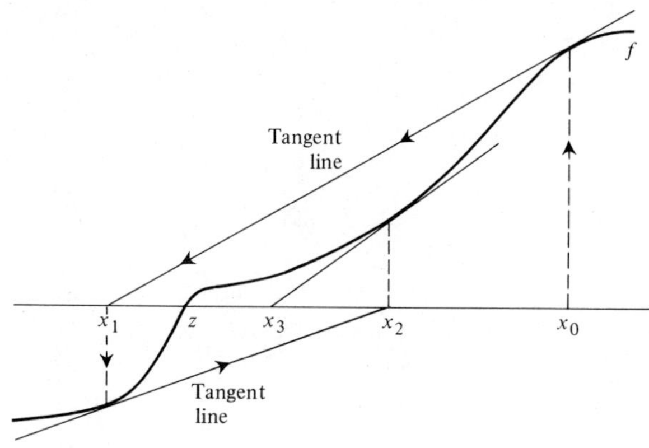

Figure 17.3

Example 17.9 Newton's Method may be used to find the value of $\sqrt{3}$. If $z = \sqrt{3}$, then $f(z) = 0$, where $f(x) = x^2 - 3$. The difference equation (17.14) is therefore

$$x_{n+1} = x_n - \frac{x_n^{\,2} - 3}{2x_n}.$$

Simplifying, this becomes

$$x_{n+1} = \frac{x_n}{2} + \frac{3}{2x_n}.$$

If $x_0 = 1$, then

$$x_1 = \tfrac{1}{2} \quad + \tfrac{3}{2} \quad = 2;$$

$$x_2 = \tfrac{2}{2} \quad + \tfrac{3}{4} \quad = 1.75;$$

$$x_3 = \frac{1.75}{2} + \frac{3}{3.5} = 1.7321.$$

The value of $\sqrt{3}$ is 1.7320 to four places, and hence x_3 is accurate to three decimal places. ◀

Example 17.10 Every cubic polynomial has at least one real root. Consider the polynomial

$$f(x) = x^3 - 3x^2 + x - 2.$$

Since the coefficient of the x^3 term is $+1$, the graph of f will have the form illustrated in Fig. 17.4. To determine a point z where the graph of f crosses the x-axis, we may use Newton's Method.

To start the method, we must choose an initial value, x_0. While theoretically any value of x_0 will eventually lead to a close approximation of z, we can make an "educated guess," which will give a reasonable approximation after only a few

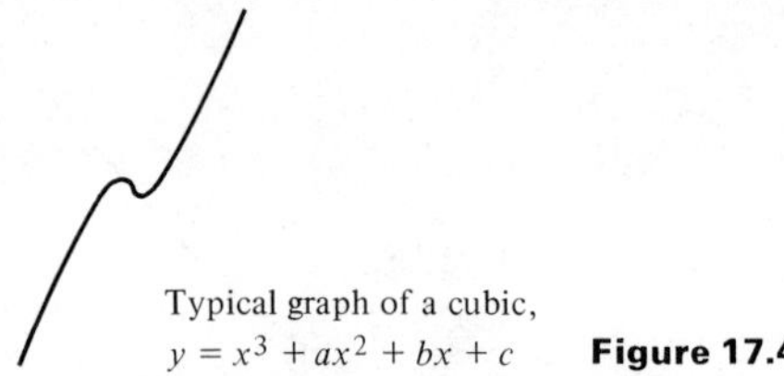

Typical graph of a cubic,
$y = x^3 + ax^2 + bx + c$ **Figure 17.4**

iterations, say x_1, x_2. In order to guess a value of x_0 that will be close to z, we first evaluate $f(x)$ at several x values:

$$f(0) = -2, \qquad f(1) = -3, \qquad f(2) = -4, \qquad f(3) = +1.$$

Since the function f changes from a negative value at $x = 2$ to a positive value at $x = 3$, we know that f must have a zero, z, in the interval (2, 3). Since $f(3)$ is closer to zero than $f(2)$, we pick

$$x_0 = 3.$$

Newton's formula is

$$x_{n+1} = x_n - \frac{f(x_n)}{f'(x_n)} = x_n - \left[\frac{x_n^3 - 3x_n^2 + x_n - 2}{3x_n^2 - 6x_n + 1}\right],$$

since $f'(x) = 3x^2 - 6x + 1$. The approximation x_1 is thus

$$x_1 = 3 - \left(\frac{f(3)}{f'(3)}\right) = 3 - \frac{1}{10} = 2.9.$$

The value of x_2 is then

$$x_2 = 2.9 - \frac{f(2.9)}{f'(2.9)} = 2.9 - \frac{0.059}{8.83} = 2.89331.$$

The value of x_3 computed as

$$x_3 = x_2 - \frac{f(x_2)}{f'(x_2)} = 2.89328.$$

As x_3 and x_2 agree to three places, the error in stating that $z = 2.893$ will be less than 0.001. The value of $f(x_2)$, used to evaluate x_3, is

$$f(2.893) \simeq 0.00025.$$

◀

17.2 IMPLICIT DIFFERENTIATION

When computing a derivative, $y'(x)$, we have always assumed that the dependent variable, y, was explicitly expressed as a function of x in the form

$$y = f(x). \tag{17.15}$$

Sometimes we are interested in examining the derivative $y'(x)$ when the variables x and y are *implicitly related*. Examples of implicit relations are as follows.

1. The equation of a circle: $x^2 + y^2 = 9$.
2. The equation of a parabola opening horizontally: $y^2 = x$.
3. The equation of a hyperbola: $xy = 4$.
4. The relation: $e^{xy} = x$.

Each of these implicit relations can be solved for y in terms of x. However, the first two do not lead to functions of x since the square root of a number may be positive or negative. We thus find the explicit forms of the above equations are as follows.

1. Explicit: $y = \sqrt{9 - x^2}$ if $y \geq 0$; $y = -\sqrt{9 - x^2}$ if $y \leq 0$.
2. Explicit: $y = \sqrt{x}$ if $y > 0$; $y = -\sqrt{x}$ if $y < 0$.
3. Explicit: $y = 4/x$ (x cannot equal zero).
4. Explicit: $y = \ln(x)/x$ since $e^{xy} = x$ corresponds to $xy = \ln(x)$.

Some relations cannot be solved explicitly. For instance, the following equations:

5. $ye^{xy} = 2$.
6. $y \sin(y) = x$.

If we are interested in the rate of change of the y-variable as a function of the x-variable, we must evaluate $y'(x)$. For relations such as (1) to (5) above, this may be accomplished by solving explicitly for y as a function of x and then differentiating. This method does not work when we consider relations (5) and (6). To determine $D_x y$ from equations such as (5) and (6), we utilize a technique known as *implicit differentiation*. This is a technique that utilizes the definition of the differential of the function y. Formula (17.6) with $f = y$; hence $df = dy$ and $f'(x) = y'$ becomes

$$\underset{\text{the differential of } y}{dy} = \underset{y' \cdot \text{the differential of } x.}{(D_x y) \cdot dx}$$

Solving for $D_x y$, we obtain the formula

$$D_x y = dy \div dx.$$

The right side of this equation is a ratio of differentials. Writing the division as a fraction, we find

$$y' = \frac{(dy)}{(dx)} = \frac{\text{differential of } y}{\text{differential of } x}. \tag{17.16}$$

It is from this equation that the Leibniz notation for the derivative,

$$\frac{dy}{dx} \quad \text{or} \quad \frac{d}{dx}f(x)$$

was developed.

To utilize the formula (17.16) to implicitly determine y', we need to first determine the differential, dy. This is done using the "calculus of differentials."

Consider a sum of two functions,

$$y = f(x) + g(x).$$

By the definition of the differential,

$$\begin{aligned} dy &= y'(x)dx \\ &= (f'(x) + g'(x))dx \qquad \text{Sum rule for derivatives} \\ &= f'(x)dx + g'(x)dx \\ &= df + dg. \end{aligned}$$

Thus we find

$$\text{differential}(f + g) = \text{differential}(f) + \text{differential}(g);$$
$$d(f + g) = df + dg.$$

In a similar manner, we may use the corresponding differentiation properties to establish a complete "calculus of differentials." The basic rules for evaluating differentials include the following.

Differential Properties

Sum rule: $d(f + g) = df + dg.$

Constant rule: $d(k \cdot f) = k \cdot df.$

Product rule: $d(f \cdot g) = f \cdot dg + g \cdot df.$

Power rule: $d(f^r) = rf^{r-1}df.$

Chain rule: $d(f \circ g) = df(g) \cdot dg.$

If f = constant, then $df = 0.$

If $y = e^f$, then $dy = e^f \cdot df.$

If $y = \ln(f)$, then $dy = \frac{1}{f}\,df$.

If $y = \sin(f)$, then $dy = \cos(f)df$, and similarly for the other trigonometric functions.

Example 17.11 The *calculus of differentials* is illustrated by the following examples, where each variable is treated as a function.

a) $d(u + v) = du + dv$

b) $d(u \cdot v) = u\,dv + v\,du$

c) $d(u^5) = 5u^4\,du$

d) $d(e^u) = e^u\,du$

e) $d(xe^y) = xd(e^y) + e^y\,dx = xe^y\,dy + e^y\,dx$

f) $d(\cos(u^2)) = -\sin(u^2)d(u^2) = -\sin(u^2) \cdot 2u \cdot du$

g) $d(y^2x^3) = y^2\,d(x^3) + x^3\,d(y^2) = y^2 \cdot 3x^2\,dx + x^3 \cdot 2y\,dy$

h) $d(e^{uv}) = e^{uv}\,d(u \cdot v) = e^{uv}(u\,dv + v\,du)$ ◀

The differential calculus as illustrated in the above example may be used to implicitly differentiate equations such as (1) to (6) at the beginning of this subsection. The process of implicit differentiation consists of three steps.

Method of Implicit Differentiation

I. Compute the differential of both sides of the given equation involving x and y using the differential properties listed above.

II. Gather all terms involving the differential dx and factor out dx. Do the same for dy.

III. Solve for the quotient dy/dx. This yields the derivative y'.

Example 17.12 We apply the process of implicit differentiation to obtain $D_x y$, where y is considered a function of x.

a) Computing the differential of both sides of the equation

$$x^2 + y^2 = 9,$$

we obtain

$$d(x^2 + y^2) = d(9),$$

or

$$d(x^2) + d(y^2) = 0, \text{ (the differential of a constant is zero)}$$

or

$$2x\,dx + 2y\,dy = 0.$$

We next transfer the dx term to the opposite side of the equation

$$2y\,dy = -2x\,dx.$$

Dividing by $2y$ and dx, we obtain the derivative

$$y' = dy/dx = -2x/2y = -x/y.$$

b) Consider the equation

$$y^2 = x.$$

Taking differentials, we have

$$d(y^2) = dx \qquad \text{or} \qquad 2y\,dy = dx.$$

Solving for the fraction dy/dx gives us

$$y' = dy/dx = 1/2y.$$

c) If $xy = 4$, then by taking differentials, we find

$$d(xy) = d(4) \qquad \text{or} \qquad x\,dy + y\,dx = 0.$$

Solving for dy/dx, we obtain the derivative

$$y' = dy/dx = -y/x.$$

d) If $e^{xy} = x$, then $d(e^{xy}) = dx$. The differential $d(e^{xy})$ is computed as illustrated in Example 17.11(d). Thus we have

$$e^{xy}x\,dy + e^{xy}y\,dx = dx,$$

or

$$e^{xy}x\,dy = dx - e^{xy}y\,dx = (1 - e^{xy}y)dx.$$

Dividing by dx and $e^{xy} \cdot x$, we obtain

$$y' = (1 - ye^{xy})/xe^{xy}.$$

e) Let $y\sin(y) = x$. Then taking differentials, we find

$$d(y\sin(y)) = dx;$$

$$yd\sin(y) + \sin(y)dy = dx;$$

$$y\cos(y)dy + \sin(y)dy = dx.$$

Hence,

$$y' = dy/dx = 1/(y\cos(y) + \sin(y)).$$ ◀

Observe that each derivative, y', evaluated in the above example is in fact a function of both x and y. Previously, the function y' has always been expressed as a function of x only; $y' = f'(x)$. Fortunately, in many instances, we do not need an explicit representation of the derivative y' as a function of x. When we wish to compute the rate of change at a given point on a graph, (x_0, y_0), or to give the equation of the tangent line at (x_0, y_0), we only need a formula by which we can evaluate $y'(x_0)$. Implicit differentiation results in such a formula even though we cannot solve explicitly for y!

Example 17.13

a) The graph of $y^2 - x^2 = e^{xy}$ passes through the point $(x_0, y_0) = (0, 1)$. To give the equation of the tangent line to the graph at this point, we must evaluate $D_x y(x_0)$. We compute $D_x y$ using implicit differentiation:

$$d(x^2) - d(y^2) = d(e^{xy});$$

$$2x\,dx - 2y\,dy = e^{xy}(x\,dy + y\,dx);$$

$$(-2y - xe^{xy})dy = (ye^{xy} - 2x)dx;$$

$$D_x y = \frac{ye^{xy} - 2x}{-2y - xe^{xy}}.$$

We evaluate $D_x y$ at the point (0, 1) by setting $x = 0$ and $y = 1$:

$$D_x y(0) = \frac{1 \cdot e^{0 \cdot 1} - 2 \cdot 0}{-2 \cdot 1 - 0 \cdot e^{0 \cdot 1}} = -\tfrac{1}{2}.$$

Thus the tangent line is $y = (-\frac{1}{2})x + 1$.

b) We evaluate the derivative $D_x y$, at $x_0 = 1$, for the function implicitly specified by $2xy + \ln(x) = x^2y^3$. The y-coordinate of this point is the value y_0 which satisfies the equation with $x = 1$:

$$2 \cdot 1 \cdot y_0 + \ln(1) = 1^2 \cdot {y_0}^3,$$

or $2y_0 = {y_0}^3$. Thus $y_0 = \sqrt{2}$ or $y_0 = -\sqrt{2}$. We use implicit differentiation to evaluate $D_x y$:

$$d(2xy) + d(\ln(x)) = d(x^2y^3);$$

$$2x\,dy + 2y\,dx + \frac{dx}{x} = 2xy^3\,dx + 3x^2y^2\,dy;$$

$$(2x - 3x^2y^2)dy = (2xy^3 - 2y - 1/x)dx;$$

$$D_x y = \frac{2xy^3 - 2y - 1/x}{2x - 3x^2y^2}.$$

Evaluating $D_x y$ at $(1, \sqrt{2})$, we obtain

$$D_x y = \frac{2 \cdot 1 \cdot (\sqrt{2^3}) - 2\sqrt{2} - 1/1}{2 \cdot 1 - 3 \cdot 1^2 \cdot (\sqrt{2})^2}$$

$$= \frac{4\sqrt{2} - 2\sqrt{2} - 1}{2 - 6} = \frac{2\sqrt{2} - 1}{-4} \simeq -0.46.$$

Evaluating $D_x y$ at $(1, -\sqrt{2})$, we obtain

$$D_x y = \frac{2\cdot 1\cdot(-\sqrt{2})^3 - 2(-\sqrt{2}) - 1/1}{2\cdot 1 - 3\cdot 1^2\cdot(-\sqrt{2})^2}$$

$$= \frac{-4\sqrt{2} + 2\sqrt{2} - 1}{2 - 6} = \frac{2\sqrt{2} + 1}{4} \simeq 0.96.$$

However, $y(x)$ is not a single-valued function. We can represent y as two functions when x is near $x_0 = 1$. (We saw this situation when we considered the two branches of a parabola, $y = \sqrt{x}$ and $y = -\sqrt{x}$.) Without being able to solve for the functions $y(x)$, we can estimate their graphs by the tangent lines obtained from the above values for y'. These are illustrated in Fig. 17.5.

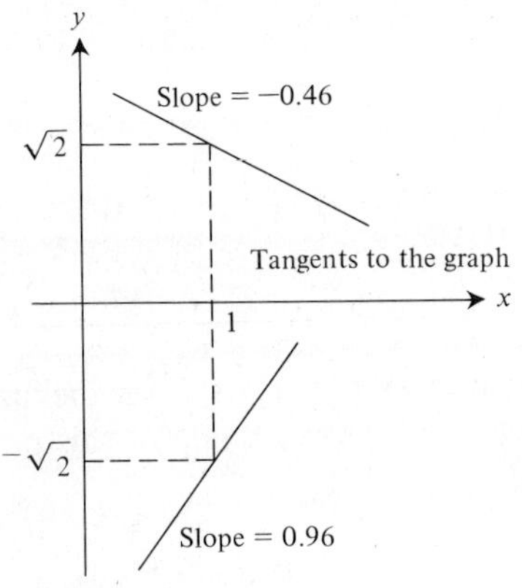

Figure 17.5

◀

SUMMARY

1. *The differential of an independent variable* x is another independent variable, dx, which is used as an increment of the variable x.

   The differential of a function f is a function of x and dx defined by

$$df = f'(x)dx.$$

2. Applications of the differential approximation formula

$$f(x_0 + dx) \simeq f(x_0) + f'(x_0)dx.$$

   a) To evaluate numbers such as $\sqrt{4.5}$, we set

$$x = 4.5, \qquad f(x) = \sqrt{x}, \qquad f'(x) = 1/(2\sqrt{x}), \qquad \text{and} \qquad x_0 = 4,$$

   since $\sqrt{4} = 2$. Then

$$dx = x - x_0 = 4.5 - 5 = 0.5$$

and

$$\sqrt{4.5} = f(x_0 + dx) \simeq f(x_0) + f'(x_0)dx$$

$$= \sqrt{4} + \frac{1}{2\sqrt{4}} \cdot (0.5) = 2 + \tfrac{1}{8} = 2\tfrac{1}{8}.$$

b) To simplify formulas (as in Example 17.4), for instance, for small values of x,

$$\ln(1 + x) \simeq x.$$

c) Using Newton's Method for finding a root of $f(z) = 0$;
   i) pick x_0 as an estimate of z;
   ii) use the formula

$$x_{n+1} = x_n - \frac{f(x_n)}{f'(x_n)}$$

   to generate terms $x_1, x_2, x_3, \ldots$;
   iii) then $x_n \to z$ as n becomes large.

3. *Calculus of differentials*: Treat all variables like functions of an independent variable. Compute the differential of a complex expression like you would compute its derivative. For example,

$$d(ye^u) = y\,d(e^u) + e^u\,dy = ye^u\,du + e^u\,dy.$$

(Leave the differentials du and dy.)

4. *Implicit differentiation*: Given an implicit relationship, such as

$$x + \ln(y) = xy,$$

take the differential of both sides of the equation:

$$dx + d(\ln(y)) = d(xy).$$

Using differential calculus, simplify and factor the differentials, dy and dx,

$$dx + \frac{1}{y}dy = x\,dy + y\,dx \qquad \text{or} \qquad \left(\frac{1}{y} - x\right)dy = (y - 1)dx.$$

Solve for $y' = dy \div dx$:

$$y' = (y - 1)/(1/y - x).$$

EXERCISE SET 17

17.1 Evaluate the differential df, the difference Δf, and the error $E = |\Delta f - df|$ at the point $x_0 = 2$ for $dx = 0.1, -0.1$, and 0.5.

a) $f(x) = x^3$
b) $f(x) = e^x$
c) $f(x) = \ln(x + 3)$
d) $f(x) = x^2 - 2x + 3$
e) $f(x) = xe^x$
f) $f(x) = (2x^3 - x)^2$
g) $f(x) = \sin(\pi x/4)$
h) $f(x) = \tan(\pi x)$
i) $f(x) = e^{\cos(x-2)}$

17.2 Using differentials, approximate the following numbers.

a) $\sqrt{48}$ b) $0.9^{1/3}$ c) $\sqrt{4.5}$ d) $e^{0.1}$
e) $(\frac{1}{5})^{1/2}$ f) $\ln(3)$ g) $\sqrt{17}$ h) $7^{2/3}$
i) $\ln(1.1)$ j) $28^{1/3}$ k) $\sqrt{3.75}$ l) $15^{1/4}$

17.3 Verify the approximation formulas of Example 17.4.

17.4 A box, which is cubic with an edge length 2 metres, is being constructed. An error in measurement resulted in each edge being 2.5 cm too short. What is the resulting error in the volume of the book? What is the estimated error using differentials?

17.5 Assume that a cell is rod-shaped. If the diameter of the cell is measured to be $d = 100$ Å and the length of the cell is measured to be $l = 500$ Å. What would be the error in computing the cell volume if there was a 10 percent error in measuring the diameter and the length? If both the diameter and the length were in error, give an estimate of the possible error in volume. Is your answer accurate?

17.6 To determine an approximation of a zero of

$$x^3 - 5.5x^2 + 8.5x - 3 = 0,$$

use Newton's method for the given value of x_0 to compute x_2.

a) 1 b) 2.5 c) 2.6 d) 3.5 e) 10

17.7 Use Newton's Method to find a root of $x^4 - 2 = 0$.

17.8 Use Newton's Method to find a root of $x^3 + 2 = 0$.

17.9 Use Newton's Method to find a root of $x \ln(x) - 1 = 0$.

17.10 Find a root of the polynomial $x^3 + 2x + 1$.

17.11 Develop general difference equations that provide an approximation of the following values. Determine the values when $\alpha = 1.5$.

a) $\sqrt{\alpha}$ b) $\sqrt[3]{\alpha}$ c) $\alpha^{1/5}$

17.12 Solve the equations to two decimal places by using Newton's Method.

a) $2\sin(x) = x$ b) $xe^{-x} = 0.1$
c) $\sin(x) = \cos(x)$ d) $x\ln(x) = 1$

17.13 Use the differential calculus formulas to compute the differential of the following.

a) $u^2 - v$ b) u^2v c) u^2/v d) $xy - x$
e) $x + x^2y^3$ f) e^{-x} g) ye^x h) $x\ln(y)$
i) $\cos(xy)$ j) $x\tan(y)$ k) $\sin^2(x)$ l) $y\sin^2(x)$

17.14 Determine $D_x y$ when x and y satisfy the specified equation.

a) $3x + 2y + 6 = 0$ b) $x^2 - y^2 = 10$
c) $xy - x = y$ d) $x^3 + xy^2 = y$
e) $e^{xy} = x$ f) $\ln(x - y) = 3$
g) $(x^2 - y^2)/xy = 5$ h) $x^{2/3} + y^{2/3} = 8$
i) $x - \ln(x) = e^y$ j) $(x + y)y^2 = x - y$
k) $\cos(x) = \sin(y)$ l) $\tan(y) = xy$
m) $\cos^2(x) + \sin^2(y) = 1$ n) $\sin(xy) = \cos(x + y)$

17.15 Determine the tangent line to the graph of the given relation at the point indicated.

a) $y^4 = 4x^4 + 6xy$ at the point $(1, 2)$
b) $x^3 + y^2 + 2x - 6 = 0$ at the point (x_0, y_0), where $y_0 = 3$
c) $x^2 + 3xy + y^2 = 2$ at the point (x_0, y_0), where $y_0 = -\sqrt{2}$
d) $x^5 - 2xy + y^5 = 0$ at the point $(1, 1)$
e) $(x - y)/(x - 2y) = 2$ at the point $(3, 1)$
f) $\cos^2(x) + \cos^2(y) = 1$ at the point $(\pi/4, -\pi/4)$
g) $\tan(x/y) = 1$ at the point $(\pi, 4)$

17.16 Determine y' two ways: (1) using implicit differentiation and (2) "solving" for y explicitly as a function of x. Verify that your results are the same at the indicated point.

a) $x^2 + y^2 = 25$ at $(3, -4)$.
b) $x^2y^2 = 9$ at $(1, -3)$.
c) $y^2 = (x - 2)(x + 1)$ at $(-1, 0)$.
d) $x^2 + 3xy + y^2 = 2$ at $(3\sqrt{2}, -\sqrt{2})$.

17.17 The volume flux of a fluid through a tube such as a capillary is given by Poiseuille's formula:

$$Q = \frac{\pi}{8}\left(\frac{P_1 - P_2}{\mu L}\right)a^4,$$

where a is the radius of the tube, L is its length, μ is a density dependent constant, and P_1 and P_2 are the fluid pressures at the two ends of the tube.

a) What is the relative error in estimating Q due to a 3 percent error in estimating the radius, a?
b) What is the relative error in estimating Q due to a 3 percent error in estimating the length, L?

17.18 In radioactive dating of archaeological artifacts, the age t is estimated by evaluating the relative amount of a radioactive substance that has not decayed, $N(t)/N(0)$, and solving the equation $N(t) = N(0)e^{-kt}$ for t. What is the error in estimating t corresponding to an error, dN, in estimating $N(t)/N(0)$?

17.19 The area of a rectangle is $A = lw$, where l is the length and w the width of the rectangle. By sketching a rectangle, show that the actual change in the area, A, due to a change, dl, in length and a change, dw, in width is not equal to the sum of the differentials:

$$dA_w = \left(\frac{dA}{dw}\right)\cdot dw \qquad \text{and} \qquad dA_l = \left(\frac{dA}{dl}\right)\cdot dl.$$

17.20 Show that the functions $\theta(t) = \sin(\lambda t)$ and $\theta(t) = \cos(-\lambda t)$ are solutions of equation (17.7) when $G = 0$ and $\lambda = \sqrt{g/l}$.

17.21 Verify that the function given by (17.10) is a solution of the differential equation (17.9).

SECTION 18

THE MEAN VALUE THEOREM; TAYLOR'S THEOREM AND TAYLOR'S POLYNOMIAL APPROXIMATIONS

18.1 ROLLE'S THEOREM AND THE MEAN VALUE THEOREM

In this subsection, we present two theorems of differential calculus. These theorems are of interest mathematically because they assert the existence of particular properties. While these theorems are not frequently used explicitly in applications, some very important and useful results concerning the representation and approximation of functions may be derived from them.

The reader whose primary interest is in applications of the derivative may proceed directly to Section 18.2, where we discuss the polynomial approximation of functions.

The first of the two theorems to be presented is known as *Rolle's Theorem.* This theorem combines two earlier results concerning the graph of a function f on a closed interval $[a, b]$. In Section 12, Theorem 12.11, we stated that a continuous function must assume its maximum and minimum values on a closed interval. Then in Section 16, Theorem 16.2, we stated that the derivative of a function is zero at any extremum in the open interval, (a, b). Rolle's Theorem guarantees that there is an extremum in the open interval (a, b) by requiring the value of the function at the two endpoints, a and b, to be equal. Then, at this extremum, we know that the derivative is zero.

Theorem 18.1 *Rolle's Theorem*
Assume that the function f is continuous on $[a, b]$, differentiable on the open interval (a, b), $f(a) = 0$ and $f(b) = 0$. Then there exists a point $\xi \in (a, b)$, such that $f'(\xi) = 0$. (A more general form assumes only that $f(a) = f(b)$.)

The sketch in Fig. 18.1(a) illustrates Rolle's Theorem. Note that there are two points ξ which satisfy the conclusion $f'(\xi) = 0$ in this illustration. The hypothesis of Rolle's Theorem requires $f(a) = 0$ and $f(b) = 0$. The reason for this assumption is to ensure that the extremum occurs on the open interval, (a, b), and not at one of the endpoints, a or b. Rolle's Theorem may be generalized by only requiring the value of the function at the endpoints to be the same, $f(a) = f(b)$. This will also ensure the existence of an extremum on the open interval, (a, b). (See Fig. 18.1(b), in which the function assumes its minimum value at the endpoints, but has an interior maximum at ξ.)

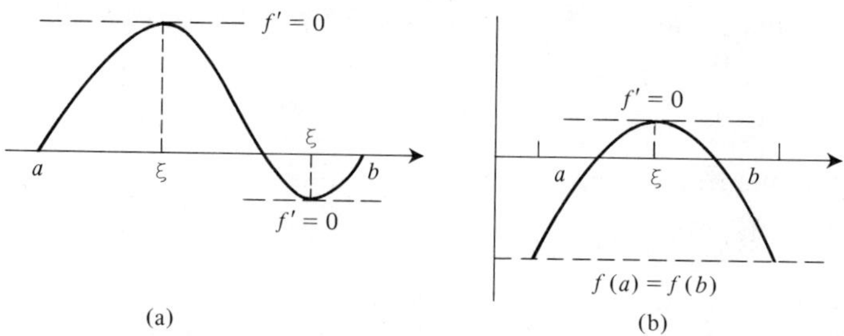

Figure 18.1

Example 18.1

a) The function $f(x) = (x - 1)(x - 4)$ satisfies the hypothesis of Rolle's Theorem on the interval $[1, 4]$; $f(1) = 0$, $f(4) = 0$, f is continuous, and $f'(x) = 2x - 5$ exists for all values of x. Consequently, we conclude from Rolle's Theorem that there exists a point $\xi \in (1, 4)$ such that $f'(\xi) = 0$. The exact value of ξ is easily determined by solving $f'(\xi) = 0$:

$$2\xi - 5 = 0 \qquad \text{or} \qquad \xi = \tfrac{5}{2}.$$

b) The function $f(x) = (e^x - e^2) \cdot \ln(x)$ satisfies the hypothesis of Rolle's Theorem on $[1, 2]$:

$$f(1) = (e^1 - e^2) \cdot \ln(1) = 0,$$

since $\ln(1) = 0$, and

$$f(2) = (e^2 - e^2) \cdot \ln(2) = 0.$$

As f is the product of two continuous functions, it is continuous for $x > 0$:

$$f'(x) = \frac{(e^x - e^2)}{x} + e^x \ln(x).$$

Consequently, we conclude from Rolle's Theorem that there exists a point $\xi \in (1, 2)$ such that $f'(\xi) = 0$. The value of ξ is difficult to determine since the equation $f'(\xi) = 0$ is

$$\frac{(e^\xi - e^2)}{\xi} + e^\xi \ln(\xi) = 0,$$

and this equation is not easily solved.

c) The function $f(x) = x^{-2} - 1$ does not satisfy the hypothesis of Rolle's Theorem on the interval $[-1, 1]$; $f(-1) = 0$ and $f(1) = 0$. But f is not continuous or even defined at $x = 0$. There is no number $\xi \in (-1, 1)$ such that $f'(\xi) = 0$.

d) The function

$$g(x) = 1 + 2x + x^2 - x^3$$

satisfies the more general hypothesis of Rolle's Theorem on the interval $[0, 2]$;

$$g(0) = 1 \qquad \text{and} \qquad g(2) = 1.$$

To determine a number ξ that satisfies the conclusion of Rolle's Theorem, we set $g'(\xi) = 0$ and solve:

$$g'(\xi) = 2 + 2\xi - 3\xi^2 = 0.$$

Using the quadratic formula, we find two values of ξ that satisfy $g'(\xi) = 0$:

$$\xi_1 = \tfrac{1}{3}(1 - \sqrt{7}) \simeq -0.55 \qquad \text{and} \qquad \xi_2 = \tfrac{1}{3}(1 + \sqrt{7}) \simeq 1.22.$$

The number ξ_1 is not in the interval (0, 2) and hence is not the desired number. Here ξ_2 is the number that Rolle's Theorem specifies as it is in the interval (0, 2), and $g'(\xi_2) = 0$. ◀

Is it possible to generalize Rolle's Theorem to functions where $f(a) \neq f(b)$? If $f(a) \neq f(b)$, we have no assurance that the function f has an extremum in the open interval (a, b). For instance, the linear function $f(x) = 2x + 3$ considered on the interval [1, 5] clearly has its maximum at $x = 5$ and has no extremum in (1, 5). Consequently, we cannot assert that f' is zero at some point ξ in (a, b). What then can we assert? To answer the question, we examine how the graph of a function f for which $f(a) \neq f(b)$ might be related to a function satisfying Rolle's Theorem.

Assume that $f(b) > f(a)$, then the graph of f over $[a, b]$ may have a form similar to the function sketched in Fig. 18.2(a). Consider the graph obtained by rotating the graph of f in a clockwise direction, keeping the point $(a, f(a))$ fixed, until the point $(b, f(b))$ is at the same level as $(a, f(a))$. (See Fig. 18.2(b).) The curve now appears to be the graph of another function $\hat{f}$ over a larger interval, $[a, \hat{b}]$. By the rotation process, we find that $\hat{f}(a) = \hat{f}(\hat{b})$. If this new function $\hat{f}$ is differentiable, it satisfies the generalized hypothesis of Rolle's Theorem. Consequently, there exists a point denoted by $\hat{\xi}$, in the interval $(a, \hat{b})$, such that $\hat{f}'(\hat{\xi}) = 0$.

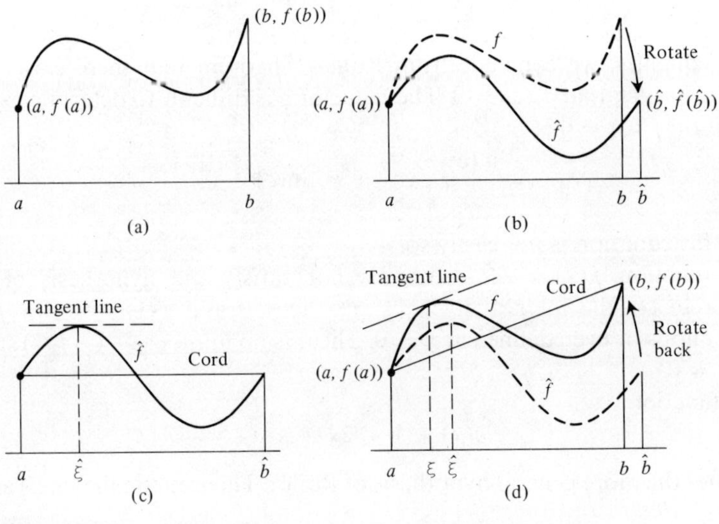

Figure 18.2

This implies that the tangent line to the graph of $\hat{f}$ at $(\hat{\xi}, \hat{f}(\hat{\xi}))$ is horizontal and hence parallel to the cord joining the endpoints $(a, \hat{f}(a))$ and $(\hat{b}, \hat{f}(\hat{b}))$. This is illustrated in Fig. 18.2(c).

Reversing the above rotation process, the point $(\hat{\xi}, \hat{f}(\hat{\xi}))$ is rotated to a point $(\xi, f(\xi))$ on our original graph. The tangent line to the graph of $\hat{f}$ at this point is also rotated. It is now the tangent to the graph of f at $(\xi, f(\xi))$ and is parallel to the cord joining $(a, f(a))$ and $(b, f(b))$. (See Fig. 18.2(d).) The conclusion that may be drawn from this geometrical discussion is that there exists a point $\xi \in (a, b)$ such that the tangent line through $(\xi, f(\xi))$ is parallel to the cord joining the endpoints, $(a, f(a))$ and $(b, f(b))$. Since two lines are parallel if and only if they have the same slope, we can phrase our conclusion in a slightly different form, known as the *Mean Value Theorem* (MVT). Recall that the slope of the tangent line is $f'(\xi)$ and the slope of the line through $(a, f(a))$ and $(b, f(b))$ is $m = [f(b) - f(a)]/(b - a)$.

Theorem 18.2 *The Mean Value Theorem*
If f is continuous on the closed interval $[a, b]$ and differentiable on the open interval (a, b), then there exists a point $\xi \in (a, b)$ such that

$$f'(\xi) = \frac{f(b) - f(a)}{b - a}. \tag{18.1}$$

Example 18.2

a) The function $f(x) = x^2 - 3x + 2$ is continuous and differentiable for all x. Hence it satisfies the hypothesis of the MVT on the interval $[1, 3]$. Consequently, we conclude that there exists a number, $\xi \in (1, 3)$, such that

$$f'(\xi) = \frac{f(3) - f(1)}{3 - 1};$$

$$2\xi - 3 = \frac{2 - 0}{3 - 1}, \qquad \xi = 2.$$

Fig. 18.3 illustrates the point ξ and the conclusion of the MVT.

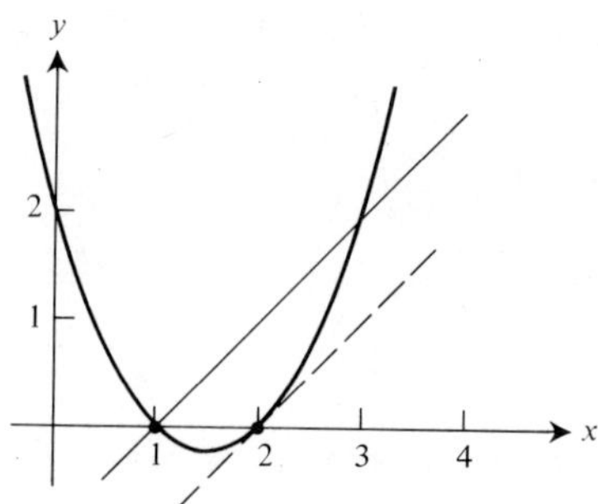

Figure 18.3

b) The function $f(x) = e^{-2x}$ is continuous and differentiable everywhere. Consequently, f satisfies the hypothesis of the MVT on the interval $[2, 5]$, and we know that there exists a number $\xi \in (2, 5)$ such that

$$f'(\xi) = \frac{f(5) - f(2)}{5 - 2}; \qquad -2e^{-2\xi} = \frac{e^{-10} - e^{-4}}{3}.$$

Solving, we obtain

$$\xi = -\tfrac{1}{2}\ln[(e^{-4} - e^{-10})/6] \simeq 2.0002.$$

Illustrating this example with a graph is difficult since $f(2) = e^{-4} \simeq 0.018316$ and $f(5) = e^{-10} \simeq 0.000045$; as a result, the value of $f(2) \simeq 400 \cdot f(5)$.

c) The function $f(x) = x^3 - x^2 - 4x + 2$ is differentiable for all x and hence satisfies the hypothesis of the MVT on the interval $[-2, 3]$. Setting

$$f'(\xi) = \frac{f(3) - f(-2)}{3 - (-2)},$$

we obtain the quadratic equation $3\xi^2 - 2\xi - 4 = 2$. This equation has two roots, which we find using the quadratic formula:

$$\xi_1 = \frac{2 - \sqrt{76}}{6} \simeq -1.1 \qquad \text{and} \qquad \xi_2 = \frac{2 + \sqrt{76}}{4} \simeq 1.8.$$

These roots are illustrated in Fig. 18.4.

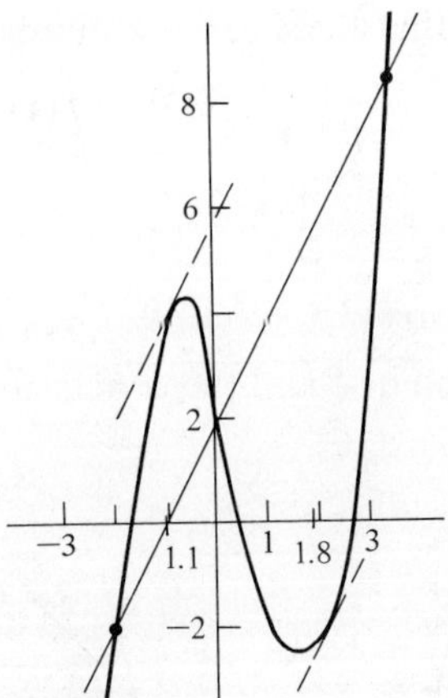

Figure 18.4

◀

There are several ways to interpret the MVT. One is to visualize a particle traveling from the point $(a, f(a))$ to the point $(b, f(b))$ along a path that is specified by the graph of f over $[a, b]$. Then, for any value $x_0 \in (a, b)$, the particle is traveling in the direction of the tangent line, $y = f'(x_0)(x - x_0) + f(x_0)$, when it is at the point $(x_0, f(x_0))$. That is, if the particle were to leave the path f—for example, due to a sudden loss of the friction that holds it to the path, then its momentum would carry it along the tangent line at the point of departure. The direction of the tangent line is its slope $f'(x_0)$. The Mean Value Theorem states that for a particle to travel from point $(a, f(a))$ to point $(b, f(b))$, at some intermediate point, $(\xi, f(\xi))$, it must be traveling in the direction of the point $(b, f(b))$ from $(a, f(a))$.

18.2 TAYLOR'S THEOREM

In this section, we introduce one of the most useful theorems of differential calculus, Taylor's theorem. We have remarked several times in the preceding text that almost any *nice* function can be represented or approximated by a polynomial. Taylor's Theorem provides the link between general functions and their approximation by polynomials. The *Taylor expansion* or *Taylor polynomial* representation of a function will involve a fixed reference point, x_0, and powers of the form $(x - x_0)^n$. We have seen the advantage of such forms when graphing polynomials. For example, the graph of

$$y = x^3 - 6x^2 + 12x - 8$$

is much more apparent when we express the function as

$$y = (x - 2)^3.$$

For simplicity, we will first consider the special case where the fixed point $x_0 = 0$. We assume that the function f is defined and has derivatives of every order on the interval $[0, T]$ for some $T > 0$. This means that $f(x), f'(x), f''(x), f^{(3)}(x), f^{(4)}(x), \ldots$, all exist and are continuous on the interval $[0, T]$. We use the following notation for derivatives

$$f^{(n)}(x) = \frac{d^n f(x)}{dx^n} \qquad \text{and} \qquad f^{(0)}(x) \equiv f(x).$$

In Section 17, we used the differential $df = f'(x)dx$ to approximate functions. Setting $x_0 = 0$ and $dx = x$ in equation (17.6), we obtain the linear, or first-order approximation of $f(x)$:

$$f(x) \simeq f(0) + f'(0)x \qquad \textbf{Linear approximation of } f \textbf{ at } x_0 = 0. \tag{18.2}$$

This formula approximates $f(x)$ by a polynomial of degree one, whose coefficients are obtained from the function f and its derivative f' evaluated at the reference point $x_0 = 0$.

A second-order or quadratic approximation of $f(x)$ is obtained by adding the term $(\frac{1}{2})f''(0)x^2$ to the approximation (18.2):

$$f(x) \simeq f(0) + f'(0)x + \frac{f''(0)x^2}{2}. \tag{18.3}$$

The reason for dividing the x^2 term by two is so that the second derivative of the polynomial

$$P_2(x) = f(0) + f'(0)x + \frac{f''(0)x^2}{2}$$

will be identical to $f''(0)$:

$$P_2'(x) = f'(0) + \frac{2f''(0)x}{2} = f'(0) + f''(0)x;$$

$$P_2''(x) = f''(0).$$

Therefore, $P_2(0) = f(0)$; $P_2'(0) = f'(0)$; and $P_2''(0) = f''(0)$. This means that the function, $P_2(x)$, agrees with $f(x)$ at $x = 0$ up to the second derivative.

The nth derivative of x^n is $n!$ Recall that

$$n! = n(n-1)(n-2)\ldots\cdot 3\cdot 2\cdot 1\ldots$$

For instance, if $y = x^3$, then $y''' = 3! = 6$. Therefore, to generalize the second-order polynomial approximation of $f(x)$ given by (18.3) to an *nth order approximation*, we simply add similar additional powers of x. The coefficient of each term x^i is

$$\frac{f^{(i)}(0)}{i!},$$

the ith derivative of f at $x_0 = 0$ divided by $i!$. This results in the approximation

$$\boxed{f(x) \simeq f(0) + f'(0)x + \frac{f''(0)x^2}{2!} + \cdots + \frac{f^{(n)}(0)x}{n!}.} \tag{18.4}$$

The polynomial on the right of (18.4) is an nth degree polynomial. We can express this polynomial using summation notation, as

$$P_n(x) = \sum_{i=0}^{n} \frac{f^{(i)}(0)x^i}{i!} \qquad \textbf{Taylor polynomial at } \boldsymbol{x_0 = 0} \textbf{ of order } \boldsymbol{n}. \tag{18.5}$$

The polynomial (18.5) is called an *nth order Taylor polynomial for the function f about the point* $x_0 = 0$. Equation (18.4) gives the nth order Taylor approximation of $f(x)$ expanded about $x_0 = 0$. Equation (18.3) is therefore the second-order Taylor approximation of $f(x)$ about $x_o = 0$, since it has the form of (18.4) with $n = 2$.

Example 18.3

a) The 4th order Taylor approximation of the function $f(x) = e^{2x}$ about $x_0 = 0$ is computed as follows. First we determine the derivatives $f^{(i)}(0)$ for $i = 0, 1, 2, 3, 4$:

$$\begin{aligned} f(x) &= e^{2x}; & f(0) &= e^0 &&= 1 \\ f'(x) &= 2e^{2x}; & f'(0) &= 2 \cdot e^0 &&= 2 \\ f''(x) &= 4e^{2x}; & f''(0) &= 4e^0 &&= 4 \\ f^{(3)}(x) &= 8e^{2x}; & f^{(3)}(0) &= 8e^0 &&= 8 \\ f^{(4)}(x) &= 16e^{2x}; & f^{(4)}(0) &= 16e^0 &&= 16 \end{aligned}$$

Next, we use (18.5) with $n = 4$ to determine the Taylor polynomial, $P_4(x)$:

$$P_4(x) = f(0) + f'(0)x + \frac{f''(0)^2}{2!}x^2 + \frac{f^{(3)}(0)x^3}{3!} + \frac{f^{(4)}(0)x^4}{4!}.$$

Substituting the above values of the derivatives, $f^{(i)}(0)$, we obtain the polynomial

$$P_4(x) = 1 + 2x + \frac{4x^2}{2} + \frac{8x^3}{6} + \frac{16x^4}{24}$$

or

$$P_4(x) = 1 + 2x + 2x^2 + (\tfrac{4}{3})x^3 + (\tfrac{2}{3})x^4.$$

To illustrate the accuracy of the approximation,

$$e^{2x} \simeq P_4(x),$$

we have computed e^{2x}, $P_4(x)$ and the error, $E = |e^{2x} - P_4(x)|$, for several values of x. These are indicated in Table 18.1. The values of e^{2x} are accurate to as many decimal places as indicated. Observe that the accuracy of the approximation decreases as x becomes larger.

Table 18.1

x	$P_4(x)$	e^{2x}	Error (approximate)
0.01	1.02020134	1.2020134	0 (to 8 places)
0.1	1.2214	1.221402	0 (to 5 places)
0.5	2.70833...	2.718	0.01
1.0	7	7.38	0.38
2.0	34.333...	54.598	20

For those wishing to verify the tabulated value of $P_4(x)$ in Table 18.1 using a hand calculator, the simplest way to evaluate $P_4(x)$ is to first express $P_4(x)$ in the following form:

$$P_4(x) = \tfrac{2}{3}x^4 + \tfrac{4}{3}x^3 + 2x^2 + 2x + 1.$$

Then, factoring this, we obtain

$$P_4(x) = \{[(\tfrac{2}{3}x + \tfrac{4}{3})x + 2]x + 2\}x + 1.$$

The numbers are the coefficients in reverse order. To evaluate $P_4(x)$, the operations of multiplying by x and adding the constants are performed from left to right. The values for $x = 0.5$ are as follows.

	Calculator Display
$\tfrac{2}{3}x$	0.333...
$\tfrac{2}{3}x + \tfrac{4}{3}$	1.666...
$(\tfrac{2}{3}x + \tfrac{4}{3})x$	0.833...
$(\tfrac{2}{3}x + \tfrac{4}{3})x + 2$	2.833...
$[(\tfrac{2}{3}x + \tfrac{4}{3})x + 2]x$	1.416...
⋮	⋮
$\{[(\tfrac{2}{3}x + \tfrac{4}{3})x + 2]x + 2\}x + 1$	2.708333...

b) The third-order Taylor approximation of the function $f(x) = \sin(x)$ about $x_0 = 0$ is computed as follows. We first evaluate the coefficients, $f^{(i)}(0)$:

$$\begin{aligned} f(x) &= \sin(x), & f(0) &= 0 \\ f'(x) &= \cos(x), & f'(0) &= 1 \\ f''(x) &= -\sin(x), & f''(0) &= 0 \\ f^{(3)}(x) &= -\cos(x), & f^{(3)}(0) &= -1 \end{aligned}$$

Then by formula (18.4), with $n = 3$, we find

$$\sin(x) \simeq f(0) + f'(0)x + \frac{f''(0)x^2}{2!} + \frac{f''(0)x^3}{3!}.$$

If we substitute the above values into this equation it yields

$$\sin(x) \simeq x - \frac{x^3}{6}.$$

Using this approximation, we obtain the values indicated in Table 18.2 with the indicated errors.

Table 18.2

x	Approximate $\sin(x)$	True $\sin(x)$	Approximate error
0.01	0.0099998	0.0099998	0
0.1	0.0998333	0.0998334	0.000001
0.5	0.4791666	0.4794255	0.00026
1.0	0.8333333	0.84147	0.008
2.0	0.6666	0.9092	0.24

c) The error in a Taylor approximation of a function, $f(x)$, depends on the distance of x from $x_0 = 0$ and on the order of the approximating polynomial. Consider the function

$$f(x) = xe^{-x}.$$

Its first three derivatives are as follows.

$$f'(x) = (1 - x)e^{-x}, \qquad f'(0) = 1$$
$$f''(x) = (-2 + x)e^{-x}, \qquad f''(0) = -2$$
$$f^{(3)}(x) = (3 - x)e^{-x}, \qquad f^{(3)}(0) = 3$$

The Taylor approximations of f about $x_0 = 0$ of order 1, 2, and 3 are given by the following.

$n = 1$.

$$f(x) \simeq P_1(x) = 0 + x$$

Therefore, the first-order approximation is simply the line, $P_1(x) = x$, indicated in Fig. 18.5.

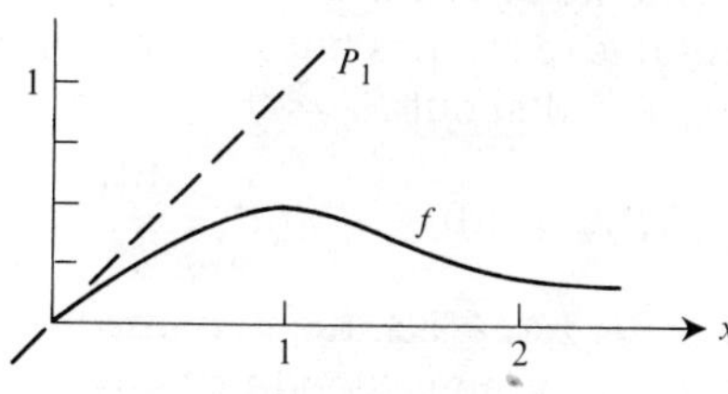

Figure 18.5

$n = 2$.

$$f(x) \simeq P_2(x) = 0 + x - \frac{2x^2}{2}$$

Therefore, the second-order approximation is the parabola, $P_2(x) = x - x^2$, indicated in Fig. 18.6.

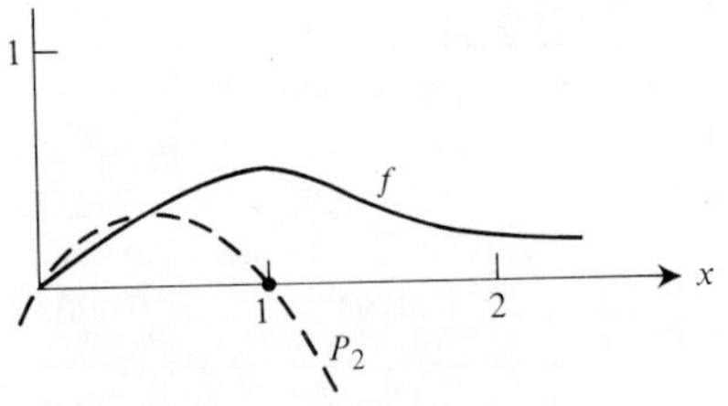

Figure 18.6

$\underline{n = 3.}$

$$f(x) \simeq P_3(x) = 0 + x - \frac{2x^2}{2} + \frac{3x^3}{3!}$$

Therefore, the third-order approximation is the cubic, $P_3(x) = x - x^2 + (x^3/2)$, indicated in Fig. 18.7.

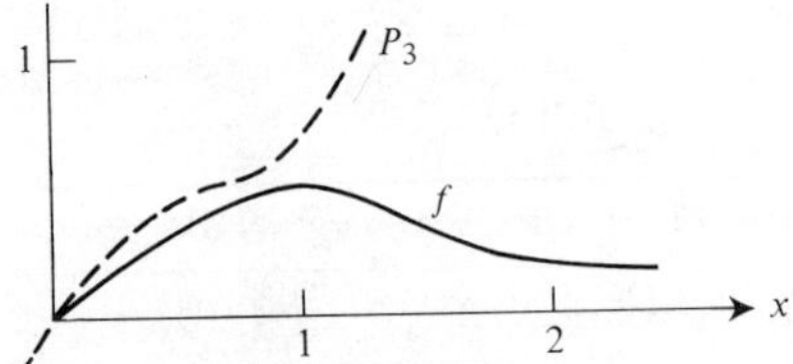

Figure 18.7

To illustrate the improved accuracy of a Taylor approximation as the order n increases, we have computed $P_n(1)$ for $n = 1, 2, \ldots, 9$. The actual value of f at $x = 1$ is $f(1) = e^{-1} = 0.3678$:

$$P_1(1) = 1 \qquad P_2(1) = 0 \qquad P_3(1) = 0.5$$

$$P_4(1) = 0.3333 \qquad P_5(1) = 0.375 \qquad P_6(1) = 0.3666$$

$$P_7(1) = 0.3680 \qquad P_8(1) = 0.3678 \qquad P_9(1) = 0.3678$$

◀

Example 18.4 The volume, V, of the aorta is a function of the internal blood pressure, P. The form of this relation is not known, but may be approximated by a Taylor polynomial expanded about $P_0 = 0$:

$$V(P) \simeq V(0) + V'(0)P + \frac{V''(0)}{2} P^2.$$

The reference pressure P_0 is zero—i.e., the pressure that causes no expansion of the aorta. (This is actually the venous blood pressure.) The coefficients in this approximation can be determined experimentally. $V(0)$ is the undistorted volume of the aorta. $V'(0)$ is the rate at which the volume of the aorta begins to expand as blood pressure begins to increase. $V''(0)$ is the initial rate at which the expansion rate $V'(P)$ changes. These values will depend on the elasticity of the aorta and are affected by various heart diseases. ◀

Not every function can be approximated by a Taylor polynomial of the form (18.5). For instance, the function $f(x) = \ln(x)$ is not defined at $x_0 = 0$. To represent the function $f(x) = \ln(x)$ by a Taylor approximation, we must choose the reference point x_0 to be a positive number.

To represent a function about an arbitrary reference point x_0, we generalize equations (18.4) and (18.5) to provide an "expansion" about the point x_0. The differential approximation formula (17.6) with $dx = x - x_0$ is

$$f(x) \simeq f(x_0) + f'(x_0)(x - x_0) \qquad \textbf{Linear approximation of } \boldsymbol{f} \textbf{ at } \boldsymbol{x_0}. \tag{18.6}$$

This is the first-order, or linear, approximation of $f(x)$ at x_0. It has the same form as equation (18.2), which is the linear approximation at $x_0 = 0$. The difference between the two approximations is that in (18.6) the derivative is evaluated at x_0 and the term x has been replaced by $x - x_0$. The generalization of equations (18.4) and (18.5) are obtained by replacing x^i by $(x - x_0)^i$ and by replacing the derivative $f^{(i)}(0)$ by $f^{(i)}(x_0)$.

Definition 18.1 Let $f(x)$ and all of its derivatives $f^{(i)}(x)$ be defined on an interval (a, b) containing the point x_0.

I. *The nth-order (or degree) Taylor polynomial* for the function f expanded about the point x_0 is

$$P_n(x) = \sum_{i=0}^{n} \frac{f^{(i)}(x_0)(x - x_0)^i}{i!} \qquad \textbf{nth order Taylor polynomial.} \tag{18.7}$$

II. *The nth-order Taylor approximation of $f(x)$ expanded about the point x_0* is

$$f(x) \simeq P_n(x), \tag{18.8}$$

where $P_n(x)$ is defined in part I.

The equations (18.7) and (18.8) are general. The following are expanded forms of the approximation (18.8) for $n = 0, 1, \ldots, 4$.

$$n = 0;\ f(x) \simeq f(x_0)$$

$$n = 1;\ f(x) \simeq f(x_0) + f'(x_0)(x - x_0)$$

$$n = 2;\ f(x) \simeq f(x_0) + f'(x_0)(x - x_0) + \frac{f''(x_0)(x - x_0)^2}{2}$$

$$n = 3;\ f(x) \simeq f(x_0) + f'(x_0)(x - x_0) + \frac{f''(x_0)(x - x_0)^2}{2} + \frac{f^{(3)}(x_0)(x - x_0)^3}{6}$$

$$n = 4;\ f(x) \simeq f(x_0) + f'(x_0)(x - x_0) + \frac{f''(x_0)(x - x_0)^2}{2} + \frac{f^{(3)}(x_0)(x - x_0)^3}{6} + \frac{f^{(4)}(x_0)(x - x_0)^4}{24}$$

Example 18.5

a) Consider the function $f(x) = \ln(x)$; $f(0)$ is not defined. Hence, we cannot use (18.6) to approximate $\ln(x)$. For $x > 0$, the derivatives of $f(x)$ are

$$f'(x) = x^{-1}; \qquad f^{(2)}(x) = -x^{-2}; \qquad f^{(3)}(x) = 2x^{-3};$$

$$f^{(4)}(x) = -3!x^{-4}; \ldots; \qquad f^{(n)}(x) = (-1)^{n+1}(n-1)!x^{-n}.$$

If we choose $x_0 = 1$, all of these derivatives can easily be evaluated at x_0. For $x \in (0, 2)$, we can approximate $\ln(x)$ with its Taylor expansion about $x_0 = 1$ of order 4:

$$\ln(x) \simeq \sum_{i=0}^{4} \frac{f^{(i)}(1)(x-1)^i}{i!}.$$

The values of the coefficients are

$$f(1) = 0; \qquad f'(1) = 1; \qquad f''(1) = -1; \qquad f^{(3)}(1) = 2; \qquad f^{(4)}(1) = -6.$$

The approximation in expanded form becomes

$$\ln(x) \simeq 0 + \frac{1 \cdot (x-1)}{1!} + \frac{(-1)(x-1)^2}{2!} + \frac{(+2)(x-1)^3}{3!} + \frac{(-6)(x-1)^4}{4!};$$

$$\ln(x) \simeq (x-1) - \tfrac{1}{2}(x-1)^2 + \tfrac{1}{3}(x-1)^3 - \tfrac{1}{4}(x-1)^4.$$

Using this approximation, we obtain $\ln(1.5) \simeq 0.4010$; while from a table of ln values, we find $\ln(1.5) = 0.4054$. Hence, the fourth-order Taylor expansion is good to two significant figures in computing $\ln(1.5)$.

b) Consider the function $f(x) = x^3 - 3x + 2$. We determine the fourth-order Taylor expansion of this function about the point $x_0 = 2$ as follows. The derivatives of f are

$$f'(x) = 3x^2 - 3; \qquad f''(x) = 6x, \qquad f^{(3)}(x) = 6; \qquad \text{and} \qquad f^{(4)}(x) = 0.$$

Hence,

$$f(2) = 4; \qquad f'(2) = 9; \qquad f''(2) = 12; \qquad f^{(3)}(2) = 6; \qquad f^{(4)}(2) = 0.$$

The fourth-order Taylor expansion is thus

$$\begin{aligned} f(x) &\simeq \sum_{i=0}^{4} \frac{f^{(i)}(2)(x-2)^i}{i!} \\ &= f(2) + f'(2)(x-2) + \frac{f''(2)(x-2)^2}{2} + \frac{f^{(3)}(2)(x-2)^3}{6} \\ &\quad + \frac{f^{(4)}(2)(x-2)^4}{24} \\ &= 4 + 9(x-2) + \frac{12(x-2)^2}{2} + \frac{6(x-2)^3}{6} + \frac{0(x-2)^4}{24} \\ &= 4 + 9(x-2) + 6(x-2)^2 + (x-2)^3. \end{aligned}$$

How accurate is this approximation? By expanding the right side, we see that it is exact. Note that the $(x - 2)^4$ term is not present, since a third-degree polynomial is always exactly represented by a third-order Taylor expansion and cannot be better approximated by a fourth or higher degree Taylor polynomial. ◀

Example 18.6 *Growth equations*

The basic equation used to describe the population dynamics of a single species is a differential equation that specifies the rate of growth of the species:

$$\frac{dN}{dt} = F(N, t).$$

When the rate of growth is independent of time, the function $F = F(N)$ only depends on the size of the population, N. If we approximate the function F by its second-order Taylor polynomial, we obtain the equation

$$\frac{dN}{dt} = F(N_0) + F'(N_0)(N - N_0) + F''(N_0)\frac{(N - N_0)^2}{2}, \tag{18.9}$$

where we have used an equal sign instead of an approximation symbol. The reference value N_0 is usually taken to be either zero or an *equilibrium population size*, N_E. An equilibrium size, N_E, is defined as a size of population at which

$$\left.\frac{dN}{dt}\right|_{N=N_E} = 0.$$

and thus $F(N_E) = 0$. Substituting $N_0 = N_E$ in equation (18.9), we obtain

$$\frac{dN}{dt} = F'(N_E)(N - N_E) + F''(N_E)\frac{(N - N_E)^2}{2}.$$

This equation is used when experimental measurements provide the difference, $N - N_E$, between the size of a population and an equilibrium value, rather than the actual population size, N.

When working with actual population sizes, the reference value N_0 is chosen to be zero. In such models, we assume that $F(0) = 0$, since a population of size $N = 0$ cannot reproduce (there is no population). Substituting $N_0 = 0$ into equation (18.9), we obtain the Pearl–Verhulst growth equation

$$\frac{dN}{dt} = F'(0)N + \frac{F''(0)}{2}N^2.$$

Setting $K = F''(0)/2$ and $M = F'(0)/K$, we can express this equation in the form

$$\frac{dN}{dt} = K(M - N)N.$$

This equation states that the growth rate is proportional to the size N times the difference between the size N and a maximum size M. This equation is also referred to as a logistic growth equation as its solution is given by the logistic function discussed in Section 8; it will be solved in Section 26. ◀

Example 18.7 We compute the third-order Taylor polynomial for the given function about $x_0 = 2$ and compare the value of $P_3(1)$ to the value of $f(1)$.

a) Let

$$f(x) = 1 + x + x^5.$$

Then

$$f(2) = 35,$$

and the derivatives of $f(x)$ are

$$f'(x) = 1 + 5x^4, \qquad f'(2) = 81;$$

$$f''(x) = 20x^3, \qquad f''(2) = 160;$$

$$f^{(3)}(x) = 60x^2, \qquad f^{(3)}(2) = 240.$$

Thus

$$P_3(x) = 35 + 81(x - 2) + \frac{160}{2}(x - 2)^2 + \frac{240}{6}(x - 2)^3$$

$$= 35 + 81(x - 2) + 80(x - 2)^2 + 40(x - 2)^3;$$

$$P_3(1) = 35 + 81(-1) + 80(-1)^2 + 40(-1)^3 = -6,$$

while $f(1) = 3$.

b) Let

$$f(x) = \sin\left(\frac{\pi}{2}x\right).$$

Then

$$f(2) = \sin(\pi) = 0.$$

The derivatives of $f(x)$ are

$$f'(x) = \frac{\pi}{2}\cos\left(\frac{\pi}{2}x\right), \qquad f'(2) = \frac{\pi}{2}\cos(\pi) = -\frac{\pi}{2};$$

$$f''(x) = -\left(\frac{\pi}{2}\right)^2 \sin\left(\frac{\pi}{2}x\right), \qquad f''(2) = -\left(\frac{\pi}{2}\right)^2 \sin(\pi) = 0;$$

$$f^{(3)}(x) = -\left(\frac{\pi}{2}\right)^3 \sin\left(\frac{\pi}{2}x\right), \qquad f^{(3)}(2) = -\left(\frac{\pi}{2}\right)^3 \cos(\pi) = \left(\frac{\pi}{2}\right)^3.$$

Thus for this function f,

$$P_3(x) = -\frac{\pi}{2}(x-2) + \left(\frac{\pi}{2}\right)^3 \frac{(x-2)^3}{6};$$

$$P_3(1) = -\left(\frac{\pi}{2}\right)(-1) + \left(\frac{\pi}{2}\right)^3 \frac{(-1)^3}{6} \simeq 0.9248,$$

while $\sin(\pi/2) = 1$.

c) Let

$$f(x) = \ln(x).$$

Then

$$f(2) = \ln(2) \simeq 0.6931.$$

The first three derivatives of $f(x)$ are

$$f'(x) = \frac{1}{x}, \qquad f'(2) = \frac{1}{2};$$

$$f''(x) = \frac{-1}{x^2}, \qquad f''(2) = -\frac{1}{4};$$

$$f^{(3)} = \frac{2}{x^3}, \qquad f^{(3)}(2) = \frac{1}{4}.$$

The third-order Taylor polynomial is thus

$$P_3(x) = \ln(2) + \frac{1}{2}(x-2) - \frac{1}{4}\frac{(x-2)^2}{2} + \frac{1}{4}\frac{(x-2)^3}{6}$$

$$\simeq 0.6931 + 0.5(x-2) - 0.125(x-2)^2 + 0.0417(x-2)^3.$$

$P_3(1) \simeq 0.0265$, while $\ln(1) = 0$. ◀

From the above examples it is clear that the error resulting from approximating a function by its Taylor polynomials can be quite large. To make the use of Taylor approximations reasonable, especially when approximating functions whose exact values are not known, we need to be able to predict the associated error. The following theorem, named after the English mathematician Brook Taylor (1635–1781), states the relationship between a function, f, and its associated polynomials, P_n.

Theorem 18.3 *Taylor's Theorem*
Assume that the $(n+1)$st derivative of f, $f^{(n+1)}(x)$, is continuous on an open interval containing the point x_0. Let $P_n(x)$ be defined as in (18.7). Then for every x in the open interval,

$$f(x) = P_n(x) + R_n(x),$$

where the remainder, $R_n(x)$, is the error in the approximation $f(x) \simeq P_n(x)$. The *Lagrange* form of the remainder function states that there exists a number ξ between x and x_0 such that

$$R_n(x) = \frac{f^{(n+1)}(\xi)(x - x_0)^{n+1}}{(n + 1)!}.$$

The number ξ in the remainder formula arises in the proof of Taylor's Theorem through an application of the Mean Value Theorem discussed in the previous subsection. The value of ξ corresponding to each value of x will be different and in practice cannot be determined. Consequently, a bound for the error in a Taylor polynomial approximation is taken as

$$E_n(x) = \underset{\xi \in [x_0, x]}{\text{maximum}} \left| f^{(n+1)}(\xi) \cdot \frac{(x - x_0)^{n+1}}{(n + 1)!} \right|. \tag{18.10}$$

The ability to use the above error formula generally requires considerable mathematical experience. For completeness, we illustrate its computation in the next example.

Example 18.8 We calculate the error bound (18.10) for each approximation $P_3(1) \simeq f(1)$ in Example 18.7. Here $x = 1$, $x_0 = 2$, and $n = 3$; thus we use the following formula for the error bound, $E_3(1)$:

$$E_3(1) = \underset{1 \le \xi \le 2}{\text{maximum}} \left| f^{(4)}(\xi) \frac{(1 - 2)^4}{4!} \right| = \frac{1}{24} \underset{1 \le \xi \le 2}{\text{maximum}} \left| f^{(4)}(\xi) \right|.$$

a) Let $f(x) = 1 + x + x^5$, then $f^{(4)}(x) = 120x$.

$$\underset{1 \le \xi \le 2}{\text{maximum}} |120\xi| = 120 \cdot 2 = 240 \text{ (when } \xi = 2).$$

Thus

$$E_3(1) = \frac{1}{24} \cdot 240 = 10.$$

Note that the actual error was

$$|P_3(1) - f(1)| = |-6 - 3| = 9.$$

b) Let $f(x) = \sin((\pi/2)x)$, then $f^4(x) = (\pi/2)^4 \sin((\pi/2)x)$.
The maximum of $|\sin((\pi/2)\xi)|$ for $1 \le \xi \le 2$ occurs at $\xi = 1$ where $\sin(\pi/2) = 1$. Thus the bound for the error is

$$E_3(1) = \frac{1}{24} \underset{1 \le \xi \le 2}{\text{maximum}} \left| \left(\frac{\pi}{2}\right)^4 \sin\left(\frac{\pi}{2} \xi\right) \right| = \frac{1}{24} \left(\frac{\pi}{2}\right)^4 \simeq 0.2537.$$

The actual error was

$$\left|P_3(1) - \sin\left(\frac{\pi}{2}\right)\right| = |0.9248 - 1| = 0.0752.$$

c) Let $f(x) = \ln((x)$, then

$$f^{(4)}(x) = \frac{-6}{x^4}.$$

Since the function $y = x^{-4}$ is decreasing, the error bound is

$$E_3(1) = \frac{1}{24} \underset{1 \le \xi \le 2}{\text{maximum}} \left|\frac{6}{\xi^4}\right| = \frac{6}{24} = 0.25 \text{ (when } \xi = 1).$$

The actual error was

$$|P_3(1) - \ln(1)| = |0.0265 - 0| = 0.0265.$$

◀

If the index n in formula (18.7) is allowed to increase and approach $+\infty$, we obtain an *infinite Taylor series.*

Definition 18.2 *The Taylor series expansion of a function f about a point x_0 is* the infinite series

$$\sum_{n=0}^{\infty} \frac{f^{(n)}(x_0)(x - x_0)^n}{n!}.$$

When the reference point $x_0 = 0$, the Taylor series is sometimes referred to as a *Maclaurin series* or a *Maclaurin expansion.* Not all functions can be represented by their Taylor series expansion even though they may be infinitely differentiable. Functions that can be represented by their Taylor series are called *analytic functions.*

Definition 18.3 *A function f is analytic* on an interval $I = (x_0 - \delta, x_0 + \delta)$ if, for each $x \in I$,

$$f(x) = \sum_{n=0}^{\infty} \frac{f^{(n)}(x_0)(x - x_0)^n}{n!}. \tag{18.11}$$

To say that equation (18.11) is valid, we mean that the Taylor series converges (in the sense discussed in Section 10, Definition 10.3), when a particular value of x is used to evaluate each term of the series and that it converges to the value $f(x)$. One of the important properties of analytic functions is that they may be differentiated, just like polynomials, by differentiating each term of their Taylor series expansion. We explore this property in Exercise 18.12.

Example 18.9 We indicate several infinite Taylor series expansions about $x_0 = 0$. The details of computing the exact derivatives and verifying these expansions are left as exercises.

a) $\sin(x) = x - \dfrac{x^3}{3!} + \dfrac{x^5}{5!} - \dfrac{x^7}{7!} + \cdots$

b) $\cos(x) = 1 - \dfrac{x^2}{2!} + \dfrac{x^4}{4!} - \dfrac{x^6}{6!} + \cdots$

c) $e^x = 1 + x + \dfrac{x^2}{2!} + \dfrac{x^3}{3!} + \dfrac{x^4}{4!} + \cdots$

d) $\dfrac{1}{1 + x} = 1 - x + x^2 - x^3 + x^4 - \cdots \qquad \text{for } |x| < 1$

e) $e^{x^2} = x^2 - \dfrac{x^6}{3!} + \dfrac{x^{10}}{5!} - \dfrac{x^{14}}{7!} + \cdots$

f) $\ln(1 + x) = x - \dfrac{x^2}{2} + \dfrac{x^3}{3} - \dfrac{x^4}{4} + \cdots \qquad \text{for } |x| < 1$ ◀

Example 18.10 In studying species diversity in ecology, the following situation is encountered. Let the probability that a particular species will be found in a complete sampling of one unit of area be p. Then the probability that the species will not be found in an area of one unit is $q = 1 - p$. The probability that the species will not be found in an area of A units is then q^A. Thus the probability $P(A)$ that the species will be found in an area of size A is given by

$$P(A) = 1 - q^A = 1 - (1 - p)^A.$$

For ease in manipulating $P(A)$, and to facilitate various ecological interpretations of species diversity, expected number of species, and species-area, the function $P(A)$ is approximated using Taylor expansions as follows. First, using the fact that

$$(1 - p)^A = e^{A \ln(1 - p)},$$

we can express

$$P(A) = 1 - e^{A \ln(1 - p)}.$$

Taylor's expansion of $\ln(1 - p)$ is

$$\ln(1 - p) = -p - \frac{p^2}{2} - \frac{p^3}{3} - \cdots$$

Using only the first term of this expansion (which can be justified if p is very small), we obtain

$$P(A) \simeq 1 - e^{-pA}.$$

To further manipulate $P(A)$, we can now approximately represent $P(A)$ by using the Taylor's series expansion of e^{-pA}. Using the expansion (c) of Example 18.9 with $x = -pA$, we obtain

$$P(A) \simeq 1 - [1 + (-pA) + \tfrac{1}{2}(-pA)^2 + \tfrac{1}{6}(-pA)^3 + \cdots];$$
$$P(A) \simeq pA - \tfrac{1}{2}(pA)^2 + \tfrac{1}{6}(pA)^3 - \cdots$$

Each of the different approximations of $P(A)$ given above is used to give theoretical interpretations of species diversity in ecology texts. ◀

SUMMARY

1. The first two theorems are about properties of a function $f(x)$ that is assumed to be continuous on an interval $[a, b]$ and differentiable on (a, b).
   Rolle's Theorem: If $f(a) = f(b) = 0$ (or just $f(a) = f(b)$), then for some point
   $$\xi \in (a, b), \qquad f'(\xi) = 0.$$
   (See Fig. 18.8.)

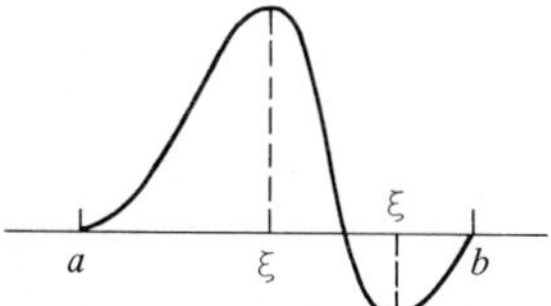

Figure 18.8

Mean Value Theorem (MVT): There exists a point $\xi \in (a, b)$ such that

$$f'(\xi) = \frac{f(b) - f(a)}{b - a}.$$

(See Fig. 18.9.)

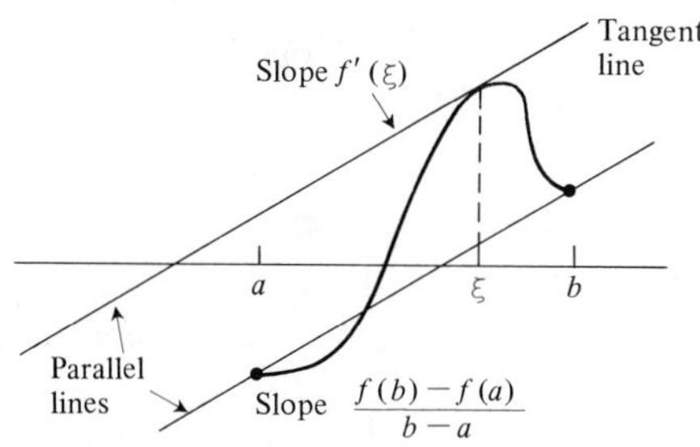

Figure 18.9

2. *Taylor polynomials*:

About $x_0 = 0$.	About x_0 arbitrary
$P_n(x) = \sum_{i=0}^{n} \frac{f^{(i)}(0)x^i}{i!}$	$P_n(x) = \sum_{i=1}^{n} \frac{f^{(i)}(x_0)(x - x_0)^i}{i!}$
$n = 1, P_1(x) = f(0) + f'(0)x$	$P_1(x) = f(x_0) + f'(x_0)(x - x_0)$
$n = 2, P_2(x) = f(0) + f'(0)x + \frac{f''(0)x^2}{2}$	$P_2(x) = f(x_0) + f'(x_0)(x - x_0) + f''(x_0)\frac{(x - x_0)^2}{2}$
$n = 3, P_3(x) = f(0) + f'(0)x + \frac{f''(0)x^2}{2} + \frac{f^{(3)}(0)^{x^3}}{6}$	$P_3(x) = f(x_0) + f'(x_0)(x - x_0) + \frac{f''(x_0)(x - x_0)^2}{2} + \frac{f^{(3)}(x_0)(x - x_0)^3}{6}$

Recall that $n! = 1 \cdot 2 \cdot 3 \cdot 4 \ldots \cdot n$, and that $0! = 1$ by definition.

$$1! = 1; 2! = 2; 3! = 6; 4! = 24; 5! = 120.$$

3. *Taylor's Theorem*:
A function, $f(x)$, can be approximated by its Taylor polynomials. The exact relationship is

$$f(x) = P_n(x) + R_n(x),$$

where the remainder term is

$$R_n(x) = \frac{f^{(n+1)}(\xi)(x - x_0)^{n+1}}{(n + 1)!},$$

for some value ξ between x_0 and x. Since R_n has $(n + 1)!$ in its denominator, its value will usually become small as n increases.

A function f is called *analytic* if it can be represented by its *infinite Taylor Series*:

$$f(x) = \sum_{i=0}^{\infty} \frac{f^{(i)}(x_0)(x - x_0)^i}{i!}.$$

Particularly important Taylor Series include the following:

$$e^x = 1 + x + \frac{x^2}{2!} + \frac{x^3}{3!} + \frac{x^4}{4!} + \cdots$$

$$\sin(x) = x - \frac{x^3}{3!} + \frac{x^5}{5!} - \frac{x^7}{7!} + \cdots \quad \text{(alternates)}$$

$$\cos(x) = 1 - \frac{x^2}{2!} + \frac{x^4}{4!} - \frac{x^6}{6!} + \cdots \quad \text{(alternates)}$$

EXERCISE SET 18

18.1 Which of the given functions satisfies the hypothesis of Rolle's Theorem on the given interval? Indicate why the other functions do not satisfy the hypothesis.

a) $f(x) = 4 - x^2$, $[-2, 2]$
b) $f(x) = 3 - x^2$, $[-2, 2]$
c) $f(x) = (x + 7)(x - 3)$, $[3, 7]$
d) $f(x) = 1 - 1/x$, $[-1, 1]$
e) $f(x) = 1 - 2/(1 + x^2)$, $[-1, 1]$
f) $f(x) = (1 - e^x)\ln(x - 1)$, $[0, 2]$

18.2 For each of the given functions and intervals, determine all points ξ that satisfy the conclusion of the MVT. If no such point ξ exists, explain why.

a) $f(x) = x^2 + 3x - 2$, $[0, 3]$
b) $g(x) = e^x - 1$, $[0, 3]$
c) $f(x) = x(x - 2)^2$, $[0, 4]$
d) $f(x) = (x - 1)^{1/3}$, $[-1, 1]$
e) $f(x) = \sqrt{(x - 1)^2}$, $[-2, 2]$
f) $f(x) = (x - 1)^{1/3}(x + 3)^{5/3}$, $[-3, +1]$
g) $f(x) = (x - 2)^2(x + 2)^2$, $[-2, 2]$

18.3 For each of the following functions, derive the third-order Taylor polynomial, $P_3(x)$, with $x_0 = 0$.

a) $f(x) = 2x^2 - 3x + 1$
b) $f(x) = x^3 - 2x + 1$
c) $f(x) = x^4 + 5x^2 - 2$
d) $f(x) = e^{2x}$
e) $f(x) = e^{x+1}$
f) $f(x) = \ln(x + 1)$
g) $f(x) = e^{x^2}$

h) $f(x) = (x - 1)^5$
i) $f(x) = e^{-x}$
j) $f(x) = xe^{-x}$

18.4 For each function indicated in Exercise 18.3, evaluate $f(x)$ and the value of the associated third-order Taylor polynomial at x and indicate the error in the approximation, where

a) $x = 1$ b) $x = 0.1$ c) $x = 0.01$ d) $x = -0.1$

18.5 Repeat Exercise 18.3 with the reference point $x_0 = 1$.

18.6 Repeat Exercise 18.4 using the approximations in Exercise 18.5 and

a) $x = 2$ b) $x = 1.5$ c) $x = 1.1$ d) $x = 0.9$

18.7 a) Compute the first, second, third, and fourth-order Taylor approximations of the function $f(x) = x^4 - 1$ about $x_0 = 1$.
b) Evaluate each approximation at $x = 2$ and compare their accuracy.

18.8 Construct a table similar to Table 18.2 for $f(x) = \cos(x)$ using the fourth-order Taylor approximation of $\cos(x)$.

18.9 a) Verify the values, $P_n(1)$, given in Example 18.3(c) for $n = 1, 2, 3, 4, 5$.
b) Evaluate these polynomials at $x = 0.5$.
c) Evaluate these polynomials at $x = 2$.

18.10 Show that the logistics function $N(t) = M(1 + ce^{-KMt})^{-1}$ satisfies the equation, $N' = K(M - N)N$, derived in Example 18.6.

18.11 a) What is the differential equation obtained in Example 18.6 by representing $F(N)$ by its first-orderTaylor approximation?
b) What is the solution of this equation when $N_0 = 0$? When $N_0 = N_E$?

18.12 Determine the first ten powers of the Taylor series expansion of the indicated function about x_0.

a) $f(x) = \sin(x), x_0 = 0$
b) $f(x) = \cos(x), x_0 = 0$
c) $f(x) = e^x, x_0 = 0$
d) $f(x) = xe^x, x_0 = 0$
e) $f(x) = \ln(x), x_0 = 1$
f) $f(x) = 1/x, x_0 = 1$
g) $f(x) = \ln(1 + x), x_0 = 0$
h) $f(x) = 1/(1 + x), x_0 = 0$

18.13 If $f(x)$ is an analytic function, then it can be expressed as an infinite series,

$$f(x) = \sum_{n=0}^{\infty} a_n(x - x_0)^n.$$

The derivative of $f(x)$ can be obtained by differentiating this series term by term. Therefore, the infinite series representation of $f'(x)$ is

$$\frac{df(x)}{dx} = \sum_{n=0}^{\infty} na_n(x - x_0)^{n-1}. \tag{18.12}$$

Use equation (18.12) to demonstrate the following. Use $x_0 = 0$ unless indicated otherwise.

a) Show that $d/dx\, e^x = e^x$.
b) Show that $d/dx \sin(x) = \cos(x)$.
c) Show that $d/dx \ln(x + 1) = 1/(x + 1)$.
d) Show that $d/dx\, e^{2x} = 2e^{2x}$.
e) Show that $d/dx \ln(x) = 1/x$, using $x_0 = 1$.

18.14 Substitute $x = 1$ into the fourth-order Taylor polynomial for e^x to obtain an approximation of the number e.

18.15 Show that the fourth-order Taylor polynomial for $f(x) = e^{x^2}$ is the same as the second-order Taylor polynomial for $f(t) = e^t$ with $t = x^2$. (Use $x_0 = 0$ and $t_0 = 0$.)

18.16 In Example 18.10, we approximate the function $P(A) = 1 - [1 - p]^A$ in a round-about manner (because the intermediate steps are useful in ecological discussions). Compute directly the Taylor's expansion of $P(A)$ about $A = 0$. Use $P(A) = 1 - e^{A \ln(1-p)}$.

CHAPTER FIVE

INTEGRAL CALCULUS

The antiderivative is introduced in Section 19 as the reverse of the derivative process. The duality between derivatives and antiderivatives is utilized to develop the standard antiderivative or indefinite integral formulas. In Section 20, we examine the discrete solution of three probelms; each solution leads to the same type of finite series approximation. This section introduces the fundamental concept of integral calculus. In Section 21, the definite integral is introduced as the limit of a series of the form present in each solution of Section 20. The remainder of Section 21 is devoted to developing the basic aspects of integral calculus using the *fundamental theorem of calculus* to evaluate integrals. In Section 22, the definition of the integral is examined more closely and the concept of area, as defined by an integral, is also examined. The numerical approximation of an integral is examined in the last part of Section 22.

SECTION 19

THE ANTIDERIVATIVE—THE INTEGRAL

19.1 THE ANTIDERIVATIVE

Taking the derivative of a function, $F(x)$, can be thought of as an operation that transforms the function $F(x)$ into a new function, $F'(x)$. The objective of this section is to examine the reverse of the differentiation process. To visualize the derivative as a transformation, we use the D_x notation:

$$D_x : F(x) \to D_x F(x) = F'(x).$$

For instance,

$$D_x : 3x^2 - 1 \to D_x(3x^2 - 1) = 6x.$$

The reverse of the differentiation process can be visualized as an operation that transforms a derivative, $F'(x)$, into a function, $F(x)$. To denote this reverse differentiation process we use the notation D_x^{-1}, which is similar to the inverse notation for ordinary functions. Thus the reverse differentiation process can be described by

$$D_x^{-1} : F'(x) \to D_x^{-1} F'(x) = F(x).$$

For instance, using the above illustration, we find that

$$D_x^{-1} : 6x \to D_x^{-1}(6x) = 3x^2 - 1.$$

However, if $G(x) = 3x^2 + 9$, the derivative of $G(x)$ is

$$D_x G(x) = D_x(3x^2 + 9) = 6x.$$

To reverse this differentiation, we must have

$$D_x^{-1} G'(x) = D_x^{-1}(6x) = 3x^2 + 9.$$

The fact that two different functions, $F(x) = 3x^2 - 1$ and $G(x) = 3x^2 + 9$, have the same derivative creates a problem when we try to reverse the differentiation process. Should $D_x^{-1}(6x) = 3x^2 - 1$ or should $D_x^{-1}(6x) = 3x^2 + 9$? In general, any given function $f(x)$, such as $f(x) = 6x$, will be the derivative of more than one function. Therefore, to discuss the reverse process of differentiation, we must identify all functions whose derivatives equal $f(x)$; these functions are called *antiderivatives* of $f(x)$.

Definition 19.1 A function $F(x)$ is *an antiderivative* of $f(x)$ if $F'(x) = f(x)$.

The following phrases,

"$F(x)$ is an antiderivative of $f(x)$"

and

"$f(x)$ is the derivative of $F(x)$,"

are synonymous.

An important observation about antiderivatives is that although a function $f(x)$ will have more than one antiderivative, any two antiderivatives of $f(x)$ are closely related. For instance, both of the functions

$$F_1(x) = 2x^3 + 3x - 5 \qquad \text{and} \qquad F_2(x) = 2x^3 + 3x + 67$$

are antiderivatives of $f(x) = 6x^2 + 3$. To verify this, we compute their derivatives:

$$F_1'(x) = 6x^2 + 3 \qquad \text{and} \qquad F_2'(x) = 6x^2 + 3.$$

How do F_1 and F_2 differ or agree with each other? They are the same except for their constant terms. In fact, $F_1(x) = F_2(x) - 73$. This type of relationship exists between all antiderivatives of the same function.

Theorem 19.1 If F_1 and F_2 are two antiderivatives of the same function f, then $F_1(x) = F_2(x) + C$, where C is an appropriate constant.

Consequently, if we know one antiderivative of a function $f(x)$, say $F(x)$, then all other antiderivatives of this function $f(x)$ will have the form

$$F_1(x) = F(x) + C.$$

Thus, to specify the general form of an antiderivative of $f(x)$, we use the inverse differential operator notation D_x^{-1} and write

$$D_x^{-1} f(x) = F(x) + C \qquad \textbf{General antiderivative,} \tag{19.1}$$

where $F(x)$ is any particular antiderivative of $f(x)$, and C is an arbitrary constant. Equation (19.1) is read "The antiderivative of $f(x)$ is capital $F(x)$ plus C." For each choice of the arbitrary constant, C, equation (19.1) specifies a particular antiderivative of $f(x)$. Consequently, the symbol $D_x^{-1} f(x)$ is frequently referred to in the singular form as "an antiderivative of $f(x)$." The following example illustrates situations that require the determination of an antiderivative.

Example 19.1

a) A drug is excreted in a patient's urine; the urine is monitored continuously using a catheter. The rate at which the drug is excreted is a measurable function, $f(t)$. The amount of drug present in the patient at any time t (assuming that no additional drug is administered) is therefore specified by an antiderivative of $f(t)$,

$$F(t) = D_t^{-1}(f(t)).$$

b) *Paleoclimatology* is the study of past climates. One method for analyzing the climate in glacial regions is to take core samples of snow-ice, which reveal the year by year snow accumulation. From this data, it is possible to estimate the rate of change in the climatic conditions such as temperature. If the temperature change is estimated by a function $r(t)$, then the actual temperature would be an antiderivative of $r(t)$, $T(t) = D_t^{-1}r(t)$.

c) A simple type of flight recorder records the inclination and speed of an aircraft. These data result in a function $g(x)$, which gives the slope of the tangent to the path that the aircraft flies. The altitude of the aircraft is therefore an antiderivative of this slope function, $A(x) = D_x^{-1}g(x)$. ◀

In the above examples the exact forms of the functions were not specified. Consequently, the antiderivatives could not be determined explicitly. How would we determine an antiderivative of a specific function, say $f(x) = 5x^3 + 2$? One approach to finding an antiderivative, $F(x) = D_x^{-1}f(x)$, is to guess. Three different guesses would be as follows.

$$F_1(x) = 15x^2; \qquad F_2(x) = 5x^4 + 2; \qquad \text{and} \qquad F_3(x) = 5x^4 + 2x^2 + 1.$$

These choices are not derived in any logical way. Are any of these guesses correct? How can we check them? The purpose of this paragraph is not to encourage wild guessing, but to examine how the guesses may either be verified as valid or discarded. When attempting to guess an antiderivative, the procedure we use for checking each guess is to compute the derivative $F'(x)$ of a guessed antiderivative $F(x)$. If $F'(x)$ equals $f(x)$, then the guessed function $F(x)$ is an antiderivative of $f(x)$. Since $F_1'(x) = 30x$, $F_2'(x) = 20x^3$, and $F_3'(x) = 20x^3 + 4x$, none of these guesses are correct. The following is obvious, but worth stating:

> Wild guessing is useless and should be avoided. However, *intelligent guessing* is very useful. An intelligent guess is a guess based upon specific knowledge about the proposed problem, experience from solving similar problems, and utilization of "known" rules, which mathematicians call theorems.

When we seek an antiderivative of $f(x) = 5x^3 + 2$, our guess should be based upon our knowledge of the power rule for differentiating,

$$D_x(x^r) = rx^{r-1}.$$

Choosing $r = 4$, so that $r - 1 = 3$, we *guess* that an antiderivative of $5x^3$ would have the form ax^4 for some constant a. Since $D_x(x) = 1$, we *guess* that an antiderivative of 2 would have the form bx for some constant b. Our guess for $D_x^{-1}(f(x))$ is thus a function of the form

$$F(x) = ax^4 + bx.$$

We can determine the constants a and b for the identity $F'(x) = f(x)$, which must hold if our guess is correct. Two polynomials are equal only if the coefficients of like powers of x are equal. Since $F'(x) = 4ax^3 + b$, the identity $F'(x) = f(x)$ will hold only if the two equations,

$$\underset{\text{coefficients of } x^3}{4a = 5} \qquad \text{and} \qquad \underset{\text{constants}}{b = 2,}$$

are satisfied. Thus, $a = \frac{5}{4}$, $b = 2$, and an antiderivative of $f(x)$ is

$$F(x) = (\tfrac{5}{4})x^4 + 2x.$$

However, another antiderivative is $F_1(x) = (\frac{5}{4})x^4 + 2x + 3$, since $F_1'(x) = 5x^3 + 2$. We observe that $F_1(x) = F(x) + 3$, and thus the antiderivative $F_1(x)$ is determined from the general formula

$$D_x^{-1}(5x^3 + 2) = (\tfrac{5}{4})x^3 + 2x + C,$$

by setting $C = 3$.

The algebraic properties of the antiderivative operator, D_x^{-1}, parallel the properties of the differential operator, D_x. This is because of the complete duality between derivatives and antiderivatives. Any statement or equation involving the derivative has a *dual statement* that refers to the antiderivative:

$D_x(f) = g \Leftrightarrow f = D_x^{-1}g$	**Duality between derivatives and antiderivatives.**

(19.2)

Consequently, each of the properties of derivatives given in Section 14 has a corresponding property for antiderivatives. The antiderivative rule corresponding to the power rule for derivatives states the reverse of the formula, $D_x(x^r) = rx^{r-1}$. The actual formula for antiderivatives is slightly different in that it gives an antiderivative for x^r rather than for rx^{r-1}.

Theorem 19.2 If r is a real number, then the antiderivative of x^r is

$D_x^{-1}(x^r) = \dfrac{x^{r+1}}{r+1} + c$, for $r \neq 1$	**Power rule for antiderivative.**

(19.3)

Example 19.2 The following antiderivatives are evaluated using formula (19.3).

a) $D_x^{-1}x^3 = \dfrac{x^{3+1}}{3+1} + c = \dfrac{x^4}{4} + c$

b) $D_x^{-1}x^\pi = \dfrac{x^{\pi+1}}{\pi+1} + c$

c) $D_x^{-1}x^{1/2} = \dfrac{x^{3/2}}{(\frac{3}{2})} + c = (\frac{2}{3})x^{3/2} + c$

d) $D_x^{-1}\dfrac{1}{\sqrt{x}} = D_x^{-1}x^{-1/2} = 2x^{1/2} + c$

e) $D_x^{-1}\dfrac{1}{x^2} = D_x^{-1}x^{-2} = -x^{-1} + c$ ◀

Since $D_x(kf(x)) = kD_x f(x)$ for any constant k, the same rule holds for antiderivatives.

Theorem 19.3 For any constant k

$$D_x^{-1}(kf(x)) = kD_x^{-1}f(x). \tag{19.4}$$

Example 19.3

a) the antiderivative of $f(x) = 3x^2$ is

$$D_x^{-1}(3x^2) = 3D_x^{-1}x^2 = 3\left(\frac{x^3}{3} + c\right) = x^3 + 3c.$$

Since the constant c represents an arbitrary constant, $3c$ is simply another arbitrary constant. If we replace $3c$ by a constant C we obtain the equation

$$D_x^{-1}(3x^2) = x^3 + C$$

In general practice, the new constant C is not represented by a capital letter but by the same small c. Furthermore, the switch from $3c$ to c will simply be made without mention.

b) $D_x^{-1}(5x^2) = 5D_x^{-1}x^2 = \frac{5}{3}x^3 + c.$

c) $D_x^{-1}(4x) = 4D_x^{-1}x = \frac{4}{2}x^2 + c = 2x^2 + c.$

d) $D_t^{-1}(\pi t^{-5}) = \pi D_t^{-1}t^{-5} = \dfrac{-\pi}{4}t^{-4} + c.$ ◀

Theorem 19.4 The antiderivative of a sum of two functions is the sum of the antiderivatives of each of the functions:

$$D_x^{-1}(f(x) + g(x)) = D_x^{-1}f(x) + D_x^{-1}g(x) \quad \textbf{Sum rule.} \tag{19.5}$$

Example 19.4

a) The antiderivative of $3x^4 - 2x$ is

$$\begin{aligned} D_x^{-1}(3x^4 - 2x) &= D_x^{-1}(3x^4) + D_x^{-1}(-2x) \\ &= 3D_x^{-1}x^4 - 2D_x^{-1}x \\ &= \tfrac{3}{5}x^5 + c_1 - \frac{2x^2}{2} + c_2 \\ &= \tfrac{3}{5}x^5 - x^2 + c. \end{aligned}$$

Note that each antiderivative contains an arbitrary constant; however, when the two antiderivatives are added together, the two constants, c_1 and c_2, are added to form a new arbitrary constant, c. As a general rule, when the antiderivative of a sum is evaluated, the individual constants associated with each term are not indicated.

b) $$\begin{aligned} D_x^{-1}(x^3 + x^2 + 1) &= D_x^{-1}x^3 + D_x^{-1}x^2 + D_x^{-1}1 \\ &= \frac{x^4}{4} + \frac{x^3}{3} + x + c. \end{aligned}$$

c) $$\begin{aligned} D_x^{-1}(x^{1/2} - 3x^{-2}) &= D_x^{-1}x^{1/2} - 3D_x^{-1}x^{-2} \\ &= \frac{x^{3/2}}{(\frac{3}{2})} - 3\frac{x^{-1}}{(-1)} + c \\ &= (\tfrac{2}{3})x^{3/2} + 3x^{-1} + c. \end{aligned}$$ ◀

Combining the last three Theorems, we obtain a general formula for the antiderivative of any polynomial function.

Theorem 19.5 The antiderivative of the polynomial

$$f(x) = a_0 + a_1x + a_2x^2 + \cdots + a_nx^n$$

is

$$D_x^{-1}f(x) = c + a_0x + \frac{a_1x^2}{2} + \frac{a_2x^3}{3} + \cdots + \frac{a_nx^{n+1}}{n+1}.$$

Examples of the polynomial rule, Theorem 19.5, are given by parts (a) and (b) of the above example.

In the foregoing, we have evaluated antiderivatives and stated properties of the antidifferentiation process without mentioning the existence of antiderivatives. While there are functions that do not have an antiderivative, most of the functions you encounter will have antiderivatives. Although some continuous functions are not differentiable (e.g., $f(x) = |x|$ at $x = 0$), it can be shown that every continuous function has antiderivatives.

Theorem 19.6 If f is a continuous function, then the antiderivative of f, $D_x^{-1}f$, exists.

As a consequence of Theorem 19.6, we know that $D_x^{-1}f(x)$ exists for each of the continuous functions listed in Theorem 13.9. Knowing that $D_x^{-1}f(x)$ exists and being able to specifically describe the function $D_x^{-1}f(x)$ are unfortunately not the same. Some functions have antiderivatives that cannot be expressed in terms of any elementary functions. Such a function is $f(x) = e^{-x^2}$.

Initially, the only functions whose antiderivatives we will be able to evaluate are those that can be expressed as sums or multiples of terms that we can "recognize" as derivatives. The derivatives of the trigonometric functions are listed in Theorem 14.6. Using the duality principle (19.2), we see that each trigonometric derivative formula has a corresponding antiderivative formula.

Theorem 19.7 The antiderivatives of the selected trigonometric functions are specified as follows, where c denotes an arbitrary constant.

$$\begin{aligned}
D_x^{-1} \cos(x) &= \sin(x) + c \\
D_x^{-1} \sin(x) &= -\cos(x) + c \\
D_x^{-1} \sec^2(x) &= \tan(x) + c \\
D_x^{-1} \csc^2(x) &= -\cot(x) + c \\
D_x^{-1}[\sec(x)\tan(x)] &= \sec(x) + c \\
D_x^{-1}[\csc(x)\cot(x)] &= -\csc(x) + c
\end{aligned}$$

Note that the above list does not include all trigonometric functions; for instance, there is no formula for $D_x^{-1}\tan(x)$. This is because $\tan(x)$ is not the derivative of one of the basic trigonometric functions. Since $\tan(x)$ is continuous (when it is defined), Theorem 19.6 guarantees the existence of $D_x^{-1}\tan(x)$. A formula for this antiderivative will be developed later.

Example 19.5 The formulas of Theorem 19.7 are used to evaluate the following antiderivatives.

a) $f(x) = \cos(x)$; $\quad D_x^{-1}f(x) = \sin(x) + c$

b) $f(x) = 3\sin(x)$; $\quad D_x^{-1}f(x) = -3\cos(x) + c$

c) $f(x) = 2\sec(x)\tan(x)$; $\quad D_x^{-1}f(x) = 2\tan(x) + c$

d) $f(t) = t^2 - \sec^2(t)$; $\quad D_t^{-1}f(t) = \dfrac{t^3}{3} - \tan(t) + c$

e) $f(\theta) = \csc(\theta)\cot(\theta)$; $\quad D_\theta^{-1}f(\theta) = -\csc(\theta) + c$ ◀

Example 19.6 The constant c appearing in each antiderivative formula allows for the representation of every antiderivative of the given function. If we require a specific antiderivative, we must determine the value of c. When the value of the antiderivative $F(x) = D_x^{-1}f(x)$ is specified at a number x_0, then the constant c can be evaluated. In (a), (b), and (c), we specify a function f, a number x_0, and a value $y_0 = F(x_0)$. We then determine the appropriate value of c.

a) Let $f(x) = 2x^4 - 3$, $x_0 = 1$, and $y_0 = 6$. Then

$$D_x^{-1}f(x) = F(x) = (\tfrac{2}{5})x^5 - 3x + c.$$

Setting $F(1) = 6$, we have $(\frac{2}{5})(1)^5 - 3(1) + c = 6$ or $c = 8.6$.

b) Let $f(x) = x^2 + 2x - 4$, $x_0 = 2$, and $y_0 = 8$. Then

$$D_x^{-1}f(x) = F(x) = x^3/3 + x^2 - 4x + c.$$

Setting $F(2) = 8$, we have $\frac{8}{3} + 4 - 8 + c = 8$ or $c = \frac{28}{3}$.

c) Let $f(x) = \sin(x) - 2\cos(x)$, $x_0 = \pi/4$ and $y_0 = 5$. Then

$$D_x^{-1}f(x) = F(x) = -\cos(x) + 2\sin(x) + c.$$

Setting $F(\pi/4) = 5$, we have $-\cos(\pi/4) - 2\sin(\pi/4) + c = 5$ or $c = 5 + 3/\sqrt{2}$. ◀

Example 19.7

a) If the instantaneous rate of change in a population is $R(t) = 3t + \sin(t)$, and the population size is 100 at $t = 0$, what is the size of the population at $t = 5$? *Solution*: The population is described by the antiderivative

$$N(t) = D_t^{-1}(3t + \sin(t)) = 3t^2/2 - \cos(t) + c.$$

Since $N(0) = -1 + c$, $N(0) = 100$ gives $c = 101$. Hence $N(5) = 3(5)^2/2 - \cos(5) + 101 \simeq 138.8$.

b) A dog chases a rabbit and the distance between them is changing at a rate of $t^{-1/2}$ metres/sec^{-1}. If they were 30 metres apart at $t = 1$, how far apart will they be at $t = 8$? *Solution*: The distance $f(t)$ between them is

$$f(t) = D_t^{-1}(t^{-1/2}) = \frac{t^{-(1/2)+1}}{-(\frac{1}{2})+1} + c = 2t^{1/2} + c.$$

Setting $f(1) = 30$, we have $c = 28$. Hence at $t = 8$, $f(8) = 2(8)^{1/2} + 28$ or $f(8) \simeq 33.66$. Will the dog catch the rabbit? ◀

The antiderivative formulas involving the exponential function e^x and the natural logarithm function $\ln(x)$ are obtained by reversing the derivative formulas (14.8) and (14.9).

Theorem 19.8 For an arbitrary constant c,

$$\boxed{D_x^{-1}(e^x) = e^x + c,} \tag{19.6}$$

and, for all $x > 0$,

$$\boxed{D_x^{-1}\left(\frac{1}{x}\right) = \ln(x) + c.} \tag{19.7}$$

These formulas are illustrated in the next subsection after we present their general forms.

19.2 THE ANTIDERIVATIVE AND THE CHAIN RULE

The Chain Rule relates the derivative of a composite function $f \circ g$ to the derivatives of the individual functions, f and g:

$$D_x[f \circ g)(x)] = f'(g(x)) \cdot g'(x) \qquad \textbf{Chain Rule.}$$

Reversing this equation according to the duality principle (19.2), we obtain the equation

$$(f \circ g)(x) = D_x^{-1}[f'(g(x)) \cdot g'(x)].$$

Consequently, when a function $y(x)$ is expressed in the form

$$y = f'(g(x)) \cdot g'(x),$$

the antiderivative of y is expressed by

$$D_x^{-1}(y) = (f \circ g)(x) + c.$$

Combining these expressions, we obtain the *Inverse Chain Rule* for evaluating antiderivatives:

$$\boxed{D_x^{-1}[f'(g(x))g'(x)] = f(g(x)) + c \qquad \textbf{Inverse Chain Rule for antiderivatives.}} \tag{19.8}$$

This rule is general in form and, consequently, difficult to use. In the following, we will present more specific formulas, which correspond to particular choices of the function $f(x)$. It is suggested that you *memorize* the *specific formulas* rather than the general formula.

Example 19.8

a) Let $f(x) = x^3$ and $g(x) = 2x + 1$. The function

$$y = 3(2x + 1)^2 \cdot 2$$

has the form

$$y = f'(g(x)) \cdot g'(x).$$

To verify this, note that

$$g'(x) = 2 \qquad \text{and} \qquad f'(x) = 3x^2,$$

so

$$f'(g(x)) = 3(2x + 1)^2.$$

By the Inverse Chain Rule, the antiderivative of y is

$$D_x^{-1}(y) = (f \circ g)(x) + c = (2x + 1)^3 + c.$$

b) Let $f(x) = \sqrt{x}$ and $g(x) = \sin(x)$. Then the function,

$$y = \frac{\cos(x)}{2\sqrt{\sin(x)}},$$

has the form $y = f'(g(x))g'(x)$. To verify this, we evaluate

$$g'(x) = \cos(x)$$

and

$$f'(x) = (\tfrac{1}{2})x^{-1/2} = \frac{1}{2\sqrt{x}}.$$

Then

$$f'(g(x)) = \frac{1}{2\sqrt{\sin(x)}}.$$

Consequently, the antiderivative of y is

$$D_x^{-1}(y) = (f \circ g)(x) + c = \sqrt{\sin(x)} + c.$$ ◀

In practice, a function is seldom expressed in the form $f'(g(x))g'(x)$ with the functions f and g given. Usually we are simply presented a function,

$$y = (\text{an expression in } x),$$

and asked to compute $D_x^{-1}(y)$. If y is not a polynomial and not one of the trigonometric functions listed in Theorem 19.7 its antiderivative may not be apparent. However, we may be able to express y in the form $f'(g(x))g'(x)$ and thereby obtain $D_x^{-1}(y)$ using the above formula. The main problem will be to discern the particular functions $f(x)$ and $g(x)$ since these will not be given.

One type of function frequently encountered in calculus has the form

$$y = [g(x)]^r \cdot g'(x).$$

The antiderivative formula (19.8) may be used to evaluate $D_x^{-1}y$. The derivative of the function $f(x)$ in this case is

$$f'(x) = x^r, \text{ a power function.}$$

Hence,

$$f(x) = \frac{x^{r+1}}{r+1}.$$

The antiderivative of y is therefore

$$D_x^{-1}(y) = f(g(x)) + c = \frac{[g(x)]^{r+1}}{r+1} + c.$$

Since this formula is so frequently used, we state it in a slightly more general form with $u(x) = g(x)$ and $u^r(x) = [u(x)]^r$.

Theorem 19.9 If $r \neq -1$ and k is a constant, then

$$D_x^{-1}[ku^r(x)u'(x)] = \frac{ku^{r+1}(x)}{r+1} + c \qquad \textbf{Generalized Power Rule for antiderivatives,} \tag{19.9}$$

where c is an arbitrary constant.

The equation (19.9) *should be memorized* as it is frequently encountered. In a specific problem, the function $u(x)$ must be determined. If a problem is suspected to be suitable for applying equation (19.9), the choice of $u(x)$ should be apparent; $u(x)$ will be the quantity raised to a power in the given function.

Example 19.9 The use of the generalized power rule for antiderivatives, (19.9), is illustrated as follows.

a) If $f(x) = 3 \cdot (3x + 1)^5$, then the "obvious" choice for $u(x)$ is $u(x) = 3x + 1$. Since $u'(x) = 3$,

$$f(x) = u^5(x)u'(x)$$

and formula (19.9) with $r = 5$ and $k = 1$ yield

$$D_x^{-1}f(x) = \frac{u^6(x)}{6} + c = \frac{(3x+1)^6}{6} + c.$$

b) If $f(x) = (x^2 + 1)^4 x$, then the choice for $u(x)$ is $u(x) = x^2 + 1$. For this choice of $u(x)$, $u'(x) = 2x$. Hence $f(x) \neq u^4(x)u'(x)$. Instead, since $x = \frac{1}{2}u'(x)$,

$$f(x) = (\tfrac{1}{2})u^4(x)u'(x)$$

and formula (19.9) with $r = 4$ and $k = \frac{1}{2}$ yields

$$D_x^{-1}f(x) = (\tfrac{1}{2})\frac{u^5(x)}{5} + c = \frac{(x^2+1)^5}{10} + c.$$

c) If $f(x) = 2(2x^2 - x + 1)^{-3}(4x - 1)$, then the "obvious" choice for $u(x)$ is $u(x) = 2x^2 - x + 1$ with $r = -3$. Since $u'(x) = 4x - 1$, formula (19.9) with $k = 2$ yields

$$D_x^{-1}f(x) = \frac{2u^{-2}(x)}{-2} + c = -(2x^2 - x + 1)^{-2} + c.$$

d) If $f(x) = (x^2 + 3x)^{10}(x + 1)$, the "obvious" choice for $u(x)$ is $u(x) = x^2 + 3x$. Then $u'(x) = 2x + 3$. Since $u'(x)$ does not equal $x + 1$, or a multiple of $x + 1$, we cannot use formula (19.9) to compute $D_x^{-1}f(x)$. Since $f(x)$ is a continuous function, we know that its antiderivative exists; in fact, by expanding $f(x)$ as a polynomial of degree 21, we can find $D_x^{-1}f(x)$ using Theorem 19.5. ◀

The procedure used to apply formula (19.9) for evaluating antiderivatives is always the same. The same steps indicated below must be followed. First, you should carefully write out the computation involved in each step. Later, when you have more experience with such problems, you will simply perform the sequence of steps mentally.

Procedure for evaluation *$D_x^{-1}f(x)$ when $f(x)$ involves an expression raised to a power.*

i) If $f(x)$ is the sum of expressions, consider each term of the sum separately.

ii) Look for the most conspicuous subterm of an expression: This will be the term raised to a power other than -1. Call this term $u(x)$. Then $f(x) = u^r(x) \cdot v(x)$ for some function $v(x)$ and constant $r \neq -1$.

iii) Compute the derivative $u'(x)$. The function $v(x)$ must be a multiple of $u'(x)$ if equation (19.9) is to be of use. Set $v(x) = k \cdot u'(x)$ and solve for the constant k. If you can solve for the constant k, then applying the generalized Inverse Power Rule (19.9), we find

$$D_x^{-1}f(x) = \frac{k[u(x)]^{r+1}}{r+1} + c,$$

where c is an arbitrary constant.

iv) If you cannot solve the equation $v(x) = k \cdot u'(x)$ for a constant k, then either $u(x)$ was incorrectly chosen or equation (19.9) is not appropriate, assuming that you have not made an error. You should check your work for an error in differentiating.

Example 19.10 To illustrate the above procedure, we determine an antiderivative of the indicated function.

a) Let $f(x) = x(x^2 + 1)^{1/2}$.
   i) f is not the sum of two terms.
   ii) Let $u(x) = x^2 + 1$ and $r = \frac{1}{2}$; then $v(x) = x$.
   iii) $u'(x) = 2x$; hence $v(x) = ku'(x)$ implies that $x = k \cdot 2x$ or $k = \frac{1}{2}$.
   iv) Therefore,

$$D_x^{-1}f(x) = \frac{(\frac{1}{2})[u(x)]^{3/2}}{\frac{3}{2}} = \frac{[x^2 + 1]^{3/2}}{3} + c.$$

b) Let $f(x) = (2x - 1)^3 - 3x^4(x^5 - 1)^{-3}$.
   i) Write $f = f_1 + f_2$, where $f_1(x) = (2x - 1)^3$ and $f_2(x) = -3x^4(x^5 - 1)^{-3}$. Consider $f_1(x)$ first.

ii) Let $u(x) = 2x - 1$ and $r = 3$, then $v(x) = 1$.
iii) $u'(x) = 2$; hence $v(x) = ku'(x)$ yields $k = \frac{1}{2}$.
iv) Therefore

$$D_x^{-1}f_1(x) = \frac{(\frac{1}{2})(2x-1)^4}{4} = \frac{(2x-1)^4}{8} + c_1.$$

Now consider $f_2(x)$.
ii) Let $u(x) = x^5 - 1$ and $r = -3$; then $v(x) = -3x^4$.
iii) $u'(x) = 5x^4$; hence $v(x) = ku'(x)$ yields $k = -\frac{3}{5}$.
iv) Therefore,

$$D_x^{-1}f_2(x) = \frac{(-\frac{3}{5})(x^5-1)^{-2}}{-2} = \frac{3(x^5-1)^{-2}}{10} + c_2.$$

Combining the above results, we find

$$D_x^{-1}f(x) = D_x^{-1}f_1(x) + D_x^{-1}f_2(x);$$

$$D_x^{-1}f(x) = \frac{(2x-1)^4}{8} + \frac{3(x^5+1)^{-2}}{10} + c,$$

where $c = c_1 + c_2$.

c) Let $f(x) = \sin(x)\cos^2(x)$.
   i) f is not the sum of two terms.
   ii) Let $u(x) = \cos(x)$ and $r = 2$; then $v(x) = \sin(x)$.
   iii) $u'(x) = -\sin(x)$; hence $v(x) = ku'(x)$ yields $k = -1$.
   iv) $D_x^{-1}f(x) = -\cos^3(x)/3 + c$.

d) Let $f(x) = (x^4 - 2x + 1)^2(x^3 - 1)$.
   i) f is not the sum of two terms.
   ii) Let $u(x) = x^4 - 2x + 1$ and $r = 2$; then $v(x) = x^3 - 1$.
   iii) $u'(x) = 4x^3 - 2$; hence $v(x) = ku'(x)$ requires $x^3 - 1 = k(4x^3 - 2)$. Equating the coefficients of like powers of x, we require k to satisfy $1 = 4k$ and $-1 = -2k$. There is no value of k that satisfies both of these equations, thus $v(x)$ is not a constant multiple of $u'(x)$.
   iv) Checking our work, we conclude that formula (19.9) cannot be used in this case to compute an antiderivative of $f(x)$. This does not mean that the antiderivative cannot be determined. By multiplying the terms and expanding, we can express $f(x)$ as a polynomial of degree 11 and the antiderivative, $F(x) = D_x^{-1}f(x)$, can then be evaluated. The resulting function, which you should verify, is

$$F(x) = \frac{x^{12}}{12} - \frac{5x^9}{9} + \frac{x^8}{4} + \frac{4x^6}{3} - \frac{6x^5}{5} + \frac{x^4}{4} - \frac{4x^3}{3} + 2x^2 - x + c.$$ ◀

The antiderivative formulas for the trigonometric functions given in Theorem 19.7 may also be generalized. Combining the Inverse Chain Rule with Theorem 19.7 results in the following generalized antiderivative formulas.

Theorem 19.10 The generalized antiderivative formulas for trigonometric functions include, for any differentiable function $u(x)$, the following.

$$\begin{aligned}
D_x^{-1}[\cos(u(x))u'(x)] &= \sin(u(x)) + c,\\
D_x^{-1}[\sin(u(x))u'(x)] &= -\cos(u(x)) + c,\\
D_x^{-1}[\sec^2(u(x))u'(x)] &= \tan(u(x)) + c,\\
D_x^{-1}[\csc^2(u(x))u'(x)] &= -\cot(u(x)) + c,\\
D_x^{-1}[\sec(u(x))\tan(u(x))u'(x)] &= \sec(u(x)) + c,\\
D_x^{-1}[\csc(u(x))\cot(u(x))u'(x)] &= -\csc(u(x)) + c.
\end{aligned}$$

The application of the formulas in Theorem 19.9 should be dictated by the presence of the trigonometric functions with an argument other than the independent variable. If one of the above equations is thought applicable in evaluating $D_x^{-1}f(x)$, the procedure used to apply it is directly analogous to the procedure used for the power formula.

Example 19.11 The following antiderivatives are obtained using the generalized formulas for Theorem 19.10.

a) To evaluate $D_x^{-1}[\cos(3x + 1)]$, we let $u(x) = 3x + 1$. Since $u'(x) = 3$, the function $f(x) = \cos(3x + 1)$ has the form $(\frac{1}{3})\cos(u(x))u'(x)$. Hence,

$$D_x^{-1}[\cos(3x + 1)] = \tfrac{1}{3}\sin(3x + 1) + c.$$

b) If $f(x) = (4x^3 - 1)\sec^2(x^4 - x)$, then the antiderivative of $f(x)$ is obtained by setting $u(x) = x^4 - x$. Thus $u'(x) = 4x^3 - 1$ and $f(x) - \sec^2(u(x))u'(x)$. Hence

$$D_x^{-1}f(x) = \tan(u(x)) = \tan(x^4 - x) + c.$$

c) Let $g(t) = t\sec(t^2)\tan(t^2)$. To evaluate $D_t^{-1}g(t)$, we first set $u(t) = t^2$. Then $u'(t) = 2t$. Since $g(t)$ has a factor t, we set $t = ku'(t)$ and determine the constant $k = \frac{1}{2}$ as was done in Example 19.10. Then

$$g(t) = \tfrac{1}{2}\sec(u(t))\tan(u(t))u'(t).$$

Consequently,

$$D_t^{-1}g(t) = \tfrac{1}{2}\sec(u(t)) = \tfrac{1}{2}\sec(t^2) + c.$$

d) Let $f(\theta) = \sin(\sin(\theta))\cos(\theta)$. To find the antiderivative $D_\theta^{-1}f(\theta)$, we first set $u(\theta) = \sin(\theta)$. Then $u'(\theta) = \cos(\theta)$ and $f(\theta) = \sin(u(\theta))u'(\theta)$. Thus the antiderivative

$$D_\theta^{-1}f(\theta) = -\cos(u(\theta)) = -\cos(\sin(\theta)) + c.$$

e) To find an antiderivative of

$$\frac{\cos(\sqrt{x-1})}{\sqrt{x-1}},$$

we let

$$u(x) = \sqrt{x-1}; \qquad \text{then} \qquad u'(x) = \frac{1}{2\sqrt{x-1}}.$$

Determining the constant $k = 2$, we obtain the desired antiderivative:

$$D_x^{-1}[2\cos(u(x))u'(x)] = 2\sin(u(x)) = 2\sin(\sqrt{x-1}). \quad ◀$$

The generalized antiderivative formulas involving e^x and $\ln(x)$ are similar to the formulas of Theorem 19.8. These are obtained from the general formula (19.8) with $g(x) = u(x)$.

Theorem 19.11 For any differentiable function $u(x)$,

$$\boxed{D_x^{-1}(e^{u(x)}u'(x)) = e^{u(x)} + c,} \tag{19.10}$$

and, if $u(x) > 0$,

$$\boxed{D_x^{-1}\left[\frac{u'(x)}{u(x)}\right] = \ln(u(x)) + c.} \tag{19.11}$$

Example 19.12 The antiderivative $D_x^{-1}f(x)$ is obtained using either formula (19.10) or (19.11).

a) Let $f(x) = 2xe^{x^2}$. Then $f(x) = e^{u(x)} \cdot u'(x)$, where $u(x) = x^2$. By formula (19.10), we find

$$D_x^{-1}(2xe^{x^2}) = e^{x^2} + c.$$

b) Let $f(x) = (x+1)e^{(x^2+2x-1)}$. To use formula (19.10) to determine an antiderivative of $f(x)$, we choose $u(x)$ as the function in the exponent:

$$u(x) = x^2 + 2x - 1; \qquad u'(x) = 2x + 2.$$

The function $f(x)$ does not have the exact form $e^{u(x)}u'(x)$. To see if $f(x)$ is a multiple of this form, we set the coefficient term, $x + 1$, equal to $k \cdot u'(x)$ and solve for k:

$$x + 1 = k(2x + 2); \qquad k = \tfrac{1}{2}.$$

Therefore,

$$f(x) = \tfrac{1}{2}e^{u(x)}u'(x),$$

and the antiderivative of $f(x)$ is

$$D_x^{-1}f(x) = \tfrac{1}{2}e^{u(x)} + c = \tfrac{1}{2}e^{(x^2+2x-1)} + c.$$

c) To determine a function whose derivative is

$$f(x) = \frac{3x^2 - 1}{x^3 - x + 6},$$

we have only one possible formula to use. As $f(x)$ is a rational expression—i.e., a fraction, the only formula which may give us $D_x^{-1}f(x)$ is the generalized formula (19.11). To apply this formula, we first set $u(x)$ equal to the denominator:

$$u(x) = x^3 - x + 6.$$

Since the derivative $u'(x) = 3x^2 - 1$ is exactly the numerator of f,

$$f(x) = \frac{u'(x)}{u(x)}.$$

Applying (19.11), we find

$$D_x^{-1} \frac{3x^2 - 1}{x^3 - x + 6} = \ln(x^3 - x + 6) + c.$$

d) Let

$$f(x) = \frac{x^2 + 1}{x^3 - 3x + 2}.$$

To determine an antiderivative of $f(x)$, we set

$$u(x) = x^3 - 3x + 2; \qquad \text{then} \qquad u'(x) = 3x^2 - 3.$$

As $f(x)$ does not have exactly the form $u'(x)/u(x)$, we set the numerator of $f(x)$ equal to a multiple of $u'(x)$ and solve for k;

$$x^2 - 1 = k(3x^2 - 3); \qquad k = \tfrac{1}{3}.$$

Therefore, $f(x) = (\frac{1}{3})u'(x)/u(x)$, and hence

$$D_x^{-1}f(x) = \tfrac{1}{3}\ln(x^3 - 3x + 2) + c.$$

e) Formulas (19.10) and (19.11) cannot be used to evaluate the antiderivative of every exponential function or every function that involves a rational expression. Instances in which these formulas are not applicable occur when

$$f(x) = e^{x^2} \qquad \text{or} \qquad f(x) = \frac{1}{x^2 + 3x + 2}.$$

In the first case, $f(x)$ cannot be put into the form $e^{u(x)}u'(x)$ as $u(x)$ must equal x^2 and then $u'(x) = 2x$. Setting the coefficient of e^{x^2} (which is 1) equal to $k \cdot u'(x)$, we find that

$$1 = k\,2x \qquad \text{or} \qquad k = \frac{1}{2x}, \qquad \text{not a constant!}$$

The antiderivative of e^{x^2} cannot be evaluated in terms of elementary functions. In the second case, if we try to use formula (19.11), we must set $u(x) = x^2 + 3x + 2$. In this case, the numerator of $f(x)$ is not a *constant* multiple of $u'(x) = 2x + 3$. Consequently, formula (19.11) cannot be used to evaluate an antiderivative of $f(x)$. In this case, the antiderivative can be found using a slightly more advanced technique that is discussed in Section 24. ◀

Example 19.13 *Exponential growth*
The function $N(t)$, describing a population which satisfies the proportional growth model, must satisfy the equation

$$N'(t) = k \cdot N(t) \qquad \text{(rate of growth} \propto \text{population size).} \tag{19.12}$$

The only function that satisfies equation (19.12) is the exponential function

$$N(t) = Ce^{kt}. \tag{19.13}$$

To derive this equation for $N(t)$ from the differential equation (19.12), we argue that the function $N(t)$ must satisfy

$$\frac{N'(t)}{N(t)} = k.$$

The left side of this equation has exactly the same form as in formula (19.11), except that the function $u(x)$ is now $N(t)$. If we argue that the antiderivative (with respect to t) of both sides of this equation are equal, we have

$$D_t^{-1}\left[\frac{N'(t)}{N(t)}\right] = D_t^{-1}(k),$$

or

$$\underset{\text{Formula (19.11)}}{\ln[N(t)]} = \underset{\text{Power formula (19.9) with } r = 0}{kt + c}$$

Taking the exponential of both sides of this equation gives

$$e^{\ln[N(t)]} = e^{kt+c}.$$

Since $e^{\ln(y)} = y$ for any y, this becomes

$$N(t) = e^{kt+c}.$$

To obtain equation (19.13), we use the fact that

$$e^{kt+c} = e^{kt}e^{c} = Ce^{kt},$$

where $C = e^c$. Combining the last two equations, we have equation (19.13).

Because the "solution" of the growth equation (19.12) is obtained by expressing it as a derivative of $\ln(N(t))$, many ecology and population dynamics texts also refer to the exponential growth model as a *logarithmic growth model.* ◀

19.3 THE INDEFINITE INTEGRAL

Another very common name for the antiderivative is the *integral.* A complete set of *integral notation* is used to discuss antiderivatives using the term "integral."

Definition 19.3

i) *The integral* or *indefinite integral* of a function $f(x)$, with respect to the variable x, is any function $F(x)$ such that

$$\frac{dF(x)}{dx} = f(x).$$

ii) The integral $F(x)$, of a function $f(x)$, is expressed in *integral notation* by

$$F(x) = \int f(x)dx,$$

which is read "The function capital F of x is equal to the integral (or indefinite integral) of the function small f of x with respect to the variable x." This may be shortened to read "Capital F of x is the integral of small f of x, dx."

iii) The symbol to denote an integral consists of an *integral sign*, $\int$ or $\displaystyle\int$, followed by a function $f(x)$, and terminated by a term "dx," which indicates that the independent variable is denoted by "x."

iv) When considering an integral, $\int f(x)dx$, the function $f(x)$ is called the *integrand.*

v) The process of determining $\int f(x)dx$ as a function $F(x)$ is called *integrating the function* $f(x)$.

The indefinite integral, $\int f(x)dx$, is a function of x; it is another symbol for the antiderivative, $D_x^{-1}f(x)$. Since the terms "integral" and "antiderivative" are synonymous, every theorem concerning antiderivatives may be restated using integrals and integral notation. (We use this notation in the summary at the end of the section.) The general integral of a function $f(x)$ will be determined only up to an arbitrary additive constant,

$$\int f(x)dx = F(x) + c.$$

If the independent variable is other than x, say t, then the term dx must be replaced by dt. The process of integrating or evaluating integrals is identical to the process of determining the antiderivative of the integrand.

The integral formulas are indicated at the end of this Section.

Example 19.14 The following equations illustrate the evaluation of indefinite integrals. The constant c is arbitrary.

a) $\int 3x\,dx = \frac{3}{2}x^2 + c$

b $\int 2x^3 - x^{1/2}\,dx = \frac{1}{2}x^4 - \frac{3}{2}x^{3/2} + c$

c) $\int (t+1)^2\,dt = \dfrac{(t+1)^3}{3} + c$

d) $\int (z^5 + 2)dz = \dfrac{z^6}{6} + 2z + c$

e) $\int w(w^2 - 1)^{2/3}\,dw = \frac{3}{10}(w^2 - 1)^{5/3} + c$

f) $\int \dfrac{7}{(2x+1)^2}\,dx = \dfrac{-\frac{7}{2}}{(2x+1)} + c$

g) $\int \sin(2x)dx = -\frac{1}{2}\cos(2x) + c$

h) $\int 2x\sec^2(x^2+1)dx = \tan(x^2+1) + c$

i) $\int \sin^2(x)\cos(x)dx = \frac{1}{3}\sin^3(x) + c$

j) $\int e^{2x}\,dx = \frac{1}{2}e^{2x} + c$ ◀

Integral Curves

In the above example each integral is represented by a sum of a function and a constant in the form

$$\int f(x)dx = F_0(x) + c.$$

The function $F_0(x)$ represents one particular integral corresponding to the choice of zero for the constant. Any other choice of the constant leads to another particular integral of $f(x)$. Each particular integral $F(x)$ of $f(x)$ corresponds to a particular choice of the constant c. To indicate this correspondence, we may *index* or *parameterize* the integrals of $f(x)$ by employing a subscript to indicate the particular constant c.

If $F'(x) = f(x)$—that is, $F(x)$ is an integral of $f(x)$—and $F(x) = F_0(x) + c$, then we denote $F(x)$ by $F_c(x)$.

For instance, every integral of $f(x) = x^5 + x - 2$ is of the form

$$F_c(x) = \frac{x^6}{6} + \frac{x^2}{2} - 2x + c.$$

The set of all integrals of a function $f(x)$ is therefore a *one-parameter* family of functions:

$$\{F_c(x)\} \qquad c \text{ a real number,}$$

which is similar to a sequence, but for which the index or parameter c assumes all real values. The graphs of the functions of this family are called *integral curves.* As the parameter c varies, the graphs of the associated integral curves vary. It can be shown that for reasonable functions, $f(x)$, the associated family of integral curves covers the entire Euclidean plane. By this, we mean that through any given point, (x_0, y_0), there passes one and only one integral curve of a given family.

Example 19.15 To illustrate the concept of a one-parameter family of integral curves and how a different family of curves is associated with each integrand, we consider three integrands, $f(x) = 3$, $f(x) = 3x$, and $f(x) = x^2 - 1$.

a) Associated with the integrand $f(x) = 3$ is the one-parameter family of integral curves, which are the graphs of the functions

$$F_c(x) = \int 3\, dx = 3x + c.$$

The family of integral curves $\{F_c(x)\}$ consists of all lines with slope three. (See Fig. 19.1(a).) Given any point in the plane, (x_0, y_0), there is exactly one line with slope three which passes through the point:

$$y = 3(x - x_0) + y_0.$$

b) Associated with the integrand $f(x) = 3x$ is a one-parameter family of integrals,

$$F_c(x) = \int 3x\, dx = (\tfrac{3}{2})x^2 + c.$$

The family $\{F_c(x)\}$ consists of all vertical translations of the parabola $y = \frac{3}{2}x^2$. (See Fig. 19.1(b).) Through any point (x_0, y_0), there passes exactly one such parabola. For instance, through the point (2, 3) passes the curve

$$F_{-3}(x) = \tfrac{3}{2}x^2 - 3.$$

c) Associated with the integrand $f(x) = x^2 - 1$ is the family of integrals

$$F_c(x) = \int x^2 - 1\, dx = \frac{x^3}{3} - x + c.$$

The family of integral curves $\{F_c(x)\}$ is a family of cubic curves, which are shifted vertically. (See Fig. 19.1(c).) The unique member of this family passing

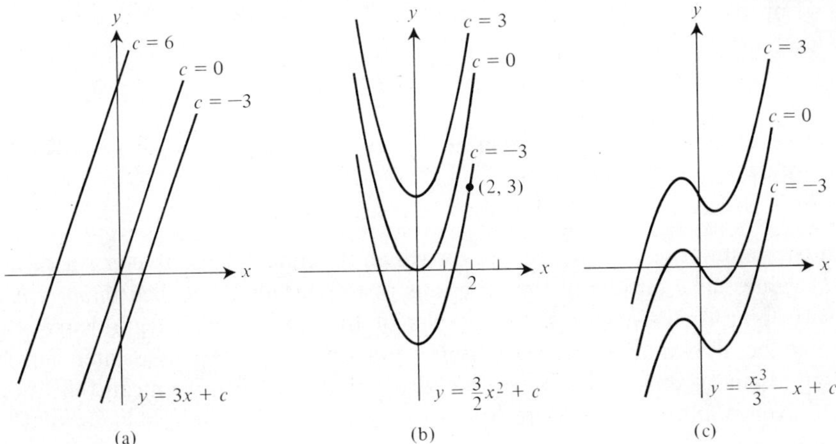

Figure 19.1

through the point (2, 1) is $F_c(x)$, where the value of c is obtained by solving $1 = F_c(2)$:

$$1 = \frac{2^3}{3} - 2 + c, \; c = \tfrac{1}{3}.$$

Therefore the curve is

$$F_{1/3}(x) = \frac{x^3}{3} - x + \tfrac{1}{3}.$$

d) In each of the examples, we observe that the value of the constant c determines the vertical displacement of the *basic form of the integral curves*; c is not necessarily the y-intercept. Consider the family of integrals,

$$F_c(t) = \int \sin(t)dt + c.$$

The integral curves of this family are the graphs of the function $F_c(t) = -\cos(t) + c$. Hence, the y-intercept of each curve $y = F_c(t)$ is in fact $c - 1$. (See Fig. 19.2.)

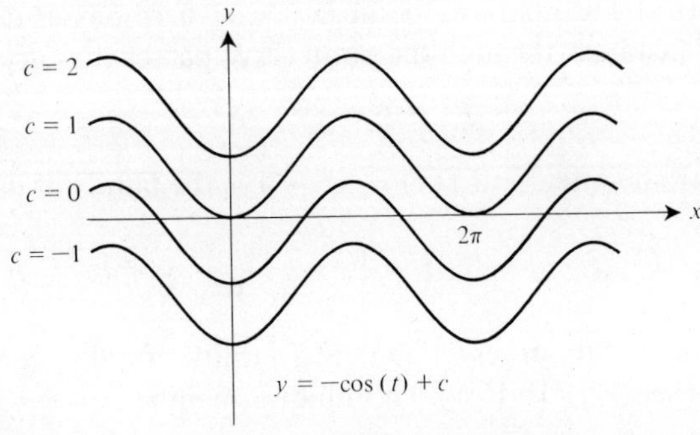

Figure 19.2

◀

In general, if an experimental observation is depicted as an integral curve, the choice of the constant c is dictated by knowing the value of one point on the curve.

Example 19.16 A common apparatus, which is used in virtually every scientific discipline, is the mechanical recorder. This is an instrument that produces a continuous graph, which depicts a measurable quantity over an interval of time. Such recorders require a second interface instrument, which translates the change in an observable quantity into an electrical current. The recorder then *integrates* the current and mechanically produces a graph. The graph is produced by a pen which moves vertically as the graph paper moves horizontally past the pen. (See Fig. 19.3(a).) The physical dimensions of the paper will limit the *amplitude* of the

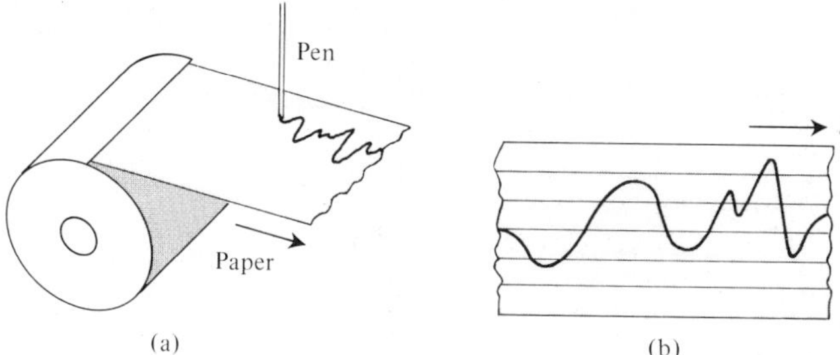

Figure 19.3 (a) (b)

graph. The speed of the paper will affect the *period* of the graph. However, these are scaling factors, which we do not consider here. What we do wish to consider is the scale associated with the y-coordinate of the graph paper. The graph produced is *an integral curve*, $y = F_{c_0}(t)$, as indicated in Fig. 19.3(b).

The scientist must decide which parameter c_0 fits the true graph of the quantity being measured. This is done by specifying the value y_1 corresponding to a specific time t_1. This corresponds to algebraically determining the constant c_0 so that

$$F_{c_0}(t_1) = y_1.$$

In Fig. 19.4, we illustrate two different scales for the same integral curve, corresponding to the values $y(t_1) = 10$ and $y(t_1) = 51$. The curve depicted is a basic sine curve, $y = \sin(\pi t/2) + c_0$, and $t_1 = 1$. In Fig. 19.4(a), $c_0 = 9$, while in Fig. 19.4(b), $c_0 = 50$.

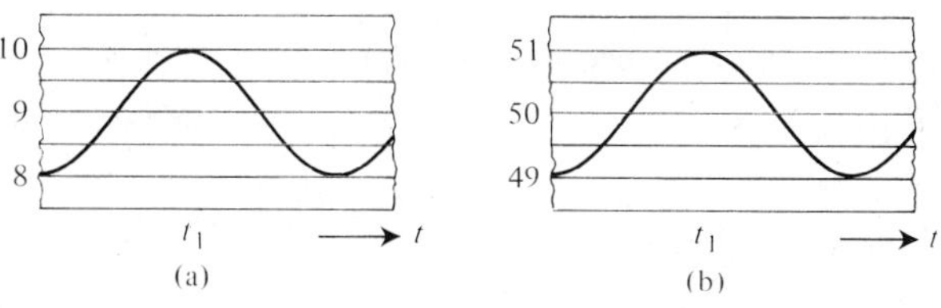

Figure 19.4 (a) (b)

◀

SUMMARY

We summarize the formulas and results of this section using integral notation.

1. *Algebraic properties*:

   a) *Sum Rule*

$$\int f(x) + g(x)dx = \int f(x)dx + \int g(x)dx$$

   b) *Constant Rule*

$$\int kf(x)dx = k \cdot \int f(x)dx, \qquad k \text{ a constant}$$

2. *Theory*:

   a) $F(x) = \int f(x)dx$ implies that $F'(x) = f(x)$.

   b) $F'(x) = f(x)$ implies that $F(x) = \int f(x)dx + c$.

   c) If $f(x)$ is continuous, then $\int f(x)dx$ exists.

3. *Basic integrals*:

   a) *Power Rule*

$$\int x^r\,dx = \frac{x^{r+1}}{r+1} + c, \text{ for } r \neq -1$$

   b) *Trigonometric formulas*

$$\int \sin(x)dx = -\cos x + c$$

$$\int \cos(x)dx = \sin(x) + c$$

$$\int \sec^2(x)dx = \tan(x) + c$$

$$\int \csc^2(x)dx = -\cot(x) + c$$

$$\int \sec(x)\tan(x)dx = \sec(x) + c$$

$$\int \csc(x)\cot(x)dx = -\csc(x) + c$$

c) *Exponential*

$$\int e^x\,dx = e^x + c$$

Natural logarithm

$$\int \frac{1}{x}\,dx = \ln(x) + c$$

4. *Generalized formulas*:

a) *Inverse Chain Rule*

$$\int f(u(x))u'(x)dx = f(u(x)) + c$$

b) *Generalized Power Rule*

$$\int [u(x)]^r u'(x)dx = \frac{[u(x)]^{r+1}}{r+1} + c, r \neq -1.$$

c) For trigonometric functions formulas corresponding to 3(b) above, replace x by $u(x)$ and replace dx by $u'(x)dx$. For example,

$$\int \cos(u(x))u'(x)dx = \sin(u(x)) + c.$$

d) *Generalized exponential*

$$\int e^{u(x)}u'(x)dx = e^{u(x)} + c$$

e) *Generalized logarithm*

$$\int \frac{u'(x)}{u(x)}\,dx = \ln(u(x)) + c$$

EXERCISE SET 19

19.1 Verify that each of the given functions is the antiderivative of the same function and determine the constant as in Theorem 19.1.

a) $F_1(x) = x^2 - 6x + 3$; $F_2(x) = x^2 - 6x + 8$
b) $F_1(x) = x^{1/2} - x^{3/2}$; $F_2(x) = x^{1/2}(1 - x) + 8$
c) $F_1(x) = (x + 3)^2$; $F_2(x) = x^2 + 6x - 12$
d) $F_1(x) = (x + 1)/(x - 1) + 3$; $F_2(x) = (x + 1)^2/(x^2 - 1)$
e) $F_1(x) = (x^2 + 3x + 2)/(x + 2)$; $F_2(x) = x - 6$

19.2 Compute the general antiderivative of the indicated function.

a) $f(x) = x^2$
b) $f(x) = 2x^{-5}$
c) $f(x) = 3x^5 + 2$
d) $f(x) = x^2 - x^{-2}$
e) $f(x) = 5x^3 - 2x^5$
f) $f(x) = 6(x + 1)(x + 7)$
g) $f(x) = 6x^{-7} - 7x^6$
h) $f(x) = 3x^2 - \frac{2}{3}x^4$
i) $f(x) = x^{1/2} + x^{3/4}$
j) $f(x) = 2x^{-1/2} + x^{-2}$

19.3 Determine the antiderivative $F(x)$ of the given function $f(x)$ satisfying the specified value.

a) $f(x) = x^2 - 2x - 2;\ F(3) = 2$
b) $f(x) = x^{1/2} - x;\ F(1) = 1$
c) $f(x) = 5x^5 + 2;\ F(0) = 1$
d) $f(x) = (x - 2)^2;\ F(2) = 8$
e) $f(x) = x^{-1/2} - 1;\ F(0) = 3$
f) $f(x) = \sin(x);\ F(\pi) = 0$
g) $f(x) = \sec^2(x);\ F(\pi/4) = 0$
h) $f(x) = 3e^x;\ F(0) = 1$
i) $f(x) = 3/x;\ F(1) = 2$

19.4 Compute the antiderivative of the given function, if possible, using the generalized power formula.

a) $y = (x + 3)^4$
b) $y = 6x(x^2 + 1)^{-3}$
c) $y = (x^2 + x + 2)^2(6x + 3)$
d) $y = (x^2 + x + 2)6x$
e) $y = (2x^3 - 1)^5x^2$
f) $y = (2x + 3)^4 - (2x + 3)^{-2}$
g) $y = (x^2 + 7x - 2)^{10}(x + 7)$
h) $y = (5x^{1/5} + 1)^4/x^{4/5}$
i) $y = x(3x^2 + 6)^{20}$
j) $y = (3x + 2)(x^3 + 2x + 7)^{-1/2}$

19.5 Represent the following functions in the form $u^r(x)v(x)$, indicating the functions u and v. Then determine the antiderivative, $D_x^{-1}(y)$, if possible.

a) $y = x(x^2 - 3)^{1/2}$
b) $y = -6(x^2 + 7x - 2)^4(2x + 7)$
c) $y = (x^2 + 3x - 2)[x^3 + 4.5x^2 - 6x]^5$
d) $y = (2x + 1)^{11} - (2x + 1) + (2x + 1)^{-3}$.

19.6 Determine the antiderivative of the following functions.

a) $2e^{2x}$
b) e^{2x}
c) e^{2x+1}
d) e^{-x}
e) xe^{x^2}
f) xe^{-x^2}
g) xe^{3x^2+1}
h) $\cos(x)e^{\sin(x)}$
i) $(x^2 + x + 1)e^{2x^3+3x^2+6x-2}$

19.7 Determine the antiderivatives of the following functions.

a) $2/x$
b) $1/2x$
c) $2x/x^2$
d) $2x/(x^2 + 3)$
e) $x/(x^2 - 4)$
f) $x^2/(x^3 + 1)$
g) $(x - 2)/(x^2 - 4x + 7)$
h) $(2x - 2)/(x^2 - 2x + 1)$
i) $(1 + \cos(x))/(x + \sin(x))$

19.8 Evaluate the given integrals.

a) $\int 5x\,dx$
b) $\int 3x^2 - 2x + 1\,dx$
c) $\int (2x + 1)^2\,dx$
d) $\int x^{-7} + 5x^2 - 2\,dx$
e) $\int 2t^{1/2}\,dt$
f) $\int \frac{(3t^{1/2} - 2)}{t^{1/2}}\,dt$
g) $\int 2x(x^2 - 3)^4\,dx$
h) $\int x^3(x^4 + 5)^{-2}\,dx$
i) $\int (x^2 + 2x - 1)^2\,dx$
j) $\int (x^2 - 1)^3x\,dx$

19.9 Integrate the following integrands.

a) $f(x) = \cos(3x + \pi)$
b) $f(x) = \sin(x) - \cos(2x)$
c) $f(x) = \sec(x)[\sec(x) - \tan(x)]$
d) $f(x) = x\sin(x^2 + 3)$
e) $f(x) = 3\sin(\pi x)$
f) $f(x) = \tan^2(x - x^2)(1 - 2x)$
g) $f(x) = \sin^3(x)\cos(x)$
h) $f(x) = \tan(x)\sec^2(x)$
i) $f(x) = \sin^3(x)$
i) $f(x) = \sin(\sin(x))\cos(x)$

19.10 Evaluate the following integrals.

a) $\int 2e^t\,dt$
b) $\int e^{2x}\,dx$
c) $\int xe^{2x^2}\,dx$
d) $\int e^{-0.01x}\,dx$
e) $\int (x^2 + 2)e^{x^3+6x-2}\,dx$
f) $\int 5te^{t^2}\,dt$
g) $\int \sin(\theta)e^{\cos(\theta)}\,d\theta$
h) $\int \frac{1}{t^2}e^{(1/t)}\,dt$

19.11 Evaluate the given integral.

a) $\int \frac{2x}{x^2 + 1}\,dx$
b) $\int \frac{t - 1}{t^2 - 2t + 1}\,dt.$
c) $\int \frac{x - 1}{2x^2 - 4x + 1}\,dx$
d) $\int \frac{2t - \cos(t)}{\sin(t) + t^2}\,dt$
e) $\int \frac{20x^3 + 15x^2 - 2}{5x^4 + 5x^3 - 2x}\,dx$
f) $\int \frac{\sin(\theta) + \theta\cos(\theta)}{\theta\sin(\theta)}\,d\theta$

19.12 A population of size $N(t)$ has an instantaneous rate of change $R(t) = -t(t - 100)$.

a) Represent $N(t)$ as an integral.
b) If $N(0) = 50$, what is the equation for $N(t)$?
c) If $N(0) = 50$, what is $N(50)$?

19.13 a) If a population $N(t)$ has an instantaneous rate of growth of $3N(t)$, what is the equation for $N(t)$?

b) If the population in (a) has a size of $3.2 \cdot 10^4$ at $t_0 = 0$, what will be its size at $t_1 = 10$?
c) If the population in (a) has a size of $5 \cdot 10^4$ at $t_0 = 2$, what will be its size at $t_1 = 10$?

19.14 a) Assume that a population, $N(t)$, grows at a rate $N'(t) = 3t \cdot N(t)$. Use the technique employed in Example 19.13 to determine the form of $N(t)$.

b) Repeat parts (b) and (c) of Exercise 19.13 for this population.

19.15 In reference to Example 19.1(a), assume that a patient at time $t = 0$ is administered 10 g of a drug, which is excreted at a rate of $3t^{-1/2}$ mg/hr. What will be the amount of drug in the patient at time $t > 0$? When will the patient be drug free?

19.16 In reference to Example 19.1(b), assume that a scientist has estimated that the average annual temperature 12,000 years ago was varying according to the sine function, $r(t) = 0.03\sin(\pi t/100)$. If t is measured in years, indicate the period of fluctuation and give a function indicating the temperature, $T(t)$. What would be a suitable choice for the constant c if T is in degrees Celsius and $T(-10{,}000) = -4$?

19.17 If a patient's blood pressure changes at a rate $r(t) = \frac{1}{2}\cos(t/4)\sin(t/4)$ mm Hg/hr in response to a test drug, what would be the patient's blood pressure in two hours if it was 96 mm Hg when the drug was administered? When will it return to normal?

19.18 For each of the integrands, $f(x)$, perform the following.

a) Determine the associated one-parameter family of integral curves.
b) Sketch the graph of the integral curve associated with $C = 0$, $C = 3$, and $C = -3$.
c) On the same graph, sketch the integral curve passing through (1, 3). (Give the equation of this curve.)

i) $f(x) = -6$ ii) $f(x) = -2x + 2$ iii) $f(x) = 4x^3 - 4x$

19.19 The flow rate of a stream over the spillway of a dam is given by $\mathring{V}(t) = e^{-Kt}$, where the constant K is related to the size of the spillway. Express the flow volume of the stream as an integral and then integrate to establish the explicit function $V(t)$. What interpretation could you give to the constant of integration?

SECTION 20

THE METHOD OF DISCRETE APPROXIMATION

20.1 THREE PROBLEMS OF MEASUREMENT

In this section, we introduce three problems of measurement, each of which presents the same difficulty—that is, measuring the amount of a quantity that accumulates over an interval of time.

We then examine how the total change in the quantity can be approximated using a discrete modeling process. This process consists of representing the change in a quantity over a time interval by the sum of its changes over shorter intervals of time. The shorter intervals of time are obtained by dividing up the original time period. Then, the actual change over a short time interval is approximated. Adding up the approximate change over each of the short time periods leads to an approximation of the total change.

This discrete modeling process is examined for two reasons. First, it illustrates the basic approach that underlies all mathematical models. It is the method used in computer models or computer simulations. Consequently, it has great practical use. Secondly, it is the process by which the concept of *integration* was historically developed. It provides the theoretical bases for both the abstract mathematical concept as well as the practical application and interpretation of the *definite integral*. These connections are discussed in the following Sections.

Problem A

A microbial population grows in a complete medium containing all the nutrients required for growth. The rate of growth of the microbial culture is a function, $f(t)$, which depends on the particular organism, its interdivision time, and the mean numbers of progeny of the dividing cell. For most microorganisms, the growth rate is exponential and the culture grows exponentially. However, for some organisms, nonexponential growth has been demonstrated. The problem is to determine how many cells are produced over a time interval, $[0, T]$, when the form of the growth-rate function $f(t)$ is known. (Assume that the death rate is incorporated in the growth rate.)

Problem B

A solute, S, enters and leaves a mixing compartment, C. Let the rate at which S enters the compartment be denoted by $\mathring{Q}_{\text{in}}(t)$ g/min., and let the rate of S leaving the compartment be denoted by $\mathring{Q}_{\text{out}}(t)$ g/min. Fick's first principle states that, during any interval of time, the quantity of solute entering the compartment must equal the quantity of solute leaving the compartment plus the quantity of solute accumulated in the compartment. We denote the quantity of solute S in the compartment at time t by $Q(t)$ g (see Fig. 20.1). The problem is to determine the amount $Q(T)$ of solute in the compartment at time T, assuming that the inflow and outflow rates, $\mathring{Q}_{\text{in}}(t)$ and $\mathring{Q}_{\text{out}}(t)$, are known, as well as the initial amount of solute in the compartment, $Q(0)$.

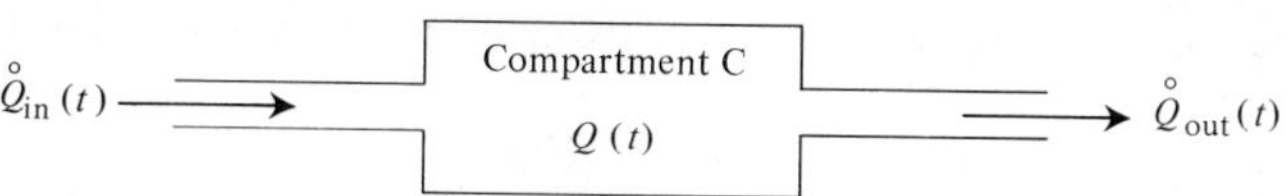

Figure 20.1

Problem C

An industrial plant discharges a relatively constant flow of effluent into a river. Unfortunately, the effluent contains mercury compounds. The amount of mercury in the effluent varies considerably over a 24-hour period. The problem is to determine how much mercury is introduced into the river in a one-day period. The only information about the amount of mercury being discharged is obtained by sampling the effluent.

In each of these problems we encounter the same situation. We are asked to determine the amount of a quantity that accumulates over an interval of time. The difficulty is that the quantity cannot easily be measured. But we assume that we know how the quantity accumulates and, from this, we will estimate the total amount accumulated.

20.2 DISCRETE SOLUTIONS: APPROXIMATE SOLUTIONS

In this section, we examine the discrete solution of the problems presented. In each problem, the discrete solution is obtained by considering how the quantity being examined accumulates or changes over successive short intervals of time. Combining these changes leads us to a solution that is expressed in terms of a finite series.

The reader encountering such arguments for the first time may find a general examination of each problem a bit formidable. If this is the case, we suggest reading concurrently the more computational examples following each general discussion. In any event, rereading the following solutions, after working some of the exercises, will be well worth the effort in terms of understanding the approximation method.

In the following, we will use the sigma or summation symbol $\sum$ to denote a sum. For a review of this notation, see Section 10.1.

Problem A Solution

We consider the total number of individual cells produced over the interval $[0, T]$ as a function, $P(T)$, of T. Denote the number of cells at time $t = T$ by $F(T)$. Then $P(T)$ is $F(T)$ minus $F(0)$, the number of cells at time $t = 0$,

$$P(T) = F(T) - F(0).$$

Alternatively, the number of cells produced over the interval $[0, T]$ can be obtained as the sum of the number of cells produced over a sequence of subintervals of the interval $[0, T]$.

For instance, if T is 1 hour, the time interval $[0, 1]$ could be divided into three time periods, each 20 minutes long. In units of hours, this would be the intervals

$$[0, \tfrac{1}{3}], \qquad [\tfrac{1}{3}, \tfrac{2}{3}], \qquad \text{and} \qquad [\tfrac{2}{3}, 1].$$

We denoted the number of cells produced over each subinterval by P_i; P_1 for the first 20 minutes, P_2 for the second 20 minutes, and P_3 for the last $\frac{1}{3}$ hour. The total production over the one-hour period would be

$$P(1) = P_1 + P_2 + P_3 = \sum_{i=1}^{3} P_i.$$

To develop a general formula to represent the production $P(T)$ using an arbitrary subdivision of the interval $[0, T]$, we introduce the notion of a *partition* of the interval. The idea of a *partition* is to state the endpoints of the subintervals—i.e., where the original interval will be divided or partitioned. In general terms, a partition of $[0, T]$ is a linearly ordered set of points, $\{t_0, t_1, t_2, \ldots, t_N\}$, with the first point being $t_0 = 0$, and the last point being $t_N = T$:

$$0 = t_0 < t_1 < t_2 < \cdots < t_N = T \qquad \textbf{Partition of } [0, T].$$

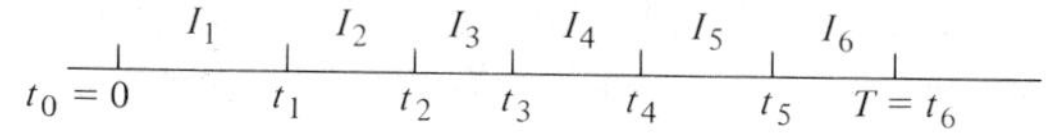

I_1 I_2 I_3 I_4 I_5 I_6

$t_0 = 0$ t_1 t_2 t_3 t_4 t_5 $T = t_6$

A partition of size $N = 6$

Figure 20.2

The set of points $\{t_i\}_{i=1}^N$ partitions the interval $I = [0, T]$ into subintervals, $I_1, I_2, \ldots, I_N$, where $I_1 = [t_0, t_1]$, $I_2 = [t_1, t_2]$, and the nth subinterval is

$$I_n = [t_{n-1}, t_n] \qquad \textbf{nth subinterval.}$$

(See Fig. 20.2.) The length of each subinterval is the difference

$$\Delta t_n \equiv t_n - t_{n-1} \qquad \textbf{Length of nth subinterval.}$$

Thus $\Delta t_5 = t_5 - t_4$ and $\Delta t_1 = t_1 - t_0$.

The number of cells produced over the interval $[0, T]$ may be expressed as a sum of the number produced over the subintervals, $I_1, I_2, \ldots, I_N$. If

$$P_n = \text{number of cells produced in the interval, } I_n,$$

then

$$P(T) = \sum_{n=1}^{N} P_n. \tag{20.1}$$

How can we use equation (20.1) to solve our problem? What are we trying to determine and what is known? We want to find $P(T)$. We know the rate of production $f(t)$, at any given time t. Trying to determine the production P_n over the subinterval $I_n = [t_{n-1}, t_n]$ is the same as trying to determine the production over $[0, T]$. Apparently, we have replaced our original problem with N-problems of equal difficulty, trying to determine $P_1, P_2, \ldots, P_N$. Are we solving the original problem or compounding it? The answer is both. The problem is compounded by increasing the number of intervals over which we wish to compute the production. But for the extra work, we gain access to an approximate solution if it is possible to accurately estimate the production over small time intervals.

Estimating Production Over an Interval

Consider the problem of estimating the production over an interval $[a, b]$. The average rate of production over the interval $[a, b]$ is defined by the property that the total production, P, over the interval $[a, b]$ is the product of the average rate of production and the length of the interval:

$$P = (\text{average rate}) \times (b - a).$$

However, we do not know the average rate of production, but simply the instantaneous rate of production, $f(t)$. To estimate the production P we approximate the average of the function f over an interval $[a, b]$. The estimate is taken to be $f(t^*)$, where t^* is an arbitrary point in the interval $[a, b]$. Thus we pick any point $t^* \in [a, b]$ and approximate the average rate by $f(t^*)$. Then the production P is approximated by the following formula:

$$P \simeq f(t^*)(b - a). \tag{20.2}$$

The accuracy of this approximation depends on the function, f, and the length of the interval $[a, b]$. For continuous functions, it can be argued that the maximum possible error in approximation (20.2) will become smaller as the length of the interval $[a, b]$ becomes smaller.

Approximating the Total Production

To estimate the total production, we use the approximation formula (20.2) to estimate the production P_n over the subinterval I_n for $n = 1, 2, \ldots, N$. These estimates are then used in equation (20.1). Since the error in estimating the production will be smaller as the interval becomes shorter, we see that the advantage of using the partitioning method is to reduce the error of estimation.

The estimation procedure consists of the following steps.

I. Partition the interval $[0, T]$. This means that we must choose (at our convenience) N points $\{t_n\}$, with

$$t_0 = 0 < t_1 < t_2 < \cdots < t_N = T.$$

This creates N-subintervals,

$$I_n = [t_{n-1}, t_n], \qquad n = 1, 2, 3, \ldots, N.$$

The length of each subinterval is

$$\Delta t_n = t_n - t_{n-1}, \qquad n = 1, 2, 3, \ldots, N.$$

II. Pick a sequence of evaluation points $\{t_n^*\}$; for $n = 1, 2, \ldots, N$, t_n^* must be in the closed interval $I_n = [t_{n-1}, t_n]$. Thus t_1^* is in the first subinterval, t_2^* is in the second subinterval, and so forth.
Note: The function f is to be evaluated at the points t_n^*. Therefore, in practice, choose t_n^* so that $f(t_n^*)$ is easy to evaluate.
III. Approximate the production P_n over the interval I_n, for $n = 1, 2, 3, \ldots, N$, by

$$P_n \simeq f(t_n^*)(t_n - t_{n-1}) = f(t_n^*)\Delta t_n.$$

Then approximate $P(T)$ by the sum obtained by substituting the above approximations into (20.1):

$$P(T) \simeq \sum_{n=1}^{N} f(t_n^*)\Delta t_n \qquad \textbf{Approximate solution.} \tag{20.3}$$

Example 20.1 To illustrate the approximate solution of Problem A given by equation (20.3), we assume that the function $f(t) = t + 1$ and $T = 2$. We will obtain different approximations for $P(2)$ by evaluating (20.3) for different choices of the partition points, $\{t_n\}$, and the evaluation points, $\{t_n^*\}$.

a) Consider the partition of $[0, 2]$ given by $t_0 = 0, t_1 = \frac{1}{2}, t_2 = 1, t_3 = 1\frac{1}{4}, t_4 = 2$. Here $N = 4$. The subintervals and their lengths are as follows:

$$I_1 = [0, \tfrac{1}{2}], \quad \Delta t_1 = \tfrac{1}{2},$$

$$I_2 = [\tfrac{1}{2}, 1], \quad \Delta t_2 = \tfrac{1}{2},$$

$$I_3 = [1, 1\tfrac{1}{4}], \quad \Delta t_3 = \tfrac{1}{4},$$

$$I_4 = [1\tfrac{1}{4}, 2], \quad \Delta t_4 = \tfrac{3}{4}.$$

t_0 t_1 t_2 t_3 t_4

0 $\frac{1}{2}$ 1 $\frac{5}{4}$ 2

We choose the following sequence of evaluation points:

$$t_1^* = 0, \qquad t_2^* = 1, \qquad t_3^* = 1, \qquad t_4^* = 1\tfrac{1}{2}.$$

Substituting these points into equation (20.3), we obtain the approximate number of cells produced:

$$\begin{aligned} P(2) &\simeq \sum_{n=1}^{4} f(t_n^*)\Delta t_n \\ &= f(0)\Delta t_1 + f(1)\Delta t_2 + f(1)\Delta t_3 + f(1\tfrac{1}{2})\Delta t_4 \\ &= 1 \cdot \tfrac{1}{2} + 2 \cdot \tfrac{1}{2} + 2 \cdot \tfrac{1}{4} + 2\tfrac{1}{2} \cdot \tfrac{3}{4} \\ &= 0.5 + 1 + 0.5 + 1.875 = 3.875; \end{aligned}$$

$$P(2) \simeq 3.875$$

b) We consider the same partition points $\{t_n\}$ used in part (a). Making a second choice of the evaluation points $\{t_n^*\}$, we obtain a second approximation with the same partition. Let

$$t_1^* = 0.25, \qquad t_2^* = 0.75, \qquad t_3^* = 1.25, \qquad t_4^* = 1.75.$$

Substituting these values into equation (20.3), we obtain the following approximation of $P(2)$:

$$\begin{aligned} P(2) &\simeq \sum_{n=1}^{4} f(t_n^*)\Delta t_n \\ &= f(0.25)\Delta t_1 + f(0.75)\Delta t_2 + f(1.25)\Delta t_3 + f(1.75)\Delta t_4 \\ &= (1.25)(0.5) + (1.75)(0.5) + (2.25)(0.25) + (2.75)(0.75) \end{aligned}$$

$$P(2) \simeq 4.125.$$

c) We take a *uniform* partition of $[0, 2]$, which has five equal subintervals, by making the length of each subinterval one-fifth of the total length—i.e.,

$$\Delta t_n = h \qquad \text{where} \qquad h = \frac{2 - 0}{5} = 0.4.$$

The corresponding partition points are given by the following formula:

$$t_n = n \cdot h = n(0.4);$$

$$t_0 = 0, \quad t_1 = 0.4, \quad t_2 = 0.8, \quad t_3 = 1.2, \quad t_4 = 1.6, \quad t_5 = 2.0,$$

We select the set of evaluation points as the left endpoint of each subinterval:

$$t_n^* = t_{n-1};$$

$$t_1^* = 0, \quad t_2^* = 0.4, \quad t_3^* = 0.8, \quad t_4^* = 1.2, \quad \text{and} \quad t_5^* = 1.6.$$

The approximation of $P(2)$ with these points is as follows:

$$\begin{aligned} P(2) &\simeq \sum_{n=1}^{5} f(t_n^*)\Delta t_n \\ &= \sum_{n=1}^{5} [f(t_{n-1}) \cdot h] = h\left[\sum_{n=1}^{5} f(t_{n-1})\right] \\ &= (0.4)[f(0) + f(0.4) + f(0.8) + f(1.2) + f(1.6)]. \end{aligned}$$

$$P(2) \simeq (0.4)(9) = 3.6.$$

The true value of $P(2)$ for this problem is $P(2) = 4.0$. ◀

Problem B Solution

We approach the solution to this problem in the same manner that was used to approximate the solution of Problem A. We do not know the function $Q(t)$, we only know its rate of change, which is the difference $\mathring{Q}_{\text{in}}(t) - \mathring{Q}_{\text{out}}(t)$. Since the accumulation in the compartment over an interval $[a, b]$ is the difference $Q(b) - Q(a)$, we can approximate $Q(T)$ by approximating the accumulation over the interval $[0, T]$. As in Problem A, this quantity may be approximated by the sum of the approximate accumulation over a collection of subintervals; the subintervals are obtained from a partition of the interval $[0, T]$.

More precisely, let

$$\{t_n\}_{n=0}^{N}$$

be a partition of the interval $[0, T]$. Then the nth subinterval determined by this partition is

$$I_n = [t_{n-1}, t_n],$$

and its length is

$$\Delta t_n = t_n - t_{n-1}.$$

We denote the accumulation over the interval I_n by A_n. Then

$$A_n = Q(t_n) - Q(t_{n-1}),$$

and the accumulation over the interval $[0, T]$ is

$$Q(T) - Q(0) = \sum_{n=1}^{N} A_n.$$

(Note: The series $\sum A_n = \sum[Q(t_n) - Q(t_{n-1})]$ is a collapsing series.) We approximate A_n by approximating the average rate of accumulation,

$$A_n = (\text{average rate of accumulation}) \times (\text{length of interval}).$$

Thus

$$A_n \simeq [\mathring{Q}_{\text{in}}(t_n{}^*) - \mathring{Q}_{\text{out}}(t_n{}^*)]\Delta t_n,$$

where $t_n{}^*$ is an arbitrary point in the interval $[t_{n-1}, t_n]$. Using these approximations for $n = 1, 2, \ldots, N$, the accumulation over $[0, T]$ may be approximated by the following formula:

$$Q(T) - Q(0) \simeq \sum_{n-1}^{N} [\mathring{Q}_{\text{in}}(t_n{}^*) - \mathring{Q}_{\text{out}}(t_n{}^*)]\Delta t_n \qquad \textbf{Approximate accumulation.}$$

(20.4)

Formula (20.4) gives the net change of the amount of solute present from time $t = 0$ to time $t = T$. The amount of solute present at time T is thus approximated by

$$Q(T) \simeq Q(0) + \sum_{n=1}^{N} [\mathring{Q}_{\text{in}}(t_n{}^*) - \mathring{Q}_{\text{out}}(t_n{}^*)]\Delta t_n. \tag{20.5}$$

Equation (20.5) is the discrete approximation of the solution of Problem B.

Example 20.2 We illustrate the approximation (20.4) for specific flow functions. Let $\mathring{Q}_{\text{in}}(t) = 2t$ and $\mathring{Q}_{\text{out}}(t) = t^2$ over the interval $[0, 2]$—i.e., $T = 2$. The accumulation of the substance S is approximated for the partition, $t_0 = 0$, $t_1 = 1$, $t_2 = 1.5$, and $t_3 = 2$. With this partition, the subintervals and their lengths are as follows.

$I_1 = [0, 1]$, $\Delta t_1 = 1$,

$I_2 = [1, 1.5]$, $\Delta t_2 = 0.5$,

$I_3 = [1.5, 2]$, $\Delta t_3 = 0.5$.

t_0 t_1 t_2 t_3
0 1 1.5 2
$N = 3$

a) We first choose the evaluation points, $t_1{}^* = 0.5$, $t_2{}^* = 1.25$, and $t_3{}^* = 1.75$ ($t_n{}^*$ is the midpoint of the interval I_n). Then an approximation of the accumulation as given by equation (20.4) is as follows.

$$\begin{aligned} Q(T) - Q(0) &\simeq \sum_{n=1}^{3} [\mathring{Q}_{\text{in}}(t_n{}^*) - \mathring{Q}_{\text{out}}(t_n{}^*)]\Delta t_n \\ &= [\mathring{Q}_{\text{in}}(0.5) - \mathring{Q}_{\text{out}}(0.5)](1) + [\mathring{Q}_{\text{in}}(1.25) - \mathring{Q}_{\text{out}}(1.25)](0.5) \\ &\quad + [\mathring{Q}_{\text{in}}(1.75) - \mathring{Q}_{\text{out}}(1.75)](0.5) \\ &= (0.75)(1) + (0.9375)(0.5) + (0.4375)(0.5) \\ &= 1.4375. \end{aligned}$$

This term is

$$[\mathring{Q}_{\text{in}}(0.5) - \mathring{Q}_{\text{out}}(0.5)] = [2(0.5) - (0.5)^2] = [1 - 0.25].$$

b) With the same partition, we choose the evaluation points to be the right endpoints of the intervals, $t_n^* = t_n$. Then $t_1^* = 1$, $t_2^* = 1.5$, and $t_3^* = 2$. The resulting approximation is

$$
\begin{aligned}
Q(T) - Q(0) &\simeq \sum_{n=1}^{3} [\mathring{Q}_{\text{in}}(t_n^*) - \mathring{Q}_{\text{out}}(t_n^*)]\Delta t_n \\
&= [\mathring{Q}_{\text{in}}(1) - \mathring{Q}_{\text{out}}(1)](1) + [\mathring{Q}_{\text{in}}(1.5) - \mathring{Q}_{\text{out}}(1.5)](0.5) \\
&\quad + [\mathring{Q}_{\text{in}}(2) - \mathring{Q}_{\text{out}}(2)](0.5) \\
&= (1)(1) + (0.75)(0.5) + (0)(0.5) \\
&= 1.375.
\end{aligned}
$$

The true solution for this example is $Q(T) - Q(0) = 1.333\ldots$ ◀

Problem C Solution

Problem C is essentially the same as Problems A and B. Since we cannot directly measure the total amount of mercury discharged over a 24-hour period, we will approximate this quantity. Our approximation technique will be the experimental counterpart of the partitioning method employed in the solutions to Problems A and B. We will sample the effluent periodically over the time interval $[a, b]$. The times at which we take a sample correspond to the evaluation times, t_n^*. The amount of mercury in each sample will provide the approximation of the average amount of mercury being discharged over the time interval in which the sample is taken.

If we were to take a sample once each hour for a 24-hour period, then N would equal 24. Assume that the flow rate of the effluent is F(cubic metres/hour) and the amount of mercury found in a sample in the nth hourly period is $m_n(\text{g}/10^3\text{m}^3)$. The amount of mercury being discharged in the nth time period is

$$M_n = (\text{flow rate}) \times (\text{average amount of mercury}/\text{m}^3).$$

The amount of mercury being discharged is then approximated by

$$M_n \simeq F \cdot m_n \cdot 10^{-3}.$$

The total amount of mercury discharged over the 24-hour period is

$$M = \sum_{n=1}^{24} M_n.$$

Using the above approximations for M_n, we approximate M by

$$M \simeq 10^{-3} F \sum_{n=1}^{24} m_n.$$

If sampling periods were taken to be other than hourly intervals, the above process would be modified as follows. Let $t_0, t_1, t_2, \ldots, t_N$ be the endpoints of the sampling intervals ($t_0 = 0$ and $t_N = 24$). The length of the nth interval, $I_n = [t_{n-1}, t_n]$, would be $\Delta t_n \equiv t_n - t_{n-1}$. Consequently, the approximate amount of mercury discharged in the time period $[t_{n-1}, t_n]$ would be

$$M_n \simeq (F \cdot \Delta t_n) m_n \cdot 10^{-3},$$

where again m_n is the amount of mercury in the nth sample measured in $g/10^3 m^3$. The term $F \cdot \Delta t_n$ gives the total volume of effluent discharged in the nth time interval. The total mercury discharged would be estimated by the following formula:

$$M = \sum_{n=1}^{N} M_n \simeq 10^{-3} F \cdot \sum_{n=1}^{M} m_n \cdot \Delta t_n \qquad \textbf{Approximate solution.} \tag{20.6}$$

Note: The sample m_n can be taken at any time in the nth time interval.

Example 20.3 To illustrate the dependence of the estimate (20.6) on the choice of the interval periods as well as on the recorded sample levels of mercury, consider the two approximations obtained by using the same sample data but different sample periods. Assume that samples are taken at 3 A.M., 7 A.M., 1 P.M., and 11 P.M. To make the computations simple, we assume that $F = 10^3$ and the amounts of mercury discharged at these times were

$$m_1 = 10, \qquad m_2 = 80, \qquad m_3 = 10, \qquad \text{and} \qquad m_4 = 120.$$

a) Assume that the sample periods are taken to be 6-hour intervals:

[12 A.M., 6 A.M.], [6 A.M., 12 noon], [12 noon, 6 P.M.], [6 P.M., 12 A.M.],

Then $\Delta t_n = 6$ for $n = 1, 2, 3$, and 4 and equation (20.6) yields the approximation

$$\begin{aligned} M &\simeq 10^{-3} \cdot 10^3 \sum_{n=1}^{4} m_n \cdot 6 \\ &= 6[10 + 80 + 10 + 120] = 1440 \text{ g}. \end{aligned}$$

b) Assume that the sample periods are chosen to be

[12 A.M., 6 A.M.], [6 A.M., 8 A.M.], [8 A.M., 10 P.M.], [10 P.M., 12 A.M.].

Then the lengths of these periods are

$$\Delta t_1 = 6, \qquad \Delta t_2 = 2, \qquad \Delta t_3 = 14, \qquad \Delta t_4 = 2.$$

The approximation (20.6) now becomes

$$\begin{aligned} M &\simeq 10^{-3} \cdot 10^3 \sum_{n=1}^{4} m_n \cdot \Delta t_n \\ &= (10 \cdot 6) + (80 \cdot 2) + (10 \cdot 14) + (120 \cdot 2) \\ &= 600 \text{ g}. \end{aligned}$$

The difference between the two approximations is considerable. Yet both approximations were made with the same data; we cannot say which is more accurate. For instance, the approximation of 600 g could be most accurate if the high sample values, m_2 and m_4, corresponded with the start and end of a day's work, when changes in operation procedures accounted for brief increases in the discharge levels. On the other hand, the values of m_1 and m_3 could reflect low levels due to lunch breaks. To improve the accuracy of these estimates, it would be best to take more samples. A scientist would actually want to take repeated samples on different days and statistically analyze the estimates obtained. ◀

To summarize, the procedure used in each problem of this section was to partition the time interval into many shorter subintervals. Then the original problem, considered over each subinterval, was approximated. It was assumed that the rate of change in the quantity being measured was a constant value over the subinterval. The constant value is chosen as the value of the instantaneous rate of change at an arbitrary evaluation time, t^*. The approximate quantities over the subintervals are taken to be the product of the instantaneous rate at t^* and the lengths of the respective intervals. The total quantity was then approximated by the sum of the approximations over all of the subintervals.

EXERCISE SET 20

20.1 Indicate a partition of the given interval, I, into N subintervals; list the subintervals and their lengths, Δt_n, which are restricted as indicated. The answers to (a), (d), and (e) are not unique.

a) $I = [0, 1]$, $N = 4$, $\max \Delta t_n \leq \frac{1}{2}$
b) $I = [2, 5]$, $N = 3$, uniform partition
c) $I = [2, 5]$, $N = 4$, uniform partition
d) $I = [-1, 1]$, $N = 3$, $\max \Delta t_n = 1$
e) $I = [0, 3]$, $N = 5$, $\max \Delta t_n \leq 1$

20.2 For each partition in Exercise 20.1 list the following:

a) the right endpoint of each subinterval;
b) the left endpoint of each subinterval;
c) the midpoint of each subinterval.

20.3 Evaluate the approximation of $P(2)$ using the partition of Example 20.1(c), but taking the right endpoints for the evaluation points, t_n^*.

20.4 Evaluate the approximation of $Q(T) - Q(0)$ using the data of Example 20.2, but taking the left endpoints as approximation points.

20.5 What would be the approximate mercury discharge in Example 20.3 if samples were taken every four hours and the corresponding m_n values were found to be 2, 10, 3, 5, 2, and 4? Assume that the sample periods are uniform four-hour intervals.

20.6 Let the rate of production in Problem A be $f(t) = 2t$. Estimate the production $P(T)$ for $T = 3$ when using a uniform partition of size $N = 3$ and t_n^* chosen as (a) the right endpoint; (b) the left endpoint; and (c) the midpoint or the subinterval, I_n.

20.7 Repeat Exercise 20.6 with $N = 5$.

20.8 Repeat Exercise 20.6 with the partition $t_0 = 0, t_1 = 1, t_2 = 1.5, t_3 = 2, t_4 = 3$.

20.9 Repeat Exercise 20.6 with the value of N unspecified—i.e., N will be arbitrary and the resulting sums will depend on N. In this problem, specify explicitly the forms of t_n and t_n^*.

20.10 Let $f(t) = t^2$ and with this value of f repeat the following exercises:

a) Exercise 20.6. b) Exercise 20.7. c) Exercise 20.8. d) Exercise 20.9.

20.11 In Exercises 20.6 and 20.10(a), the approximation obtained using the left endpoints was less than that obtained using the midpoints, which was less than that obtained using the right endpoints. Is this always true? If so, explain why; and, if not, give an example in which it is not true.

20.12 In Problem B, assume that

$$\mathring{Q}_{\text{in}}(t) = 1 + t^2 \qquad \text{and} \qquad \mathring{Q}_{\text{out}}(t) = 2t^2.$$

a) When will the outflow equal the inflow? Take this as the value T.
b) When will the accumulated solute, S, reach its maximum value? Give a logical answer without using calculus.
c) Using T, as determined in (a), approximate $Q(T)$ using a uniform partition of size $N = 2$ and t_n^* as the right endpoint of the interval, I_n.
d) Repeat (c) with $N = 3$ and $N = 4$.
e) Approximate $Q(T)$ with a uniform partition of length $\Delta t = 0.1$.

20.13 Repeat Exercise 20.12 with the functions

$$\mathring{Q}_{\text{in}}(t) = t + 2 \qquad \text{and} \qquad \mathring{Q}_{\text{out}}(t) = t^2.$$

20.14 The number of revolutions per minute (rpm) in an ultracentrifuge increases at a rate $f(t)$ over a period $[0, T]$.

a) If rpm(0) = 0, develop an approximate formula for the rpm(T) at time, T.
b) Use your formula to approximate rpm (T) when $f(t) = 10e^{t-1}$ and $T = 5$. Use a uniform partition with $N = 5$ and $t_n^* = t_n$.
c) Repeat parts (b) using $t_n^* = t_{n-1}$.

20.15 The rate of growth of most plants is a periodic function with period of one day. Assume that a corn plant grows at a rate $\mathring{g}(t) = k \sin^2(\pi t)$, where t is days and k is a constant.

a) Construct a discrete approximation to the height of corn, $g(T)$.
b) If $T = 10$ and $g(0) = 0$, what is an upper limit for the value of $g(T)$? (*Hint*: Sketch a graph of $\mathring{g}(t)$.)
c) Approximate $g(10)$ if $k = 1$. (The exact value is 5.)

SECTION 21

THE DEFINITE INTEGRAL AND THE FUNDAMENTAL THEOREM OF CALCULUS

21.1 DEFINING THE DEFINITE INTEGRAL

In the previous section, we examined three problems and found an approximate solution to each in the form of a *finite series*. Using an arbitrary function, $f(x)$, with the independent variable x rather than t, the general form of the series developed to solve each problem was

$$\sum_{n=1}^{N} f(x_n{}^*)\Delta x_n. \tag{21.1}$$

The same technique used in Section 20 to develop this type of series can be applied to any continuous function, $f(x)$. Considering the resulting series in a more abstract mathematical setting, in particular the value of the series as the number N of subintervals is increased, we obtain the definition of the *definite integral*. Consequently, the approximation method illustrated in Section 20 forms the basis for the branch of mathematics called *Integral Calculus*.

Historically, integral calculus was developed before the derivative and differential calculus. The basic concepts of integral calculus were known as early as 440 B.C., when the Greek mathematician Democritus used it to develop the formula for the volume of a pyramid. The derivative and differential notation were not developed until the seventeenth century, when the German mathematician Gottfried Leibniz (1646–1716) introduced the dy/dx notation for the derivative, and the English scientist Isaac Newton (1642–1727) introduced the $\mathring{y}$-notation. One reason why calculus texts still introduce the subject in the reverse order of its discovery can be traced to the fact that, when the French mathematician Marquis G. L'Hospital published the first calculus text in 1696, he presented the new ideas of differential calculus and added only as an appendix the more well-known integral calculus.

In this subsection, we present the definition of the definite integral as the limit of a series of the form (21.1). The definition is presented here without an extensive discussion. The problems and solutions of the preceding Section 20 are intended to demonstrate both the technique and motivation for developing the definite-integral concept. In Section 22, we will examine in greater detail the definition presented below. At that time, we will also present the more traditional mathematical development of the definite integral to define *area* under a curve.

The definite integral of a function, $f(x)$, over an interval, $[a, b]$, is a number. This number is defined as the limit of the series (21.1) as N approaches $+\infty$ and Δx_n approaches 0. This limit can be thought of as the limit associated with an infinite series as discussed in Section 10. However, mathematically, this is a much more general limit. The symbol used to denote the definite integral utilizes the same symbol used to denote the antiderivative; but in order to denote the definite integral, the endpoints a and b of the interval are indicated in the form

$$\int_a^b f(x)dx.$$

The integral symbol $\int$ was originally a long S and was used to denote a sum. The formal definition of the definite integral uses the terminology used in solving the problems of Section 20.

Definition 21.1 *The definite integral of* $f(x)$ *over the interval* $[a, b]$ is defined as follows:

Notation:

1. Let $\{x_n\}_{n=1}^N$ partition $[a, b]$ into N subintervals, $I_n = [x_{n-1}, x_n]$,
$$a = x_0 < x_1 < x_2 < \cdots < x_N = b.$$
2. Denote the lengths of the subintervals by $\Delta x_n \equiv x_n - x_{n-1}$ and denote the maximum $\{\Delta x_n | n = 1, 2, \ldots, N\}$ by Δx.
3. Let $\{x_N^*\}_{n=1}^N$ be an arbitrary sequence of evaluation points, with $x_n^* \in I_n$, $n = 1, 2, \ldots, N$.

Hypothesis: If the limit as $\Delta x \to 0$ and $N \to \infty$ of

$$\sum_{n=1}^{N} f(x_n^*)\Delta x_n$$

exists as a unique number L, independent of the choice of the evaluation points, $\{x_n^*\}$, then the conclusion is as follows.

Conclusion:

I. The function f *is* said to be *integrable over the interval* $[a, b]$,

II. The *definite integral of* f *over* $[a, b]$ is the number

$$L = \lim_{\substack{N \to \infty \\ \Delta x \to 0}} \sum_{n=1}^{N} f(x_n^*)\Delta x_n.$$

III. The *definite integral of* f *over* $[a, b]$ is expressed using *integral notation* by

$$L = \int_a^b f(x)dx,$$

read "L equals the (definite) integral from a to b of $f(x)dx$."

IV. $[a, b]$ is called the *interval of integration*; a and b are called the *lower* and *upper limits of integration*, and the function $f(x)$ is called the *integrand*.

Table 21.1

Function	Common approximating sum	Definite integral	Interpretation of integral
$f(t)$: rate of change at time t	$\sum_{n=1}^{N} f(t_n^*)\Delta t_n$	$\int_a^b f(t)dt$	Total change over time $[a, b]$
$f(t)$: flow rate of a fluid through a vessel of cross-section A	$A \cdot \sum_{n=1}^{N} f(t_n^*)\Delta t_n$	$A \int_a^b f(t)dt$	Volume of flow over time interval $[a, b]$
$f(x)$ density of a linear string or bar at point x	$\sum_{n=1}^{N} f(x_n^*)\Delta x_n$	$\int_0^l f(x)dx$	Total mass of bar of length l
$N(t)$: size of population at time t	$\frac{1}{T} \sum_{n=1}^{N} N(t_n^*)\Delta t_n$	$\frac{1}{T} \int_0^T N(t)dt$	Average size of population over time interval $[0, T]$
$S(t)$: speed at time t	$\sum_{n=1}^{N} S(t_n^*)\Delta t_n$	$\int_a^b S(t)dt$	Distance covered in time interval $[a, b]$
$P(x)$: probability density function	$\sum_{n=1}^{N} P(x_n^*)\Delta x_n$	$\int_a^b P(x)dx$	Probability that the random variable x is between a and b
$f(x)$: a positive function	$\sum_{n=1}^{N} f(x_n^*)\Delta x_n$	$\int_a^b f(x)dx$	Area under the graph of f over $[a, b]$
$F(x)$: a "force" acting on an object at position x	$\sum_{n=1}^{N} F(x_n^*)\Delta x_n$	$\int_a^b F(x)dx$	"Work" in moving the object through the interval $[a, b]$
$f(x)$: any function	$\frac{1}{b-a} \sum_{n=1}^{N} f(x^*)\Delta x$	$\frac{\int_a^b f(x)dx}{b-a}$	The average value of f over $[a, b]$

The definite integral

$$\int_a^b f(x)dx \qquad \textbf{Definite integral.}$$

is frequently simply referred to as "the integral of f over $[a, b]$." This should not be confused with "the integral of f,"

$$\int f(x)dx \qquad \textbf{Indefinite integral.}$$

which is an antiderivative of $f(x)$. *Remember: The definite integral is a number, whereas the antiderivative is a function.*

Mathematically, the definite integral, $\int_a^b f(x)dx$, is a number. In the pure mathematical world, numbers are dimensionless and convey no special meaning. In the real world, where mathematics is used, this is not the case. Functions are used to describe particular events and have dimensions such as $f(x)$ metres/sec to describe velocity, or $V(t)$ m^3 to describe volume. The application of the definite integral to real problems requires specific interpretation of the number $\int_a^b f(x)dx$. The interpretation, in any particular application, depends on how the integrand, $f(x)$, is interpreted. Table 21.1 provides some of the more common interpretations of the definite integral. These and other interpretations will be illustrated in the remaining Examples and Exercises of the text.

21.2 THE FUNDAMENTAL THEOREM OF CALCULUS

Using Definition 21.1, the value of a specific integral, $\int_a^b f(x)dx$, can be calculated (or at least approximated). The procedure for evaluating integrals using the definition is theoretically straightforward, but in practice it is difficult. It requires either representing a series in closed form (see Section 10) or performing an enormous amount of calculations. This approach is sometimes the only practical way to evaluate an integral, especially if the function $f(x)$ is specified by experimental data rather than by an algebraic expression. In such cases, we turn to the computer to perform the extensive calculations. However, for many functions, $f(x)$, there is a much simpler way to evaluate

$$\int_a^b f(x)dx.$$

This simpler method is so important and useful that it is referred to as the *Fundamental Theorem of Calculus*. We present it now and defer until Section 22 the evaluation of integrals using Definition 21.1.

For the theoretical manipulation of integrals and to interpret the biological significance associated with integrals, it is desirable that we be able to represent the value of a definite integral in a functional form, which displays the dependence on both the integrand, f, and the interval of integration, $[a, b]$. This is exactly what the Fundamental Theorem of Calculus provides. Assume that $f(x)$ is the instantaneous rate of change in a quantity whose value is denoted by $F(x)$ for $x \in (a, b)$. Then the total change in the quantity over the interval $[a, b]$ is

$$F(b) - F(a).$$

By using the same arguments employed in Section 20, this change can be approximated by a series

$$F(b) - F(a) \simeq \sum_{n=1}^{N} f(x_n^*)\Delta x_n.$$

If we then let $N \to +\infty$ and $\Delta x_n \to 0$, this approximation should improve and, in the limit, give an equality. But the limit of the series is by definition the integral of f over $[a, b]$:

$$F(b) - F(a) = \int_a^b f(x)dx. \tag{21.2}$$

This gives a simple form for the integral; but, to use it, we must know the function $F(x)$. The key which makes this formula practical is the term "instantaneous rate of change." In Section 14, the instantaneous rate of change of a function $F(x)$ was defined to be the derivative $F'(x)$. Thus our assumption that "$f(x)$ is the instantaneous rate of change of $F(x)$" is the same as stating that

$$F'(x) = f(x) \qquad \text{or} \qquad F(x) \text{ is an antiderivative of } f(x). \tag{21.3}$$

The relationship between equation (21.2) and (21.3) forms the basis for the Fundamental Theorem of Calculus (FTC).

Theorem 21.1 *The Fundamental Theorem of Calculus*
If f is continuous on $[a, b]$, then the definite integral of f over $[a, b]$,

$$\int_a^b f(x)dx,$$

exists. If F is an antiderivative of f—that is, $F'(x) = f(x)$, then

$$\int_a^b f(x)dx = F(b) - F(a) \qquad \textbf{Fundamental Theorem of Calculus.}$$

A notation that is widely used to denote the difference $F(b) - F(a)$ is the symbol $F(x)|_b^a$. Using this notation, we can express equation (21.2) as

$$\int_a^b f(x)dx = F(x)\Big|_a^b \equiv F(b) - F(a).$$

We illustrate the Fundamental Theorem of Calculus in the following example.

Example 21.1

a) If $f(x) = x^2$, an antiderivative of $f(x)$ is $F(x) = x^3/3$. Thus

$$\int_a^b x^2\,dx = F(x)\Big|_a^b = \frac{b^3}{3} - \frac{a^3}{3}.$$

b) If $f(x) = 2x - 3$, an antiderivative of $f(x)$ is $F(x) = x^2 - 3x + 6$. Hence, we evaluate the integral of $f(x)$ over the interval $[1, 2]$ as

$$\int_1^2 2x - 3\,dx = (x^2 - 3x + 6)\Big|_1^2$$

$$= [2^2 - 3(2) + 6] - [1^2 - 3(1) + 6)]$$

$$= 4 - 4 = 0.$$

c) To evaluate $\int_{-1}^3 2x + x^2 - 3\,dx$, we let $F(x) = D_x^{-1}(2x + x^2 - 3)$. Then $F(x) = x^2 - x^3/3 - 3x + c$ for some constant, c. By the FTC,

$$\int_{-1}^3 2x + x^2 - 3\,dx = F(x)\Big|_{-1}^3 = F(3) - F(-1)$$

$$= \left[(3)^2 + \frac{(3)^3}{3} - 3(3) + c\right]$$

$$- \left[(-1)^2 + \frac{(-1)^3}{3} - 3(-1) + c\right]$$

$$= [9 + c] - [4 - \tfrac{1}{3} + c] = \tfrac{14}{3}.$$

Note that the value of c does not affect the value of the integral.

d) Using the Chain Rule, we may verify that the derivative of $F(x) = (x^2 + 5)^4$ is $F'(x) = 8x(x^2 + 5)^3$. Consequently,

$$\int_0^2 8x(x^2 + 5)^3\,dx = (x^2 + 5)^4\Big|_0^2 = ((2)^2 + 5)^4 - ((0)^2 + 5)^4$$

$$= 9^4 - 5^4 = 5936.$$

e) To evaluate $I = \int_0^6 (3x + 2)(3x^2 + 4x + 1)^4\,dx$, using the FTC, we first compute the antiderivative,

$$F(x) = D_x^{-1}[(3x + 2)(3x^2 + 4x + 1)^4] = \tfrac{1}{10}(3x^2 + 4x + 1)^5.$$

The value of the integral is thus

$$I = F(x)\Big|_0^6 = \tfrac{1}{10}[3(6)^2 + 4(6) + 1]^5 - \tfrac{1}{10}[3(0)^2 + 4(0) + 1]^5$$

$$= \tfrac{1}{10}[(61)^5 - 1] = 84{,}459{,}630.$$

f) As $D_\theta \sec(\theta) = \sec(\theta)\tan(\theta)$, by the FTC,

$$\int_0^{\pi/4} \sec(\theta)\tan(\theta)d\theta = \sec(\theta)\Big|_0^{\pi/4} = \sec\left(\frac{\pi}{4}\right) - \sec(0) = \sqrt{2} - 1 \simeq 0.2929.$$

◀

Example 21.2 *Basal metabolism*
Is a term used to describe the normal chemical activity in an organism not subject to stress—for instance, a plant growing under ideal conditions or an animal resting over a period of time.

The metabolic rate of an animal will vary in response to environmental changes (temperature, humidity, air quality) and changes in physical activity. As air temperature fluctuates on a daily basis, the basal metabolic rate (BMR) of an animal will very over a *diurnal cycle*; the BMR increases at night to compensate for the lower temperature and decreases during the day. Fig. 21.1 shows the graph of such a cycle.

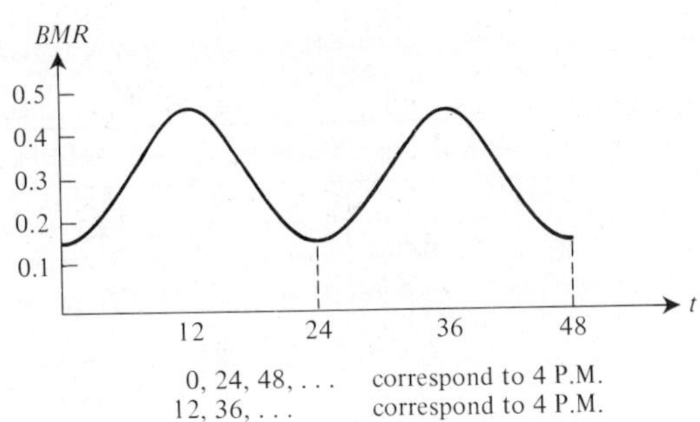

Figure 21.1

Metabolic rates are expressed in several equivalent ways—as a measure of heat produced in kilocalories per unit (kcal/hr), as a measure of oxygen consumption per unit of body weight (cm^3 O_2/g), and as a measure of carbon dioxide expelled per unit of time (cm^3 CO_2/hr). In all cases, the total basal metabolism, BM, over a period is obtained as the integral of the basal metabolic rate over the time interval

$$BM = \int_{t_1}^{t_2} BMR(t)dt. \tag{21.4}$$

For the rate sketched in Fig. 21.1, the BMR is given by the following function:

$$BMR(t) = -(0.15)\cos\left(\frac{\pi t}{24}\right) + 0.3 \text{ kcal/hr}.$$

Consequently, the BM value for a one-day period would be

$$BM = \int_0^{24} \left[-0.15 \cos\left(\frac{\pi t}{24}\right) + 0.3 \right] dt$$

$$= -(0.15)\left(\frac{24}{\pi}\right) \sin\left(\frac{\pi t}{24}\right) + 0.3t \Big|_0^{24}$$

$$= \frac{-3.6}{\pi} [\sin(\pi) - \sin(0)] + (7.6 - 0)$$

$$= 7.6 \text{ kcal/day}.$$

This would correspond to the BM of a mouse, whereas the value for an adult human would be approximately 2000 kcal/day. ◀

Example 21.3 *Cardiac Output*
In measuring cardiac output, one method, known as the *Dye dilution method*, is performed as follows. A fixed amount of a dye is injected into a vein or the right side of the heart. This dye then is circulated with the blood through the heart to the lungs, back to the heart, and into the arterial system. At a peripheral artery, the blood is continuously monitored for the presence of the dye for 30 seconds from the time of injection. The concentration of dye passing the monitored artery is then plotted as a function, $c(t)$, of time. (After about 15 seconds, recirculation of the dye occurs and care in monitoring the blood circulation must be exercised.) The cardiac output is defined to be the volume of blood pumped *per minute*. This is obtained as the ratio of the amount of dye injected to the average concentration monitored over the 30-second period, multiplied by two, so that it corresponds to one minute:

$$\text{Cardiac output} = \frac{2\,[M_g \text{ of injected dye}]}{\frac{1}{30}\int_0^{30} c(t)dt}. \tag{21.5}$$

The dye-dilution method is used in experiments in basic physiology laboratories. The integral,

$$\int_0^{30} c(t)dt,$$

is estimated by drawing a continuous curve through dye-concentration values which are plotted on standard graph paper over a 30-second interval. The integral is then approximated by counting the squares of the graph paper under the curve. This corresponds to interpretating the integral as an area.

Assume that in an experiment in which 5 mg of dye was injected at time $t = 0$, the concentration curve was found to be

$$C(t) = 0 \quad \text{if} \quad 0 \le t \le 3 \quad \text{or} \quad 18 \le t \le 30,$$

and

$$C(t) = (t^3 - 40t^2 + 453t - 1026)10^{-3} \quad \text{if} \quad 3 \le t \le 18.$$

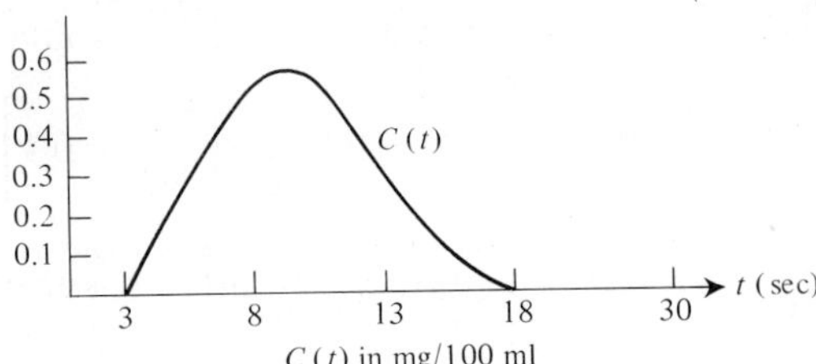

Figure 21.2

The graph of $C(t)$ has the shape of the curve sketched in Fig. 21.2. No dye passes the observation artery for three seconds, and then a large quantity of dye passes. After 18 seconds, all measurable amounts of dye have passed. To compute the cardiac output determined by this experiment, we evaluate the average

$$A = \frac{1}{30}\int_0^{30} c(t)dt.$$

Since

$$\int_0^3 c(t)dt = 0 \qquad \text{and} \qquad \int_{18}^{30} c(t)dt = 0,$$

the average, A, is given by

$$A = \frac{10^{-3}}{30}\int_3^{18} [t^3 - 40t^2 + 453t - 1026]dt.$$

Using the FTC to evaluate this integral, and substituting into equation (21.5), we obtain the cardiac output for this experiment, which is approximately 6.275 litres/min. ◀

The evaluation of definite integrals can often be simplified by using the following algebraic properties of definite integrals. These correspond directly with the algebraic properties of the antiderivative or indefinite integral, which are listed at the end of Section 19. It should be noted that these properties hold for all definite integrals, even those that cannot be evaluated by the FTC.

Theorem 21.2 Assume that f and g are integrable over $[a, b]$. Then

$$\int_a^b [f(t) + g(t)]dt = \int_a^b f(t)dt + \int_a^b g(t)dt \qquad \textbf{Sum rule.} \tag{21.6}$$

The integral of the sum is the sum of the integrals.

$$\int_a^b kf(t)dt = k\int_a^b f(t)dt, \; k \text{ a constant} \qquad \textbf{Constant } \boldsymbol{R} \textbf{ rule.} \tag{21.7}$$

The integral of a constant times a function is the constant times the integral of the function.

If $f(t) \leq g(t)$ for $t \in [a, b]$, then

$$\int_a^b f(t)dt \leq \int_b^a g(t)dt \qquad \textbf{Inequality rule.} \tag{21.8}$$

The last property, (21.8), is very useful when approximating integrals. For instance, if we know that a function g satisfies

$$f_1(t) \leq g(t) \leq f_2(t), \qquad \text{for} \qquad t \in [a, b],$$

and we know the definite integrals of the functions f_1 and f_2 over $[a, b]$, then the integral of g over $[a, b]$ will lie between these two numbers:

$$\int_a^b f_1(t)dt \leq \int_a^b g(t)dt \leq \int_a^b f_2(t)dt.$$

When the integrand, $g(t)$, is held fixed and the limits of integration are altered, the value of the integral will vary. The basic relationships that result from changing the limits of integration are given by the next theorem.

Theorem 21.3 If g is integrable over $[a, b]$ and $c \in [a, b]$, then

$$\int_a^b g(t)dt = \int_a^c g(t)dt + \int_c^b g(t)dt; \tag{21.9}$$

$$\int_b^b g(t)dt = 0; \tag{21.10}$$

$$\int_b^c g(t)dt = -\int_c^b g(t)dt. \qquad \textbf{Interchange the limits of integration.} \tag{21.11}$$

Equation (21.9) corresponds to the fact that the amount of a quantity, which accumulates over an interval $[a, b]$, is the sum of the amount that accumulates over the first portion of the interval, $[a, c]$, plus the amount that accumulates over the remaining portion of the interval, $[c, b]$. The equation (21.10) reflects the fact that no additional quantity accumulates when the time is held fixed.

Equation (21.11) will prove useful in many theoretical manipulations; it states that, by interchanging the limits, we obtain the negative of the original integral's value. Visualize this by thinking of the integral with its limits interchanged as reversing the process described by the original integral. For instance, if $\int_4^6 g(t)dt$ represented the amount of fluid flowing from one tank to another tank over a two-hour period, then $\int_6^4 g(t)dt$ would represent the change in the fluid in the second tank if we reversed the flow over the last two hours. The change would be negative, indicating that the flow was out of the tank.

Example 21.4 The properties of integrals given in the above theorems are illustrated by the following equations. (The evaluation of the integrals is left as an exercise.)

a) $\int_2^5 t^2 - 2t\,dt = \int_2^5 t^2\,dt - 2\int_2^5 t\,dt$

b) $\int_{-2}^1 5t^3 + t\,dt = 5\int_{-2}^1 t^3\,dt + \int_{-2}^1 t\,dt$

c) $\int_0^{2\pi} \cos(\theta)d\theta = \int_0^{\pi} \cos(\theta)d\theta + \int_{\pi}^{2\pi} \cos(\theta)d\theta$

d) $\int_3^6 t\,dt \le \int_3^6 t + 1\,dt$, since $t \le t + 1$ for all t

e) $\int_0^{10} e^t - \dfrac{1}{(t+1)}\,dt = \int_0^{10} e^t\,dt - \int_0^{10} \dfrac{1}{(t+1)}\,dt$

f) $\int_0^8 t^2 + 1\,dt = \int_0^4 t^2 + 1\,dt + \int_4^8 t^2 + 1\,dt$

g) $\int_0^3 t^2\,dt = -\int_3^0 t^2\,dt$ ◀

The following example illustrates the algebraic manipulation of integrals to obtain an equation in which one integral can be expressed in terms of other integrals whose values are experimentally measurable.

Example 21.5 CO_2 *exchange in plants*
Carbon dioxide plays an important role in the metabolic process of plants. Plants consume carbon dioxide from the atmosphere, where it normally constitutes 0.03 percent of the air. This carbon dioxide is utilized in the process of photosynthesis, during which a plant utilizes solar energy (light) and chloroplast to convert the carbon dioxide into carbohydrate:

Photosynthesis

$$\underset{\left(\substack{\text{consumed}\\ \text{from the air}}\right)}{CO_2} + 2H_2O \xrightarrow[\text{chloroplast}]{\text{solar energy}} \underset{\text{(stored)}}{(CH_2O)} + H_2O + \underset{\left(\substack{\text{released}\\ \text{to the air}}\right)}{O_2}$$

A second metabolic process of plants is respiration, which releases carbon dioxide to the atmosphere. In this process, the plant combines the stored carbohydrates with oxygen to yield energy for growth and in the process liberates carbon dioxide and water vapor into the surrounding atmosphere:

$$CH_2O + O_2 \xrightarrow{\text{Respiration}} CO_2 + H_2O + \text{Energy}$$

The carbon dioxide used by a plant is typically monitored over 24-hour periods. The net consumption of carbon dioxide over a one-day period is given by the integral:

$$\text{Net photosynthesis} = \int_0^{24} \mathring{P}_n(t)dt, \tag{21.12}$$

where $\mathring{P}_n(t)$ is the instantaneous flux, or rate of change, of carbon dioxide between the plant and its environment. Net photosynthesis is actually the difference between the carbon dioxide consumed in photosynthesis and liberated in respiration. If $\mathring{P}(t)$ is the rate at which the plant consumes carbon dioxide and $\mathring{R}(t)$ is the rate at which the plant respires carbon dioxide, then the net flux, $\mathring{P}_n$, is given by

$$\mathring{P}_n(t) = \mathring{P}(t) - \mathring{R}(t).$$

Consequently, the net photosynthesis can be expressed as the difference of two integrals:

$$\text{Net photosynthesis} = \int_0^{24} \mathring{P}(t)dt - \int_0^{24} \mathring{R}(t)dt. \tag{21.13}$$

The first of these integrals measures the total or gross photosynthetic consumption of carbon dioxide:

$$\text{Gross photosynthesis} = \int_0^{24} \mathring{P}(t)dt. \tag{21.14}$$

The second integral measures the gross respiration function of the plant:

$$\text{Gross respiration} = \int_0^{24} \mathring{R}(t)dt.$$

One of the plant physiologist's main difficulties is separating the effects of photosynthesis and respiration in the carbon-dioxide exchange. One approach sometimes utilized is to assume that the function $\mathring{P}(t)$ is zero during the night. If we assume that our day has 12 hours of sunlight and $t = 0$ corresponds to daybreak, this means that $\mathring{P}(t) = 0$ for $12 \leq t \leq 24$. Then

$$\int_0^{24} \mathring{P}(t)dt = \int_0^{12} \mathring{P}(t)dt + \int_{12}^{24} 0\,dt = \int_0^{12} \mathring{P}(t)dt.$$

Next it is assumed that respiration is essentially the same over the daytime period and the nighttime period. Respiration is measured at night and this quantity is then also used to represent the daytime respiration. (There is no mathematical justification for this assumption and very little biological evidence to support it.) The mathematical statement of this assumption is that

$$\int_0^{12} \mathring{R}(t)dt = \int_{12}^{24} \mathring{R}(t)dt.$$

Mathematically, this means that

$$\int_0^{24} \mathring{R}(t)dt = 2\int_{12}^{24} \mathring{R}(t)dt. \tag{21.15}$$

The net photosynthesis and the nighttime respiration are measurable quantities. Substituting equation (21.15) into equation (21.13) and then equating (21.13) and (21.12), we obtain an equation for the gross photosynthesis.

$$\text{Gross photosynthesis} = \int_0^{24} \mathring{P}_n(t)dt + 2\int_{12}^{24} \mathring{R}(t)dt.$$

Both of these integrals represent measurable quantities. ◀

Example 21.6 *Availability of drugs*

When a drug is administered orally, it must be assimilated into the blood system before it is available to "act" on different parts of the body. Not all of the drug in an oral dose will be assimilated in an active form. To measure the total amount of the drug available in the blood system, it is necessary to monitor the rate at which the drug is excreted in the urine, using a standard clinical method. If this rate is $r(t)$, then the total amount of drug which passes through the body in a time interval $[0, T]$ is

$$D = \int_0^T r(t)dt.$$

In testing a drug, the upper limit T is chosen as a time after which no measurable amounts of the drug can be detected. In theory, T should be $+\infty$; we will discuss the consideration of an integral over an infinite interval in Section 25.

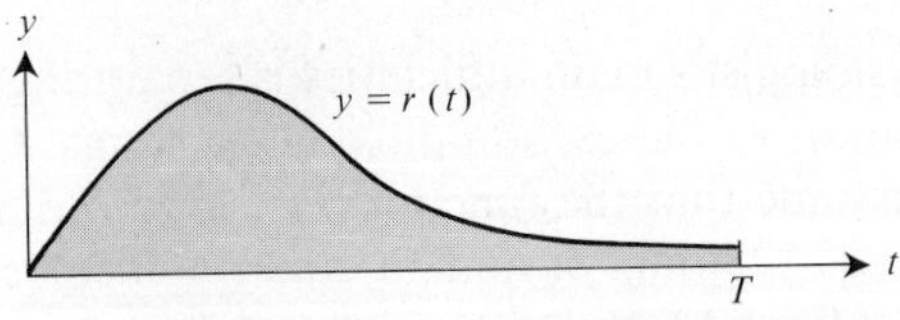

Area under curve = total drug available.

Figure 21.3

The available dose, D, may be visualized as the area under the graph of $r(t)$ between $t = 0$ and $t = T$. (See Fig. 21.3.) A typical elimination-rate function is

$$r(t) = te^{-kt}, \qquad k > 0.$$

The corresponding available drug dose is

$$D = \int_0^T te^{-kt}\,dt.$$

An antiderivative of $r(t)$ is

$$R(t) = -e^{-kt}\left[\frac{t}{k} + \frac{1}{k^2}\right].$$

To verify this, we simply differentiate $R(t)$ using the product rule. This antiderivative cannot be obtained using the techniques given in Section 19. We show how it is derived in Section 24, where we consider special integration methods.

By the FTC, we have

$$\begin{aligned} D &= \int_0^T r(t)dt = R(T) - R(0) \\ &= \left[-e^{-kT}\left(\frac{T}{k} + \frac{1}{k^2}\right)\right] - \left[-e^{-k\cdot 0}\left(\frac{0}{k} + \frac{1}{k^2}\right)\right] \\ &= \frac{1}{k^2} - e^{-kT}\left(\frac{T}{k} + \frac{1}{k^2}\right). \end{aligned}$$

If T is very large, the term

$$e^{-kT}\left(\frac{T}{k} + \frac{1}{k^2}\right)$$

will be very small. For instance, if $k = 0.1$ and $T = 100$, the value of the above term is

$$e^{-10}\left(\frac{100}{0.1} + \frac{1}{0.01}\right) \simeq 0.05;$$

whereas, when $T = 1000$ and $k = 0.1$, this value becomes

$$e^{-100}\left(\frac{1000}{0.1} + \frac{1}{0.01}\right) \simeq 3.7 \times 10^{-40}.$$

If T is minutes, $T = 1000$ corresponds to monitoring the drug for 16 hours and 40 minutes. Thus, for large T, the available drug level is

$$D \simeq \frac{1}{k^2}.$$

◀

21.3 VARIABLE LIMITS OF INTEGRATION; THE DERIVATIVE OF AN INTEGRAL

When presenting the basic properties of the definite integral, we observed that the definite integral

$$\int_a^b f(t)dt,$$

depends on two quantities: (1) the integrand, $f(t)$ and (2) the interval of integration, $[a, b]$. The Fundamental Theorem of Calculus provides a relationship which emphasizes the dependence of the definite integral on the interval of integration:

$$F(b) - F(a) = \int_a^b f(t)dt.$$

If we consider the right endpoint of the interval $[a, b]$ as a variable quantity, call it x, we can replace b with x in the above equation. Solving for $F(x)$, we obtain the equation

$$F(x) = F(a) + \int_a^x f(t)dt. \qquad (21.16)$$

To illustrate equation (21.16), we let $f(t) = 2t - 1$. An antiderivative of $f(t)$ is the function $F(t) = t^2 - t + 3$ (the 3 is an arbitrary choice). Equation (21.16) for this example is then

$$F(x) = (a^2 - a + 3) + \int_a^x (2t - 1)dt.$$

When $a = 5$, this equation reads

$$F(x) = 23 + \int_5^x (2t - 1)dt.$$

If we take the derivative with respect to x of both sides of equation (21.16), we have

$$D_x F(x) = D_x F(a) + D_x \int_a^x f(t)dt.$$

Since $F(a)$ is a constant (it was 23 in the above illustration), $D_x F(a)$ is zero. By assumption, the function $F(x)$ is an antiderivative of $f(x)$. Therefore

$$D_x F(x) = f(x).$$

Substituting these into the above equation, we obtain a very useful result concerning the derivative of an integral.

Theorem 21.4 Assume that f is a continuous function over an interval $[a, b]$. Consider the number x, $a < x < b$, as an independent variable, then $\int_a^x f(t)dt$ is a differentiable function of x and

$$\frac{d}{dx}\int_a^x f(t)dt = f(x) \qquad \textbf{Derivative of an integral.} \qquad (21.17)$$

Example 21.7 The following illustrate the application of equation (21.17) to take the derivative of an integral.

a) $D_x \int_3^x (2t + 1)dt = 2x + 1,$

b) $D_x \int_0^x (3t^2 + 2)^{-1}\, dt = (3x^2 + 2)^{-1},$

c) $D_x \int_1^x \frac{1}{t}\, dt = \frac{1}{x},$

d) $D_t \int_2^t e^{-x^2}\, dx = e^{-t^2},$

e) $D_\theta \int_0^\theta \tan(t)dt = \tan(\theta)$

f) $D_s \int_6^s \sqrt{100 - x^2}\, dx = \sqrt{100 - s^2}.$ ◀

How can we use equation (21.17) to differentiate the function $G(x)$ defined by

$$G(x) = \int_x^5 (3t^2 - 6t)dt?$$

First, we apply the change of limits property (21.11) to express $G(x)$ in the form of the integral in equation (21.17):

$$G(x) = -\int_5^x (3t^2 - 6t)dt.$$

We then use (21.17) to differentiate $G(x)$:

$$G'(x) = -\frac{d}{dx}\int_5^x (3t^2 - 6y)dt = -(3x^2 - 6x).$$

Next consider the problem of differentiating the function $G(x)$ defined by

$$G(x) = \int_5^{x^2} (2t + 3)dt.$$

One way to determine $G'(x)$ is to first evaluate $G(x)$ using the *FTC*:

$$G(x) = (t^2 + 3t)\Big|_5^{x^2} = [(x^2)^2 + 3(x^2)] - [5^2 + 3(5)];$$

$$G(x) = x^4 + 3x^2 - 10.$$

Then, we can compute $G'(x) = 4x^3 + 6x$. However, this method requires the evaluation of the integral first. Can we compute $G'(x)$ without first evaluating the integral $G(x)$? The answer is yes. We can express $G(x)$ as a composition of the functions $h(x) = x^2$ and $g(u) = \int_5^u (2t + 3)dt$:

$$G(x) = (g \circ h)(x) = g[h(x)] = g(x^2) = \int_5^{x^2} (2t + 3)dt.$$

We now use the Chain Rule to evaluate $G'(x)$; setting $u = x^2$, we have

$$\frac{d}{dx} G(x) = \frac{d}{du} g(u) \cdot \frac{du}{dx}.$$

Equation (21.17) with $f \equiv g$ and $x \equiv u$ then gives:

$$\frac{d}{dx} g(u) = \frac{d}{du} \int_5^u (2t + 3)dt = 2u + 3.$$

As $D_x u = 2x$ and $2u + 3 = 2x^2 + 3$, the above equations yield

$$G'(x) = (2x^2 + 3)(2x).$$

We can repeat this process for an arbitrary function, $h(x)$, to obtain the following generalization of equation (21.17).

Theorem 21.5 Assume that f is continuous on the interval $[a, b]$ and $h(x)$ is a differentiable function of x with $a < h(x) < b$; then the integral

$$G(x) = \int_a^{h(x)} f(t)dt$$

exists and is a differentiable function of x with the derivative

$$\boxed{G'(x) = \frac{d}{dx} \int_a^{h(x)} f(t)dt = f(h(x)) \cdot h'(x).} \qquad (21.18)$$

Example 21.8 As an example of Theorem 21.5, we evaluate the following derivatives of integrals:

a) $\displaystyle \frac{d}{dx} \int_2^{x^2} 3t - t^5 \, dt = [3(x^2) - (x^2)^5](2x) = 6x^3 - 2x^{11}$

b) $\displaystyle \frac{d}{dx} \int_0^{(2x+1)^2} (5t + 2)dt = [5(2x + 1)^2 + 2] \cdot 4(2x + 1)$

c) $\displaystyle \frac{d}{dx} \int_{-1}^{1/x} 3t^4 \, dt = (3x^{-4})(-x^{-2}) = -3x^{-6}$

d) $\displaystyle \frac{d}{ds} \int_0^{s^2 - s} 3t + t^2 \, dt = [3(s^2 - s) + (s^2 - s)^2](2s - 1)$

e) $\displaystyle \frac{d}{dx}\int_0^{\sin(x)} t^3\,dt = \sin^3(x)\cos(x)$

f) $\displaystyle \frac{d}{dt}\int_0^{t^2} 5x - \frac{1}{x+1}\,dx = \left[5(t^2) - \frac{1}{t^2+1}\right](2t)$

Formula (21.18) may also be used to evaluate the derivative of integrals in which the lower limit of integration is a function as follows:

g) $\displaystyle \frac{d}{dt}\int_{-2t}^{0} 3x\,dx = \frac{d}{dt}\left[-\int_0^{2t} 3x\,dx\right] = [-3(2t)](2) = -12t$

h) $$\begin{aligned}\frac{d}{dx}\int_x^{x^2}(3t+1)dt &= \frac{d}{dx}\left[\int_x^0 (3t+1)dt + \int_0^{x^2}(3t+1)dt\right]\\ &= \frac{d}{dx}\left\{-\int_0^x (3t+1)dt\right\} + \frac{d}{dx}\int_0^{x^2}(3t+1)dt\\ &= -(3x+1) + [3(x^2)+1](2x) = 6x^3 - x - 1\end{aligned}$$

i) $$\begin{aligned}\frac{d}{d\theta}\int_{\cos^2\theta}^{\theta^2}(x-1)dx &= \frac{d}{d\theta}\left\{\int_0^{\theta^2}(x-1)dx - \int_0^{\cos^2(\theta)}(x-1)dx\right\}\\ &= (\theta^2-1)2\theta - (\cos^2(\theta)-1)(-2\cos(\theta)\sin(\theta))\\ &= 2\theta^3 - 2\theta - 2\cos(\theta)\sin^3(\theta).\end{aligned}$$ ◀

Example 21.9 *Release of creatinine*

When a muscle is exercised, the compound creatinine is released into the blood. During an experiment involving dogs, it was found that the speed at which a dog runs on a tread mill determines the release rate of creatinine into the blood. Assume that the function describing the rate of release as a function of the speed, s, for $s < 5$, is

$$f(s) = 6s(5-s).$$

The total amount of creatinine released as a dog progresses (uniformly in time) from a standing position to its maximum speed is

$$C = \int_0^M f(s)ds,$$

where M equals maximum speed. The maximum speed, M, is a function of w, the dog's body weight. Assume that

$$M(w) = M_0\left[1 + \sin\left(\left(\frac{w}{w_0} - 1\right)\right)\right],$$

where M_0 is the average maximum speed and w_0 is the average weight of a given breed of dog. The amount of creatinine circulating in a dog's blood when the treadmill reaches maximum speed is thus a function of the dog's weight:

$$C(w) = \int_0^{M(w)} f(s)ds + C(0), \tag{21.19}$$

where $C(0)$ is the amount of circulating creatinine in the dog at the start of the experiment. The rate of change in the amount of creatinine in the blood as a function of body weight is, by equation (21.18),

$$\begin{aligned} D_w C(w) &= [f(M(w))][M'(w)] \\ &= [6M(w)(5 - M(w))]\left[\frac{M_0 \pi}{w_0} \cos\left(\pi\left(\frac{w}{w_0} - 1\right)\right)\right]. \end{aligned}$$

This rate of change can be used to ascertain abnormal metabolism. When computing $C'(w)$, the value $M(w)$ must be evaluated first. For instance, a dog of average weight, w_0, has an average maximum speed:

$$M(w_0) = M_0\left[1 + \sin\left(\pi\left(\frac{w_0}{w_0} - 1\right)\right)\right] = M_0.$$

Hence the rate of creatinine release in the average dog is

$$\begin{aligned} C'(w_0) &= 6M_0(5 - M_0)\left[\frac{M_0 \pi}{w_0} \cos\left(\pi\left(\frac{w_0}{w_0} - 1\right)\right)\right] \\ &= \frac{\pi M_0{}^2}{w_0}(30 - 6M_0). \end{aligned}$$

◀

Example 21.10 *Air-pollution standards*

Air-pollution monitoring consists in measuring the quantity of a gas or suspended particulate matter in the air. The unit of measurement is usually given in *parts per million* (ppm), but, for design of flues, chimneys, etc., it is convenient to indicate the amount in mass per unit volume. These units are related by the following formula:

$$1 \text{ ppm} \cdot \frac{\text{molecular weight} \cdot 10^6}{22{,}400} = \frac{1 \mu\text{g}}{\text{m}^3},$$

where the molecular weight of the particulate matter or gas is a fixed constant. When monitoring the emission of pollutants, it is usually easier to measure the amount being emitted per unit of time by a particular source. This is accomplished by measuring a sample of air from a chimney to obtain the concentration of the substance (mass per unit volume) and multiplying it by the flow rate (volume per unit time). This results in a mass-per-unit-time rate of emission, $C(t)$. By monitoring $C(t)$ over an interval $[t_0, t_1]$, we find that the total amount of emission over this interval is given by

$$TC = \int_{t_1}^{t_2} C(t)dt. \tag{21.20}$$

For the monitoring of abnormally high emissions, the cumulative value of the excess emission is given by the *dosage* above a threshold concentration, C_T. This dosage, D_T, is given by the integral

$$D_T = \int_{t_1}^{t_2} C(t) - C_T \, dt. \tag{21.21}$$

C_T may represent the *normal* level of the substance being monitored in the air.

Air quality standards for industries typically set levels for the average amount of emission over a period of time. These standards will vary depending on the length of the associated time period. For instance, acceptable sulphur-dioxide (SO_2) exposure levels are as follows:

0.3 ppm over 8 hr;	$D_A = 2.4$ ppm/hr
1.0 ppm over 1 hr;	$D_A = 1$ ppm/hr
1.5 ppm over 3 min;	$D_A = 0.075$ ppm/hr

The average dosage, D_A ppm/hr, is given by

$$D_A = \frac{\int_{t_1}^{t_2} C(t)dt}{t_2 - t_1}. \tag{21.22}$$

The rate of change in the average dosage as a function of the length of the time interval, $[t_1, t_2]$, can be obtained by differentiating D_A with respect to the variable t_2. To simplify the calculations, we let $t_1 = 0$ and $t_2 = l$. Then D_A is a function of the length l given by

$$D_A(l) = \frac{\int_0^l C(t)dt}{l}.$$

By the quotient rule, we see that

$$D'_A(l) = \frac{l \dfrac{d}{dl}\left[\displaystyle\int_0^l C(t)dt\right] - \displaystyle\int_0^l C(t)dt}{l^2}.$$

The derivative of the integral is $C(l)$. Upon simplifying, we find that

$$D'_A(l) = \frac{C(l) - D_A(l)}{l}.$$

Hence the rate of change in D_A is the difference between the instantaneous rate of emission, $C(l)$, and the average rate, $D_A(l)$, *averaged* over the interval $[0, l]$. Note that for the large values of l, $D'_A(l)$ will be small unless there is a very large difference between the emission rate and the average dosage. Consequently, when averaging over a large interval, physiologically detectable changes (like pungent odors) would not be easily detected by observing changes in the average levels D_A. ◀

The following example illustrates a type of integral equation in which the limit of integration is an unknown to be determined by evaluating the integral using the FTC and then solving for the unknown constant.

Example 21.11 *Energy expenditure and oxygen metabolism associated with vigorous exercise*

When a person experiences vigorous exercise, energy, which is converted into muscle action, is derived primarily from adenosine triphosphate (ATP), creatine phosphate (CP), and glycogen stored in the muscle cells. Each of these chemicals is altered in the process of releasing energy. In order for this energy to be returned to the body for future use, the altered forms of these chemicals must be transformed into their original form and then stored again in the tissue cells. This reverse process is an *aerobic process*, which utilizes oxygen to complete the reversal. To maintain a balance, the body must compensate for the energy expended in exercise with an equivalent amount of *energy gain* through *oxidative metabolism*—i.e., by using oxygen to metabolize, or reconvert, the altered energy sources to their original state. Since energy can bc consumed very quickly during strenuous exercise and since the supply of oxygen is limited by lung capacity, oxidative metabolism usually continues after exercise and energy consumption have terminated. This is illustrated in Fig. 21.4, where we have plotted the relative rates of energy expenditure, $\mathring{E}$, and oxygen metabolism, $\mathring{O}$, over a period containing an interval of exercise.

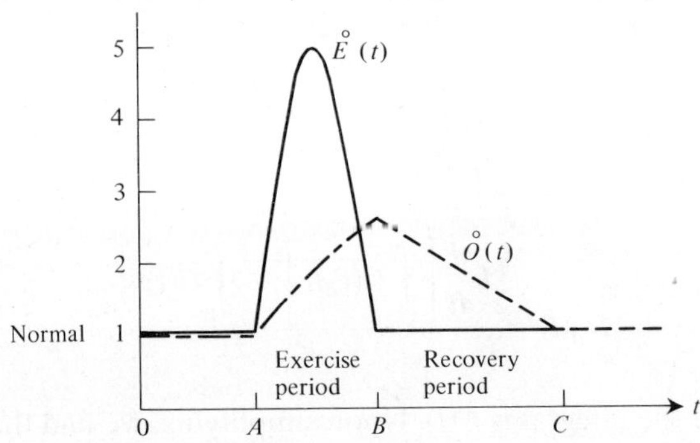

Figure 21.4

In Fig. 21.4, the rate of energy expenditure, $\mathring{E}(t)$, is constant (normal) up to a time $t = A$, at which point exercise commences. $\mathring{E}(t)$ then is greater than one over the interval $[A, B]$ of exercise, and it returns to normal for $t > B$. The oxidative metabolism rate, $\mathring{O}(t)$, is also constant (normal) up to time $t = A$. Then $\mathring{O}(t)$ increases throughout the short period of exercise, $[A, B]$; and, upon the termination of the exercise, the $\mathring{O}(t)$ curve decreases over the interval $[B, C]$.

The balance of energy requires that the total energy expelled be equal to the total energy gained by the oxidative metabolism. The total energy expelled is

$$E = \int_A^B \mathring{E}(t)dt.$$

The total oxidative metabolism is

$$O = \int_A^C \mathring{O}(t)dt.$$

One question that arises is, "How long does it take for the *oxygen debt* caused by the exercise to be made up?" That is, what is the length of the recovery interval $[B, C]$?

To answer this, we must make some assumptions concerning the form of the functions $\mathring{E}$ and $\mathring{O}$. For example, assume that a person exercises for two minutes, with $A = 2$, $B = 4$, and that $\mathring{E}(t)$ is defined by

$$\mathring{E}(t) = \begin{cases} 5 - 4(t-3)^2 & \text{for } 2 \le t \le 4; \\ 1 & \text{otherwise.} \end{cases}$$

Assume that for an unknown value, C, the function $\mathring{O}$ is as follows:

$$\mathring{O}(t) = \begin{cases} 1 & \text{if} \quad t < 2; \\ 3 - \dfrac{(t-4)^2}{2} & \text{if} \quad 2 \le t < 4; \\ \text{a linear function} & \text{if} \quad 4 \le t < C; \\ 1 & \text{if} \quad t \ge C. \end{cases}$$

By requiring the function $\mathring{O}(t)$ to be continuous, we force $\mathring{O}(4) = 3$ and $\mathring{O}(C) = 1$. Therefore the equation for $\mathring{O}(t)$ over the interval $[4, C]$ is given by the linear equation

$$\mathring{O}(t) = \left[\frac{3-1}{4-C}\right] \cdot (t-4) + 3.$$

The total energy expended by this person is

$$\begin{aligned} E &= \int_2^4 5 - 4(t-3)^2\, dt \\ &= 5t - \frac{4(t-3)^3}{3}\bigg|_2^4 \quad \text{(by the } FTC\text{)} \\ &= (20 - \tfrac{4}{3}) - (10 + \tfrac{4}{3}) = \tfrac{22}{3}. \end{aligned}$$

The total oxidative metabolism is given by

$$O = \int_2^C \mathring{O}(t)dt = \int_2^4 \mathring{O}(t)dt + \int_4^C \mathring{O}(t)dt.$$

The first integral of the sum is evaluated using the FTC as follows:

$$\int_2^4 3 - \frac{(t-4)^2}{2}\,dt = 3t - \frac{(t-4)^3}{6}\bigg|_2^4$$

$$= (12) - (6 + \tfrac{8}{6}) = 4\tfrac{2}{3}.$$

The second integral may also be evaluated using the FTC, but the result is a function of the yet to be determined constant, C:

$$\int_4^C \left(\frac{2}{4-C}\right)(t-4) + 3\,dt = \left(\frac{2}{4-C}\right)\frac{(t-4)^2}{2} + 3t\bigg|_4^C$$

$$= \left(\frac{2}{4-C}\right)\frac{(C-4)^2}{2} + 3C - 12$$

$$= -(C-4) + 3C - 12 = 2C - 8 = 2(C-4).$$

Therefore, adding the values of the two integrals,

$$O = 4\tfrac{2}{3} + 2(C-4) = 2C - \tfrac{10}{3}.$$

Setting energy, $E = 10$, equal to the oxidative metabolism, O, and solving, we are able to determine the value of C:

$$C = 5\tfrac{1}{3}.$$

Therefore it would take 80 seconds to recover from the 2 minutes of exercise if the rates $\mathring{E}$ and $\mathring{O}$ are assumed to have the above forms. ◀

SUMMARY

1. The *definite integral* of $f(x)$ over an interval $[a, b]$ is defined as a limit of a particular type of

   Approximating Series: $\sum_{n=1}^{N} f(x_n^*)\Delta x_n$.

   This series is obtained from a

   Partition of the interval $[a, b]$: $a = x_0 < x_1 < x_2 < \cdots < x_N = b$

   and from a set of

   Evaluation points: $x_1^*, x_2^*, \ldots, x_n^*$.

The partition $\{x_n\}$ divides the interval, $[a, b]$, into N

Subintervals: $I_n = [x_{n-1}, x_n]$, $n = 1, 2, \ldots, N$.

The term Δx_n is the

Length of the nth subinterval: $\Delta x_n \equiv x_n - x_{n-1}$.

The evaluation point

x_n^* *is any point in the interval* I_n: $x_{n-1} \le x_n^* \le x_n$.

The

Definite integral of $f(x)$ *over* $[a, b]$: $\int_a^b f(x)dx$,

is defined as the limit of the approximating series:

$$\int_a^b f(x)dx \equiv \lim_{\substack{\Delta x \to 0 \\ N \to +\infty}} \sum_{n=1}^{N} f(x_n^*)\Delta x.$$

2. The *Fundamental Theorem of Calculus* states that the definite integral of $f(x)$ can be expressed in terms of any antiderivative, $F(x)$, of $f(x)$.

$$\text{If } F'(x) = f(x), \qquad \text{then } \int_a^b f(x)dx = F(b) - F(a).$$

Terminology:

a is called the *lower limit of integration*.

b is called the *upper limit of integration*.

$f(x)$ is called the *integrand*.

x is called the *variable of integration*.

The evaluation of an integral is sometimes expressed using the notation:

$$\int_a^b f(x)dx = F(x)\Big|_a^b \equiv F(b) - F(a); \qquad \textbf{FTC}$$

$$\int_a^b x^3\, dx = \frac{x^4}{4}\Big|_a^b = \frac{b^4}{4} - \frac{a^4}{4}.$$

3. The derivative of an integral with respect to the upper limit of integration:

$$\frac{d}{dx}\int_a^x f(t)dt = f(x);$$

for example,

$$\frac{d}{dx}\int_1^x e^{t^2}dt = e^{x^2}.$$

Note that the derivative is with respect to x which is a different variable than the variable t of integration. The generalized derivative formula states that

$$\frac{d}{dx}\int_a^{h(x)} f(t)dt = f(h(x))\cdot h'(x) \qquad \textbf{Derivative of an Integral.}$$

for any differentiable function $h(x)$. An example with $h(x) = x^2 - x + 3$ is

$$\frac{d}{dx}\int_a^{x^2-x+3} \sin(t)dt = \sin(x^2 - x + 3)\cdot(2x - 1).$$

EXERCISE SET 21

21.1 Use the Fundamental Theorem of Calculus to evaluate the following integrals.

a) $\int_0^2 3x\,dx$ b) $\int_1^4 2 - x^2\,dx$

c) $\int_{-1}^3 2x + x\,dx$ d) $\int_0^{10} t^{1/2}\,dt$

e) $\int_5^6 (t-5)^2\,dt$ f) $\int_{-2}^1 (s+3)^{1/2}\,ds$

g) $\int_0^h x + 5x^4\,dx$ h) $\int_5^6 t^3 - 2\,dt$

i) $\int_5^x t^3 - 2\,dt$ j) $\int_0^x 2s^2 - 3\,ds$

k) $\int_1^3 2x(x^2+5)^3\,dx$ l) $\int_{-1}^3 2x(x^2-1)^4\,dx$

m) $\int_{-1}^1 (x+1)^2\,dx$ n) $\int_{-1}^1 x^2 + 1\,dx$

o) $\int_0^T t(3t^2+2)^{1/2}\,dt$ p) $\int_0^T (2t-3)(t^2-3t+6)^2\,dt$

21.2 Use the *FTC* to evaluate the indicated integrals.

a) $\int_0^{\pi} \sin(x)dx$ b) $\int_0^{\pi/4} \sec^2(x)dx$ c) $\int_{-\pi}^{\pi} \cos^2(x)\sin(x)dx$

d) $\int_{\pi/6}^{\pi/2} \csc(t)\cos(t)dt$ e) $\int_{\pi}^{\pi/2} \cos(x)dx$ f) $\int_0^{\pi} \theta\cos(\theta^2 + \pi)d\theta$

21.3 Which of the following integrals does not satisfy the hypothesis of the FTC? Indicate why.

a) $\int_0^{10} x^2 - 1\, dx$ b) $\int_0^{10} \sqrt{x^2 - 1}\, dx$ c) $\int_0^{\pi/2} \sec^2(\theta)d\theta$

d) $\int_0^{\pi/2} \sec(\theta/2)d\theta$ e) $\int_0^{5} \ln(x)dx$ f) $\int_0^{5} (1/t)dt$

21.4 Which of the two integrals is easier to evaluate? Evaluate the simpler one and indicate why the other one is more difficult.

a) $\int_0^{3} x^5(x^6 + 1)^{1/2}\, dx;\ \int_0^{3} (x^6 + 1)^{1/2}\, dx.$

b) $\int_0^{\pi} \sin(x^2)dx;\ \int_0^{\pi} x\sin(x^2)dx.$

c) $\int_0^{\pi/2} \cos(x)\sin(x)dx;\ \int_0^{\pi/2} \sin^2(x)dx.$

21.5 Express each of the following as a single definite integral.

a) $\int_0^{6} t^2\, dt + 3\int_0^{6} (t - 2)dt$ b) $\int_3^{5} (t + 1)^2\, dt - \int_3^{5} (t + 1)dt$

c) $\int_3^{4} (t + 7)dt + \int_4^{6} (t + 7)dt$ d) $\int_1^{3} x^2 - 2x\, dx + \int_{-2}^{1} x^2 - 2x\, dx$

e) $\int_3^{4} e^x\, dx - \int_4^{3} e^x\, dx$ f) $\int_3^{5} xe^x\, dx + \int_5^{4} xe^x\, dx$

21.6 Evaluate the following integrals.

a) $\int_0^{1} e^x\, dx$ b) $\int_{-1}^{0} e^x\, dx$ c) $\int_0^{-1} e^x\, dx$

d) $\int_0^{\ln(2)} e^{2x}\, dx$ e) $\int_{-1}^{1} xe^{x^2}\, dx$ f) $\int_0^{\ln(2)} e^{-3x}\, dx$

g) $\int_1^{e} \frac{1}{x}\, dx$ h) $\int_1^{e^2} \frac{1}{x}\, dx$ i) $\int_0^{10} \frac{1}{x + 1}\, dx$

j) $\int_0^{10} \frac{1}{3x + 1}\, dx$ k) $\int_2^{1} \frac{3}{x}\, dx$ l) $\int_{e^2}^{e} \frac{1}{x}\, dx$

21.7 Verify the following identities using the FTC. Also, state the integral property that is being demonstrated.

a) $\int_0^3 2t^2 + 4\,dt = 2\int_0^3 t^2 + 2\,dt$

b) $\int_1^4 x^{-3}\,dx \le \int_1^4 x^{-2}\,dx$

c) $\int_0^5 (t-3)^2\,dt = \int_0^1 (t-3)^2\,dt + \int_1^5 (t-3)^2\,dt$

d) $\int_1^1 6t^2 - t^{10}\,dt = 0$

e) $\int_{-4}^{3} (x+1)^2\,dx = \int_3^{-4} -(x+1)^2\,dx$

f) $\int_1^x 3t(t^2-6)^4\,dt = 3\int_1^2 t(t^2-6)^4\,dt - 3\int_x^2 t(t^2-6)dt$

21.8 Verify the algebraic properties of the definite integral by evaluating each integral in Example 21.4 and showing that equations (a) through (g) are valid.

21.9 An animal generates an amount of "heat" above its normal temperature depending on its activities. Assume that over a six-hour period the amount of radiated heat given off by the animal at time t is

$$h(t) = 32 - (t-2)^2(t-4)^2, \qquad t \in [0, 6].$$

a) Sketch a graph of the function, $h(t)$.
b) Estimate the total excess heat generated over the time intervals [0, 3] and [2, 4].
c) Compute the actual total excess heat generated over the intervals [0, 3] and [2, 4] by expanding the function $h(t)$ as a polynomial in t and integrating over the appropriate interval.

21.10 The integral

$$\int_3^{18} t^3 - 40t^2 + 453t - 1026\,dt.$$

was used in Example 21.3. Evaluate the integral and use it to obtain the cardiac output given in the example.

21.11 Use equation (21.4) to compute the basal metabolism of an animal whose BMR over the interval [0, 24] is given by $BMR = -0.5\cos(\pi t/12) + 1$ kcal/hr.

21.12 Use equation (21.19) to compute the circulating creatinine in a certain breed dog that weighs 10 kg, if for the same breed the average weight is $w_0 = 8$ kg and the average maximum speed is $M_0 = 2$.

21.13 a) As in Example 21.10, compute the dosage, D_T, (equation (21.21)) of a particulate matter for which the threshold value is $C_T = 5$ ppm, and the measured levels over the interval [0, 24] are given by $C(t) = 3\sin(\pi t) \pm 6$ ppm.

b) What is the average dosage D_A (equation (21.22)) of this pollutant?

21.14 Evaluate the following derivatives.

a) $\dfrac{d}{dx}\displaystyle\int_3^x (2t^2 + t^3)dt$ b) $\dfrac{d}{dt}\displaystyle\int_0^t 5e^x\,dx$ c) $\dfrac{d}{ds}\displaystyle\int_2^s (4x^2 - 2)dx$

d) $\dfrac{d}{dx}\displaystyle\int_0^{x^2} (t + 3)dt$ e) $\dfrac{d}{dx}\displaystyle\int_0^{x^2+1} (t + 3)dt$ f) $\dfrac{d}{dx}\displaystyle\int_1^{x^2-2x} \left(t + \frac{3}{t}\right)dt$

g) $\dfrac{d}{dt}\displaystyle\int_1^{(t+1)^2} (x - x^2)dx$ h) $\dfrac{d}{dx}\displaystyle\int_{2x}^{3} (1 + t)dt$ i) $\dfrac{d}{dx}\displaystyle\int_{x^2-1}^{5} (2t + t^4)dt$

j) $\dfrac{d}{dx}\displaystyle\int_{x^2}^{1+3x} t^2\,dt$ k) $\dfrac{d}{ds}\displaystyle\int_{-\sqrt{1-s^2}}^{\sqrt{1+s^2}} (t^2 + 1)dt$ l) $\dfrac{d}{ds}\displaystyle\int_1^{(s^2-2s)^6} \left(t - \frac{1}{t}\right)dt$

21.15 Find $F'(\theta)$ if $F(\theta)$ is given by the following integrals.

a) $\displaystyle\int_0^\theta \sec^2(t)dt$ b) $\displaystyle\int_0^{\sin(\theta)} 1 + t^2\,dt$ c) $\displaystyle\int_{-\theta}^{\theta} (1 + t^2)^{-1/2}\,dt$

d) $\displaystyle\int_{\sin(\theta)}^{\tan(\theta)} 1 - x^2\,dx.$ e) $\displaystyle\int_\theta^{\pi-\theta} \sin(t)dt$ f) $\displaystyle\int_{\sin(\theta)}^{\cos(\theta)} \ln(x)dx$

21.16 Compute the "period of recovery" as was done in Example 21.11 corresponding to the following functions:

$$\mathring{E}(t) = \begin{cases} 5 - (t - 4)^2 & \text{if} \quad 2 \le t \le 6; \\ 1 & \quad\text{otherwise.} \end{cases}$$

$$\mathring{O}(t) = \begin{cases} 1 & \text{if} \quad t \le 2; \\ 1 + \dfrac{(t-2)^3}{64} & \text{if} \quad 2 \le t \le 6; \\ \text{linear} & \text{if} \quad 6 \le t \le C; \\ 1 & \text{if} \quad t \ge C. \end{cases}$$

21.17 If an animal travels at a speed of $s(t) = 1 - t^{-2}$ miles per hour at time t hours, what is the distance it travels over the time interval $[2, 10]$?

21.18 Let the flow rate of a fluid into and out of a chamber be

$$\mathring{Q}_{\text{in}}(t) = t^2 + 3 \qquad \text{and} \qquad \mathring{Q}_{\text{out}}(t) = 4t + 3.$$

a) Sketch the graph of $\mathring{Q}_{\text{in}}(t)$ and $\mathring{Q}_{\text{out}}(t)$ over $[0, 6]$.
b) Compute the volume, $V(t)$, in the chamber at time t assuming $V(0) = 10$.
c) Interpret computation (b).

21.19 Assume that the growth rate of a plant is $g(x) = (t - 3)^2$ at time t. The maximum period of growth is $[0, T]$, where T is a function of rainfall given by $T = x(2 - x)$ and x is the rainfall in inches.

a) Express the total growth of the plant as an integral.
b) What is the instantaneous rate of change in the total growth with respect to variations in the rainfall, x?
c) Evaluate the integral in (a) and verify that the derivative computed in (b) is correct.

21.20 In Example 21.6, we used the rate of elimination of a drug to determine the total amount of a drug available. Another method is to use the rate at which the drug enters the blood system. This type of curve is extensively used in television commercials to "compare" different asprin-type drugs. A typical rate of assimilation function has the form

$$f(t) = kt(t - b)^2 \qquad 0 \le t \le b. \tag{21.23}$$

The graph of this function is shown in Fig. 21.5.

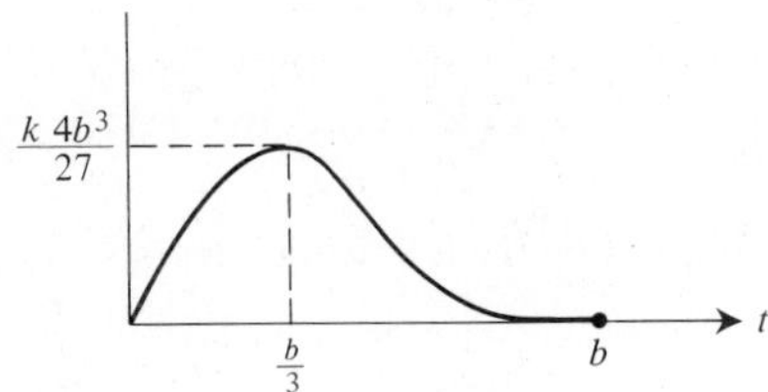

Figure 21.5

The total amount of drug assimilated is

$$A = \int_a^b f(t)dt.$$

Assume that three drugs, d_1, d_2, and d_3, are administered in equal pill sizes and are tested and found to be assimilated at the rates given by (21.23) with the following constraints:

$$k_1 = 0.15, b_1 = 3 \qquad \text{for drug } d_1;$$

$$k_2 = 0.01, b_2 = 6 \qquad \text{for drug } d_2;$$

$$k_3 = 0.001, b_3 = 9 \qquad \text{for drug } d_3.$$

a) Sketch on the same graph the assimilation rate functions, f_1, f_2, and f_3 corresponding to the above values of k and b.
b) Which drug is assimilated at a faster initial rate and which drug is assimilated over the longest period of time?
c) Which drug provides the greatest amount A, and which drug provides the least amount A?
d) From a practical point of view, does the above information tell you anything about which drug should be prescribed?

SECTION 22

THE DEFINITE INTEGRAL, AREA, AND NUMERICAL INTEGRATION

22.1 AREA AND THE DEFINITE INTEGRAL

In this section, the concept of area is examined. If we use the intuitive concept of the area of a rectangle, it is shown that the area of a nonrectangular region can be defined using the definite integral. This association between the easily visualized concept of an area and the definite integral provides not only a method for determining the area of specific regions, but insight into the approximation of integrals.

In most calculus texts, the following geometric interpretation of the definite integral is used to introduce the integral concept. The reason is the obvious advantage of using a sketch to convey concepts that are more difficult to express in prose. *One picture is worth a thousand words.* While we recognize the simplicity and advantage of introducing integrals as "areas," we have postponed this approach until now for two basic reasons. First, we wish to avoid the mistaken idea that "an integral is an area." True, as we will see shortly, areas can be computed via integrals. But, as the three problems of Section 20 illustrate, many other problems also may have their solutions expressed as definite integrals. Our second reason is that we wish to emphasize the process of deriving a definite integral through the procedure of discrete approximations to continuous processes.

In this section, we adopt the standard convention in mathematics of denoting the independent variable by x to represent linear distance. We consider a continuous function, $f(x)$, defined on the interval $[a, b]$. *For the moment, we assume that* $f(x) > 0$ *for* $x \in [a, b]$. The graph of f thus creates a region R in the $x - y$ plane. The region R is bounded above by the curve $y = f(x)$, on the sides by the vertical lines $x = a$ and $x = b$, and below by the x-axis. (See Fig. 22.1(a).)

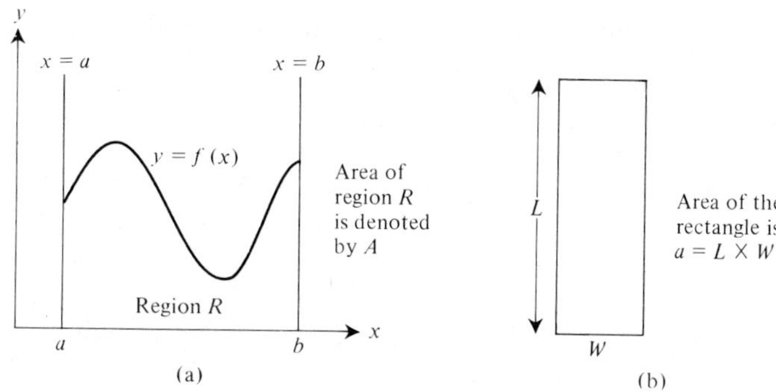

Figure 22.1

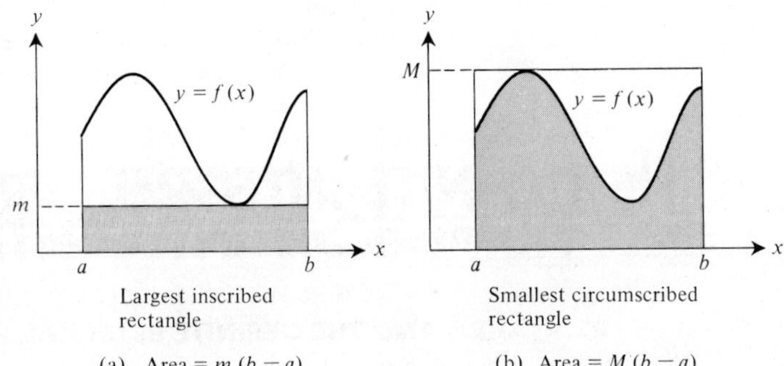

Largest inscribed rectangle

(a) Area = $m(b-a)$

Smallest circumscribed rectangle

(b) Area = $M(b-a)$

Figure 22.2

The region R has an area that we denote by A. Given the function f and the interval $[a, b]$ how can we compute the area A? To answer this, we must have some preconceived notion of area. We shall use the concept of area that arises from defining the area of a rectangle of height L and width W to be

$$\boldsymbol{a} = L \cdot W.$$

Looking at the particular function sketched in Fig. 22.1(a), we see that this definition of area does not tell us the area of the region R, since R is not a rectangular region.

We can associate various rectangles with the region R. The region R contains a largest inscribed rectangle with the interval $[a, b]$ as its base. R is also contained in a smallest circumscribed rectangle with the same base. These rectangles are depicted in Fig. 22.2. The heights of the inscribed and circumscribed rectangles are the values m and M given by Extreme Value Theorem 12.13:

$$m = \min_{x \in [a, b]} f(x) \qquad \text{and} \qquad M = \max_{x \in [a, b]} f(x).$$

From the Fig. 22.2, it is clear that the area A of the region R is bounded below and above by the areas of the inscribed and circumscribed rectangles, respectively:

$$m(b - a) \leq A \leq M(b - a). \tag{22.1}$$

Is there a rectangle whose area is the same as the area of R? To see that the answer is yes, we consider the rectangle associated with an arbitrary point, $x_1{}^* \in [a, b]$, depicted in Fig. 22.3. The area of this rectangle is $(b - a) \cdot f(x_1{}^*)$, since the height of the rectangle is the y-coordinate, $y = f(x_1{}^*)$, obtained from the graph. It should be

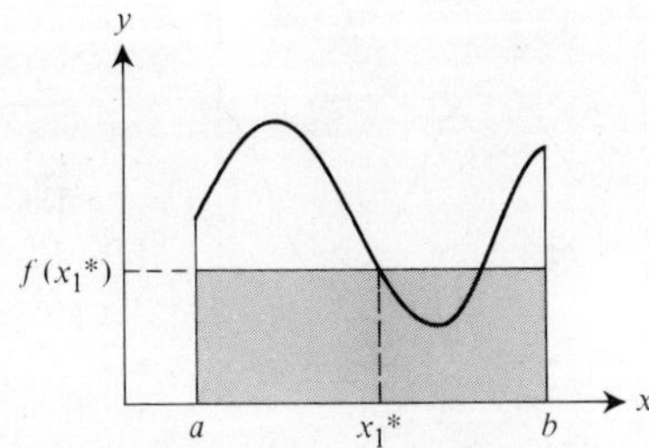

Figure 22.3

clear from the graph that this particular rectangle does not have area A. Since $x_1{}^*$ was randomly picked, we would be surprised if the area was A. Is it possible to pick such a point systematically so that the area of the associated rectangle will be equal to A? As we shall see later, the answer is yes. Intuitively, if we slowly raise the height of the inscribed rectangle, at some height $\bar{Y}$ less than M we would have the area of the rectangle exactly equal to the area A:

$$A = \bar{Y}(b - a).$$

The point that gives this value of $\bar{Y}$ is a point ξ such that $\bar{Y} = f(\xi)$. The existence of such a point ξ can be verified using the Intermediate Value Theorem 12.12. Instead, we will demonstrate it later by using the Mean Value Theorem for integrals.

What is the error in approximating the area A by the area $f(x_1{}^*)(b - a)$ of the rectangle as illustrated in Fig. 22.3? To give the specific error,

$$E = |A - f(x_1{}^*)(b - a)|,$$

we must know the area A as well as the point $x_1{}^*$ and $f(x_1{}^*)$. We can, however, give an upper bound for this error without knowing either A or $x_1{}^*$. To do this, we observe that the area of the rectangle of height $f(x_1{}^*)$ is also bounded by the areas of the two rectangles in Fig. 22.2:

$$m(b - a) \leq f(x_1{}^*)(b - a) \leq M(b - a). \tag{22.2}$$

Combining the inequalities (22.1) and (22.2), we find that the error E is not larger than the maximum area, $M(b - a)$, minus the minimum area, $m(b - a)$:

$$E \leq M(b - a) - m(b - a) = (M - m)(b - a).$$

This inequality states that the maximum possible error in evaluating the area A by the area $f(x_1{}^*) \cdot (b - a)$ of a rectangle is

1. proportional to the length $b - a$ of the interval $[a, b]$, and
2. proportional to the difference $M - m$ between the maximum and minimum values of $f(x)$ on the interval $[a, b]$.

These observations suggest that the error E would become smaller if a shorter interval $[a, b]$ were considered. The first reason is that the length $(b - a)$ would be smaller. The second reason, which is really the most important, is that the difference $(M - m)$ will never be larger and will usually be smaller when the continuous function $f(x)$ is considered over a smaller interval. It is this idea, "A shorter interval $[a, b]$ results in a smaller error in approximating the area A under the graph of f," which motivates the following derivation of the area A.

For a specific problem, we are given the function f and interval $[a, b]$ and asked to find the area of the region R. How do we obtain a reasonable estimate of this area if the interval $[a, b]$ is large? The solution is a simple and standard method. We employ the same procedure used to obtain approximate solutions in Section 20. We reduce our problem to several similar problems by partitioning the interval

$[a, b]$ into N-subintervals. To be more precise, we let $\{x_n\}_{n=0}^{N}$ be a partition of the of the interval $[a, b]$. Then

$$a = x_0 < x_1 < x_2 < \cdots < x_N = b.$$

The partition determines the subintervals

$$I_1 = [x_0, x_1] I_2 = [x_1, x_2], \ldots, I_N = [x_{N-1}, x_N].$$

It also results in the partitioning of the region R into N-subregions, $R_1, R_2, \ldots, R_N$, where the nth subregion, R_n, is bounded by the curves

$$y = f(x), \qquad x = x_{n-1}, \qquad x = x_n, \qquad \text{and} \qquad y = 0.$$

See Fig. 22.4(a), where $N = 4$.

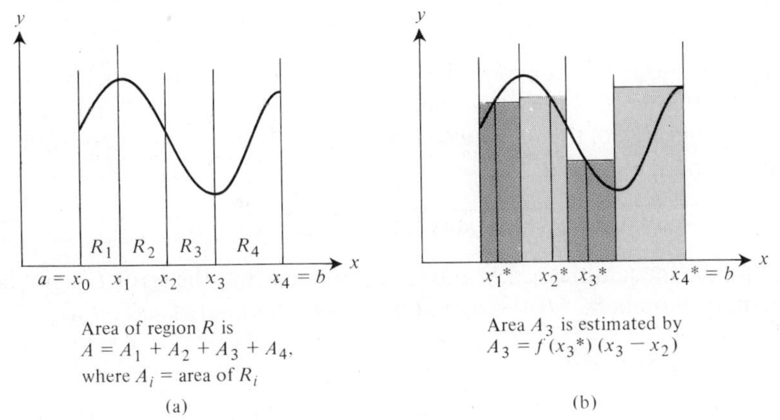

Figure 22.4

Denoting the area of the nth subregion by A_n, we obtain the area of R as the sum of the areas of the subregions, $R_1, R_2, \ldots, R_N$;

$$A = \sum_{n=1}^{N} A_n. \tag{22.3}$$

We have reduced the problem of determining the area A to N problems of determining the areas $A_1, A_2, \ldots, A_N$. However, by choosing N large, we can choose the partition $\{x_n\}$ in such a way that each subregion is determined by the graph of f over a *small* subinterval $[x_{n-1}, x_n]$—i.e., we can ensure that $\Delta x_n \equiv (x_n - x_{n-1})$ is small. Making the length of the subintervals small suggests that we should be able to approximate the areas A_n, $n = 1, 2, \ldots, N$, reasonably well. Then, if $\hat{A}_n$ is an estimate of the area A_n, we can estimate the area A by

$$A \simeq \hat{A} \equiv \sum_{n=1}^{A} \hat{A}_n. \tag{22.4}$$

For each subregion $R_1, R_2, \ldots, R_n, \ldots, R_N$, we estimate the respective area, $A_1, A_2, \ldots, A_n, \ldots, A_N$ by the procedure illustrated in Fig. 22.4(b). We arbitrarily pick an x-value in each interval, $x_1{}^*$ in the first interval, $x_2{}^*$ in the second interval, and in general $x_n{}^*$ in the nth interval. The area of the nth subregion is then estimated by

$$\hat{A}_n = f(x_n{}^*)(x_n - x_{n-1}).$$

The value $\hat{A}_n$ is the area of a rectangle having the subinterval $[x_{n-1}, x_n]$ as a base and height equal to $f(x_n{}^*)$. Adding the areas of the approximating rectangles, we obtain the estimate of A,

$$A \simeq \sum_{n=1}^{N} f(x_n{}^*)\Delta x_n. \tag{22.5}$$

Geometrically, this approximates the area under the graph of f by the area of N rectangles.

The error in the approximation (22.5) is the value

$$E = \left| A - \sum_{n=1}^{N} f(x_n{}^*)\Delta x_n \right|. \tag{22.6}$$

In Fig. 22.5, the error E is indicated by the area of the region marked horizontally, minus the area of the region marked vertically.

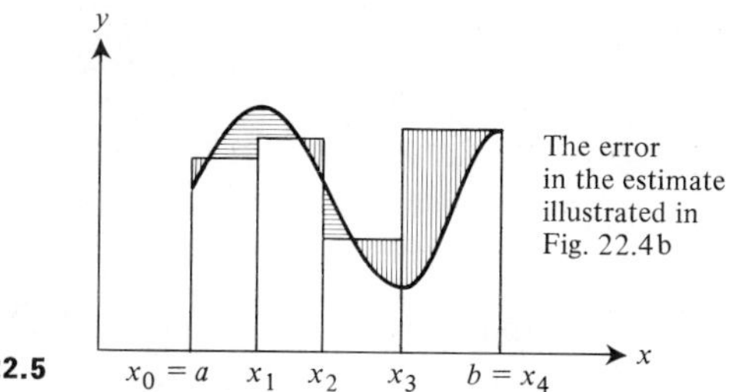

Figure 22.5

To improve upon the estimate (22.5), it seems reasonable that we should increase the value of N and simultaneously decrease the lengths Δx_n of the subintervals. One method of increasing N graphically consists of inserting more partition points between existing partition points and arbitrarily choosing additional evaluation points, $x_n{}^*$. When this is done and f is continuous, the error E given by equation (22.6) cannot increase and will in fact decrease to zero as N increases without bound. Consequently, the area A may be defined as the limiting value of the approximation (22.5) as $N \to \infty$:

$$A = \lim_{\substack{N \to \infty \\ \Delta x_n \to 0}} \sum_{n=1}^{N} f(x_n{}^*)\Delta x_n.$$

However, this limiting value was defined in Definition 21.1 to be the definite integral of f over $[a, b]$. We now define the area of the region R using the definite integral.

Definition 22.1 The area A of a region bounded by the graph of a positive continuous function, $y = f(x)$, the vertical lines, $x = a$, $x = b$, and the x-axis, $y = 0$, is equal to the definite integral of the function f over $[a, b]$—i.e.,

$$A \equiv \int_a^b f(x)dx \qquad \textbf{Area under graph of } \boldsymbol{f(x)}. \tag{22.7}$$

Example 22.1

a) The above definition may be used to evaluate the area of the triangle illustrated in Fig. 22.6. The function $f(x) = \frac{1}{2}x$, and the interval $[a, b] = [0, 2]$. Consequently, the area is given by equation (22.7) as

$$A = \int_0^2 \tfrac{1}{2}x\,dx \underset{\text{FTC}}{=} \left.\frac{x^2}{4}\right|_0^2 = 1 - 0 = 1.$$

This agrees with the standard formula for the area of a triangle:

$$A = \tfrac{1}{2}(\text{base}) \times (\text{height}) = (\tfrac{1}{2})(2)(1) = 1.$$

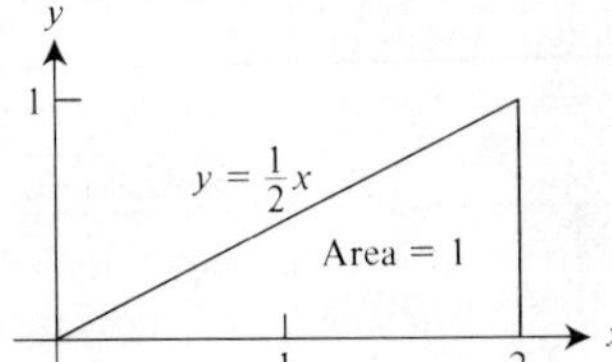

Figure 22.6

b) In Fig. 22.7, the area under the curve, $y = (x - 1)(4 - x)$, over the interval $[1, 4]$ is

$$\begin{aligned} A &= \int_1^4 (x - 1)(4 - x)dx \\ &= \int_1^4 - x^2 + 5x - 4\,dx \\ &= \left.\frac{-x^3}{3} + \frac{5x^2}{2} - 4x\right|_1^4 = 4.5. \end{aligned}$$

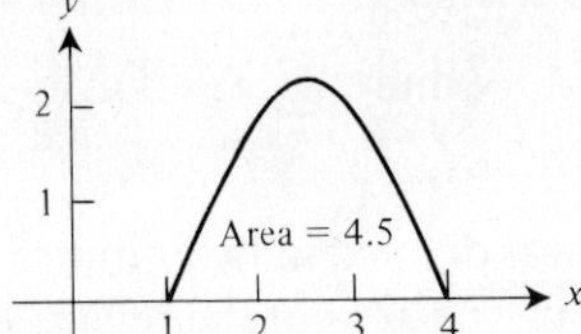

Figure 22.7

c) In Fig. 22.8, the area under the curve, $y = e^{-x^2}$, between $x = 0$ and $x = 1$ is given by the integral

$$A = \int_0^1 e^{-x^2}\,dx.$$

As we cannot express an antiderivative of e^{-x^2} in terms of elementary functions, the only way to obtain this area is to approximate it. Approximating methods are discussed in the next subsection.

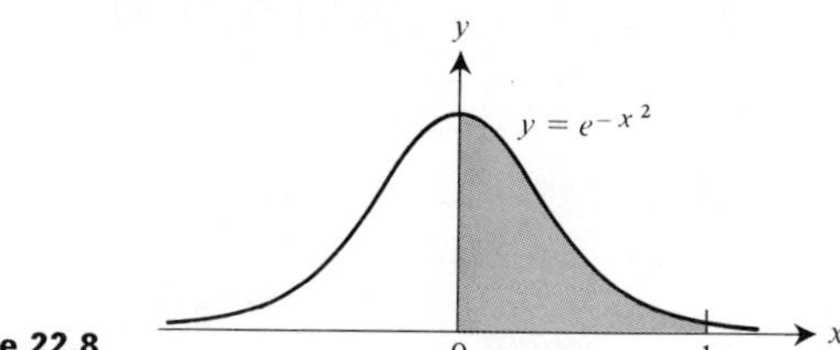

Figure 22.8

◀

To obtain the area of more complex regions, we use the procedure of representing the region in terms of several regions, each of which has the form whose area is defined in Definition 22.1.

Consider a region, R, which is bounded above by a function $g(x)$ and bounded below by a function $f(x)$, as illustrated in Fig. 22.9. To obtain the area of R, we visualize the region as the region under the graph of g minus the region under the graph of f. Then the area, A, of the region R, is as follows:

$$\begin{aligned} A &= (\text{area under } g) - (\text{area under } f). \\ &= \int_a^b g(x)dx - \int_a^b f(x)dx. \\ &= \int_a^b [g(x) - f(x)]dx. \end{aligned}$$

We state this result as a theorem, which holds even for negative functions.

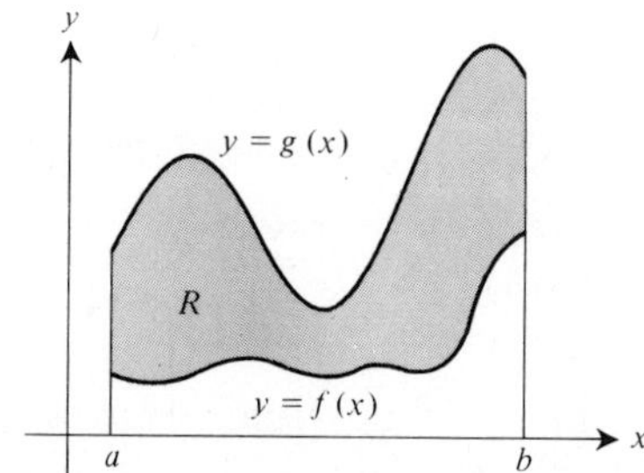

Figure 22.9

Theorem 22.1 Assume that f and g are continuous functions and $f(x) \leq g(x)$ for $x \in [a, b]$; then the area A of the region bounded by the graphs of $y = f(x)$, $y = g(x)$, $x = a$, and $x = b$ is given by

$$A = \int_a^b [g(x) - f(x)]dx \qquad \textbf{Area between two curves when } g(x) \geq f(x). \tag{22.8}$$

Example 22.2

a) Let $g(x) = 3x$ and $f(x) = x$. (See Fig. 22.10(a).) Then the area of the region between the graphs of f and g over the interval $[0, 2]$ is as follows:

$$\begin{aligned} A &= \int_0^2 g(x) - f(x)dx \\ &= \int_0^2 3x - x\,dx = \int_0^2 2x\,dx = x^2 \Big|_0^2 = 4. \end{aligned}$$

Note that $f(x) \leq g(x)$ or the interval $[0, 2]$.

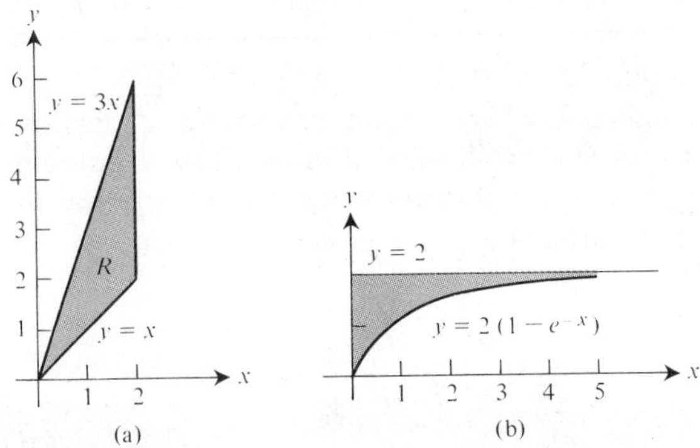

Figure 22.10

b) Let $g(x) = 2$ and $f(x) = 2 - 2e^{-x}$. (See Fig. 22.10(b).) The area bounded by the graphs of $g(x)$ and $f(x)$ over the interval $[0, 5]$ is

$$\begin{aligned} A &= \int_0^5 (2) - (2 - 2e^{-x})dx \\ &= \int_0^5 2e^{-x}\,dx = -2e^{-x}\Big|_0^5 = 2 - 2e^{-5} \simeq 1.99. \end{aligned}$$ ◀

Example 22.3 The area of the finite region between the graphs of the two functions, $f(x) = 5x - x^2$ and $g(x) = x^2 - 5x + 4.5$, is determined as follows.

First, we find the points of intersection of the two curves, $y = f(x)$ and $y = g(x)$. To do this, we set $f(x) = g(x)$ and solve algebraically for the two roots, x_1 and x_2:

$$5x - x^2 = x^2 - 5x + 4.5,$$

$$2x^2 - 10x + 4.5 = 0,$$

$$x = \tfrac{1}{4}(10 \pm \sqrt{100 - 36}),$$

$$x_1 = 0.5 \qquad \text{and} \qquad x_2 = 4.5.$$

The points x_1 and x_2 are indicated in Fig. 22.11.

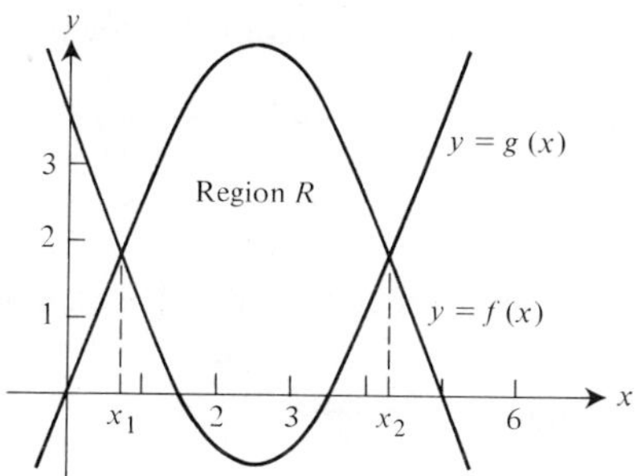

Figure 22.11

To find the desired area, we integrate the difference $f(x) - g(x)$ over $[x_1, x_2]$, since the graph of f is above the graph of g on this interval.

$$\int_{x_1}^{x_2} f(x) - g(x)dx = \int_{0.5}^{4.5} (5x - x^2) - (x^2 - 5x + 4.5)dx$$

$$= \int_{0.5}^{4.5} -2x^2 + 10x - 4.5\, dx$$

$$= -\tfrac{2}{3}x^3 + 5x^2 - 4.5x \Big|_{0.5}^{4.5}$$

$$= [-\tfrac{2}{3}(4.5)^3 + 5(4.5)^2 - (4.5)^2]$$
$$-[-\tfrac{2}{3}(0.5)^3 + 5(0.5)^2 - (4.5)(0.5)]$$
$$= 21\tfrac{1}{3}.$$

◀

A natural question is, "What is the area interpretation of $\int_a^b f(x)dx$ if $f(x)$ is not a positive function?" One interpretation is the following. Assume that the function f has a graph as illustrated in Fig. 22.12. Then the $\int_a^b f(x)dx$ can be visualized as the difference between two areas:

$$\int_a^b f(x)dx = A_+ - A_-,$$

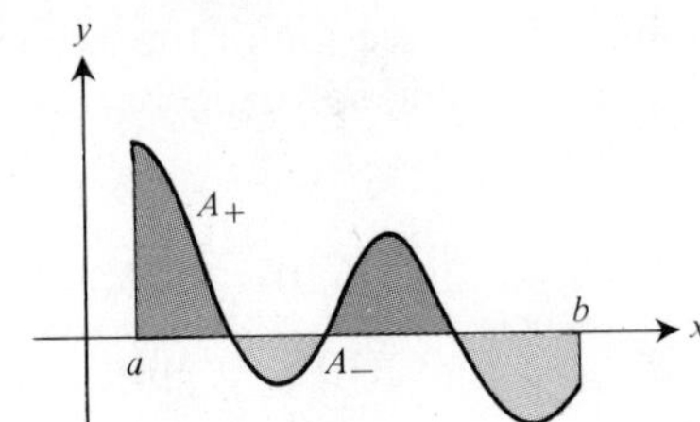

Figure 22.12

where

A_+ is the area between the x-axis and the positive $f(x)$ values,

A_- is the area between the x-axis and the negative $f(x)$ values.

In this sense the integral represents a "signed area," which may be positive or negative or zero.

Example 22.4

a) If we integrate $f(x) = \sin(x)$ over the interval $[0, 2\pi]$, we find

$$\int_0^{2\pi} \sin(x)dx = -\cos(x)\Big|_0^{2\pi} = -1 + 1 = 0.$$

This may be interpreted as stating that the area under the positive portion of the sine curve equals the area between the x-axis and the negative portion of the sine curve.

b) Assume that the rate of blood flow through an artery was found experimentally to be

$$r(t) = 0.5 - \sin(\pi t) \qquad \text{for} \qquad 0 < t \leq 2.$$

What is the total blood flow over the interval $[0, 2]$? The volume of blood flow is

$$V = \int_0^2 r(t)dt = \int_0^2 0.5 - \sin(\pi t)dt$$

$$= 0.5t + \frac{\cos(\pi t)}{\pi}\Big|_0^2 = \left(1 + \frac{1}{\pi}\right) - \left(0 + \frac{1}{\pi}\right) = 1 \text{ unit.}$$

This states that the area under the graph of $r(t)$ above the t-axis is one unit greater than the area below the t-axis over the interval $[0, 2]$. ◀

Returning to a question raised earlier in this section, let us examine the problem of finding a point ξ in the interval $[a, b]$ such that the area under the graph of $f(x)$ over $[a, b]$ is equal to the area of a rectangle having height $\overline{Y} = f(\xi)$. Thus we want to find a number ξ such that

$$\int_a^b f(x)dx = f(\xi)(b - a). \tag{22.9}$$

Area under the graph of f Area of a rectangle of height $f(\xi)$

The value ξ is obtained by applying the Mean Value Theorem 18.2 to an antiderivative of $f(x)$. Assume that F is an antiderivative of f:

$$F'(x) = f(x).$$

Then by the Fundamental Theorem of Calculus, the area under the graph of f is

$$A = \int_a^b f(x)dx = F(b) - F(a).$$

But the Mean Value Theorem tells us that there is a point ξ in the interval (a, b) such that

$$\frac{F(b) - F(a)}{b - a} = F'(\xi),$$

or

$$F(b) - F(a) = F'(\xi)(b - a).$$

Since $F'(\xi) = f(\xi)$, setting $F(b) - F(a) = f(\xi)(b - a)$, we obtain formula (22.9). We state this result as follows.

Theorem 22.2 *Mean Value Theorem of Integral Calculus*
If f is continuous on the interval $[a, b]$, then there exists a point ξ in the interval (a, b) such that

$$\int_a^b f(x)dx = f(\xi)(b - a) \qquad \textbf{Mean value theorem for integrals.} \tag{22.10}$$

We have frequently used the term "average of $f(x)$ over the interval..." in the foregoing; we now give its formal definition.

Definition 22.3 Let f be a continuous function over the interval $[a, b]$. The *average value of f over* $[a, b]$ is the number

$$\overline{Y} = \frac{\int_a^b f(x)dx}{b - a} \qquad \textbf{Average of } \boldsymbol{f} \textbf{ over } [\boldsymbol{a}, \boldsymbol{b}].$$

The average of a function $f(x)$ is therefore the value $\overline{Y} = f(\xi)$. The value of ξ is not unique. The Fig. 22.13 illustrates the average value $\overline{Y}$ for the same curve used in earlier illustrations in this section. For the function as illustrated, there are three possible values for ξ.

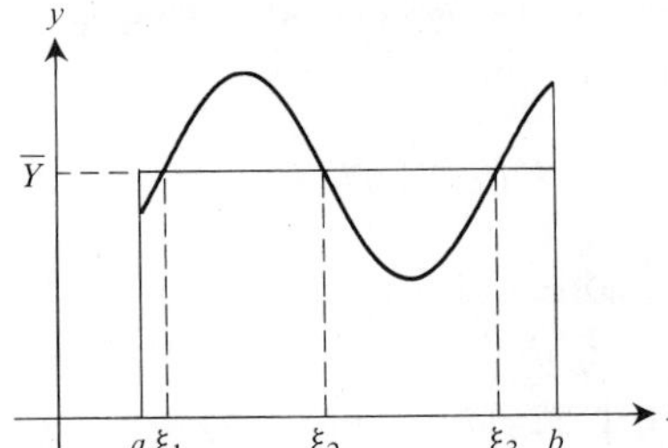

The area of the rectangle with height $\overline{Y}$ equals the area under the curve $y = f(x)$.

$f(\xi_1) = \overline{Y}$
$f(\xi_2) = \overline{Y}$
$f(\xi_3) = \overline{Y}$

Figure 22.13

Example 22.5 The amount of insulin in a normal individual's blood is affected by the amount of blood sugar present. As the blood sugar level rises, insulin is secreted by the pancreas into the bloodstream. Once in the bloodstream, the insulin degrades (becomes inactive biochemically) exponentially with a half-life of about 20 minutes.

In an experiment, a patient has fasted (to reduce her or his blood sugar level) and is given a large amount of sugar by injection. Assume that the experimentally determined blood-insulin level is described by the following function:

$$I(t) = \begin{cases} t(10 - t) & \text{for} \quad 0 \le t \le 5; \\ 25e^{-k(t-5)} & \text{for} \quad 5 \le t, \end{cases}$$

where $k = \ln(2)/20$.

What is the average insulin level over the interval [0, 60], a one hour period? The solution is

$$\bar{I} = \int_0^{60} I(t)dt/(60 - 0).$$

To evaluate the integral, we must separate it into two integrals using the property (21.9):

$$\int_0^{60} I(t)dt = \int_0^5 I(t)dt + \int_5^{60} I(t)dt.$$

The first of the integrals on the right is

$$\int_0^5 t(10 - t)dt = \int_0^5 10t - t^2\,dt = 5t^2 - \frac{t^3}{3}\Big|_0^5$$
$$= (125 - \tfrac{125}{3}) = 83.33\ldots$$

The second integral is

$$\int_5^{60} 25e^{-\ln(2)(t-5)/20}\,dt = -(25)\frac{(20)}{\ln(2)}\,e^{-\ln(2)(t-5)/20}\Big|_5^{60}$$
$$= \frac{-(25)(20)}{\ln(2)}\,[e^{-\ln(2)(55)/20} - e^0]$$
$$= -721.35[0.15 - 1] \simeq 614.12.$$

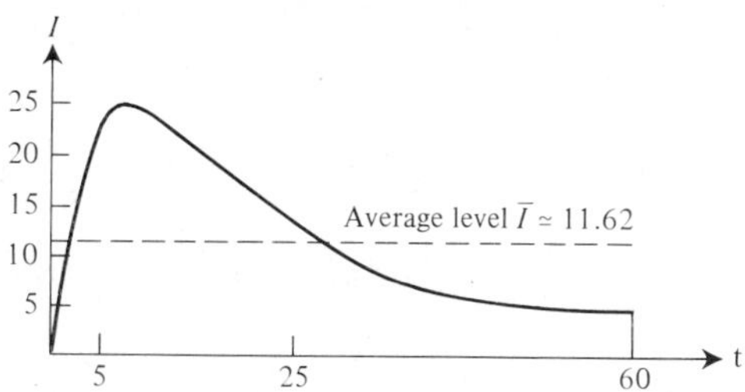

Fig. 22.14 Insulin level following the introduction of blood sugar.

Combining the value of these two integrals (see Fig. 22.14), we find

$$\bar{I} \simeq \frac{83.33 + 614.12}{60} \simeq 11.62.$$

◀

22.2 EVALUATING THE DEFINITE INTEGRAL AS A LIMIT

In this subsection, the limit definition of the integral is used to derive the value of specific definite integrals. The methods considered here will form the basis for the next subsection on numerical integration.

Using t to denote the independent variable, the integral of a function $g(t)$ is defined (see Definition 20.1) as

$$\int_a^b g(t)dt = \lim_{\substack{\Delta t \to 0 \\ t \to \infty}} \sum_{n=1}^{N} g(t_n{}^*)\Delta t_n, \tag{22.11}$$

where $\{t_n\}$ is a partition of the interval $[a, b]$, $\Delta t = \text{maximum } \{\Delta t_n = t_n - t_{n-1}, n = 1, 2, \ldots, N\}$, and the arbitrary evaluation points $\{t_n{}^*\}$ satisfy

$$t_{n-1} \leq t_n{}^* \leq t_n.$$

The above definition at first appears to be formidable, but in practice it is very convenient. Its generality at the same time provides a flexibility that allows for extreme freedom in evaluating integrals. If a function $g(t)$ is integrable, then we may evaluate its integral over the interval $[a, b]$ by considering the limit in equation (22.11) for *any* particular choice of partition and evaluation points. This means that we can simply choose the partition $\{t_n\}$ and points $\{t_n{}^*\}$ at our convenience to make our calculations easy, and then let $N \to \infty$ and max $\Delta t \to 0$.

From a mathematical point of view, the evaluation of a series $\sum g(t_n{}^*)\Delta t_n$ is greatly simplified if all of the lengths Δt_n are the same. If $\Delta t_n = h$ for $n = 1, 2, \ldots, N$, then, by the distributive law (Definition 10.3),

$$\sum_{n=1}^{N} g(t_n{}^*)\Delta t_n = h \sum_{n=1}^{N} g(t_n{}^*).$$

When this occurs the partition is said to be a *uniform partition*.

Definition 22.4 A partition, $\{t_n\}$, of the interval $[a, b]$ is a *uniform partition of size N* if

$$\Delta t_n = h \equiv \frac{b-a}{N} \qquad \text{for} \qquad n = 1, 2, \ldots, N.$$

The number $(b - a)/N$ is the *norm* of the partition and the *n*th partition point, t_n, is given by

$$t_n = a + n \cdot h.$$

Thus

$$t_0 = a, \qquad t_1 = a + h, \qquad t_2 = a + 2h, \ldots, \qquad t_N = a + Nh = b.$$

A uniform partition of size $N = 5$ of the interval $[3, 7]$ has norm $h = (7 - 3)/5 = 0.8$. The partition points would be

$$t_0 = 3, \qquad t_1 = 3.8, \qquad t_2 = 4.6, \qquad t_3 = 5.4, \qquad t_4 = 6.2, \qquad t_5 = 7.$$

Since the partition points of a uniform partition are easily described by an equation, it is often convenient to use the endpoints of the subintervals, $I_n = [t_{n-1}, t_n]$, as the evaluation points $t_n{}^*$. When this is done, it is possible to give rather straightforward approximation formulas for

$$\int_a^b g(t)dt.$$

This is accomplished by setting $t_n{}^* = t_n$ or $t_n{}^* = t_{n-1}$ in the approximation

$$\int_a^b g(t)dt \simeq h \sum_{n=1}^{N} g(t_n{}^*) \tag{22.12}$$

The approximation formula for $\int_a^b g(t)dt$ using a uniform partition of size N and the left endpoint of each subinterval as an evaluation point is obtained by setting $t_n{}^* = t_{n-1}$. This gives an approximation of the form

$$\int_a^b g(t)dt \simeq h \cdot \sum_{n=1}^{N} g(t_{n-1}); \tag{22.13}$$

as $t_{n-1} = a + (n - 1)h$, this finite sum can be expanded.

$$\int_a^b g(t)dt \simeq \frac{b-a}{N}[g(a) + g(a + h) + g(a + 2h) + \cdots + g(a + (N - 1)h)]$$

Approximating the integral using left endpoints.

(22.14)

Geometrically, the approximation (22.14) represents the area under the graph of g by the sum of the area of the rectangles illustrated in Fig. 22.15. Note that the height of each rectangle is the value $g(t_{n-1})$ of the function g at the left endpoint t_{n-1}.

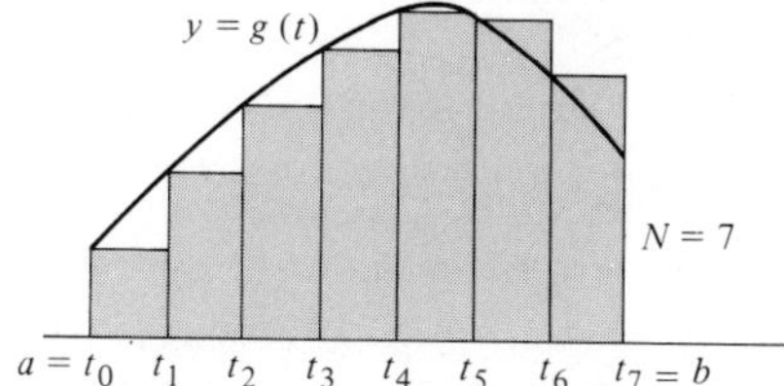

Fig. 22.15 Approximating the area under the graph of $g(t)$ using the left endpoint.

Example 22.6 The function $g(t) = 2^t$ is a continuous function and hence is integrable over the interval [0, 5]. We have yet to discuss antiderivatives of this type of function and consequently cannot use the Fundamental Theorem of Calculus to evaluate

$$\int_0^5 2^t \, dt.$$

Let us estimate the value of this integral using a uniform partition with $N = 5$, $a = 0$, and $b = 5$. Then $h = (5 - 0)/5 = 1$ and the partition points are

$$t_0 = 0, \quad t_1 = 1, \quad t_2 = 2, \quad t_3 = 3, \quad t_4 = 4, \quad \text{and} \quad t_5 = 5.$$

Formula (22.14) then gives

$$\int_0^5 2^t \, dt \simeq h[g(0) + g(1) + g(2) + g(3) + g(4)]$$
$$= (1)[2^0 + 2^1 + 2^2 + 2^3 + 2^4] = 31.$$

◀

The approximation formula for $\int_a^b g(t)dt$ using a uniform partition of size N and the right endpoint is obtained by setting $t_n^* = t_n$ for $n = 1, 2, \ldots, N$, in formula (22.12):

$$\int_a^b g(t)dt \simeq h \cdot \sum_{n=1}^{N} g(t_n). \tag{22.15}$$

As $t_n = a + nh$, the expanded form of the equation (22.14) is usually easier to remember:

$$\int_a^b g(t)dt \simeq \frac{b - a}{N} [g(a + h) + g(a + 2h) + \cdots + g(b)]$$

Approximating the integral using right endpoints.

(22.16)

The geometrical interpretation of the approximation (22.16) is illustrated in Fig. 22.16. The area under the graph of $y = g(t)$ is approximated by the sum of the areas of the rectangles. Each rectangle has height equal to $g(t_n)$, the value of g at the right endpoint of the respective subinterval.

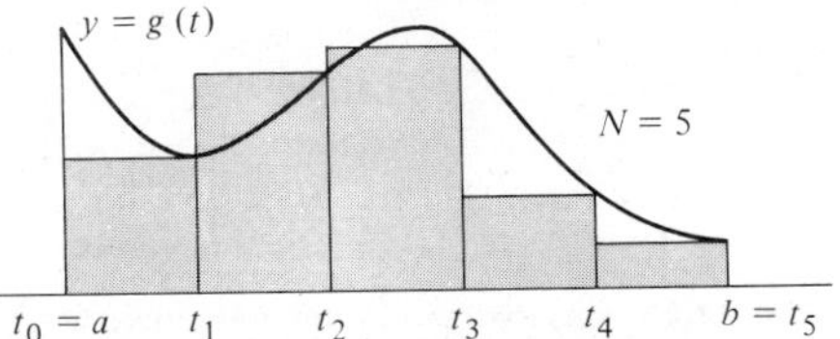

Fig. 22.16 Approximating the area under the curve using the right endpoints of a uniform partition.

Example 22.7 We approximate the definite integral of the function $g(t) = 2t$ over the interval [1, 3].

a) First, we consider a uniform partition of size $N = 4$. This gives a value of

$$h = (b - a)/N = (3 - 1)/4 = 0.5,$$

and the uniform partition points,

$$t_0 = 1, \qquad t_1 = 1 + h = 1.5, \qquad t_2 = 1 + 2h = 2.0, \qquad t_3 = 1 + 3h = 2.5,$$

and

$$t_4 = 1 + 4b = 3.$$

Using the left endpoint for $t_n{}^*$, equation (22.14) provides the following approximation

$$\int_1^3 2t\, dt \simeq (0.5)[2(1.0) + 2(1.5) + 2(2.0) + 2(2.5)] = 7.$$

(See Fig. 22.17.) Using the right endpoint for $t_n{}^*$, equation (22.16) provides the following approximation:

$$\int_1^3 2t\, dt \simeq (0.5)[2(1.5) + 2(2.0) + 2(2.5) + 2(3.0)] = 9.$$

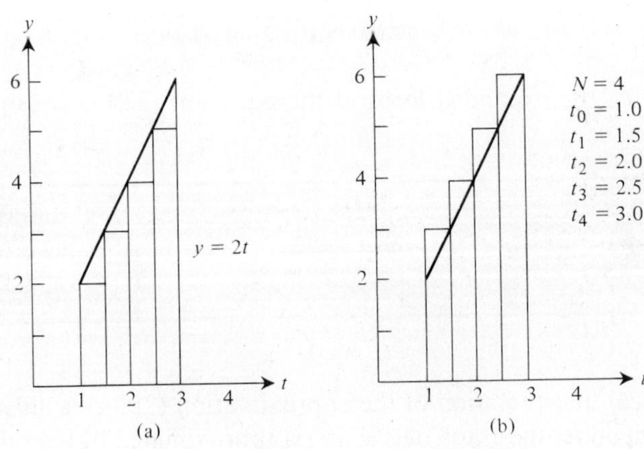

Fig. 22.17 (a) Approximation using left endpoints; (b) approximation using right endpoints.

b) Next we consider $N = 10$. Then

$$h = (b - a)/N = (3 - 1)/10 = 0.2.$$

The uniform partition points are

$$t_n = (1 + 0.2n) \qquad \text{for} \qquad n = 0, 1, 2, \ldots, 10.$$

Using the left endpoints, $t_n^* = t_{n-1} = 1 + 0.2(n - 1)$, we evaluate approximation (22.13) as follows:

$$\begin{aligned}
\int_1^3 2t\,dt &\simeq (0.2)\sum_{n=1}^{10} g(1 + 0.2(n - 1)) && \text{(Sum of } g(t_{n-1}))\\
&= (0.2)\sum_{n=1}^{10} 2(1 + (0.2)(n - 1)) && \text{(Evaluating } g(t_{n-1}) = 2t_{n-1})\\
&= (0.2)\sum_{n=1}^{10} (1.6 + (0.4)n) && \text{(Algebra)}\\
&= (0.2)[(1.6)(10) + (0.4)(10 \cdot 11)/2] && \text{(By formula (10.8))}\\
&= (0.2)[16 + 22] = 7.6.
\end{aligned}$$

Repeating the same procedure, we evaluate approximation (22.15) corresponding to choosing t_n^* as the right endpoint t_n:

$$\begin{aligned}
\int_1^3 2t\,dt &\simeq (0.2)\sum_{n=1}^{10} g(1 + (0.2)n)\\
&= (0.2)\sum_{n=1}^{10} 2(1 + (0.2)n)\\
&= (0.2)\sum_{n=1}^{10} [2 + (0.4)n]\\
&= (0.2)\left[(2)(10) + (0.4)\left(\frac{10 \cdot 11}{2}\right)\right] && \text{(By formula (10.8))}\\
&= (0.2)[20 + 22] = 8.4.
\end{aligned}$$

The approximations of the integral $\int_1^3 2t\,dt$ obtained thus far are 7.0, 7.6, 8.4, and 9.0.

c) Now, to obtain the exact value of the integral and to demonstrate that both formulas (22.14) and (22.16) lead to the same limiting value, we consider N to be arbitrary and repeat the procedures of parts (a) and (b):

$$h = (b - a)/N = (3 - 1)/N = 2/N,$$

and the partition points are

$$t_n = 1 + n \cdot h, \qquad \text{for} \qquad n = 0, 1, 2, \ldots, N.$$

Using equation (22.14), we obtain the approximation with $t_n^* = t_{n-1}$:

$$\begin{aligned}\int_1^3 2t\,dt &\simeq h\cdot\sum_{n=1}^{N} g(1 + h(n-1))\\ &= h\cdot\sum_{n=1}^{N} 2(1 + h(n-1))\\ &= h\sum_{n=1}^{N}[2(1-h) + 2\cdot h\cdot n]\\ &= h\left[2(1-h)N + 2\cdot h\cdot\frac{N(N+1)}{2}\right] \qquad \text{(By formula (10.8))}\\ &= h[2N + hN(N-1)]\\ &= \left(\frac{2}{N}\right)\left[2N + \left(\frac{2}{N}\right)N(N-1)\right] \qquad \left(\text{As } h = \frac{2}{N}\right)\\ &= 8 - \frac{4}{N}.\end{aligned}$$

Taking the limit as $N \to \infty$, we obtain the value of the integral:

$$\int_1^3 2t\,dt = \lim_{N\to\infty}\left(8 - \frac{4}{N}\right) = 8.$$

Repeating the same procedure using equation (22.16), we obtain the approximation with $t_n^* = t_n$:

$$\begin{aligned}\int_1^3 2t\,dt &\simeq h\cdot\sum_{n=1}^{N} g(1 + nh)\\ &= h\cdot\sum_{n=1}^{N} 2(1 + nh)\\ &= h\cdot\sum_{n=1}^{N}(2 + 2h\cdot n)\\ &= h\left[2N + 2h\frac{N(N+1)}{2}\right]\\ &= h[2N + hN(N+1)]\\ &= \frac{2}{N}\left[2N + \left(\frac{2}{N}\right)N(N+1)\right]\\ &= 8 + \frac{4}{N}.\end{aligned}$$

Taking the limit as $N \to \infty$, we obtain the value of the integral:

$$\int_1^3 2t\,dt = \lim_{n\to\infty}\left(8 + \frac{4}{N}\right) = 8. \qquad ◀$$

The method of evaluating an integral illustrated in Example 22.7 is more difficult than using the Fundamental Theorem of Calculus. Yet this is precisely the method used by the ancient Greeks and Egyptians to evaluate area and volume. The tricky part of performing this type of evaluation is to represent the sum

$$\sum_{n=1}^{N} g(t_n{}^*)$$

in a *closed form* as a function of N. In Section 10, we state several special types of finite sums that are expressible in closed forms. Consequently, with only these special cases, you will be very limited as to the type of integrals that you can evaluate using the limit definition. These will essentially be the integrals of polynomials of degree three or less. We give one additional example using only the right endpoint formula.

Example 22.8 We evaluate the integral

$$\int_0^4 t^2\, dt,$$

using right endpoint formula (22.15). For any N, the increment

$$h = \frac{4-0}{N} = \frac{4}{N}.$$

The partition points are

$$t_n = 0 + n \cdot h = \left(\frac{4}{N}\right)n.$$

Thus formula (22.15) gives

$$\begin{aligned} \int_0^4 t^2\, dt &\simeq \left(\frac{4}{N}\right) \sum_{n=1}^{N} \left(\frac{4}{N}\cdot n\right)^2 \\ &= \left(\frac{4}{N}\right)^3 \sum_{n=1}^{N} n^2 \\ &= \left(\frac{4}{N}\right)^3 \left[\frac{N(N+1)(2N+1)}{6}\right] \qquad \text{(Using formula (10.6) for the sum of squares)} \\ &= \frac{64}{6}\,\frac{(2N^3+3N^2+N)}{N^3} \\ &= \frac{32}{3}\left[2 + \frac{3}{N} + \frac{1}{N^2}\right]. \end{aligned}$$

If we let N approach $+\infty$, we obtain

$$\int_0^4 t^2\,dt = \lim_{N\to\infty}\left\{\frac{32}{3}\left[2+\frac{3}{N}+\frac{1}{N^2}\right]\right\} = \frac{64}{3},$$

because

$$\lim_{N\to\infty}\frac{3}{N} = 0 \qquad \text{and} \qquad \lim_{N\to\infty}\frac{1}{N^2} = 0. \qquad \blacktriangleleft$$

22.3 NUMERICAL APPROXIMATION OF INTEGRALS

Scientists must often evaluate integrals in which the integrands are given in tabulated form from experimental results. In such cases, the elaborate theory developed from the Fundamental Theorem of Calculus is of no use. Instead, the scientist must turn to numerical approximations to evaluate specific integrals. In this subsection, we present three approximation formulas that are frequently used to perform integrations on computers. Each formula is associated with a uniform partition of size N. A computer can be programmed to use such formulas in two ways. One way is to simply compute the approximation for a given value of N. A second way is to compute the approximation successively for increasing values of N, say $N = 10, 20, 40, 80, \ldots$, and, after each approximation, compare the last two computed values. If these values are the same, they are printed as the value of the integral. If they are different, the process continues until it reaches a maximum permissible value of N, at which point the computer "bombs" and provides no answer.

In the following, we present basic numerical approximation formulas for evaluating

$$\int_a^b f(x)dx.$$

These formulas are based on the representation of the function $f(x)$ over each subinterval of a uniform partition by a polynomial. The basic assumption is that when $P(x)$ is a polynomial that approximates $f(x)$ over an interval $[\alpha, \beta]$, then for all $x \in [\alpha, \beta]$,

$$f(x) = P(x) + \text{small error}.$$

Ignoring the error term, we can obtain an approximation of the integral of $f(x)$ by the integral of the polynomial $P(x)$:

$$\int_\alpha^\beta f(x)dx \simeq \int_\alpha^\beta P(x)dx.$$

Three different approximation formulas are obtained by considering the polynomial P to be the degree zero (a constant); degree one (a linear function); and degree two (a parabola).

Degree 0—the Rectangle Formula

When $P(x) = C$, a constant, then

$$\int_{\alpha}^{\beta} p(x)dx = \int_{\alpha}^{\beta} C\,dx = C\cdot(\beta - \alpha).$$

This simply states that the area of a rectangle is its height times its width. (See Fig. 22.18.) By approximating $f(x)$ by a constant over each subinterval of a uniform partition of $[a, b]$, we obtain the approximation formula of the form (22.12),

$$\int_a^b f(x)dx \simeq h\sum_{n=1}^{N} f(x_n^*),$$

where

$$h = (b - a)/N \qquad \text{and} \qquad a + (n - 1)h \leq x_n^* \leq a + nh.$$

Taking x_n^* to be the right endpoint, $x_n = a + nh$, we obtain the approximation formula

$$\int_a^b f(x)dx \simeq h\sum_{n=1}^{N} f(a + nh),$$

or

$\int_a^b f(x)dx \simeq h[f(a + h) + f(a + 2h) + \cdots + f(b)]$	**Rectangle formula using right endpoint.**

(22.17)

This is the same as (22.16) with $g(t)$ replaced by $f(x)$.

Choosing x_n^* to be the left endpoint, $x_{n-1} = a + (n - 1)h$, we obtain the approximation formula

$$\int_a^b f(x)dx \simeq h\sum_{n=1}^{N} f(a + (n - 1)h),$$

or

$\int_a^b f(x)dx \simeq h[f(a) + f(a + h) + \cdots + f(a + (N - 1)h)]$	**Rectangle formula using left endpoints.**

(22.18)

This is the same as formula (22.14) with $g(t)$ replaced by $f(x)$.

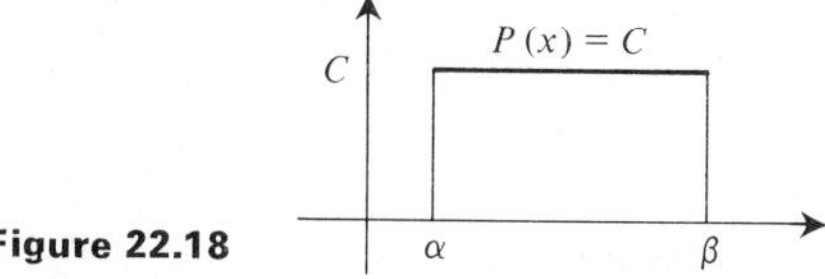

Figure 22.18

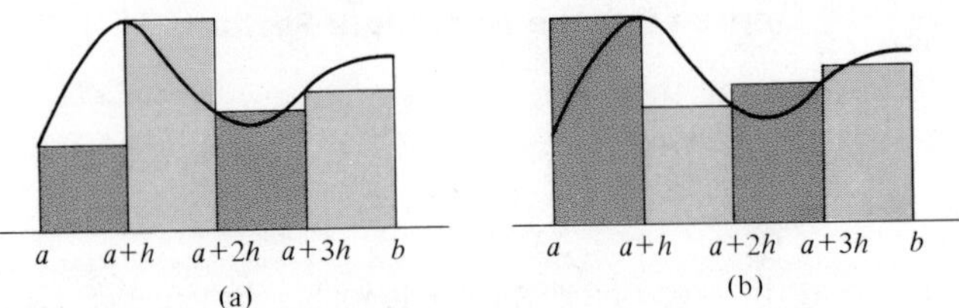

Figure 22.19

The graphical illustrations of formulas (22.17) and (22.18) are indicated in Fig. 22.19, with $N = 4$.

Formulas (22.17) and (22.18) are referred to as the *Rectangle Formula* for approximating integrals. Examples using these formulas were presented in the previous subsection. While these formulas may be found in some "package" computer programs, they are generally not used. This is because there are formulas, which result in better approximations with less error and require the same amount of computation. One such method is the *Trapezoid Formula*, which follows.

Degree 1—the Trapezoid Formula

When $P(x)$ is linear over the interval $[\alpha, \beta]$, the value of

$$\int_\alpha^\beta P(x)dx$$

can be obtained using the formula for the area of a trapezoid. In geometry, it is shown that the area of the trapezoid illustrated in Fig. 22.20 is one-half the product of the length of the base and the sum of the lengths of the two sides. This yields the formula

$$\int_\alpha^\beta P(x)dx = \tfrac{1}{2}(\beta - \alpha)[P(\alpha) + P(\beta)] \qquad \textbf{Area of a trapezoid.} \qquad (22.19)$$

Assume that the function f is approximated over each subinterval, I_n, of the partition by the linear function having the same value as $f(x)$ at the endpoints of the

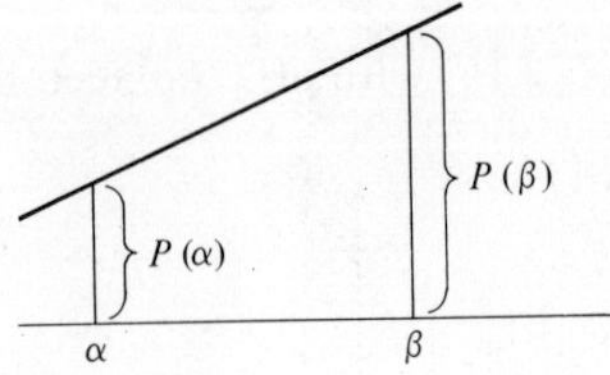

Figure 22.20

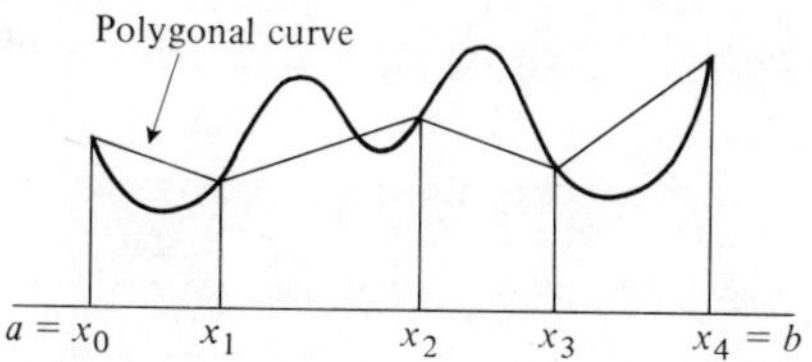

Figure 22.21

interval. Fig. 22.21 illustrates the linear approximation of a function f. The corresponding approximation of the integral of f over $[a, b]$ is the area under the polygonal curve. The approximation of the integral of f is obtained by setting

$$\int_a^b f(x)dx = \sum_{n=1}^{N} \int_{x_{n-1}}^{x_n} f(x)dx.$$

Then each integral

$$\int_{x_{n-1}}^{x_n} f(x)dx$$

is approximated using the linear polynomial $P(x)$. which agrees with f at the endpoints:

$$P(x_{n-1}) = f(x_{n-1}) \qquad \text{and} \qquad P(x_n) = f(x_n).$$

According to formula (22.19) this gives us

$$\int_{x_{n-1}}^{x_n} f(x)dx \simeq \int_{x_{n-1}}^{x_n} P(x)dx = (\tfrac{1}{2})\Delta x_n[P(x_{n-1}) + P(x_n)]$$

or

$$\int_{x_{n-1}}^{x_n} f(x)dx \simeq \frac{h}{2}[f(x_{n-1}) + f(x_n)],$$

assuming a uniform partition of norm $h = (b - a)/N$. Summing these approximations for $n = 1, 2, \ldots, N$ gives us

$$\int_a^b f(x)dx \simeq \sum_{n=1}^{N} \left\{\left(\frac{h}{2}\right)[f(x_{n-1}) + f(x_n)]\right\}.$$

Factoring out the $h/2$ term and expanding, we find that this series has the form

$$\left(\frac{h}{2}\right)\{\underbrace{[f(x_0) + f(x_1)]}_{n=1} + \underbrace{[f(x_1) + f(x_2)]}_{n=2} + \cdots + \underbrace{[f(x_{N-1}) + f(x_N)]}_{n=N}\}.$$

Gathering terms, we see that each term $f(x_n)$ appears twice, except for the first and last terms:

$$\left(\frac{h}{2}\right)\{f(x_0) + 2f(x_1) + 2f(x_2) + \cdots + 2f(x_{N-1}) + f(x_N)\}.$$

Dividing each term by two and writing $x_n = a + nh$, the above formula gives the following:

$$\int_a^b f(x)dx \simeq h\left[\frac{f(a)}{2} + f(a+h) + f(a+2h) + \cdots + f(a+(N-1)h) + \frac{f(b)}{2}\right].$$

Trapezoid Formula for integrals.

(22.20)

Observe that the Trapezoid Formula requires exactly the same information used in the Rectangle formulas (22.17) and (22.18). It does require one more addition.

Example 22.9 We illustrate the use of the Trapezoid Formula to approximate

$$\int_0^4 x(4-x)dx.$$

Using a partition of size $N = 4$, we find that the subinterval length is $h = (4-0)/4 = 1$. The formula (22.20) with $f(x) = x(4-x)$, $a = 0$, and $b = 4$ gives the approximation

$$\int_0^4 x(4-x)dx \simeq (1)\left[\frac{f(0)}{2} + f(1) + f(2) + f(3) + \frac{f(4)}{2}\right]$$

$$= \left[\frac{0}{2} + 3 + 4 + 3 + \frac{0}{2}\right] = 10.$$

(See Fig. 22.22).

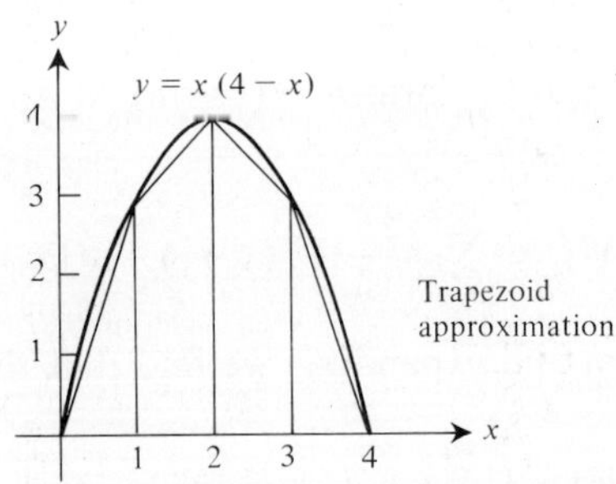

Figure 22.22

◀

Degree 2—Simpson's Formula

To approximate the function f by a polynomial of degree two is to represent it by a quadratic polynomial of the form

$$P(x) = Ax^2 + Bx + C.$$

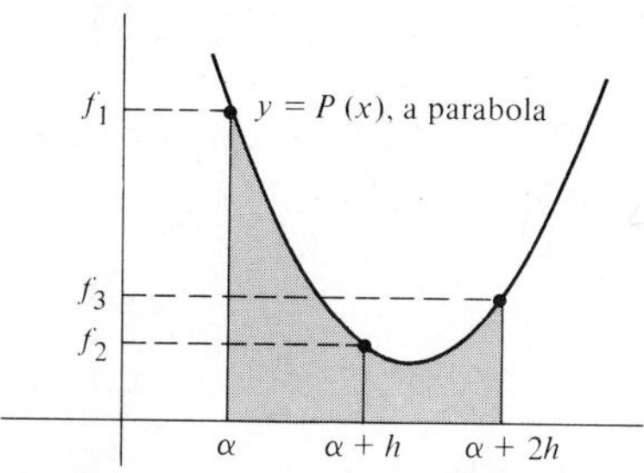

Figure 22.23

The graph of a parabola is determined by three points. To continue the practice of using the partition points as evaluation points, we choose to approximate the function f by a quadratic P over two subintervals instead of one subinterval. This provides the three points needed to determine the coefficients of $P(x)$, but limits us to considering only an even number of subintervals. Not only does this allow us to evaluate the approximating parabolas but, by making the three points equally spaced, the value of the integral of $P(x)$ can be simply represented. Assume that $P(x)$ is a quadratic with $P(\alpha) = f_1$, $P(\alpha + h) = f_2$, and $P(\alpha + 2h) = f_3$ as indicated in Fig. 22.23. It can then be shown that the area under the graph of the parabola is

$$\int_{\alpha}^{\alpha+2h} P(x)dx = \frac{h}{3}[f_1 + 4f_2 + f_3]. \qquad (22.21)$$

By choosing $P(x)$ over successive pairs of subintervals so that the values of $P(x)$ agree with the values of $f(x)$ at the partition points, it is possible to approximate

$$\int_a^b f(x)dx,$$

using the above formula. Using the sketch in Fig. 22.24 as an example, we would approximate the function $f(x)$ by four parabolas:

$$\begin{aligned}
f(x) &\simeq P_1(x) \quad &\text{over} \quad &[a, a + 2h];\\
f(x) &\simeq P_2(x) \quad &\text{over} \quad &[a + 2h, a + 4h];\\
f(x) &\simeq P_3(x) \quad &\text{over} \quad &[a + 4h, a + 6h];\\
f(x) &\simeq P_4(x) \quad &\text{over} \quad &[a + 6h, b].
\end{aligned}$$

Simpson's approximation sets

$$\int_a^b f(x)dx \simeq \int_a^{a+2h} P_1(x)dx + \int_{a+2h}^{a+4h} P_2(x)dx + \int_{a+4h}^{a+6h} P_3(x)dx + \int_{a+6h}^{b} P_4(x)dx.$$

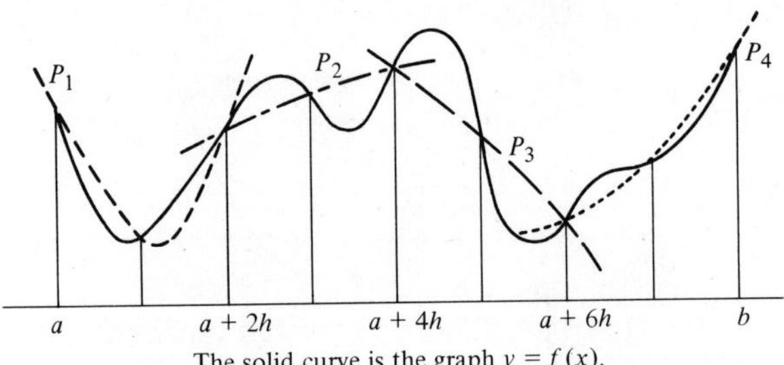

The solid curve is the graph $y = f(x)$.

Fig. 22.24 Uniform partition with $N = 8$ and $h = (b - a)/8$.

Using equation (22.21) to evaluate each of these integrals with $P(a + nh) \equiv f(a + nh)$, we obtain the approximation known as Simpson's Formula:

$$\begin{aligned}\int_a^b f(x)dx \simeq \frac{h}{3}&[f(a) + 4f(a + h) + f(a + 2h)] \\ &+ \frac{h}{3}[f(a + 2h) + 4f(a + 3h) + f(a + 4h)] \\ &+ \frac{h}{3}[f(a + 4h) + 4f(a + 5h) + f(a + 6h)] \\ &+ \frac{h}{3}[f(a + 6h) + 4f(a + 7h) + f(b)].\end{aligned}$$

Factoring out $h/3$ and adding terms such as $f(a + 2h)$, this becomes

$$\begin{aligned}\int_a^b f(x)dx \simeq \frac{h}{3}\{&f(a) + 4f(a + h) + 2f(a + 2h) + 4f(a + 3h) + 2f(a + 4h) \\ &+ 4f(a + 5h) + 2f(a + 6h) + 4f(a + 7h) + f(b)\}.\end{aligned}$$

This sum adds the value of f at the first and last points, $f(a)$ and $f(b)$. It then adds four times f evaluated at each odd partition point, $4f(a + nh)$, n odd. It then adds two times f evaluated at each even partition point, $2f(a + nh)$, n even. Expressed in compact form, this formula for arbitrary N would read as follows (remember N is assumed to be an even number):

$$\int_a^b f(x)dx \simeq \frac{h}{3}\left\{f(a) + f(b) + 4\sum_{\substack{n=1 \\ (n\text{ odd})}}^{N-1} f(a + nh) + \sum_{\substack{n=1 \\ (n\text{ even})}}^{N-2} f(a + nh)\right\}. \tag{22.22}$$

Simpson's Formula is easy to remember if we set

$$x_n = a + nh \qquad n = 0, 1, 2, \ldots, N, \qquad N \text{ even}, \qquad h = \frac{b - a}{N}.$$

It is then given by the following:

$$\int_a^b f(x)dx \simeq \frac{h}{3}\{f(x_0) + 4f(x_1) + 2f(x_2) + 4f(x_3) + 2f(x_4) + \cdots + 2f(x_{N-2}) + 4f(x_{N-1}) + f(x_N)\}. \qquad (22.23)$$

Simpson's Formula for approximating integrals.

Example 22.10 We use Simpson's Formula with $N = 6$ to approximate

$$\int_0^4 x^3\,dx;$$

$$f(x) = x^3, \qquad h = (4 - 0)/6 = \tfrac{2}{3}.$$

Formula (22.23) gives the following approximation:

$$\frac{(\frac{2}{3})}{3}\{f(0) + 4f(\tfrac{2}{3}) + 2f(\tfrac{4}{3}) + 4f(\tfrac{6}{3}) + 2f(\tfrac{8}{3}) + 4f(\tfrac{10}{3}) + f(4)\}$$

$$= (\tfrac{2}{9})\{0^3 + 4(\tfrac{2}{3})^3 + 2(\tfrac{4}{3})^3 + 4(\tfrac{6}{3})^3 + 2(\tfrac{8}{3})^3 + 4(\tfrac{10}{3})^3 + 4^3\}$$

$$= (\tfrac{2}{9})\{0 + \tfrac{32}{27} + \tfrac{128}{27} + \tfrac{864}{27} + \tfrac{1024}{27} + \tfrac{4000}{27} + 64\}$$

$$= (\tfrac{2}{9})\{\tfrac{6048}{27} + 64\} = (\tfrac{2}{9})\{288\} \simeq 64.$$

The exact value of this integral may be obtained via the Fundamental Theorem of Calculus:

$$\int_0^4 x^3\,dx = \frac{x^4}{4}\bigg|_0^4 = \frac{4^4}{4} = 64.$$

This illustrates the fact that Simpson's Formula is exact for polynomials of degree three, even though it is obtained using second-order polynomials. ◀

The error in each of the above approximation formulas depends on the length of the step size, $h = (b - a)/N$, and the function f. The following can be shown:

The error of the Rectangular Formula $\propto h$

The error of the Trapezoid Formula $\propto h^2$

The error of Simpson's Formula $\propto h^4$

Consequently, in return for the increased complexity and the amount of work with the Trapezoid and Simpson's formulas, we obtain increased accuracy in the approximations.

Example 22.11 We compute the approximations to

$$\int_2^6 x^2\,dx,$$

using formulas (22.18), (22.19), (22.20), and (22.22). Here $f(x) = x^2$ and $[a, b] = [2, 6]$. We choose $N = 8$ and therefore $h = (6 - 2)/8 = 0.5$. The values of $x_n = a + nh$ are indicated in Table 22.1 were we have set up the calculations necessary for each formula. We have indicated the coefficient of each term $f(x_n)$ in formulas (22.20) and (22.22) under the columns indicated Wt (for weight $\equiv$ coefficient). Using the actual value of the integral, which is $69\frac{1}{3}$, we also indicate the absolute value of the error. Simpson's Formula is exact in this example since the function $f(x)$ is exactly equal to its second-degree approximation.

Table 22.1

n	x_n	$f(x_n) = x_n^2$		Wt	Trapezoid	Wt	Simpson's
0	2.0	4.0		0.5	2.0	1	4
1	2.5	6.25	6.25	1	6.25	4	25
2	3.0	9.0	9.0	1	9.0	2	18
3	3.5	12.25	12.25	1	12.25	4	49
4	4.0	16.0	16.0	1	16.0	2	32
5	4.5	20.25	20.25	1	20.25	4	81
6	5.0	25.0	25.0	1	25.0	2	50
7	5.5	30.25	30.25	1	30.25	4	121
8	6.0		36.0	0.5	18.0	1	36
Sums		123	155		139		416

Formula	Approximation	$\lvert$Error$\rvert$
Rectangle (22.18)	$(0.5)\cdot(123) = 61.5$	$\simeq 7.8$
Rectangle (22.19)	$(0.5)\cdot(155) = 77.5$	$\simeq 8.2$
Trapezoid (22.20)	$(0.5)\cdot(139) = 69.5$	$\simeq 0.2$
Simpson's (22.22)	$(\frac{0.5}{3})\cdot(416) = 69.33\ldots$	0.0

◀

SUMMARY

1. The area A of the region bounded above by the curve $y = f(x)$ is the sum of the areas over subintervals of $[a, b]$:

$$A = A_1 + A_2 + \cdots + A_N.$$

(See Fig. 22.25.)

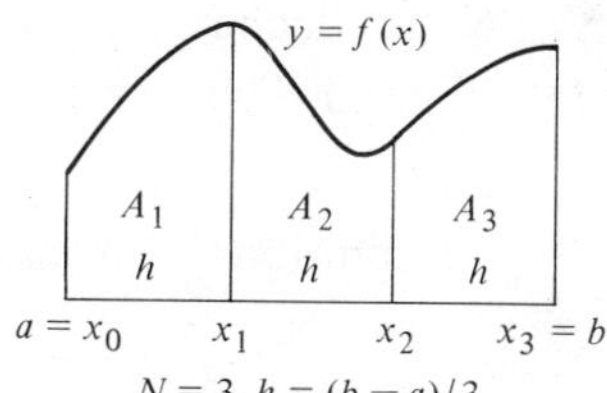

Figure 22.25 $N = 3, h = (b - a)/3$

Each area A_n, $n = 1, 2, \ldots, N$, *is approximated* by the area of a rectangle of height $f(x_n^*)$, where x_n^* is *any point* satisfying

$$x_{n-1} \leq x_n^* \leq x_n.$$

Thus

$$A \simeq \sum_{n=1}^{N} f(x_n^*)\Delta x_n.$$

(See Fig. 22.26.)

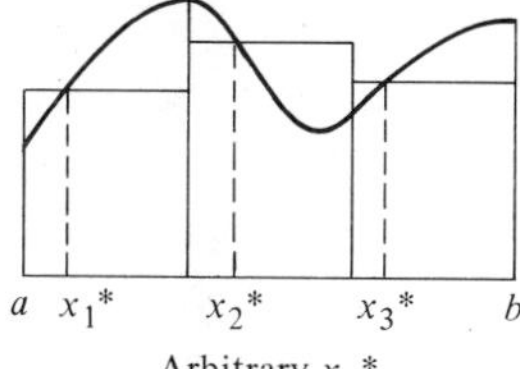

Figure 22.26 Arbitrary x_n^*

If x_n^* is the *left endpoint*, x_{n-1}, this gives

$$A \simeq h \sum_{n=1}^{N} f(a + (n - 1)h),$$

where $h = (b - a)/N$ is the length of the subinterval of a uniform partition of $[a, b]$. If x_n^* is the *right endpoint*, x_n, the formula becomes

$$A \simeq h \sum_{n=1}^{N} f(a + nh).$$

(See Figs. 22.27 and 22.28.)

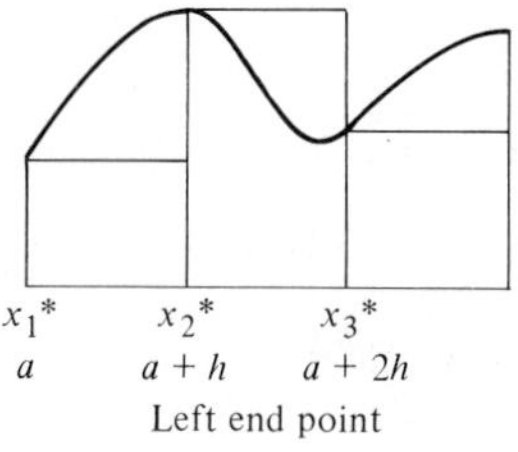

Left end point

Figure 22.27

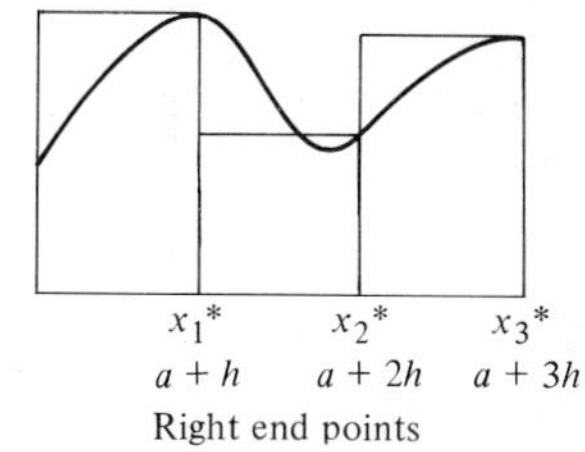

Right end points

Figure 22.28

In the limit as N approaches $+\infty$, the above approximations converge to the integral of f over $[a, b]$. Thus the area A is defined as the integral:

$$\underline{\text{Definition}} \qquad \text{Area } A \equiv \int_a^b f(x)dx.$$

The area between two curves is the integral of the difference:

$$\int_a^b \begin{pmatrix}\text{upper}\\ \text{function}\end{pmatrix} - \begin{pmatrix}\text{lower}\\ \text{function}\end{pmatrix} dx.$$

(See Fig. 22.29).

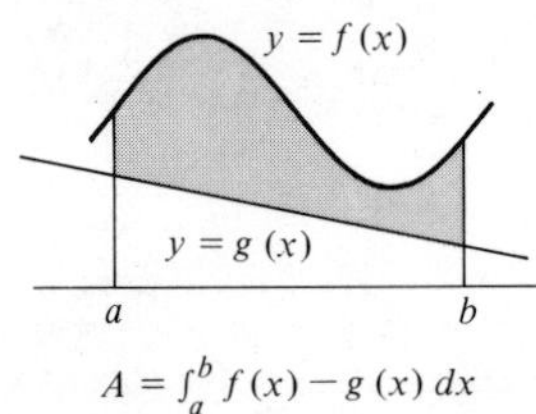

Figure 22.29

2. To obtain the value of a definite integral using the definition, assume that $f(x)$ is continuous on $[a, b]$. Then $\int_a^b f(x)dx$ exists and is given by

$$\int_a^b f(x)dx = \lim_{\substack{\Delta x \to 0 \\ N \to \infty}} \sum_{n=1}^{N} f(x_n^*)\Delta x_n$$

for any partition $\{x_n\}$ and sequence of evaluation points $\{x_n^*\}$. Therefore, choose a *uniform partition* with $h = (b - a)/N$. Let x_n^* be an endpoint of the interval $[x_{n-1}, x_n]$. Then represent the series in the definition as a function of N. Let $N \to \infty$ to obtain the value of the integral. Example: Let $[a, b] = [2, 3]$, $f(x) = 2x$, and $h = (3 - 2)/N = 1/N$. Using the right endpoint formula, we evaluate the series

$$\frac{1}{N}\sum_{n=1}^{N} f\left(2 + \frac{n}{N}\right) = \frac{1}{N}\sum_{n=1}^{N} 2\left(2 + \frac{n}{N}\right) = \frac{1}{N}\sum_{n=1}^{N}\left[4 + \left(\frac{2}{N}\right)n\right]$$

$$= \left(\frac{1}{N}\right)\left[4N + \left(\frac{2}{N}\right)\left(\frac{N(N+1)}{2}\right)\right] \qquad \text{(See (10.8))}$$

$$= 4 + \left(\frac{N+1}{N}\right) = 4 + 1 + \frac{1}{N}.$$

Thus

$$\int_2^3 2x\,dx = \lim_{N \to \infty}\left(5 + \frac{1}{N}\right) = 5.$$

The *Mean Value Theorem for Integrals* states that for some point ξ, between a and b,

$$\int_a^b f(x)dx = f(\xi)(b - a);$$

The *average of* $f(x)$ *over* $[a, b]$ is

$$\bar{Y} = \frac{\int_a^b f(x)dx}{b - a}.$$

3. *Numerical Integration*:
Rectangle Formula (left endpoint):

$$\int_a^b f(x)dx \simeq \left[\frac{b-a}{N}\right][f(a) + f(a + h) + f(a + 2h) + \cdots + f(a + (N - 1)h)].$$

Trapezoid Formula:

$$\int_a^b f(x)dx \simeq \left[\frac{b-a}{N}\right][f(a) + 2f(a + h) + 2f(a + 2h) + \cdots + 2f(a + (N - 1)h) + f(b)].$$

Simpson's Formula (N must be even):

$$\int_a^b f(x)dx \simeq \tfrac{1}{3}\left[\frac{b-a}{N}\right]\{f(a) + 4f(a + h) + 2f(a + 2h) + 4f(a + 3h) + \cdots + 4f(a + (N - 1)h) + f(b)\}.$$

Note that in Simpson's Formula the coefficients alternate between four and two, except for the first and last terms, which have the coefficient one.

EXERCISE SET 22

22.1 a) Sketch the graph of $f(x) = \frac{1}{2}x$ over $[0, 4]$.
b) Estimate the area of the region determined by f over $[0, 4]$ using a uniform partition of size $N = 4$ and the left endpoint of each interval for evaluation.
c) Sketch the rectangles used in your estimation.
d) Compute the error of your approximation using the formula for the area of a triangle.
e) Repeat (b), (c), and (d) using the right endpoints for $x_n{}^*$.

22.2 Determine the area of the indicated region R using the integral definition. Sketch the region.

a) R is the region under the curve $y = e^x - x$ over the interval $[0, 2]$.
b) R is a triangle with the interval $[3, 6]$ as a base and an apex at $(6, 3)$.
c) R is a triangle with the interval $[3, 6]$ as a base and an apex at $(3.5, 3)$.
d) R is the region inside the parabola $y^2 = x$ for $0 < x < 4$.
e) R is the diamond-shaped region with vertices at $(0, 0)$, $(1, 2)$, $(3, 0)$, and $(1, -2)$.
f) R is the region between the sine and cosine curves over the interval $[0, 2\pi]$.

22.3 Estimate the area of the regions indicated by the graphs in Fig. 22.30 by constructing approximating rectangles and measuring their heights.

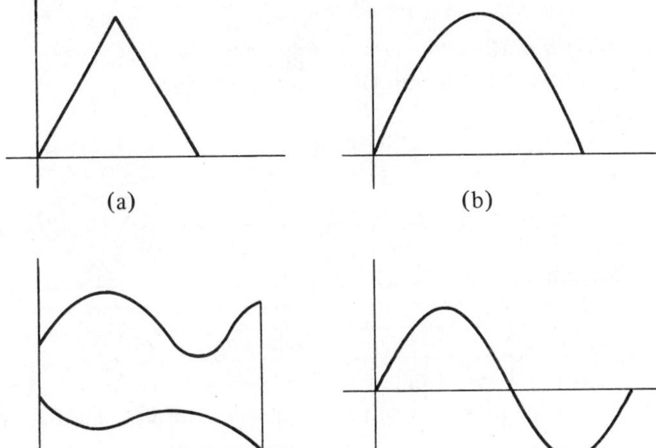

Figure 22.30

22.4 a) Sketch the graph of $y = 1/t$.
b) For $x > 1$, sketch the region whose area is $A(x) = \int_1^x 1/t\,dt$.
c) Using the area of rectangles, show that $A(2) = \int_1^2 1/t\,dt$ satisfies $\frac{1}{2} < A(2) < \frac{3}{4}$.
d) Using the area of rectangles, show that $A(3) = \int_1^3 1/t\,dt$ is less than 2.
e) Is $A(3) > 1$? How can you show this?

22.5 Let the function $F(x) = \int_1^x \frac{1}{2}[t + 1]dt$.

a) Draw a diagram to illustrate the value of $F(x)$ as an area.
b) For what value of x will this area equal three?

22.6 For each given pair of functions, (1) determine the points of intersection of their graphs and sketch their graphs; and (2) compute the finite area bounded by the graphs of the functions:

a) $f(x) = x^2 + 1; g(x) = 5$
b) $f(x) = x^2 + 1; g(x) = x + 3$
c) $f(x) = x^2 + 1; g(x) = 5 - x^2$
d) $f(x) = x; g(x) = \sin(\pi x/2)$
e) $f(x) = \cos(x); g(x) = \sin(x), |x| < \pi$
f) $f(x) = 2x^2; g(x) = 3x^2 - 1$
g) $f(x) = \sqrt{x}; g(x) = x^2$

22.7 Determine the area of the finite region R that is

a) bounded by $f(x) = 2x$, $g(x) = x/2$, and $h(x) = 4 - x$;
b) below the curves $f(x)$ and $h(x) = 1 - x$ and above the curve $g(x) = x^2$;
c) between the curves $f(x) = \cos(x)$ and $g(x) = x^2 - \pi^2/4$ and below $h(x) = 2x + 1$;
d) below the curve $f(x) = x(4 - x)$ and above the curves $g(x) = -x$ and $h(x) = 2(x - 4)$; and
e) below the curve $f(x) = x$, above the curve $g(x) = -x$, and to the left of $y^2 = -x + 4$.

22.8 Estimate the definite integral of $g(t) = 3t + 5$ over $[0, 2]$ using the following:

a) Equation (22.14) with $N = 4$,
b) Equation (22.16) with $N = 4$,
c) Equation (22.14) with $N = 6$,
d) Equation (22.16) with $N = 6$,
e) Equation (22.14) with arbitrary N.

22.9 Repeat Exercise 22.6 with $g(t) = t^2$.

22.10 a) Approximate the integral over $[0, 3]$ of the function $f(x) = [\![t]\!]$ using (22.14) and (22.16) with (a) $N = 6$; (b) $N = 9$; (c) N an arbitrary multiple of three.

22.11 Estimate $\int_0^\pi \sin(t)dt$, using a uniform partition of size $N = 6$ and formula (22.14).

22.12 If a flock of geese can fly at speed $s(t) = A \sin^2(\pi t)$ (t = days), give a formula expressing the distance they can fly in 10 days. Approximate this distance using a uniform partition of size $N = 20$ and formula (22.14).

22.13 If the size of a population is given $N(t) = \log(t + 1)$ for $t \geq 0$, estimate the average size of the population over the interval $[0, 10]$ using subintervals of length one and formula (22.16).

22.14 The density of a tree x feet above the ground is $f(x) = \frac{1}{50}(50 - x)$.

a) At what height does the tree have zero density? How would you interpret this?
b) Compute an estimate for the total mass of the tree using 10 subintervals and formula (22.14).

22.15 If the probability density function for a random variable x is $P(x) = 1/x^2$ for $1 \leq x < \infty$, estimate the probability that x is between one and three using a uniform partition of size $N = 4$ and formula (22.14). (See Table 21.1.)

22.16 Estimate the average value of the function $f(x) = 3^x$ over the interval $[0, 2]$ with a uniform partition of size $N = 4$: (a) using (22.14); and (b) using (22.16).

22.17 Rework Example 22.7 using the interval $[2, 4]$.

22.18 Water flows into a tank at the rate of $5t + 3$ litres/min. It flows out at a rate of t^2 litres/min. If the tank is empty to begin with, estimate its volume after 4 minutes with uniform partitions of size $N = 4$ and $N = 8$ using formula (22.16).

22.19 Sketch a graph to illustrate the properties of the integral given in Theorem 21.3 using areas.

22.20 a) Carefully sketch the graph of $y = x^2$ over the interval $[0, 4]$ on $\frac{1}{4}$ inch or $\frac{1}{2}$ cm graph paper. Then, by physically counting the squares under the curve and one-half of the squares that the curve passes through, estimate

$$\int_0^4 x^2\, dx.$$

b) Repeat part (a) for $y = 2 \sin(\pi x/4)$.

22.21 Assume that a flat fish is symmetric about its longitudinal axis. Represent its surface area as an integral. (Define the variables and functions involved.)

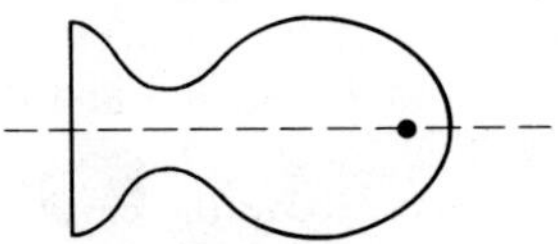

Figure 22.31

22.22 Approximate

$$\int_0^1 5e^{-x^2}\,dx,$$

a uniform partition of size $N = 6$: (a) using rectangles; (b) with the Trapezoid Formula; and (c) with Simpson's Formula.

22.23 Repeat Exercise 22.22 with the integrand $f(x) = \sin(\pi x)$.

22.24 Repeat Exercise 22.22 for the integral

$$\int_1^7 \log(x)dx.$$

22.25 If a function $f(x)$ is experimentally found to have the values indicated below, estimate

$$\int_0^4 f(x)dx.$$

x	0	0.5	1.0	1.5	2.0	2.5	3.0	3.5	4.0
$f(x)$	1.0	1.0	2	1.5	3.0	1.2	1.1	0.8	0.1

Use (a) the Rectangle Formula, (b) the Trapezoid Formula; and (c) Simpson's Formula.

22.26 Sometimes experimental values cannot be obtained at evenly spaced intervals. In this case, the formulas of Section 22.3 are not applicable.

a) Develop a formula corresponding to (22.18) when the partition is not uniform.
b) Develop a formula corresponding to (22.20) when the partition is not uniform.
c) Use the formulas developed in parts (a) and (b) to approximate

$$\int_0^2 f(x)dx,$$

when f is obtained from the following table.

x	0	0.2	0.4	0.5	0.9	1.3	1.8	2.0
$f(x)$	0	1.5	1.2	2.0	1.6	3.4	0.3	0.5

22.27 The use of Taylor's polynomial approximations provides another method of approximating integrals. This method consists in approximating the integral by

$$\int_a^b f(t)dt \simeq \int_a^b P_n(t)dt,$$

where P_n is an nth order Taylor polynomial of f.

a) Approximate

$$\int_0^1 e^x\,dx,$$

using the third-order Taylor polynomial,

$$P_3(x) = 1 + x + \frac{x^2}{2} + \frac{x^3}{3!}.$$

b) Approximate

$$\int_1^5 \frac{1}{t}\,dt,$$

using a third-order approximation of $f(t) = t^{-1}$ about $t_0 = 1$.

c) Approximate

$$\int_0^1 \frac{1}{1+t^3}\,dt$$

using a third-order approximation of $1/(1 + t^3)$. (*Hint*: $1/(1 + t^3) = 1/1 + x$, where $x = t^3$; hence you can use the expansion of $1/x$ about $x_0 = 1$ to obtain the expansion of $1/(1 + t^3)$ about $t_0 = 0$.)

22.28 Determine approximate fourth-order formulas for the indicated integrals, using Taylor approximations of the integrand $f(t)$ about the lower limit of integration.

a) $\int_0^x \sin(t)dt$ b) $\int_0^x \sin(t^2)dt$ c) $\int_0^x \cos(\pi t)dt$ d) $\int_0^x \cos(t^3)dt$

e) $\int_1^x \frac{1}{t}\,dt$ f) $\int_1^x \ln(t)dt$ g) $\int_0^x e^{2t}\,dt$ h) $\int_0^x e^{t^2}\,dt$

CHAPTER SIX

APPLICATIONS USING THE INTEGRAL

The exponential and logarithmic functions are very important in biological applications. In Section 23, the calculus of these functions is considered from a different point of view and in greater depth. This section also contains a host of additional applications of these functions. In Section 24, we present additional techniques of integration. While *integration by parts* is relatively important and is used in subsequent sections, the other special integration techniques are optional and could easily be omitted in a course for life-science students. In Section 25, we examine indeterminant limits and L'Hospital's Rule. This rule is frequently necessary in the following material on singular or infinite integrals. This topic may seem too theoretical, but it is in fact very important, since most biological models have an infinite time interval $[0, +\infty)$. Indeed, the material on infinite integrals is essential in the study of probability functions such as the normal distribution or the t-distribution.

SECTION 23

ANOTHER LOOK AT THE CALCULUS OF LOGARITHMIC AND EXPONENTIAL FUNCTIONS

INTRODUCTION

In this section, we present a more complete development of the calculus of exponential and logarithmic functions. We begin by defining the function $\ln(x)$ as a particular integral. The definition requires only the concept of the area under a curve as given by a definite integral. This definition leads to a simple definition of the number e and in turn to the definition of the exponential function e^x. Using only basic derivative and integral properties, we then develop the rules for differentiating and integrating more general logarithm and exponential functions.

We have chosen to include this "second look" at the calculus of e^x and $\ln(x)$ for two reasons. First, these functions are the most extensively used functions in biological and life sciences. Therefore they deserve a closer and more thorough examination than we have presented up to this time. In this section, we not only review the calculus of these functions, but we also expand and extend these concepts to general exponential and logarithmic functions. It is hoped that the alternative approach presented here will provide new insight into these topics, especially for the reader who is no longer struggling to grasp the more fundamental calculus concepts. The second and less important reason for presenting the following approach is to provide a mathematically sound development of the exponential and logarithm functions. This approach does not depend on the limit,

$$\lim_{\varepsilon \to 0} (1 + \varepsilon)^{1/\varepsilon} = e,$$

used in Section 14.

23.1 THE NATURAL LOGARITHM FUNCTION

The function $\ln(x)$ was defined in Section 5 as the inverse of the function $y = e^x$. However, as was noted in Section 5, the definition of the exponential function $y = a^x$ for a nonrational number a could not be provided without using calculus concepts. In this subsection, we introduce a way of defining the function $\ln(x)$ explicitly using integrals. We obtain the base of the natural logarithm, the number e, as the unique real number that satisfies an easily conceived geometric property.

We begin by examining the curve $y = 1/t$. Since the function $f(t) = 1/t$ is continuous for positive t, the definite integral

$$\int_1^x \frac{1}{t}\, dt \qquad (23.1)$$

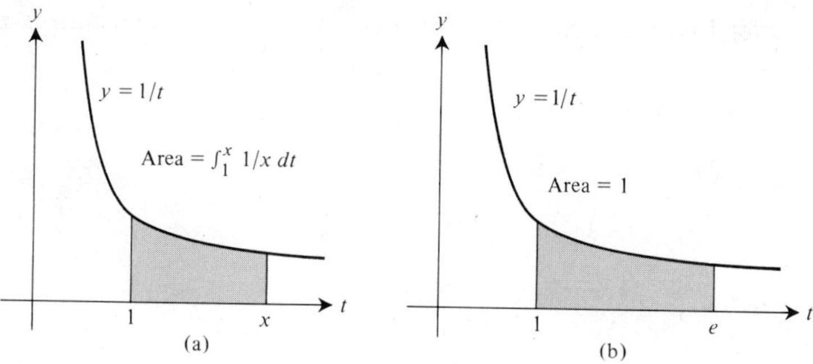

Figure 23.1

exists for any positive value of x. This integral gives the area under the graph of the curve, $y = 1/t$, over the interval $[1, x]$ if $x > 1$. (See Fig. 23.1.) Although we have integrated this integral in the past, the value was obtained using the Fundamental Theorem of Calculus and the derivative

$$\frac{d}{dx}\ln(x) = \frac{1}{x}. \tag{23.2}$$

At this time, we elect to not use this knowledge about the derivative of $\ln(x)$; instead, we will proceed to analyze the integral (23.1) directly and in so doing define the $\ln(x)$ function and obtain the derivative (23.2) directly.

We begin by defining the function $L(x)$ by

$$L(x) = \int_1^x \frac{1}{t}\,dt, \tag{23.3}$$

which is a differentiable function of x for $0 < x < +\infty$. The derivative of $L(x)$ is, by Theorem 21.3,

$$\frac{d}{dx}L(x) = \frac{1}{x}. \tag{23.4}$$

It can be shown that the function $L(x)$ satisfies all of the properties associated with a logarithm function. These are listed in Section 5. The following illustrate the verifications of these properties for the function $L(x)$.

Property L-P2 states that $\log_b(1) = 0$. To show that $L(1) = 0$, we set $x = 1$ in equation (23.3) and use equation (21.10):

$$L(1) = \int_1^1 \frac{1}{t}\,dt = 0.$$

Property L-P4 states that $\log_b(x^r) = r\log_b(x)$. To verify this property for $L(x)$, we use a technique that is often used to show that two functions are equal. We first show that they are both antiderivatives of the same function. This implies that they differ by only a constant; we then show that the constant is zero. By formula (23.4),

$$\frac{d}{dx}\, rL(x) = rL'(x) = \frac{r}{x}.$$

The derivative of

$$L(x^r) = \int_1^{x^r} \frac{1}{t}\, dt$$

is obtained by using the formula (21.18).

$$\frac{d}{dx}\, L(x^r) = \frac{1}{x^r} \cdot \frac{dx^r}{dx} = \frac{1}{x^r} \cdot rx^{r-1} = \frac{r}{x}.$$

Consequently,

$$\frac{d}{dx}\, rL(x) = \frac{d}{dx}\, L(x^r);$$

and, by Theorem 16.1,

$$rL(x) = L(x^r) + c,$$

for some constant c. Setting $x = 1$ and using the fact that $L(1) = 0$, we find $c = 0$. Thus

$$L(x^r) = rL(x).$$

Since as x increases the area under the graph of $y = 1/t$ over the interval $[1, x]$ also increases, the function $L(x)$ is monotone increasing. It therefore satisfies Property L-P5. While the domain of the function $L(x)$ is clearly $(0, +\infty)$, to show that the range of $L(x)$ is $(-\infty, +\infty)$, we are required to use more advanced techniques. Similarly, to show that

$$L(a \cdot b) = L(a) + L(b),$$

requires the use of some "tricks," which we shall not explore here.

These results support the contention that $L(x)$ is in fact a logarithm function, $\log_b(x)$, for some base $b > 1$. To decide what the number b should be, we let

$$L(x) = \log_b(x) \qquad \text{and} \qquad y = L(x).$$

(See Equation (5.9).) Then we have the relation

$$y = \log_b(x) \qquad \text{or equivalently} \qquad x = b^y.$$

Using the exponential property, we obtain

$$y = L(x) = L(b^y) = yL(b).$$

Solving for $L(b)$, we find that

$$L(b) = 1.$$

That is, if $L(x)$ is in fact the logarithm function, $y = \log_b(x)$, then the base b is the unique number such that $L(b) = 1$. Mathematicians choose to represent this unique number by the letter e.

Definition 23.1 The real number e is the unique number x greater than 1 such that the area (illustrated in Fig. 23.1(b)) bounded by the graph of $y = 1/t$, the t-axis, and the vertical lines $t = 1$ and $t = x$ is exactly 1. Thus e is defined by the equation

$$1 = \int_1^e \frac{1}{t}\, dt.$$

In light of the fact that the function $L(x)$ satisfies the properties associated with the logarithm function $\ln(x)$, we define the natural logarithm function to be $L(x)$.

Definition 23.2 The natural logarithm function is defined for all $x \in (0, \infty)$ by

$$\ln(x) \equiv \int_1^x \frac{1}{t}\, dt \qquad \textbf{Definition of the natural logarithm.}$$

The derivative of the function $\ln(x)$ is determined using Theorem 21.3:

$$\frac{d}{dx}\ln(x) = \frac{1}{x}. \tag{23.5}$$

Applying Theorem 21.4 or using the Chain Rule and equation (23.5), we can compute the derivative of $\ln(g(x))$ for any differentiable, *positive function* $g(x)$:

$$\frac{d}{dx}\ln(g(x)) = \frac{g'(x)}{g(x)}. \tag{23.6}$$

Example 23.1 The derivative formula (23.6) and the definition of $\ln(x)$ are illustrated below.

a) $\displaystyle\int_1^5 \frac{1}{t}\, dt = \ln(5).$

b) $\displaystyle\int_1^{\sqrt{x^2+1}} \frac{1}{t}\, dt = \ln(\sqrt{x^2+1}) = \tfrac{1}{2}\ln(x^2+1).$

c) $\displaystyle\int_2^x \frac{1}{t}\, dt = \int_1^x \frac{1}{t}\, dt - \int_1^2 \frac{1}{t}\, dt = \ln(x) - \ln(2);$

or, since $D_x(\ln(x)) = 1/x$, directly by the *FTC*,

$$\int_2^x \frac{1}{t}\, dt = \ln(t)\Big|_2^x = \ln(x) - \ln(2).$$

d) If $f(x) = \ln(x^2+3)$, then $f'(x) = \dfrac{2x}{x^2+3}$.

e) If $f(x) = \ln(5x + x^3)$, then $f'(x) = \dfrac{5 + 3x^2}{5x + x^3}$.

f) If $f(x) = \ln(2x^{-1})$, then $f'(x) = \dfrac{-2x^{-2}}{2x^{-1}} = \dfrac{-1}{x}$. Alternatively,

$$\frac{d}{dx} f(x) = \frac{d}{dx}(\ln(2) + \ln(x^{-1})) = \frac{d}{dx}\ln(2) - \frac{d}{dx}\ln(x) = 0 - \frac{1}{x} = \frac{-1}{x}.$$

◀

23.2 THE EXPONENTIAL e^x

In this subsection, we reverse the process used in Section 5. There we introduced the exponential function $y = e^x$ (without a complete definition) and then defined the function $\ln(x)$ as the inverse of the exponential function. In Section 23.1, we presented a complete definition of the natural logarithm function $\ln(x)$, without using an exponential function. Consequently, $\ln(x)$ is a well-defined function, which we may use to define the function $y = e^x$. We define e^x as the inverse function, with respect to composition, of the function $y = \ln(x)$.

Definition 23.3 The *exponential function, with base e,* is the inverse of the natural logarithm function, with respect to composition. The exponential function with base e is denoted e^x or $\exp_e(x)$ and, by definition,

$$y = e^x \qquad \text{if and only if} \qquad \ln(y) = x. \tag{23.7}$$

Geometrically, Definition 23.3 may be interpreted as saying that $y = e^x$ is the number such that the area under the graph of $f(t) = 1/t$ between $t = 1$ and $t = y$ is equal to x. Thus, to compute e^3, we need only determine the unique number y such that

$$\int_0^y \frac{1}{t}\, dt = 3;$$

this is depicted in Fig. 23.2.

For negative exponents, the geometric interpretation of $y = e^{-x}$ is that y is the number between 0 and 1 such that the area between y and 1 below the graph of $f(t) = 1/t$ is equal to x. (See Fig. 23.3.)

The exponential function $y = e^x$ satisfies each of the properties attributed to exponential functions in Section 5. These properties follow from the equivalence

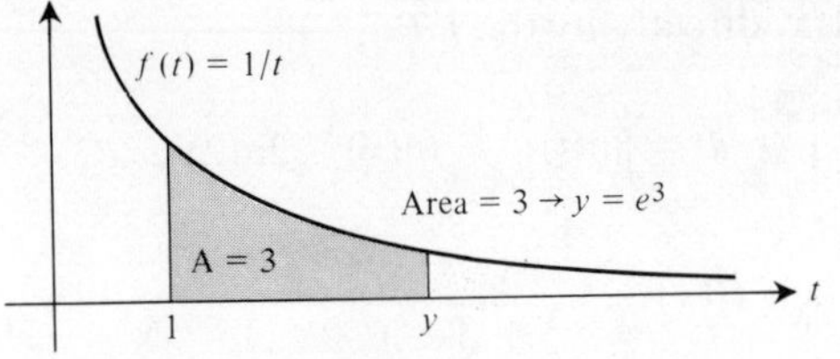

Figure 23.2

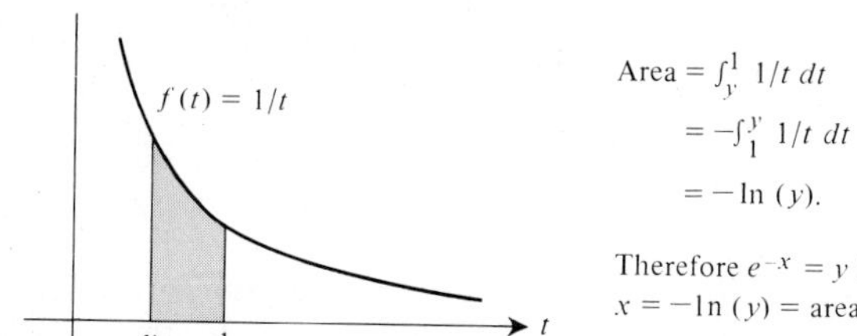

Figure 23.3

(23.7) once the corresponding properties have been established for the logarithm function $y = \ln(x)$. We restate these properties here for convenience.

E-P1 $\quad e^x > 0$ for all $x \in R$.

E-P2 $\quad e^0 = 1$.

E-P3 $\quad e^{x_1 + x_2} = e^{x_1} \cdot e^{x_2}$.

E-P4 $\quad e^{kx} = [e^x]^k$.

E-P5 $\quad e^x$ is monotone increasing on $(-\infty, +\infty)$.

The derivative of the exponential function e^x is derived using the equivalence (23.7) and the derivative of $\ln(x)$ given by equation (23.5).

$$\text{If } y = e^x \qquad \text{then} \qquad x = \ln(y).$$

Taking the derivative with respect to x, we obtain the equations

$$\frac{dy}{dx} = \frac{de^x}{dx} \qquad \text{and} \qquad \frac{dx}{dx} = \frac{d}{dx}\ln(y) = \frac{(dy/dx)}{y}.$$

The last equality follows from equation (23.6) with $g(x) = y$. Hence, as $dx/dx = 1$, it follows that

$$1 = \frac{(dy/dx)}{y}.$$

Solving for (dy/dx) and setting $y = e^x$ in the last equation, we obtain the equation $dy/dx = e^x$ or

$$\boxed{\frac{de^x}{dx} = e^x.} \tag{23.8}$$

From equation (23.8), we see that the function $y = e^x$ satisfies the first-order differential equation

$$y' = y \qquad \text{or} \qquad D_x y(x) = y(x).$$

This property of the exponential function is very important and is the primary reason why the exponential function appears in virtually every equation describing natural biological growth.

Example 23.2 *Proportional Growth*

In the discrete model of proportional growth, we assumed that the change in a population at generation n was proportional to its size X_n: $\Delta X_n = kX_n$. When we introduced the time variable into the discrete model by considering the generation length to depend on time, the model for the change in the population became

$$\frac{\Delta X(t)}{\Delta t} = kX(t),$$

where $\Delta t = (t + h) - h$, and h denoted the generation length. If we consider the effect of letting the generation length approach zero, we find that the difference equation results in a differential equation:

$$\lim_{h \to 0} \frac{\Delta X(t)}{\Delta t} = \lim_{h \to 0} \frac{X(t + h) - X(t)}{h} = \lim_{h \to 0} k \cdot X(t).$$

In the limit, this leads to the equation

$$\mathring{X}(t) = kX(t).$$

Therefore, the instantaneous rate of change at time t is proportional to the population size at time t. This is the most fundamental differential equation in the biological and physical sciences. Its solution is given by

$$X(t) = e^{kt}.$$

To verify that this is a solution, we differentiate $X(t) = e^{kt}$ using the Chain Rule and equation (23.8). Let $u = k \cdot t$, then

$$\mathring{X}(t) = \frac{de^{kt}}{dt} = \left(\frac{de^u}{du}\right)\left(\frac{du}{dt}\right) = (e^u)(k) = k \cdot e^{kt} = kX(t).$$

If $k > 0$, then $\mathring{X}(t) > 0$ and the population increases. If $k < 0$, then $\mathring{X}(t) < 0$ and the population decreases. In this case, we usually write k as $-\lambda$ to emphasize that it is negative. When this is the case, the model is called a *decay model* rather than a growth model. See in the examples and problems of Sections 5 and 8. ◀

Generalizing the technique used in Example 23.2, we can combine the derivative of e^x given in equation (23.8) with the Chain Rule to compute the derivative of the composition of the exponential function with other functions. If $g(x)$ is a differentiable function and

$$y = e^{g(x)},$$

then setting $u = g(x)$ the Chain Rule states that

$$\frac{dy}{dx} = \frac{de^u}{du} \cdot \frac{du}{dx}.$$

Since $de^u/du = e^u$ and $du/dx = g'(x)$. this equation becomes

$$\frac{d(e^{g(x)})}{dx} = e^{g(x)} \cdot g'(x) = g'(x)e^{g(x)}. \tag{23.9}$$

Example 23.3 We employ equation (23.9) to compute the following derivatives.

a) $\dfrac{de^{x^2}}{dx} = e^{x^2} \cdot 2x = 2xe^{x^2}.$

b) $\dfrac{de^{3x+1}}{dx} = 3e^{3x+1}.$

c) $\dfrac{de^{-t^3}}{dt} = -3t^2e^{-t^3}.$

d) $\dfrac{de^{-(t+1)^3}}{dt} = -3(t+1)^2e^{-(t+1)^3}.$

e) If $y = \dfrac{e^{2x}}{1+e^x}$, then, by the quotient rule,

$$y' = \frac{(1+e^x)(2e^{2x}) - (2^{2x})(e^x)}{(1+e^x)^2}$$
$$= \frac{e^{2x}(2 + 2e^x - e^x)}{(1+e^x)^2} = \frac{e^{2x}(2+e^x)}{(1+e^x)^2}.$$

f) If $y = e^{x^2} + e^{-3x+2}$, then $y' = 2xe^{x^2} - 3e^{-3x+2}$.

g) If $y = (e^{5x^2+1})(e^{-2x^2-1})$, then by the product rule

$$y' = (e^{5x^2+1})(-4xe^{-2x^2-1}) + (10xe^{5x^2+1})(e^{-2x^2-1}).$$

A simpler approach is to use Property E-P3 to express y as $y = e^{3x^2}$; then $y' = 6xe^{3x^2}$. You should verify that these are the same results.

h) $\dfrac{d}{dx} e^{\sin(x)} = \cos(x)e^{\sin(x)}.$

i) $\dfrac{d}{d\theta} e^{\cot(\theta)} = -\csc^2(\theta)e^{\cot(\theta)}.$ ◀

Example 23.4 *Oxygen supply and hematocrit levels*

The hematocrit, H, is a measure of the percentage of red blood cells in the circulating blood. Since the red blood cells carry the oxygen from the lungs to the body tissues, a natural assumption would be that an increase in hematocrit would lead to an increase in oxygen supply. This is valid up to a point, but the supply of oxygen also depends on the rate of blood flow, $\mathring{Q}$. For instance, one blood system with twice the hematocrit but only one-fourth the flow rate as compared to a second system would in fact supply only half as much oxygen per unit of time. The rate of blood flow depends on the blood pressure and the viscosity of the blood. (The viscosity of a fluid is a measure of how it flows; the higher the viscosity the slower the flow.) The viscosity of blood depends on the temperature and the hematocrit of the blood. As the hematocrit increases, the viscosity increases, and hence the flow rate decreases. The relationship between hematocrit and oxygen supply is therefore not linear, but is given by the equation

$$O_2 = mH\mathring{Q}_{pl}e^{-K_1H} \cdot (\text{ml } O_2/\text{min}). \tag{23.10}$$

$\mathring{Q}_{pl}$ is the flow rate of plasma (blood with the red cells removed) at the given temperature and pressure; m is a constant, which reflects the oxygen carrying capacity of the red cells and varies between individuals and species. In experiments with the bullfrog, *Rana catesbeiana*, m was found to be 0.257. The rate of change in the oxygen supply due to a change in hematocrit is obtained using the product rule and equation (23.9):

$$\begin{aligned}\frac{dO_2}{dH} &= m\mathring{Q}_{pl}[e^{-K_1H} - K_1He^{-K_1H}] \qquad (23.11)\\ &= [1 - K_1H]m\mathring{Q}_{pl}e^{-K_1H}.\end{aligned}$$

◀

Example 23.5 *Animal growth as a function of food intake*
The following equation has been developed to relate the weight, W, to the food intake, q, of growing animals with unlimited food access:

$$W = A[1 - (1 - q/C)^{\alpha} \exp(\alpha q/C)], \tag{23.12}$$

where A is the mature animal weight, C is mature food intake, and α is a constant that reflects the nutritional requirements and the efficiency with which the animal utilizes the food. The change in weight due to a change in food intake is obtained from equation (23.12) using the product rule and formula (23.9):

$$\begin{aligned}\frac{dW}{dq} &= A\frac{\alpha}{C}\left(1 - \frac{q}{C}\right)^{\alpha-1} e^{\alpha q/C} - A\left(1 - \frac{q}{C}\right)^{\alpha} \cdot \frac{\alpha}{c} \cdot e^{\alpha q/c}\\ &= A\frac{\alpha}{C}\left(1 - \frac{q}{C}\right)^{\alpha-1} e^{\alpha q/C}\left[1 - \left(1 - \frac{q}{C}\right)\right]\\ &= A\frac{\alpha q}{C^2}\left(1 - \frac{q}{C}\right)^{\alpha-1} e^{\alpha q/C}.\end{aligned}$$

Since $q < C$, we see that (dW/dq) is positive for all values of q. Often the food intake q is assumed to be a function of time. One such function is

$$q(t) = C(1 - e^{-t/\tau}), \tag{23.13}$$

where τ is the time when $q = C(1 - e^{-1}) \simeq 0.63C$—i.e., when the food intake is about 63 percent of the mature food intake C. With this assumed form of food consumption, the weight function (23.12) becomes a function of time. Substituting (23.13) into (23.12) leads us to an equation for W, which involves a Gompertz function:

$$W(t) = A[1 - e^{-\alpha(1-t/\tau)}e^{-\alpha e^{-t/\tau}}]. \tag{23.14}$$

The rate of weight change $\mathring{W}(t)$ is also a positive function for all values of t. (We leave the computation of $\mathring{W}(t)$ to the exercises.) ◀

Example 23.6 The tapeworm, *Hymenolepis diminuta*, is introduced to a host by way of eggs deposited in a food source. These eggs upon ingestion into an animal hatch into embryos called *hexacanths*, which then penetrate the midgut wall of the host. The number of embryos that survive to maturity depends on the time it takes for them to penetrate the gut wall and the natural death rate of the hexacanths. The basic model of the survival of the hexacanths is given by

$$\frac{dN(t)}{dt} = -\mu N(t), \tag{23.15}$$

where μ is the death rate. The solution of this equation for constant μ is

$$N(t) = N_0 e^{-\mu t}, \tag{23.16}$$

where N_0 is the initial number of hexacanths. This solution is easily verified using equation (23.9). However, when the death rate is not constant, the form of the solution is given by

$$N(t) = N_0 \exp\left\{-\int_0^t \mu(x)dx\right\}. \tag{23.17}$$

This solution may be verified using equations (23.9) and (21.17):

$$\begin{aligned}\frac{dN(t)}{dt} &= \left[N_0 \exp\left\{-\int_0^t \mu(x)dx\right\}\right]\frac{d}{dt}\left(-\int_0^t \mu(x)dx\right)\\ &= N(t)\cdot(-\mu(t)).\end{aligned}$$

The form of $\mu(t)$ has been found experimentally to be exponential:

$$\mu(t) = ae^{bt},$$

where values of the constants are $a = 0.0608$ and $b = 0.2944$. Thus $N(t)$ is a Gompertz function when $\mu(t)$ is an exponential:

$$N(t) = N_0 e^{-(a/b)(e^{bt}-1)} \tag{23.18}$$

◀

23.3 GENERAL LOGARITHMIC AND EXPONENTIAL FUNCTIONS

We have defined the natural logarithm and exponential functions with base e. The logarithm functions and exponential functions are defined for any base b, $b > 0$, in terms of $\ln(x)$ and e^x. The motivation for these definitions is provided by the reasoning used to obtain the change of base formula (5.10) for logarithms.

Definition 23.4 For any positive real number b, we define the logarithm function with base b, $\log_b(x)$, by

$$\boxed{\log_b(x) \equiv \frac{\ln(x)}{\ln(b)}} \tag{23.19}$$

Definition 23.5 For any positive real number b, we define the exponential function with base b, $\exp_b(x)$ or simply b^x as the inverse, with respect to composition, of the function $\log_b(x)$. Thus

$$\boxed{y = b^x \text{ is equivalent to } \log_b(y) = x.} \tag{23.20}$$

> *Note*: Unless specified otherwise $\log(x)$ denotes the logarithm function with base $b = 10$.

The derivative of the function $\log_b(x)$ may be computed using the identity (23.19) and the formula (23.5):

$$\boxed{\frac{d}{dx}\log_b(x) = \frac{1}{\ln(b)}\cdot\frac{1}{x}.} \tag{23.21}$$

The derivative of $y = b^x$ is obtained in the same way that the derivative of $y = e^x$ was determined. By (23.19),

$$\text{if} \qquad y = b^x \qquad \text{then} \qquad x = \log_b(y) = \frac{\ln(y)}{\ln(b)}.$$

Taking derivatives, we find $D_x(x) = D_x[\ln(y)/\ln(b)]$. Using (23.9) to compute the derivative of $\ln(y)$, we obtain the equation

$$1 = \frac{1}{\ln(b)}\cdot\frac{y'}{y} \qquad \text{or} \qquad y' = \ln(b)y.$$

Thus

$$\boxed{\frac{d}{dx}b^x = \ln(b)\cdot b^x.} \tag{23.22}$$

> **Caution:**
>
> $$x^b \neq b^x;$$
>
> therefore
>
> $$\frac{db^x}{dx} \neq bx^{b-1} \qquad \text{and} \qquad \frac{db^x}{dx} \neq xb^{x-1}.$$
>
> These are two of the most common mistakes made in computing $D_x(b^x)$. Avoid these mistakes by either memorizing formula (23.22) or learning how to derive this formula whenever it is needed. We recommend that you learn to derive the formula, as it is really the easier method.

The derivatives given by equations (23.21) and (23.22) may be generalized to encompass general types of functions by employing the Chain Rule. If $g(x)$ is any positive differentiable function, then

$$\frac{d}{dx}\log_b(g(x)) = \frac{1}{\ln(b)}\cdot\frac{g'(x)}{g(x)}. \tag{23.23}$$

For any differentiable function, $g(x)$,

$$\frac{d}{dx}b^{g(x)} = \ln(b)\cdot b^{g(x)}\cdot g'(x). \tag{23.24}$$

Example 23.7 We compute the following derivatives by employing equations (23.23) and (23.24).

a) $\frac{d}{dx}\log(x^2) = \frac{1}{\ln(10)}\frac{2x}{x^2} = \frac{2}{\ln(10)x}$

b) $\frac{d}{dx}\log_3(2x+1) = \frac{1}{\ln(3)}\cdot\frac{2}{2x+1}$

c) $D_x(4^x) = \ln(4)4^x$

d) $D_x(4^{x^2}) = \ln(4)\cdot 4^{x^2}\cdot 2x$

e) $\frac{d}{dt}(3^{2t}) = \ln(3)\cdot 3^{2t}\cdot 2$

f) $\frac{d}{dt}5^{t^2-t} = \ln(5)\cdot 5^{t^2-t}\cdot(2t-1)$

g) $\frac{d}{dt}\log(t^2+1) = \frac{1}{\ln(10)}\cdot\frac{2t}{t^2+1}$

h) $\frac{d}{dx}(x^2 - \log(x^2)) = 2x - \frac{2}{\ln(10)x}$

i) $\frac{d}{dx}\log(\sin(x)) = \frac{1}{\ln(10)}\frac{\cos(x)}{\sin(x)} = \frac{\cot(x)}{\ln(10)}$

j) $\frac{d}{dx}10^{\cot(x)} = -\ln(10)\csc^2(x)10^{\cot(x)}$

k) $\frac{d}{dx}[e^{x^2}\cdot\log(4x-2)] = (2xe^{x^2})(\log(4x-2)) + \frac{4e^{x^2}}{\ln(10)(4x-2)}$ ◀

Example 23.8 Various techniques have been developed to estimate animal sizes. One method used in fisheries research relates the length of a fish to the size of its *otolith*. An otolith is a granule of calcium carbonate in the inner ear of vertebrates; it is relatively larger in fish than in other species. The relationship between the otolith radius r(10^{-1} mm) and the length L (cm) of sole has been found to be as follows:

$$r = 2.18\log(L) - 1.3, \qquad \text{for females with } 9 < L < 48;$$
$$r = 1.91\log(L) - 0.99, \qquad \text{for males with } 8 < L < 35.$$

According to these equations, we find the rates of change in r as the length L increases are, by formula (23.21), as follows:

$$\text{Females}\quad \frac{dr}{dL} = \frac{2.18}{\ln(10)}\frac{1}{L} = \frac{2.18}{2.30}\frac{1}{L} = \frac{0.95}{L};$$

$$\text{Males}\quad \frac{dr}{dL} = \frac{1.91}{\ln(10)}\frac{1}{L} = \frac{(1.91)\log(e)}{L} = \frac{(1.91)(0.4343)}{L} = \frac{0.83}{L}.$$

Note: $\ln(10) \simeq 2.3026$, $\log(e) \simeq 0.4343$, and $1/\ln(10) = \log(e)$.

Thus, as L increases, the otolith radius also increases, but the rate of increase is inversely proportional to L. ◀

Example 23.9 The dimensions of a rod-shaped bacteria such as *Escherichia coli* (*E. coli*) or *Salmonella typhimurium* are dependent on the growth conditions and on the replication time of the bacteria's chromosomes. The relationship for the length of the bacteria is indicated by the equation

$$l = K2^{D\mu},$$

where l is cell length, K is a constant, D is the time between termination of chromosome replication and the subsequent cell division, and μ is the growth rate of the cell. The diameter of the cell, $\mathbf{d}$, is similarly related to the growth parameter,

$$\mathbf{d} = K_2 2^{C\mu/2},$$

where K_2 is a constant and $C\mu$ gives the number of replication positions of a cell. The rate of change in the bacteria's shape due to changes in the growth rate, μ, are described by the following equations:

$$\frac{dl}{d\mu} = \ln(2)K2^{D\mu}\cdot D; \qquad \text{and} \qquad \frac{d\mathbf{d}}{d\mu} = \ln(2)K_2 2^{C\mu/2}(C/2).$$

From these equations, we see that the shape of the bacteria will depend on the values of D and C. For fixed C, if the value of D increases, then the bacteria's shape will be longer and will have an increased rate of elongation as the growth rate μ increases. However, for *E. coli*, experimental evidence shows that for all growth rates μ, the replication time D and the replication position constant C are related by $C = 2D$. When this occurs, we see that both derivatives, $D_\mu(l)$ and $D_\mu(\mathbf{d})$, are equal. Thus, as the growth rate, μ, changes, the relative shape of *E. coli* remains the same. ◀

23.4 INTEGRALS OF THE LOGARITHMIC AND EXPONENTIAL FUNCTIONS

Indefinite integrals (or antiderivatives) involving the logarithmic and exponential functions are obtained directly from the formulas expressing derivatives involving these functions. Recall that each differentiation formula, $F'(x) = f(x)$, has a corresponding integration (or antiderivative) formula,

$$\int f(x)dx = F(x) + c.$$

Using this relationship, we may use the differentiation formulas derived above in this section to obtain equivalent integration formulas. We state these formulas below, indicating the corresponding differentiation formulas.

Integrals	Corresponding Derivatives	
$\int e^x \, dx = e^x + c$	$\frac{d}{dx} e^x = e^x$	(23.25)
$\int g'(x)e^{g(x)} \, dx = e^{g(x)} + c$	$\frac{d}{dx} e^{g(x)} = g'(x)e^{g(x)}$	(23.26)
$\int \frac{1}{x} \, dx = \ln(x) + c$	$\frac{d}{dx} \ln(x) = \frac{1}{x}$	(23.27)
$\int b^{g(x)}g'(x)dx = \frac{b^{g(x)}}{\ln(b)} + c;$	$\frac{d}{dx} b^{g(x)} = \ln(b) \cdot b^{g(x)}g'(x)$	(23.28)
$\int \frac{g'(x)}{g(x)} \, dx = \ln(g(x)) + c;$	$\frac{d}{dx} \ln(g(x)) = \frac{g'(x)}{g(x)}$	(23.29)
$\int b^x \, dx = \frac{b^x}{\ln(b)} + c;$	$\frac{d}{dx} b^x = \ln(b)b^x$	(23.30)

Each of these integral formulas may be extended by introducing a constant, k. For instance,

$$\int k \frac{g'(x)}{g(x)} \, dx = k \ln(g(x)) + c.$$

Example 23.10 The following illustrate the use of the above integral formulas.

a) To evaluate

$$\int e^{2x} \, dx,$$

we recognize that the integrand involves a term of the form $e^{g(x)}$. The choice for $g(x)$ is obviously $g(x) = 2x$. Then $g'(x) = 2$. The coefficient of e^{2x} in the integral is one. Consequently since $1 = \frac{1}{2}g'(x)$, the integral is of the appropriate form and may be evaluated by using (23.26).

$$\int e^{2x} \, dx = \int \tfrac{1}{2}g'(x)e^{g(x)} \, dx = \tfrac{1}{2}e^{g(x)} + c = \tfrac{1}{2}e^{2x} + c.$$

b) To evaluate

$$\int 5xe^{x^2}\,dx,$$

we again recognize that the integrand contains an exponential term, $e^{g(x)}$, where $g(x) = x^2$. We set $5x$, the coefficient of e^{x^2}, equal to a constant multiple of the derivative of $g(x)$: $5x = k \cdot g'(x)$. Since $g'(x) = 2x$, we find $k = \frac{5}{2}$. Therefore, the integral may be evaluated using formula (23.26):

$$\int 5xe^{x^2}\,dx = \int \tfrac{5}{2}g'(x)e^{g(x)}\,dx = \tfrac{5}{2}e^{g(x)} + c = \tfrac{5}{2}\,e^{x^2} + c.$$

c) To evaluate the integral

$$\int 5xe^{x^3}\,dx,$$

we recognize that the integrand contains the term, $e^{g(x)}$, where $g(x) = x^3$. In this case, $g'(x) = 3x^2$. Proceeding as above, we attempt to represent the coefficient of e^{x^3} as a constant multiple of $g'(x)$ so that we may employ formula (23.26). However, setting $5x = k \cdot g'(x)$, that is, $5x = k \cdot 3x^2$, we solve for k and find that to satisfy this equation $k = \frac{5}{3}x$, which is not a constant. The conclusion we draw is that the integral does not have the form (23.26) and, consequently, we do not know how to evaluate this integral.

d) To evaluate the integral

$$\int \frac{2x}{x^2 + 1}\,dx,$$

our attention is first called to the fact that the integrand involves a quotient. Consequently, we check to see if it has the particular form given in formula (23.28), $g'(x)/g(x)$. To this end, we set $g(x)$ equal to the denominator, $g(x) = x^2 + 1$. Then $g'(x) = 2x$, and we immediately see that the integrand does have the form $g'(x)/g(x)$ and may be evaluated as

$$\int \frac{2x}{x^2 + 1}\,dx = \int \frac{g'(x)}{g(x)}\,dx = \ln(g(x)) + c = \ln(x^2 + 1) + c.$$

e) To evaluate the integral

$$\int \frac{e^x}{3 + e^x}\,dx,$$

the key to the solution is the fact that the integrand is a quotient. Consequently, we set $g(x)$ equal to the denominator; $g(x) = 3 + e^x$. Then $g'(x) = e^x$ and the integral may be evaluated using formula (23.28):

$$\int \frac{e^x}{3 + e^x}\,dx = \int \frac{g'(x)}{g(x)}\,dx = \ln(g(x)) + c = \ln(3 + e^x) + c.$$

f) The integral $\int 3^x\,dx$ is evaluated by formula (23.29):

$$\int 3^x\,dx = \frac{3^x}{\ln(3)} + c.$$

g) To evaluate the integral

$$\int 3x2^{x^2}\,dx,$$

we chose $g(x) = x^2$. Then setting the coefficient of $2^{g(x)}$ equal to a multiple of $g'(x)$, we have $3x = k \cdot g'(x) = k \cdot 2x$, and thus $k = \frac{3}{2}$. Consequently, by formula (23.30) the integral may be evaluated:

$$\int 3x2^{x^2}\,dx = \int \frac{3}{2} g'(x)2^{g(x)}\,dx = \frac{3}{2}\frac{2^{g(x)}}{\ln(2)} + c = \frac{3}{2}\frac{2^{x^2}}{\ln(2)} + c.$$ ◀

Example 23.11 The formulas for integrals (23.25) to (23.30) may likewise be used to evaluate definite integrals via the *FTC*.

a) $\displaystyle\int_3^5 e^x\,dx = e^x\Big|_3^5 = e^5 - e^3 \simeq 128.33$

b) $\displaystyle\int_2^7 \frac{1}{x}\,dt = \ln(x)\Big|_2^7 = \ln(7) - \ln(2) = \ln(\tfrac{7}{2}) \simeq 1.25$

c) $\displaystyle\int_0^2 xe^{x^2}\,dx = \tfrac{1}{2}e^{x^2}\Big|_0^2 = \tfrac{1}{2}e^{2^2} - \tfrac{1}{2}e^{0^2} = \tfrac{1}{2}(e^4 - 1) \simeq 26.80.$

d) $$\int_{-2}^2 \frac{x}{x^2+1}\,dx = \tfrac{1}{2}\ln(x^2+1)\Big|_{-2}^2$$
$$= \tfrac{1}{2}\ln(2^2+1) - \tfrac{1}{2}\ln((-2)^2+1)$$
$$= \tfrac{1}{2}\ln(5) - \tfrac{1}{2}\ln(5) = 0$$

e) $\displaystyle\int_3^5 2^x\,dx = \frac{2^x}{\ln(2)}\Big|_3^5 = \frac{2^5}{\ln(2)} - \frac{2^3}{\ln(2)} = \frac{24}{\ln(2)} \simeq 34.62$

f) $\displaystyle\int_0^3 2x5^{x^2+1}\,dx = \frac{5^{x^2+1}}{\ln(5)}\Big|_0^3 = \frac{5^{10}-5}{\ln(5)} \simeq 6{,}067{,}720.87$

g) $\displaystyle\int_0^{\pi/2} \sin(\theta)e^{\cos(\theta)}\,d\theta = -e^{\cos(\theta)}\Big|_0^{\pi/2} = -e^0 + e^1 = e - 1 \simeq 1.72$ ◀

Example 23.12 *Passage of digesta through sheep*
Sheep, as ruminants, have a complicated stomach that has a storage compartment called the rumen, in which unchewed food may be partially digested and then rechewed as "cud." This extra stomach activity allows ruminants to feed on very low grade feed such as grass or corn silage. For this reason ruminants form a

vital link in the food chain for human consumption. (Nonruminants, like swine or fowl, require food that humans can also utilize.) To examine how ruminants utilize their food, studies are made of the way ingested food passes through the digestive tract of different animals. One such investigation led to the mathematical model describing the passage of material through sheep by the following equation:

$$y(t) = \begin{cases} 0 & \text{if} \quad t < T; \\ A[e^{-k_1(t-T)} - e^{-k_2(t-T)}] & \text{if} \quad t \geq T. \end{cases} \tag{23.31}$$

In this equation, A is a constant, which represents the quantity of a substance introduced to the rumen at time $t = 0$. T is a delay time, which describes the minimum time required for a substance to pass through the digestive tract; k_1 and k_2 are rate constants, which have a physiological interpretation that we shall not discuss. And y represents the concentration of monitored substance that is excreted in the faeces. Experimentally, y is measured using a radioactive marker. The cumulative amount of marker that has been excreted up to time $t > T$ is given by the following integral:

$$\begin{aligned} E(t) &= \int_T^t y(s)ds \\ &= A \int_T^t e^{-k_1(s-T)} - e^{-k_2(s-T)}\, ds \\ &= A\left\{\frac{e^{-k_1(s-T)}}{-k_1} - \frac{e^{-k_2(s-T)}}{-k_2}\Bigg|_T^t\right\} \\ &= A\left\{\left(\frac{e^{-k_1(t-T)}}{-k_1} - \frac{e^{-k_2(t-T)}}{-k_2}\right) - \left(\frac{e^{-k_1(T-T)}}{-k_1} - \frac{e^{-k_2(T-T)}}{-k_2}\right)\right\}. \end{aligned}$$

This equation simplifies to

$$E(t) = \frac{-A}{k_1 k_2}[k_2 e^{k_1(t-T)} - k_1 e^{-k_2(t-T)} - k_2 + k_1].$$

If the total amount of marker injected into the sheep is $A(k_2 - k_1)/k_1 k_2$, then the percentage of recovered marker at time t is

$$PC(t) = \frac{E(t)}{A(k_2 - k_1)/k_1 k_2} \cdot 100 \text{ percent},$$

or

$$PC(t) = 100\left[1 + \frac{k_1 e^{-k_2(t-T)} - k_2 e^{-k_1(t-T)}}{k_2 - k_1}\right].$$

The time necessary for 50 percent of the marker to be excreted is a standard clinical value used to describe an animal's digestive process. This time is the difference $t_d - T$, where $PC(t_d) = 50$. Mathematically, it can be shown that this is not a very good indicator of the passage curve given by formula (23.31). (We investigate this in the exercises.) From a practical point of view, the time t_d is difficult to determine since the equation $PC(t_d) = 50$ cannot be solved explicitly for t_d. ◀

23.5 LOGARITHMIC DIFFERENTIATION

When confronted with a problem which requires the differentiation of the product of several functions, it is always possible to obtain the derivative through repeated application of the product rule. For example, let

$$y = (x + 1)^{1/2}(x^2 - 2x + 3)(x^3 - x).$$

Then, treating $f(x) = (x + 1)^{1/2}$ and $g(x) = (x^2 - 2x + 3)(x^3 - x)$, we compute

$$\begin{aligned} y' &= f'(x)g(x) + f(x)g'(x) \\ &= \tfrac{1}{2}(x + 1)^{-1/2}(x^2 - 2x + 3)(x^3 - x) + (x + 1)^{1/2} \cdot g'(x). \end{aligned}$$

To evaluate the derivative $g'(x)$, we again use the product rule:

$$g'(x) = (2x - 2)(x^3 - x) + (x^2 - 2x + 3)(3x^2 - 1).$$

Substituting this value of $g'(x)$ into the expression for y', we obtain the following equation:

$$\begin{aligned} y' = \tfrac{1}{2}&(x + 1)^{-1/2}(x^2 - 2x + 3)(x^3 - x) \\ &+ (x + 1)^{1/2}[(2x - 2)(x^3 - x) + (x^2 - 2x + 3)(3x^2 - 1)]. \end{aligned}$$

The procedure used works; however, it is a long process and, as a consequence, would be prone to errors. What we would like to have is a simpler method of computing the derivative of a product, which might also result in a more compact or organized expression for y'. There is such a method, called *logarithmic differentiation.*

Assume that $y(x)$ is the product of three functions:

$$y(x) = f_1(x) \cdot f_2(x) \cdot f_3(x). \tag{23.32}$$

Then, taking the natural logarithm of $y(x)$, we obtain the equation

$$\begin{aligned} \ln(y(x)) &= \ln(f_1(x) \cdot f_2(x) \cdot f_3(x)) \\ &= \ln(f_1(x) + \ln(f_2(x)) + \ln(f_3(x)). \end{aligned}$$

Next we differentiate this equation with respect to the variable x. For each term, we apply the formula (23.6) to obtain the equation

$$\frac{y'(x)}{y(x)} = \frac{f_1'(x)}{f_1(x)} + \frac{f_2'(x)}{f_2(x)} + \frac{f_3'(x)}{f_3(x)}.$$

Solving for $y'(x)$, we obtain the equation

$$y'(x) = y(x)\left[\frac{f_1'(x)}{f_1(x)} + \frac{f_2'(x)}{f_2(x)} + \frac{f_3'(x)}{f_3(x)}\right]. \tag{23.33}$$

Equation (23.33) yields the derivative of the function y, but cannot be used when $f_1(x)$, $f_2(x)$, or $f_3(x)$ are zero. If one of these functions is zero, it is necessary to expand equation (23.33), using equation (23.32), in the equivalent form

$$y'(x) = f_1'(x)f_2(x)f_3(x) + f_1(x)f_2'(x)f_3(x) + f_1(x)f_2(x)f_3'(x). \qquad (23.34)$$

The results derived for the product of three functions easily generalize to products of N-functions.

Theorem 23.1 If $\{f_n\}_{n=1}^{N}$ is a sequence of N differentiable functions and $y(x)$ is a product of the N-functions,

$$y(x) = f_1(x) \cdot f_2(x) \cdot \cdots \cdot f_N(x),$$

then the derivative of y is given by

$$\boxed{y'(x) = y(x) \sum_{n=1}^{N} \frac{f_n'(x)}{f_n(x)}.} \qquad (23.35)$$

Example 23.13 We employ equation (23.35) to evaluate the following derivatives.

a) $\dfrac{d}{dx} x(x+1)(x+2) = x(x+1)(x+2)\left[\dfrac{1}{x} + \dfrac{1}{x+1} + \dfrac{1}{x+2}\right].$

b) $D_x(2x)(x^2 - 1)(3x^4 - x)$

$$= (2x)(x^2-1)(3x^4-x)\left[\frac{2}{2x} + \frac{2x}{x^2-1} + \frac{12x^3-1}{3x^4-x}\right].$$

c) If $y(x) = x^2(x+3)(e^{2x})$, then

$$y'(x) = y(x)\left[\frac{2x}{x^2} + \frac{1}{x+3} + \frac{2e^{2x}}{e^{2x}}\right]$$

$$= y(x)\left[\frac{2}{x} + \frac{1}{x+3} + 2\right].$$

d) If $y(x) = \ln(2x) \cdot e^{x^2}(x+2)^5(x-3)^2$, then

$$y'(x) = y(x)\left[\frac{2/2x}{\ln(2x)} + \frac{2xe^{x^2}}{e^{x^2}} + \frac{5(x+2)^4}{(x+2)^5} + \frac{2(x-3)}{(x-3)^2}\right]$$

$$y'(x) = y(x)\left[\frac{1}{x\ln(2x)} + 2x + \frac{5}{x+2} + \frac{2}{x-3}\right].$$

e) If $y(x) = x \cdot \sin(x) \cdot \cos(x)$, then

$$y'(x) = y(x)\left[\frac{1}{x} + \frac{\cos(x)}{\sin(x)} - \frac{\sin(x)}{\cos(x)}\right]$$

$$= y(x)[x^{-1} + \cot(x) - \tan(x)]. \qquad ◀$$

Example 23.14 *Maximum likelihood estimate in genetics*
The number of individuals in a randomly mating population having a specific genotype can be predicted by a genetic model of the population. When a particular population is sampled, the geneticist uses statistics to decide whether or not her or his model of the genetic make-up of the population fits the observed frequencies in the population. For instance, we consider a model of the gene frequency in a two-allele loci, where heterozygotes are not distinguishable. Let the alleles be denoted by A and a with the probability of A occurring being p and the probability of a occurring being $(1 - p)$. Then the expected number of individuals of each genotype can be computed and compared to the observed numbers. Assume that the observed numbers having genotype AA, Aa and aa are A, B and C, respectively, with a total $A + B + C = N$ as indicated in Table 23.1. Then the probability, P, of the observed frequencies occurring under the hypothesis of the model is a function of the probability parameter, p, in the model. This probability function has the form

$$P(p) = \frac{N!}{A!B!C!} \cdot p^{2A} \cdot (2p(1 - p))^B \cdot (1 - p)^{2C}.$$

When the probability p is not known theoretically, it is estimated from observed frequencies. When trying to estimate the probability p of the model, it is necessary to compute $D_p(P)$. The estimate for p, denoted $\hat{p}$, is taken to be the value of p for which $P'(\hat{p}) = 0$. To compute P', we use formula (23.35) with $N = 3$ and $f_1(p) = p^{2A}$, $f_2(p) = (2p(1 - p))^B$, and $f_3(p) = (1 - p)^{2C}$. The resulting formula for the derivative is

$$\frac{d}{dp} P(p) = P' = P\left[\frac{2A}{p} + \frac{(1 - 2p)B}{p(1 - p)} + \frac{-2C}{1 - p}\right].$$

We have evaluated and simplified each term f_i'/f_i. You should verify this formula. Setting $P' = 0$, we obtain the estimate $\hat{p}$ as the value of p for which the sum on the right of this equation is zero. Solving, we find

$$\hat{p} = \frac{2A + B}{2N};$$

$\hat{p}$ is called a *maximum likelihood estimate* of the probability p.

Table 23.1

Genotype	Expected proportion	Observed number
AA	p^2	A
Aa	$2p(1 - p)$	B
aa	$(1 - p)^2$	C
		Total N

EXERCISE SET 23

23.1 Sketch a graph to illustrate the area represented by the given number.

a) $\ln(3)$ b) $\ln(5)$ c) $\ln(1)$ d) $\ln(\frac{1}{2})$ e) $\ln(\frac{1}{3})$ f) $\ln(0.2)$

23.2 Compute the derivative y' for the indicated function.

a) $y(x) = e^{3x}$ b) $y(x) = e^{x^3}$
c) $y(x) = e^{x^2+2}$ d) $y(x) = e^{-x}$
e) $y(x) = e^{(2x-1)^5}$ f) $y(x) = e^{(2x^5-1)}$
g) $y(x) = e^{1/x}$ h) $y(x) = xe^x$
i) $y(x) = x^2e^{x^2}$ j) $y(x) = \frac{1}{2}(e^{(1/2)x} - e^{-(1/2)x})$
k) $y(x) = e^{x^2}[1 - e^{3x}]$ l) $y(x) = e^{2x}/(e^{5x} - 1)$

23.3 Compute the indicated derivative.

a) $D_x(\ln(x^2 + 1))$ b) $D_x(\ln(4x))$

c) $\dfrac{d}{dt}\ln(4t - t^2)$ d) $\dfrac{d}{ds}\ln(\sqrt{1 - s^2})$

e) $\dfrac{d}{dt}t\ln(t)$ f) $\dfrac{d}{dt}e^t - \ln(t^{-1})$

g) $\dfrac{d}{dx}\dfrac{4\ln(x)}{\ln(x + 1)}$ h) $\dfrac{d}{dx}(\ln(x))^2$

i) $\dfrac{d}{dx}3\ln(x) - (\ln(x))^{1/2}$ j) $\dfrac{d}{dx}\ln(\ln(x))$

23.4 Compute $f'(\theta)$ for the specified function f.

a) $f(\theta) = e^{\tan(\theta)}$ b) $f(\theta) = 2^{\sin(\theta)}$
c) $f(\theta) = \ln(\cos(\theta))$ d) $f(\theta) = \log(\tan(\theta))$
e) $f(\theta) = \theta e^{\sin(\theta)}$ f) $f(\theta) = \log(\theta)e^{\cos(\pi\theta/2)}$
g) $f(\theta) = e^{\theta}/\ln(\theta)$ h) $f(\theta) = \ln(1 - \cos^2(\theta))$

23.5 Compute the indicated derivatives.

a) $D_x(10^x)$ b) $D_x(4^{(x-2)})$ c) $D_t(2^t)$
d) $D_x(2^{x^2})$ e) $D_x(10^{x^2-2x+3})$ f) $D_x(e^{\log(x)})$
g) $D_t(t^5)$ h) $D_t(t \cdot 5^t)$ i) $D_t(t^2 \cdot 3^{4t})$
j) $D_x(2^x \cdot 3^x)$ k) $D_x \ln(3^x)$ l) $D_x \ln(3^x + 1)$

23.6 Compute the derivative as indicated.

a) $\dfrac{d}{dx}\log(3x)$ b) $\dfrac{d}{dx}\log_3(x)$ c) $\dfrac{d}{dx}\log(x^3)$

d) $\dfrac{d}{dx}\log(x + 3)$ e) $\dfrac{d}{dt}\log_2(t)$ f) $\dfrac{d}{dt}\log(t^2 - t + 3)$

g) $\dfrac{d}{dx}\log(xe^x)$ h) $\dfrac{d}{dx}\log\left(\dfrac{x + 1}{x - 1}\right)$ i) $\dfrac{d}{dx}x\log(x)$

j) $\dfrac{d}{dx}2^x\log(x)$

23.7 The following problem involves a trick. Think carefully about what you are doing. Compute $D_x y$ if $y(x) = x^x$.

23.8 A population is assumed to have size $N(t) = e^{t^2} - t$.

a) Compute the growth rate of the population.
b) Determine when the growth rate equals 1000 units/time unit

23.9 Two animals, A_1 and A_2, have a total mass of $M_1 = e^{l-2}$ and $M_2 = \log(l^5)$, respectively, where l denotes the body length of the animal. Which animal is growing at a faster rate when $l = \frac{1}{2}$, $l = 1$, and $l = 2$? Determine when each animal has a growth rate of 100.

23.10 In Example 23.4, the oxygen supply, O_2, was given as a function of the hematocrit level, H, in equation (23.10). Assume that an animal has a constant blood plasma flow of $\mathring{Q} = 4$ ml/min, $m = 0.257$, and $K_1 = 300$.

a) At what value of H will $D_H O_2 = 0$?
b) If H is a function of time, $H(t) = \frac{1}{4}\sin(\pi t/24) + \frac{1}{2}$, compute $D_t O_2$.
c) When will $D_t O_2 = 0$ in part (b)?

23.11 Compute $\mathring{W}(t)$ for the function $W(t)$ given in equation (23.14).

23.12 In a study of the biological changes in a membrane (such as a cell wall), it was determined that the pressure, P, on a membrane being stressed is

$$P(\delta) = Y \int_{\delta_0}^{\delta} \frac{1}{x}\,dx,$$

where Y is a constant called an *elastic modulus*, δ_0 is the unstrained thickness of the membrane, and δ is the stressed thickness.

a) Express $P(\delta)$ without using an integral.
b) The stress energy W is defined by the integral equation

$$W = -\int_{t=\delta_0}^{t=\delta} P(t)dt.$$

Show that the function $W(\delta) = Y[\delta \ln(\delta_0/\delta) + \delta - \delta_0]$ satisfies this equation by differentiating W and showing that this is the same as the derivative of the integral.

23.13 The water retention of terrapins (turtles) depends on their environment. Studies in which terrapins were changed from fresh water to salt water led to the following set of equations to describe the rate of water flux between an animal and its environment.

$$\text{flux} = k_1 W_1; \quad k_1 = \lambda/(1 + (W_1/W_2));$$

$$\lambda = \frac{2.3}{t}\log\left[\frac{W_2/(W_1 + W_2)}{(C_t/C_0) - W_1(W_1 + W_2)}\right]$$

W_1 = radioactive labeled water exterior (ml)
W_2 = radioactive labeled water interior (ml)
C_t/C_0 = relative radioactivity of exterior water (counts per minute)

a) Compute the rate of change in λ as a function of W_2.
b) Compute the rate of change in k_1 as a function of W_2.
c) Compute the rate of change in the flux due to a change in W_2.

23.14 a) Solve the two equations in Example 23.8 for L as a function of r.
b) Use the results of part (a) to compute the rate of change in the fishes' lengths, L, with respect to the radius, r.
c) Are the derivatives dr/dL and dL/dr related in this example?

23.15 In Example 23.6, show that the Gompertz function (23.18) is obtained from formula (23.15) with $\mu(t) = ae^{bt}$.

23.16 The average age-length relationship for commercial-sized male rock sole are found to satisfy the *von Bertalanffy growth relationship*:

$$l(t) = l_\infty[1 - e^{-K(t-t_0)}],$$

where $t_0 = -1.56$ years, $K = 0.26$, and $l_\infty = 402$ mm. (a) What is the rate of increase in the length, $l(t)$, when the fish is eight years old? (b) What is the rate of increase in the length, $l(t)$, for a fish with length $l = 376$ mm. (c) Repeat (a) and (b) for female sole that satisfy the same von Bertalanffy equation with $t_0 = -2.5$ years, $K = 0.146$, and $l_\infty = 516$ mm.

23.17 Compute the second derivative, y'', of the given function.

a) $y = 3^{-t}$ b) $y = \log(t)$
c) $y = B/(1 + ce^{-\lambda t})$ d) $y = e^{-ke^{-\lambda t}}$
e) $y = A3^{-k_1t} - B3^{-k_2t}$ f) $y = 2^{\sin(x)}$
g) $y = t2^{-kt}$ h) $y = 3[1 - e^{3(t-1)}]$
i) $y = \log(x^2)$ j) $y = \log_2(\sin(t))$
k) $y = \log(1 + t)$ l) $y = x\ln(x) - x$
m) $y = 2^{e^t}$ n) $y = e^{x^2}$

23.18 Evaluate each of the following integrals.

a) $\int e^t\,dt$ b) $\int e^{3t+1}\,dt$

c) $\int xe^{x^2-2}\,dx$ d) $\int x^2e^{x^3}\,dx$

e) $\int (x+3)e^{(x+3)^2}\,dx$ f) $\int e^{-x}\,dx$

g) $\int xe^{3x^2-2}\,dx$ h) $\int \frac{e^{\sqrt{x+1}}}{\sqrt{x+1}}\,dx$

23.19 Evaluate each of the following integrals.

a) $\int \frac{3}{x}\,dx$ b) $\int \frac{1}{x^3}\,dx$

c) $\int \frac{x^2}{x^3-1}\,dx$ d) $\int \frac{x-1}{x^2-2x+6}\,dx$

e) $\int \frac{e^x}{1+e^x}\,dx$ f) $\int \frac{1+e^x}{x+e^x}\,dx$

g) $\int \frac{1+2x^{-3}}{-x+x^{-2}}\,dx$ h) $\int \frac{1}{x(1+\ln(x))}\,dx$

i) $\int \frac{1}{x^2(1-x^{-1})}\,dx$ j) $\int \frac{2x}{(x^2+1)\ln(x^2+1)}\,dx$

23.20 Evaluate each of the following integrals.

a) $\int 2^x \, dx$　　b) $\int 4^{x/2} \, dx$

c) $\int 3^{x+1} \, dx$　　d) $\int x3^{x^2+1} \, dx$

e) $\int e^x 2^{e^x} \, dx$　　f) $\int (5x^2+4)3^{5x^3+12x-2} \, dx$

g) $\int \frac{1}{2^x} \, dx$　　h) $\int \left(\frac{1}{3}\right)^x dx$

23.21 Evaluate the definite integrals.

a) $\int_0^5 e^x \, dx$　　b) $\int_0^{\ln(5)} e^x \, dx$

c) $\int_1^2 xe^{x^2} \, dx$　　d) $\int_2^5 (x-3)e^{(x-3)^2 dx}$

e) $\int_1^6 \frac{1}{t} \, dx$　　f) $\int_1^4 \frac{t}{t^2+1} \, dt$

g) $\int_{-1}^2 \frac{t}{t^2+1} \, dt$　　h) $\int_{-2}^3 \frac{(x-3)^2}{1+(x-3)^3} \, dx$

i) $\int_0^2 3^x \, dx$　　j) $\int_{-1}^1 2^x \, dx$

k) $\int_{-1}^1 x5^{x^2} \, dx$　　l) $\int_0^{10} \frac{1}{4^x} \, dx$

23.22 Derive formulas for $\int \tan(x)dx$ and $\int \cot(x)dx$.

23.23 Evaluate the following integrals.

a) $\int \sec^2(x)e^{\tan(x)} \, dx$　　b) $\int \cos(\theta)e^{\sin(\theta)} \, d\theta$

c) $\int \sec^2(\theta)\cot^2(\theta)e^{\cot(\theta)} \, d\theta$　　d) $\int \frac{\cos(\theta)}{\sin(\theta)+1} \, d\theta$

e) $\int \sin(\theta)\cot(\theta)2^{\sin(\theta)} \, d\theta$　　f) $\int_{-\pi}^{\pi} \sin(\theta)e^{\cos(\theta)} \, d\theta$

g) $\int_{-\pi}^{\pi} \sin(\theta)\cos(\theta)e^{\cos^2(\theta)} \, d\theta$　　h) $\int \frac{\sec^2\theta}{1-\tan(\theta)} \, d\theta$

i) $\int \csc(\theta)\cos(\theta)d\theta$　　j) $\int \frac{\cos(\theta)\sin(\theta)}{1+\cos^2(\theta)} \, d\theta$

k) $\int \frac{1}{(1+\tan(\theta))\cos^2\theta} \, d\theta$　　l) $\int \frac{\cos(\theta)}{\sin^2(\theta)} \, d\theta$

23.24 Derive the derivative of $y = 3^x$ directly without using formula (23.12).

23.25 Derive formulas (23.23) and (23.24) using the Chain Rule.

23.26 In Example 23.12, the equation (23.31) was used as the basis of the model. Assume that $T = 676$, $A = 1000$, and k_1 and k_2 are

i) $k_1 = \ln(2)/1425$; $k_2 = \ln(2)/50$.
ii) $k_1 = \ln(2)/1104$; $k_2 = \ln(2)/250$.
iii) $k_1 = \ln/880$; $k_2 = \ln(2)/400$.

a) Evaluate

$$E(t) = \int_T^t y(t)dt$$

for each choice of parameters.

b) Show that each set of parameters leads to the same t_d value, $t_d = 2176$. (Show $PC(t_d) = 50$.)

c) Compute $\mathring{Y}(t)$ at $t = 2176$ for each choice of the parameters.

23.27 For each function in Exercise 8.9 (Section 8), determine the value of t for which $y'(t) = 0$.

23.28 For each function in Exercise 8.13 (Section 8), determine the value of t for which $y'(t) = 0$.

23.29 Compute $y'(x)$ for the indicated functions $y(x)$.

a) $y(x) = (3x + 1)(2x)(-x + 6)$
b) $y(x) = (4x^2 + 2)(2x - 1)(x^3)$
c) $y(x) = (x^2 + 1)(x^2 - 1)(2x + 3)$
d) $y(x) = e^x(x + 2)(\ln(x))$
e) $y(x) = (x + 1)(x + 2)(x + 3)(x + 4)$
f) $y(x) = x^2(x + 1)^2(x + 6)^3$
g) $y(x) = (2x^2 - 1)^{1/2}(x^2 + 3x - 7)(x^{5/4})$
h) $y(x) = e^{\ln(x^2)}(3x + 2) \cdot \ln(x^2)$
i) $y(x) = (x^2 + 1)(2x^2 + 1)^2(3x^2 + 1)^3(4x^2 + 1)^4$

23.30 In Example 23.14, we derive the value $\hat{p}$. Complete the details of the calculations of this example.

23.31 If in Table 23.1 of Example 23.14, the number of individuals with the three genotypes was AA – 30, Aa – 24, and aa, 6, what is the maximum likelihood estimate for the probability of allele a?

23.32 Color blindness is a sex-linked phenomenon. If p is the probability of an individual being color blind, then the theoretical probability of a population having a given distribution of color-blind people depends on p and the proportion, m, of males in the population. This function is

$$\text{Prob.} = k(mp)^A[m(1 - p)]^B[(1 - m)p^2]^C[(1 - m)(1 - p^2)]^D.$$

The constants A, B, C, and D represent the number of color-blind males, normal males, color-blind females, and normal females, respectively.

a) Consider m constant and compute $D_p(\text{Prob.})$.
b) Consider p constant and compute $D_m(\text{Prob.})$.
c) Determine the maximum likelihood estimates of m and p, as the values $\hat{m}$ and $\hat{p}$ for which the two derivatives in (a) and (b) are simultaneously zero.

23.33 Show that equations (23.33) and (23.34) are equivalent.

SECTION 24

ADDITIONAL INTEGRATION TECHNIQUES

This section provides additional techniques for evaluating antiderivatives. The integration by parts technique is presented first as the integral counterpart to the product rule for differentiation. This technique is important and is used extensively in the remaining sections of this text. The partial-fraction technique is easily illustrated and, in Section 24.2, we present its simplest form. While this technique can become complicated, we feel the reader wishing to use the method will be best able to understand the simple cases and should refer to integral tables for more complicated integrals. Subsection 24.3 is devoted to special substitution methods. Again, we have only provided the simplest substitutions, omitting some of the more complex substitution methods. While these methods are traditionally presented in calculus courses, it is our intention to minimize these tricks of integration in favor of directing the reader to integral tables. Section 24.3 is not required in the remaining sections and is therefore optional.

24.1 INTEGRATION BY PARTS

The technique commonly called *integration by parts* is the integral counterpart of the product rule for derivatives. If f and g are differentiable functions, the product rule states:

$$D_x(fg) = fD_xg + gD_xf.$$

Integrating this equation, we have

$$\int [f(x)g(x)]'\,dx = \int f(x)g'(x)dx + \int g(x)f'(x)dx.$$

The integral (or antiderivative) of a derivative $y'(x)$ is simply the function $y(x)$, (setting the constant $c = 0$),

$$\int y'(x)dx = y(x).$$

Therefore

$$\int [f(x)g(x)]'\,dx = f(x)g(x).$$

Substituting this into the left side of the integrated product rule and transferring one term from the right to the left side of the equation, we obtain the

Integration by parts (IP) formula:

$$\int f(x)g'(x)dx = f(x)g(x) - \int g(x)f'(x)dx. \tag{24.1}$$

When considering definite integrals, the IP formula becomes

$$\int_a^b f(x)g'(x)dx = f(x)g(x)\Big|_a^b - \int_a^b g(x)f'(x)dx. \tag{24.2}$$

The integration by parts formula is used to evaluate certain integrals in which the integrand is a product of two functions. Previously, we have been able to integrate integrals of this type using the Chain Rule if the integrands had the form $f(u(x))u'(x)$. However, this method would not provide integrals of terms such as

$$xe^x, \qquad x\cos(x), \qquad \text{or} \qquad x(3x+2)^{1/5}.$$

Each of these terms is a product that is not of a form suitable for applying the Chain Rule. As we shall see, each term may be integrated using the IP formula.

The IP formula does not give the integral directly, rather it represents one integral in terms of a second integral, which hopefully is more amenable to integration. When presented with an integral involving a product of terms, it is difficult to decide which term in the integrand should be $f(x)$ and which should be $g'(x)$. Our choice for f and g' is guided by the knowledge that we must be able to integrate g' to obtain g. Then we must be able to integrate the resulting integral,

$$\int f'(x)g(x)dx.$$

This suggests that f would be chosen, so the $f'(x)$ is less complicated than $f(x)$, if possible.

Example 24.1 We illustrate the IP technique for evaluating the following integrals.

a) Consider the integral

$$\int xe^x\, dx.$$

The integrand is the product of two terms, x and e^x. As both terms are easily integrated, we choose $f(x) = x$ on the basis that $f'(x) = 1$ is a simpler function than $f(x)$. The value of $g'(x)$ is then $g'(x) = e^x$ and consequently, $g(x) = e^x$.

With this choice of f and g', the IP formula (24.1) becomes

$$\int xe^x\,dx = x \cdot e^x - \int 1 \cdot e^x\,dx$$

$$= xe^x - \int e^x = xe^x - e^x.$$

(We choose the constant of integration to be $c = 0$, unless otherwise specified.)

What would happen if we chose the functions f and g' as $f(x) = e^x$ and $g'(x) = x$? Then $f'(x) = e^x$ and $g(x) = x^2/2$. The IP formula for this choice has the form

$$\int xe^x\,dx = e^x \cdot x - \int e^x \cdot \frac{x^2}{2}\,dx.$$

This equation expresses the original integral in terms of a more complicated integral and thus presents us with a more difficult integration problem than we started with. This illustrates that the choice of the functions f and g' is crucial when using equation (24.1).

b) To evaluate

$$\int x(3x + 2)^{1/5}\,dx$$

using the IP formula, we must choose one of the terms x or $(3x + 2)^{1/5}$ to be $f(x)$ and the other to be $g'(x)$. We let

$$f(x) = x \qquad \text{and} \qquad g'(x) = (3x + 2)^{1/5},$$

then

$$f'(x) = 1 \qquad \text{and} \qquad g(x) = \tfrac{5}{18}(3x + 2)^{6/5}.$$

We used the power rule to evaluate

$$g(x) = \int (3x + 2)^{1/5}\,dx.$$

Substituting these functions into formula (24.1), we find that the IP formula becomes

$$\int x(3x + 2)^{1/5}\,dx = x \cdot \tfrac{5}{18}(3x + 2)^{6/5} - \int 1 \cdot \tfrac{5}{18}(3x + 2)^{6/5}\,dx$$

$$= \tfrac{5}{18}x(3x + 2)^{6/5} - \tfrac{25}{594}(3x + 2)^{11/5}.$$

c) $\int \ln(x + 2)dx$ does represent the integral of a product: $1 \cdot \ln(x + 2)$. Since we cannot integrate $\ln(x + 2)$, we let $f(x) = \ln(x + 2)$ and $g'(x) = 1$. Then $f'(x) = 1/(x + 2)$ and $g(x) = x + 2$. Recall that the antiderivative of one is

$x + c$; we have chosen $c = 2$ to make the arithmetic easier. Using the IP formula, we find that

$$\int \ln(x+2)dx = (x+2)\ln(x+2) - \int (x+2)\left(\frac{1}{x+2}\right)dx$$

$$= (x+2)\ln(x+2) - \int 1\,dx$$

$$= (x+2)\ln(x+2) - x.$$

d) $\int x^2 e^{-x}\,dx$. The choice of $f(x) = x^2$ and $g'(x) = e^{-x}$ leads to $f'(x) = 2x$ and $g(x) = -e^{-x}$. The IP formula then gives

$$\int x^2 e^{-x}\,dx = -x^2 e^{-x} - \int 2x(-e^{-x})dx$$

$$= -x^2 e^{-x} + 2\int x e^{-x}\,dx.$$

To evaluate the integral

$$\int x e^{-x}\,dx,$$

we make a second application of the IP formula. Let $f_1(x) = x$ and $g_1'(x) = e^{-x}$. Then $f'_1(x) = 1$ and $g_1(x) = -e^{-x}$. Using the IP formula,

$$\int x e^{-x}\,dx = -xe^{-x} - \int 1\cdot(-e^{-x})dx = -xe^{-x} - e^{-x}.$$

Combining the above equations, we obtain the complete evaluation:

$$\int x^2 e^{-x} = -x^2 e^{-x} - 2xe^{-x} - 2e^{-x}$$

$$= -e^{-x}(x^2 + 2x + 2).$$ ◀

Part (d) of the above example illustrates the process of repeated application of the IP formula. When this is done, often the original integral reappears. This may not be a bad occurrence. If the coefficient of the original integral when it reappears is not one, then the integral can be obtained algebraically without further evaluation of integrals. We illustrate this in the next example.

Example 24.2

a) To evaluate

$$\int \frac{1}{x}\ln(x)dx$$

using integration by parts, we let $f(x) = \ln(x)$ and $g'(x) = 1/x$. Then $f'(x) = 1/x$ and $g(x) = \ln(x)$. The IP formula then results in the equation

$$\int \frac{1}{x}\ln(x)dx = \ln(x)\cdot\ln(x) - \int \ln(x)\frac{1}{x}\,dx. \tag{24.3}$$

Do not despair that we have run into the same integral. Equation (24.3) has the form $I = A - I$. Solving for I, we have $2I = A$ or $I = A/2$. Thus we find the value of the integral is

$$\int \frac{1}{x} \ln(x)dx = \tfrac{1}{2}(\ln(x))^2.$$

Note: We could also use the Chain Rule for integrals to obtain this result.

b) $\int \cos(x)e^x\,dx$ is evaluated by letting $f(x) = \cos(x)$ and $g'(x) = e^x$. Then $f'(x) = -\sin(x)$ and $g(x) = e^x$. Thus the IP formula gives

$$\int \cos(x)e^x\,dx = \cos(x)e^x + \int \sin(x)e^x\,dx. \tag{24.4}$$

To evaluate

$$\int \sin(x)e^x\,dx,$$

we again apply the IP formula. Let $f(x) = \sin(x)$ and $g'(x) = e^x$. Then $f'(x) = \cos(x)$ and $g(x) = e^x$. The IP formula yields

$$\int \sin(x)e^x\,dx = \sin(x)e^x - \int \cos(x)e^x. \tag{24.5}$$

Substituting formula (24.5) into equation (24.4), we have

$$\int \cos(x)e^x\,dx = \cos(x)e^x + \sin(x)e^x - \int \cos(x)e^x\,dx,$$

or

$$2\int \cos(x)e^x\,dx = (\cos(x) + \sin(x))e^x.$$

Thus

$$\int \cos(x)e^x\,dx = (\cos(x) + \sin(x))e^x/2.$$ ◀

Another common way to formulate the integration by parts method is to use differentials. (Review Section 17.2.) If we let $u = f(x)$ and $v = g(x)$, then $du = f'(x)dx$ and $dv = g'(x)dx$. With this notation, the integration by parts formula reads as follows:

$$\int u\,dv = uv - \int v\,du \qquad \textbf{Integration by parts using differentials.} \tag{24.6}$$

This equation is shorter than formula (24.1) and is thus easier to remember.

Example 24.3

a) To evaluate the integral $\int x \ln(x)dx$, we let

$$u = \ln(x) \qquad \text{and} \qquad dv = x\,dx.$$

Then

$$du = \frac{1}{x}\,dx \qquad \text{and} \qquad v = \frac{x^2}{2}.$$

Thus the above version of the IP formula yields

$$\begin{aligned}\int x\ln(x)dx &= \ln(x)\frac{x^2}{2} - \int \frac{x^2}{2}\frac{1}{x}\,dx \\ &= \ln(x)\frac{x^2}{2} - \int \frac{x}{2}\,dx \\ &= \ln(x)\frac{x^2}{2} - \frac{x^2}{4}.\end{aligned}$$

b) To evaluate $\int x\cos(2x)dx$, using IP formula (24.6), we let

$$u = x \qquad \text{and} \qquad dv = \cos(2x)dx.$$

Then

$$du = dx \qquad \text{and} \qquad v = \tfrac{1}{2}\sin(2x).$$

Equation (24.6) gives

$$\begin{aligned}\int x\cos(2x)dx &= \frac{x}{2}\sin(2x) - \int \tfrac{1}{2}\sin(2x)dx \\ &= \frac{x}{2}\sin(2x) + \tfrac{1}{4}\cos(2x).\end{aligned}$$

c) $\int \sin^2(x)dx$ is evaluated by writing $\sin^2(x) = \sin(x)\sin(x)$. Setting

$$u = \sin(x) \qquad \text{and} \qquad dv = \sin(x)dx,$$

we find that

$$du = \cos(x)dx \qquad \text{and} \qquad v = -\cos(x).$$

Therefore formula (24.6) gives us

$$\begin{aligned}\int \sin^2(x)dx &= -\sin(x)\cos(x) - \int -\cos(x)\cos(x)dx \\ &= -\sin(x)\cos(x) + \int \cos^2(x)dx.\end{aligned}$$

At this point, we use the Pythagorean identity

$$\cos^2(x) = 1 - \sin^2(x).$$

Substituting into the above equation gives us

$$\begin{aligned}\int \sin^2(x)dx &= -\sin(x)\cos(x) + \int [1 - \sin^2(x)]dx \\ &= -\sin(x)\cos(x) + x - \int \sin^2(x)dx.\end{aligned}$$

Solving for $\int \sin^2(x)dx$, we find

$$\int \sin^2(x)dx = \tfrac{1}{2}[x - \sin(x)\cos(x)].$$

◀

In each of the above examples, we evaluated the indefinite integrals with the arbitrary constant $c = 0$. This was done for convenience only and the most general form of each integral would be obtained by adding an arbitrary constant to each integral. The same procedure used in the above examples could also be used to evaluate definite integrals. For instance, Example 24.3(a) provides an antiderivative of $x \ln(x)$. Thus, by the Fundamental Theorem of Calculus,

$$\begin{aligned}\int_1^e x \ln(x)dx &= \ln(x)\cdot\frac{x^2}{2} - \frac{x^2}{4}\bigg|_1^e \\ &= \left(\ln(e)\cdot\frac{e^2}{2} - \frac{e^2}{4}\right) - \left(\ln(1)\cdot\frac{1^2}{2} - \frac{1^2}{4}\right) \\ &= \frac{e^2}{4} + \frac{1}{4} = \frac{e^2+1}{4}.\end{aligned}$$

24.2 PARTIAL FRACTIONS

The *partial-fractions method* is an integration technique used to evaluate integrals whose integrands are rational expressions; these integrals have the form

$$\int \frac{f(x)}{g(x)}\,dx, \tag{24.7}$$

where $f(x)$ and $g(x)$ are polynomials and degree$(f) <$ degree(g). The basic principle underlying this technique is the rule for adding fractions. The sum of $\frac{2}{5}$ and $\frac{1}{3}$,

$$\frac{2}{5} + \frac{1}{3} = \frac{3\cdot 2 + 5\cdot 1}{5\cdot 3} = \frac{11}{15},$$

is straightforward. However, if you were given the number $\frac{11}{15}$, would you immediately associate it with the sum of $\frac{2}{5}$ and $\frac{1}{3}$? The answer is most likely not, yet this is precisely the basis of the partial fraction technique. We are given a fraction $f(x)/g(x)$ and must find two fractions $P_1(x)/Q_1(x)$ and $P_2(x)/Q_2(x)$, such that

$$\frac{f(x)}{g(x)} = \frac{P_1(x)}{Q_1(x)} + \frac{P_2(x)}{Q_2(x)}. \tag{24.8}$$

If we can represent f/g by equation (24.8), then the integral (24.7) is given by

$$\int \frac{f(x)}{g(x)}\,dx = \int \frac{P_1(x)}{Q_1(x)}\,dx + \int \frac{P_2(x)}{Q_2(x)}\,dx. \tag{24.9}$$

Just as both $\frac{2}{5}$ and $\frac{1}{3}$ are simpler than $\frac{11}{15}$, both of the integrals on the right side of the above equation should be simpler to evaluate than the original integral.

The essential feature of the quotient $f(x)/g(x)$ which must be met if we are to represent it in the form (24.8), is that the denominator $g(x)$ be a product of the polynomials $Q_1(x)$ and $Q_2(x)$:

$$g(x) = Q_1(x) \cdot Q_2(x).$$

Therefore, the choice of Q_1 and Q_2 should be apparent from factoring g. If $g(x) = x^2 - x$, then $g(x) = x(x - 1)$, and we must choose $Q_1(x) = x$ and $Q_2(x) = x - 1$ (the order does not matter). If $g(x) = (x^2 - 2)(x + 1)$, we must choose $Q_1(x) = x^2 - 2$ and $Q_2(x) = x + 1$. Once Q_1 and Q_2 have been chosen, the functions P_1 and P_2 are determined algebraically as polynomials with

$$\text{degree}(P_1) < \text{degree}(Q_1) \qquad \text{and} \qquad \text{degree}(P_2) < \text{degree}(Q_2),$$

such that equation (24.8) is valid. This requires

$$f(x) = Q_2(x)P_1(x) + Q_1(x)P_2(x).$$

The algebra involved in determining P_1 and P_2, together with the factoring of $g(x)$, are the difficult steps in using partial fractions. To avoid an extensive discussion of the algebraic theory of representing polynomials, we do not present an analysis of the general procedure. Instead, we will illustrate the above method in the special case where the function $g(x)$ is factorable into the product of linear functions.

Example 24.4

a) To evaluate

$$\int \frac{2x}{x^2 - x - 2}\,dx,$$

we represent the integrand as a quotient f/g, where

$$f(x) = 2x \qquad \text{and} \qquad g(x) = x^2 - x - 2.$$

Since $g(x)$ factors, $g(x) = (x - 2)(x + 1)$, we set

$$Q_1(x) = x - 2 \qquad \text{and} \qquad Q_2(x) = x + 1.$$

Since Q_1 and Q_2 are of degree one, we seek constants P_1 and P_2 such that

$$\frac{f(x)}{g(x)} = \frac{P_1}{Q_1} + \frac{P_2}{Q_2} \qquad \text{or} \qquad \frac{2x}{x^2 - x - 2} = \frac{P_1}{x - 2} + \frac{P_2}{x + 1}.$$

This requires that

$$2x = P_1(x + 1) + P_2(x - 2);$$

$$2x = [P_1x + P_1 + P_2x - 2P_2] = (P_1 + P_2)x + (P_1 - 2P_2).$$

This is an equation involving polynomials and it is valid only if the coefficient of like powers of x are identical. We obtain two equations by setting the coefficients of the x terms equal and the constants equal

$$2 = P_1 + P_2 \qquad \text{and} \qquad 0 = P_1 - 2P_2.$$

Solving these equations simultaneously, we find

$$P_1 = \tfrac{4}{3} \qquad \text{and} \qquad P_2 = \tfrac{2}{3}.$$

Consequently, corresponding to equation (24.9), we find

$$\begin{aligned}\int \frac{2x}{x^2 - x - 2}\,dx &= \int \frac{4/3}{x - 2}\,dx + \int \frac{2/3}{x + 1}\,dx \\ &= (\tfrac{4}{3})\ln(x - 2) + (\tfrac{2}{3})\ln(x + 1) \\ &= \tfrac{2}{3}\ln[(x - 2)^2(x + 1)].\end{aligned}$$

b) We apply the same technique when the denominator, $g(x)$, factors into three linear terms. To evaluate

$$I = \int \frac{2x^2 - 3x + 2}{x^3 - 6x^2 + 11x - 6}\,dx,$$

we set

$$f(x) = 2x^2 - 3x + 2$$

and observe that the denominator factors into the form

$$g(x) = (x - 1)(x - 2)(x - 3).$$

Consequently, we express the integrand in the form

$$\frac{f(x)}{g(x)} = \frac{P_1}{x - 1} + \frac{P_2}{x - 2} + \frac{P_3}{x - 3},$$

where the constants P_1, P_2, and P_3 are obtained by adding the three fractions and setting the numerator equal to $f(x)$:

$$f(x) = P_1(x - 2)(x - 3) + P_2(x - 1)(x - 3) + P_3(x - 1)(x - 2).$$

Multiplying the terms and gathering this gives us

$$f(x) = (P_1 + P_2 + P_3)x^2 + (-5P_1 - 4P_2 - 3P_3)x + (6P_1 + 3P_2 + 2P_3).$$

Equating the coefficients of x^2, x, and the constants of this equation, we obtain three simultaneous equations:

$$\begin{aligned} P_1 + P_2 + P_3 &= 2, && x^2 \text{ coefficients;} \\ -5P_1 - 4P_2 - 3P_3 &= -3, && x \text{ coefficients;} \\ 6P_1 + 3P_2 + 2P_3 &= 2, && \text{constants.} \end{aligned}$$

The solution of this set of equations is

$$P_1 = -\tfrac{1}{2}, \qquad P_2 = -3, \qquad P_3 = \tfrac{11}{2}.$$

The integral I is therefore equal to the sum of three integrals:

$$\begin{aligned} I &= \int \frac{-1/2}{x-1}\,dx + \int \frac{-3}{x-2}\,dx + \int \frac{11/2}{x-3}\,dx \\ &= (-\tfrac{1}{2})\ln(x-1) - 3\ln(x-2) + (\tfrac{11}{2})\ln(x-3) \\ &= \tfrac{1}{2}\ln\left[\frac{(x-3)^{11}}{(x-1)(x-2)^6}\right]. \end{aligned}$$

◀

To apply the partial fraction techniques to an integrand f/g it is assumed that degree(f) < degree(g). When this is not the case, we may reduce the integral by division so that it is valid. This is the same concept used to write $\frac{32}{5}$ as $6 + (\frac{2}{5})$. By dividing $g(x)$ into $f(x)$ (see Appendix I for division of polynomials), we can express

$$\frac{f(x)}{g(x)} = h(x) + \frac{R(x)}{g(x)},$$

where $R(x)$ is the remainder polynomial and hence degree(R) < degree(g). If we divide $f(x) = x^2 - 1$ into $g(x) = 2x^3 - x^2 + 2$, we find that $h(x) = 2x - 1$ and the remainder is $R(x) = 2x + 1$:

$$\frac{2x^3 - x^2 + 2}{x^2 - 1} = (2x - 1) + \frac{2x + 1}{x^2 - 1}.$$

Consequently,

$$\begin{aligned} \int \frac{2x^3 - x^2 + 2}{x^2 - 1}\,dx &= \int 2x - 1\,dx + \int \frac{2x+1}{x^2-1}\,dx \\ &= x^2 - x + \int \frac{2x+1}{x^2-1}\,dx. \end{aligned}$$

The integral

$$\int \frac{2x+1}{x^2-1}\,dx$$

is of the form suited for partial fractions and may be evaluated using the method illustrated in Example 24.4; $x^2 - 1 = (x - 1)(x + 1)$.

The partial fraction technique is also applicable when the denominator $g(x)$ contains quadratic factors, such as $x^2 + 1$, that cannot be factored into linear terms. In these cases, the numerator P, corresponding to a quadratic factor Q, must be a linear term, $P = Ax + B$. For instance,

$$\frac{3x + 2}{(x-1)(x^2+1)} = \frac{Ax + B}{x^2 + 1} + \frac{C}{x + 1},$$

where the constants A, B, and C are obtained by adding the two fractions and setting the numerator equal to $3x + 2$:

$$\begin{aligned} 3x + 2 &= (Ax + B)(x - 1) + C(x^2 + 1) \\ &= (A + C)x^2 - Ax + B - C. \end{aligned}$$

This equation is equivalent to the following three equations:

$$\begin{aligned} 0 &= A + C, && \text{coefficients of } x^2; \\ 3 &= -A, && \text{coefficients of } x. \\ 2 &= B - C, && \text{constants.} \end{aligned}$$

The values of A, B, and C satisfying these equations are

$$A = -3, \qquad B = 5, \qquad C = 3.$$

Thus

$$\int \frac{3x + 2}{(x + 1)(x^2 + 1)}\, dx = \int \frac{-3x + 5}{x^2 + 1}\, dx + \int \frac{3}{x - 1}\, dx.$$

The integral

$$\int \frac{-3x + 5}{x^2 + 1}\, dx$$

will be considered in the next subsection.

Another special situation arises when the denominator $g(x)$ has multiple factors, such as $g(x) = (x - 2)(x + 1)^2$ or $g(x) = (x + 2)^3$. When g has a factor $(x - a)^n$, the fraction expansion of g must include terms of the form

$$\frac{P_1}{x - a} + \frac{P_2}{(x - a)^2} + \cdots + \frac{P_n}{(x - a)^n} + \text{terms with other factors of } g.$$

For instance,

$$\frac{3x^2 + 2x + 1}{x^3} = \frac{3}{x} + \frac{2}{x^2} + \frac{1}{x^3};$$

$$\frac{2x^2 - 4x - 5}{(x - 3)(x + 2)^2} = \frac{1/25}{x - 3} + \frac{49/25}{x + 2} + \frac{-11/5}{(x + 2)^2}.$$

The integral of a rational expression, f/g, is therefore always represented by the sum of n integrals, where n is the number of factors of $g(x)$, counting repeated factors (assuming no $P_i = 0$).

24.3 SUBSTITUTION METHODS

Substitution methods for evaluating integrals are used to transform a complicated integral into a simpler form, which can be evaluated using the basic integral formulas. We first experienced a substitution procedure in Section 19 when we discussed the use of the Chain Rule for integrals (or antiderivatives) to develop generalized integral formulas. To recognize that an integral has the form of one of the basic integrals (i.e., the form of a power function, exponential function, etc.), we substitute $u(x)$ for an expression involving x. We then compute $u'(x)$ and check to see whether or not the integrand is a function of u and u' that we could easily integrate. For instance, each of the following integrals can be expressed in the simplified form indicated by choosing u as an appropriate function of x. We use the differential du in place of $u'(x)dx$:

$$\int 2x(x^2 + 5)^6\, dx \to \int u^6\, du;$$

$$\int \frac{1}{3x - 1}\, dx \to k \int \frac{1}{u}\, du;$$

$$\int (2x - 1)\cos(x^2 - x + 5)dx \to \int \cos(u)du.$$

In each of the above integrals, the substitution $u(x)$ is the "obvious" term that appears to a power, in the denominator, or as the argument of the trigonometric function. We present in this subsection some special substitutions, which will enable us to evaluate a few additional types of integrals that are not obviously in one of the basic forms as discussed in Section 19.

Forms Involving $\sqrt{ax + b}$ or $(ax + b)^r$

Integrals involving a fractional power of a linear term $ax + b$ and the variable x do not have one of the basic integral forms, since x is not the derivative of $(ax + b)^r$. With our previous integration formulas, we have been unable to evaluate integrals such as

$$\int x\sqrt{x - 1}\, dx, \quad \int \frac{x^2}{\sqrt{3x + 4}}\, dx, \quad \int xe^{\sqrt{2 - x}}\, dx,$$

since these integrals involve a term $\sqrt{ax + b}$. To evaluate each of these integrals, we make the substitution

$$u = \sqrt{ax + b}. \tag{24.10}$$

Then, squaring both sides and solving for x, we find

$$x = \frac{u^2 - b}{a}. \tag{24.11}$$

Taking the differential of this equation, we have

$$dx = \frac{2u}{a}\, du. \tag{24.12}$$

These values are then substituted into the integral to obtain a rational integrand of the new variable u.

Example 24.5 We use equations (24.10), (24.11), and (24.12) to evaluate the three integrals listed above.

a) $\int x\sqrt{x-1}\,dx$. Set $u = \sqrt{x-1}$. Then

$$x = u^2 + 1 \qquad \text{and} \qquad dx = 2u\,du.$$

Substituting these values into the integral, we obtain a new integral with u as the variable of integration, which we integrate:

$$\begin{aligned}\int x\sqrt{x-1}\,dx &= \int (u^2+1)\cdot u \cdot 2u\,du \\ &= \int 2u^4 + 2u^2\,du \\ &= \frac{2u^5}{5} + \frac{2u^3}{3}.\end{aligned}$$

At this point, we must substitute back $\sqrt{x-1} = u$:

$$\int x\sqrt{x-1}\,dx = \frac{2(x-1)^{5/2}}{5} + \frac{2(x-1)^{3/2}}{3}.$$

We have not indicated the arbitrary constant, c.

b) To evaluate

$$\int \frac{x^2}{\sqrt{3x+4}}\,dx,$$

we set

$$u = \sqrt{3x+4}, \qquad x = \frac{u^2-4}{3}, \qquad \text{and} \qquad dx = \frac{2u}{3}\,du.$$

Then

$$\begin{aligned}\int \frac{x^2}{\sqrt{3x+4}}\,dx &= \int \frac{[(u^2-4)/3]^2}{u}\cdot\frac{2u}{3}\,du \\ &= \int \frac{2}{27}(u^2-4)^2\,du \\ &= \frac{2}{27}\int u^4 - 8u^2 + 16\,du \\ &= \frac{2}{27}\left[\frac{u^5}{5} - \frac{8u^3}{3} + 16u\right] \\ &= \frac{2}{27}\left[\frac{(3x+4)^{5/2}}{5} - \frac{8(3x+4)^{3/2}}{3} + 16(3x+4)^{1/2}\right].\end{aligned}$$

c) To evaluate

$$\int xe^{\sqrt{2-x}}\,dx,$$

we set

$$u = \sqrt{2-x}, \qquad x = 2 - u^2 \qquad \text{and} \qquad dx = -2u\,du.$$

Then

$$\int xe^{\sqrt{2-x}}\,dx = \int (2 - u^2)e^u(-2u)du$$

$$= -4\int ue^u\,du + 2\int u^3e^u\,du.$$

Both of these integrals may be evaluated using integration by parts. Leaving the details to exercises, we find that

$$\int xe^{\sqrt{2-x}} = 2e^u[u^3 - 3u^2 + 4u + 4]$$

$$= 2e^{\sqrt{2-x}}[(2-x)^{3/2} + 4(2-x)^{1/2} + 3x - 2]. \quad ◀$$

Trigonometric Substitutions for Integrals Involving $a^2 + x^2$ or $a^2 - x^2$

To evaluate integrals containing quadratic forms, usually in the denominator or in a square root, the substitution of a trigonometric function will reduce the integral to one involving only trigonometric functions.

If the integral has terms	The substitution is
$a^2 - x^2$	$x = a\sin(\theta)$ or $\theta = \sin^{-1}(x/a)$
$a^2 + x^2$,	$x = a\tan(\theta)$ or $\theta = \tan^{-1}(x/a)$

These substitutions are illustrated in Fig. 24.1. The choice of the substitution in each case is made with the idea of utilizing the following Pythagorean identities:

$$\cos^2(\theta) = 1 - \sin^2(\theta) \qquad \text{and} \qquad \sec^2(\theta) = 1 + \tan^2\theta.$$

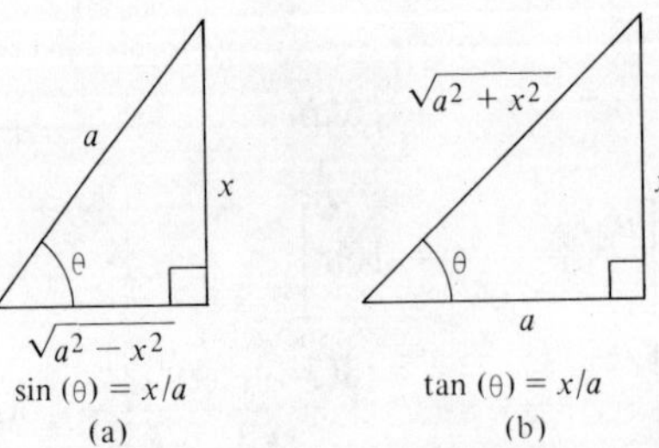

Figure 14.1

Example 24.6 We illustrate the method of trigonometric substitution to evaluate the following integrals.

a) To evaluate

$$\int \frac{1}{\sqrt{1-x^2}}\,dx,$$

we set

$$x = \sin(\theta); \qquad \text{then} \qquad dx = \cos(\theta)d\theta.$$

Then

$$\begin{aligned}\int \frac{1}{\sqrt{1-x^2}}\,dx &= \int \frac{1}{\sqrt{1-\sin^2\theta}}\cos(\theta)d\theta\\ &= \int \frac{\cos(\theta)}{\sqrt{\cos^2\theta}}\,d\theta\\ &= \int \frac{\cos(\theta)}{\cos\theta}\,d\theta\\ &= \int 1\,d\theta = \theta.\end{aligned}$$

To substitute back, we must solve $x = \sin(\theta)$ for θ; the solution is denoted $\theta = \sin^{-1}(x)$ or $\theta = \arcsin(\theta)$, where $\sin^{-1}$ (or arcsin) is the inve. ` of the sine function. Thus

$$\int \frac{1}{\sqrt{1-x^2}}\,dx = \sin^{-1}(x).$$

Compare this with Exercise 15.27(c).

b) To evaluate

$$\int \frac{1}{x\sqrt{4+x^2}}\,dx,$$

we set

$$x = 2\tan(\theta); \qquad dx = 2\sec^2(\theta)d\theta.$$

Then

$$\begin{aligned}\int \frac{1}{x\sqrt{4+x^2}}\,dx &= \int \frac{1}{2\tan(\theta)\sqrt{4+4\tan^2(\theta)}}\cdot 2\sec^2(\theta)d\theta\\ &= \int \frac{2\sec^2(\theta)}{2\tan(\theta)\cdot 2\sqrt{1+\tan^2(\theta)}}\,d\theta = \frac{1}{2}\int \frac{\sec^2(\theta)}{\tan(\theta)\sqrt{\sec^2(\theta)}}\,d\theta\\ &= \frac{1}{2}\int \frac{\sec^2(\theta)}{\tan(\theta)\sec(\theta)}\,d\theta = \frac{1}{2}\int \frac{\sec(\theta)}{\tan(\theta)}\,d\theta = \frac{1}{2}\int \frac{1}{\sin(\theta)}\,d\theta.\end{aligned}$$

To evaluate $\int (1/\sin(\theta))d\theta$, we use a "trick":

$$\int \frac{1}{\sin(\theta)}\,d\theta = \int \frac{\sin(\theta)}{\sin^2(\theta)}\,d\theta = \int \frac{\sin(\theta)}{1 - \cos^2(\theta)}\,d\theta$$

$$= \int \frac{\sin(\theta)}{(1 - \cos(\theta))(1 + \cos(\theta))}\,d\theta.$$

We now use the partial fraction technique,

$$\int \frac{1}{\sin(\theta)}\,d\theta = \int \frac{(\frac{1}{2})\sin(\theta)}{1 - \cos(\theta)}\,d\theta + \int \frac{(\frac{1}{2})\sin(\theta)}{1 + \cos(\theta)}\,d\theta.$$

Each of these integrals can be evaluated as a natural logarithm:

$$\int \frac{1}{\sin(\theta)}\,d\theta = (\tfrac{1}{2})\ln(1 - \cos(\theta)) - (\tfrac{1}{2})\ln(1 + \cos(\theta))$$

$$= (\tfrac{1}{2})\ln\left[\frac{1 - \cos(\theta)}{1 + \cos(\theta)}\right].$$

From the diagram in Fig. 24.1(b), with $a = 2$, we see that $\cos(\theta) = 2/\sqrt{4 + x^2}$. Substituting this value into the above equation and simplifying, we reach the value of the integral:

$$\int \frac{1}{x\sqrt{4 + x^2}}\,dx = \left(\frac{1}{4}\right)\ln\left[1 + \frac{8}{x^2} - \frac{4\sqrt{4 + x^2}}{x^2}\right]. \qquad ◀$$

The problems which require the use of the above trigonometric substitutions are usually not easily solved. If you come upon such integrals, it is often quicker and more reliable to seek their values in a book of tables. However, the tables were constructed by going through the process illustrated above. Also, many of these integrals must be expressed using inverse trigonometric functions. We have not discussed these in detail and have not required a familiarity of the reader with these functions. If you are not familiar with inverse trigonometric functions, it will suffice to work the exercises to simply denote the solution of $x = \tan(\theta)$ by $\theta = \tan^{-1}(x)$ and the solution of $x = \sin(\theta)$ by $\theta = \sin^{-1}(x)$.

EXERCISE SET 24

24.1 Use integration by parts to explicitly evaluate the following indefinite integrals.

a) $\int \ln(x)dx$ b) $\int x\sqrt{x + 4}\,dx$ c) $\int x^3 \cdot e^{x^2}\,dx$

d) $\int x^3 \cdot e^x\,dx$ e) $\int xe^{-kx}\,dx$ f) $\int x^2 \cdot e^{-kx}\,dx$

g) $\int \ln(\sqrt{5 + x})dx$ h) $\int \ln(x) \cdot \ln(x)dx$ i) $\int 4xe^{-6x}\,dx$

j) $\int x^5 \cdot e^{x^2}\,dx$ k) $\int [\ln(x)]^3\,dx$ l) $\int x(\ln(x))^2\,dx$

m) $\int x^{1/3}\ln(x)dx$ n) $\int \frac{\log(\log(x))}{x}\,dx$

24.2 Use integration by parts to find an antiderivative, $D_x^{-1}f$, for the given function f.

a) $f(x) = x \cdot \cos(x)$ b) $f(x) = x \cdot \sin(x)$ c) $f(x) = x \cdot \sec^2(3x)$
d) $f(x) = x^2\sin(x)$ e) $f(x) = x^3 \cdot \cos(x^2)$ f) $f(x) = \cos(x)\cos(x)$
g) $f(x) = \sin(x)e^x$ h) $f(x) = \cos^3(x)$ i) $f(x) = \sin(2x)e^x$
j) $f(x) = \cos(x)e^{-2x}$

24.3 Evaluate the following integrals.

a) $\int \sin^2(3x)dx$ b) $\int \sin^3(x)\cos(x)dx$ c) $\int \sin^3(x)dx$

d) $\int \sin^4(x)dx$ e) $\int \tan^2(x)dx$ f) $\int \frac{1}{\cos(x)}\,dx$

24.4 Evaluate the following integrals using the partial-fractions technique.

a) $\int \frac{2x+1}{x^2-1}\,dx$ b) $\int \frac{1}{x^2-1}\,dx$

c) $\int \frac{x+4}{x^2+x-2}\,dx$ d) $\int \frac{x+3}{x^2-2x}\,dx$

e) $\int \frac{3x^2+2}{(x-1)(x+2)(x+3)}\,dx$ f) $\int \frac{3x^2+x+2}{(x^2-1)(x+4)}\,dx$

g) $\int \frac{3x^2-2x}{(x-1)(x^2-3)}\,dx$ h) $\int \frac{x^2}{x^3-3x^2+2x}\,dx$

24.5 Evaluate the following integrals.

a) $\int \frac{x^2+2}{x}\,dx$ b) $\int \frac{x^2+2}{x+2}\,dx$ c) $\int \frac{x^3-x+1}{x-2}\,dx$

d) $\int \frac{x^3+1}{x+2}\,dx$ e) $\int \frac{x^2+1}{x^2-1}\,dx$ f) $\int \frac{x^3+1}{x^2-1}\,dx$

g) $\int \frac{x^3+1}{x^2+3x+2}\,dx$ h) $\int \frac{x^{10}-1}{x-1}\,dx$

24.6 Evaluate the following integrals.

a) $\int \frac{x}{(x-2)^2}\,dx$ b) $\int \frac{x^2+2}{x^3}\,dx$ c) $\int \frac{x+1}{(x-1)^2}$

d) $\int \frac{x-1}{x^2(x+1)}\,dx$ e) $\int \frac{x^3}{(x-1)^2}\,dx$ f) $\int \frac{x^2+x+1}{(x-1)^2(x+2)^2}\,dx$

g) $\int \frac{x^2+1}{(x-1)^2(x+3)^2}\,dx$ h) $\int \frac{x+2}{x(x-1)^3}\,dx$

24.7 If $g(x)$ has a quadratic factor $Q(x) = ax^2 + bx + c$, the fraction P/Q in the representation of $f(x)/g(x)$ has a numerator of the form $P(x) = Ax + B$. Represent each of the following quotients as a sum of partial fractions, and integrate the quotients using the fractions. (Some integrals may not be evaluated with the methods we have covered thus far.)

a) $\dfrac{3x^2 + 1}{(x^2 + 1)x}$ b) $\dfrac{3x^2 - 6x + 3}{(x^2 - x + 1)(x - 2)}$

c) $\dfrac{3x^2}{(x^2 - x + 1)(x + 1)}$ d) $\dfrac{3}{(x^2 + 1)x}$

e) $\dfrac{x^2 - 3x - 2}{(x^2 + x + 1)(x)}$ f) $\dfrac{7x^3 - 1}{(x^3 - 1)(2x)}$

24.8 Evaluate the following integrals.

a) $\displaystyle\int \frac{1}{\sqrt{x + 3}}\,dx$ b) $\displaystyle\int \frac{x}{\sqrt{x + 3}}\,dx$

c) $\displaystyle\int \frac{\sqrt{x + 3}}{x}\,dx$ d) $\displaystyle\int x\sqrt{x + 3}\,dx$

e) $\displaystyle\int \frac{x^2}{\sqrt{1 - 2x}}\,dx$ f) $\displaystyle\int \frac{x - 2}{\sqrt{x - 2}}\,dx$

g) $\displaystyle\int \sin(\sqrt{2 + x})dx$ h) $\displaystyle\int e^{\sqrt{3 - 2x}}\,dx$

i) $\displaystyle\int \frac{x}{(x + 3)^{1/3}}\,dx$ j) $\displaystyle\int x(3x - 2)^{1/3}\,dx$

24.9 Verify the omitted calculations in Example 24.5(c).

24.10 Determine an antiderivative of the indicated function.

a) $f(x) = \sqrt{1 - x^2}$ b) $f(x) = \sqrt{1 + x^2}$

c) $f(x) = \dfrac{1}{1 + x^2}$ d) $f(x) = \dfrac{1}{\sqrt{1 + x^2}}$

e) $f(x) = \dfrac{1}{\sqrt{4 - x^2}}$ f) $f(x) = \dfrac{1}{3 - 2x^2}$

g) $f(x) = \dfrac{1}{x\sqrt{1 - x^2}}$ h) $f(x) = \dfrac{x^2}{\sqrt{4 + x^2}}$

24.11 Verify that

$$\int \frac{1}{\sqrt{x^2 - a^2}}\,dx = \ln(x + \sqrt{x^2 - a^2})$$

by differentiating both sides of the equation.

24.12 If we wish to apply a trigonometric substitution to an integrand involving a term

$$\sqrt{Ax^2 + Bx + C},$$

we must first complete the square of the quadratic so that it has the form

$$\sqrt{A(x - b)^2 + c}.$$

(See Section 3.3) Then we set $u = x - b$ and factor out the coefficient A to obtain a term of the form

$$(\sqrt{|A|})(\sqrt{u^2 + (c/A)}).$$

Use this technique to evaluate the following integrals.

a) $\displaystyle\int \frac{1}{\sqrt{2 - x - x^2}}\,dx$ b) $\displaystyle\int \sqrt{3x^2 - 6x + 9}\,dx$

c) $\displaystyle\int \frac{x}{\sqrt{x^2 + 2x}}\,dx$ d) $\displaystyle\int \frac{1}{\sqrt{x^2 - 6x}}\,dx$

SECTION 25

INDETERMINANT LIMITS AND SINGULAR INTEGRALS

25.1 INDETERMINANT FORMS; L'HOSPITAL'S RULE

The limit properties presented in Section 12.2 for the limit of a quotient or product are as follows:

$$\lim_{x \to b} \left[\frac{f(x)}{g(x)}\right] = \frac{\lim_{x \to b} f(x)}{\lim_{x \to b} g(x)}, \tag{25.1}$$

and

$$\lim_{x \to b} [f(x) \cdot g(x)] = \left[\lim_{x \to b} f(x)\right] \cdot \left[\lim_{x \to b} g(x)\right]. \tag{25.2}$$

These rules assume that the limits

$$A = \lim_{x \to b} f(x) \qquad \text{and} \qquad B = \lim_{x \to b} g(x)$$

are both finite, i.e., they are not equal to $+\infty$ or $-\infty$. Furthermore, the property (25.1) is valid only when $B = \lim_{x \to b} g(x)$ is not zero. When these conditions are not satisfied, the limits of f/g or $f \cdot g$ will be called *indeterminant* when the limits on the right of equation (25.1) or (25.2) assume one of the following *indeterminant forms*;

$$\frac{\text{“}0\text{”}}{0}; \quad \frac{\text{“}\infty\text{”}}{\infty}; \quad \text{“}\infty \cdot 0\text{”}.$$

For instance, the limit $\lim_{x\to 1}(\ln(x)/(1-x))$ assumes the form "0/0" if we consider

$$\frac{\lim_{x\to 1}\ln(x)}{\lim_{x\to 1}(1-x)}.$$

An example of the indeterminant form "∞/∞" is given by the limit

$$\lim_{x\to\infty}\frac{x^2}{e^x} = \frac{\lim_{x\to\infty} x^2}{\lim_{x\to\infty} e^x}.$$

The third indeterminant form, "$0 \cdot \infty$", arises in a limit such as

$$\lim_{x\to+\infty}(xe^{-x}) = \left(\lim_{x\to+\infty} x\right)\left(\lim_{x\to+\infty} e^{-x}\right).$$

The forms "0/0", "∞/∞", and "$0 \cdot \infty$" are said to be indeterminant since *they do not represent any real number*. Hence, they can not be used to evaluate the limits illustrated above. These limits are called *indeterminant limits.*

Limits of the Form "0/0"

In this subsection, we introduce a special technique, which sometimes allows us to determine the value of a limit that results in an indeterminant form. The technique for general functions employs a theorem known as *L'Hospital's Rule*. The basic idea underlying L'Hospital's Rule and the evaluation of indeterminant forms is easily illustrated by considering the limit of a rational function. What is

$$\lim_{x\to 2}\frac{x^2-6x+8}{x-2}?$$

Since

$$\frac{\lim_{x\to 2} x^2 - 6x + 8}{\lim_{x\to 2} x^x - 2} = \frac{\text{“}0\text{”}}{0},$$

this limit results in an indeterminant form. However, by factoring the numerator, we see that

$$\lim_{x\to 2}\frac{x^2-6x+8}{x-2} = \lim_{x\to 2}\frac{(x-2)(x-4)}{(x-2)}$$

$$(\text{cancelling}) = \lim_{x\to 2}(x-4) = -2.$$

Another way to *factor the numerator* is to represent it by its Taylor expansion about $x = 2$ (see Section 18.2):

$$x^2 - 6x + 8 = -2(x-2) + (x-2)^2.$$

Using this expansion, we obtain the same conclusion:

$$\lim_{x\to 2}\frac{x^2-6x+8}{x-2}=\lim_{x\to 2}\left[\frac{-2(x-2)+(x-2)^2}{x-2}\right]$$

$$=\lim_{x\to 2}[-2+(x-2)]=-2.$$

While the idea of expressing the numerator, $f(x)=x^2-6x+8$ in this instance, by its Taylor expansion seems more difficult than factoring when f is a polynomial, it provides a *method of factoring* that enables us to consider more general functions.

To examine a limit,

$$\lim_{x\to b}\frac{f(x)}{g(x)}$$

that results in the indeterminant form "0/0", we assume that f and g are *analytic functions.* This means that $f(x)$ and $g(x)$ can be represented by their Taylor series expansions about the limit point $x = b$. Assume that

$$\lim_{x\to b} f(x)=0 \qquad \text{and} \qquad \lim_{x\to b} g(x)=0.$$

This requires $f(b) = 0$ and $g(b) = 0$ so that the Taylor series for $f(x)$ and $g(x)$ about $x = b$ will have the forms

$$f(x)=f'(b)(x-b)+\frac{f''(b)(x-b)^2}{2!}+\frac{f^{(3)}(b)(x-b)^3}{3!}+\cdots, \tag{25.3}$$

and

$$g(x)=g'(b)(x-b)+\frac{g''(b)(x-b)^2}{2!}+\frac{g^{(3)}(b)(x-b)^3}{3!}+\cdots. \tag{25.4}$$

Consequently, the ratio $f(x)/g(x)$ has the form:

$$\frac{f(x)}{g(x)}=\frac{f'(b)(x-b)+\frac{1}{2}f''(b)(x-b)^2+\frac{1}{6}f^{(3)}(b)(x-b)^3+\cdots}{g'(b)(x-b)+\frac{1}{2}g''(b)(x-b)^2+\frac{1}{6}g^{(3)}(b)(x-b)^3+\cdots}. \tag{25.5}$$

Thus it appears that we can factor a power of $(x - b)$ out of both the numerator and the denominator of formula (25.5) (which corresponds to cancelling) to obtain

$$\frac{f(x)}{g(x)}=\frac{f'(b)+\frac{1}{2}f''(b)(x-b)+\frac{1}{6}f^{(3)}(b)(x-b)^2+\cdots}{g'(b)+\frac{1}{2}g''(b)(x-b)+\frac{1}{6}g^{(3)}(b)(x-b)^2+\cdots}. \tag{25.6}$$

If we let x approach b in equation (25.6) and $g'(b) \neq 0$, then we may suspect that

$$\lim_{x\to b}\frac{f(x)}{g(x)}=\frac{f'(b)}{g'(b)}, \tag{25.7}$$

since in the limit each term $(x - b)^n$ becomes zero. This conclusion holds under less stringent assumptions on f and g, even when the limit of f'/g' is infinite. We state this result as L'Hospital's Rule.

Theorem 25.1 *L'Hospital's Rule*
Let f and g be differentiable in an open interval containing $x = b$ and assume that

$$\lim_{x \to b} f(x) = 0 \quad \text{and} \quad \lim_{x \to b} g(x) = 0.$$

If

$$\lim_{x \to b} \frac{f'(x)}{g'(x)} = A,$$

i.e., the limit exists for A a finite number, $A = +\infty$, or $A = -\infty$, then

$$\lim_{x \to b} \frac{f(x)}{g(x)} = \lim_{x \to b} \frac{f'(x)}{g'(x)} = A \qquad \textbf{L'Hospital's Rule.}$$

When f' and g' are continuous and $g'(b) \neq 0$, then

$$\lim_{x \to b} \frac{f'(x)}{g'(x)} = \frac{f'(b)}{g'(b)}.$$

Also note that the hypothesis of L'Hospital's Rule contains the stipulation that

$$\lim_{x \to b} \frac{f'(x)}{g'(x)}$$

exists. This need not be the case as both $f'(b)$ and $g'(b)$ may be zero or the limit may not exist for other reasons. If $f'(b) = 0$ and $g'(b) = 0$, then, by examining the equation (25.6), we can conjecture that the limit of f/g may be $f''(a)/g''(b)$. This would correspond to employing L'Hospital's Rule twice. This is demonstrated in the examples, but we leave the formal statement as an exercise.

Example 25.1 *Applications of L'Hospital's Rule*

a) To evaluate

$$\lim_{x \to 0} \frac{e^x - 1}{x},$$

we let $f(x) = e^x - 1$ and $g(x) = x$. Then $f(0) = 0$, $g(0) = 0$, and $f'(x) = e^x$, $g'(x) = 1$. Since f' and g' are continuous,

$$\lim_{x \to 0} \frac{f'(x)}{g'(x)} = \lim_{x \to 0} \frac{e^x}{1} = \frac{1}{1} = 1.$$

Consequently, by L'Hospital's Rule, we conclude that

$$\lim_{x \to 0} \frac{e^x - 1}{x} = 1.$$

b) To evaluate

$$\lim_{x \to 3} \frac{\ln(4 - x)}{x - 3},$$

we let $f(x) = \ln(4 - x)$ and $g(x) = x - 3$. Then $f(3) = \ln(1) = 0$ and $g(3) = 0$; $f'(x) = -1/(4 - x)$ and $g'(x) = 1$. Thus

$$\lim_{x \to 3} \frac{f'(x)}{g'(x)} = \lim_{x \to 3} \frac{-1/(4 - x)}{1} = -1.$$

Consequently, by L'Hospital's Rule, we find that

$$\lim_{x \to 3} \frac{\ln(4 - x)}{x - 3} = -1.$$

c) To evaluate

$$\lim_{x \to 0} \frac{e^x - x - 1}{x^2},$$

we let $f(x) = e^x - x - 1$ and $g(x) = x^2$; $f(0) = 0$, $g(0) = 0$, $f'(x) = e^x - 1$, and $g'(x) = 2x$:

$$\lim_{x \to 0} \frac{f'(x)}{g'(x)} = \lim_{x \to 0} \frac{e^x - 1}{2x} = \frac{\text{“}0\text{”}}{0}.$$

Since the limit of f'/g' results in the indeterminant form, "0/0", we apply L'Hospital's Rule a second time. That is, we compute $(f'(x))' = f''(x) = e^x$ and $(g'(x))' = g''(x) = 2$. Then

$$\lim_{x \to 0} \frac{f''(x)}{g''(x)} = \lim_{x \to 0} \frac{e^x}{2} = \frac{1}{2}.$$

By L'Hospital's Rule, we conclude that

$$\lim_{x \to 0} \frac{f'(x)}{g'(x)} = \frac{1}{2} \quad \text{and hence} \quad \lim_{x \to 0} \frac{f(x)}{g(x)} = \frac{1}{2}.$$ ◀

The limit point b in L'Hospital's Rule need not be a finite number, it can be replaced by either $+\infty$ or $-\infty$. In so doing, the assumption that the derivatives exist in an interval containing b must be altered. The necessary condition would be for the derivative to exist in a neighborhood of $+\infty$ or $-\infty$. Such a neighborhood would be an infinite interval $(a, +\infty)$ or $(-\infty, a)$, respectively. The limit as $x \to +\infty$ or $x \to -\infty$ is necessarily a one-sided limit.

Even for a finite limit point, the limit in L'Hospital's Rule can be replaced by the one-sided limits, $x \to b^+$ or $x \to b^-$, with corresponding changes in the assumption concerning the existence of derivatives. One-sided limits are necessary when considering limits of functions which are not defined on one side of the limit point. For instance, we can only consider the limit of $x \log(x)$ as $x \to 0^+$, since $\log(x)$ is not defined for $x \le 0$. In this case, the function $\log(x)$ is not even defined at the limit point. The hypothesis of Theorem 25.1 can actually be extended so that $f(b)$ and $g(b)$ need not exist and the derivatives, f' and g', need only exist on the puncture neighborhood of b, $(b - \delta, b) \cup (b, b + \delta)$.

Limits of the Form "∞/∞"

The limits resulting in the indeterminant form "∞/∞" are analytically very similar to limits resulting in the form "0/0". This is because if

$$\lim_{x\to b} h(x) = +\infty \text{ (or } -\infty) \qquad \text{then} \qquad \lim_{x\to b} \frac{1}{h(x)} = 0.$$

In fact, it can be shown that L'Hospital's Rule also applies to the limit f/g when f and g both become unbounded. The symbol "∞/∞" is used to represent one of the four possible combinations, "$+\infty/+\infty$", "$+\infty/-\infty$", "$-\infty/+\infty$", and "$-\infty/-\infty$". When used as a limit point or as a limit of integration, ∞ will simply represent $+\infty$, while $\pm\infty$ will mean either $+\infty$ or $-\infty$.

Theorem 25.2 *L'Hospital's Rule*
Assume that f and g are differentiable in a neighborhood of a limit point b (b may be finite, $+\infty$, or $-\infty$.) Assume that

$$\lim_{x\to b} f(x) = \pm\infty \qquad \text{and} \qquad \lim_{x\to b} g(x) = \pm\infty.$$

If

$$\lim_{x\to b} \frac{f'(x)}{g'(x)} = A \text{ (a finite number or } \pm\infty),$$

then

$$\lim_{x\to b} \frac{f(x)}{g(x)} = \lim_{x\to b} \frac{f'(x)}{g'(x)} = A.$$

Example 25.2

a) To evaluate $\lim_{x\to\infty} x/e^x$, we set $f(x) = x, g(x) = e^x$. Then $f'(x) = 1, g'(x) = e^x$ and

$$\lim_{x\to\infty} \frac{f'(x)}{g'(x)} = \lim_{x\to\infty} \frac{1}{e^x} = 0.$$

Hence, by L'Hospital's Rule (Theorem 25.2),

$$\lim_{x\to\infty} \frac{x}{e^x} = 0.$$

b) To evaluate

$$\lim_{x\to 0^+} \frac{\ln(x)}{1/x},$$

we let $f(x) = \ln(x)$ and $g(x) = 1/x$. Then $f'(x) = 1/x$ and $g'(x) = -1/x^2$;

$$\lim_{x\to 0^+} \frac{f'(x)}{g'(x)} = \lim_{x\to 0^+} -x = 0.$$

Consequently, by L'Hospital's Rule,

$$\lim_{x\to 0^+} \frac{\ln(x)}{1/x} = 0.$$

c) The

$$\lim_{x\to\infty} \frac{2^x}{3^x + 1}$$

is evaluated by letting $f(x) = 2^x$ and $g(x) = 3^x + 1$. Since $\lim_{x\to\infty} f(x) = \lim_{x\to\infty} g(x) = \infty$, we compute $f'(x) = \ln(2)2^x$ and $g'(x) = \ln(3)\cdot 3^x$:

$$\lim_{x\to\infty} \frac{f'(x)}{g'(x)} = \lim_{x\to\infty} \frac{\ln(2)2^x}{\ln(3)3^x}$$

$$= \lim_{x\to\infty} \frac{\ln(2)}{\ln(3)}\left(\frac{2}{3}\right)x$$

$$= \frac{\ln(2)}{\ln(3)} \lim_{x\to\infty} \left(\frac{2}{3}\right)^x = 0,$$

since $0 < \frac{2}{3} < 1$. Consequently, by L'Hospital's Rule,

$$\lim_{x\to\infty} \frac{2^x}{3^x + 1} = 0.$$

d) Since $\lim_{x\to\infty} P(x) = \infty$ for any polynomial, we need to apply L'Hospital's Rule three times to evaluate

$$\lim_{x\to\infty} \frac{2x - x^3}{1 + x^2 + 3x^3}.$$

The conclusion is that

$$\lim_{x\to\infty} \frac{f(x)}{g(x)} = \lim_{x\to\infty} \frac{f'(x)}{g'(x)} = \lim_{x\to\infty} \frac{f''(x)}{g''(x)} = \lim_{x\to\infty} \frac{f'''(x)}{g'''(x)},$$

or

$$\lim_{x\to\infty} \frac{2x - x^3}{1 + x^2 + 3x^3} = \lim_{x\to\infty} \frac{2 - 3x^2}{2x + 9x^2}$$

$$= \lim_{x\to\infty} \frac{-6x}{2 + 18x}$$

$$= \lim_{x\to\infty} \frac{-6}{18} = \frac{-1}{3}.$$

When considering the limit of rational functions at ∞, repeated applications of L'Hospital's Rule will always be necessary. There is, however, a *short-cut method*, which involves factoring the highest power of x in the rational expression from both the numerator and the denominator. This is illustrated for the above quotient:

$$\frac{2x - x^3}{1 + x^2 + 3x^3} = \frac{x^3\left(\frac{2}{x^2} - 1\right)}{x^3\left(\frac{1}{x^3} + \frac{1}{x} + 3\right)} = \frac{\frac{2}{x^2} - 1}{\frac{1}{x^3} + \frac{1}{x} + 3}.$$

Using the fact that

$$\lim_{x \to \infty} \frac{1}{x^n} = 0$$

for $n > 0$, the limit of the factored expression is easily seen to be $-\frac{1}{3}$. ◀

Limits of the Form "0·∞"

Indeterminant limits of the form "$0 \cdot \infty$" are evaluated by first translating them into one of the two forms, "0/0" or "∞/∞". This is accomplished by dividing by the reciprocal of one of the functions.

If $\lim_{x \to b} f(x) = 0$ and $\lim_{x \to b} g(x) = \infty$, we can consider the

$$\lim_{x \to b} [f(x) \cdot g(x)]$$

in two ways:

$$\lim_{x \to b} f(x) \cdot g(x) = \lim_{x \to b} \left[\frac{f(x)}{\left(\dfrac{1}{g(x)} \right)} \right]; \tag{25.8}$$

or

$$\lim_{x \to b} f(x) \cdot g(x) = \lim_{x \to b} \left[\frac{g(x)}{\left(\dfrac{1}{f(x)} \right)} \right]. \tag{25.9}$$

Equation (25.8) results in the indeterminant form "0/0" and equation (25.9) leads to the indeterminant for "∞/∞." Both are evaluated using L'Hospital's Rule.

Given a specific limit problem, which form should you use, equation (25.8) or equation (25.9)? The answer depends upon the particular functions f and g. Using equation (25.8) and L'Hospital's Rule, it will be necessary to evaluate the limit of the derivatives:

$$\lim_{x \to b} \left[\frac{f'(x)}{\left(\dfrac{-g'(x)}{g^2(x)} \right)} \right]. \tag{25.10}$$

Using the form (25.9) and L'Hospital's Rule, it will be necessary to evaluate the limit of the derivatives:

$$\lim_{x \to b} \left[\frac{g'(x)}{\left(\dfrac{-f'(x)}{f^2(x)} \right)} \right]. \tag{25.11}$$

For particular functions f and g, one of the limits, (25.10) or (25.11), may be easily evaluated, while the other may prove more difficult to evaluate than the original problem.

Example 25.3

a) Consider $\lim_{x\to\infty} x^{-2}e^x$. Let $f(x) = x^{-2}$ and $g(x) = e^x$. Then $\lim_{x\to\infty} f(x) = 0$ and $\lim_{x\to\infty} g(x) = \infty$. If we use form (25.9), we consider

$$\lim_{x\to\infty} \frac{e^x}{x^2}.$$

We can apply L'Hospital's Rule twice to obtain

$$\lim_{x\to\infty} \frac{e^x}{x^2} = \lim_{x\to\infty} \frac{e^x}{2x} = \lim_{x\to\infty} \frac{e^x}{2} = \infty.$$

However if we use form (25.8), we must evaluate

$$\lim_{x\to\infty} \frac{x^{-2}}{e^{-x}}.$$

To apply L'Hospital's Rule, we take derivatives and obtain

$$\lim_{x\to\infty} \frac{-2x^{-3}}{-e^{-x}}.$$

This limit is no easier to evaluate than the original limit.

b) Consider

$$\lim_{x\to 3^+} [(x-3)\cdot \ln(x-3)].$$

Let $f(x) = x - 3$, and $g(x) = \ln(x - 3)$; $\lim_{x\to 3^+} f(x) = 0$ and $\lim_{x\to 3^+} g(x) = -\infty$. If we consider the limit (25.9), we obtain

$$\lim_{x\to 3^+} \frac{\ln(x-3)}{\dfrac{1}{x-3}} = \lim_{x\to 3^+} \left[\frac{\dfrac{1}{x-3}}{\dfrac{-1}{(x-3)^2}}\right] = \lim_{x\to 3^+} -(x-3) = 0.$$

If on the other hand we consider the limit (25.8) and try to apply L'Hospital's Rule, we must evaluate the limit of form (25.10):

$$\lim_{x\to 3^+} \frac{1}{\left(\dfrac{-1}{(x-3)(\ln(x-3))^2}\right)} = \lim_{x\to 3^+} -(x-3)(\ln(x-3))^2.$$

This limit is more difficult to evaluate than the original limit. ◀

There are several indeterminant forms, which are obtained as the limit of an exponential function

$$\lim_{x \to b} [f(x)]^{g(x)}.$$

These are evaluated using the fact that the exponential and logarithm functions are continuous. Therefore,

$$\lim_{x \to b} e^{f(x)} = \exp\left[\lim_{x \to b} f(x)\right] \qquad \text{and} \qquad \lim_{x \to b} \ln(f(x)) = \ln\left[\lim_{x \to b} f(x)\right].$$

Combining these with the identity

$$[f(x)]^{g(x)} = \exp\{\ln[[f(x)]^{g(x)}]\} = \exp\{g(x) \cdot \ln(f(x))\},$$

we obtain the identity

$$\lim_{x \to b} (f(x))^{g(x)} = \exp\left\{\lim_{x \to b} g(x) \cdot \ln(f(x))\right\}. \tag{25.12}$$

The limit of the exponential, f^g, is therefore translated into the limit of a product that, in the indeterminant case, can be evaluated by L'Hospital's Rule.

Example 25.4

a) As x approaches zero, any positive power of x also approaches zero:

$$\lim_{x \to 0^+} x^k = 0, \qquad \text{for all} \qquad k > 0.$$

But, as x approaches zero, for any number $r \in (0, 1)$, r^x approaches one:

$$\lim_{x \to 0^+} r^x = 1, \qquad \text{for all} \qquad 0 < r < 1.$$

What then is

$$\lim_{x \to 0^+} x^x?$$

Is it zero, is it one, or does it exist? Using equation (25.12), we evaluate this limit as follows:

$$\begin{aligned} \lim_{x \to 0^+} x^x &= \exp\left\{\lim_{x \to 0^+} [x \cdot \ln(x)]\right\} \\ &= \exp\left\{\lim_{x \to 0^+} \frac{\ln(x)}{1/x}\right\} \\ &= e^{(0)} = +1. \qquad \text{(By Example 24.2(b))} \end{aligned}$$

b) What is the limit as $x \to +\infty$ of $(1 + (1/x))^x$? Using arguments similar to those in part (a), we could argue that this limit should be either $+\infty$ or 1. To evaluate the limit, we use equation (25.12):

$$\lim_{x \to +\infty} \left(1 + \frac{1}{x}\right)^x = \exp\left\{ \lim_{x \to +\infty} \left[x \cdot \ln\left(1 + \frac{1}{x}\right)\right]\right\}$$

$$= \exp\left\{ \lim_{x \to +\infty} \left[\frac{\ln\left(1 + \frac{1}{x}\right)}{\frac{1}{x}} \right]\right\}$$

(By L'Hospital's Rule.)
$$= \exp\left\{ \lim_{x \to +\infty} \left[\frac{\frac{-1}{x^2}}{1 + \frac{1}{x}} \div \frac{-1}{x^2} \right]\right\}$$

$$= \exp\left\{ \lim_{x \to +\infty} \left[\frac{1}{1 + \frac{1}{x}} \right]\right\}$$

$$= e^1 = e.$$

Consequently, an alternative definition of the number e is to define

$$e = \lim_{x \to +\infty} \left(1 + \frac{1}{x}\right)^x.$$

c) In Sections 13 and 14, we used the following limit to derive the derivative of e^x and $\ln(x)$:

$$\lim_{\varepsilon \to 0} (1 + \varepsilon)^{1/\varepsilon}.$$

To evaluate this limit, we use (25.12).

$$\lim_{\varepsilon \to 0} (1 + \varepsilon)^{1/\varepsilon} = \exp\left\{ \lim_{\varepsilon \to 0} \frac{1}{\varepsilon} \ln(1 + \varepsilon)\right\}$$

$$= \exp\left\{ \lim_{\varepsilon \to 0} \left[\frac{\frac{d}{d\varepsilon} \ln(1 + \varepsilon)}{\frac{d\varepsilon}{d\varepsilon}} \right]\right\}$$

$$= \exp\left\{ \lim_{\varepsilon \to 0} \frac{\left(\frac{1}{1 + \varepsilon}\right)}{1} \right\} = e^1 = e.$$

Note that this required the knowledge of the derivative of $\ln(1 + \varepsilon)$ to apply L'Hospital's Rule. ◀

25.2 SINGULAR OR IMPROPER INTEGRALS

The definite integral $\int_a^b f(x)dx$ was defined in Section 21 as the limit of a particular type of series, $\sum f(x_n^*)\Delta x_n$. The development of the integral required (1) that the interval $[a, b]$ be finite and (2) that $f(x)$ be defined on the closed interval $[a, b]$. Theorem 21.1 stated that the $\int_a^b f(x)dx$ exists when f is continuous on $[a, b]$. Consequently, we have only considered integrals with continuous integrands over finite intervals of integration. In this section, we examine the integral $\int_a^b f(x)dx$, when either f is unbounded on (a, b) or when the endpoints, a and b, are not both finite.

Infinite Intervals of Integration

When f is continuous on every finite interval, it is possible to consider the integral of f over an unbounded interval. The basic concept is easily visualized when f is positive over $[a, \infty)$ for some finite value a. Then for any number $t > a$, $\int_a^t f(x)dx$ represents the area under the graph of f over $[a, t]$. (See Fig. 25.1.) The function

$$F(t) = \int_a^t f(x)dx$$

is monotone increasing in t (we have assumed $f > 0$). Thus, as $t \to \infty$, the value of $F(t)$ either approaches a finite limit value, A, or becomes unbounded. If it becomes unbounded, we attach no meaning to the symbol $\int_a^{+\infty} f(x)dx$. If the limit is a finite number A, then we attribute the total area under the graph of f over the infinite interval $[a, \infty)$ as the value A and represent this by the *singular* or *improper integral*

$$\int_a^{+\infty} f(x)dx.$$

The area under the graph is the limit, as $t \to \infty$, of the area under the finite portion of the graph from a to t. We formalize this concept and extend it to the infinite intervals, $(-\infty, b]$ and $(-\infty, \infty)$, as follows.

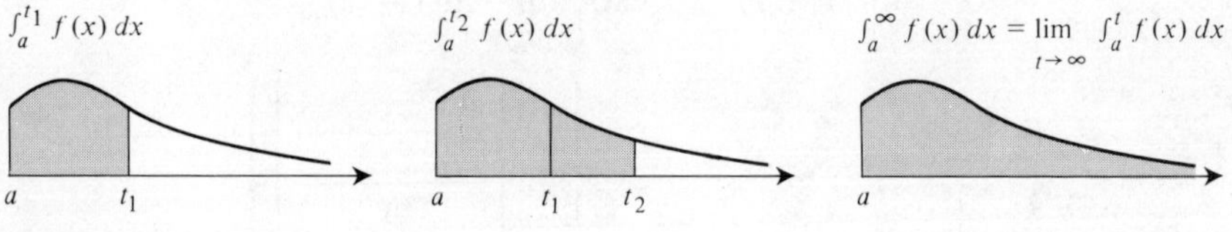

Figure 15.1

Definition 25.1 Assume that f is continuous on every finite interval.

i) If

$$\lim_{t\to+\infty} \int_a^t f(x)dx = A, \qquad \text{a finite number,}$$

then we say that $\int_a^{+\infty} f(x)dx$ exists and is equal to the number A.

ii) If

$$\lim_{t \to -\infty} \int_t^b f(x)dx = B, \qquad \text{a finite number,}$$

then we say that $\int_{-\infty}^{b} f(x)dx$ exists and is equal to the number B.

iii) If

$$\int_{-\infty}^{0} f(x)dx \qquad \text{exists and} \qquad \int_{0}^{+\infty} f(x)dx \qquad \text{exists,}$$

then we say that $\int_{-\infty}^{+\infty} f(x)dx$ exists, and we define

$$\int_{-\infty}^{+\infty} f(x)dx \equiv \int_{-\infty}^{0} f(x)dx + \int_{0}^{+\infty} f(x)dx.$$

iv) The $\int_a^b f(x)dx$, where $a = -\infty$ or $b = +\infty$, is an *improper integral* or a *singular integral.* When a singular integral exists as defined in (i), (ii), or (iii), it is said to *converge*; when it does not exist, it is said to *diverge.*

Example 25.5 We examine several integrals over unbounded intervals. In each case, we employ the appropriate clause of Definition 25.1 to determine whether or not the integral exists.

a) $\int_1^{+\infty} \frac{1}{x^2}\, dx$. We examine

$$\lim_{t \to +\infty} \int_1^t \frac{1}{x^2}\, dx = \lim_{t \to +\infty} \left.\frac{-1}{x}\right|_1^t = \lim_{t \to +\infty} \left(1 - \frac{1}{t}\right) = 1.$$

Consequently, the improper integral exists and

$$\int_1^{+\infty} \frac{1}{x^2}\, dx = 1.$$

b) $\int_3^{+\infty} \frac{1}{x}\, dx$. We examine

$$\lim_{t \to +\infty} \int_3^t \frac{1}{x}\, dx = \lim_{t \to +\infty} \ln(x)\Big|_3^t$$

$$= \lim_{t \to +\infty} \ln\left(\frac{t}{3}\right) = +\infty.$$

Hence $\int_3^{+\infty} 1/x\, dx$ does not exist.

c) $\int_{-\infty}^{4} e^x\, dx$. We examine

$$\lim_{t \to -\infty} \int_t^4 e^x\, dx = \lim_{t \to -\infty} e^x\Big|_t^4$$

$$= \lim_{t \to -\infty} (e^4 - e^t) = e^4.$$

Thus $\int_{-\infty}^{4} e^x\, dx$ exists and equals $e^4 \simeq 54.6$.

d) $\int_{-\infty}^{+\infty} xe^{-x^2}\,dx$. To evaluate this integral, we must examine

$$\int_0^{+\infty} xe^{-x^2} \qquad \text{and} \qquad \int_{-\infty}^{0} xe^{-x^2}\,dx.$$

First,
$$\lim_{t\to+\infty}\int_0^t xe^{-x^2}\,dx = \lim_{t\to+\infty}\left.\frac{-e^{-x^2}}{2}\right|_0^t$$
$$= \lim_{t\to+\infty}\left(\frac{1}{2}-\frac{e^{-t^2}}{2}\right)=\frac{1}{2}.$$

Thus
$$\int_0^{+\infty} xe^{-x^2}\,dx = \tfrac{1}{2}.$$

Similarly, we have
$$\lim_{t\to-\infty}\int_t^0 xe^{-x^2}\,dx = \lim_{t\to-\infty}\left.\frac{-e^{-x^2}}{2}\right|_t^0$$
$$= \lim_{t\to-\infty}\left(\frac{-1}{2}+\frac{e^{-t^2}}{2}\right)=\frac{-1}{2}.$$

Thus
$$\int_{-\infty}^{0} xe^{-x^2}\,dx = \frac{-1}{2}.$$

Combining these, we find
$$\int_{-\infty}^{+\infty} xe^{-x^2} = \int_{-\infty}^{0} xe^{-x^2}\,dx + \int_0^{+\infty} xe^{-x^2}\,dx = \frac{-1}{2}+\frac{1}{2}=0.$$

e) $\int_1^{+\infty} xe^{-x}\,dx$. To evaluate this improper integral, we consider
$$\lim_{t\to+\infty}\int_1^t xe^{-x}\,dx.$$

To evaluate this finite integral, we obtain an antiderivative of xe^{-x} using integration by parts. This was done in Example 24.1(c). Using the result obtained there, we find that this limit becomes
$$\lim_{t\to+\infty}\left. -(1+x)e^{-x}\right|_1^t = \lim_{t\to\infty}\,[2e^{-1}-(1+t)e^{-t}].$$

To evaluate this limit, we must employ L'Hospital's Rule as we have an indeterminate form "$\infty\cdot 0$". We write $(1+t)e^{-t}$ as $(1+t)/e^t$. Then we evaluate
$$\lim_{t\to+\infty}\frac{1+t}{e^t}=\lim_{t\to\infty}\frac{1}{e^t}=0.$$

Consequently, $\int_1^{+\infty} xe^{-x}\,dx$ exists and equals $2/e \simeq 0.736$. ◀

Unbounded Integrands

Let $[a, b]$ be a finite interval. Assume f is continuous on the open interval (a, b) but becomes unbounded as x approaches one of the end points, a or b. In this case the integral $\int_a^b f(x)dx$ is not defined according to our previous definition of the definite integral. To give a definition which will apply in this situation we mimic the procedure used above for infinite intervals. The situation is the same in principle. The integral of f is defined on any interior interval not including the end point where the function becomes unbounded. Thus we define the integral as the limit of the integrals over interior intervals approaching the desired interval $[a, b]$.

Definition 25.2 Let $[a, b]$ be a finite interval.

i) Assume that f is continuous on $[a, b)$ and that $\lim_{x \to b^-} f(x) = \infty$. If

$$\lim_{t \to b^-} \int_a^t f(x)dx = B, \text{ a finite number,}$$

then $\int_a^b f(x)dx$ exists and is equal to B.

ii) Assume that f is continuous on $(a, b]$ and that $\lim_{x \to a^+} f(x) = \infty$. If

$$\lim_{t \to a^+} \int_t^b f(x)dx = A, \text{ a finite number,}$$

then $\int_a^b f(x)dx$ exists and is equal to A.

iii) Assume that f is continuous on (a, b) and that $\lim_{x \to a^+} f(x) = \infty$ and $\lim_{x \to b^-} f(x) = \infty$. For any point $c \in (a, b)$, if

$$\int_a^c f(x)dx \qquad \text{and} \qquad \int_c^b f(x)dx$$

both exist, then $\int_a^b f(x)dx$ exists and

$$\int_a^b f(x)dx \equiv \int_a^c f(x)dx + \int_c^b f(x)dx.$$

iv) The integrals $\int_a^b f(x)dx$ discussed in (i), (ii), and (iii) are called *singular* or *improper integrals*. If they exist they are said to *converge* and if they do not exist they are said to *diverge*.

Example 25.6

a) The integral

$$\int_0^1 \frac{1}{x} dx$$

is an improper integral as it has a singularity at $x = 0$. We examine

$$\lim_{t \to 0^+} \int_t^1 \frac{1}{x} dx = \lim_{t \to 0^+} \ln(x)\Big|_t^1 = \lim_{t \to 0^+} - \ln(t) = \infty.$$

Consequently, $\int_0^1 1/x\, dx$ does not exist.

b) The singular integral

$$\int_0^1 \frac{1}{x^{1/2}}\, dx$$

does exist. To verify this, we examine

$$\lim_{t \to 0^+} \int_t^1 \frac{1}{x^{1/2}}\, dx = \lim_{t \to 0^+} 2x^{1/2}\Big|_t^1 = \lim_{t \to 0^+} 2 - 2\sqrt{t} = 2.$$

Thus

$$\int_0^1 \frac{1}{x^{1/2}}\, dx = 2.$$

EXERCISE SET 25

25.1 Evaluate the following limits.

a) $\lim_{x \to 1} \dfrac{x^4 + 3x^2 - 6x + 2}{x^3 - 2x^2 + 5x - 4}$

b) $\lim_{x \to 2} \dfrac{x^4 - 4x^3 + 16x - 16}{x^4 - 4x^3 + 5x^2 - 4x + 4}$

c) $\lim_{x \to 3} \dfrac{(x-3)^5(x^2 + 2x - 1)}{(x-3)^6(x+2)}$

d) $\lim_{x \to -1} \dfrac{x^3 + 3x^2 + 3x + 1}{x^4 + 5x^3 + 9x^2 + 7x + 2}$

25.2 Evaluate the following limits.

a) $\lim_{x \to 0^+} \dfrac{\sqrt{x}}{x^2 - x}$

b) $\lim_{x \to 2^+} \dfrac{\sqrt{x^2 - 4}}{\ln(x-1)}$

c) $\lim_{x \to \infty} \dfrac{e^{-x}}{\left(\dfrac{1}{x^3}\right)}$

d) $\lim_{x \to 1} \dfrac{x^3 - 1}{1 - e^{x-1}}$

e) $\lim_{x \to 5} \dfrac{e^{5-x} - 1}{(x-5)^2}$

f) $\lim_{x \to 0^+} \sqrt{x}\, \ln(x)$

25.3 Evaluate the following limits

a) $\lim_{x \to \infty} \dfrac{4x^2 - 2x + 7}{x^3 - 6x + 3}$

b) $\lim_{x \to \infty} \dfrac{5x^3 - 2x + 1}{(x+1)^3}$

c) $\lim_{x \to \infty} \dfrac{(x+2)(x-3)^2}{(x+5)^4}$

d) $\lim_{x \to \infty} \dfrac{5x^5 - 4x^6 + 2}{6x^4 + 7x^5 + 8x^6}$

25.4 Evaluate the following limits.

a) $\lim_{x \to \infty} \dfrac{\ln(x)}{x}$

b) $\lim_{x \to \infty} \dfrac{x^2}{e^x}$

c) $\lim_{x \to \infty} \dfrac{e^x}{x^2}$

d) $\lim_{x \to \infty} \dfrac{e^x + e^{-x}}{e^{x^2}}$

e) $\lim_{x \to \infty} \dfrac{e^{\sqrt{x}}}{e^x - 1}$

f) $\lim_{x \to \infty} \dfrac{e^{e^x}}{x^2}$

25.5 Evaluate the following limits.

a) $\lim_{x \to 0} xe^{-x}$ b) $\lim_{x \to 0} xe^{1/x}$

c) $\lim_{x \to 0^+} \sqrt{x} \ln(x)$ d) $\lim_{x \to 3}(x^2 - 9)/(e^{x/3} - e)$

e) $\lim_{x \to 0^+} (1 - \cos(x)) \ln(x)$ f) $\lim_{x \to \infty} \dfrac{\ln(x+1)}{x-1}$

25.6 Evaluate the following limits.

a) $\lim_{x \to \infty} x^{-x}$ b) $\lim_{x \to 0^+} x^{-x}$

c) $\lim_{x \to 0^+} x^{\sqrt{x}}$ d) $\lim_{x \to 0}(1 - x)^x$

e) $\lim_{x \to 1} (\ln(x))^{x-1}$ f) $\lim_{x \to \infty} x^{1/x}$

g) $\lim_{x \to \infty}(1 + 2/x)^x$ h) $\lim_{x \to 0} (x^2 + 1)^{x^2}$

25.7 Compute the following limits.

a) $\lim_{x \to 0} \dfrac{3^x - 5^x}{x}$ b) $\lim_{x \to 0} \dfrac{5^x - 3^x}{x^2}$

c) $\lim_{x \to \infty} \dfrac{3^x - 5^x}{x}$ d) $\lim_{x \to \infty} \dfrac{5^x - 3^x}{x^2}$

25.8 State L'Hospital's Rule and the conditions under which it holds for the one-sided limits (a) $\lim_{x \to b^+}$ and (b) $\lim_{x \to b^-}$.

25.9 Which of the following limits may not be evaluated using L'Hospital's Rule? For these, indicate the actual limit and the erroneous limit, which is obtained by incorrectly using L'Hospital's Rule.

a) $\lim_{x \to 3^+} \dfrac{\ln(x-3)}{x-3}$ b) $\lim_{x \to 2} \dfrac{e^{-x}}{(x-2)^2}$

c) $\lim_{x \to \infty} \dfrac{\cos(x)}{x}$ d) $\lim_{x \to 0} \dfrac{\cos(x)}{x}$

e) $\lim_{x \to 0} \dfrac{\sin(\sqrt{x^2})}{x^2}$ f) $\lim_{x \to 0} x \sin(x^{-2})$

g) $\lim_{x \to \infty} e^x - x$ h) $\lim_{x \to \infty} (2x)^{\sin(x)}$

25.10 State L'Hospital's Rule when $f(b) = f'(b) = g(b) = g'(b) = 0$.

25.11 Limits of the form "$\infty - \infty$" arise in limits such as $\lim_{x \to 0^+}((1/x) + \ln(x))$. Such limits can be reduced to one of the product forms by factoring one of the terms. Thus

$$\lim_{x \to 0^+} \left(\left(\frac{1}{x}\right) + \ln(x)\right) = \lim_{x \to 0^+} \left[\frac{1}{x}(1 + x \ln(x))\right].$$

In this form we see that $\lim_{x \to 0^+} 1 + x\ln(x) = 1$ and hence $\lim_{x \to 0^+}[(1/x) + \ln(x)] = +\infty$. Use this method to evaluate the following limits.

a) $\lim_{x \to 0^+}\left[\frac{1}{x} - e^{-x}\right]$ b) $\lim_{x \to \infty}[e^x - x^2]$

c) $\lim_{x \to 1^+}\left[\frac{1}{\ln(x)} - \frac{1}{x-1}\right]$ d) $\lim_{x \to 0^+}\left[\frac{1}{x} - \frac{1}{x^2}\right]$

e) $\lim_{x \to 0^+}\left[\frac{1}{x^2} - \frac{1}{x}\right]$ f) $\lim_{x \to \pi/2}\left[\tan(x) - \frac{1}{x - (\pi/2)}\right]$

25.12 Determine which of the following integrals exist. Give their value when they converge.

a) $\int_4^{+\infty} x^{-3/4}\,dx$ b) $\int_3^{\infty} xe^{-x^2}\,dx$

c) $\int_2^{\infty} x^3e^{-x^2}\,dx$ d) $\int_0^1 x^{-2/3}\,dx$

e) $\int_0^1 x^{-2}\,dx$ f) $\int_1^{10} \log(x)dx$

g) $\int_0^3 x^3\log_3(x)dx$ h) $\int_1^{10} \log_3(t-1)dt$

i) $\int_0^5 \frac{e^{-1/x}}{x^2}\,dx$ j) $\int_0^{\infty} te^{-t}\,dt$

k) $\int_0^{10} \frac{1}{(x-5)^{1/3}}\,dx$ l) $\int_1^{\infty} \frac{1}{x\ln(x)}\,dx$

m) $\int_{-\infty}^{\infty} \frac{x}{(1+x^2)^2}\,dx$ n) $\int_{-\infty}^{\infty} \frac{x}{1+x^2}\,dx$

o) $\int_{-\infty}^{\infty} x^2e^{-x^3}\,dx$ p) $\int_{-\infty}^{\infty} x^3e^{-x^4}\,dx$

25.13 Evaluate the following integrals if they exist.

a) $\int_0^{\pi/2} \tan(x)dx$ b) $\int_0^{\pi/2} \frac{\cos\sqrt{x}}{\sqrt{x}}$

c) $\int_0^{\infty} \sin(x)e^{-x}\,dx$ d) $\int_0^{\infty} \cos(t)e^{-t}\,dt$

e) $\int_0^{\pi} \frac{-\sin(1/t)}{t^2}\,dt$ f) $\int_0^{\pi/2} \sec^2(x)dx$

g) $\int_0^{\infty} \cos(t)e^{\sin(t)}\,dt$ h) $\int_0^{\infty} (\sin(t) + t\cos(t))e^{-t\sin(t)}\,dt$

i) $\int_0^{\infty} (1 + t\cos(t))e^{\sin(t)}\,dt$ j) $\int_0^{\pi} \tan(\theta)d\theta$

k) $\int_1^{\infty} \left(\frac{-1}{t^2} + \frac{\cos(t)}{t}\right)e^{\sin(t)}\, dt$ l) $\int_0^{\pi} \frac{\sin(x)}{1 - \cos^2(x)}\, dx$

m) $\int_0^{\infty} (\cos(t) - 1)e^{\sin(t) - t}\, dt$ n) $\int_0^{\infty} \sin(2\theta)d\theta$

o) $\int_{-2}^{1} \frac{x^2 + 2x}{\sqrt{4 - 3x^2 - x^3}}\, dx$ p) $\int_{-2}^{1} \frac{1}{\sqrt{2 - x - x^2}}\, dx$

25.14 The volume of an infinite *cone* obtained by rotating the graph of $y = f(x)$ about the x-axis for $x \geq 1$ is

$$V = \pi \int_1^{\infty} (f(x))^2\, dx.$$

The surface area of this cone is

$$A = 2\pi \int_1^{\infty} f(x)dx.$$

Show that the cone obtained by rotating the graph of $y = (1/x)$ has infinite surface area but finite volume. (Thus it would take an infinite amount of paint to paint its interior but it could be filled up with a finite amount of paint.)

25.15 If the amount of a drug being eliminated from an animal at time t is the indicated function $C(t)$, what is the total amount of drug administered to the animal, assuming that all of the drug will eventually be eliminated?

a) $C(t) = Ae^{-kt}$ b) $C(t) = Ate^{-kt}$ c) $C(t) = A[e^{-2t} - e^{-6t}]$
d) $C(t) = At^2e^{-kt}$ e) $C(t) = -35[e^{-t} + e^{-0.5t} - 2e^{-0.1t}]$

CHAPTER SEVEN

DIFFERENTIAL EQUATIONS

SECTION 26 AN INTRODUCTION TO DIFFERENTIAL EQUATIONS

The study of differential equations forms one of the major areas of mathematics. Traditionally, an introductory course provides a careful exhaustive examination of special methods for solving differential equations. However, for the practicing life scientists, the most important concepts are frequently the qualitative aspects of solution of differential equations. Consequently, we have avoided the standard plethora of solution methods in favor of a quick introduction to the qualitative aspects of differential equations. We only examine the direct separation of variable methods and the solution of second-order constant-coefficient homogeneous equations. (These methods may be compared to those for the second-order difference equations in Section 11.) We then examine the solution and qualitative aspects of the standard growth equations, which are used to model everything from animal populations to physiological aspects of vision. Since, in the real world of applications, most differential equations can only be solved numerically, we examine Euler's method for numerically solving initial-value problems.

SECTION 26

AN INTRODUCTION TO DIFFERENTIAL EQUATIONS

26.1 INTRODUCTION: BASIC CONCEPTS

The most fundamental aspect of any biological entity is the way in which it changes with time and how it interacts with its environment, grows, functions, and dies. The mathematical description, or model, of any biological process therefore originates in equations which describe the change that occurs. Continuous models use the derivative with respect to time to describe rates of change and are consequently described by *differential equations.*

This section is intended to be an introduction to the broad theory of differential equations. Only the simplest methods of solution are presented, but these suffice to solve many classical equations arising in biological models. The qualitative aspects of differential equations are discussed in reference to particular examples. The differential equations arising in most biological models will be fairly complicated and their solution or analysis will require techniques beyond those discussed here.

The theory of differential equations is based upon the study of *first-order equations*, which can be expressed in the general form

$$y'(t) = f(t, y), \tag{26.1}$$

where f is a function of t and y only. The simplest first-order equation is the *linear equation*:

$y'(t) = a(t)y(t) + b(t)$	**Nonhomogeneous linear differential equation.**

(26.2)

If $b(t)$ is not identically zero, then this equation is said to be *nonhomogeneous.* The *homogeneous linear first-order differential equation* has the form

$y'(t) = a(t)y(t)$	**Homogeneous first-order linear differential equation.**

(26.3)

(These equations should be compared to the corresponding difference equations (11.5), (11.6), and (11.7).)

By a solution of the differential equation (26.1), we will mean a function $y(t)$ that is differentiable and hence continuous and for which equation (26.1) is satisfied for all values of t, (or for all t in a specified set). For instance, the function $y_1(t) = \sin(t)$ is not a solution of the equation $y'(t) = y(t)$, since $y_1'(t) = \cos(t) \neq y_1(t)$. However, y_1 is a solution of the equation $y'(t) = 1 - y^2(t)$ on the interval $[-\pi/2, \pi/2]$, since for $-\pi/2 \leq t \leq \pi/2$,

$$y_1'(t) = \cos(t) = \sqrt{1 - \sin^2(t)} = \sqrt{1 - y_1^2(t)}.$$

Although there are a few tricks and shortcuts, the only way to explicitly solve a differential equation is to transform the equation into an integral equation that can be integrated. The equivalence between integrals and derivatives,

$$y = \int f(t)dt \quad \rightleftarrows \quad \frac{dy}{dt} = f(t),$$

has been repeatedly used in this text to develop both differentiation and integration formulas. The *integral equation* corresponding to the differential equation (26.1) is

$$y(t) = \int f(t, y)dt + c \tag{26.4}$$

Before discussing general methods for solving differential equations, we give a physiological model to illustrate the correspondence between differential and integral equations.

Example 26.1 *A two-chamber model of the aorta*
The aorta is a major blood vessel leaving the heart. In humans, the aorta leaves the top of the left ventricle and turns downward. (See Fig. 26.1.)

In this model, we consider the aorta to be composed of two chambers, an increasing chamber and a decreasing chamber. We will examine the blood flow through these two chambers. We denote the flow rate of the blood leaving the heart, the ascending chamber, and the descending chamber by $W(t)$, $F_1(t)$, and $F_2(t)$,

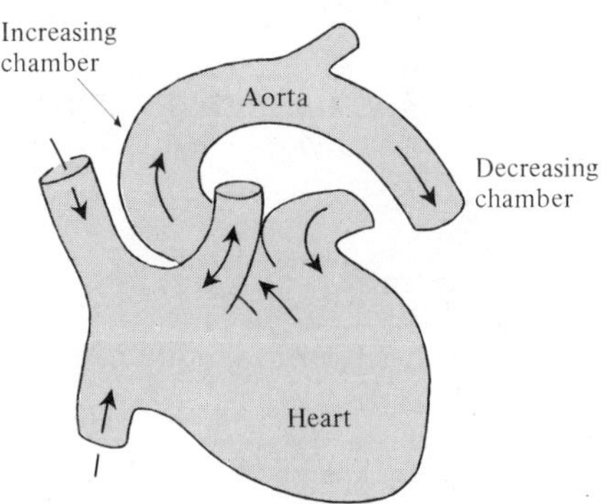

Figure 26.1

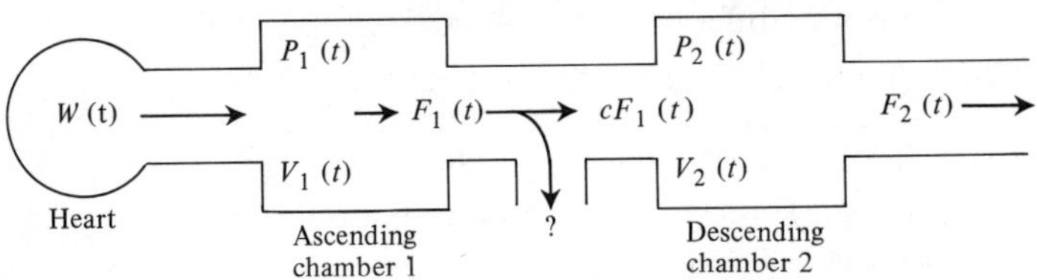

Figure 26.2

respectively. Figure 26.2 illustrates the model, where $P_1(t)$ and $P_2(t)$ denote blood pressure and $V_1(t)$ and $V_2(t)$ denote blood volume in the first and second chambers, respectively. As there are major vessels between chamber 1 and chamber 2, the flow rate of blood entering chamber 2 is specified by $c \cdot F_1(t)$, where c is a constant less than one.

We assume the pressures in the chambers remain constant and that the chambers expand or contract as their volume changes. The rate of change in their volumes is the difference between the flow rate of the blood entering and leaving the respective chambers. We express the rate of change in the volumes by differential equations:

$$\frac{dV_1(t)}{dt} = W(t) - F_1(t),$$

$$\frac{dV_2(t)}{dt} = cF_1(t) - F_2(t).$$

The *integral equations*, which correspond to these differential equations, are

$$V_1(t) = \int [W(t) - F_1(t)]dt + c_1$$

and

$$V_2(t) = \int [cF_1(t) - F_2(t)]dt + c_2.$$

The constants c_1 and c_2 are arbitrary, since an indefinite integral is only determined up to an additive constant. ◀

Separation of Variables

At first glance, the solution of the differential equation

$$y' = f(t, y)$$

appears to be given by equation (26.4). However, since the function $y(t)$ is presumably unknown, we do not know how to evaluate the integral in equation (26.4) to find $y(t)$, unless $f(t, y)$ is in fact a function of t only. The basic method of solving a differential equation will be to manipulate it into the form

$$U' = f(t), \qquad \text{whence} \qquad U = \int f(t)dt + c. \tag{26.5}$$

We employed a capital U to emphasize that in the form (26.5) U will not generally be the desired solution, $y(t)$, but a function of $y(t)$. Examples of equation (26.5) are given by

$$y' = 3t, \qquad \text{whence} \qquad y = \tfrac{3}{2}t^2 + c,$$

where $U = y$ and $U' = y'$; and by

$$e^y y' = \cos(t), \qquad \text{whence} \qquad e^y = \sin(t) + c_3$$

where $U = e^y$ and by the Chain Rule, $D_t U = e^y y'$. The process of transforming a given differential equation into the form (26.5) is called *separation of variables.* Not all first-order differential equations can be put into the form (26.5); those which can be expressed in this form are called *separable equations.*

Example 26.2 We illustrate the solution of separable equations. When no confusion is likely to result, we denote $y(t)$ and $y'(t)$ by y and y' respectively.

a) The equation $y' = t^2/y$ is separable. We can separate the t and y terms to obtain the nonlinear first-order equation

$$yy' = t^2.$$

Recognizing that

$$yy' = \frac{d}{dt}\left(\frac{y^2}{2}\right),$$

we set $U = y^2/2$. With this substitution, the equation assumes the form (26.5):

$$U' = t^2, \qquad \text{whence} \qquad U = t^3/3 + c.$$

Substituting $y^2/2$ for U, we find that the solution of our original equation is

$$\frac{y^2(t)}{2} = \left(\frac{t^3}{3}\right) + c \qquad \text{or} \qquad y(t) = \pm\sqrt{\frac{2t^3}{3} + c}.$$

Observe that the solution is only defined when $(2t^3/3) + c > 0$ and that there are two possible solutions depending on the sign of the square root.

b) We illustrate a slightly different separating procedure, utilizing differentials, to solve the homogeneous linear equation (26.3). We replace y' by the quotient of differentials $y' = dy \div dt$, as discussed in Section 17;

$$\frac{dy}{dt} = a(t)y.$$

As the differential dt is a number, we can algebraically manipulate the equation into the *separable form* by multiplying by dt and dividing by y:

$$\frac{dy}{y} = a(t)dt.$$

This equation can now be *integrated*:

$$\int \frac{dy}{y} = \int a(t)dt,$$

or

$$\ln(y) = \int a(t)dt + c.$$

Solving for y, we obtain the general solution of the equation (26.3):

General solution of $y'(t) = a(t)y(t)$,

$$y(t) = Ce^{\int a(t)dt},$$

where C is related to the constant of integration c by

$$C = e^c.$$

For instance, the solution of

$$y' = ty$$

is obtained by setting $a(t) = t$:

$$y(t) = Ce^{\int t\,dt} = Ce^{t^2/2}.$$

As another example, consider the equation

$$y' = \cos(t)y;$$

here $a(t) = \cos(t)$ and the solution is

$$y(t) = Ce^{\int \cos(t)dt} = Ce^{\sin(t)}.$$

c) Sometimes it is possible to put an equation in separable form but not be able to perform the integration. The differential equation

$$\frac{dy}{dt} = \frac{y+1}{\sqrt{2-\sin^2(t)}}$$

is separable. We determine its solution y as follows. First multiply by the differential dt, then

Divide by $y + 1$: $\quad \dfrac{dy}{y+1} = \dfrac{dt}{\sqrt{2-\sin^2(t)}}$

Integrate: $\quad \ln(y+1) = \displaystyle\int \frac{1}{\sqrt{2-\sin^2(t)}}\,dt + c$

Solve for y: $\quad y(t) = -1 + C\exp\left[\displaystyle\int \frac{1}{\sqrt{2-\sin^2(t)}}\,dt\right]$

The above solution $y(t)$ cannot be expressed in simpler terms. The integral in the exponential is an *elliptic integral* and cannot be expressed with elementary functions. Values of elliptic integrals are tabulated just as the values of $\sin(t)$ are tabulated. In this sense, the above solution is just as exact and useful as the solution in part (b) with $a(t) = \cos(t)$.

d) The equation $y' = \sin(yt) + (t/y)$ is not separable. It is impossible to express this equation in the form (26.5):

$$(\text{a function of } y \text{ only})y' = (\text{a function of } t \text{ only}).$$

Its solution therefore cannot be determined using the separation of variables method. ◀

Each solution in the above example contained an *arbitrary constant of integration*. This reflects the fact that a first-order differential equation has one degree of freedom. To determine a particular solution of a first-order differential equation, we must specify the desired solution by giving its value (or the value of its derivative) at a specific point t_0. This is called an *initial-value problem* and is posed in the following form:

$$y' = f(t, y), \quad y(t_0) = y_0 \qquad \textbf{Initial-value problem.}$$

The solution of an initial value problem is obtained from the general solution of the equation $y' = f(t, y)$ by specifying a particular value for the arbitrary constant. The value of the constant is found by substituting y_0 for y and t_0 for t in the general solution and solving for C.

Example 26.3 We solve the following initial value problems using the given general solution of the differential equation and the specified point (t_0, y_0) to determine the constant corresponding to the desired specific solution.

a) $y' = 3y$, $y(2) = 5$

General solution:	$y(t) = Ce^{3t}$
Substituting:	$5 = y(2) = Ce^6$
Solving:	$C = 5e^{-6}$
Particular solution:	$y(t) = 5e^{-6}e^{3t}$,
	$y(t) = 5e^{3(t-2)}$

b) $y^2y' = t + \sin(t)$, $y(0) = 2$.

General solution:	$y(t) = [3(t^2/2) - 3\cos(t) + c]^{1/3}$
Substituting:	$2 = [3(0^2/2) - 3\cos(0) + c]^{1/3}$
Solving	$c = 11$
Particular solution:	$y(t) = [3(t^2/2) - 3\cos(t) + 11]^{1/3}$

c) $y' = 2y + 1,\ y(0) = 3$

General solution: $y(t) = Ce^{2t} - (\frac{1}{2})$

Substituting: $3 = Ce^0 - \frac{1}{2}$

Solving: $C = 3.5$

Particular solution: $y(t) = (3.5)e^{2t} - (1/2)$ ◀

Solving $y' = ay + b$

The nonhomogeneous linear first-order equation cannot be expressed in a separable form by algebraic manipulation. The solution of the equation,

$$y'(t) = a(t)y(t) + b(t), \tag{26.2}$$

is obtained from the general solution of the associated homogeneous equation

$$y'(t) = a(t)y(t). \tag{26.3}$$

The heuristic argument is as follows: If $b(t)$ is very small, then equation (26.2) is very similar to equation (26.3). Consequently, we would expect the solution of the nonhomogeneous equation (26.2) to be essentially similar to the solution of the associated homogeneous equation (26.3). In Example 26.2 (b), we determined the solution of the homogeneous equation. For convenience, we denote this solution in the form

$$y_h(t) = Cy_1(t) \qquad (h \text{ for homogeneous}), \tag{26.6}$$

where

$$y_1(t) = e^{\int a(t)dt}$$

is a particular solution of (26.3). To obtain a solution of the nonhomogeneous equation (26.2) that is similar to the solution $y_h(t) = Cy_1(t)$ of the homogeneous equation, we replace the constant C by a function of time, $C(t)$. Thus we assume that the general solution of equation (26.2) can be expressed in the form

$$y(t) = C(t)y_1(t). \tag{26.7}$$

To determine the exact form of the function $C(t)$, we substitute the function $y(t)$, and its derivative (obtained using the Product Rule),

$$y'(t) = C'(t)y_1(t) + C(t)y_1'(t),$$

into equation (26.2). The resulting equation is as follows:

$$\underbrace{C'(t)y_1(t) + C(t)y_1'(t)}_{y'(t)} = a(t) \cdot \underbrace{C(t)y_1(t)}_{y(t)} + b(t).$$

Rearranging this equation leads to

$$C(t)[y_1'(t) - a(t)y_1(t)] + y_1(t)C'(t) = b(t).$$

However, since y_1 is a solution of the homogeneous equation, $y_1' = ay_1$, the term in the brackets is zero. Thus we are left with the following equation:

$$y_1(t)C'(t) = b(t) \qquad \text{or} \qquad C'(t) = \frac{b(t)}{y_1(t)}.$$

However, this equation is a separable differential equation, which we can solve. The left side is the derivative $C'(t)$ and the right side is a known function of t. Integrating, we obtain the function

$$C(t) = \int \frac{b(t)}{y_1(t)}\, dt + c_0,$$

where c_0 is a constant of integration. Substituting this value of $C(t)$ into equation (26.7) gives us the general solution of the nonhomogeneous equation (26.2):

$$y(t) = \left[\int \frac{b(t)}{y_1(t)}\, dt + c_0\right] y_1(t). \tag{26.8}$$

Substituting the function

$$y_1(t) = \exp \int a(t)dt$$

into (26.8), we arrive at the following:

The general solution of equation $y' = ay + b$:

$$y(t) = \left[c_0 + \int b(t)e^{-\int a(t)dt}\, dt\right] e^{\int a(t)dt}. \tag{26.9}$$

Example 26.4

a) To solve the nonhomogeneous equation

$$y' = 2y - 3,$$

we set $a(t) = 2$ and $b(t) = -3$. As

$$y_1(t) = \int a(t)dt = \int 2\, dt = 2t,$$

equation (26.9) gives the solution

$$y(t) = \left[c_0 + \int (-3)e^{-2t}\, dt\right] e^{2t}$$
$$= [c_0 + (\tfrac{3}{2})e^{-2t}]e^{2t},$$

or

$$y(t) = c_0 e^{2t} + (\tfrac{3}{2}).$$

b) To solve the nonhomogeneous equation

$$y'(t) = 3y(t) + t$$

without just substituting $a(t) = 3$, $b(t) = t$ into the formula (26.9), we can go through the process that led to the formula. We consider the particular solution of the associated homogeneous equation $y'(t) = 3y(t)$,

$$y_1(t) = e^{3t}.$$

We then set

$$y(t) = C(t)y_1(t) = C(t)e^{3t}.$$

The derivative of this function $y(t)$ is

$$y'(t) = C'(t)e^{3t} + C(t) \cdot 3e^{3t}.$$

Substituting these expressions for y and y' into the original equation gives us

$$C'(t)e^{3t} + C(t)3e^{3t} = 3C(t)e^{3t} + t.$$

Canceling the terms $C(t)3e^{3t}$ (which must cancel because of the way $y_1(t) = e^{3t}$ was chosen) results in an equation for $C'(t)$:

$$C'(t)e^{3t} = t \qquad \text{or} \qquad C'(t) = te^{-3t}.$$

Integrating this equation (using integration by parts), we find that the function $C(t)$ is determined:

$$C(t) = \int te^{-3t}\,dt = \frac{-te^{-3t}}{3} - \frac{e^{-3t}}{9} + c_0.$$

Therefore the general solution of $y'(t) = 3y(t) + t$ is as follows:

$$y(t) = \left[\frac{-te^{-3t}}{3} - \frac{e^{-3t}}{9} + c_0\right]e^{3t}$$

or

$$y(t) = c_0 e^{3t} + \left(\frac{-t}{3} - \frac{1}{9}\right). \qquad \blacktriangleleft$$

The method of solution employed to solve the nonhomogeneous equation $y' = ay + b$ is called the *variation of parameter method*, since the "trick" of the method was to allow the constant parameter C in equation (26.6) to vary with time. In biological models, the constants appearing in equations are associated with particular biological entities, which are being considered fixed. To replace a constant c by a function $c(t)$ in such cases corresponds to generalizing the model by allowing the fixed value (for instance, the thickness of a cell wall, the rate of a reaction, etc.) to vary with time; this can often result in a more realistic model.

An important observation concerning the solution of the nonhomogeneous equation (26.2) is that the solution, as expressed in (26.8), is in fact the sum of two functions:

$$y(t) = [c_0 y_1(t)] + \left[y_1(t) \int \frac{b(t)}{y_1(t)}\, dt\right].$$

Thus,

$$y(t) = \begin{bmatrix} \text{General solution of} \\ y' = ay \\ \text{Homogeneous} \end{bmatrix} + \begin{bmatrix} \text{Specific solution of} \\ y' = ay + b \\ \text{Nonhomogeneous} \end{bmatrix}.$$

This is exactly the same situation encountered when we found the general solution of the nonhomogeneous difference equation (11.6) in Section 11 (see equation (11.13)).

Example 26.5 In the single compartment model illustrated in Fig. 26.3, the amount, $Q(t)$, of a substance in the compartment is given by a first-order differential equation,

$$\mathring{Q}(t) = \mathring{Q}_{in}(t) - \mathring{Q}_{out}(t).$$

Here $\mathring{Q}_{in}$ is the rate at which the substance enters the compartment and $\mathring{Q}_{out}$ is the rate at which the substance leaves the compartment. (Recall $\mathring{Q} = (dQ/dt)$.) The standard model assumes that

$$\mathring{Q}_{out}(t) = kQ(t);$$

the quantity leaves the compartment at a rate proportional to the amount present. This would be the case if the compartment were an organ or body tissue and $Q(t)$ denoted the amount of a radioactive substance that decays at rate k. It would

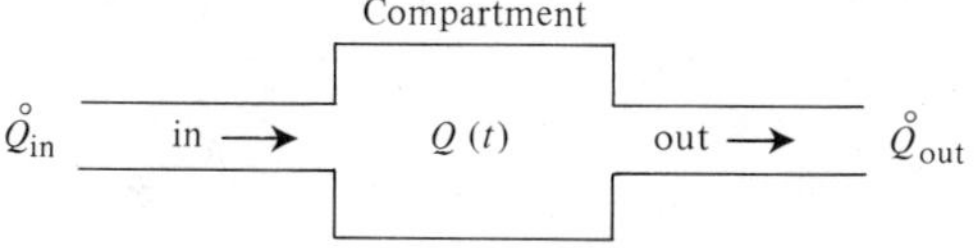

Figure 26.3

also be a reasonable assumption if the compartment were a volume of liquid such as the circulating blood and the substance $Q(t)$ is removed by uniform filtration or removal of the fluid. For convenience, we denote the rate at which the substance enters the compartment by the *input function*,

$$I(t) = \mathring{Q}_{in}(t).$$

The corresponding differential equation describing the quantity Q is

$$\mathring{Q}(t) = I(t) - kQ(t).$$

This is of the form (26.2) with $a(t) = -k$ and $b(t) = I(t)$. As $\int -k\,dt = -kt$, the quantity Q is given by the solution of form (26.9):

$$Q(t) = c_0 e^{-kt} + e^{-kt} \int I(t) e^{kt}\,dt. \tag{26.10}$$

Note that c_0 is the amount of substance in the compartment at time $t = 0$. The exact form of $Q(t)$ will depend on the input function $I(t)$.

a) Constant rate of input. If $I(t) = K$ a constant, then

$$\int I(t)e^{kt}\,dt = \int Ke^{kt}\,dt = \frac{K}{k}\,e^{kt}.$$

Consequently,

$$\begin{aligned} Q(t) &= c_0 e^{-kt} + e^{-kt}[(K/k)e^{kt}] \\ &= c_0 e^{-kt} + (K/k). \end{aligned}$$

The term $c_0 e^{-kt}$ describes the natural decay of the quantity Q in the absence of an input. When the input is at a uniform rate, the net effect is to raise the level of Q by the ratio, K/k, of the input-to-outflow rates. Consequently, as t increases towards $+\infty$, the amount $Q(t)$ approaches the steady-state K/k, regardless of the initial amount c_0.

b) If the input function is periodic, say a sinusoidal function $I(t) = 1 + \sin(\lambda t)$, then

$$\begin{aligned} \int I(t)e^{kt}\,dt &= \int [1 + \sin(\lambda t)]e^{kt}\,dt, \\ &= \int e^{kt}\,dt + \int \sin(\lambda t)e^{kt}\,dt. \end{aligned}$$

The first integral is simply e^{kt}/k. The second integral is obtained using integration by parts. We let $u = \sin(\lambda t)$ and $dv = e^{kt}\,dt$; then $du = \lambda\cos(\lambda t)dt$ and $v - c^{kt}/k$. Hence

$$\begin{aligned} \int \sin(\lambda t)\frac{e^{kt}}{k}\,dt &= \int u\,dv = uv - \int v\,du \\ &= \sin(\lambda t)\frac{e^{kt}}{k} - \int \lambda\cos(\lambda t)\frac{e^{kt}}{k}\,dt. \end{aligned}$$

To evaluate the last integral, a second application of integrations by parts is necessary. We let $u_1 = \lambda\cos(\lambda t)$ and $dv_1 = (e^{kt}/k)dt$, then $du_1 = -\lambda^2\sin(\lambda t)dt$ and $v_1 = e^{kt}/k^2$. Hence

$$\begin{aligned} \int \lambda\cos(\lambda t)\frac{e^{kt}}{k}\,dt &= \int u_1\,dv_1 = u_1 v_1 - \int v_1\,du_1 \\ &= \lambda\cos(\lambda t)\frac{e^{kt}}{k^2} - \int -\lambda^2\sin(\lambda t)\frac{e^{kt}}{k^2}\,dt. \end{aligned}$$

Combining the last two equations gives us

$$\int \sin(\lambda t)e^{kt}\,dt = \sin(\lambda t)\frac{e^{kt}}{k} - \lambda\cos(\lambda t)\frac{e^{kt}}{k^2} - \frac{\lambda^2}{k^2}\int \sin(\lambda t)e^{kt}\,dt.$$

Hence, solving this equation for the desired integral, we find that

$$\int \sin(\lambda t)e^{kt}\,dt = \frac{1}{1 + (\lambda^2/k^2)}\left[\sin(\lambda t)\frac{e^{kt}}{k} - \lambda\cos(\lambda t)\frac{e^{kt}}{k^2}\right].$$

Substituting this into the expression for $\int I(t)e^{kt}\,dt$ and the resulting expression in equation (26.10) gives us, upon simplification,

$$Q(t) = c_0 e^{-kt} + \left[\frac{1}{k} + \frac{k\sin(\lambda t) - \lambda\cos(\lambda t)}{k^2 + \lambda^2}\right].$$

The effect of having a periodic input $I(t)$ is thus seen to result in the natural decay plus a periodic component having the same period, $2\pi/\lambda$. ◀

26.2 EQUATIONS OF GROWTH AND QUALITATIVE ASPECTS OF DIFFERENTIAL EQUATIONS

One aspect of biology that has been extensively modeled and analyzed mathematically is the growth and dynamics of populations and organisms. We have repeatedly referred to specific equations used to describe particular populations. In this section, we demonstrate how the various continuous-growth functions introduced in Section 8 are arrived at as solutions of differential equations.

The Fundamental Equation of Population Dynamics

All continuous population models make the basic assumption that the dynamics of a closed population, whose size is denoted by $N(t)$, are described by a differential equation of the form

$$\boxed{\frac{dN(t)}{dt} = R(t, N)N,} \tag{26.11}$$

where $R(t, N)$ is a function that incorporates all reproductive and death factors affecting individuals of the population. The product $R \cdot N$ is then a reflection of how the total population is affected.

Exponential Growth

Perhaps the simplest and most basic form of the "reproduction" term R is obtained by setting R equal to the birth rate of the population minus the death rate of the population:

$$R = b - d \equiv r, \qquad \text{the growth rate.}$$

When these rates are constant, equation (26.11) is the model of exponential growth:

$$\frac{dN}{dt} = (b - d)N, \qquad \text{hence} \qquad N(t) = N_0 e^{(b-d)t}.$$

Malthus, in 1798, observed that, for a population with a greater birth rate than death rate ($b > d$), this equation predicts an ever-increasing population, which will become infinitely large. This qualitative analysis, $\lim_{t\to\infty} N(t) = \infty$, was in obvious contradiction to the real world in which organs, organisms, and the human population did not grow indefinitely, but after an initial exponential-growth phase tapered off to an apparent maximum size. Malthus' observations on this are said to have influenced Darwin in his studies of species.

If the birth and death rates are functions of time—for instance, sinusoidal corresponding to seasons of mating and severe weather, the solution of equation (26.11) is still exponential in form:

$$N(t) = N_0 e^{\int b(t) - d(t) dt}.$$

For instance, if

$$b(t) - d(t) = \sin\left(\frac{t + 90}{365} 2\pi\right),$$

then

$$N(t) = N_0 e^{(365/2\pi)\cos((t+90/365)2\pi)} \qquad (t \text{ days})$$

is a periodic function whose period is 1 year; it therefore describes a finite population whose size oscillates, reaching a peak size in the fall when $t = 275$.

Equations Describing Limited Growth

To describe populations that exhibit limited growth, we assume that the growth rate R of equation (26.11) is a function of the population size N. To have limited growth, $R(N)$ must approach zero as N approaches a limiting size, M.

Example 26.6 *The Logistic Equation*
The first model of this type, examined by Verhulst in 1839, is the *logistic differential equation*, which we write in the form

$$\frac{dN(t)}{dt} = r(M - N(t))N(t), \tag{26.12}$$

where M is the maximum population size. Equation (26.12) is solved using the partial-fraction method (Section 24.2) as follows. Separating variables and multiplying by $-M$, we find that the equation becomes

$$\frac{-M}{(M - N)N} dN = -Mr\, dt.$$

To integrate the term on the left, we express it as a partial fraction:

$$\int \frac{-M}{(M-N)N}\,dN = \int \frac{-1}{M-N} + \frac{-1}{N}\,dN$$

$$= \ln(M-N) - \ln(N) = \ln\left(\frac{M-N}{N}\right).$$

As

$$\int -Mr\,dt = -Mrt + C,$$

we have

$$\ln\left(\frac{M-N}{N}\right) = -Mrt + C.$$

Solving, we find that

$$\frac{M-N}{N} = ce^{-Mrt} \qquad \text{or} \qquad M = N + Nce^{-Mrt} = (1 + ce^{-Mrt})N.$$

Hence, we obtain the logistic equation (see equation (8.1)):

$$N(t) = \frac{M}{1 + ce^{-rMt}}. \tag{26.13}$$

The above derivation of the solution $N(t)$ (or a minor variant of it) appears continually in biological literature, as the basic differential equation (26.12) is used to model everything from forest growth, the spread of epidemics, and the production of wool to the generation of cell membranes. (The only tricky part of the solution is keeping the minus signs straight.)

A common alternative form of the *Pearl-Verhulst* or *logistic differential equation* is

$$\frac{dN}{dt}(t) = r\left(\frac{K - N(t)}{K}\right) N(t). \tag{26.14}$$

In this form, K is called the *carrying capacity of the population* and $(K - N)/K$ is a relative measure of how close the population is to its carrying capacity. ◀

Example 26.7 *The Gompertz equation*
A second sigmoid function, used to model limited growth, is the *Gompertz function,* which is the solution of the differential equation

$$\frac{dN(t)}{dt} = r\ln\left(\frac{M}{N(t)}\right)N(t),$$

where M is the maximum population size. The choice of the growth-rate function,

$$R = r\ln\left(\frac{M}{N}\right),$$

is not as esoteric as it might first appear. The object is to have the growth rate, R, be a positive multiple of r, the instrinsic growth rate, and to become negligible as the population size, N, approaches the maximum size, M. Since

$$\lim_{N\to M^-} \ln\left(\frac{M}{N}\right) = \ln(1) = 0,$$

this choice of R has the desired growth-limiting property. But, how can we solve the equation (26.15)? This equation is separable and assumes the form

$$\frac{-1}{\ln(M/N)N}\frac{dN}{dt} = -r.$$

The left side of this equation appears to be too difficult to evaluate, since it is more complicated than standard "textbook" equations. However, if we try to integrate it by substituting u for the most complicated part, $u = \ln(M/N)$, we see that

$$\frac{du}{dt} = \frac{du}{dN}\cdot\frac{dN}{dt} = \left[\frac{1}{(M/N)}\left(-\frac{M}{N^2}\right)\right]\cdot\frac{dN}{dt} = \frac{-1}{N}\frac{dN}{dt}.$$

Consequently, the differential equation has the form

$$\frac{1}{u}\frac{du}{dt} = -r,$$

which upon integrating becomes

$$\ln(u) = -rt + C.$$

Substituting for u and taking exponentials, we have

$$\ln\left(\frac{M}{N}\right) = ce^{-rt} \quad \text{or} \quad \ln\left(\frac{N}{M}\right) = -ce^{-rt},$$

as $\ln(N/M) = -\ln(M/N)$. The solution for $N(t)$ is given by the *Gompertz equation* (see equation (8.4)):

$$N(t) = Me^{-ce^{-rt}}. \tag{26.16}$$

◀

Qualitative Analysis of Growth Models

We were able to solve the differential equations in the growth models considered above; however, it is not difficult to construct models whose differential equations we cannot solve explicitly. In many situations, the exact form of the population function $N(t)$ need not be known. Instead, what is often required from a model is qualitative information about the growth of the population. (Keep in mind that "population" is a very general term, which can refer to any biological entity.) We consider two qualitative aspects of a population, $N(t)$, satisfying a differential equation of the form

$$N' = F(N). \tag{26.17}$$

The first qualitative aspect that we examine is the *steady state* of the solution subject to an initial value, N_0. The steady state $\bar{N}$ of an initial value problem

$$N' = F(N), \qquad N(0) = N_0,$$

is defined as the limit of the population size $N(t)$, as $t \to \infty$, if it exists.

$$\bar{N} \equiv \lim_{t \to \infty} N(t) \qquad \textbf{Steady state.}$$

The second qualitative aspect that we investigate is the *equilibrium states* of the solution and their *stability*. The equilibrium states are the values of N at which the derivative N' is zero:

N_E is an equilibrium state if $F(N_E) = 0$.

Equilibrium values are also called the *stationary values* of the solution N, since these are the values corresponding to no change in the size of the population—i.e., $N' = 0$.

$$\text{If } N(t_1) = N_E \qquad \text{then} \qquad N'(t_1) = F(N(t_1)) = F(N_E) = 0.$$

The concepts of steady-state and equilibrium values can be visualized with the aid of the graph of $F(N)$. This is usually sketched on an N versus N' coordinate system as illustrated in Fig. 26.4. If the function $F(N)$ is determined by the graph

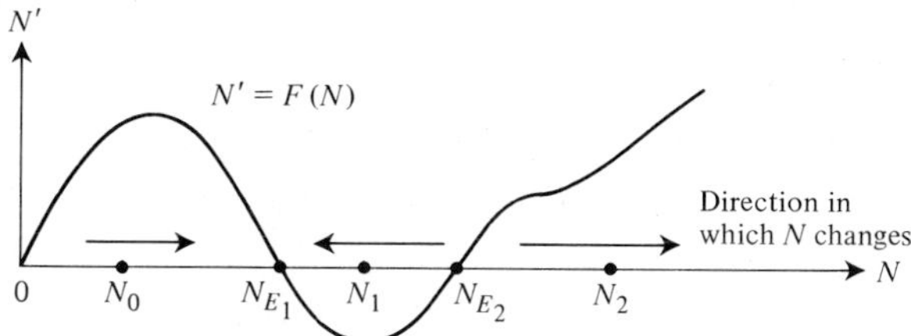

Figure 26.4

in Fig. 26.4, then the population N that satisfies the differential equation $N' = F(N)$ has three equilibrium values, which correspond to the values of N where the graph of F crosses the N-axis. These are the points $N = 0$, N_{E_1}, and N_{E_2}.

The steady state, or limiting size, of the population N depends on the initial state of the population. If at time $t = 0$, $N(0) = N_0$, as indicated on the graph, then the population will initially increase in size since its rate of change, $N'(0)$, is positive—i.e., $F(N_0)$ is positive. Since $F(N)$ is positive for all N in the interval (N_0, N_{E_1}), the size of the population increases throughout this interval. Consequently, the population must approach the equilibrium value N_{E_1} as $t \to \infty$.

$$\text{If } N(0) = N_0 \qquad \text{then} \qquad \bar{N} = \lim_{t \to \infty} N(t) = N_{E_1}.$$

If $N(0) = N_1$, as indicated on the graph, then the rate of change in the population is negative. Since $F(N) < 0$ for $N_{E_1} < N < N_1$, the population must steadily decrease in size from the value N_1 until it reaches the equilibrium value N_{E_1}.

$$\text{If } N(0) = N_1 \qquad \text{then} \qquad \bar{N} = \lim_{t \to \infty} N(t) = N_{E_1}.$$

If $N(0) = N_2$, as indicated on the graph, then the population size increases. Assuming that the function F remains positive for all $N > N_2$, the size N will increase without bound.

$$\text{If } N(0) = N_2 \qquad \text{then} \qquad \bar{N} = \lim_{t \to \infty} N(t) = +\infty.$$

From the above discussion, we see that the population N has two possible steady states (assuming that $N(0)$ is not an equilibrium value), $\bar{N} = N_{E_1}$ and $\bar{N} = +\infty$. Consequently, the model given by the graph in Fig. 26.1 and equation (26.17) predicts that a population that is not at a stationary value will either approach the stationary value N_{E_1}, or grow infinitely large. This type of qualitative analysis is often of more value than finding a complicated equation to describe $N(t)$.

The model described by the graph in Fig. 26.4 also illustrates the concept of *stability. The equilibrium value N_{E_1} is stable.* If the population size were shifted slightly from the value N_{E_1}, say by an oil spill, a forest fire, or an abnormal physiological state, the size $N(t)$ would return to the value N_{E_1}. *The equilibrium value, N_{E_2}, is unstable.* If the population size were to shift slightly from the value N_{E_2}, it would not return to the value N_{E_2}; it would either approach N_{E_1} or $+\infty$. A third possibility exists, which is not illustrated in Fig. 26.4; this is for an equilibrium value N_E to be *semistable*. In this case, a population that is slightly perturbed from the value N_E will return to the value N_E if it is on one side of the point N_E and will approach a different value if it is on the other side of N_E.

Example 26.8 The steady state, equilibrium, and stability of a population can be examined by graphing the derivative $N' = F(N)$. (See Fig. 26.5.)

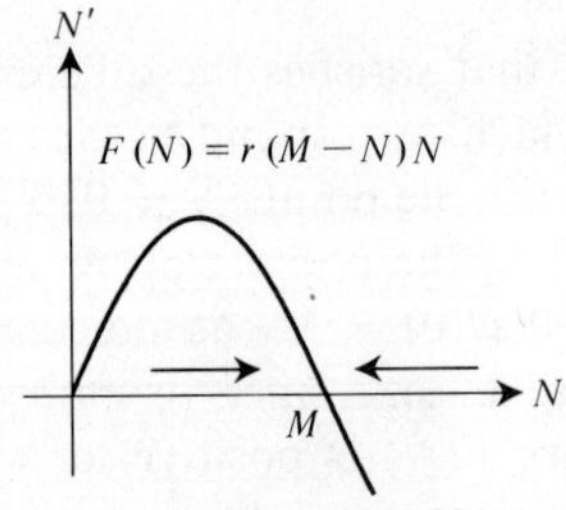

Figure 26.5

a) The logistic equation (26.12) is determined by the function

$$F(N) = r(M - N)N.$$

This function is sketched in Fig. 26.5. From the graph, we see that the population has two stationary values:

$$N = 0 \text{ (no population)} \qquad \text{and} \qquad N = M \text{ its maximum value.}$$

Since the graph of F is decreasing as it crosses the N-axis at M, we see that the equilibrium value $N_E = M$ is a stable equilibrium. Also, for $N_0 > 0$, the steady state

$$\bar{N} = \lim_{n \to \infty} N(t) = M.$$

Thus a population described by a logistic curve always approaches its maximum value.

b) The Gompertz equation is determined by the function

$$F(N) = r \ln\left(\frac{M}{N}\right) \cdot N.$$

As $F(M) = 0$, and by L'Hospital's Rule, $\lim_{N \to 0^+} F(N) = 0$, the graph of F has the form indicated in Fig. 26.6. Consequently, in this case, the population again has a stable equilibrium at $N_E = M$.

c) An example of a model that has a semi-stable equilibrium is given by the function

$$F(N) = -16N - 20N^2 + 8N^3 - N^4$$

As F factors into

$$F(N) = -N(N-2)^2(N-4),$$

the graph of F is as indicated in Fig. 26.7. Since F is positive on the intervals (0, 2) and (2, 4) and negative for $N > 4$, we see that the corresponding population has two nonzero equilibrium values:

$N_{E_1} = 2$ is a semistable equilibrium, and $N_{E_2} = 4$ is a stable equilibrium.

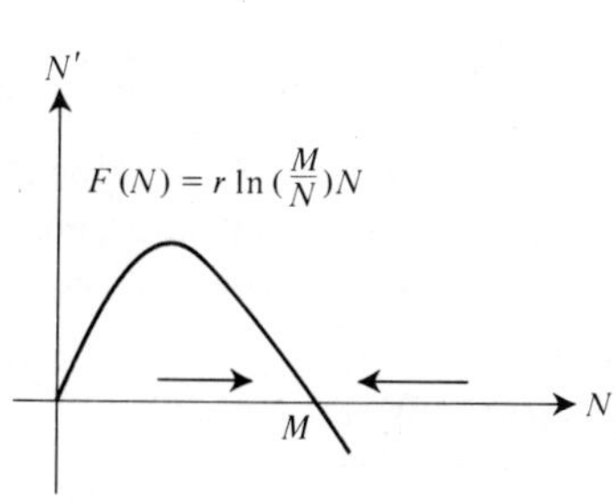

Figure 26.6

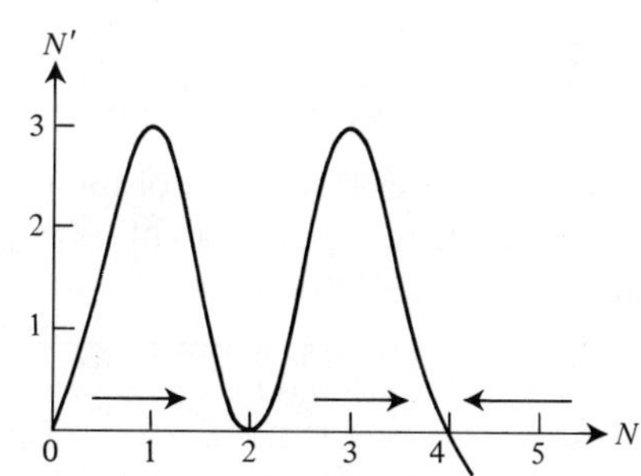

Figure 26.7

◀

26.3 SECOND-ORDER DIFFERENTIAL EQUATIONS

The solutions of higher-order differential equations (order $n \geq 2$)* in most instances cannot be explicitly obtained, especially when the equations are nonlinear. Theoretically, such differential equations can be solved by reducing them to systems of first-order differential equations; we will not explore this concept here.

There is one particular type of higher-order differential equation whose solution can be directly obtained. The most frequently encountered equation of this type is the *homogeneous second-order linear equation with constant coefficients*,

$$y'' + a_1 y' + a_0 y = 0 \qquad \textbf{Second-order equation.} \tag{26.18}$$

The solution of this equation can be derived algebraically using the fact that all derivatives of the exponential function $ce^{\lambda t}$ (c a constant) are simply multiples of $ce^{\lambda t}$. If

$$y = ce^{\lambda t} \qquad \text{then } y' = \lambda ce^{\lambda t} \qquad \text{and} \qquad y'' = \lambda^2 ce^{\lambda t}.$$

Consequently, if we substitute the function $y = ce^{\lambda t}$ and its derivatives into equation (26.18), we can factor the exponential term $e^{\lambda t}$ from each term of the equation:

$$\lambda^2 ce^{\lambda t} + a_1(\lambda ce^{\lambda t}) + a_0(ce^{\lambda t}) = 0;$$

$$ce^{\lambda t}[\lambda^2 + a_1\lambda + a_0] = 0. \tag{26.19}$$

This equation is valid when either $c = 0$, an undesirable result as it makes $y \equiv 0$, or the quadratic term is zero:

$$\lambda^2 + a_1\lambda + a_0 = 0. \tag{26.20}$$

Equation (26.20) is the *auxiliary* or *characteristic equation* associated with the differential equation (26.18). For each of the two solutions, denoted λ_1 and λ_2, of the quadratic equation (26.20), the function $y = ce^{\lambda t}$ is a solution of equation (26.18) for any choice of the constant c. We thus have two *independent* solutions of the second-order equation (26.18):

$$y_1(t) = c_1 e^{\lambda_1 t} \qquad \text{and} \qquad y_2(t) = c_2 e^{\lambda_2 t},$$

where

$$\lambda_1 = (\tfrac{1}{2})(-a_1 + \sqrt{a_1{}^2 - 4a_0}) \qquad \text{and} \qquad \lambda_2 = (\tfrac{1}{2})(-a_1 - \sqrt{a_1{}^2 - 4a_0}).$$

The most general solution of the second-order differential equation (26.18) is the sum of these two independent solutions and contains two arbitrary parameters, c_1 and c_2:

$$y(t) = c_1 e^{\lambda_1 t} + c_2 e^{\lambda_2 t} \qquad \textbf{General solution.} \tag{26.21}$$

---

* The order of a differential equation is the highest-order derivative appearing in the equation; $y''' = 3y' - t$ is of order 3; $y'' + y = 0$ is of order 2.

The fact that we can add two solutions and still have a solution is due to the linearity of the differential equation. (This is usually not possible with nonlinear equations.) The two parameters, c_1 and c_2, reflect the fact that there must be two *integrations* to transform a second derivative into the desired function,

$$y = \int y' = \iint y''.$$

The two constants, c_1 and c_2, also indicate that there are two *degrees of freedom* in the solutions.

To specify a particular solution of a second-order equation, we must specify both constants. Therefore two "bits" of information must be supplied to determine the desired specific solution. This information is usually given as an *initial-value problem*:

$$\boxed{\begin{gathered} y'' + a_1 y' + a_0 y = 0, \\ y(t_0) = y_0, \qquad y'(t_0) = y_1, \end{gathered}}$$

in which the value of the function and its derivative are specified at a particular reference point ($t_0 = 0$ in most cases). A second way to specify a specific solution is as a *boundary-value problem*:

$$\boxed{\begin{gathered} y'' + a_1 y' + a_0 y = 0, \\ y(t_0) = y_0, \qquad y(t_1) = y_1, \end{gathered}}$$

in which the value of the function is specified at the endpoints (boundary) of an interval, $[t_0, t_1]$.

Example 26.9 The general solution of the equation

$$y'' + 4y' + 3y = 0$$

is obtained by first solving the characteristic equation,

$$\lambda^2 + 4\lambda + 3 = 0; \qquad (\lambda + 3)(\lambda + 1) = 0.$$

The roots are $\lambda_1 = -3$ and $\lambda_2 = -1$ (the order in which they are chosen is not important). The general solution is therefore

$$y(t) = c_1 e^{-3t} + c_2 e^{-t}.$$

a) A specific solution is specified by the initial values

$$y(0) = 2 \qquad \text{and} \qquad y'(0) = -1.$$

To determine this specific solution, we solve the simultaneous system of two equations in the two unknowns, c_1 and c_2, obtained by evaluating the general solution and its derivative at $t_0 = 0$.

$$y(0) = 2: \qquad c_1 e^{-0} + c_2 e^{-0} = 2; \qquad c_1 + c_2 = 2.$$
$$y'(0) = -1: \qquad -3c_1 e^{-0} - c_2 e^{-0} = -1; \qquad -3c_1 - c_2 = -1.$$

Solving, we find $c_1 = -\frac{1}{2}$ and $c_2 = \frac{5}{2}$. The particular solution is thus

$$y(t) = -(\tfrac{1}{2})e^{-3t} + (\tfrac{5}{2})e^{-t}.$$

b) Another specific solution is specified by the boundary conditions $y(0) = 5$ and $y(1) = 1$. These boundary conditions, in a similar manner, yield two simultaneous equations in the two unknown constants, c_1 and c_2:

$$y(0) = c_1 e^{-0} + c_2 e^{-0}; \qquad c_1 + c_2 = 5$$
$$y(1) = c_1 e^{-3} + c_2 e^{-1}; \qquad c_1 e^{-3} + c_2 e^{-1} = 1.$$

The solutions of these equations are obtained by substituting $c_1 = 5 - c_2$ into the second equation and solving for c_2:

$$c_1 = \frac{5 - e}{1 - e^{-2}} \simeq 2.64 \qquad \text{and} \qquad c_2 = \frac{5 - e^3}{1 - e^2} \simeq 2.36.$$

The specific solution satisfying these boundary conditions is (approximately)

$$y(t) = 2.64e^{-3t} + 2.36e^{-t}.$$ ◀

If we attempt to use the general solution (26.21) to solve the equation

$$y'' + 2y' + y = 0,$$

we experience difficulties. For this equation, we find the characteristic equation has repeated roots:

$$\lambda^2 + 2\lambda + 1 = 0, \qquad \lambda_1 = -1, \qquad \lambda_2 = -1.$$

Substituting these into equation (26.21), we obtain a solution that involves only one constant, $C = c_1 + c_2$:

$$y(t) = c_1 e^{-t} + c_2 e^{-t} = (c_1 + c_2)e^{-t} = Ce^{-t}.$$

We therefore cannot have the most general solution. If we use a variation of parameter method similar to that used in Section 26.1, it can be shown that, whenever the roots λ_1 and λ_2 are the same, the general solution of the equation

$$y'' + a_1 y' + a_0 y = 0$$

has the form

$$y(t) = (c_1 + c_2 t)e^{\lambda t}, \tag{26.22}$$

where

$$\lambda_1 = \lambda_2 = \lambda = -\frac{a_1}{2}.$$

Thus the general solution of the equation $y'' + 2y' + y = 0$ is

$$y(t) = (c_1 + c_2 t)e^{-t}.$$

Using the general solution (26.21) to solve the equation $y'' + 4y = 0$, we again encounter difficulties. The characteristic equation in this case has no real solutions. It has two "imaginary" roots:

$$\lambda^2 + 4 = 0; \qquad \lambda_1 = 2i, \qquad \lambda_2 = -2i,$$

where

$$i^2 = -1.$$

To a theoretical mathematician, engineer, or physicist, the resulting general solution,

$$y(t) = c_1 e^{2it} + c_2 e^{-2it},$$

is perfectly acceptable. However, this form of solution is not practical when describing real phenomenon. It can be shown that, when the roots of the characteristic equation are "complex," the general solution of equation (26.18) can be expressed in terms of trigonometric functions.

If $\lambda_1 = \alpha + \beta i$ and $\lambda_2 = \alpha - \beta i$, then the general solution of equation (26.18) has the form

$$y(t) = [c_1 \cos(\beta t) + c_2 \sin(\beta t)]e^{\alpha t}. \tag{26.23}$$

Thus the general solution of the equation $y'' + 4y = 0$ is ($\alpha = 0, \beta = 2$):

$$y(t) = c_1 \cos(2t) + c_2 \sin(2t).$$

Observe that when the roots, λ, are complex numbers, the resulting equation involving trigonometric functions will oscillate. When the roots are "purely imaginary," $\alpha = 0$, as in the equation $y'' + 4y = 0$, the solution is strictly oscillating and a periodic function since it contains no exponential terms.

The general solution of equation (26.18) has one of the three forms given by (26.21), (26.22), and (26.23), depending on the roots of the characteristic equation (26.20). We restate these results as a theorem.

Theorem 26.1 The general solution of the equation

$$y'' + a_1 y' + a_0 y = 0$$

is determined by the characteristic roots:

$$\lambda_1 = (\tfrac{1}{2})(-a_1 + \sqrt{a_1{}^2 - 4a_0})$$

and

$$\lambda_2 = (\tfrac{1}{2})(-a_1 - \sqrt{a_1{}^2 - 4a_0}).$$

i) If λ_1 and λ_2 are real and distinct,

$$y(t) = C_1 e^{\lambda_1 t} + C_2 e^{\lambda_2 t}.$$

ii) If $\lambda_1 = \lambda_2$, $(a_1{}^2 - 4a_0 = 0)$,

$$y(t) = [c_1 + c_2 t]e^{-\lambda_1 t}.$$

iii) If λ_1 and λ_2 are complex,

$$\lambda_1 = \alpha + \beta i \quad \text{and} \quad \lambda_2 = \alpha - \beta i,$$

$$y(t) = [c_1 \cos(\beta t) + c_2 \sin(\beta t)]e^{\alpha t},$$

where

$$\alpha = -a_1/2 \quad \text{and} \quad \beta = \sqrt{4a_0 - a_1{}^2}.$$

Example 26.10 We use the formulas (26.21), (26.22), and (26.23) to express the general solution of the following differential equations according to whether the roots of their associated characteristic equations (26.20) are real and distinct, repeated, or complex numbers, respectively.

a) $y'' + 3y' - 10y = 0$: $\quad \lambda^2 + 3\lambda - 10 = 0; (\lambda + 5)(\lambda - 2) = 0;$

$\lambda_1 = -5, \lambda_2 = 2.$

$y(t) = c_1 e^{-5t} + c_2 e^{2t}.$

b) $y'' + 2y' + 0.1y = 0$: $\quad \lambda^2 + 2\lambda + 0.1 = 0;$

$\lambda_1 = -1 + \sqrt{3.6}/2; \lambda_2 = -1 - \sqrt{3.6}/2.$

$y(t) = c_1 e^{\lambda_1 t} + c_2 e^{\lambda_2 t}$

c) $y'' + 2y' + 4y = 0$: $\quad \lambda^2 + 2\lambda + 4 = 0;$

$\lambda_1 = -1 + \sqrt{3}i; \lambda_2 = -1 - \sqrt{3}i.$

$y(t) = [c_1 \cos(\sqrt{3}t) + c_2 \sin(\sqrt{3}t)]e^{-t}.$

d) $4y'' + 12y' + 9y = 0$; $\quad 4\lambda^2 + 12\lambda + 9 = 0$, or

$y'' + 3y' + \frac{9}{4}y = 0$; $\quad \lambda^2 + 3\lambda + \frac{9}{4} = 0;$

$$\lambda_1 = \lambda_2 = \frac{-3}{2}.$$

$y(t) = (c_1 + c_2 t)e^{-3t/2}.$ ◀

The general *nth-order linear constant-coefficient homogeneous differential equation* has the form

$$a_n y^{(n)} + a_{n-1} y^{(n-1)} + \cdots + a_2 y'' + a_1 y' + a_0 y = 0,$$

where $a_0, \ldots, a_n$ are constants and $a_n \neq 0$.

Using the same approach employed to solve the second-order equation (26.18), the general nth-order equation can be shown to have a solution of the form

$$y(t) = c_1 e^{\lambda_1 t} + c_2 e^{\lambda_2 t} + \cdots + c_n e^{\lambda_n t}.$$

This equation must be modified slightly if there are either repeated or complex roots of the associated characteristic equation,

$$a_n \lambda^n + a_{n-1} \lambda^{n-1} + \cdots + a_2 \lambda^2 + a_1 \lambda + a_0 = 0.$$

The essential result is that the solution of a linear constant coefficient differential equation is a sum of exponential terms if all roots of its characteristic equation are distinct. The roots $\lambda_1, \lambda_2, \ldots$ are called *characteristic values* or *eigenvalues* of the differential equation. The study of the qualitative aspects of a higher-order equation centers around the nature of the eigenvalues $\lambda_1, \lambda_2, \ldots$ We refer you to a differential equations text for a more detailed discussion of these topics. We only note that, if all of the eigenvalues are negative (or have negative real parts, in the case of complex roots), then the solution $y(t)$ will be a sum of exponentially decaying terms; consequently, as t approaches $+\infty$, the solution $y(t)$ will approach zero.

26.4 NUMERICAL SOLUTION OF DIFFERENTIAL EQUATIONS

In this section, we examine numerical solutions of first-order initial value problems of the form

$$y'(t) = F(t, y(t)), \qquad y(t_0) = y_0; \tag{26.24}$$

t_0 represents an initial time and y_0 represents the value of the solution y at time t_0. In some instances, when the term $F(t, y)$ has a particularly simple form, it is possible to obtain an exact solution $y(t)$ of problem (26.24). However, as your experience may have indicated, when the equation $y' = F(t, y)$ is not separable, the exact form of a solution $y(t)$ may be very difficult if not impossible to obtain. Even the most experienced specialists in differential equations can solve explicitly only a small class of differential equations of the form (26.24). Consequently, many initial-value problems that arise naturally in various models cannot be explicitly solved. In such cases, since real problems must be solved and not just discarded like a difficult textbook problem, the solution of the problem is approximated numerically. Thus, if the value of the solution, at a point $t = b$ is desired, a computational algorithm is employed, which will result in an approximation of $y(b)$.

In the remainder of this section, we present two basic numerical methods for approximating the solution of the initial-value problem (26.24). Each is based upon the following general concept.

The initial-value problem (26.24) is equivalent on an interval $[t_0, b]$ to the integral equation:

$$y(b) - y(t_0) = \int_{t_0}^{b} y'(t)dt$$

or, since $y' = F(t, y)$ and $y(t_0) = y_0$,

$$y(b) = y_0 + \int_{t_0}^{b} F(t, y(t))dt. \tag{26.25}$$

This suggests that the numerical solution of the initial-value problem (26.24) can be obtained by numerically approximating the integral in (26.25) using the methods discussed in Section 22. The difficulty that arises is that the approximation methods of Section 22.3 are only applicable to integrals whose integrands are completely known. In the present problem, the integrand $F(t, y(t))$ is not known since $y(t)$ is the unknown solution that we are trying to determine. Consequently, to approximate

$$\int_{t_0}^{b} F(t, y(t))dt,$$

we approximate the unknown integrand $F(t, y(t))$ over the interval $[t_0, b]$ by a function $\hat{F}(t)$ and then integrate. If

$$F(t, y(t)) \simeq \hat{F}(t)$$

then

$$\int_{t_0}^{b} F(t, y(t))dt \simeq \int_{t_0}^{b} \hat{F}(t)dt.$$

The simplest approximation of $F(t, y(t))$ on the interval $[t_0, b]$ is the constant $F(t_0, y(t_0))$. This leads to the approximation

$$y(b) \simeq y_0 + \int_{t_0}^{b} y_0'\, dt,$$

where we use the symbol y_0' to denote the constant $F(t_0, y_0)$. Since y_0' is a constant, the above integral is easily evaluated to obtain the approximation

$$y(b) \simeq y_0 + y_0'(b - t_0).$$

This is simply the differential approximation to $y(b)$ and is equivalent to equation (17.6) or the Taylor approximation (18.6). As discussed in Sections 17 and 18, the error in such approximations is proportional to the difference $b - t_0$. Consequently, to improve this method of approximating $y(b)$, we utilize the partitioning technique on which the entire theory of calculus is based.

To approximate a solution of the problem $y' = F(t, y)$, $y(t_0) = y_0$ at a point $b > t_0$, we first partition the interval $[t_0, b]$ into N subintervals having endpoints

$$t_0 < t_1 < t_2 < \cdots < t_N = b.$$

We then apply the method used above to sequentially estimate the solution y at the points $t_1, t_2, \ldots, t_N = b$. To start, we consider the interval $[t_0, t_1]$. If we approximate $y'(t) = F(t, y(t))$ on this interval by the constant

$$y_0' \equiv F(t_0, y_0),$$

then the value of $y(t_1)$ is approximated by

$$y(t_1) \simeq y_1 \equiv y_0 + \int_{t_0}^{t_1} y_0'\, dt.$$

Integrating, we obtain the approximation of $y(t_1)$,

$$y_1 = y_0 + y_0'(t_1 - t_0). \tag{26.26}$$

Graphically, this approximation is illustrated in Fig. 26.8.

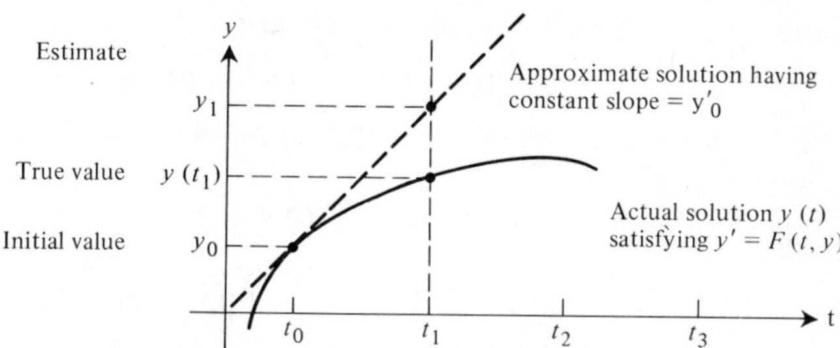

Figure 26.8

Next consider the interval $[t_1, t_2]$. To estimate $y(t_2)$, we approximate $y' = F(t, y(t))$ on this interval by the constant

$$y_1' \equiv F(t_1, y_1),$$

where y_1 is given by (26.26). We then substitute the approximation y_1 for $y(t_1)$ and y_1' for $F(t, y(t))$ into the identity

$$y(t_2) = y(t_1) + \int_{t_1}^{t_2} F(t, y(t))dt.$$

This results in the approximation

$$y(t_2) \simeq y_2 \equiv y_1 + \int_{t_1}^{t_2} y_1' \, dt.$$

Integrating, we obtain the approximation of $y(t_2)$,

$$y_2 = y_1 + y_1'(t_2 - t_1).$$

This is simply a repetition of the process used to obtain y_1.

Continuing the above process, we may obtain estimates for $y(t_i)$, $i = 2, 3, \ldots, N - 1$, and finally an estimate for $y(b) = y(t_N)$. The entire process is known as Euler's Method and may be summarized as follows.

Euler's Method

To estimate $y(b)$, where $y(t)$ is a solution of the initial-value problem, $y' = F(t, y)$, $y(t_0) = y_0$, $(b > t_0)$, complete the following steps.

I. Construct a partition of the interval $[t_0, b]$, $\{t_i\}$ $i = 0, 1, 2, \ldots, N$. (A uniform partition will simplify the arithmetic.)

II. Set

$$y_0' = F(t_0, y_0)$$

(y_0 is the initial value) and inductively, for $n = 0, 1, 2, \ldots, N - 1$, set

$$y_{n+1} = y_n + y_n'(t_{n+1} - t_n) \tag{26.27}$$

and

$$y_{n+1}' = F(t_{n+1}, y_{n+1}) \tag{26.28}$$

III. An estimate for $y(b)$ is y_N.

Example 26.11 The solution of the logistic equation,

$$y' = 10^{-2}y(10 - y), \qquad y(0) = 1,$$

using Euler's Method to determine $y(2)$, $b = 2$, is illustrated using a uniform partition of size $N = 4$. Thus

$$\Delta t_n = (2 - 0)/4 = 0.5$$

and

$$t_0 = 0, \qquad t_1 = 0.5, \qquad t_2 = 1.0, \qquad t_3 = 1.5, \qquad t_4 = 2.0.$$

$$y_0 = 1, \qquad \text{given initial condition,} \qquad y_0' = 10^2 \cdot y_0(10 - 1) = 0.09$$

$$y_1 = y_0 + \Delta \cdot y_0' \qquad (\text{where } \Delta = \Delta t_n = 0.5)$$

$$= 1 + (0.5)[0.09] = 1.045$$

$$y_2 = y_1 + \Delta \cdot y_1'$$

$$= (1.045) + (0.5)[10^{-2} \cdot 1.045(10 - 1.045)] \simeq 1.0918$$

$$y_3 = y_2 + \Delta \cdot y_2'$$

$$= 1.0918 + (0.5)[10^{-2}(1.0918)(10 - 1.0918)] \simeq 1.1404$$

$$y_4 = y_3 + \Delta \cdot y_3'$$

$$= 1.1404 + (0.5)[10^{-2}(1.1404)(10 - 1.1404)] \simeq 1.1909$$

The actual solution of this equation is

$$y(t) = \frac{10}{1 + 9e^{-t/10}}.$$

To four-figure accuracy, we find that the true value $y(2)$ is $y(2) = 1.1949$, while the above approximation is $y(2) \simeq y_4 = 1.1909$. The error in the approximation is $E \simeq 0.004$. ◀

Euler's Method is known as a *predictor method* since it simply predicts the value of the solution $y(t_n)$, utilizing information about the solution at values of $t < t_n$. (See Fig. 26.9.) Euler's method uses tangent lines to "shoot" from one grid point to the next.

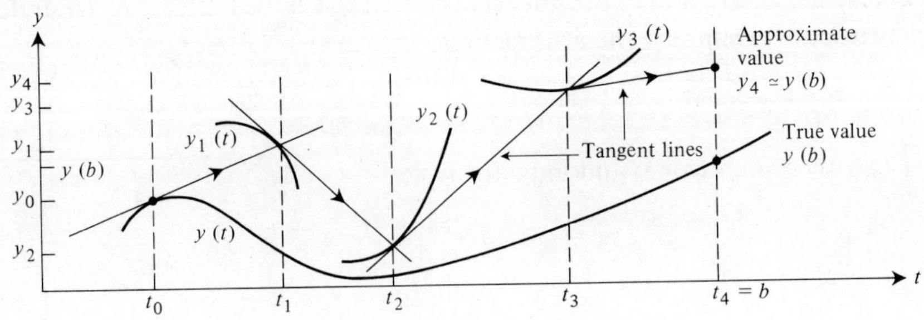

Figure 26.9

The curved lines in Fig. 26.9 represent the solutions of the initial value problems:

$$y'(t) = F(t, y(t)),\ y(t_0) = y_0;$$

$$y_1'(t) = F(t, y_1(t)),\ y_1(t_1) = y_1;$$

$$y_2'(t) = F(t, y_2(t)),\ y_2(t_2) = y_2;$$

$$y_3'(t) = F(t, y_3(t)),\ y_3(t_3) = y_3.$$

The values y_1, y_2, y_3, and y_4 are the y-coordinates, where the tangent lines meet the vertical line $t = t_1$, $t = t_2$, $t = t_3$, and $t = t_4$.

Example 26.12 Euler's method is used to estimate $y(3)$, where y is a solution of the initial value problem

$$y'(t) = \sqrt{y} \qquad y(1) = 0.25.$$

Since this equation is separable, its solution is easily found to be $y(t) = (t/2)^2$ and hence $y(3) = 2.25$. We approximate this solution using different partitions of the interval [1, 3].

a) Consider the partition of size $N = 4$ given by $t_0 = 1$, $t_2 = 1.5$, $t_3 = 2.0$, $t_4 = 3.0$. Then Euler's Method is used to complete Table 26.1 where $y_n' = \sqrt{y_n}$.

Table 26.1

t_n	y_n	y_n'	Δt_n	$y_{n+1} = y_n + y_n'\Delta t_n$
1	0.25	0.5	0.5	0.5
1.5	0.5	0.707	0.5	0.854
2.0	0.854	0.924	1	1.778
3.0	1.778 (Estimate)			

b) Consider the uniform partition of size $N = 8$,

$$t_0 = 1.0, \quad t_2 = 1.25, \quad t_3 = 1.5, \quad \ldots, t_8 = 3.0.$$

Using Euler's Method, we complete Table 26.2, where $y_n' = \sqrt{y_n}$, and $\Delta = (3 - 1)/8 = 0.25$.

Table 26.2

t_n	y_n	y_n'	$y_{n+1} = y_n + y_n' \cdot \Delta$
1	0.25	0.5	0.375
1.25	0.375	0.612	0.528
1.5	0.528	0.727	0.709
1.75	0.709	0.842	0.920
2.0	0.920	0.959	1.160
2.25	1.160	1.077	1.430
2.5	1.430	1.196	1.728
2.75	1.728	1.315	2.057
3.0	2.057		
	Estimate		

Euler's Method may be improved by utilizing a better estimate of the derivative, $y'(t) = F(t, y(t))$, over each subinterval $[t_n, t_{n+1}]$. Euler's Method only uses information concerning the derivative at the left endpoint, t_n. It seems reasonable that a better approximation could be obtained by using information about the derivative at both endpoints, t_n and t_{n+1}. The Modified Euler Method is a *predictor–corrector method*, which at stage n uses the estimate y_n and Euler's Method to obtain a first estimate to $y(t_{n+1})$, denoted $\hat{y}_{n+1}$. Then, this first "prediction," $\hat{y}_{n+1}$ is used to "correct" the estimate of $y' = F(t, y)$ over $[t_n, t_{n+1}]$. The corrected estimate of y' is the average of the slopes at the endpoints t_n and t_{n+1}; we denote the corrected slope by

$$\bar{y}_n' = \tfrac{1}{2}[F(t_n, y_n) + F(t_{n+1}, \hat{y}_{n+1})].$$

This averaged slope is then used to obtain a second estimate of $y(t_{n+1})$, which is denoted y_{n+1}. Using the estimate y_{n+1}, we repeat the procedure on the interval $[t_{n+1}, t_{n+2}]$ as the next stage of the procedure. Fig. 26.10 illustrates the basic idea of the Modified Euler Method.

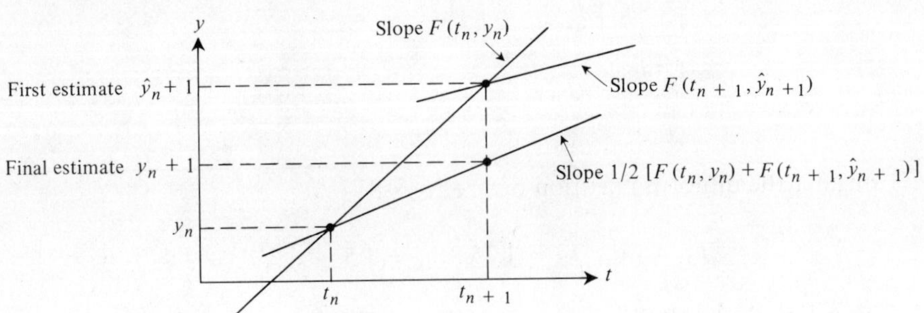

Figure 26.10

Modified Euler Method

To estimate $y(b)$ where $y(t)$ is a solution of the initial-value problem, $y'(t) = F(t, y(t))$, $Y(t_0) = y_0$ $(b > t_0)$, complete the following steps.

I. Construct a partition of the interval, $[t_0, b]$, $\{t_n\}$ $n = 0, 1, \ldots, N$. (A uniform partition is most convenient.) Let $\Delta t_n = t_{n+1} - t_n$.

II. Construct a sequence $\{y_n\}$ of approximations to the values $\{y(t_n)\}$ as follows, for $n = 0, 1, \ldots, N - 1$:

$$\begin{aligned} &1) \quad \text{Set } \hat{y}_{n+1} = y_n + F(t_n, y_n)\Delta t_n. \\ &2) \quad \text{Set } \bar{y}_n' = \tfrac{1}{2}[F(t_n, y_n) + F(t_{n+1}, \hat{y}_{n+1})] \\ &3) \quad \text{Set } y_{n+1} = y_n + \bar{y}_n' \cdot \Delta t_n. \end{aligned} \tag{26.29}$$

III. The estimate of $y(b)$ is y_N.

Example 26.13 Using the Modified Euler Method we approximate the value $y(3)$, where $y' = \sqrt{y}$, $y(1) = 0.25$, using the same partitions used in Example 26.12.

a) We complete Table 26.3 using formulas (1), (2), and (3) of Step II.

Table 26.3

t_n	Δt_n	y_n	$F(t_n, y_n)$	$\hat{y}_{n+1}$	$F(t_{n+1}, \hat{y}_{n+1})$	$\bar{y}_n'$	y_{n+1}
1	0.5	0.25	0.5	0.5	0.707	0.603	0.552
1.5	0.5	0.552	0.743	0.924	0.961	0.852	0.978
2.0	1.0	0.978	0.989	1.967	1.402	1.19	2.168
3.0	2.168						

b) For the uniform partition of size $N = 8$, $\Delta t_n = 0.25$, the Modified Euler Method is used to complete Table 26.4 (using three decimal places).

Table 26.4

t_n	y_n	$F(t_n, y_n)$	$\hat{y}_{n+1}$	$F(t_{n+1}, \hat{y}_{n+1})$	$\bar{y}_n'$	y_{n+1}
1	0.25	0.5	0.375	0.612	0.556	0.389
1.25	0.389	0.624	0.545	0.738	0.681	0.559
1.5	0.559	0.748	0.746	0.864	0.806	0.761
1.75	0.761	0.872	0.979	0.989	0.931	0.994
2.0	0.994	0.997	1.243	1.115	1.056	1.258
2.25	1.258	1.122	1.539	1.241	1.182	1.554
2.5	1.454	1.247	1.866	1.366	1.307	1.881
2.75	1.881	1.371	2.224	1.491	1.431	2.239
3.0	2.239					

Recall that the true solution is $y(3) = 2.25$. ◀

The error associated with the Modified Euler Method is consistently less than that of the regular Euler Method. To further reduce the error, more complicated algorithms have been developed, most of which follow the same format as the Modified Euler Method, in that they estimate $y' = F(t, y)$ over an interval $[t_n, t_{n+1}]$ by an average of y' values at several partition points. The most frequently used method of this type is known as the *Runge–Kutta Method.* This method was developed around 1900 and involves relatively few steps with greatly improved error control. The five-step Runge–Kutta method (averaging using five partition points) is the basic numerical routine used in packaged computer programs to solve ordinary differential equations. Due to their complexity, we do not present the Runge–Kutta Method or other more complicated types of numerical routines; these routines may be found in numerical methods or differential equations texts.

EXERCISE SET 26

26.1 The following equations are separable; determine the general solution of the given equation.

a) $y' = 2t - e^t$ b) $y' = 2ty$ c) $y' = \cos(2t)$
d) $y' = \tan(y) \cdot 5t$ e) $y' = ye^{3t}$ f) $y' = te^{3y}$
g) $y' = t - ty$ h) $yy' = \sin(t)$ i) $(t + 1)y' = y$
j) $t \ln(t)y' = y \ln(y)$

26.2 Find the general solution of the following equations.

a) $y' = 3y$ b) $y' = 3y + 2$ c) $y' = 3ty$
d) $y' = 3ty + 2$ e) $y' = 5y/t$ f) $y' = 5y/t + 2$
g) $y' = \sin(t)y$ h) $y' = \sin(t)y - e^{-\cos(t)}$ i) $y' = t\sin(t)/y$
j) $y' - 3y = t^2$

26.3 Find the specific solution of the given equation satisfying the indicated condition.

a) $y' = 7y$, $y(1) = 1$ b) $y' = 7y$, $y(1) = 0$
c) $y' = 3y$, $y'(5) = 1$ d) $y' = \sin(t)y^2$, $y(0) = 2$
e) $y' = t/y$, $y(1) = 2$ f) $y' = t/y$, $y'(1) = 2$
g) $y' = 5y$, $y(2) + y'(2) = 3$

26.4 Determine the specific solution y satisfying the indicated condition.

a) $y' = 3y + 2$, $y(0) = 3$ b) $y' = 3y + 2$, $y(1) = 3$
c) $y' = 3y + t$, $y(0) = 2$ d) $y' = 3y + t$, $y(1) = 2$
e) $y' = -y + \cos(t)$, $y(0) = 1$ f) $y' = -y + \cos(t)$, $y'(0) = 1$
g) $y' = (y/t) + t$, $y(1) = 0$ h) $y' = (y/t) + t$, $y(1) = 3$
i) $y' = (-y/t) + e^t$, $y(1) = 1$

26.5 Newton's Law of cooling states that the change in temperature of a body is proportional to the difference between the temperature T, and the surrounding temperature, $\hat{T}$.

a) State Newton's Law as a differential equation
b) Solve this differential equation.
c) How long will it take a body of temperature $T_0 = 37°C$ immersed in a 15°C bath to reach a temperature of 26°C if the proportionality constant is $-k = -2°/\text{hr}$?

26.6 A wet porous material with adequate ventilation will loose its moisture at a rate proportional to its moisture content. If a wet blanket that is being used to protect a seal in transit looses 30 percent of its moisture in one hour, when will only 20 percent of its original moisture remain?

26.7 A tank contains 1000 litres of brine containing 10 kilos of salt. The tank water is diluted by adding fresh water at a rate of 25 litres/min, which presumably mixes instantaneously with the water in the tank. The tank overspills at a rate equal to the amount of water added so that its volume remains constant.

a) Set up a differential equation to describe the amount of salt in the tank.
b) How much salt will be in the tank after 15, 30, and 60 minutes?
c) When will the tank contain 2 kilos of salt?

26.8 One pessimistic model of human population growth used an equation of the form $N'(t) = \lambda N(t)/(D - t)$; $N(0) = D^{-\lambda}$.

a) Solve this equation (k, λ, D constants.)
b) If $k = 1.79 \times 10^{11}$, $\lambda = 0.99$, and $D = 2026.87$, how close is this estimate to the 1975 world population of 3.97 billion people. (Use $t = 1975$.)
c) Using the parameters of part (b), why would this model be titled "Doomsday: Friday, 13 November 2026 A.D."?

26.9 In the study of epidemics, a usual assumption is that the rate of new infections is proportional to the number of contacts between infectives, I, and susceptibles, S.

a) Express this as a differential equation (assume that no one gets over the disease).
b) In a finite population of size N, the number of infectives plus susceptibles equals N. Use this to describe the change in the number of susceptibles as a function of S (and the constant N).
c) Solve the equation derived in part (b).

26.10 The equation $N(t) = 450/[1 + \exp(5.041 - 1.022t)]$ was found to fit experimental data for the growth of the bacteria *Paramecium aurelia*. Determine the differential equation describing this growth function.

26.11 In studies of energy flow through forest ecosystems, the accumulation of litter, mainly leaves and other plant matter, is modeled by the equation

$$\frac{dx}{dt} = L - kx,$$

where L is the rate of litter formation and k is the decomposition rate.

a) What is the equilibrium state described by this equation? Most forests are very close to the equilibrium state, but, at the time when the Carboniferous forest created the present coal beds, this was not the case.
b) In most real forests, the litter rate L is not constant over a one-year period; most litter is deposited in the fall. Model a forest over a three-year period in which a total of l units of litter are added to the accumulation $x(t)$, at the start of each year and $L = 0$ throughout the year. Assume that $x(0) = x_0 + l$.
c) Sketch a graph to illustrate the solution of part (b), where $x_0 = 2$ g/(cm)^2 and $l = 5$ g/(cm)^2.

26.12 The yield of a healthy crop is predicted by

$$\frac{dw}{dt} = \lambda(t)\left[1 - \frac{w(t)}{W}\right]w(t),$$

where W equals the dry weight at maximum yield, w equals the dry weight at time t, and λ equals the coefficient of growth as a function of time.

a) Solve for $w(t)$ if λ equals a constant.
b) Solve for $w(t)$ if λ equals t.
c) Solve for $w(t)$ if λ equals $t(T - t)$, where T is a fixed constant.

26.13 In predicting epidemics of leaf rust in plants, crop scientists have used the equation

$$\frac{dx}{dt} = kx(1 - x),$$

where x is the rusted fraction of the leaf area and k relates the susceptibility of the plant to the blight fungi. k has been found to be sensitive to temperature, T, with

$$k = k_{\max} - c(T - 20)^2,$$

c and $k_{\max}$ constants. In a simple model, the mean daily maximum temperature is assumed to have the form

$$T = a + b\sin(\pi t/120), \ (t \text{ days}).$$

Determine the function of $x(t)$ under these conditions.

26.14 A population is assumed to grow according to equation

$$N'(t) = -N(N - 2)(N - 3).$$

a) What are the equilibrium and steady-state values of this population? (Indicate the initial values, N_0, which lead to the steady states.)
b) Discuss the stability of the equilibrium values. Use a graph if necessary.
c) Solve this equation. Can you solve explicitly for $N(t)$? How does the integrated equation reflect the stability properties determined in part (c)?

26.15 Discuss the equilibrium values and their stability for the equations.

a) $N' = \sin(N)$
b) $N' = (1 - N)^2$
c) $N' = \sin^2(N)$
d) $N' = 1 - N^2$
e) $N' = N - 2\sin(N)$
f) $N' = N(N - 1)^2(N - 2)^2$

26.16 Carefully sketch on the same graph the rate functions, R, of the logistic (26.12) and Gompertz (26.15) models with $r = 0.1$ and $M = 10$. Indicate the maximum rates. How can these graphs be used to compare the resulting logistic and Gompertz functions?

26.17 Determine the general solution of the following equations.

a) $y'' + 3y' - 4y = 0$
b) $y'' - 3y' - 4y = 0$
c) $y'' + y' - 2y = 0$
d) $y'' - y' - 2y = 0$
e) $y'' - 4y = 0$
f) $y'' + 4y = 0$
g) $2y'' - y' + y = 0$
h) $y'' - y' = 0$
i) $y'' + 3y' - 3y = 0$
j) $y'' + 2y = 0$

26.18 Solve the following equations subject to the initial conditions $y(0) = 1$, $y'(0) = 2$.

a) $y'' - 2y' - 3y = 0$
b) $y'' - 2y' + y = 0$
c) $y'' + 5y = 0$
d) $y'' - 5y = 0$
e) $2y'' - 3y' + y = 0$
f) $2y'' + 4y' - 7y = 0$

26.19 Solve the following boundary-value problems.

a) $y'' + 9y = 0,\quad y(0) = 1, y'(0) = 2$
b) $y'' - 9y = 0,\quad y(0) = 0, y(\pi) = 0$
c) $y'' + 2y' + y = 0,\quad y(0) = 1, y(2) = -1$
d) $y'' + 2y' + 2y = 0,\quad y(0) = 1, y(2) = -1$
e) $y'' + \lambda^2 y = 0,\quad y(0) = 0, y(\pi/\lambda) = 0$
f) $y'' + \lambda^2 y = 0,\quad y(0) = 0, y(\pi/2\lambda) = 1$
g) $y'' + \lambda^2 y = 0,\quad y(0) = 1, y(\pi/2\lambda) = 0$
h) $y'' + \lambda^2 y = 0,\quad y(0) = 1, y(\pi/2\lambda) = 1$

26.20 Find a second-order differential equation having both of the indicated functions as solutions.

a) e^{-t}, e^{-2t}
b) e^{3t}, e^{-3t}
c) e^{2t}, e^{-t}
d) e^{2t}, te^{2t}
e) $\cos(3t), \sin(3t)$
f) $e^t \cos(3t), e^t \sin(3t)$
g) $e^{-k_1 t}, e^{-k_2 t}$
h) te^{-kt}, e^{-kt}

26.21 The diffusion of a compound between two compartments separated by a membrane of thickness δ depends on the concentrations of the substance in each compartment and the diffusion constant D, which is characteristic of the substance and membrane. If A is the area of the membrane, V is the volume of compartment 1, c_1 and c_2 are the concentrations in compartments 1 and 2, respectively, then from Fick's Law it follows that

$$\frac{dc_1}{dt} = \frac{DA}{\delta V}(c_1 - c_2).$$

a) Solve this equation with the initial value $c_1(0) = 0$.
b) If c_2 is held constant, when will $c_1(t) = (\frac{1}{2})c_2$?
c) Theoretically will c_1 ever equal c_2? Consider c_2 constant and $c_2(t) = k[1 + \sin^2(t)]$.

26.22 The law of mass action leads to the equation $x_1' = -k_1x_1 + k_2x_2$, where x_1 and x_2 are the concentrations of chemicals A and B that satisfy the reversible monomolecular reaction with reaction rates k_1 and k_2:

$$A \underset{k_2}{\overset{k_1}{\rightleftharpoons}} B.$$

Assume that the total amount of A and B in the system is fixed, $x_1 + x_2 = 5$. Solve this equation and discuss the equilibrium and steady states of the solution in terms of the rate constants, k_1 and k_2.

26.23 Consider the two-compartment system illustrated in Fig. 26.11 A substance, X, having concentrations x_1 and x_2 in compartments 1 and 2, respectively, diffuses across the membrane separating the compartments. Assume that the diffusion is not (necessarily)

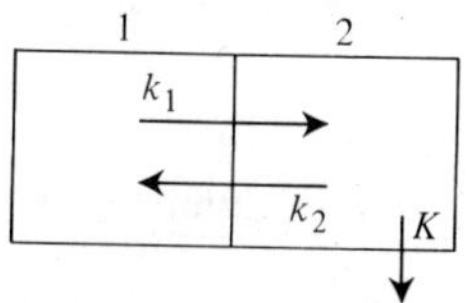

Figure 26.11

equal in both directions with diffusitinity constants, k_1 and k_2, as indicated in the diagram. Also assume that the substance diffuses from compartment 2 at a rate K. Then, if A is the area of the membrane separating the compartments and v_1 and v_2 are their volumes, respectively, the equations describing the concentrations x_1 and x_2 are

$$\frac{dx_1}{dt} = \frac{-k_1 A}{v_1} x_1 + \frac{k_2 A}{v_2} x_2; \tag{26.30}$$

$$\frac{dx_2}{dt} = \frac{k_1 A}{v_1} x_1 - \frac{k_2 A}{v_2} x_2 - \frac{K}{v_2} x_2. \tag{26.31}$$

Solve this system of equations using the following steps.

a) Differentiate equation (26.30) to obtain a second-order equation, $x_1'' = \ldots$
b) Substitute x_2' from equation (26.31) into the equation of part (a)
c) Solve equation (26.30) for x_2 in terms of x_1 and x_1'; $x_2 = \ldots$
d) Substitute the expression of x_2 (part (c)) into the equation of part (b).
e) You should now have a second-order equation of the form (26.18) (after rearranging and simplifying). Solve this equation for x_1.
f) Substitute the value of x_1 (and its derivative x_1') as obtained in part (e) into the equation obtained in part (c) for x_2. You now have the solution: $x_1 = \ldots, x_2 = \ldots$

If the constants and algebra of this method seem too difficult and general, first try to use this method to solve the specific equations in Exercise 26.24.

26.24 Use the technique described in Exercise 26.23 to solve the following systems.

a) $x_1' = -2x_1 + x_2$, $x_2' = +2x_1 - x_2$
b) $x_1' = -2x_1 + 2x_2$, $x_2' = +2x_1 - 2x_2$
c) $x_1' = -3x_1 + x_2$, $x_2' = 3x_1 - x_2$
d) $x_1' = -3x_1 + x_2$, $x_2' = 3x_1 - 2x_2$
e) $x_1' = -3x_1 + x_2$, $x_2' = 3x_1 - 3x_2$
f) $x_1' = -x_1 + 3x_2$, $x_2' = +x_1 - 4x_2$
g) $x_1' = x_1 + x_2$, $x_2' = -x_1 + x_2$
h) $x_1' = x_1 + 2x_2$, $x_2' = -2x_1 + x_2$

26.25 Use Euler's Method to approximate $y(2)$, where $y(t)$ satisfies the given initial value problem. Use a partition of size $N = 4$.

a) $y' = 2y$, $y(0) = 0.1$
b) $y' = t/y$, $y(0) = 10$
c) $y' = t + y$, $y(0) = 0$
d) $y' = ty$, $y(0) = 1$
e) $y' = \cos(y)$, $y(0) = 0$

26.26 Repeat Exercise 26.25 with $N = 8$ and compare the results with the values obtained when $N = 4$.

26.27 Use the modified Euler Method to solve the following equations for $y(1)$ using $N = 5$.

a) $y' = 2x + y$, $y(0) = 0$
b) $y' = 2xy$, $y(0) = 1$
c) $y' = 10^{-2}y^2$, $y(0) = 1$
d) $y' = y$, $y(0) = 1$

26.28 Solve the equations of Exercise 26.26.

a) With Euler's Method, $N = 5$. Compare the results to those of the Modified Euler Method.
b) With the Modified Euler Method, $N = 10$. Compare the results to those obtained with $N = 5$.

26.29 The existence and uniqueness of a solution to an initial-value problem, $y' = f(t, y)$, $y(t_0) = y_0$ can be demonstrated when the function f is not too complicated by an iteration procedure known as *Picard's Method*. This method also provides another way to approximate the solution $y(t)$. It consists of constructing a sequence of functions $\{y_n(t)\}$ such that $\lim_{n\to\infty} y_n(t) = y(t)$, the solution. Picard's Iteration Method is to set

$$y_0(t) \equiv y_0, \text{ the initial constant,}$$

and inductively for $n \geq 0$, set

$$y_{n+1}(t) = y_0 + \int_{t_0}^{t} f(x, y_n(x))dx.$$

a) Construct an approximate solution to the problem $y' = 1 + y^2$, $y(0) = 0 = y_0$, by evaluating $y_1(t)$, $y_2(t)$, and $y_3(t)$ by Picard's iteration.
b) Compare the function $y_3(t)$ to the Taylor series expansion of $y(t) = \tan(t)$ about $t_0 = 0$.
c) Apply Picard's Method to the following problems (evaluate $y_3(t)$).
   i) $y' = 3y$, $y(0) = 2$
   ii) $y' = xy$, $y(0) = -1$
   iii) $y' = xy + 1$, $y(0) = 0$
d) Compare the function $y_3(t)$ to the true solution $y(t)$ in part (c) when $t = 1$.

CHAPTER EIGHT

AN INTRODUCTION TO PROBABILITY AND STATISTICS

This Chapter is intended to introduce the basic aspects of probability and statistics. Section 27 considers only finite sample space and can be covered without any calculus background, preferably after Section 10, where the sigma notation is introduced. Section 28 examines continuous probability spaces and requires as background Chapter 5 for subsections 28.1 and 28.2 and Section 25 of Chapter 6 for the last subsection 28.3.

The objective of this chapter is to provide an introduction to the basic concepts of probability and statistics and, in particular, to the role and use of *calculus* to describe continuous probability spaces. We have therefore avoided the computational stumbling blocks of combinations and permutations and their use in obtaining probabilities via counting arguments.

SECTION 27

AN INTRODUCTION TO PROBABILITY AND STATISTICS

27.1 INTRODUCTION

There is an anonymous quote that goes something like this: "Of one thing you can be certain, you will die, all the rest is left to chance." This quote embodies the idea that we are continually presented with situations whose exact outcome cannot be known in advance. However, except for accidents, the chance of the outcome in most situations can be predicted. One of the primary objectives of science is to be able to predict what will occur in given physical and biological situations. In the following, we shall employ the more scientific term "experiment" to refer to an observable situation that results in a recorded "outcome." When all possible outcomes of an experiment are known, our ability to predict what outcome will actually occur depends on two things. First, the actual structure of the experiment, i.e., how will possible outcomes occur, since this determines the actual chance of each possible outcome occurring. And, second, our ability to recognize the underlying structure of the experiment, utilizing data obtained by repeating the experiment several times and observing the resulting outcomes.

Example 27.1 An experiment consists of flipping a coin and recording as the outcome the side that lands facing upward. Our ability to predict the outcome depends first on the coin itself. Is it a "fair" coin, having a "head" and a "tail" side, each of which is equally likely to occur? Or, is it a coin with "tails" on two sides or otherwise "unfair"? Our chance of predicting the outcome also depends on our ability to recognize the type of coin being flipped. Suppose that we observed two successive flips of the coin, each resulting in "tails." Could we deduce the type of coin being utilized and predict the outcome of the next flip from these observations? Probably not. What if we observed 30 successive flips, all resulting in "tails"; could we determine if a fair coin was being used? ◀

Probability is the scientific term used to describe the concept of chance or likelihood. The origin of probability theory was a mathematical attempt in the early 1700's to analyze fashionable gambling games and to decide which bets were likely to win more often than they were to lose. The study of probability concerns the structure of experiments and the theoretical chance of outcomes based upon the assumed structure. *Statistics* is the study of data generated by repeating experiments. The objectives of statistics are to represent the data concisely and to ascertain the fundamental probability structure of the experiment that generated the data. Individual values, which are used to represent the data or the probability structure, are called *statistics*. Statistics can be used to predict the nature of an experiment's outcome and to test proposed probability models for the experiment.

A probability statement about the flipping of a coin (Example 27.1) would be, "The outcome heads will occur with probability $\frac{1}{2}$ if the coin is fair." A statistical statement concerning the observation of 30 successive "tails" would be that this observation is "highly unlikely" with a fair coin, since it would have a probability of less than 0.005 of occurring. (As we shall see, the chance that a fair coin would land "tails" 30 successive times is $(\frac{1}{2})^{30} \simeq 9 \times 10^{-10}$.)

There are two basic concepts of probability. The *classical concept* is the concept of "equilikely" or "equiprobable" events. If a ball is randomly drawn from a bag containing three balls that are identical except for color, and the balls are red, blue, and green, then the chance of selecting a red ball is $\frac{1}{3}$ since there are three equilikely outcomes, and only one results in a red ball. The *frequency concept* is the concept used to predict the probability of an event occurring based upon how frequently the event has occurred previously. This is the concept of probability used by a weather forecaster when predicting the chance that it will rain.

The intuitive concept of probability is based upon experience with simple events. When we encounter more complex situations, our intuitive notions of probability are frequently challenged. For example, consider a game in which one of three cards is drawn from a hat and placed on a table before it is examined. Suppose one card is red on both sides, one is blue on both sides and one is red on one side and blue on the other side. Now, if the card on the table has a red side facing upwards, what is the probability that the side face down is also red? Since the side down is either red or blue, you might say that the probability is $\frac{1}{2}$, one out of two possible events. But this is not the correct answer.

To analyze the probability of events in more complex situations we must turn to more precise mathematical terminology to describe the concepts of probability. The following is a very basic and brief introduction to the terminology of probability theory. The most fundamental concept in statistics and probability is the concept of an experiment.

Definition 27.1 An *experiment* is a procedure

1. which results in the recording of an "outcome," and
2. in which the "outcome" cannot be determined before the experiment is performed.

The set of all possible "outcomes" of an experiment is called the *sample space* of the experiment.

Each individual possible recorded "outcome" of an experiment is called an *elementary event*.

Example 27.2 Consider the experiment that consists of determining a person's ABO blood group. In this experiment, a person's blood sample is exposed to two different antigens called *anti-A* and *anti-B*. There are four possible reactions that can occur based upon whether the blood sample reacts or does not react to each of the antigens.

Reaction	Classification type
Does not react to either antigen	Type O
Reacts to anti-A but not anti-B	Type A
Reacts to anti-B but not anti-A	Type B
Reacts to both anti-A and anti-B	Type AB

Thus the sample space of this experiment is as follows:

$$S = \{\text{Type O, Type A, Type B, Type AB}\},$$

The elementary events are the individual blood types.

To avoid adverse reactions, people with one blood type should not be given blood transfusions with another blood type. An immediate consequence of this is that a blood bank must keep a sufficient supply of each type of blood to cover emergencies. This requires a blood bank to be able to estimate the proportion of people having each of the different blood types. To obtain such an estimate a record was kept of blood donors at a blood center in Great Britain. The proportion of the general population having a particular blood type was then taken to be the number observed with this type divided by the total number observed. This data is summarized in Table 27.1.

Table 27.1

Blood type	Observed	Proportion
O	88,700	0.467
A	79,300	0.417
B	16,300	0.086
AB	5,700	0.030
Total	190,000	1.000

Based upon this data, what is the probability that a person, randomly selected from the English population, would have type B blood? The answer would be 0.086, the proportion of people having type B blood. A probability of 0.086 can be interpreted to mean that, on the average, 86 out of every 1000 people randomly selected will have type B blood.

To make the probability of an event equal to the relative frequency of its occurrence—i.e., the proportion of times it occurs—when the experiment is repeated a large number of times is called the *frequency concept* of probability.

◀

Intuitively, the probability of any particular outcome of an experiment is a number that indicates the chance that the outcome will occur when the experiment is performed. Since a fair coin has a head and a tail side, each of which is equally likely to occur, the probability of their occurrence should be the same:

$$\text{Probability (head)} = \text{probability (tail)}.$$

To standardize the numerical value assigned to each of these probabilities, a probability function is required to satisfy some basic properties. These properties are motivated by the frequency concept of probability illustrated in Example 27.2. Note that the proportion of times an event occurs (that is, its relative frequency) is a number between zero and one. Furthermore, observe that the sum of the proportions associated with all possible outcomes is equal to one. If the relative frequency is to represent the probability of an event, these observations would require that the probability associated with the flip of a fair coin be given by the following.

$$P(\text{"tail"}) = 0.5$$
$$P(\text{"head"}) = 0.5$$
$$\text{Total Probability} = 1.0$$

Definition 27.2 Let S denote a finite sample space $S = \{X_1, X_2, X_3, \ldots, X_n\}$. A *probability function*, or *distribution*, defined on S is any real valued function P that satisfies the following:

$0 \leq P(X_i) \leq 1$ for $i = 1, 2, 3, \ldots, n$	**All probabilities are between 0 and 1.**

(27.1)

$P(X_1) + P(X_2) + \cdots + P(X_n) = 1$	**The sum of the probabilities is 1.**

The sample space S together with a probability function P is called a *probability space*.

Example 27.3 An experiment consists of randomly selecting a ball from a sack and recording the color of the ball. If the sack contains only red, blue, and green balls, then the sample space will be

$$S = \{\text{red, blue, green}\}.$$

a) The function P defined on S by

$$P(\text{red}) = 0.5,\ P(\text{blue}) = 0.3,\ P(\text{green}) = 0.2$$

is a probability function on S.

b) The function P defined on S by

$$P(\text{red}) = 0.5,\ P(\text{blue}) = 0.4,\ P(\text{green}) = 0.2$$

is not a probability function, since the sum of the probabilities is greater than one.

c) The function P defined on S by

$$P(\text{red}) = 0.6,\ P(\text{blue}) = 0.6,\ P(\text{green}) = -0.2$$

is not a probability function since $P(\text{green})$ is negative. ◀

27.2 FUNDAMENTALS OF PROBABILITY

In Example 27.3, the elementary events are the outcomes red, blue, and green. We can also consider compound events, which consist of more than one elementary event. Consider the event, denoted A, that the ball selected will be either red or blue. Using the probability function of part (a) of Example 27.3, what would be the probability of event A? We can answer this question in two ways. Directly, event A consists of the two elementary events, red and blue. We can thus express event A as a subset of sample space S:

$$A = \{\text{red, blue}\}.$$

The chance of event A occurring is the chance of selecting a red ball plus the chance of selecting a blue ball. Using $P(A)$ to denote the probability of the event A,

$$P(A) = P(\text{red}) + P(\text{blue}) = 0.5 + 0.3 = 0.8.$$

A second approach is to note that whenever the experiment is performed either event A occurs or event A does not occur.

> We denote the event "A does not occur" by $\tilde{A}$.

Therefore, we should have

$$P(A) + P(\tilde{A}) = 1.$$

Solving for $P(A)$, we have

$$P(A) = 1 - P(\tilde{A}).$$

Since the only way event A cannot occur is for the ball to be green, the event $\tilde{A}$ is equivalent to the elementary event, green. This means that

$$P(\tilde{A}) = P(\text{green}) = 0.2,$$

and

$$P(A) = 1 - P(\tilde{A}) = 1 - 0.2 = 0.8.$$

The same principles apply in general when event A is any event associated with a given probability space.

Definition 27.3 Let S be a (finite) probability space with distribution P. Let A be an event that consists of a subset of the set S, $A \subseteq S$. Then, the probability of event A occurring is defined to be $P(A) = \text{Sum } P(X_i)$ for all elementary events X_i in A.

For instance, if $S = \{R, B, G, Y\}$ and $A = \{R, G\}$, then

$$P(A) = P(R) + P(G).$$

Definition 27.4 Let S be a probability space with distribution P. For any event A which is a subset of S, the *complimentary event*, denoted $\tilde{A}$, is the set of all elementary events in S that are not in A. Thus the complimentary event $\tilde{A}$ is the event that A does not occur. The probabilities of the events A and $\tilde{A}$ are related by the equation

$$\boxed{P(A) = 1 - P(\tilde{A}) \qquad \text{or} \qquad P(\tilde{A}) = 1 - P(A).} \tag{27.2}$$

Example 27.4 An experiment consists of rolling a single die with six sides. There are two sides marked 1 and the remaining sides are marked 2, 3, 4, and 5, respectively. The recorded outcome of the experiment is the number on the top face of the die. The sample space of the experiment is

$$S = \{1, 2, 3, 4, 5\}.$$

To assign a probability distribution to this experiment, we assume that each face has an equal chance of ending "up." Since there are six faces, each one has one chance in six of occurring. Thus we assume that each number 2, 3, 4, 5 has a probability of $\frac{1}{6}$, while the number 1, being on two sides, has a probability of $\frac{2}{6}$.

The probability distribution is therefore

$$P(1) = \tfrac{2}{6}, \quad P(2) = P(3) = P(4) = P(5) = \tfrac{1}{6}.$$

Let A denote the event that the outcome is either 1 or 4 and let B denote the event that the outcome is a number greater than 3. As subsets of the sample space S, these events are

$$A = \{1, 4\} \qquad \text{and} \qquad B = \{4, 5\}.$$

The complements of these events are

$$\tilde{A} = \{2, 3, 5\} \qquad \text{and} \qquad \tilde{B} = \{1, 2, 3\}.$$

What is the probability of A occurring? Using Definition 27.3, this value is

$$P(A) = P(1) + P(4) = \tfrac{2}{6} + \tfrac{1}{6} = \tfrac{1}{2}.$$

What is the probability of $\tilde{B}$ occurring? Since

$$P(B) = P(4) + P(5) = \tfrac{1}{3},$$

by (27.2),

$$P(\tilde{B}) = 1 - P(B) = 1 - \tfrac{1}{3} = \tfrac{2}{3}.$$

What is the probability that either A or B will occur? Since A or B occurring is the event

$$A \cup B = \{1, 4, 5\},$$

we have, by formula (27.1),

$$P(A \cup B) = P(1) + P(4) + P(5) = \tfrac{2}{6} + \tfrac{1}{6} + \tfrac{1}{6} = \tfrac{2}{3}.$$

What is the probability that both events A and B will occur? For both events to occur, the die must come up 4, since it is the only elementary event in common to both A and B;

$$A \cap B = \{4\}.$$

Thus

$$P(A \cap B) = P(4) = \tfrac{1}{6}.$$

Does the probability of either A or B occurring equal the sum of the probability that A will occur plus the probability that B will occur? The answer is no, as can be seen from the above calculations:

$$P(A) + P(B) = \tfrac{1}{2} + \tfrac{1}{3} = \tfrac{5}{6} \neq P(A \cup B).$$

The reason for this is because the event 4, which is in common to both events A and B, is "counted" twice when we simply add $P(A)$ to $P(B)$. In the Venn diagram in

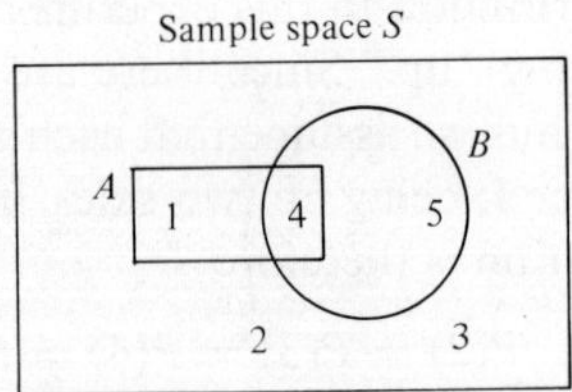

Figure 27.1

Fig. 27.1, event A is represented by the rectangle and event B by the circle. The intersection of the two events contains the elementary event 4. As indicated above,

$$P(A) + P(B) = [P(1) + P(4)] + [P(4) + P(5)]$$

counts $P(4)$ twice. By subtracting this probability, which must equal $P(A \cap B)$, we obtain

$$\begin{aligned} P(A) + P(B) - P(A \cap B) &= [P(1) + P(4) + P(4) + P(5)] - P(4) \\ &= P(1) + P(4) + P(5) \\ &= P(A \cup B). \end{aligned}$$

Thus the probability that event A or event B occurs is equal to the probability that A occurs plus the probability that B occurs, minus the probability that both A and B occur. ◀

The situation illustrated in the above example holds for any sample space. Let A and B be two events of a sample space S with distribution P. Then the probabilities of the event "A and B," denoted $A \cap B$, and the event "A or B," denoted $A \cup B$, are related by the formula

$$\boxed{P(A \cup B) = P(A) + P(B) - P(A \cap B).} \tag{27.3}$$

Two events A and B are said to be *mutually exclusive* if they cannot both occur simultaneously. In terms of sets, this means that $A \cap B = \phi$, the empty set. In terms of probability theory, since an event that can never occur has probability zero, we conclude that

$$\boxed{A \text{ and } B \text{ are mutually exclusive if } P(A \cap B) = 0.}$$

Example 27.5 A survey of farm-land utilization found that 74 percent of the farms in a given region grow crops of some type, 52 percent raise some commercial animals, and 31 percent do both. What percentage of the farms are not producing either livestock or crops? *Solution*: Think of the percentage as a probability. Then letting C represent crops and L represent livestock, we can determine the percentage of producing farms, $C \cup L$, from the following equation:

$$\begin{aligned} P(C \cup L) &= P(C) + P(L) - P(C \cap L) \\ &= 0.74 + 0.52 - 0.31 = 0.95. \end{aligned}$$

The percentage of nonproducing farms is thus determined using formula (27.2);

$$P(\widetilde{C \cup L}) = 1 - P(C \cup L) = 0.05.$$

Hence 5 percent of the farms are nonproducing.

What is the chance that a farm in the survey area would produce livestock only? The livestock producing farms can be split into two groups, the ones who also produce crops, denoted $L \cap C$, and the ones who do not produce crops, denoted $L \cap \tilde{C}$:

$$L = \underbrace{(L \cap C)}_{\text{Produces crops}} \cup \underbrace{(L \cap \tilde{C})}_{\text{Does not produce crops}}.$$

Since the two groups, $L \cap C$ and $L \cap \tilde{C}$, are mutually exclusive,

$$P((L \cap C) \cap (L \cap \tilde{C})) = 0.$$

By (27.3),

$$P(L) = P(L \cap C) + P(L \cap \tilde{C}) - P((L \cap C) \cap (L \cap \tilde{C})).$$

Solving for $P(L \cap \tilde{C})$ using the mutual exclusion of the two groups leads to the following formula:

$$P(L \cap \tilde{C}) = P(L) - P(L \cap C).$$

Using this equation, we determine the probability that a farm produces only livestock as $0.52 - 0.31 = 0.21$. ◀

27.3 SEQUENTIAL EVENTS, CONDITIONAL PROBABILITY, AND BAYES' THEOREM

Sequential Events

A *sequential event* or *experiment* is one that considers two or more separate events in order and *assumes the events to be unrelated*. The most frequent sequential event is the repetition of an experiment. The probability of a sequence of events occurring is obtained by multiplying the probabilities of each individual event occurring. The multiplication principle can be motivated by considering the following illustration.

Suppose you wish to track an animal that travels from point A to point B by one of three trails, and then proceeds to a further destination C by one of two trails. (See Fig. 27.2.) The total number of trails that the animal can take is $3 \times 2 = 6$. Assuming that all trails are equally likely to be used by the animal, your

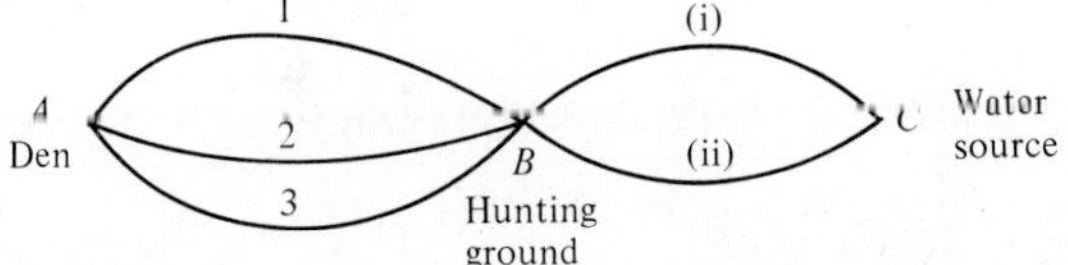

Figure 27.2

chance of taking the same trail chosen by the animal is one in six. This chance can also be calculated as follows:

the chance of being on the same first part of trail, $\frac{1}{3}$,

the chance of being on the same second part of trail, $\frac{1}{2}$,

the chance of being on the same total trail, $\frac{1}{3} \times \frac{1}{2} = \frac{1}{6}$.

We formalize this concept by the following equation:

$$\text{Probability of two sequential events, } A \text{ and } B = P(A) \cdot P(B). \tag{27.4}$$

For example, if we flip a fair coin twice, what will be the probability that it comes up heads on the first flip and tails on the second flip? The answer is

$$\text{Probability of heads then tails} = P(\text{heads}) \cdot P(\text{tails}) = \tfrac{1}{2} \times \tfrac{1}{2} = \tfrac{1}{4}.$$

Similarly, if we consider more than two events, we multiply the probabilities.

Consider the probability that a fair coin, flipped three times, will land on heads each time.

$$\text{Probability of three heads} = P(\text{heads}) \cdot P(\text{heads}) \cdot P(\text{heads}) = (\tfrac{1}{2})^3.$$

The probability that the special die in Example 27.4 would result in the outcomes 5, 1, 2, 1 on four successive tosses would be

$$\text{Probability of } 5, 1, 2, 1 = P(5) \cdot P(1) \cdot P(2) \cdot P(1) = \frac{4}{6^4}.$$

Conditional Probability and Bayes' Theorem

In the preceding, we have primarily considered experiments with simple outcomes. Most real experiments are more complicated and their outcomes consist of several bits of data. For instance, a medical examination would record several aspects of a person's health, perhaps height, weight, blood pressure, pulse rate, and sex. A laboratory test may determine the pH and optical density of a urine sample. A study may examine the level of PCB's (polychlorinated biphenyls, a toxic chemical) found in lake fish and the incidence of nervous disorders in communities surrounding the lake.

When considering experiments that record more than one bit of data, the first scientific question that must be asked is, "Are the different data components independent of each other?"

Using an intuitive concept of independence, the height and weight of a healthy individual would not be independent of each other. On the other hand, the number of brothers of an individual would be independent of the individual's sex. In most laboratory experiments, none of the individual data components will be independent of any other component. This is because an experiment is designed to study a particular phenomenon and each bit of recorded data is selected (hopefully) to monitor the same phenomenon.

The next obvious question which arises is "If two aspects of an experiment are not independent, what does it mean to say that they are dependent?"

The question of *dependency* is difficult to answer. Smoking and the incidence of cancer have been shown not to be independent events. The evidence that indicates that smoking induces cancer is overwhelming. However, the same statistical test that shows that smoking and the incidence of cancer are related can also be applied to historical data to show that the number of breweries constructed and the number of churches built in eighteenth century England are not independent. Does this mean that more breweries result in more churches, or that more churches lead to more breweries? Neither can be concluded because the association between the two events can be simply explained in terms of the general increase in the population. The implications of stating that two components of an experiment or survey are not independent, in particular the ideas of cause and effect, cannot be covered in an introductory course.

Two events are considered independent if the occurrence of one event does not influence the occurrence of the other event, and conversely. In terms of probability, this can be stated in the following way. Let A and B denote two events concerning the outcome of an experiment. For instance, event A could be that a person has high blood pressure and event B could be that a person has an abnormally fast pulse rate.

Event A is said to be independent of event B if the probability of event A occurring is not changed by the knowledge that event B also occurs.

For instance, high blood pressure would be considered independent of fast pulse rates if the chance that a person with a fast pulse rate will have high blood pressure is the same as the chance that any individual will have high blood pressure.

Definition 27.5

i) *The conditional probability that event A occurs, given the fact that event B is known to occur*, is denoted by the symbol $P(A|B)$ and is read "The probability of A given B."

ii) *Event A is independent of event B* if the probability of event A is the same as the probability of A given B:

$$A \text{ is independent of } B \text{ if } P(A) = P(A|B).$$

iii) *Events A and B are independent* if event A is independent of event B and event B is independent of event A.

Example 27.6 Consider the experiment in which a fair coin is flipped twice. The sample space is

$$S = \{HH, HT, TH, TT\},$$

where HT denotes a heads on the first flip and a tails on the second flip. Since the coin is fair, each outcome is equally likely to occur and hence will have probability of $\frac{1}{4}$. Intuitively, the result of the second flip should be independent of the result of the first flip. To examine this mathematically, we pose two events; one involves only the outcome of the second flip and one only involves the first flip. For instance,

Let A be the event that the second flip is tails, and

let B be the event that the first flip is heads.

As a subset of the sample space, S, the event A consists of two events, TT and HT (similarly for B):

$$A = \{TT, HT\} \quad \text{and} \quad B = \{HH, HT\}.$$

The probability of A is computed with the aid of formula (27.1)

$$P(A) = P(TT) + P(HT) = \tfrac{1}{4} + \tfrac{1}{4} = \tfrac{1}{2}.$$

Similarly, we find that

$$P(B) = P(HH) + P(HT) = \tfrac{1}{4} + \tfrac{1}{4} = \tfrac{1}{2}.$$

Event A given B is the event that the second flip is tails, given that the first flip is heads. If we know that event B occurs, the possible outcomes are limited to *two* elementary events, HT and HH. Since only *one* of these two events results in tails on the second flip, the probability of this occurring is $\tfrac{1}{2}$ (one way of two possible ways). Thus we have argued that

$$P(A \mid B) = \tfrac{1}{2}.$$

Consequently,

$$P(A) = P(A \mid B)$$

and event A is independent of event B. Note that the probability of A and B occurring is

$$P(A \cap B) = P(HT) = \tfrac{1}{4},$$

and

$$P(A \cap B) \neq P(A \mid B).$$

◀

Computing the conditional probability $P(A \mid B)$ is often a stumbling block. Observe that in general

$$P(A \mid B) \neq P(A \cap B).$$

However, these two probabilities are related as may be visualized using the Venn diagram in Fig. 27.3. Figure 27.3(a) indicates the sample space S and events A and B. If we assume that event B occurs, then we are not really looking at the sample space S but only at part B of the sample space, as indicated in Figure 27.3(b).

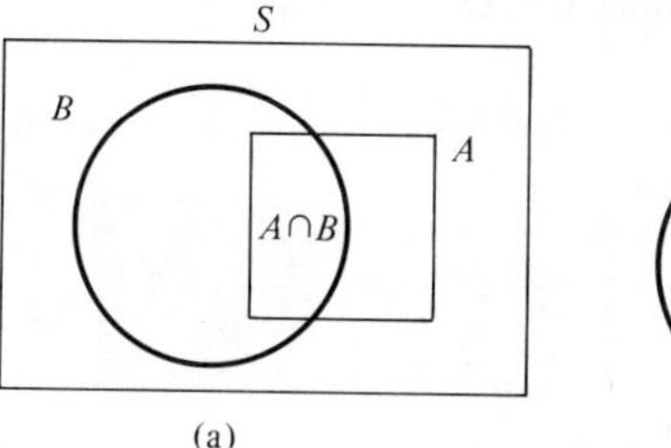

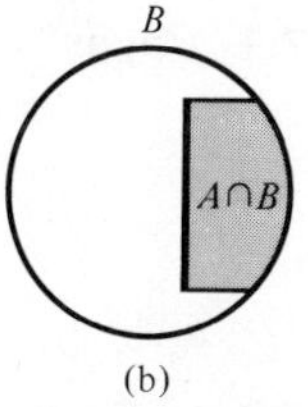

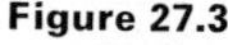

Figure 27.3 (a) (b)

The only way event A can occur, given that event B occurs, is to have the outcome represented in the region denoted $A \cap B$. Thus the event A and B can be visualized as a sequence of two events: first, event B occurs and then the event A given B occurs. Using the notion that the probability of a sequence of events occurring is the product of the probabilities of each event occurring, we have

$$P(A \cap B) = P(B) \cdot P(A \mid B). \tag{27.5}$$

The conditional probability formula (27.5) can also be visualized using counting arguments. *In an equiprobable space* S the probability of any event is given as follows:

$$P(\text{event}) = \frac{\text{number of ways the event may occur}}{\text{total number of elementary events in } S}.$$

Abbreviating this, we would have for event B,

$$P(B) = \frac{\# B}{\# S},$$

and similarly,

$$P(A \cap B) = \frac{\#(A \cap B)}{\# S}.$$

But the conditional event A given B restricts the total number of possible events to those for which B occurs, as in Fig. 27.3(b). Thus, in the same equiprobable setting,

$$P(A|B) = \frac{\#(A \cap B)}{\# B} \left(\frac{\text{the ways in which } A \text{ can occur given } B}{\text{the total possible ways for } B \text{ to occur}} \right).$$

With the above formulas, we have

$$\begin{aligned} P(B) \cdot P(A|B) &= \frac{\# B}{\# S} \cdot \frac{\#(A \cap B)}{\# B} \\ &= \frac{\#(A \cap B)}{\# S} = P(A \cap B). \qquad \text{(canceling)} \end{aligned}$$

When the equation (27.5) is solved for the conditional probability, $P(A|B)$, the resulting formula is known as Bayes' Theorem.

Theorem 27.1 *Bayes' Theorem*

$$P(A|B) = \frac{P(A \cap B)}{P(B)}. \tag{27.6}$$

Example 27.7 In Example 27.2, ABO blood types were described in terms of reactions with antigens. What is the probability that a person's blood will react with the A-antigen given that it has been tested and found to react to the B-antigen? The answer requires us to compute the conditional probability $P(A|B)$, where

A is the event that an A-antigen reaction occurs

and

B is the event that a B-antigen reaction occurs.

In terms of the elementary events of Example 27.2,

$$A = \{\text{Type A, Type AB}\},$$
$$B = \{\text{Type B, Type AB}\},$$
$$A \cap B = \{\text{Type AB}\}.$$

Consequently, using the probabilities derived from the relative frequencies in Table 27.1,

$$P(A) = P(\text{Type A}) + P(\text{Type AB})$$
$$= 0.417 + 0.030 = 0.447.$$

Similarly,

$$P(B) = 0.116 \qquad \text{and} \qquad P(A \cap B) = 0.030.$$

By Bayes' Theorem, we conclude that

$$P(A \mid B) = \frac{P(A \cap B)}{P(B)} = \frac{0.030}{0.116} \simeq 0.257.$$

Thus, while a typical person has a chance of about 3 in 100 of having blood type AB, if we know that a person's blood reacts to the B-antigen, his or her chances of having type AB blood increases to 1 in 4. ◀

Example 27.8 A (hypothetical) survey determined the following probabilities for a 50 year old male American:

$$P(C) = 0.25,\ P(S) = 0.40,\ P(C \cap S) = 0.20,$$

where C denotes "will contact cancer" and S denotes "smokes regularly." What is the probability that a person who smokes regularly will have cancer (based on the above figures)? *Solution*: Using Bayes' Theorem,

$$P(C \mid S) = \frac{P(C \cap S)}{P(S)} = \frac{0.20}{0.40} = 0.50.$$

What is the probability that a person who does not smoke will have cancer? Again we use Bayes' Theorem, but first we must compute the probability that a person does not smoke:

$$P(\tilde{S}) = 1 - P(S) = 1 - 0.40 = 0.60.$$

Next we must compute the probability that a person does not smoke and will have cancer, denoted $\tilde{S} \cap C$. To compute this probability, we observe that people with cancer consist of two groups, those who smoke and those who do not smoke. Therefore,

$$C = \underset{\text{Smokers}}{(C \cap S)} \quad \cup \quad \underset{\text{Nonsmokers}}{(C \cap \tilde{S})}.$$

As the smokers and nonsmokers are mutually exclusive,

$$P(C) = P(C \cap S) + P(C \cap \tilde{S}).$$

Solving, we have

$$P(C \cap \tilde{S}) = P(C) - P(C \cap S) = 0.25 - 0.20 = 0.05.$$

(See Fig. 27.4.)

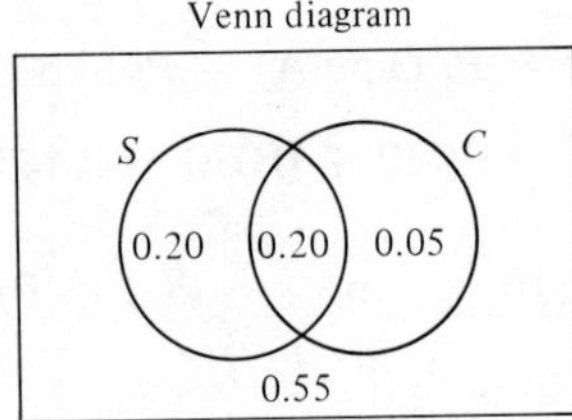

Figure 27.4

Now, using Bayes' Theorem,

$$P(C \mid \tilde{S}) = \frac{P(C \cap \tilde{S})}{P(\tilde{S})} = \frac{0.05}{0.60} \simeq 0.08.$$

Conclusion: Using the above data, a nonsmoker would have a much lower chance of having cancer than would a smoker. ◀

Using Bayes' Theorem, we can now give a simple test to check whether two events A and B are independent. By definition, A is independent of B if $P(A \mid B) = P(A)$. Replacing the conditional probability $P(A \mid B)$ by $P(A \cap B)/P(B)$, we obtain:

A and B are independent if $P(A \cap B) = P(A) \cdot P(B)$.

27.4 INTRODUCTION TO STATISTICS

Statistics can be divided into two basic areas, descriptive statistics and predictive or inference statistics.

Descriptive statistics concerns the representation of data for the purpose of clear and concise presentation. This is the most familiar aspect of statistics and is introduced early in the educational process. It includes such things as bar graphs or histograms, pie charts, and proportional representation charts as illustrated in Fig. 27.5. It also includes such numerical simplifications as specifying the range of data values, the mode or most frequent value in a list of data, and the median or midpoint value, which is one-half the sum of the largest and the smallest values. These topics are relatively well understood and will not be pursued further except for some exercises.

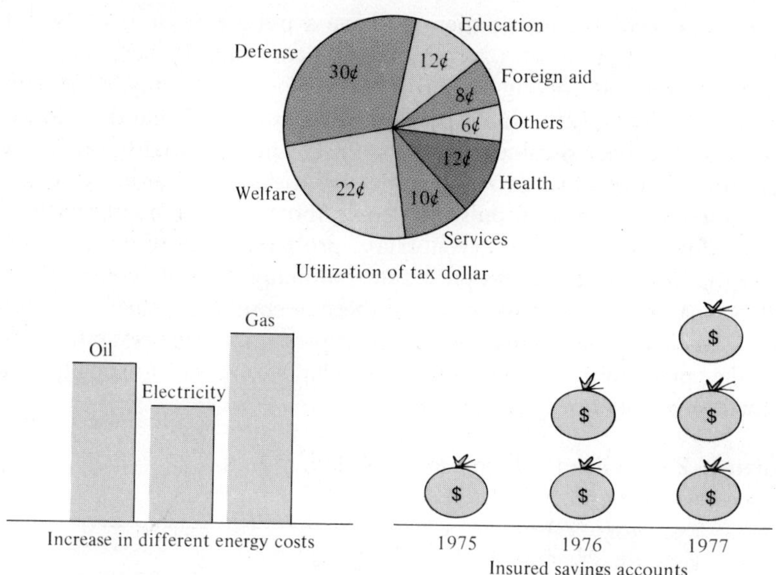

Figure 27.5

Predictive statistics examines data which are assumed to be generated by performing an experiment. Its objective is to provide information about the probabilistic mechanisms of the experiment. This information can then be utilized to predict the outcome if the experiment is repeated, to ascertain differences or similarities between different experiments or sets of data, and to test the validity of proposed relationships or dependencies. In the following, we will only introduce the rudiments of predictive statistics, the mean and standard deviation. We begin with an example.

Example 27.9 A nurse in training took the pulse rates of five patients. The values were 68, 72, 88, 76, and 88. What would be the pulse rate which he or she would expect of the next patient? To answer this, we first make some basic assumptions. For instance, we assume that the patients all fall into some general class, so that the next patient will be similar to the first five. This assumes that the nurse has not moved from the postoperative room to the labor room. This is the basic type of assumption made in all statistical analyses even though such assumptions are usually not specifically mentioned. Next we give the simplest estimate of the expected pulse rate of the next patient, the mean or average of the pulse rates previously recorded. We denote this mean value by $\bar{x}$ and define it to be the sum of the data divided by the number of data:

$$\bar{x} = \frac{68 + 72 + 88 + 76 + 88}{5} = 78.4.$$

Do we really expect the next patient to have a pulse rate of 78.4? No! This is unlikely since most pulse rates are obtained by counting the heart beats in a 15-second period and multiplying by four. This method results only in rates divisible by four such as 76 or 80 and would preclude the value 78.4. What then do we mean when we say the expected value will be $\bar{x} = 78.4$? There are many good answers to this question, none of which proves completely satisfactory. For instance, we could say that the chance of the next pulse rate being above 78.4 is the same as the chance that it is below 78.4—i.e., both events have probability 0.5 of occurring. Or, we could say that if not just the next pulse rate, but many more are considered, (say 5, 10, 100, or so) and the average of all of these is computed, then we expect this average to be close to 78.4. This is not an exact prediction because statistics involve chance—i.e., probability distributions—and what we expect statistically we can be certain will probably not occur exactly. ◀

Definition 27.6 Given a finite set of N data values, $\{x_1, x_2, x_3, \ldots, x_N\}$, the *sample mean* is the number $\bar{x}$, defined by

$$\bar{x} = \frac{x_1 + x_2 + \cdots + x_N}{N} = \frac{1}{N}\sum_{i=1}^{N} x_i \qquad \textbf{Sample mean.} \tag{27.7}$$

As illustrated in Example 27.9, should the experiment that generated the data, x_1, $x_2, \ldots, x_N$ be repeated, the "expected value" of the outcome is approximated by the sample mean $\bar{x}$. To explore further, the concept of an expected value, we consider the following example in which the actual probabilities associated with an experiment are known.

Example 27.10 A particular bird species is known to lay one, two, or three eggs. Based upon observations over a long period of time, it has been determined that the probability of each of these events occurring is given by

$$P(1) = \tfrac{1}{3},\ P(2) = \tfrac{1}{2},\ P(3) = \tfrac{1}{6}.$$

Using the frequency concept of probability, this would suggest that, on the average, out of six birds, two would lay one egg, three would lay two eggs, and only one would lay three eggs. Thus the expected average number of eggs to be laid by the six birds would be the total number of expected eggs divided by six. The total number of expected eggs would be the sum of the possible number of eggs times the number of birds expected to lay that number of eggs:

$$\begin{matrix}\text{Expected}\\ \text{total number}\\ \text{of eggs}\end{matrix} = 1\cdot\begin{bmatrix}\text{expected}\\ \text{number}\\ \text{laying 1 egg}\end{bmatrix} + 2\cdot\begin{bmatrix}\text{expected}\\ \text{number}\\ \text{laying 2 eggs}\end{bmatrix} + 3\cdot\begin{bmatrix}\text{expected}\\ \text{number}\\ \text{laying 3 eggs}\end{bmatrix}.$$

Therefore the expected average number of eggs laid would be

$$\mu = \frac{1\cdot(2) + 2\cdot(3) + 3\cdot(1)}{6},$$

where we have used the Greek letter, μ (pronounced mu), to represent this expected average. If we rearrange the above equation into the form

$$\mu = 1 \cdot (\tfrac{2}{6}) + 2 \cdot (\tfrac{3}{6}) + 3 \cdot (\tfrac{1}{6}),$$

we see that it can also be visualized as the sum of each possible outcome (one, two, or three eggs) multiplied by the probability of the outcome:

$$\mu = \sum (\text{possible outcomes}) \times (\text{probability of the outcome}).$$

Thus in this example, the expected number of eggs would be

$$\mu = 1.5.$$

Of course, a bird cannot lay 1.5 eggs. This is not what the expected value gives. The expected value μ gives the theoretical average number of eggs laid by birds of this species. One interpretation of this would be that the species increases each year at a rate of 1.5 per (female) bird. This concept would tell us that a flock containing 120 female birds would be expected to increase by

$$120 \times \mu = 120 \times 1.5 = 180 \text{ new birds},$$

assuming that all the eggs survive and no mature birds die. ◀

Definition 27.7 Given a discrete probability space S with distribution P, the *mean value* μ is defined to be the sum of each elementary event multiplied by the probability of its occurrence. If $S = \{x_1, x_2, \ldots, x_n\}$, then

$$\mu = \sum_{i=1}^{n} x_i \cdot P(x_i) \qquad \textbf{Mean } \boldsymbol{\mu} \textbf{ (theoretical).} \tag{27.8}$$

The term μ is also called the *expected value* of the distribution.

Example 27.11 We compute the mean μ for several different probability spaces.

a) Let $S = \{-1, 0, 1\}$ with $P(-1) = 0.6$, $P(0) = 0.1$, and $P(1) = 0.3$. Then

$$\begin{aligned}\mu &= (-1) \cdot P(-1) + 0 \cdot P(0) + 1 \cdot P(1)\\ &= (-1) \cdot (0.6) + 0 \cdot (0.1) + 1 \cdot (0.3) = -0.3.\end{aligned}$$

b) Let $S = \{1, 2, 3, 4, 5, 6\}$ and $P(x) = \frac{1}{6}$ for each $x \in S$. This is an equiprobable space. Then

$$\begin{aligned}\mu &= 1 \cdot \tfrac{1}{6} + 2 \cdot \tfrac{1}{6} + 3 \cdot \tfrac{1}{6} + 4 \cdot \tfrac{1}{6} + 5 \cdot \tfrac{1}{6} + 6 \cdot \tfrac{1}{6}\\ &= \tfrac{1}{6} \cdot (1 + 2 + 3 + 4 + 5 + 6)\\ &= \tfrac{1}{6} \cdot 21 = 3.5.\end{aligned}$$

c) Again let $S = \{1, 2, 3, 4, 5, 6\}$, but choose the distribution P with

$$P(1) = P(2) = P(3) = 0.1, \qquad P(4) = P(5) = 0.15, \qquad \text{and} \qquad P(6) = 0.4.$$

The expected value is

$$\mu = 1 \cdot (0.1) + 2 \cdot (0.1) + 3 \cdot (0.1) + 4 \cdot (0.15) + 5 \cdot (0.15) + 6 \cdot (0.4)$$
$$= 4.35.$$

◀

The existence of two means, μ and $\bar{x}$, may seem confusing at first glance; however, the difference between them should be clear. The distribution mean μ, or simply the mean μ, is the theoretical expected value and can only be calculated when the sample space and its probability distribution are completely known. The sample mean $\bar{x}$ is an estimated expected value based on observed data. The sample mean $\bar{x}$ is an estimate of the theoretical mean μ. In many experiments, the true distribution of outcomes is not known; therefore, the theoretical mean μ cannot be calculated. In such situations, the sample mean $\bar{x}$ is calculated on the basis of repeated experiments.

One of the basic results of statistics is the fact that the value of $\bar{x}$ will approach the value of the theoretical mean in the limit as the number of sample data values becomes larger. For instance, the distribution in Example 27.11(b) corresponds to the roll of a fair die. If we were to roll a die 10 times, we might obtain the following set of values:

$$\{2, 1, 4, 2, 1, 3, 4, 6, 6, 4\}.$$

The sample mean would be $\bar{x} = 3.3$. If we continued rolling the die, say an additional 90 times, then we would expect the value of $\bar{x}$ to be closer to the value $\mu = 3.5$.

Example 27.12 Consider the sample space $S = \{1, 2, 3\}$ with the three distributions, P_a, P_b, and P_c, given by:

a) $P_a(1) = P_a(2) = P_a(3) = \frac{1}{3}$;
b) $P_b(1) = 0.45$, $P_b(2) = 0.1$, $P_b(3) = 0.45$;
c) $P_c(1) = 0.05$, $P_c(2) = 0.9$, $P_c(3) = 0.05$.

These distributions are illustrated in Fig. 27.6. The mean μ of each distribution is the value $\mu = 2$.

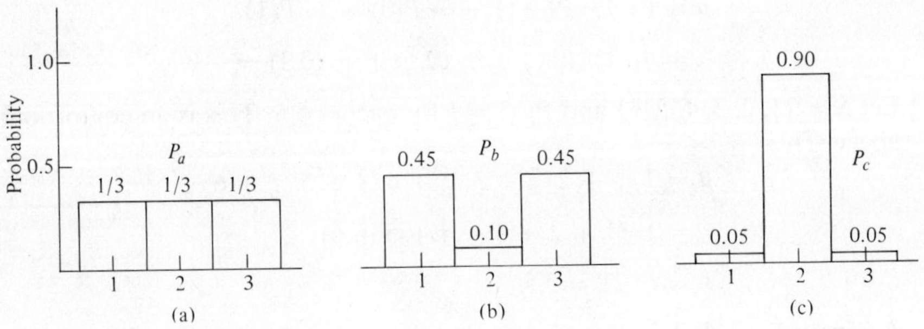

Figure 27.6

◀

Obviously, the three distributions in Example 27.12 are different. In the first, all outcomes are equally likely; in the second, outcomes 1 and 3 are much more likely to occur than outcome 2; and, in the third distribution 95 percent of the time the outcome will be 2, with 1 and 3 much less likely to occur. These differences are not reflected by the mean μ, as it is the same for each distribution; in each experiment, the expected value is 2. The difference between the distributions is not their mean value, but rather how close repeated outcomes are to the mean, $\mu = 2$. For instance, with the distribution P_b, we would seldom expect the outcome to be 2, while with the distribution P_c the outcome will be 2 most of the time.

To distinguish between distributions with the same mean, a second statistic is utilized. This statistic is called the *standard deviation* of the distribution and is defined to measure the degree to which the outcome of an experiment deviates from the mean μ. The lower case Greek letter sigma, σ, is used to represent the theoretical standard deviation. The following definition involves the square root of a sum of squares. The reason for this is to measure the distance between an outcome and the mean μ. If a term such as $x - \mu$ were used it would give a *signed distance* between x and μ; $x - \mu$ is positive if $x > \mu$, and negative if $x < \mu$. The absolute distance between x and μ is

$$|x - \mu| = \sqrt{(x - \mu)^2}.$$

Definition 27.8 Given a discrete probability space S with distribution P and mean μ, the *standard deviation* of the distribution is the number σ, defined by

$$\sigma = \left[\operatorname*{sum}_{x \in S} (x - \mu)^2 \cdot P(x)\right]^{1/2}.$$

If $S = \{x_1, x_2, \ldots, x_n\}$, then

$$\sigma = \left[\sum_{i=1}^{n} (x_i - \mu)^2 P(x_i)\right]^{1/2} \qquad \textbf{Standard deviation (theoretical).} \qquad (27.9)$$

Example 27.13 The standard deviation is an indication of the expected difference between the outcomes and the mean value μ. A small value of σ indicates that most outcomes will be close to the mean μ. A large value of σ indicates that the outcomes are not as likely to be close to μ. This interpretation of the standard deviation is illustrated by the distributions in Example 27.12, $\mu = 2$ for each distribution.

a) The distribution P_a has standard deviation:

$$\sigma_a = [(1-2)^2 \cdot P_a(1) + (2-2)^2 \cdot P_a(2) + (3-2)^2 \cdot P_a(3)]^{1/2}$$
$$= [\tfrac{1}{3} + 0 + \tfrac{1}{3}]^{1/2} = \sqrt{\tfrac{2}{3}} \simeq 0.816.$$

b) The distribution P_b has standard deviation:

$$\sigma_b = [(1-2)^2 \cdot P_b(1) + (2-2)^2 \cdot P_b(2) + (3-2)^2 \cdot P_b(3)]^{1/2}$$
$$= [0.45 + 0 + 0.45]^{1/2} = \sqrt{0.9} \simeq 0.949.$$

c) The distribution P_c has standard deviation:

$$\sigma_c = [(1-2)^2 \cdot P_c(1) + (2-2)^2 \cdot P_c(2) + (3-2)^2 \cdot P_c(3)]^{1/2}$$
$$= [0.05 + 0 + 0.05]^{1/2} = \sqrt{0.1} \simeq 0.316.$$

Observe that $\sigma_c < \sigma_a < \sigma_b$ in conjunction with the graphical representations of the distributions in Fig. 27.6. ◀

Just as we defined the sample mean $\bar{x}$ to be an estimate of the theoretical mean μ, we define a sample standard deviation, which is obtained from experimental data and estimates the theoretical standard deviation σ.

Definition 27.9 Given a finite set of data, $\{x_1, x_2, x_3, \ldots, x_N\}$, the *sample standard deviation* is the number

$$s = \left[\frac{\sum_{i=1}^{N}(x_i - \bar{x})^2}{N-1}\right]^{1/2}, \quad \textbf{Sample standard deviation,} \qquad (27.10)$$

where $\bar{x}$ is the sample mean.

The sample mean is a sum divided by N, the number of data values. The sample standard deviation involves a sum of squares divided by $N - 1$, the number of data values minus one. The term $N - 1$ in the formula for the sample standard deviation is referred to as the number of *degrees of freedom* of the statistic s. The reason for having $N - 1$ in the formula for s rather than N, as in the formula for $\bar{x}$, is because the formula for s, equation (27.10), includes the sample mean $\bar{x}$. If we know the value for $\bar{x}$ and only $N - 1$ of the data values, then we can determine the Nth data value algebraically; this is why there is only $N - 1$ degrees of freedom. For instance, if we have four data values, $x_1 = 2, x_2 = 1.5, x_3 = 5$, and $x_4 = 5.5$, the mean is $\bar{x} = (2 + 1.5 + 5 + 5.5)/4 = 3.5$. Assume that we were only told the values x_1, x_2, and x_4 and also told the mean $\bar{x}$. The value of x_3 would then be obtained by solving the equation

$$\bar{x} = \frac{x_1 + x_2 + x_3 + x_4}{4}$$

for x_3 and substituting in the known values:

$$x_3 = 4 \cdot \bar{x} - (x_1 + x_2 + x_4),$$
$$x_3 = 4 \cdot (3.5) - (2 + 1.5 + 5.5) = 5.$$

Formula (27.10) for the sample standard deviation is very useful theoretically; but, for the purpose of calculations, it is not the best one to use. The following formula is algebraically equivalent to formula (27.10) and is much easier to evaluate:

$$s = \left[\frac{N\sum(x_i^2) - (\sum x_i)^2}{N(N-1)}\right]^{1/2} \quad \textbf{Computational form of Sample standard deviation.} \qquad (27.11)$$

Observe that formula (27.11) does not require the prior evaluation of the sample mean $\bar{x}$; in essence, the value of $\bar{x}$ is determined simultaneously as s is computed using formula (27.11).

Example 27.14 We illustrate the two formulas for calculating s for the set of data

$$\{12, 8, 7, 2, 10, 3\}.$$

a) To use formula (27.10), we must first calculate $\bar{x}$ using (27.7) with $N = 6$:

$$\bar{x} = (12 + 8 + 7 + 2 + 10 + 3) \div 6 = 7.$$

Next, to calculate the sum in (27.10), we must evaluate the differences $x_i - \bar{x}$ and their squares $(x_i - \bar{x})^2$.

	x_i	$(x_i - \bar{x})$	$(x_i - \bar{x})^2$
	12	5	25
	8	1	1
	7	0	0
	2	−5	25
	10	3	9
	3	−4	16
Sums	42	0	76

Substituting the sum of $(x_i - \bar{x})^2$ and $N - 1 = 5$ into (27.10), we obtain

$$s = \sqrt{\tfrac{76}{5}} = \sqrt{15.2} \simeq 3.9.$$

b) To calculate s and simultaneously $\bar{x}$ for the given data set using formula (27.11), we arrange the data as follows

							sum
x_i	12	8	7	2	10	3	$42 = \sum x_i$
x_i^2	144	64	49	4	100	9	$370 = \sum x_i^2$

Thus $\bar{x} = \frac{42}{6} = 7$, and substituting into formula (27.11) gives us

$$S = \sqrt{\frac{6(370) - (42)^2}{6(5)}} = \sqrt{\frac{2220 - 1764}{30}}$$

$$= \sqrt{\frac{456}{30}} = \sqrt{15.2} \simeq 3.9. \qquad ◀$$

The computations illustrated in the above example are theoretically equivalent; but for larger data sets, the method using formula (27.11) involves fewer steps. Calculators that automatically evaluate $\bar{x}$ and s use formula (27.11) to compute the sample standard deviation.

Example 27.15 In many laboratory experiments, the results are biased by the person performing the experiment. The following data indicate numerical values obtained by three technicians repeating a clinical assay using the same blood sample.

Technician	A	B	C
	22.1	19.2	26.2
	21.0	20.7	21.0
Sample data	23.2	21.2	22.0
	22.2	19.7	23.2
	22.0	20.0	18.1
	26.5	20.1	26.5

The sample mean and standard deviation of these three data sets are as follows.

	$\bar{x}$	s
Technician A	22.83	1.9
Technician B	20.15	0.7
Technician C	22.83	3.2

Based upon this data, Technician B is the most consistent of the three technicians. This is observed from the relatively low standard deviation compared to the other two technicians' results. In many clinical tests, the actual values are not of interest so much as detecting values that are abnormal. Thus Technician B would be preferable to the other two technicians in the sense that abnormally high or low values reported by this technician would be less likely to result from inconsistent techniques. ◀

The Example 27.15 illustrates the simplest type of comparison between different data sets. The observations made in the example concerned the equality of the sample means and the relative sample standard deviations. A more complete statistical analysis of these data sets would investigate questions such as the following:

Are the sample means (or standard deviations) for Technicians A and B significantly different?

Are the samples likely to be representative of the same theoretical distribution?

If the samples are all from the same theoretical distribution, can we determine the value of the theoretical mean μ?

Questions such as these are answered using statistical tests. The development and methods of these tests are too involved to be examined in this text but may be found in any elementary statistics text. We will mention some of the special distributions connected with these statistical tests in Section 28.

EXERCISE SET 27

27.1 A sack contains three red balls and two white balls. What is the probability that a ball selected at random from the sack will be

a) red? b) white? c) not red?

27.2 If a lake contains approximately 10,000 fish and 500 fish are tagged, what is the probability that a tagged fish will be caught if every fish has an equal chance of being caught?

27.3 In a survey of 476 dogs admitted to a Veterinary Clinic, 52 suffered from diabetes; 10 suffered from urinary tract infections; 73 suffered kidney problems; 157 were not ill but were admitted for vaccinations; 37 had been hit by automobiles; 9 had hip displasia problems; 20 needed their teeth cleaned; 11 suffered bleeding presumably due to warfrin poisoning; and the remainder were admitted to be spayed or castrated.

a) What is the probability that a dog admitted to the clinic will have been hit by a car?
b) What is the probability that a dog will be admitted suffering from a urinary tract infection or kidney problem?
c) What is the incidence of bleeding problems encountered at the clinic? How many bleeding problems would likely be encountered in the next 200 dog admissions to the clinic?

27.4 Indicate the sample spaces for the following experiments.

a) Quail nests are examined and the number of eggs in each are recorded. (Assume that one quail lays at most eight eggs.)
b) A patient's blood pressure is recorded as high, low, or normal.
c) The genotype of an offspring is recorded when the parents' genotypes are *Aa* and *AA*.
d) A person's A-B-O blood type is indicated.
e) A person is asked to choose a number between one and nine, inclusively.

27.5 For the following "experiments" indicate (1) the sample space and (2) the probability of each outcome using the "frequency" concept of probability.

a) In a closed lake, a fisheries agent catches fish and records the species. The results over a one month period are given in Table 27.2.

Table 27.2

Bass	Pike	Trout
38	41	371

b) A coin is flipped 1000 times. Heads comes up 410 times and tails comes up 590 times. (Is the coin fair?)
c) A laboratory test for pregnancy resulted in the frequencies listed in Table 27.3.

Table 27.3

Positive	Negative	Doubtful
21	138	38

27.6 For the following experiments, list the sample space and define a probability function on the sample space.

a) The experiment consists of selecting and recording the color of one ball from a sack containing three red balls, two blue balls, and five yellow balls.
b) An experiment consists of randomly selecting an animal and recording its sex when the population from which the selection is made has 250 males and 950 females.
c) An experiment consists of selecting a number (recorded on ping-pong balls) from a box containing seven number 1's, three number 2's, eight number 3's, and one each of the numbers 4, 5, 6, 7, and 8.

27.7 Which of the following specify probability functions for the set $S = \{1, 2, 3, 4\}$?

a) $P(1) = P(2) = P(3) = P(4) = \frac{1}{4}$
b) $P(1) = 0.1, P(2) = 0.2, P(3) = 0.3, P(4) = 0.4$
c) $P(1) = 0.1, P(2) = 0.4, P(3) = 0.4, P(4) = 0.2$
d) $P(x) = x/10$ for $x \in S$
e) $P(x) = (1 - x)/6$ for $x \in S$

27.8 An experiment consists of drawing a ball from a sack containing 10 similar balls (3 red, 3 blue, and 4 white) and recording its color.

a) What is the sample space S?
b) If each ball is equally likely to be drawn, what is the probability function P?
c) Let A be the event the ball is red or blue. Give $P(A)$ and $P(\tilde{A})$.
d) Let B be the event that the ball is either red or white. Give $P(B)$ and $P(\tilde{B})$.
e) What are the events $A \cap B$, $A \cup B$, $A \cap \tilde{B}$, and $\tilde{A} \cap \tilde{B}$? Give their probabilities.

27.9 Assume that $S = \{1, 2, 3, 4, 5\}$ and $P(x) = x/15$ for $x \in S$.

a) Show that P is a probability function for S.
b) Compute the probabilities of the following events:

i) $\{1, 2\}$ ii) $\{1, 3, 5\}$ iii) $\{2, 4\}$

c) What is the probability that x will be less than 4?

27.10 The following Venn diagram illustrates a probability space with events A and B. Evaluate the following.

a) $P(A)$ b) $P(B)$ c) $P(A \cap B)$
d) $P(A \cup B)$ e) $P(\tilde{A})$ f) $P(\tilde{A} \cap \tilde{B})$

S
0.3
A B
0.4 0.2 0.1

27.11 A pair of fair dice with sides numbered one through six is tossed and the sum of their upward face values is recorded as x.

a) List the sample space S of this experiment.
b) To compute the probability distribution for this experiment, we think of the two dice as being distinguishable, say one is red and the other is green. How many different ways can the dice land? How many ways will the outcome be 7? Assuming that each outcome is equally probable, what is the probability that a 7 is rolled?
c) Determine the probability of each elementary event.

d) List the following events as subsets of the sample space S and compute their probabilities of occurring.

A: The number x is less than 5

B: The number x is greater than 7

C: The number x is 7 or 11

e) List the elementary events in the following events and the probability of the events:

i) $A \cup B$,	ii) $B \cap C$,	iii) $A \cup C$,
iv) $A \cap B$,	v) $B \cap \tilde{C}$,	vi) $A \cap \tilde{B}$,
vii) $A \cup \tilde{B}$,	viii) $\tilde{B} \cap \tilde{C}$.	ix) $\tilde{A} \cap B \cap C$.

27.12 In Example 27.4, what is the probability that the outcome will be either 1 or 3?

27.13 If the events A and B have $P(A) = 0.4$, $P(B) = 0.4$, and $P(A \cap B) = 0.2$, what is $P(A \cup B)$?

27.14 In Example 27.5, what is the probability that a farm produces crops only?

27.15 A person flips a fair coin three times, recording the outcome as either H or T after each flip.

a) List the set of all possible outcomes.
b) What is the probability that the third flip results in an H?
c) What is the probability that the first two outcomes are both T?
d) What is the probability that the outcomes in order are TTH?

27.16 A couple has a recessive sex-linked genetic defect. The male has the defect and the woman is a carrier of the defect. If they are told the chance of an offspring having the defect is $\frac{1}{4}$ and one of their two children has the defect, can they safely have two more children without fear of them having the defect? If not, explain why in probability terms.

27.17 If the experiment in Example 27.3 (a) is repeated, what is the probability

a) that two red balls will be drawn?
b) that a red and a blue ball will be drawn?
c) that neither ball will be green?

27.18 Assume that the probability of it raining in a particular location is 0.03.

a) What is the probability that it will not rain?
b) What is the probability that it will rain two days and then not rain on the third day?
c) What is the probability that it will not rain for one year (not a leap year)?
d) What is the longest period that has a probability of 0.5 that it will not rain?

27.19 In Example 27.6, let the event $C = \{TH, HT\}$ and $A = \{TT, HT\}$.

a) Compute $P(C)$ b) Compute $P(C|A)$. c) Compute $P(A|\tilde{C})$.

27.20 Assume that the probability that a person drinks alcohol, denoted D, is $P(D) = 0.35$. Assume that the probability that a person has liver problems, denoted L, is $P(L) = 0.005$, and the probability that a person drinks and has liver problems is $P(D \cap L) = 0.003$.

a) What is the probability that a person who drinks will have liver problems?
b) What is the probability that a person who does not drink will have liver problems?
c) Does drinking increase your chances of having liver problems (based on this fictitious data)?

27.21 The 1976 Swine Flu immunization program in the United States was suspended because of a statistical association between the vaccination and a rare disease known as the Guillain–Barré syndrome. The following questions lead to the statistical evidence upon which the decision was made to discontinue the vaccination program.

a) Assuming that the United States has a population of 200 million, what is the probability of the event (denoted G) of contacting the Guillain–Barré disease, if 500 cases in total were reported in a two-month period?
b) If one-half of the Guillain–Barré cases reported also received a vaccination, what is the probability that an individual both contacts the disease and is vaccinated?
c) Assume that 6.666 percent of the population was vaccinated. Let V denote the event of being vaccinated. Using the results of (a) and (b), show that events G and V are not independent. How much more likely were vaccinated individuals to contact the disease than the general population?
d) Was the decision to halt the immunization program justified by the above statistics or should other statistical considerations be included?

27.22 Using the Venn diagram, compute the following:

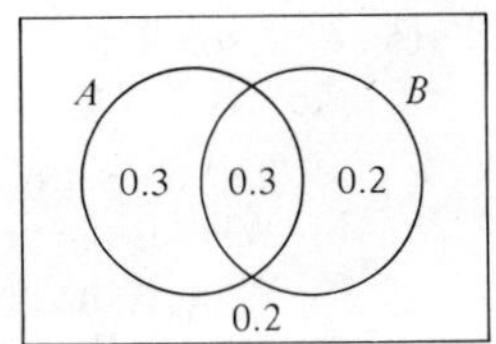

a) $P(A|B)$
b) $P(B|A)$

Are A and B independent?

27.23 If A and B independent events with $P(A) = 0.4$ and $P(B) = 0.2$, evaluate the following:

a) $P(A \cap B)$ b) $P(A \cap \tilde{B})$ c) $P(A \cup B)$

27.24 A survey of 420 students examined the incidence of being left-handed. The results of the survey are indicated in Table 27.4.

Table 27.4

	Male = M	Female = F	Total
L = Left-handed	28	32	60
R = Right-handed	168	192	360
Total	196	224	420

Answer the following questions, using the information given in Table 27.4.

a) Using the frequency concept of probability, determine the probability of the events M, F, L, and R.
b) Are the events of being a male and being left-handed independent events?
c) Are the events of being a female and being left-handed independent events?
d) What is the probability that a female is not left-handed?
e) What is the probability that a left-handed person is female?
f) If a person is right-handed, what is the probability that the person is a male?

27.25 Assume that the probability of having a boy or girl baby is equal. Consider a family with two children.

a) What is the probability that they are both of the same sex?
b) If one of the children is a girl, what is the probability that the other is a boy?
c) What is the probability that the children are of opposite sexes?
d) What is the probability that the first child is a boy when the second child is a girl?

27.26 Assume that two genes A and a have probability $p = 0.7$ and $q = 0.3$ of being passed on at each generation.

a) What is the probability of the four genotypes, AA, Aa, aA, and aa?
b) If one of an individual's genes is A, what is the probability that the other is also A?
c) When an individual is not of type AA, what is the probability that it is not of type aa?

27.27 If three cards have sides colored red-red, white-white, and red-white respectively, what is the probability that a randomly selected card, which is known to have one red side, will be red on both sides?

27.28 A survey of the sex distribution in three spider populations resulted in the data given in Table 27.5.

Table 27.5

Sex \ Species	A	B	C
F	42	70	38
M	8	28	14

a) Give the probabilities of the events, A, B, C, F, and M.
b) Give the following conditional probabilities: $P(A|F)$, $P(B|M)$, $P(C|M)$, $P(F|B)$, and $P(F|C)$.
c) Are events M and C independent?
d) Are events F and B independent?
e) Are events A and C independent?

27.29 Assuming that the events of being happy and being wealthy are independent, what are your chances of being both when only 1 percent of the population is wealthy and 34 percent of the population is happy?

27.30 Assume that there are 1000 Monarch butterflies of each sex. If the probability that a female butterfly will encounter a male is 0.05 and the probability that their mating will produce 50 butterflies is 0.45, how many butterflies will be produced?

27.31 Compute the sample mean and sample standard deviations of the given data sets.

a) $\{1, 2, 3, 4, 5, 6\}$ b) $\{1, 1, 2, 4, 6, 6\}$ c) $\{1, 3, 3, 3, 5, 6\}$
d) $\{1.1, 2.2, 3.3, 4.4\}$ e) $\{0, 2, -1, 3, -4\}$ f) $\{0, 2, 0, 2, -1\}$

27.32 Compute the mean, μ, for the probability spaces in Example 27.4.

27.33 For the sample space $S = \{-1, 0, 2, 3\}$, compute the mean, μ, and standard deviation, σ, for the following probability functions.

a) $P(-1) = 0.5$, $P(0) = 0.1$, $P(2) = 0.1$, $P(3) = 0.3$
b) $P(x) = \frac{1}{4}$ for $x \in S$
c) $P(x) = (x + 1)/8$ for $x \in S$
d) $P(-1) = P(0) = P(2) = 0.2$, $P(3) = 0.4$

27.34 A laboratory purchases rabbits that are supposed to weigh 5 lb. Ten rabbits were received, weighing as follows: 4.2 lb, 4.6 lb, 3.2 lb, 3.6 lb, 7.2 lb, 9.1 lb, 8.2 lb, 4.1 lb, 2.8 lb, and 3 lb. Compute the mean and variance of these weights. Does the laboratory have grounds to complain about the rabbits received?

27.35 Compute the values $\bar{x}$ and s given in Example 27.15.

27.36 Table 27.6 gives the air pollution index for two cities over a two-week period.

Table 27.6

	Hamilton		Toronto	
Saturday	18	19	11	15
Sunday	9	11	12	13
Monday	12	11	20	19
Tuesday	14	15	21	17
Wednesday	27	24	22	16
Thursday	29	26	23	22
Friday	34	29	21	24

a) Give the mean and standard deviation of the pollution index for each city.
b) Which city is less polluted?

27.37 A laboratory colony of drosophila is allowed to produce over a long period of time so that the development of genetic mutations introduced into the colony can be studied. Assume that the mutation has a theoretical probability of 0.05 of occurring.

a) What is the expected number of mutants to be found in a sample of 25 flies from the colony?
b) Assume that 10 sample populations of 25 flies are taken and the number of mutants in each sample is counted, resulting in the following data:

$$1, 0, 0, 3, 2, 2, 0, 0, 0, 3.$$

What is the average number of mutants found in a sample? What is the standard deviation of the above data?

27.38 Assume that 27 percent of a dairy herd have a parasite and 41 percent of the herd have received an antiparasitic drug in their food. If 20 percent of the herd received the drug and have the parasite, decide how effective the drug is by comparing the chances of having parasites for those receiving the drug and those not receiving the drug.

27.39 Which of the following data sets has the highest mean and which has the least variability (as indicated by the standard deviation)?

Data Set 1: 22, 28, 27, 14, 26, 25.
Data Set 2: 22, 23, 24, 25, 32.
Data Set 3: 18, 19, 25, 25, 26, 27.
Data Set 4: 21, 19, 32, 26, 20, 20.

SECTION 28

CALCULUS OF CONTINUOUS PROBABILITY SPACES

28.1 PROBABILITY, DENSITY, AND DISTRIBUTION FUNCTIONS

When the outcome of an experiment is a number that can assume any value in a finite interval $[\alpha, \beta]$ or an infinite interval such as $[0, +\infty)$ or $(-\infty, \infty)$, the associated interval is called a *continuous sample space*. In this situation, the outcome (usually denoted X) is called a *continuous random variable*. In this section, we introduce the basic concepts of continuous probability theory.

Given a continuous sample space I, any subset $A \subseteq I$ represents a possible event. Intuitively, a probability function for a continuous sample space I is a function $P(A)$, which indicates the probability that event A occurs:

$$P(A) = \text{probability that } X \in A.$$

A probability function P for a sample space I must satisfy the following properties:

i) $P(I) = 1$. (This guarantees that the random variable X is in the interval I.)
ii) $P(\phi) = 0$. (ϕ is the empty set.)
iii) $0 \leq P(A) \leq 1$ for all subsets $A \subseteq I$. (28.1)
iv) If $A \subseteq I$, $B \subseteq I$, and $A \cap B = \phi$, then $P(A \cup B) = P(A) + P(B)$.
v) If $\tilde{A} = \{x \in I \mid x \notin A\}$, then $P(A) = 1 - P(\tilde{A})$.

(*Note*: Capital X denotes the random variable and small x denotes numbers in the sample space I of the random variable X.)

A probability function P is defined on all subsets of the sample space and is consequently almost impossible to use. To describe the probability of a continuous random variable X, two additional probability functions are used. One function is called a *probability distribution function* and is denoted by $F(x)$. This function gives the probability that the random variable X will be less than or equal to x. Using a subscript 0 to emphasize that the small x is the argument of the function F, the distribution

$$\boxed{F(x_0) = \text{probability } (X \leq x_0).}$$

For instance, if X denotes the weight in kilograms of a lamb at birth, then $F(3)$ is the probability that a newborn lamb will weigh 3 kg or less.

The second function used to describe the probability structure of a continuous random variable is called a *probability density function* and is generally denoted by $f(x)$. A probability density function $f(x)$ gives an indication of the probability that the random variable X will be "close to x." The exact meaning of this will be discussed in the following material.

Let us state clearly at the outset that $f(x)$ is not the probability that x will occur In fact, for a continuous density function $f(x)$, the probability that any particular outcome, x_0, will occur is zero,

$$P(X = x_0) = 0.$$

In reading the following, you should be very careful to distinguish between the terms "distribution" and "density." As we shall see, the distribution $F(x)$ and the density $f(x)$ associated with a random variable X are closely related. Before we examine this relation, let us define each function separately and investigate their properties.

In the remainder of this subsection, we assume that the sample space is a finite interval $I = [\alpha, \beta]$. The infinite interval case will be examined in subsection 28.3.

Continuous Probability Distributions

Definition 28.1 Let $I = [\alpha, \beta]$ be a finite interval; a *probability distribution* $F(x)$, defined on I, is a function which satisfies the following.

i) $F(\alpha) = 0$ and $F(\beta) = 1$.

ii) If x_1 and $x_2 \in I$ with $x_1 < x_2$, then $F(x_1) \leq F(x_2)$.

Each distribution $F(x)$ is uniquely associated with a probability function P by the equation

$$\boxed{P(X \leq x) = F(x).} \tag{28.2}$$

Example 28.1 Consider the sample space $I = [0, 2]$.

a) The function

$$F(x) = \frac{x}{2}$$

is a distribution on the interval I. Condition (i) of Definition 28.1 is satisfied as $F(0) = 0$ and $F(2) = 1$. F satisfies condition (ii) of the definition since $x_1 < x_2$ implies that

$$F(x_1) = \frac{x_1}{2} < \frac{x_2}{2} = F(x_2).$$

b) The function

$$F(x) = \frac{x^2}{2}$$

is not a distribution for the interval I since $F(2) \neq 1$.

c) The "step" function given by

$$F(x) = \begin{cases} 0 & \text{if } 0 \leq x < 1.5; \\ 1 & \text{if } 1.5 \leq x \leq 2, \end{cases}$$

is a distribution on the interval I.

d) The function given by

$$F(x) = \begin{cases} x & \text{if } 0 \leq x < 1; \\ x - 1 & \text{if } 1 \leq x \leq 2, \end{cases}$$

is not a distribution on the interval I because it does not satisfy condition (ii) of the definition. For instance, $0.9 < 1.5$, but $F(0.9) = 0.9 > 0.5 = F(1.5)$. The above functions are sketched in Fig. 28.1.

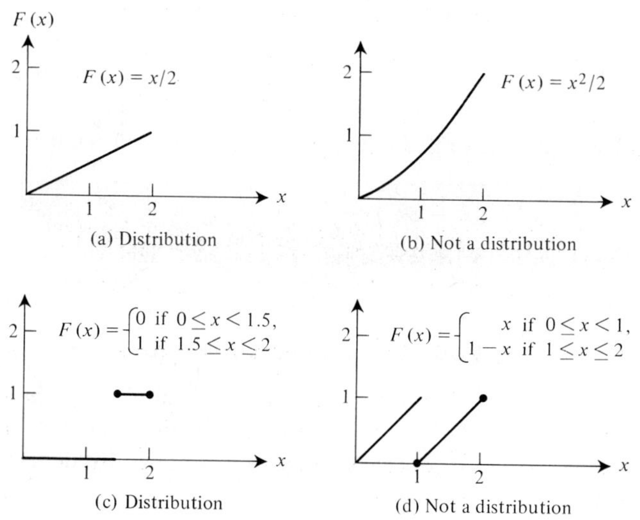

Figure 28.1

◀

The identity (28.2) and the probability properties (28.1) can be used to show that the probability that the random variable X will be in any given interval $[a, b]$, contained in I, is

$$P(a < X < b) = F(b) - F(a) \qquad \textbf{Probability that } X \in [a, b]. \tag{28.3}$$

For example, using the distribution $F(x) = x/2$ on the interval $I = [0, 2]$, we find that the probability that the random variable X will be between 1 and 1.5 is

$$P(1 < X < 1.5) = F(1.5) - F(1) = \frac{1.5}{2} - \frac{1}{2} = \frac{0.5}{2} = 0.25.$$

Similarly, the probability that X will be in the interval $[0.75, 1.35]$ is

$$P(0.75 < X < 1.35) = F(1.35) - F(0.75) = \frac{1.35}{2} - \frac{0.75}{2} = \frac{0.6}{2} = 0.30.$$

The distribution in part (a) of Example 28.1 is called a *uniform distribution*. It is a generalization of the *equiprobable distribution* in finite sample spaces. A uniform distribution has the property that the probability of X being in any interval $[a, b]$ is proportional to the length of the interval:

$$P(a < X < b) = \text{constant}(b - a).$$

In part (a) of the above example, the probability that X will be in any subinterval of length 0.2 is $P = 0.1$, since

$$\begin{aligned} P(x_0 < X < x_0 + 0.2) &= F(x_0 + 0.2) - F(x_0) \\ &= \frac{x_0 + 0.2}{2} - \frac{x_0}{2} = \frac{0.2}{2} = 0.1. \end{aligned}$$

It can be shown that the uniform distribution on a sample space $I = [\alpha, \beta]$ is

$$F(x) = \frac{x - \alpha}{\beta - \alpha} \qquad \textbf{Uniform distribution function.}$$

Probability Density Functions

The definition of a probability density function is similar to the definition of a distribution function for finite sample spaces. The difference is that the sum used with finite sample spaces is replaced by an integral when we deal with continuous sample spaces.

Definition 28.2 Let $I = [\alpha, \beta]$ be a finite interval. A *probability density function* $f(x)$ on the interval I is a function which satisfies the following.

i) f is continuous on $[\alpha, \beta]$,
ii) f is nonnegative on $[\alpha, \beta]$, $f(x) \geq 0$ for all x,

iii) $\int_{\alpha}^{\beta} f(x)dx = 1.$

Each density function is uniquely associated with a probability function P by the following equation:

$$P(a < X < b) = \int_{a}^{b} f(x)dx \quad \textbf{Probability that } X \in [a, b], \tag{28.4}$$

where $[a, b]$ is any subinterval of I. (See Fig. 28.2.)

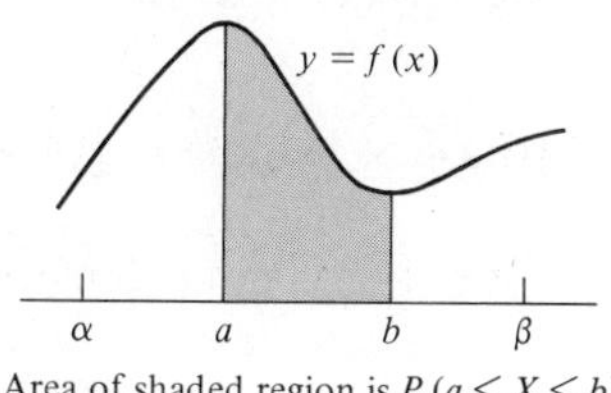

Area of shaded region is $P(a < X < b)$

Figure 28.2

Example 28.2 Let $I = [1, 5]$.

a) The constant function $f(x) = 0.25$ is a density function on I. To verify this observe that $f(x) \geq 0$ for all x and

$$\int_{1}^{5} f(x)dx = \int_{1}^{5} 0.25\, dx = (0.25)x\Big|_{1}^{5} = (0.25)\ (5 - 1) = 1.$$

b) The function,

$$f(x) = \begin{cases} \frac{1}{4}(x - 1) & \text{if } 1 \leq x < 3; \\ \frac{1}{4}(5 - x) & \text{if } 3 \leq x \leq 5, \end{cases}$$

is a density function on I. Clearly $f(x) \geq 0$; f is continuous on the intervals $[1, 3]$ and $[3, 5]$, since it is linear on these intervals. At $x = 3$, f is also continuous, since

$$\lim_{x \to 3^-} f(x) = \lim_{x \to 3^-} \tfrac{1}{4}(x - 1) = \tfrac{1}{2},$$

$$\lim_{x \to 3^+} f(x) = \lim_{x \to 3^+} \tfrac{1}{4}(5 - x) = \tfrac{1}{2},$$

and hence

$$\lim_{x \to 3} f(x) = \tfrac{1}{2} = f(3).$$

(See Fig. 28.3(b).) The integral $\int_1^5 f(x)dx$ can be evaluated by separating it into two integrals:

$$\begin{aligned}\int_1^5 f(x)dx &= \int_1^3 f(x)dx + \int_3^5 f(x)dx \\ &= \int_1^3 \tfrac{1}{4}(x-1)dx + \int_3^5 \tfrac{1}{4}(5-x)dx \\ &= \frac{(x-1)^2}{8}\bigg|_1^3 - \frac{(5-x)^2}{8}\bigg|_3^5 \\ &= (\tfrac{1}{2}-0)-(0-\tfrac{1}{2}) = 1.\end{aligned}$$

However, the same value could also be obtained using the formula for the area of a triangle and interpreting $\int_1^5 f(x)dx$ as the area of the triangle illustrated in Fig. 28.3(b):

$$\text{Area} = \tfrac{1}{2}(\text{base} \times \text{height}) = \tfrac{1}{2}(4 \times \tfrac{1}{2}) = 1.$$

c) The function $f(x) = (1-x)(x-4)$ is not a density on I since $f(x)$ is negative when $4 < x < 5$.

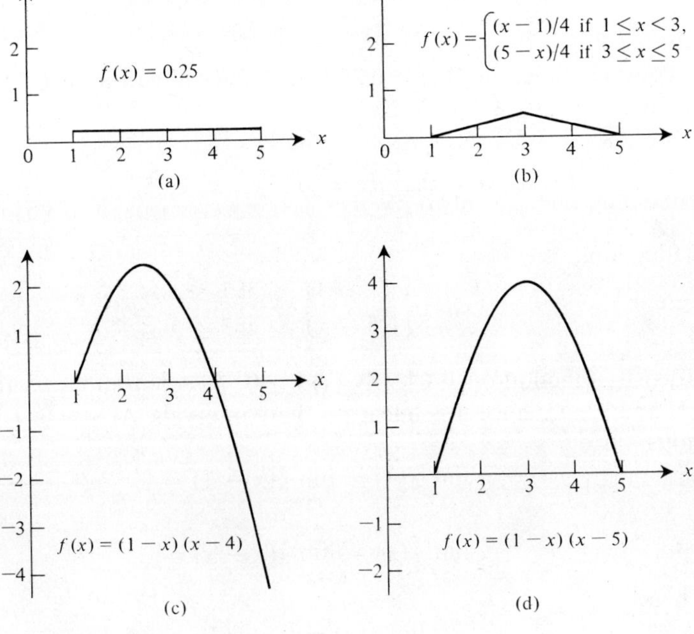

Figure 28.3

d) The function $f(x) = (1 - x)(x - 5)$ is not a density function since by direct calculation we find that

$$\int_1^5 f(x)dx = 20\tfrac{2}{3}.$$

The above functions f are sketched in Fig. 28.3. ◀

The density function in part (a) of Example 28.2 is a *uniform density*. Like the uniform distribution, the uniform density corresponds to a probability function that assigns equal probabilities to all subintervals of equal length. (See Fig. 28.4.)

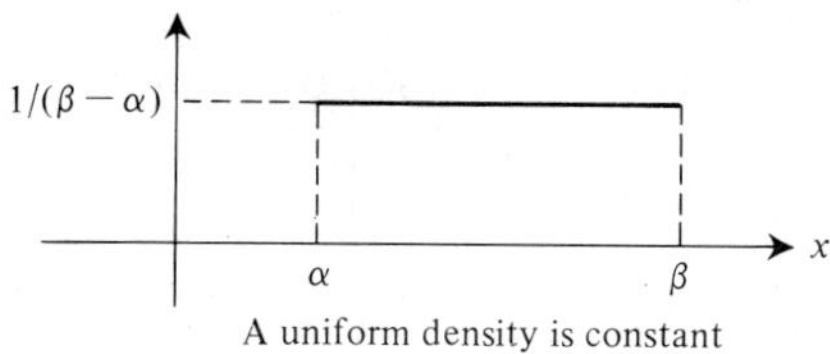

Figure 28.4

A uniform density for a sample space $[\alpha, \beta]$ is the constant function

$$f(x) = \frac{1}{\beta - \alpha} \qquad \textbf{Uniform density function.}$$

In Definition 28.2, the continuity requirement (i) imposed on a density function is made to ensure that $\int_a^b f(x)dx$ exists for all intervals $[a, b] \subseteq I$. The positivity condition (ii) is made to ensure that the probability of any event is never negative. The third condition, (iii), is a normalizing condition. It is imposed to standardize the notions of probability by ensuring that the maximum probability of any event is one. Any continuous nonnegative function defined on the interval I can be normalized to form a probability density function. The normalization is accomplished by dividing the function by the value of its integral over the interval $[\alpha, \beta]$. If g is continuous and nonnegative on $[\alpha, \beta]$, then

$$f(x) = \frac{g(x)}{\int_\alpha^\beta g(x)dx}$$

is a probability density function on $[\alpha, \beta]$. If $[\alpha, \beta] = [0, 10]$ and $g(x) = x^2$, then the corresponding density function is

$$f(x) = \frac{x^2}{\int_0^{10} x^2\,dx} = \frac{x^2}{\frac{10^3}{3}} = 0.003x^2.$$

The Relationship Between Density, Distribution, and Probability Functions

We have defined three types of functions, each associated with the chance that the numerical outcome of an experiment, called a *random variable*, X, will be in particular subintervals of a sample space $I = [\alpha, \beta]$. These are the density function $f(x)$, the distribution function $F(x)$, and a probability function P defined on subintervals of the sample space $[\alpha, \beta]$. In the preceding discussion each of these functions has been examined individually; at this point, let us see how the three functions are related.

Assume that an experiment results in a random variable X having sample space $I = [\alpha, \beta]$. Then associated with this random variable will be particular density, distribution, and probability functions: f, F, and P. The relationship between these three functions is determined by equations (28.3) and (28.4), which we repeat here:

$$P(a < X < b) = F(b) - F(a); \tag{28.3}$$

$$P(a < X < b) = \int_a^b f(x)dx. \tag{28.4}$$

Given any one of the functions f, F, or P, we can use these equations to determine the other two.

Assume that the density function $f(x)$ is known. Then the probability function P is given by equation (28.4). The corresponding distribution function is defined by

$$F(x) = \int_\alpha^x f(t)dt. \tag{28.5}$$

Thus $F(x)$ is the antiderivative of $f(x)$, which satisfies the condition $F(\alpha) = 0$. (See Fig. 28.5.)

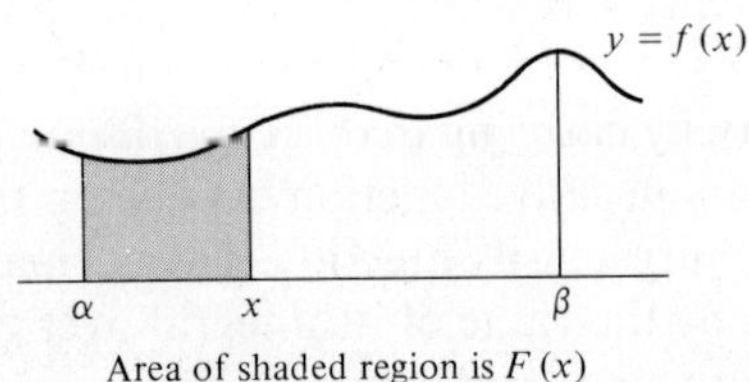

Area of shaded region is $F(x)$

Figure 28.5

Example 28.3 For the sample space $I = [0, 2]$ and the given density function $f(x)$, we compute the associated distribution function $F(x)$ using equation (28.5).

a) A uniform density: $f(x) = 1/2$ for all x. Then

$$F(x) = \int_0^x \tfrac{1}{2}\, dt = \tfrac{1}{2}x,$$

which is the same distribution in Example 28.1(a).

b) The linear density $f(x) = \frac{1}{2}x$ has the distribution

$$F(x) = \int_0^x \tfrac{1}{2}t\,dt = \frac{t^2}{4}\bigg|_0^x = \frac{x^2}{4}.$$

These functions are sketched in Fig. 28.6.

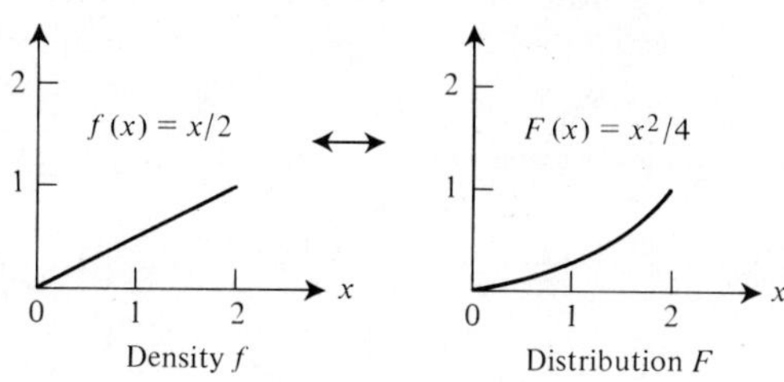

Figure 28.6

c) The symmetric triangle density,

$$f(x) = \begin{cases} x & \text{if } 0 \le x < 1; \\ 2 - x & \text{if } 1 \le x \le 2, \end{cases}$$

has the distribution

$$F(x) = \int_0^x f(t)dt.$$

To evaluate this integral, we must consider whether x is less than or greater than one.

If $x < 1$, then

$$F(x) = \int_0^x t\,dt = \frac{t^2}{2}\bigg|_0^x = \frac{x^2}{2}.$$

If $x \ge 1$, then

$$\begin{aligned} F(x) &= \int_0^x f(t)dt = \int_0^1 f(t)dt + \int_1^x f(t)dt = \int_0^1 t\,dt + \int_1^x 2 - t\,dt \\ &= \frac{t^2}{2}\bigg|_0^1 - \frac{(2-t)^2}{2}\bigg|_1^x \\ &= \frac{1}{2} - \frac{(2-x)^2}{2} + \frac{1}{2} \\ &= 1 - \frac{(2-x)^2}{2}. \end{aligned}$$

Thus the corresponding distribution is

$$F(x) = \begin{cases} \dfrac{x^2}{2} & \text{if } 0 \leq x < 1; \\ 1 - \dfrac{(2-x)^2}{2} & \text{if } 1 \leq x \leq 2. \end{cases}$$

The graph of F is a sigmoid or s-shaped curve as sketched in Fig. 28.7.

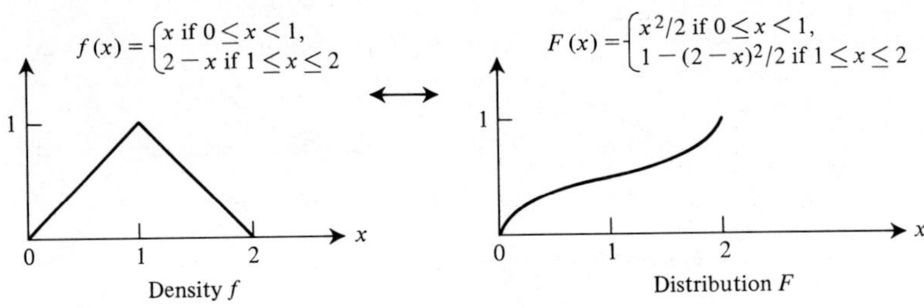

Figure 28.7

◀

Assume that the distribution function $F(x)$ is known. Then the probability function P is given by (28.3). The corresponding density function can be derived as follows. For a fixed point x_0, $\alpha < x_0 < \beta$, and small increment h, the probability that X will be in the interval $[x_0, x_0 + h]$ is given by

$$P(x_0 < X < x_0 + h) = F(x_0 + h) - F(x_0),$$

and also by

$$P(x_0 < X < x_0 + h) = \int_{x_0}^{x_0+h} f(x)dx.$$

Equating the right side of these equations gives us

$$F(x_0 + h) - F(x_0) = \int_{x_0}^{x_0+h} f(x)dx.$$

The integral

$$\int_{x_0}^{x_0+h} f(x)dx$$

may be approximated by the area of the rectangle of height $f(x_0)$ and base length $h = (x_0 + h) - x_0$ as illustrated in Fig. 28.8.

If we use the approximation

$$\int_{x_0}^{x_0+h} f(x)dx \simeq f(x_0) \cdot h \quad \text{(the area of the rectangle)} \tag{28.6}$$

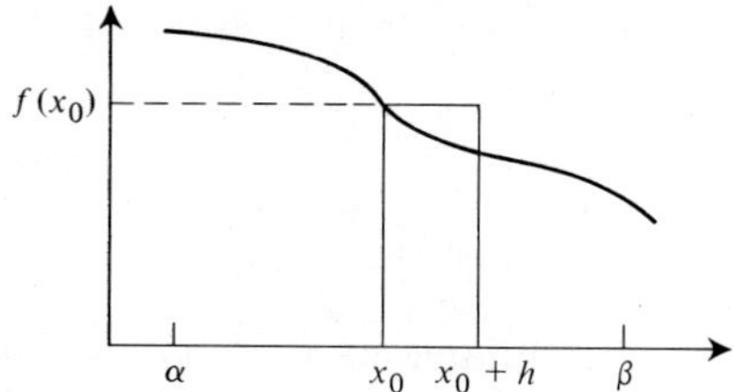

Figure 28.8

in the above identity, the resulting approximation is

$$F(x_0 + h) - F(x_0) \simeq f(x_0) \cdot h.$$

Dividing by h gives

$$\frac{F(x_0 + h) - F(x_0)}{h} \simeq f(x_0). \tag{28.7}$$

The left side of this equation is the difference quotient used to define the derivative $F'(x_0)$. If we let h approach zero, the error in the approximation (28.6) approaches zero and the resulting approximation (28.7) will become a true equality:

$$\lim_{h \to 0} \frac{F(x_0 + h) - F(x_0)}{h} = f(x_0).$$

If this limit exists, it is by definition the derivative

$$F'(x_0) = f(x_0). \tag{28.8}$$

Thus the density corresponding to a differentiable distribution $F(x)$ is simply the derivative: $f(x) = F'(x)$.

Example 28.4 For the sample space $I = [-1, 1]$, we specify the distribution function $F(x)$ and compute its derivative $F'(x) = f(x)$ to obtain the corresponding density $f(x)$.

a) Let $F(x) = (3 + 2x - x^2)/4$. Then the associated density is $f(x) = \frac{1}{2}(1 - x)$. (See Fig. 28.9.)

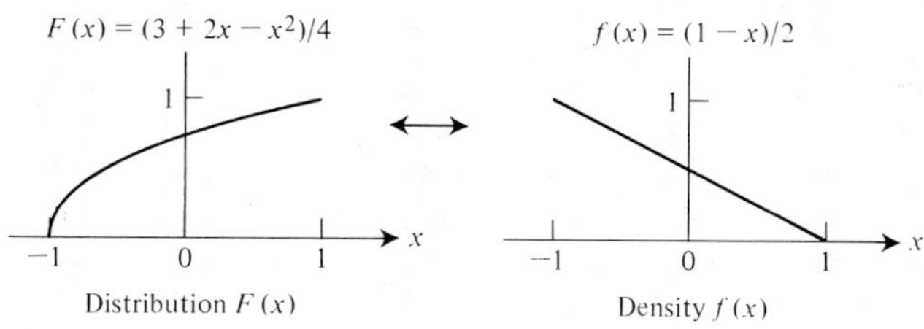

Figure 28.9

b) Let $F(x) = 1 - x^2$ if $-1 \leq x < 0$ and $F(x) = 1$ if $x \geq 0$. Then the corresponding density function is

$$f(x) = \begin{cases} -2x & \text{if } -1 \leq x < 0 \\ 0 & \text{if } 0 \leq x \leq 1. \end{cases}$$

The function $F(x)$ is differentiable at $x_0 = 0$, if we use the limit definition, $F'(x_0) = \lim_{h \to 0}(F(x_0 + h) - F(x_0))/h$, and consider the limit for $h > 0$ and $h < 0$ separately. (See Fig. 28.10.)

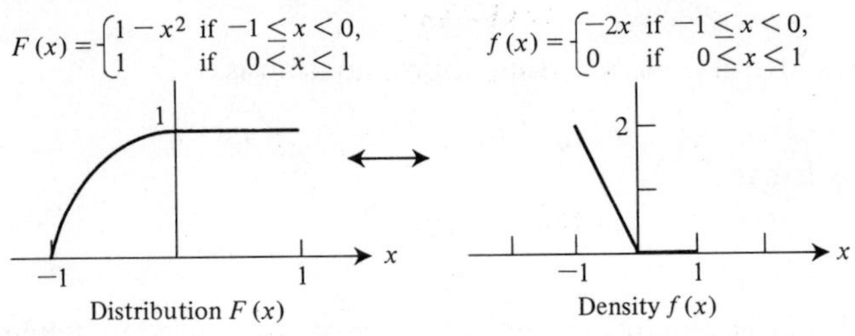

Figure 28.10 ◀

Determining a Density Function from Experimental Data

Given a sample space I, the associated probability function P is defined for all subsets of I. In general, probability functions are impossible to define explicitly without using a density or distribution function. When an experiment is performed, the probability function P associated with its outcome X is determined by the biological or physical mechanisms of the experiment leading to the value X. The main problem facing a scientist is to ascertain the actual probability structure of an experiment by observing the random variable X for many replications of the experiment. To describe the probability structure, we must determine either the associated density function $f(x)$ or the distribution function $F(x)$. The following example illustrates the basic process used to determine the density function $f(x)$ from a set of observed outcomes.

Table 28.1 Data set for example.

0.08	0.11	0.13	0.14	0.16	0.16	0.19	0.20	0.21	0.21	0.24
0.26	0.27	0.27	0.27	0.31	0.31	0.31	0.34	0.34	0.34	0.36
0.36	0.36	0.37	0.38	0.38	0.38	0.38	0.39	0.40	0.41	0.41
0.42	0.42	0.42	0.43	0.44	0.44	0.46	0.46	0.46	0.46	0.49
0.49	0.49	0.49	0.49	0.49	0.49	0.51	0.51	0.52	0.52	0.53
0.53	0.53	0.54	0.54	0.54	0.58	0.58	0.58	0.58	0.59	0.59
0.59	0.59	0.60	0.61	0.61	0.62	0.63	0.63	0.64	0.64	0.65
0.66	0.66	0.67	0.67	0.68	0.69	0.70	0.71	0.73	0.74	0.74
0.78	0.78	0.79	0.81	0.82	0.83	0.84	0.87	0.87	0.90	0.94
0.98										

An experiment results in a real value X between zero and one. To ascertain the corresponding probability density $f(x)$, the experiment was performed 100 times and the resulting outcomes were recorded as indicated in Table 28.1. The true density function $f(x)$ corresponding to this experiment is approximated by a step function, $\hat{f}$, obtained by constructing a histogram using the given data.

General Method for Approximating *f*(*x*)

The approximation function $\hat{f}$ is constructed as follows.

Step 1 Partition the sample space into subintervals—i.e., divide the interval I into nonoverlapping subintervals $I_1, I_2, \ldots, I_N$. If the sample space is not known, let $I = [\alpha, \beta]$, where α = smallest data value and β = largest data value.

Step 2 Determine the frequency of each subinterval. The frequency is the number

$$Y_n = \text{number of data values in } I_n.$$

Data that equal endpoints of the subintervals are arbitrarily assigned to the left interval.

Step 3 The approximate density function $\hat{f}$ is defined to be a constant value on each subinterval. This value is the estimated probability that X will be in the subinterval, divided by the length of the subinterval. Using the frequency concept of probability, if T denotes the total number of data values. The estimated

$$\text{Probability } (x \in I_n) = \frac{Y_n}{T} \equiv p_n.$$

Using the notation

$$\Delta_n = \text{length of interval } I_n,$$

the function $\hat{f}$ is defined as follows:

$$\text{if } x \in I_n, \hat{f}(x) = \frac{(Y_n/T)}{\Delta_n} = \frac{\text{probability } (x \in I_n)}{\text{length of } I_n}.$$

For the given set of data, $T = 100$. We construct an approximation $\hat{f}$ using a partition of $I = [0, 1]$ into four subintervals:

$$I_1 = [0, 0.25], I_2 = [0.25, 0.5], I_3 = [0.5, 0.75], I_4 = [0.75, 1.0].$$

The frequencies for these intervals are

$$y_1 = 11, \qquad y_2 = 39, \qquad y_3 = 38, \qquad y_4 = 12.$$

The estimated probabilities that X will be in the particular intervals are:

$$p_1 = 0.11, \qquad p_2 = 0.39, \qquad p_3 = 0.38, \qquad p_4 = 0.12.$$

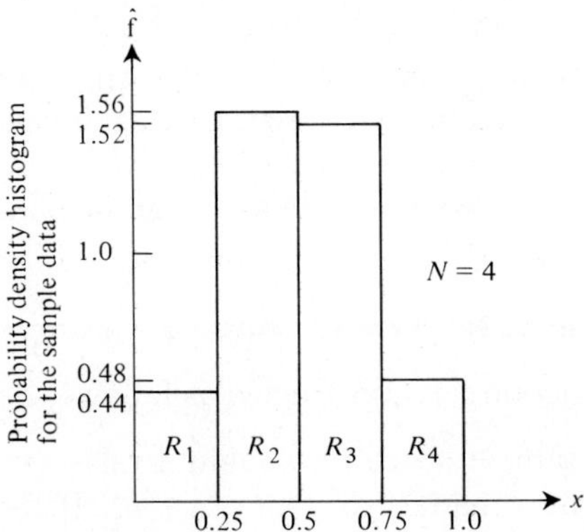

Figure 28.11

As the length of each subinterval is $\Delta_n = 0.25$, the function $\hat{f}$ is then given by:

$$\hat{f}(x) = \begin{cases} \dfrac{0.11}{0.25} = 0.44 & \text{on } [0, 0.25]; \\[2ex] \dfrac{0.39}{0.25} = 1.56 & \text{on } (0.25, 0.5]; \\[2ex] \dfrac{0.38}{0.25} = 1.52 & \text{on } (0.5, 0.75]; \\[2ex] \dfrac{0.12}{0.25} = 0.48 & \text{on } (0.75, 1.0]. \end{cases}$$

The graph of the function $\hat{f}$ is sketched in Fig. 28.11. This graph is sometimes called a *probability density histogram.* Observe that the total area under the graph is equal to one.

$$\begin{aligned} \text{Area} &= \text{sum of areas of the rectangles } R_1, R_2, R_3, \text{ and } R_4 \\ &= (0.44 \times 0.25) + (1.56 \times 0.25) + (1.52 \times 0.25) + (0.48 \times 0.25) \\ &= 0.11 + 0.39 + 0.38 + 0.12 = 1.0. \end{aligned}$$

We repeat the above procedure, this time using ten subintervals of equal length:

$$I_1 = [0, 0.1], I_2 = [0.1, 0.2], I_3 = [0.2, 0.3], \ldots, I_{10} = [0.9, 1.0].$$

The frequencies for these intervals are as follows:

$$\begin{array}{lllll} y_1 = 1, & y_2 = 6, & y_3 = 8, & y_4 = 15, & y_5 = 20, \\ y_6 = 18, & y_7 = 15, & y_8 = 8, & y_9 = 6, & y_{10} = 3. \end{array}$$

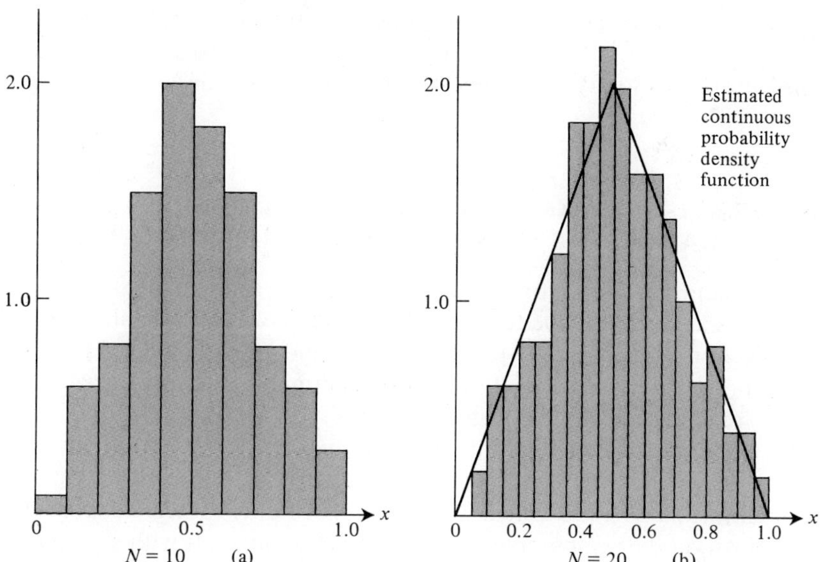

Figure 28.12

The resulting function $\hat{f}$ is defined on the nth interval by

$$\hat{f} = \frac{(y_n/100)}{0.1} = y_n \times 0.1.$$

The graph of the resulting function $\hat{f}$ is sketched in Fig. 28.12(a). In Fig. 28.12(b), we have sketched the function $\hat{f}$ obtained by considering 20 subintervals $I_1 = [0, 0.05]$, $I_2 = [0.05, 0.1], \ldots,$ etc. Observe that the shape of the histograms are similar. Generally, the number of data values must be much larger than the number of subintervals. If we continued to increase the number of subintervals, we would have to obtain additional data. Observe that as the number of subintervals is increased the resulting function $\hat{f}$ has smaller jumps at the endpoints of the subintervals. In the limit, as the number of sample data becomes infinite and the length of the subintervals approaches zero, these jumps should disappear. The corresponding approximate functions $\hat{f}$ should then approach the true density f:

$$f(x) = \lim_{\substack{T \to \infty \\ \Delta_n \to 0}} \hat{f}(x).$$

In practice, the density function f is chosen as a continuous function whose graph is similar to the density distribution histogram obtained by the above procedure. For the data of Table 28.1, one choice for the density $f(x)$ is

$$f(x) = \begin{cases} 4x & \text{if } 0 \le x < 0.5, \\ 4(1 - x) & \text{if } 0.5 \le x \le 1.0. \end{cases}$$

The graph of this function f is superimposed on the histogram in Fig. 28.12(*b*). Another choice for $f(x)$ might be $f(x) = 8x(1 - x)$. Once a function f has been assumed, more advanced statistical tests may be utilized to decide how correct it is in a probabilistic sense.

28.2 STATISTICAL INDICATORS OF A CONTINUOUS RANDOM VARIABLE: THE MEAN AND STANDARD DEVIATION

Let X be a continuous random variable with sample space $I = [\alpha, \beta]$. The same two statistics used to describe the central tendencies of finite probability distributions in Section 27 can be used to describe the random variable X. These are the mean μ and the standard deviation σ. For continuous random variables, these statistics are defined using the probability density function $f(x)$.

The motivation for the definition of the mean μ for a continuous random variable is that it should generalize the expected value concept of the mean μ and the sample mean $\bar{x}$ associated with finite probability spaces. Suppose that we were to partition the sample space $[\alpha, \beta]$ into N subintervals with endpoints

$$\alpha = x_0 < x_1 < x_2 < \cdots < x_N = \beta.$$

The subintervals would be

$$I_1 = [x_0, x_1], I_2 = [x_1, x_2], \ldots, I_N = [x_{N-1}, x_N].$$

Next let us randomly select an x value in each subinterval; we denote the selected value in the nth subinterval, I_n, by $x_n{}^*$. This gives a set of N numbers, $\{x_1{}^*, x_2{}^*, \ldots, x_N{}^*\}$. The mean μ of the random variable X should represent the *average* or *expected value* of X. Thus, intuitively, we might expect μ to equal (or be close to) the expected average of the selected points $\{x_n{}^*\}$. Thus it seems reasonable that the mean μ should be approximately equal to the value

$$\hat{\mu} = \sum_{i=1}^{N} x_i{}^* \times \text{Probability of choosing } x_i{}^*. \tag{28.9}$$

As mentioned in the previous subsection, the probability that a continuous random variable X will equal a particular number $x_i{}^*$ is zero, $P(X = x_i{}^*) = 0$. Thus, for equation (28.9) to make sense, the probability term must be interpreted differently —namely, as the probability of any point in the ith subinterval occurring:

$$\hat{\mu} = \sum_{i=1}^{N} x_i{}^* P(x_{i-1} < X < x_i).$$

The probability that X will be in the ith subinterval is

$$P(x_{i-1} < X < x_i) = \int_{x_{i-1}}^{x_i} f(x)dx.$$

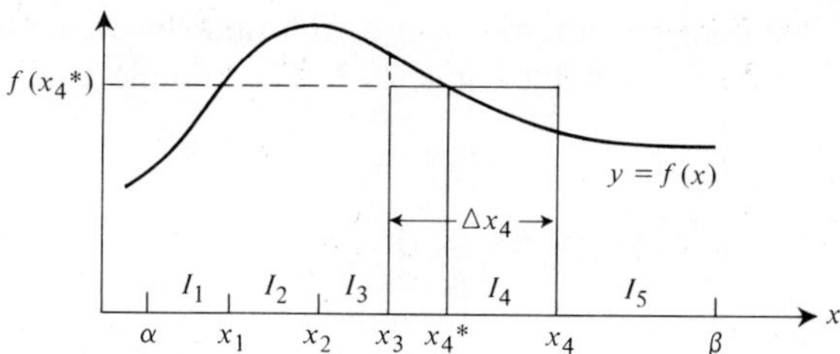

Figure 28.13

This probability, considered as an integral, can be approximated by the value $f(x_i^*) \cdot \Delta x_i$, where $\Delta x_i = x_i - x_{i-1}$ is the length of the interval I_i (see Fig. 28.13, where $N = 5$ and $I_i = I_4$.) Thus

$$P(x_{i-1} < X < x_i) \simeq f(x_i^*) \cdot \Delta x_i.$$

Using the above approximation for the probability terms in equation (28.9), we obtain the approximation

$$\mu \simeq \sum_{i=1}^{N} x_i^* f(x_i^*)\Delta x_i. \tag{28.10}$$

This approximation should improve as the number of subintervals increases while their lengths approach zero. In the limit, we obtain the identity

$$\mu = \lim_{\substack{N \to \infty \\ \Delta x \to 0}} \sum_{i=1}^{N} x_i^* f(x_i^*)\Delta x_i. \tag{28.11}$$

This limit is the limit of a Riemann sum, which by definition is the integral,

$$\mu = \int_{\alpha}^{\beta} xf(x)dx.$$

Compare this development with the examples in Section 20 and the discussion of Sections 21 and 22. We formalize the above intuitive result in the following definition.

Definition 28.3 Let X be a continuous random variable on the interval $I = [\alpha, \beta]$ with probability density function $f(x)$. The *mean of the random variable* X is the number

$$\mu = \int_{\alpha}^{\beta} xf(x)dx \quad \textbf{Mean.} \tag{28.12}$$

The mean μ is the *expected value* of X and is sometimes denoted by $E[X]$.

Example 28.5 We compute the mean μ of the random variable X having the various densities $f(x)$ given in Example 28.3. The sample space is the interval $I = [0, 2]$.

a) If $f(x) = \frac{1}{2}$, then

$$\mu = \int_0^2 xf(x)dx = \int_0^2 \frac{x}{2}\,dx = \left.\frac{x^2}{4}\right|_0^2 = 1.$$

b) If $f(x) = x/2$, then

$$\begin{aligned}\mu &= \int_0^2 xf(x)dx = \int_0^2 x \cdot \frac{x}{2}\,dx \\ &= \int_0^2 \frac{x^2}{2}\,dx = \left.\frac{x^3}{6}\right|_0^2 = 1\tfrac{1}{3}.\end{aligned}$$

c) If

$$f(x) = \begin{cases} x & \text{if } 0 \le x < 1; \\ 2 - x & \text{if } 1 \le x \le 2, \end{cases}$$

then

$$\begin{aligned}\mu = \int_0^2 xf(x)dx &= \int_0^1 xf(x)dx + \int_1^2 xf(x)dx \\ &= \int_0^1 x \cdot x\,dx + \int_1^2 x(2 - x)dx \\ &= \int_0^1 x^2\,dx + \int_1^2 2x - x^2\,dx \\ &= \left.\frac{x^3}{3}\right|_0^1 + \left. x^2 - \frac{x^3}{3}\right|_1^2 \\ &= \tfrac{1}{3} + [(4 - \tfrac{8}{3}) - (1 - \tfrac{1}{3})] = 1.\end{aligned}$$

◀

Observe that the densities in part (a) and (c) of the above example are symmetric about their mean value $\mu = 1$. In general, if a density $f(x)$ is symmetric about a number x_0, then the mean μ of the density $f(x)$ will be equal to x_0. In part (b) of the example, the density f is not symmetric about its mean. The midpoint of the interval [0, 2] is to the left of the mean μ. When this occurs, the density $f(x)$ is said to be *skewed to the left* or *negatively skewed*. (See Fig. 28.14.) This means that the random variable X tends to deviate from the mean μ to the left more than to the right.

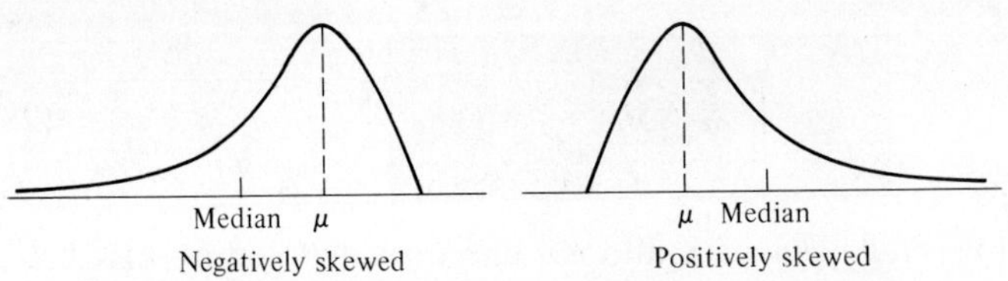

Figure 28.14

Since the two densities used in parts (a) and (c) of the above example both have mean $\mu = 1$, additional statistics are necessary to distinguish their central tendencies. The statistic most commonly used for this purpose is the standard deviation, σ. The standard deviation is defined to measure the expected deviation of the random variable X from the mean μ. The following definition parallels the definition of the mean μ in the sense that the sum over all possible x values in the sample space is replaced by an integral.

Definition 28.4 Let the random variable X with probability density $f(x)$ on the interval $[\alpha, \beta]$ have mean μ. The *standard deviation* of X is the number

$$\sigma = \left[\int_\alpha^\beta (x-\mu)^2 f(x)dx\right]^{1/2} \qquad \textbf{Standard deviation.} \tag{28.13}$$

The *variance* of X is $V = \sigma^2$;

$$V = \int_\alpha^\beta (x-\mu)^2 f(x)dx. \qquad \textbf{Variance.} \tag{28.14}$$

The variance is simply the square of the standard deviation and is frequently used to avoid taking the square root in (28.13). In more complicated statistical analysis, the variance is more suitable for manipulation than the standard deviation.

Example 28.6 We compute the variance and standard deviation of the densities in Example 28.5 on the interval [0, 2].

a) $f(x) = 1/2$, $\mu = 1$.

$$\sigma^2 = V = \int_0^2 (x-1)^2 f(x)dx = \int_0^2 \frac{(x-1)^2}{2}dx$$

$$= \frac{(x-1)^3}{6}\bigg|_0^2 = \frac{8}{6} - \left(\frac{-1}{6}\right) = 1\tfrac{1}{2}.$$

Thus $\sigma = \sqrt{V} = \sqrt{1.5} \simeq 1.22$.

b) $f(x) = x/2$, $\mu = \frac{4}{3}$.

$$\sigma^2 = V = \int_0^2 \left(x - \frac{4}{3}\right)^2 f(x)dx = \int_0^2 \left(x - \frac{4}{3}\right)^2 \frac{x}{2}dx$$

$$= \int_0^2 \frac{x^3}{2} - \frac{4}{3}x^2 + \frac{8}{9}x\,dx$$

$$= \frac{x^4}{8} - \frac{4x^3}{9} + \frac{4x^2}{9}\bigg|_0^2 = \frac{16}{8} - \frac{32}{9} + \frac{16}{9} = \frac{2}{9} \simeq 0.222.$$

Thus $\sigma = \sqrt{V} = \sqrt{2/9} \simeq 0.471$.

c) $f(x) = \begin{cases} x & \text{if } 0 < x \leq 1; \\ 2 - x & \text{if } 1 \leq x \leq 2, \end{cases} \quad \mu = 1.$

$$\begin{aligned}
\sigma^2 = V &= \int_0^2 (x-1)^2 f(x)dx \\
&= \int_0^1 (x-1)^2 f(x)dx + \int_1^2 (x-1)^2 f(x)dx \\
&= \int_0^1 (x-1)^2 \cdot x\,dx + \int_1^2 (x-1)^2(2-x)dx \\
&= \int_3^1 x^3 - 2x^2 + x\,dx + \int_1^2 -x^3 + 4x^2 - 5x + 2\,dx \\
&= 1/6 \text{ (after some arithmetic).}
\end{aligned}$$

Thus $\sigma = \sqrt{V} = \sqrt{1/6} \simeq 0.408$. ◀

The standard deviation in part (c) is less than the standard deviation in part (a) of the above example. This illustrates that the smaller the standard deviation, the more likely the random variable will be near the mean. A very general result, which indicates the usefulness of the standard deviation as a measure of how a random variable X is distributed, is

Chebyshev's Inequality

For *any* random variable X with mean μ and standard deviation σ and any real number $k > 1$,

$$P\left(\frac{|X-\mu|}{\sigma} \geq k\right) \leq \frac{1}{k^2}, \tag{28.15}$$

or

$$P(-k\sigma < X - \mu < k\sigma) \geq 1 - \frac{1}{k^2}. \tag{28.16}$$

These inequalities are illustrated in Fig. 28.15. Although they may seem complex, the first time you see them they are really not difficult to visualize. Assume that the

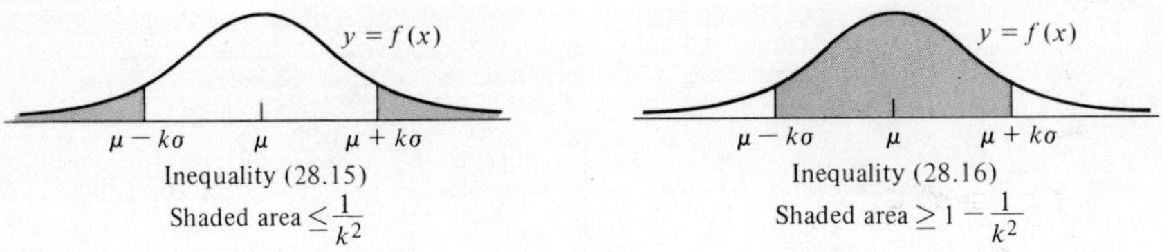

Figure 28.15

units on the x-axis are measured in terms of standard deviations—i.e., steps of length σ, just as you might measure in inches, miles, or centimeters. Then inequality (28.15) states that the probability that X is further than k standard-deviation units from the mean μ is always less than $1/k^2$. For example, if $k = 3$, inequality (28.15) states that the probability that X is greater than $3 \times \sigma$ units from μ is less than $\frac{1}{9}$.

Similarly, inequality (28.16) states that the probability that X is less than k standard-deviation units from μ is at least $1 - 1/k^2$. If $k = 3$, this probability is $1 - \frac{1}{9} = \frac{8}{9} \simeq 0.888$. Thus almost 90 percent of all X values will be within three standard deviations of the mean. For this reason, data are frequently indicated together with standard-deviation units to give an indication of how concentrated the data are about the mean. If the actual distribution of X is known, inequalities (28.15) and (28.16) can usually be improved.

28.3 INFINITE PROBABILITY SPACES; THE NORMAL DISTRIBUTION

Virtually any quantity represented by a random variable X is limited either biologically or physically. Consequently, the sample space of a random variable X will in reality be a finite interval. However, it is frequently desirable and sometimes necessary to assume that the random variable X can assume any value in an infinite interval. For instance, when studying the survival of a genetic mutation or the reliability of a heart pacemaker, the random variable X might denote the time at which the mutation ceases to exist or the pacemaker fails. Since the upper limits of these time values are not known, we must assume that X can be any value in the infinite interval $[0, \infty)$.

In the following, we briefly introduce some of the most important probability density functions defined on infinite intervals. The fundamental probability theory for these infinite sample spaces is directly analogous to that presented in the preceding subsections for finite interval sample spaces. The only difference is that for an infinite sample space the integrals are frequently "improper integrals" and thus must be evaluated using limits as discussed in Section 25.

For an infinite interval $[0, \infty)$, we require a density function $f(x)$ to satisfy (in addition to being nonnegative),

$$\int_0^\infty f(t)dt = 1.$$

This requires the distribution function

$$F(x) = \int_0^x f(t)dt$$

to satisfy

$$\lim_{x \to +\infty} F(x) = 1.$$

For an infinite interval $(-\infty, +\infty)$, the density function $f(x)$ must satisfy

$$\int_{-\infty}^{+\infty} f(t)dt = 1.$$

The associated distribution function

$$F(x) = \int_{-\infty}^{x} f(t)dt$$

must satisfy

$$\lim_{x \to -\infty} F(x) = 0 \quad \text{and} \quad \lim_{x \to +\infty} F(x) = 1.$$

The Exponential Probability Distribution

The exponential density function is defined on the half line $[0, +\infty)$ and is frequently used to model *survival times* in populations having *constant failure rates.*

Let the random variable T represent the *survival time* of an individual in a given population—i.e., T denotes the length of time that an individual lives. Measuring from time $t = 0$, T is the time at which the individual dies. The (*death*) *density function* $f(t)$ for the random variable T gives the probability that an individual will die close to time t. (In an interval $[t, t + \Delta t]$ for Δt small.) The associated distribution function

$$F(t) = \int_{0}^{t} f(x)dx$$

gives the probability that an individual will die in the time interval $[0, t]$. In reliability theory and actuarial studies, the *survival function* $S(t)$ is defined as the probability that an individual does not die in the interval $[0, t]$. Consequently,

$$S(t) = 1 - F(t) \qquad \textbf{Survival function.}$$

The relation between the distribution F and survivorship function S is illustrated in Fig. 28.16.

The *hazard function* $\lambda(t)$ is defined to measure the probability that an individual who has survived to time t will then die immediately. The hazard function is also

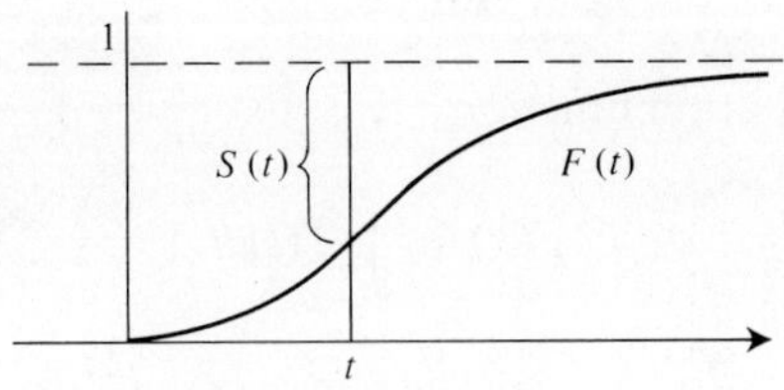

Figure 28.16

called the *instantaneous death rate* and the *force of mortality*. This function is actually a condition probability function:

$$\lambda(t) = P(\text{individual dies near time } t \mid \text{individual survives until time } t).$$

Using conditional probability theory, it follows that the hazard function is

$$\lambda(t) = \frac{f(t)}{S(t)} \qquad \textbf{Hazard function.} \tag{28.17}$$

A population that experiences no change in death rate has a constant hazard function. Using the fact that

$$F'(t) = f(t) \qquad \text{and} \qquad S(t) = 1 - F(t)$$

in the above equation, we find that the constant hazard population distribution $F(t)$ satisfies the separable differential equation

$$\lambda = \frac{F'(t)}{1 - F(t)}.$$

We solve this equation for $F(t)$ as follows:

$$\ln(1 - F(t)) = -\lambda t + c; \qquad \text{(integrating)}$$

$$F(t) = 1 - e^{-\lambda t + c}. \qquad \text{(taking exponentials)}$$

Using the requirement that $F(0) = 0$, we determine the integration constant $c = 0$ and hence

$$F(t) = 1 - e^{-\lambda t}.$$

Thus with a constant hazard rate, the distribution function $F(t)$ is one minus an exponential function. The corresponding density function is the following:

$$\text{Exponential density function: } f(t) = \lambda e^{-\lambda t},\ \lambda \text{ a constant.} \tag{28.18}$$

Examples of random variables which have been shown to have exponential distributions include the following:

a) Survival time of a population exposed to excessive radiation at the same time (data from Hiroshima survivors).
b) Survival time of patients treated for an acute disease such as myelogenous leukemia.
c) Relief times of arthritic patients receiving an analgesic.
d) The time between breakdowns of computers that are routinely maintained.

The mean of an exponential distribution is

$$\mu = \int_0^\infty x \cdot \lambda e^{-\lambda x}\, dx.$$

The singular integral is evaluated using Definition 25.1:

$$\mu = \lim_{t \to +\infty} \int_0^t x \cdot \lambda e^{-\lambda x}\, dx.$$

Using integration by parts similar to that in Example 24.1 we find

$$\int_0^t x \cdot \lambda e^{-\lambda x}\, dx = -\left(x + \frac{1}{\lambda}\right)e^{-\lambda x}\Big|_0^t.$$

Hence, applying L'Hospital's Rule, we find that

$$\mu = \lim_{t \to +\infty} \left[-\left(t + \frac{1}{\lambda}\right)e^{-\lambda t} + \left(\frac{1}{\lambda}\right)\right] = \frac{1}{\lambda}.$$

Observe that as the hazard rate λ increases, the expected survival time $\mu = 1/\lambda$ decreases.

The variance of the exponential distribution is

$$V = \int_0^\infty \left(x - \frac{1}{\lambda}\right)^2 e^{-\lambda x}\, dx.$$

The evaluation of this integral is left as an exercise.

Normal Distribution

One of the most important probability distributions is the *normal* or *Gaussian distribution.* Normal distributions are used to describe random variables whose probability density histograms, or frequency histograms, approximate a symmetric "bell-shaped" curve as indicated in Fig. 28.17.

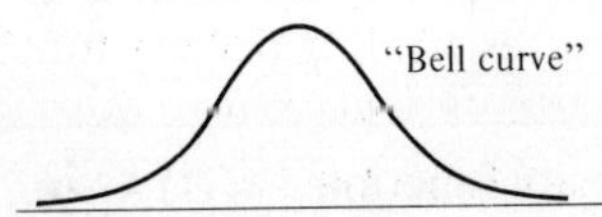

Figure 28.17

Normal distributions are characterized by their mean, μ, and their standard deviation, σ. A normal distribution is symmetric about its mean, μ. The standard deviation, σ, indicates the shape of the associated "bell curve." If σ is small, the curve is "peaked," whereas for large values of σ the curve is "flat." Figure 28.18(a) indicates several normal density functions having the same standard deviation and varying means. Figure 28.18(b) shows three normal density functions having the same mean and different standard deviations.

One of the most useful aspects of normal density functions is that they are all related by a simple algebraic identity to the *standard normal density function,* having mean $\mu = 0$ and standard deviation $\sigma = 1$.

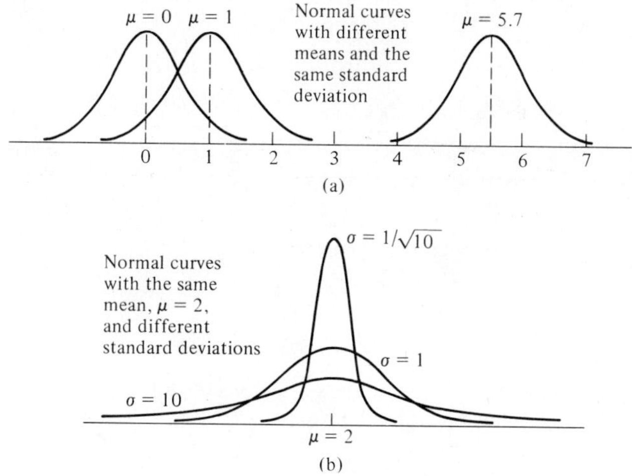

Figure 28.18

Because of its importance, the normal random variable having mean $\mu = 0$ and standard deviation $\sigma = 1$ is denoted by Z to distinguish it from all other normally distributed random variables. The density function of the random variable Z is defined on the interval $(-\infty, +\infty)$ to be as follows:

$$f(z) = \frac{1}{\sqrt{2\pi}} e^{-z^2/2} \qquad \textbf{Standard normal density function.} \tag{28.19}$$

The function $f(z)$ is symmetric about the point $z = 0$ and is positive for all values of z. Since its derivative

$$f'(z) = \frac{-z}{\sqrt{2\pi}} e^{-z^2/2}$$

is negative for $z > 0$, the graph of f is decreasing on $[0, +\infty)$. Computing the second derivative

$$f''(z) = (z^2 - 1)\frac{e^{-z^2/2}}{\sqrt{2\pi}},$$

we see that the graph of $f(z)$ has inflection points at $z = 1$ and $z = -1$. The graph is concave downward on the interval $(-1, 1)$ and is concave upward on the intervals $(-\infty, -1)$ and $(1, +\infty)$. The graph of $f(z)$ is sketched in Fig. 28.19.

To verify that $f(z)$ is a probability density function on the infinite interval $(-\infty, +\infty)$, we must show that

$$\int_{-\infty}^{+\infty} f(z)dz = 1.$$

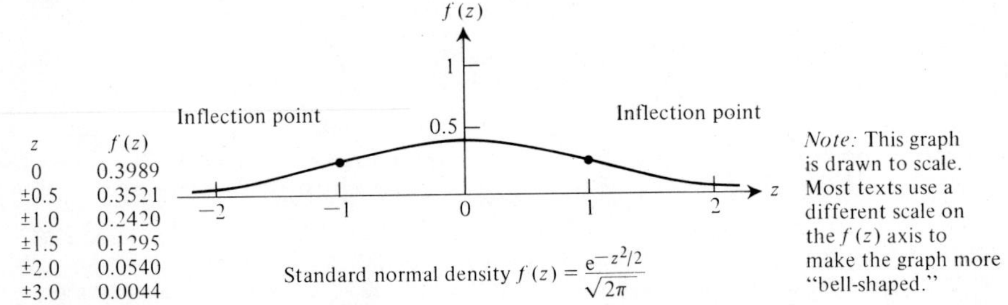

z	$f(z)$
0	0.3989
±0.5	0.3521
±1.0	0.2420
±1.5	0.1295
±2.0	0.0540
±3.0	0.0044

Figure 28.19

However, the integral

$$\frac{1}{\sqrt{2\pi}}\int_{-\infty}^{\infty} e^{-z^2/2}\,dz$$

is not easily evaluated, since the antiderivative of $e^{-z^2/2}$ cannot be represented using elementary functions. In advanced calculus, it is shown that indeed

$$\int_{-\infty}^{\infty} e^{-z^2/2}\,dz = \sqrt{2\pi}, \tag{28.20}$$

and hence the area under the graph of $f(z)$ is one. The mean value of the distribution $f(z)$ is

$$\mu = \int_{-\infty}^{+\infty} z \cdot f(z)dz = \int_{-\infty}^{+\infty} \frac{ze^{-z^2/2}}{\sqrt{2\pi}}\,dz.$$

Since the integrand is an odd function of z, it follows that this integral will be zero. However, since it is a singular integral, its evaluation requires a more detailed analysis. By definition,

$$\int_{-\infty}^{+\infty} \frac{ze^{-z^2/2}}{\sqrt{2\pi}}\,dz = \frac{1}{\sqrt{2\pi}}\left[\int_{0}^{+\infty} ze^{-z^2/2}\,dz + \int_{-\infty}^{0} ze^{-z^2/2}\,dz\right].$$

The first integral is

$$\begin{aligned}\int_{0}^{+\infty} ze^{-z^2/2}\,dz &= \lim_{t\to+\infty}\int_{0}^{t} ze^{-z^2/2}\,dz\\ &= \lim_{t\to\infty}[-e^{-t^2/2}+1] = 1.\end{aligned}$$

The second integral is

$$\begin{aligned}\int_{-\infty}^{0} ze^{-z^2/2}\,dz &= \lim_{t\to-\infty}\int_{t}^{0} ze^{-z^2/2}\,dz\\ &= \lim_{t\to-\infty}[-1+e^{-t^2/2}] = -1.\end{aligned}$$

Consequently, $\mu = 0$.

The determination of the variance of $f(z)$ requires the use of integration by parts to evaluate

$$V = \int_{-\infty}^{\infty} \frac{z^2 e^{-z^2/2}}{\sqrt{2\pi}}\, dz.$$

Formally,

$$\int_{-\infty}^{\infty} \frac{z^2 e^{-z^2/2}}{\sqrt{2\pi}}\, dz = \frac{-z}{\sqrt{2\pi}} e^{-z^2/2} \Bigg|_{-\infty}^{\infty} + \int_{-\infty}^{\infty} \frac{e^{-z^2/2}}{\sqrt{2\pi}}\, dz.$$

By examining the appropriate limits the first term on the right is found to be zero, whereas the remaining integral is again the area under the graph of $f(z)$, which by equation (28.20) is equal to one. Therefore $V = 1$ and, consequently, the standard deviation $\sigma = 1$.

The "*standard*" *normal probability-distribution function* for the random variable Z is given by

$$F(z_0) = \int_{-\infty}^{z_0} \frac{e^{-z^2/2}}{\sqrt{2\pi}}\, dz \qquad \textbf{Standard normal distribution.} \qquad (28.21)$$

$F(z_0)$ is the probability that the random variable Z will be less than the number z_0,

$$P(z \leq z_0) = F(z_0),$$

and is represented by the shaded area in Fig. 28.20(a). The only difficulty is that the value of $F(z)$ is not easily evaluated, since we do not know an antiderivative

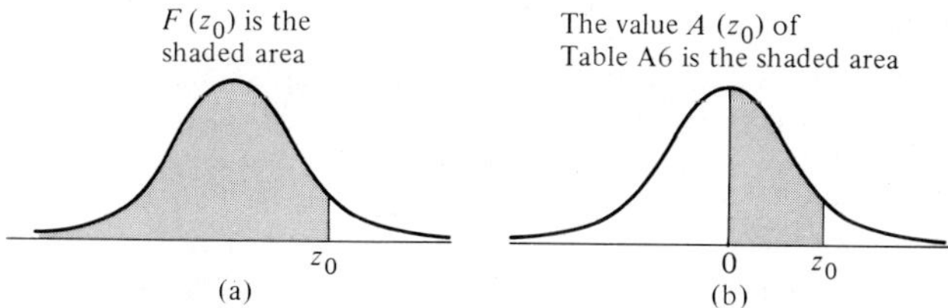

Figure 28.20

of $e^{-z^2/2}$. To circumvent this difficulty, tables of the normal distribution function have been compiled and are found in most statistics texts and scientific reference books. We have included in Appendix II, a standard table used to evaluate $F(z)$, Table A-6. The use of this table requires some explanation. To reduce the size of such tables, we utilize the symmetry of the normal density function. The probability that Z is less than zero is

$$F(0) = \int_{-\infty}^{0} \frac{e^{-z^2/2}}{\sqrt{2\pi}}\, dz = \tfrac{1}{2}.$$

The tabulated values $A(z_0)$ in Table A-6 are for the area between $z = 0$ and $z = z_0$. Thus the table value corresponding to $z_0 = 2$ is

$$A(2) = P(0 < Z < 2);$$

$$A(2) = \int_0^2 \frac{e^{-z^2/2}}{\sqrt{2\pi}}\, dz \simeq 0.4772.$$

This area is sketched in Fig. 28.20(b). To compute

$$F(z) = P(Z \leq 2),$$

we add $\frac{1}{2}$ to $A(2)$, since

$$F(z) = \int_{-\infty}^{0} f(z)dz + \int_0^2 f(z)dz$$
$$= \tfrac{1}{2} + A(2) = 0.9442.$$

To evaluate

$$F(-2) = P(Z < -2),$$

we use the symmetry of the normal density $f(z)$. Since $f(-z) = f(z)$, the area between $-z_0$ and 0 is the same as the area between 0 and z_0 (as indicated in Fig. 28.21(a)). Therefore

$$\int_{-2}^{0} f(z)dz = A(2)$$

and

$$F(-2) = \int_{-\infty}^{-2} f(z)dz = \int_{-\infty}^{0} f(z)dz - \int_{-2}^{0} f(z)dz$$
$$= \tfrac{1}{2} - A(2) = 0.0228.$$

Thus, for positive values z_0, using the table value $A(z_0)$, we find that

$$F(z_0) = \tfrac{1}{2} + A(z_0)$$
$$F(-z_0) = \tfrac{1}{2} - A(z_0).$$

The probability that Z is between -1 and 1.5 (see Fig. 28.21(b)) is

$$\begin{aligned} P(-1 < Z < 1.5) &= F(1.5) - F(-1) \\ &= [\tfrac{1}{2} + A(1.5)] - [\tfrac{1}{2} - A(1)] \\ &= A(1.5) + A(1) = 0.4332 + 0.3413 = 0.7745. \end{aligned}$$

The probability that Z is between 1 and 2 is

$$\begin{aligned} P(1 < Z < 2) &= F(2) - F(1) = (\tfrac{1}{2} + A(2)) - (\tfrac{1}{2} + A(1)) \\ &= A(2) - A(1) = 0.4772 - 0.3413 = 0.1359. \end{aligned}$$

A random variable X, which is normally distributed with mean μ and standard deviation σ, is referred to as a $N(\mu, \sigma^2)$ *random variable*; read "X is a normal μ,

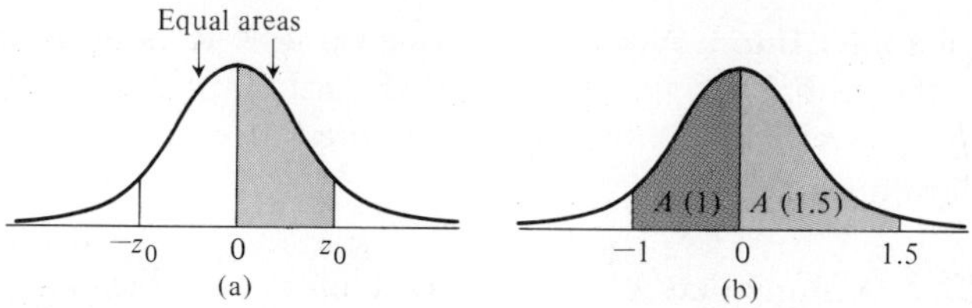

Figure 28.21

σ^2 random variable." The standard normal variable Z is thus a $N(0, 1)$ random variable. In general, a $N(\mu, \sigma^2)$ random variable X is related to the standard $N(0, 1)$ random variable Z by the equation

$$Z = \frac{X - \mu}{\sigma}. \tag{28.22}$$

The probability density function of a $N(\mu, \sigma^2)$ random variable X is therefore

$$f(x) = \frac{1}{\sigma\sqrt{2\pi}} e^{-\frac{1}{2}((x-\mu)/\sigma)^2} \qquad \textbf{Normal } N(\mu, \sigma^2) \textbf{ density function.} \tag{28.23}$$

The graph of the $N(\mu, \sigma^2)$ density function $f(x)$ is essentially the same as the graph of the $N(0, 1)$ density function $f(z)$. The difference being that the graph of $f(z)$ is shifted horizontally due to the term $X - \mu$, so that it is symmetric about the mean μ. The spread of the graph is then changed due to the standard deviation σ.

Theoretically, if X is a $N(\mu, \sigma^2)$ random variable, for any interval $[a, b]$,

$$P(a < X < b) = \int_a^b f(x)dx = \int_a^b \frac{e^{-\frac{1}{2}((x-\mu)/\sigma)2}}{\sigma\sqrt{2\pi}} dx.$$

However, the inequality $a < X < b$ is algebraically equivalent to

$$\frac{a - \mu}{\sigma} < \frac{X - \mu}{\sigma} < \frac{b - \mu}{\sigma},$$

which, using identity (28.22), is the same as

$$\frac{a - \mu}{\sigma} < Z < \frac{b - \mu}{\sigma}.$$

Therefore

$$P(a < X < b) = P\left(\frac{a - \mu}{\sigma} < Z < \frac{b - \mu}{\sigma}\right).$$

Thus the probability that a $N(\mu, \sigma^2)$ random variable X is in an interval $[a, b]$ is the same as the probability that the standard normal variable Z is in the interval $[(a - \mu)/\sigma, (b - \mu)/\sigma]$. This enables us to compute $P(a < X < b)$, using Table A-6 for the standard normal distribution.

Example 28.7 Assume that X is a $N(3, 4)$ random variable. Then X has mean $\mu = 3$ and standard deviation $\sigma = \sqrt{4} = 2$.

a) The probability that X is between zero and four is

$$P(0 < X < 4) = P\left(\frac{0-3}{2} < Z < \frac{4-3}{2}\right)$$

$$= P(-\tfrac{3}{2} < Z < \tfrac{1}{2})$$

$$\text{(From Table A-6)} \;= A(\tfrac{3}{2}) + A(\tfrac{1}{2}) = 0.4332 + 0.1915 = 0.6247.$$

b) The probability that X is greater than six is

$$P(X > 6) = P\left(Z > \frac{6-3}{2}\right) = P(Z > \tfrac{3}{2}).$$

Since

$$\int_0^{+\infty} f(z)dz = \int_0^{3/2} f(z)dz + \int_{3/2}^{+\infty} f(z)dz,$$

that is,

$$\tfrac{1}{2} = A(\tfrac{3}{2}) + P(Z > \tfrac{3}{2}),$$

we solve for $P(Z > \tfrac{3}{2})$ as

$$P(X > 6) = P(Z > \tfrac{3}{2}) = \tfrac{1}{2} - A(\tfrac{3}{2}) = 0.0668.$$

c) What is the probability that X is within two standard deviations of its mean? This question asks for

$$P(\mu - 2\sigma < X < \mu + 2\sigma).$$

As in the above examples, we determine this probability by converting it to a standard $N(0, 1)$ probability:

$$P(\mu - 2\sigma < X < \mu + 2\sigma) = P\left[\frac{(\mu - 2\sigma) - \mu}{\sigma} < Z < \frac{(\mu - 2\sigma) - \mu}{\sigma}\right].$$

As

$$\frac{(\mu - 2\sigma) - \mu}{\sigma} = \frac{-2\sigma}{\sigma} = -2,$$

this becomes

$$P(-2 < Z < 2) = A(2) + A(2) = 0.9544.$$ ◀

Note that we really did not need to know the value of μ or σ to answer the question in part (c) of the above example.

For any $N(\mu, \sigma^2)$ random variable X, the probability that X is within two standard deviations of the mean is $P(\mu - 2\sigma < X < \mu + 2\sigma) = 0.9544$.

Since this means that over 95 percent of all observations (of normal random variables) will be within two standard deviations of the mean, values outside of the interval $[\mu - 2\sigma, \mu + 2\sigma]$ are considered unlikely to occur.

EXERCISE SET 28

28.1 Which of the following are probability distribution functions on the interval [0, 4]?

a) $F(x) = x/2$ b) $F(x) = x/4$
c) $F(x) = 1 - (x/4)$ d) $F(x) = x^2/4$
e) $F(x) = x^2/16$ f) $F(x) = x(8 - x)/16$
g) $F(x) = 2x/(x + 4)$ h) $F(x) = x/(4 + (4 - x)^2)$
i) $F(x) = 1 - e^{-x}$ j) $F(x) = xe^{(4-x)}/4$

28.2 Which of the following are probability density functions on the interval [0, 4]?

a) $f(x) = x/2$ b) $f(x) = x/4$ c) $f(x) = 0.25$
d) $f(x) = 2x - 4$ e) $f(x) = x + \frac{9}{4}$ f) $f(x) = (x - 4)^2$

g) $f(x) = \begin{cases} \frac{1}{2} & \text{if} \quad 0 \le x \le 2; \\ 0 & \text{if} \quad 2 < x \le 4. \end{cases}$

h) $f(x) = \begin{cases} x & \text{if} \quad 0 \le x \le 1; \\ 1 - \frac{1}{3}(x - 1) & \text{if} \quad 1 < x \le 4. \end{cases}$

28.3 For the probability density function $f(x) = \frac{2}{9}x$, $0 \le x \le 3$, compute the following probabilities and illustrate their values as areas under the graph of $f(x)$.

a) $P(X < 1)$ b) $P(1 < X)$ c) $P(X < 2)$
d) $P(1 < X < 2)$ e) $P(X < 3)$ f) $P(1 < X < 1.5)$

28.4 For the probability distribution function $F(x) = x^2, 0 \le x \le 1$, compute the following probabilities and illustrate their values using the graph of $F(x)$.

a) $P(X < \frac{1}{2})$ b) $P(X < \frac{1}{4})$ c) $P(X > \frac{1}{2})$
d) $P(X > \frac{1}{4})$ e) $P(\frac{1}{4} < X < \frac{1}{2})$ f) $P(\frac{1}{2} < X < \frac{3}{4})$

28.5 The function $F(x) = \sin(x)$ is a probability distribution function on the interval $[0, \pi/2]$. Evaluate the following probabilities where the random variable X is assumed to have $F(x)$ as a distribution function.

a) $P(X < 0.3)$ b) $P(X < 0.4)$ c) $P(X > 0.3)$
d) $P(0.3 < X < 0.4)$ e) $P(X > \pi/4)$ f) $P(X < \pi/3)$

28.6 Determine the probability density function obtained by normalizing the given function over the interval [0, 2] and compute $P(\frac{1}{2} < X < \frac{3}{2})$ for each density function.

a) $f(x) = x$ b) $f(x) = x^2$ c) $f(x) = x^3$
d) $f(x) = x^2 - 2x + 1$ e) $f(x) = e^x$ f) $f(x) = e^{-x}$
g) $f(x) = xe^{x^2}$ h) $f(x) = xe^{-x^2}$ i) $f(x) = \sin(\pi x/2)$
j) $f(x) = \cos(\pi x/4)$ k) $f(x) = x(2 - x)$ l) $f(x) = x^3(2 - x)$

28.7 For the random variable X having probability density function $f(x) = (\frac{3}{2})(x - 1)^2$ on the interval [0, 2], calculate the following probabilities and illustrate their values as areas.

a) $P(X < 2)$ b) $P(X < 1)$ c) $P(X > 1)$
d) $P(X < 0.5)$ e) $P(0.5 < X < 1)$ f) $P(1 < X < 1.5)$

28.8 Determine the density function $f(x)$ corresponding to the specified distribution function $F(x)$ on the interval [0, 1]. Graph both $F(x)$ and $f(x)$ over the interval [0, 1].

a) $F(x) = x$ b) $F(x) = \sin(x/\pi)$
c) $F(x) = x^2$ d) $F(x) = (x^2 + x)/2$

e) $F(x) = \begin{cases} 0 & \text{if} \quad 0 \le x \le \frac{1}{2}; \\ 2(x - \frac{1}{2}) & \text{if} \quad \frac{1}{2} < x \le 1 \end{cases}$ f) $F(x) = xe^{x-1}$

g) $F(x) = (2x - 1)x$

28.9 Determine the distribution function $F(x)$ corresponding to the given density function $f(x)$ on the interval [0, 1]. Graph both $F(x)$ and $f(x)$ for $0 \le x \le 1$.

a) $f(x) = 2 - 2x$ b) $f(x) = \pi/2 \cos(\pi x/2)$
c) $f(x) = x^2/2 + 5/6$ d) $f(x) = x^2 + 4x/3$

e) $f(x) = \begin{cases} 4x & \text{if} \quad 0 \le x < \frac{1}{2}; \\ 4 - 4x & \text{if} \quad \frac{1}{2} \le x \le 1 \end{cases}$ f) $f(x) = (e/(e - 1))e^{-x}$

g) $f(x) = \begin{cases} 7(0.5 - x) & \text{if} \quad 0 \le x < 0.5; \\ 3(0.5 - x)^2 & \text{if} \quad 0.5 \le x \le 1 \end{cases}$

28.10 Assume that the length of newborn babies is uniformly distributed between 16 and 24 inches. What is the probability that a baby will be born with length between 20 and 22 inches?

28.11 Assume that the amount T of Testosterone (a hormone) produced by a male in a one-hour period is distributed between 0 and $M = (e^2 - 1)^{1/2}$ mg with a density function

$$f(t) = \frac{t}{1 + t^2}.$$

a) Show that $f(t)$ is a probability density function on the interval $[0, M]$ and calculate M.
b) What is the probability that T will be less than $M/2$?
c) What is the probability that T will be between 1 and 2?

28.12 Using the method illustrated in the text, consider the experiment resulting in the following 20 data values: 0.1, 0.1, 0.2, 0.3, 0.3, 0.3, 0.4, 0.4, 0.5, 0.5, 0.5, 0.5, 0.6, 0.6, 0.6, 0.7, 0.7, 0.8, 0.9, 1.0.

a) Construct probability density histograms $\hat{f}$ for this data using (i) $N = 4$ subintervals and (ii) $N = 10$ subintervals of the interval [0, 1].
b) Sketch the functions $\hat{f}$ of part (a) on the same graph and draw a smooth curve to approximate the true density function.

28.13 Repeat Exercise 28.12 using the following data set:

0.05	0.07	0.08	0.09	0.11	0.15
0.18	0.24	0.27	0.27	0.28	0.32
0.36	0.37	0.44	0.55	0.57	0.58
0.62	0.63	0.64	0.64	0.70	0.71
0.75	0.76	0.77	0.77	0.78	0.79
0.91	0.95	0.96	0.97	0.98	0.99

28.14 Determine the mean and standard deviation of the given density function.

a) $f(x) = \frac{1}{3}$, $3 \le x \le 6$
b) $f(x) = x - 2$, $2 \le x \le 2 + \sqrt{2}$
c) $f(x) = 3x^2$, $0 \le x \le 1$
d) $f(x) = \sqrt{x}$, $0 \le x < (\frac{3}{2})^{2/3}$

e) $f(x) = 6x(10 - x)10^{-3}$, $0 \le x \le 10$
f) $f(x) = \begin{cases} 4x & \text{if} \quad 0 \le x < \frac{1}{2}; \\ 4 - 4x & \text{if} \quad \frac{1}{2} \le x \le 1 \end{cases}$

g) $f(x) = \begin{cases} 0.2(1 - 0.2x) & \text{if} \quad 0 \le x \le 5; \\ x - 5 & \text{if} \quad 5 < x \le 6 \end{cases}$

28.15 If a random variable X has mean $\mu = 10$ and standard deviation $\sigma = 0.5$, use Chebyshev's Inequality to estimate the following.

a) $P(|X - 10|/0.5 \ge 3)$
b) $P(-1.5 \le X - 10 \le 1.5)$
c) $P(6 \le X \le 14)$
d) $P(X > 12 \text{ or } X < 8)$

28.16 a) Plot the exponential density, $f(t) = \lambda e^{-\lambda t}$, with $\lambda = \frac{1}{2}$.

b) Compute the associated distribution function $F(t)$ and survival function $S(t)$ and sketch their graphs.

c) If a population is dying due to asbestosis with this survival distribution, what is its hazard rate?

d) What is the probability that an individual of this population will die in the next two years?

e) What is the probability that an individual will survive two years but die before the third year is up?

f) Repeat parts (a) through (e) with $\lambda = \frac{1}{10}$.

g) Repeat parts (a) through (e) with $\lambda = 0.9$.

28.17 Determine the variance of the exponential distribution with $f(x) = \lambda e^{-\lambda x}$, $x \ge 0$.

28.18 A population of patients with malignant melanoma (a cancer) is found to have an exponential survival distribution. If the probability of survival for 3 years, $S(3)$, is found to be 0.1, what is the death density function $f(t)$ and the hazard rate λ?

28.19 Assume that a population with kidney disease has a death density function $f(t) = 2\lambda t e^{-\lambda t^2}$ (Weibull density).

a) Compute the associated distribution $F(t)$ and survival function $S(t)$.

b) Plot $f(t)$, $S(t)$, and $F(t)$ when $\lambda = \frac{1}{2}$.

c) What is the hazard rate $\lambda(t)$ of this population?

d) What is the probability that an individual in this population will survive at least 3 years?

e) Repeat parts (a) through (d) with $f(t) = \frac{1}{2}\lambda t^{-1/2}e^{-\lambda t^{1/2}}$.

28.20 Using Table A6 in Appendix II, evaluate the following probabilities where Z is the standard normal random variable. Draw a graph to illustrate each probability.

a) $P(0 < Z < 2)$ b) $P(0 < Z < 3)$ c) $P(0 < Z < 2.65)$
d) $P(-2 < Z < 0)$ e) $P(-2.65 < Z < 0)$ f) $P(3 < Z)$
g) $P(-2 < Z < 2)$ h) $P(-2 < Z < 3)$ i) $P(1 < Z < 2)$
j) $P(-2 < Z < -1)$ k) $P(-1 < Z)$ l) $P(Z > 2.5)$
m) $P(Z < -3.02)$ n) $P(0.6 < Z)$ o) $P(|Z| < 0.8)$
p) $P(|Z| > 0.8)$ q) $P(|Z| < 3.02)$ r) $P(|Z| > 1.2)$

28.21 Using Table A-6 in Appendix II, determine the values z_0 that satisfy the following.

a) $P(0 < Z < z_0) = 0.1$ b) $P(0 < Z < z_0) = 0.05$
c) $P(0 < Z < z_0) = 0.475$ d) $P(|Z| < z_0) = 0.95$
e) $P(|Z| > z_0) = 0.05$ f) $P(|Z| > z_0) = 0.01$
g) $P(|Z| < z_0) = 0.90$

28.22 Assume that X is normally distributed with mean $\mu = 5$ and standard deviation $\sigma = 1$. Evaluate the following and draw a graph to illustrate the probability.

a) $P(5 < X < 6)$ b) $P(5 < X < 7.7)$
c) $P(3.4 < X < 5)$ d) $P(3.4 < X < 6)$
e) $P(X < 4.1)$ f) $P(X > 6.2)$
g) $P(X > 3.7)$ h) $P(X < 6.5)$
i) $P(3 < X < 7)$ j) $P(X < 2 \text{ or } X > 8)$

28.23 Repeat Exercise 28.22 with (a) $\sigma = 2$ and (b) $\sigma = \frac{1}{2}$.

28.24 The heights of 18-year-old males are normally distributed with a mean of 70 in. and a standard deviation of 2.6 in. What is the probability that an 18-year-old male will be taller than 6 ft 4 in.?

28.25 The length of salmon is normally distributed with $\mu = 650$ mm and $\sigma = 25$ mm. What is the expected percentage of the salmon run that will be caught if gillnets catch all salmon of length 600 mm or larger? What size mesh should be used to ensure that 75 percent of the salmon will not be caught?

28.26 A holiday brochure indicates that the average daily temperature in Funsun City is 20°C ± 10°C. Assume that the temperature is $N(20, 10)$ distributed and calculate the following probabilities.

a) You will only swim with $T > 30°$.
b) You will "freeze" with $T < 10°$.
c) You will have a pleasant time with $20° < T < 25°$.

28.27 The weight of three-month-old infants is normally distributed with mean $\mu = 12.4$ lb. Assume that 95 percent of all infants weigh between 9.85 and 14.95 lbs.

a) What is the standard deviation σ?
b) What is the probability that an infant will weigh more than 13.2 lb?
c) What will be the interval containing the middle 50 percent of all infants' weights?

28.28 The acidity of saliva differs between children and adults. Assume that the pH values for adults are normally distribution with mean $\mu = 5.97$ and standard deviation 0.8. Assume that for children, the pH values are $N(7.22, 0.92)$.

a) What is the probability that an adult's saliva will have pH > 6.5?
b) What is the probability that a child's saliva will have pH < 6.5?
c) If a saliva sample was found to have a pH value between 6.4 and 6.7, would it more likely be that of a child or of an adult?

28.29 The amount of potassium in cerebrospinal fluid is normally distributed with $\mu =$ 11.57 mg/100 ml and $\sigma = 1.76$ mg/100 ml. What is the probability that an individual potassium level is between 11 and 12 mg/100 ml?

28.30 Assume that the radioactivity of an urine sample obtained from an animal that was given a radioactive labeled drug is known to be normally distributed with a $N(284, 20)$ distribution.

a) What is the probability that a sample will have a radioactive count greater than 300 units/min.?
b) What is the probability that the count will be within the interval $[250, 300]$?
c) What is the probability that the count will be less than 150 units/min.?

28.31 A medical laboratory processes a large volume of samples. It is found that their machine which gives white blood cell counts has an error which is normally distributed with mean $\mu = 0$ and variance $V = 200$ per cc.

a) If the normal count is 7500/cc, in what range will 90 percent of all normal samples fall?
b) If an individual sample is given a count of 6280/cc, what is the range of values in which the actual sample count will be with a 90 percent probability?
c) What is the probability that the sample in part (b) will lie within 90 percent of the normal count?

28.32 If the lung volume of an adult person is $N(4500, 50)$ distributed, what percentage of the population will have lung volume greater than 6000 ml?

APPENDIX I

BASIC ALGEBRA FORMULAS

1. Types of real numbers:

Natural numbers: $\{1, 2, 3, 4, \ldots\}$.

Integers: $I = \{\ldots, -3, -2, -1, 0, 1, 2, 3, \ldots\}$.

Rational numbers are numbers that can be expressed as a ratio or fraction of integers. A rational number has the form $Q = M/N$, where M and N are integers. Rational numbers can also be represented as either a finite decimal number, such as $\frac{3}{4} = 0.75$, or as an infinite repeated decimal number, such as

$$\tfrac{4}{3} = 1.3333\ldots \quad \text{or} \quad \tfrac{1}{7} = 0.142857142857\ldots,$$

where the sequence 142857 is repeated infinitely.

Irrational numbers are real numbers that cannot be represented in the form M/N, where M and N are integers. Irrational numbers have infinite decimal expansions, which never repeat. An example is

$$\sqrt{2} \simeq 1.414213562\ldots.$$

Special irrational numbers are

$$\pi = 3.141592654\ldots$$

$$e = 2.7182818275\ldots.$$

2. Fractions:

The sum of two fractions:

$$\frac{A}{B} + \frac{C}{D} = \frac{A \cdot D + B \cdot C}{B \cdot D}.$$

Multiplication of fractions:

$$\left(\frac{A}{B}\right) \times \left(\frac{C}{D}\right) = \frac{A \cdot C}{B \cdot D}.$$

Compound fractions:

$$\frac{\left(\frac{A}{B}\right)}{\left(\frac{C}{D}\right)} = \frac{A \cdot D}{B \cdot C}.$$

Negatives:

$$\frac{-A}{B} = \frac{A}{-B} = -\left(\frac{A}{B}\right).$$

3. Exponents:

For positive numbers x and y and any real numbers a and b, the following applies:
Multiplication with the same base:

$$(x^a) \cdot (x^b) = x^{a+b}$$

Multiplication of terms to the same power:

$$x^a \cdot y^a = (x \cdot y)^a$$

Negative exponents:

$$x^{-a} = \frac{1}{x^a}, \qquad x^{-1} = \frac{1}{x}.$$

Division with the same base:

$$\frac{x^a}{x^b} = x^{(a-b)}.$$

Division of terms to the same power:

$$\frac{(x^a)}{(y^a)} = \left(\frac{x}{y}\right)^a.$$

Double exponents:

$$(x^a)^b = x^{a \cdot b}.$$

Examples:

$$2^5 \cdot 2^3 = 2^8; \quad 2^3 \cdot 5^3 = 10^3; \quad 10^{-2} = \frac{1}{10^2};$$

$$\frac{5^3}{5^1} = 5^2; \quad \frac{5^3}{2^3} = \left(\frac{5}{2}\right)^3; \quad (2^3)^2 = 2^{2 \cdot 3} = 2^6.$$

4. Inequalities:

The letters a, b, c represent real numbers.

Symbol	Meaning
$a > b$	a is greater than b
$a \geq b$	a is greater than or equal to b
$a < b$	a is less than b
$a \leq b$	a is less than or equal to b
$a > 0$	a is positive
$a \geq 0$	a is nonnegative
$a < 0$	a is negative
$a \leq 0$	a is nonpositive

Adding: If $a < b$, then $a + c < b + c$ for all c. This also holds for the other inequalities: $\leq$, $>$, and $\geq$.

Multiplying: Multiplication by a positive number c leaves the inequality unchanged—i.e., if $a < b$ and $c > 0$, then $a \cdot c < b \cdot c$.

Multiplication by a negative number c reverses the inequality—i.e., if $a < b$ and $c < 0$, then $a \cdot c > b \cdot c$.

Solving inequalities: Perform operations to both sides just like solving equations, remembering the above multiplication principle.

Examples to solve the inequality:

	$-2 < 3x - 1 < 2$	$-3 < -2x + 1 < 4$
Add	$-2 + 1 < 3x < 2 + 1$	$-3 - 1 < -2x < 4 - 1$
Divide	$\frac{-1}{3} < x < \frac{3}{3}$	$\frac{-4}{-2} > x > \frac{3}{-2}$ *
	$\frac{-1}{3} < x < 1$	$2 > x > \frac{-3}{2}$

*(Change inequality because $-2 < 0$)

5. Polynomials

A polynomial of degree n has the form

$$a_0 + a_1x + a_2x^2 + \cdots + a_nx^n,$$

where $a_0, a_1, \ldots, a_n$ are numbers and $a_n \neq 0$.

Adding: Add the coefficients of like powers of x:

$$(a_0 + a_1x + a_2x^2 + a_3x^3) + (b_0 + b_1x + b_2x^2)$$
$$= (a_0 + b_0) + (a_1 + b)x + (a_2 + b_2)x^2 + a_3x^3 (\text{since } b_3 \equiv 0).$$

Multiplication: Multiply by each term:

$$(a_0 + a_1 x + a_2 x^2)(b_n x^n) = a_0 b_n x^n + a_1 b_n x^{n+1} + a_2 b_n x^{n+2}.$$

For more than one term, the standard method is as illustrated:

Multiply this	$1 + 2x + 3x^2$
By	$(4 + 5x + 7x^2)$
Times 4	$4 + 8x + 12x^2$
Times $5x$	$5x + 10x^2 + 15x^3$
Times $7x^2$	$7x^2 + 14x^3 + 21x^4$
Add (like powers)	$4 + 13x + 29x^2 + 29x^3 + 21x^4$

Division: Division of polynomials is similar to division of real numbers. As a rule, the polynomials are expressed in decreasing order of the exponents for the purpose of division.

Example:

$$\frac{1 + x^2 - 2x^3}{3 + 5x + x^2} = \frac{-2x^3 + x^2 + 1}{x^2 + 5x + 3}.$$

Express, in the following form, where zero coefficients are introduced for terms that do not appear:

$$\begin{array}{r|l} & -2x + 11 \\ x^2 + 5x + 3 & -2x^3 + x^2 + 0x + 1 \end{array}$$

Multiply by $-2x$	$-2x^3 - 10x^2 - 6x$
Subtract	$11x^2 \quad 6x \quad 1$
Multiply by 11	$11x^2 + 55x + 33$
Subtract	$-49x - 32$ Remainder

The above division results in the following identity, which is similar to the identity $\frac{47}{4} = 11 + \frac{3}{4}$:

$$\frac{1 + x^2 - 2x^3}{3 + 5x + x^2} = -2x + 11 + \frac{-49x - 32}{3 + 5x + x^2}.$$

Quadratic formula: If $ax^2 + bx + c = 0$, then

$$x = \frac{-b \pm \sqrt{b^2 - 4ac}}{2a},$$

provided that the quantity $b^2 - 4ac$ is not negative. If $b^2 - 4ac < 0$, then there are no real solutions.

Factoring quadratics: Since $(x - a)(x - b) = x^2 - (a + b)x + ab$, to factor a quadratic $x^2 + Ax + B$ into the form $(x - a)(x - b)$, we must have

$$a \cdot b = B \qquad \text{and} \qquad a + b = -A.$$

If the values of a and b are not obvious, they can be determined using the quadratic formula:

$$x^2 + Ax + B = \left[x - \frac{(-A + \sqrt{A^2 - 4B})}{2}\right]\left[x - \frac{(-A - \sqrt{A^2 - 4B})}{2}\right]$$

provided that $A^2 - 4B \geq 0$. If $A^2 - 4B < 0$, the quadratic cannot be factored.
Completing the square: Any quadratic $y = a_0 + a_1x + a_2x^2$ can be expressed as a squared term plus a constant. The formula is

$$y = a(x - b)^2 + c,$$

where

$$a = a_2, \qquad b = \frac{-a_1}{2a_2}, \qquad c = a_0 - \frac{a_1{}^2}{4a_2}.$$

APPENDIX II

TABLES

Table A-1 Common Logarithms; log (N)

N	0	1	2	3	4	5	6	7	8	9
1.0	.0000	.0043	.0086	.0128	.0170	.0212	.0253	.0294	.0334	.0374
1.1	.0414	.0453	.0492	.0531	.0569	.0607	.0645	.0682	.0719	.0755
1.2	.0792	.0828	.0864	.0899	.0934	.0969	.1004	.1038	.1072	.1106
1.3	.1139	.1173	.1206	.1239	.1271	.1303	.1335	.1367	.1399	.1430
1.4	.1461	.1492	.1523	.1553	.1584	.1614	.1644	.1673	.1703	.1732
1.5	.1761	.1790	.1818	.1847	.1875	.1903	.1931	.1959	.1987	.2014
1.6	.2041	.2068	.2095	.2122	.2148	.2175	.2201	.2227	.2253	.2279
1.7	.2304	.2330	.2355	.2380	.2405	.2430	.2455	.2480	.2504	.2529
1.8	.2553	.2577	.2601	.2625	.2648	.2672	.2695	.2718	.2742	.2765
1.9	.2788	.2810	.2833	.2856	.2878	.2900	.2923	.2945	.2967	.2989
2.0	.3010	.3032	.3054	.3075	.3096	.3118	.3139	.3160	.3181	.3201
2.1	.3222	.3243	.3263	.3284	.3304	.3324	.3345	.3365	.3385	.3404
2.2	.3424	.3444	.3464	.3483	.3502	.3522	.3541	.3560	.3579	.3598
2.3	.3617	.3636	.3655	.3674	.3692	.3711	.3729	.3747	.3766	.3784
2.4	.3802	.3820	.3838	.3856	.3874	.3892	.3909	.3927	.3945	.3962
2.5	.3979	.3997	.4014	.4031	.4048	.4065	.4082	.4099	.4116	.4133
2.6	.4150	.4166	.4183	.4200	.4216	.4232	.4249	.4265	.4281	.4298
2.7	.4314	.4330	.4356	.4362	.4378	.4393	.4409	.4425	.4440	.4456
2.8	.4472	.4487	.4502	.4518	.4533	.4548	.4564	.4579	.4594	.4609
2.9	.4624	.4639	.4654	.4669	.4683	.4698	.4713	.4728	.4742	.4757
3.0	.4771	.4786	.4800	.4814	.4829	.4843	.4857	.4871	.4886	.4900
3.1	.4914	.4928	.4942	.4955	.4969	.4983	.4997	.5011	.5024	.5038
3.2	.5051	.5065	.5079	.5092	.5105	.5119	.5132	.5145	.5159	.5172
3.3	.5185	.5198	.5211	.5224	.5237	.5250	.5263	.5276	.5289	.5302
3.4	.5315	.5328	.5340	.5353	.5366	.5378	.5391	.5403	.5416	.5428
3.5	.5441	.5453	.5465	.5478	.5490	.5502	.5514	.5527	.5539	.5551
3.6	.5563	.5575	.5587	.5599	.5611	.5623	.5635	.5647	.5658	.5670
3.7	.5682	.5694	.5705	.5717	.5729	.5740	.5752	.5763	.5775	.5786
3.8	.5798	.5809	.5821	.5832	.5843	.5855	.5866	.5877	.5888	.5899
3.9	.5911	.5922	.5933	.5944	.5955	.5966	.5977	.5988	.5999	.6010
4.0	.6021	.6031	.6042	.6053	.6064	.6075	.6085	.6096	.6107	.6117

Table A-1 (continued)

N	0	1	2	3	4	5	6	7	8	9
4.0	.6021	.6031	.6042	.6053	.6064	.6075	.6085	.6096	.6107	.6117
4.1	.6128	.6138	.6149	.6160	.6170	.6180	.6191	.6201	.6212	.6222
4.2	.6232	.6243	.6253	.6263	.6274	.6284	.6294	.6304	.6314	.6325
4.3	.6335	.6345	.6355	.6365	.6375	.6385	.6395	.6405	.6415	.6425
4.4	.6435	.6444	.6454	.6464	.6474	.6484	.6493	.6503	.6513	.6522
4.5	.6532	.6542	.6551	.6561	.6571	.6580	.6590	.6599	.6609	.6618
4.6	.6628	.6637	.6646	.6656	.6665	.6675	.6684	.6693	.6702	.6712
4.7	.6721	.6730	.6739	.6749	.6758	.6767	.6776	.6785	.6794	.6803
4.8	.6812	.6821	.6830	.6839	.6848	.6857	.6866	.6857	.6884	.6893
4.9	.6902	.6911	.6920	.6928	.6937	.6946	.6955	.6964	.6972	.6981
5.0	.6990	.6998	.7007	.7016	.7024	.7033	.7042	.7050	.7059	.7067
5.1	.7076	.7084	.7093	.7101	.7110	.7118	.7126	.7135	.7143	.7152
5.2	.7160	.7168	.7177	.7185	.7193	.7202	.7210	.7218	.7226	.7235
5.3	.7243	.7251	.7259	.7267	.7275	.7284	.7292	.7300	.7308	.7316
5.4	.7324	.7332	.7340	.7348	.7356	.7264	.7372	.7380	.7388	.7396
5.5	.7404	.7412	.7419	.7427	.7435	.7443	.7451	.7459	.7466	.7474
5.6	.7482	.7490	.7497	.7505	.7513	.7520	.7528	.7536	.7543	.7551
5.7	.7559	.7566	.7574	.7582	.7589	.7597	.7604	.7612	.7619	.7627
5.8	.7634	.7642	.7649	.7657	.7664	.7672	.7679	.7686	.7694	.7701
5.9	.7709	.7716	.7723	.7731	.7738	.7745	.7752	.7760	.7767	.7774
6.0	.7782	.7789	.7796	.7803	.7810	.7818	.7825	.7832	.7839	.7846
6.1	.7853	.7860	.7868	.7875	.7882	.7889	.7896	.7903	.7910	.7917
6.2	.7924	.7931	.7938	.7945	.7952	.7959	.7966	.7973	.7980	.7987
6.3	.7993	.8000	.8007	.8014	.8021	.8028	.8035	.8041	.8048	.8055
6.4	.8062	.8069	.8075	.8082	.8089	.8096	.8102	.8109	.8116	.8122
6.5	.8129	.8136	.8142	.8149	.8156	.8162	.8169	.8176	.8182	.8189
6.6	.8195	.8202	.8209	.8215	.8222	.8228	.8235	.8241	.8248	.8254
6.7	.8261	.8267	.8274	.8280	.8287	.8293	.8299	.8306	.8312	.8319
6.8	.8325	.8331	.8338	.8344	.8351	.8357	.8363	.8370	.8376	.8382
6.9	.8388	.8395	.8401	.8407	.8414	.8420	.8426	.8432	.8439	.8445
7.0	.8451	.8457	.8463	.8470	.8476	.8482	.8488	.8494	.8500	.8506

Table A-1 (continued)

N	0	1	2	3	4	5	6	7	8	9
7.0	.8451	.8457	.8463	.8470	.8476	.8482	.8488	.8494	.8500	.8506
7.1	.8513	.8519	.8525	.8531	.8537	.8543	.8549	.8555	.8561	.8567
7.2	.8573	.8579	.8585	.8591	.8597	.8603	.8609	.8615	.8621	.8627
7.3	.8633	.8639	.8645	.8651	.8657	.8663	.8669	.8675	.8681	.8686
7.4	.8692	.8698	.8704	.8710	.8716	.8722	.8727	.8733	.8739	.8745
7.5	.8751	.8756	.8762	.8768	.8774	.8779	.8785	.8791	.8797	.8802
7.6	.8808	.8814	.8820	.8825	.8831	.8837	.8842	.8848	.8854	.8859
7.7	.8865	.8871	.8876	.8882	.8887	.8893	.8899	.8904	.8910	.8915
7.8	.8921	.8927	.8932	.8938	.8943	.8949	.8954	.8960	.8965	.8971
7.9	.8976	.8982	.8987	.8993	.8998	.9004	.9009	.9015	.9020	.9025
8.0	.9031	.9036	.9042	.9047	.9053	.9058	.9063	.9069	.9074	.9079
8.1	.9085	.9090	.9096	.9101	.9106	.9112	.9117	.9122	.9128	.9133
8.2	.9138	.9143	.9149	.9154	.9159	.9165	.9170	.9175	.9180	.9186
8.3	.9191	.9196	.9201	.9206	.9212	.9217	.9222	.9227	.9232	.9238
8.4	.9243	.9248	.9253	.9258	.9263	.9269	.9274	.9279	.9284	.9289
8.5	.9294	.9299	.9304	.9309	.9315	.9320	.9325	.9330	.9335	.9340
8.6	.9345	.9350	.9355	.9360	.9365	.9370	.9375	.9380	.9385	.9390
8.7	.9395	.9400	.9405	.9410	.9415	.9420	.9425	.9430	.9435	.9440
8.8	.9445	.9450	.9455	.9460	.9465	.9469	.9474	.9479	.9484	.9489
8.9	.9494	.9499	.9504	.9509	.9513	.9518	.9523	.9528	.9533	.9538
9.0	.9542	.9547	.9552	.9557	.9562	.9566	.9571	.9576	.9581	.9586
9.1	.9590	.9595	.9600	.9605	.9609	.9614	.9619	.9624	.9628	.9633
9.2	.9638	.9643	.9647	.9652	.9657	.9661	.9666	.9671	.9675	.9680
9.3	.9685	.9689	.9694	.9699	.9703	.9708	.9713	.9717	.9722	.9727
9.4	.9731	.9736	.9741	.9745	.9750	.9754	.9759	.9736	.9768	.9773
9.5	.9777	.9782	.9786	.9791	.9795	.9800	.9805	.9809	.9814	.9818
9.6	.9823	.9827	.9832	.9836	.9841	.9845	.9850	.9854	.9859	.9863
9.7	.9868	.9872	.9877	.9881	.9886	.9890	.9894	.9899	.9903	.9908
9.8	.9912	.9917	.9921	.9926	.9930	.9934	.9939	.9943	.9948	.9952
9.9	.9956	.9961	.9965	.9969	.9974	.9978	.9983	.9987	.9991	.9996

Table A-2 Natural logarithms; ln(x)

x	0.0	0.1	0.2	0.3	0.4	0.5	0.6	0.7	0.8	0.9
1	0.0000	0.0953	0.1823	0.2624	0.3365	0.4055	0.4700	0.5306	0.5878	0.6419
2	0.6931	0.7419	0.7885	0.8329	0.8755	0.9163	0.9555	0.9933	1.0296	1.0647
3	1.0986	1.1314	1.1631	1.1939	1.2238	1.2528	1.2809	1.3083	1.3350	1.3610
4	1.3863	1.4110	1.4351	1.4586	1.4816	1.5041	1.5261	1.5476	1.5686	1.5892
5	1.6094	1.6292	1.6487	1.6677	1.6864	1.7047	1.7228	1.7405	1.7579	1.7750
6	1.7918	1.8083	1.8245	1.8405	1.8563	1.8718	1.8871	1.9021	1.9169	1.9315
7	1.9459	1.9601	1.9741	1.9879	2.0015	2.0149	2.0281	2.0412	2.0541	2.0669
8	2.0794	2.0919	2.1041	2.1163	2.1282	2.1401	2.1518	2.1633	2.1748	2.1861
9	2.1972	2.2083	2.2192	2.2300	2.2407	2.2513	2.2618	2.2721	2.2824	2.2925
10	2.3026	2.3125	2.3224	2.3321	2.3418	2.3514	2.3609	2.3702	2.3795	2.3888
11	2.3979	2.4069	2.4159	2.4248	2.4336	2.4423	2.4510	2.4596	2.4681	2.4765
12	2.4849	2.4932	2.5014	2.5096	2.5177	2.5257	2.5337	2.5416	2.5494	2.5572
13	2.5649	2.5726	2.5802	2.5878	2.5953	2.6027	2.6101	2.6174	2.6247	2.6319
14	2.6391	2.6462	2.6532	2.6603	2.6672	2.6741	2.6810	2.6878	2.6946	2.7014
15	2.7080	2.7147	2.7213	2.7279	2.7344	2.7408	2.7473	2.7537	2.7600	2.7663
16	2.7726	2.7788	2.7850	2.7912	2.7973	2.8034	2.8094	2.8154	2.8214	2.8273
17	2.8332	2.8391	2.8449	2.8507	2.8565	2.8622	2.8679	2.8736	2.8792	2.8848
18	2.8904	2.8959	2.9014	2.9069	2.9123	2.9178	2.9232	2.9285	2.9339	2.9392
19	2.9444	2.9497	2.9549	2.9601	2.9653	2.9704	2.9755	2.9806	2.9857	2.9907
20	2.9957	3.0007	3.0057	3.0106	3.0155	3.0204	3.0253	3.0301	3.0350	3.0397
21	3.0445	3.0493	3.0540	3.0587	3.0634	3.0681	3.0727	3.0773	3.0819	3.0865
22	3.0910	3.0956	3.1001	3.1046	3.1091	3.1135	3.1179	3.1224	3.1268	3.1311
23	3.1355	3.1398	3.1442	3.1485	3.1527	3.1570	3.1612	3.1655	3.1697	3.1739
24	3.1781	3.1822	3.1864	3.1905	3.1946	3.1987	3.2027	3.2068	3.2108	3.2149
25	3.2189	3.2229	3.2268	3.2308	3.2347	3.2387	3.2426	3.2465	3.2504	3.2542
26	3.2581	3.2619	3.2658	3.2696	3.2734	3.2771	3.2809	3.2847	3.2884	3.2921
27	3.2958	3.2995	3.3032	3.3069	3.3105	3.3142	3.3178	3.3214	3.3250	3.3286
28	3.3322	3.3358	3.3393	3.3429	3.3464	3.3499	3.3534	3.3569	3.3604	3.3638
29	3.3673	3.3707	3.3742	3.3776	3.3810	3.3844	3.3878	3.3911	3.3945	3.3979
30	3.4012	3.4045	3.4078	3.4111	3.4144	3.4177	3.4210	3.4243	3.4275	3.4308
31	3.4340	3.4372	3.4404	3.4436	3.4468	3.4500	3.4532	3.4563	3.4595	3.4626
32	3.4657	3.4689	3.4720	3.4751	3.4782	3.4812	3.4843	3.4874	3.4904	3.4935
33	3.4965	3.4995	3.5025	3.5056	3.5086	3.5115	3.5145	3.5175	3.5205	3.5234
34	3.5264	3.5293	3.5322	3.5351	3.5381	3.5410	3.5439	3.5467	3.5496	3.5525
35	3.5553	3.5582	3.5610	3.5639	3.5667	3.5695	3.5723	3.5752	3.5779	3.5807
36	3.5835	3.5863	3.5891	3.5918	3.5946	3.5973	3.6000	3.6028	3.6055	3.6082
37	3.6109	3.6136	3.6163	3.6190	3.6217	3.6243	3.6270	3.6297	3.6323	3.6350
38	3.6376	3.6402	3.6428	3.6454	3.6481	3.6507	3.6533	3.6558	3.6584	3.6610
39	3.6636	3.6661	3.6687	3.6712	3.6738	3.6763	3.6788	3.6814	3.6839	3.6864
40	3.6889	3.6914	3.6939	3.6964	3.6988	3.7013	3.7038	3.7062	3.7087	3.7111
41	3.7136	3.7160	3.7184	3.7209	3.7233	3.7257	3.7281	3.7305	3.7329	3.7353
42	3.7377	3.7400	3.7424	3.7448	3.7471	3.7495	3.7519	3.7542	3.7565	3.7589
43	3.7612	3.7635	3.7658	3.7682	3.7705	3.7728	3.7751	3.7773	3.7796	3.7819
44	3.7842	3.7865	3.7887	3.7910	3.7932	3.7955	3.7977	3.8000	3.8022	3.8044
45	3.8067	3.8089	3.8111	3.8133	3.8155	3.8177	3.8199	3.8221	3.8243	3.8265
46	3.8286	3.8308	3.8330	3.8351	3.8373	3.8395	3.8416	3.8437	3.8459	3.8480
47	3.8501	3.8523	3.8544	3.8565	3.8586	3.8607	3.8628	3.8649	3.8670	3.8691
48	3.8712	3.8733	3.8754	3.8774	3.8795	3.8816	3.8836	3.8857	3.8877	3.8898
49	3.8918	3.8939	3.8959	3.8979	3.9000	3.9020	3.9040	3.9060	3.9080	3.9100
50	3.9120	3.9140	3.9160	3.9180	3.9200	3.9220	3.9240	3.9259	3.9279	3.9299

Table A-2 (continued)

x	0.0	0.1	0.2	0.3	0.4	0.5	0.6	0.7	0.8	0.9
50	3.9120	3.9140	3.9160	3.9180	3.9200	3.9220	3.9240	3.9259	3.9279	3.9299
51	3.9318	3.9338	3.9357	3.9377	3.9396	3.9416	3.9435	3.9455	3.9474	3.9493
52	3.9512	3.9532	3.9551	3.9570	3.9589	3.9608	3.9627	3.9646	3.9665	3.9684
53	3.9703	3.9722	3.9741	3.9759	3.9778	3.9797	3.9815	3.9834	3.9853	3.9871
54	3.9890	3.9908	3.9927	3.9945	3.9964	3.9982	4.0000	4.0019	4.0037	4.0055
55	4.0073	4.0091	4.0110	4.0128	4.0146	4.0164	4.0182	4.0200	4.0218	4.0236
56	4.0254	4.0271	4.0289	4.0307	4.0325	4.0342	4.0360	4.0378	4.0395	4.0413
57	4.0431	4.0448	4.0466	4.0483	4.0500	4.0518	4.0535	4.0553	4.0570	4.0587
58	4.0604	4.0622	4.0639	4.0656	4.0673	4.0690	4.0707	4.0724	4.074!	4.0758
59	4.0775	4.0792	4.0809	4.0826	4.0843	4.0860	4.0877	4.0893	4.0910	4.0927
60	4.0943	4.0960	4.0977	4.0993	4.1010	4.1026	4.1043	4.1059	4.1076	4.1092
61	4.1109	4.1125	4.1141	4.1158	4.1174	4.1190	4.1207	4.1223	4.1239	4.1255
62	4.1271	4.1287	4.1304	4.1320	4.1336	4.1352	4.1368	4.1384	4.1400	4.1415
63	4.1431	4.1447	4.1463	4.1479	4.1495	4.1510	4.1526	4.1542	4.1558	4.1573
64	4.1589	4.1604	4.1620	4.1636	4.1651	4.1667	4.1682	4.1698	4.1713	4.1728
65	4.1744	4.1759	4.1775	4.1790	4.1805	4.1820	4.1836	4.1851	4.1866	4.1881
66	4.1897	4.1912	4.1927	4.1942	4.1957	4.1972	4.1987	4.2002	4.2017	4.2032
67	4.2047	4.2062	4.2077	4.2092	4.2106	4.2121	4.2136	4.2151	4.2166	4.2180
68	4.2195	4.2210	4.2224	4.2239	4.2254	4.2268	4.2293	4.2297	4.2312	4.2327
69	4.2341	4.2356	4.2370	4.2384	4.2399	4.2413	4.2428	4.2442	4.2456	4.2471
70	4.2485	4.2499	4.2513	4.2528	4.2542	4.2556	4.2570	4.2584	4.2599	4.2613
71	4.2627	4.2641	4.2655	4.2669	4.2683	4.2697	4.2711	4.2725	4.2739	4.2753
72	4.2767	4.2781	4.2794	4.2808	4.2822	4.2836	4.2850	4.2863	4.2877	4.2891
73	4.2905	4.2918	4.2932	4.2946	4.2959	4.2973	4.2986	4.3000	4.3014	4.3027
74	4.3041	4.3054	4.3068	4.3081	4.3095	4.3108	4.3121	4.3135	4.3148	4.3162
75	4.3175	4.3188	4.3202	4.3215	4.3228	4.3241	4.3255	4.3268	4.3281	4.3294
76	4.3307	4.3320	4.3334	4.3347	4.3360	4.3373	4.3386	4.3399	4.3412	4.3425
77	4.3438	4.3451	4.3464	4.3477	4.3490	4.3503	4.3516	4.3529	4.3541	4.3554
78	4.3567	4.3580	4.3593	4.3605	4.3618	4.3631	4.3644	4.3656	4.3669	4.3682
79	4.3694	4.3707	4.3720	4.3732	4.3745	4.3758	4.3770	4.3783	4.3795	4.3808
80	4.3820	4.3833	4.3845	4.3858	4.3870	4.3883	4.3895	4.3907	4.3920	4.3932
81	4.3944	4.3957	4.3969	4.3981	4.3994	4.4006	4.4018	4.4031	4.4043	4.4055
82	4.4067	4.4079	4.4092	4.4104	4.4116	4.4128	4.4140	4.4152	4.4164	4.4176
83	4.4188	4.4200	4.4212	4.4224	4.4236	4.4248	4.4260	4.4272	4.4284	4.4296
84	4.4308	4.4320	4.4332	4.4344	4.4356	4.4368	4.4379	4.4391	4.4403	4.4415
85	4.4427	4.4438	4.4450	4.4462	4.4473	4.4485	4.4497	4.4509	4.4520	4.4532
86	4.4543	4.4555	5.4567	4.4578	4.4590	4.4601	4.4613	4.4625	4.4636	4.4648
87	4.4659	4.4671	4.4682	4.4693	4.4705	4.4716	4.4728	4.4739	4.4751	4.4762
88	4.4773	4.4785	4.4796	4.4807	4.4819	4.4830	4.4841	4.4853	4.4864	4.4875
89	4.4885	4.4898	4.4909	4.4920	4.4931	4.4942	4.4954	4.4965	4.4976	4.4987
90	4.4998	4.5009	4.5020	4.5031	4.5042	4.5053	4.5065	4.5076	4.5087	4.5098
91	4.5109	4.5120	4.5131	4.5142	4.5152	4.5163	4.5174	4.5185	4.5196	4.5207
92	4.5218	4.5229	4.5240	4.5250	4.5261	4.5272	4.5283	4.5294	4.5304	4.5315
93	4.5326	4.5337	4.5347	4.5358	4.5369	4.5380	4.5390	4.5401	4.5412	4.5422
94	4.5433	4.5444	4.5454	4.5465	4.5475	4.5486	4.5497	4.5507	4.5518	4.5528
95	4.5539	4.5549	4.5560	4.5570	4.5581	4.5591	4.5602	4.5612	4.5623	4.5633
96	4.5643	4.5654	4.5664	4.5675	4.5685	4.5695	4.5706	4.5716	4.5726	4.5737
97	4.5747	4.5757	4.5768	4.5778	4.5788	4.5799	4.5809	4.5819	4.5829	4.5839
98	4.5850	4.5860	4.5870	4.5880	4.5890	4.5901	4.5911	4.5921	4.5931	4.5941
99	4.5951	4.5961	4.5971	4.5981	4.5992	4.6002	4.6012	4.6022	4.6032	4.6042

Table A-3 Exponential functions

x	10^x	e^x	10^{-x}	e^{-x}
0.00	1.00000	1.00000	1.00000	1.00000
0.01	1.02329	1.01005	0.97724	0.99005
0.02	1.04713	1.02020	0.95499	0.98020
0.03	1.07152	1.03045	0.93325	0.97045
0.04	1.09648	1.04081	0.91201	0.96079
0.05	1.12202	1.05127	0.89125	0.95123
0.06	1.14815	1.06184	0.87096	0.94176
0.07	1.17490	1.07251	0.85114	0.93239
0.08	1.20226	1.08329	0.83176	0.92312
0.09	1.23027	1.09417	0.81283	0.91393
0.10	1.25892	1.10517	0.79433	0.90484
0.11	1.28825	1.11628	0.77625	0.89583
0.12	1.31826	1.12750	0.75858	0.88692
0.13	1.34896	1.13883	0.74131	0.87810
0.14	1.38038	1.15027	0.72444	0.86936
0.15	1.41254	1.16183	0.70795	0.86071
0.16	1.44544	1.17351	0.69183	0.85214
0.17	1.47911	1.18530	0.67608	0.84366
0.18	1.51356	1.19722	0.66069	0.83527
0.19	1.54882	1.20925	0.64565	0.82696
0.20	1.58489	1.22140	0.63096	0.81873
0.21	1.52181	1.23368	0.61660	0.81058
0.22	1.65959	1.24608	0.60256	0.80252
0.23	1.69824	1.25860	0.58884	0.79453
0.24	1.73780	1.27125	0.57544	0.78663
0.25	1.77828	1.28403	0.56234	0.77880
0.26	1.81970	1.29693	0.54954	0.77105
0.27	1.86209	1.30996	0.53703	0.76338
0.28	1.90546	1.32313	0.52481	0.75578
0.29	1.94984	1.33643	0.51286	0.74826
0.30	1.99526	1.34986	0.50119	0.74082
0.31	2.04174	1.36342	0.48978	0.73345
0.32	2.08930	1.37713	0.47863	0.72615
0.33	2.13796	1.39097	0.46774	0.71892
0.34	2.18776	1.40495	0.45709	0.71177
0.35	2.23872	1.41907	0.44668	0.70469
0.36	2.29087	1.43333	0.43652	0.69768
0.37	2.34423	1.44773	0.42658	0.69073
0.38	2.39883	1.46228	0.41687	0.68386
0.39	2.45471	1.47698	0.40738	0.67706

Table A-3 (continued)

x	10^x	e^x	10^{-x}	e^{-x}
0.40	2.51189	1.49182	0.39811	0.67032
0.41	2.57040	1.50682	0.38905	0.66365
0.42	2.63027	1.52196	0.38019	0.65705
0.43	2.69153	1.53726	0.37154	0.65051
0.44	2.75423	1.55271	0.36308	0.64404
0.45	2.81838	1.56831	0.35481	0.63763
0.46	2.88403	1.58407	0.34674	0.63128
0.47	2.95121	1.59999	0.33884	0.62500
0.48	3.01995	1.61607	0.33113	0.61878
0.49	3.09030	1.63232	0.32359	0.61263
0.50	3.16228	1.64872	0.31623	0.60653
0.51	3.23594	1.66529	0.30903	0.60050
0.52	3.31131	1.68203	0.30200	0.59452
0.53	3.38844	1.69893	0.29512	0.58861
0.54	3.46737	1.71601	0.28840	0.58275
0.55	3.54813	1.73325	0.28184	0.57695
0.56	3.63078	1.75067	0.27542	0.57121
0.57	3.71535	1.76827	0.26915	0.56553
0.58	3.80139	1.78604	0.26303	0.55990
0.59	3.89045	1.80399	0.25704	0.55433
0.60	3.98107	1.82212	0.25119	0.54881
0.61	4.07380	1.84043	0.24547	0.54335
0.62	4.16869	1.85893	0.23988	0.53794
0.63	4.26579	1.87761	0.23442	0.53259
0.64	4.36516	1.89648	0.22909	0.52729
0.65	4.46684	1.91554	0.22387	0.52205
0.66	4.57088	1.93479	0.21878	0.51685
0.67	4.67735	1.95424	0.21380	0.51171
0.68	4.78630	1.97388	0.20893	0.50662
0.69	4.89779	1.99372	0.20417	0.50158
0.70	5.01187	2.01375	0.19953	0.49659
0.71	5.12861	2.03399	0.19498	0.49164
0.72	5.24807	2.05443	0.19055	0.48675
0.73	5.37032	2.07508	0.18621	0.48191
0.74	5.49541	2.09593	0.18197	0.47711
0.75	5.62341	2.11700	0.17783	0.47237
0.76	5.75440	2.13828	0.17378	0.46767
0.77	5.88844	2.15977	0.16982	0.46301
0.78	6.02560	2.18147	0.16596	0.45841
0.79	6.16595	2.20340	0.16218	0.45384

Table A-3 (continued)

x	10^x	e^x	10^{-x}	e^{-x}
0.80	6.30957	2.22554	0.15849	0.44933
0.81	6.45654	2.24791	0.15488	0.44486
0.82	6.60693	2.27050	0.15136	0.44043
0.83	6.76083	2.29332	0.14791	0.43605
0.84	6.91831	2.31637	0.14454	0.43171
0.85	7.07946	2.33965	0.14125	0.42742
0.86	7.24436	2.36316	0.13804	0.42316
0.87	7.41310	2.38691	0.13490	0.41895
0.88	7.58578	2.41090	0.13183	0.41478
0.89	7.76247	2.43513	0.12882	0.41066
0.90	7.94328	2.45960	0.12589	0.40657
0.91	8.12831	2.48432	0.12303	0.40252
0.92	8.31764	2.50929	0.12023	0.39852
0.93	8.51138	2.53451	0.11749	0.39455
0.94	8.70964	2.55998	0.11482	0.39063
0.95	8.91251	2.58571	0.11220	0.38674
0.96	9.12011	2.61170	0.10965	0.38289
0.97	9.33254	2.63794	0.10715	0.37908
0.98	9.54993	2.66446	0.10471	0.37531
0.99	9.77237	2.69123	0.10233	0.37158
1.00	10.00000	2.71828	0.10000	0.36788
1.01	10.23291	2.74560	0.09772	0.36422
1.02	10.47127	2.77319	0.09550	0.36060
1.03	10.71519	2.80106	0.09333	0.35701
1.04	10.96478	2.82922	0.09120	0.35345
1.05	11.22016	2.85765	0.08913	0.34994
1.06	11.48152	2.88637	0.08710	0.34646
1.07	11.74897	2.91538	0.08511	0.34301
1.08	12.02264	2.94468	0.08318	0.33960
1.09	12.30266	2.97427	0.08128	0.33622
1.10	12.58924	3.00416	0.07943	0.33287
1.11	12.88248	3.03436	0.07762	0.32956
1.12	13.18256	3.06485	0.07586	0.32628
1.13	13.48960	3.09565	0.07413	0.32303
1.14	13.80382	3.12677	0.07244	0.31982
1.15	14.12536	3.15819	0.07079	0.31664
1.16	14.45439	3.18993	0.06918	0.31349
1.17	14.79105	3.22199	0.06761	0.31037
1.18	15.13559	3.25437	0.06607	0.30728
1.19	15.48815	3.28708	0.06457	0.30422

Table A-3 (continued)

x	10^x	e^x	10^{-x}	e^{-x}
1.20	15.84892	3.32012	0.06310	0.30119
1.21	16.21806	3.35348	0.06166	0.29820
1.22	16.59584	3.38719	0.06026	0.29523
1.23	16.98241	3.42123	0.05888	0.29229
1.24	17.37799	3.45561	0.05754	0.28938
1.25	17.78279	3.49034	0.05623	0.28650
1.26	18.19698	3.52542	0.05495	0.28365
1.27	18.62085	3.56085	0.05370	0.28083
1.28	19.05458	3.59664	0.05248	0.27804
1.29	19.49844	3.63279	0.05129	0.27527
1.30	19.95258	3.66929	0.05012	0.27253
1.31	20.41734	3.70617	0.04898	0.26982
1.32	20.89294	3.74342	0.04786	0.26714
1.33	21.37961	3.78104	0.04677	0.26448
1.34	21.87756	3.81904	0.04571	0.26185
1.35	22.38718	3.85742	0.04467	0.25924
1.36	22.90865	3.89619	0.04365	0.25666
1.37	23.44228	3.93535	0.04266	0.25411
1.38	23.98828	3.97490	0.04169	0.25158
1.39	24.54704	4.01485	0.04074	0.24908
1.40	25.11884	4.05520	0.03981	0.24660
1.41	25.70393	4.09595	0.03890	0.24414
1.42	26.30261	4.13712	0.03802	0.24171
1.43	26.91530	4.17870	0.03715	0.23931
1.44	27.54225	4.22069	0.03631	0.23693
1.45	28.18381	4.26311	0.03548	0.23457
1.46	28.84032	4.30596	0.03467	0.23224
1.47	29.51204	4.34923	0.03388	0.22993
1.48	30.19948	4.39294	0.03311	0.22764
1.49	30.90292	4.43709	0.03236	0.22537
1.50	31.62277	4.48169	0.03162	0.22313
1.51	32.35930	4.52673	0.03090	0.22091
1.52	33.11307	4.57222	0.03020	0.21871
1.53	33.88438	4.61818	0.02951	0.21654
1.54	34.67368	4.66459	0.02884	0.21438
1.55	35.48126	4.71147	0.02818	0.21225
1.56	36.30775	4.75882	0.02754	0.21014
1.57	37.15349	4.80665	0.02692	0.20805
1.58	38.01892	4.85495	0.02630	0.20598
1.59	38.90443	4.90374	0.02570	0.20393

Table A-3 (continued)

x	10^x	e^x	10^{-x}	e^{-x}
1.60	39.81065	4.95303	0.02512	0.20190
1.61	40.73799	5.00281	0.02455	0.19989
1.62	41.68692	5.05309	0.02399	0.19790
1.63	42.65787	5.10387	0.02344	0.19593
1.64	43.65152	5.15517	0.02291	0.19398
1.65	44.66832	5.20698	0.02239	0.19205
1.66	45.70880	5.25931	0.02188	0.19014
1.67	46.77341	5.31216	0.02138	0.18825
1.68	47.86293	5.36555	0.02089	0.18637
1.69	48.97783	5.41948	0.02042	0.18452
1.70	50.11870	5.47395	0.01995	0.18268
1.71	51.28603	5.52896	0.01950	0.18087
1.72	52.48065	5.58452	0.01905	0.17907
1.73	53.70311	5.64065	0.01862	0.17728
1.74	54.95406	5.69734	0.01820	0.17552
1.75	56.23413	5.75460	0.01778	0.17377
1.76	57.54388	5.81243	0.01738	0.17204
1.77	58.88429	5.87085	0.01698	0.17033
1.78	60.25592	5.92985	0.01660	0.16864
1.79	61.65948	5.98945	0.01622	0.16696
1.80	63.09561	6.04964	0.01585	0.16530
1.81	64.56534	6.11044	0.01549	0.16365
1.82	66.06929	6.17186	0.01514	0.16203
1.83	67.60828	6.23389	0.01479	0.16041
1.84	69.18297	6.29653	0.01445	0.15882
1.85	70.79448	6.35982	0.01413	0.15724
1.86	72.44353	6.42373	0.01380	0.15567
1.87	74.13100	6.48830	0.01349	0.15412
1.88	75.85760	6.55350	0.01318	0.15259
1.89	77.62459	6.61936	0.01288	0.15107
1.90	79.43274	6.68589	0.01259	0.14957
1.91	81.28302	6.75309	0.01230	0.14808
1.92	83.17621	6.82095	0.01202	0.14661
1.93	85.11366	6.88951	0.01175	0.14515
1.94	87.09627	6.95875	0.01148	0.14370
1.95	89.12505	7.02869	0.01122	0.14227
1.96	91.20108	7.09933	0.01096	0.14086
1.97	93.32527	7.17067	0.01072	0.13946
1.98	95.49915	7.24274	0.01047	0.13807
1.99	97.72366	7.31553	0.01023	0.13670

Table A-3 (continued)

x	10^x	e^x	10^{-x}	e^{-x}
2.00	1.000E 02	7.389E 00	1.000E-02	1.353E-01
2.05	1.122E 02	7.768E 00	8.913E-03	1.287E-01
2.10	1.259E 02	8.166E 00	7.943E-03	1.225E-01
2.15	1.413E 02	8.585E 00	7.079E-03	1.165E-01
2.20	1.585E 02	9.025E 00	6.310E-03	1.108E-01
2.25	1.778E 02	9.488E 00	5.623E-03	1.054E-01
2.30	1.995E 02	9.974E 00	5.012E-03	1.003E-01
2.35	2.239E 02	1.049E 01	4.467E-03	9.537E-02
2.40	2.512E 02	1.102E 01	3.981E-03	9.072E-02
2.45	2.818E 02	1.159E 01	3.548E-03	8.629E-02
2.50	3.162E 02	1.218E 01	3.162E-03	8.208E-02
2.55	3.548E 02	1.281E 01	2.818E-03	7.808E-02
2.60	3.981E 02	1.346E 01	2.512E-03	7.427E-02
2.65	4.467E 02	1.415E 01	2.239E-03	7.065E-02
2.70	5.012E 02	1.488E 01	1.995E-03	6.721E-02
2.75	5.623E 02	1.564E 01	1.778E-03	6.393E-02
2.80	6.310E 02	1.644E 01	1.585E-03	6.081E-02
2.85	7.079E 02	1.729E 01	1.413E-03	5.784E-02
2.90	7.943E 02	1.817E 01	1.259E-03	5.502E-02
2.95	8.913E 02	1.911E 01	1.122E-03	5.234E-02
3.00	1.000E 03	2.009E 01	1.000E-03	4.979E-02
3.05	1.122E 03	2.112E 01	8.913E-04	4.736E-02
3.10	1.259E 03	2.220E 01	7.943E-04	4.505E-02
3.15	1.413E 03	2.334E 01	7.079E-04	4.285E-02
3.20	1.585E 03	2.453E 01	6.310E-04	4.076E-02
3.25	1.778E 03	2.579E 01	5.623E-04	3.877E-02
3.30	1.995E 03	2.711E 01	5.012E-04	3.688E-02
3.35	2.239E 03	2.850E 01	4.467E-04	3.508E-02
3.40	2.512E 03	2.996E 01	3.981E-04	3.337E-02
3.45	2.818E 03	3.150E 01	3.548E-04	3.175E-02
3.50	3.162E 03	3.312E 01	3.162E-04	3.020E-02
3.55	3.548E 03	3.481E 01	2.818E-04	2.872E-02
3.60	3.981E 03	3.660E 01	2.512E-04	2.732E-02
3.65	4.467E 03	3.847E 01	2.239E-04	2.599E-02
3.70	5.012E 03	4.045E 01	1.995E-04	2.472E-02
3.75	5.623E 03	4.252E 01	1.778E-04	2.352E-02
3.80	6.310E 03	4.470E 01	1.585E-04	2.237E-02
3.85	7.079E 03	4.699E 01	1.413E-04	2.128E-02
3.90	7.943E 03	4.940E 01	1.259E-04	2.024E-02
3.95	8.913E 03	5.194E 01	1.122E-04	1.925E-02

Table A-3 (continued)

x	10^x	e^x	10^{-x}	e^{-x}
4.00	1.000E 04	5.460E 01	1.000E-04	1.832E-02
4.05	1.122E 04	5.740E 01	8.913E-05	1.742E-02
4.10	1.259E 04	6.034E 01	7.943E-05	1.657E-02
4.15	1.413E 04	6.343E 01	7.079E-05	1.576E-02
4.20	1.585E 04	6.669E 01	6.310E-05	1.500E-02
4.25	1.778E 04	7.011E 01	5.623E-05	1.426E-02
4.30	1.995E 04	7.370E 01	5.012E-05	1.357E-02
4.35	2.239E 04	7.748E 01	4.467E-05	1.291E-02
4.40	2.512E 04	8.145E 01	3.981E-05	1.228E-02
4.45	2.818E 04	8.563E 01	3.548E-05	1.168E-02
4.50	3.162E 04	9.002E 01	3.162E-05	1.111E-02
4.55	3.548E 04	9.463E 01	2.818E-05	1.057E-02
4.60	3.981E 04	9.948E 01	2.512E-05	1.005E-02
4.65	4.467E 04	1.046E 02	2.239E-05	9.562E-03
4.70	5.012E 04	1.099E 02	1.998E-05	9.095E-03
4.75	5.623E 04	1.156E 02	1.778E-05	8.652E-03
4.80	6.310E 04	1.215E 02	1.585E-05	8.230E-03
4.85	7.079E 04	1.277E 02	1.413E-05	7.828E-03
4.90	7.943E 04	1.343E 02	1.259E-05	7.447E-03
4.95	8.913E 04	1.412E 02	1.122E-05	7.083E-03
5.00	1.000E 05	1.484E 02	1.000E-05	6.738E-03
5.05	1.122E 05	1.560E 02	8.913E-06	6.409E-03
5.10	1.259E 05	1.640E 02	7.943E-06	6.097E-03
5.15	1.413E 05	1.724E 02	7.079E-06	5.799E-03
5.20	1.585E 05	1.813E 02	6.310E-06	5.517E-03
5.25	1.778E 05	1.906E 02	5.623E-06	5.248E-03
5.30	1.995E 05	2.003E 02	5.012E-06	4.992E-03
5.35	2.239E 05	2.106E 02	4.467E-06	4.748E-03
5.40	2.512E 05	2.214E 02	3.981E-06	4.517E-03
5.45	2.818E 05	2.328E 02	3.548E-06	4.296E-03
5.50	3.162E 05	2.447E 02	3.162E-06	4.087E-03
5.55	3.548E 05	2.572E 02	2.818E-06	3.887E-03
5.60	3.981E 05	2.704E 02	2.512E-06	3.698E-03
5.65	4.467E 05	2.843E 02	2.239E-06	3.518E-03
5.70	5.012E 05	2.989E 02	1.995E-06	3.346E-03
5.75	5.623E 05	3.142E 02	1.778E-06	3.183E-03
5.80	6.310E 05	3.303E 02	1.585E-06	3.028E-03
5.85	7.079E 05	3.472E 02	1.413E-06	2.880E-03
5.90	7.943E 05	3.650E 02	1.259E-06	2.739E-03
5.95	8.913E 05	3.838E 02	1.122E-06	2.606E-03

Table A-3 (continued)

x	10^x	e^x	10^{-x}	e^{-x}
6.00	1.000E 06	4.034E 02	1.000E-06	2.479E-03
6.05	1.122E 06	4.241E 02	8.913E-07	2.358E-03
6.10	1.259E 06	4.459E 02	7.943E-07	2.243E-03
6.15	1.413E 06	4.687E 02	7.079E-07	2.133E-03
6.20	1.585E 06	4.927E 02	6.310E-07	2.029E-03
6.25	1.778E 06	5.180E 02	5.623E-07	1.930E-03
6.30	1.995E 06	5.446E 02	5.012E-07	1.836E-03
6.35	2.239E 06	5.725E 02	4.467E-07	1.747E-03
6.40	2.512E 06	6.018E 02	3.981E-07	1.662E-03
6.45	2.818E 06	6.327E 02	3.548E-07	1.581E-03
6.50	3.162E 06	6.651E 02	3.162E-07	1.503E-03
6.55	3.548E 06	6.992E 02	2.818E-07	1.430E-03
6.60	3.981E 06	7.351E 02	2.512E-07	1.360E-03
6.65	4.467E 06	7.728E 02	2.239E-07	1.294E-03
6.70	5.012E 06	8.124E 02	1.995E-07	1.231E-03
6.75	5.623E 06	8.541E 02	1.778E-07	1.171E-03
6.80	6.310E 06	8.978E 02	1.585E-07	1.114E-03
6.85	7.079E 06	9.439E 02	1.413E-07	1.059E-03
6.90	7.943E 06	9.823E 02	1.259E-07	1.008E-03
6.95	8.913E 06	1.043E 03	1.122E-07	9.586E-04
7.00	1.000E 07	1.097E 03	1.000E-07	9.119E-04
7.05	1.122E 07	1.153E 03	8.913E-08	8.674E-04
7.10	1.259E 07	1.212E 03	7.943E-08	8.251E-04
7.15	1.413E 07	1.274E 03	7.079E-08	7.849E-04
7.20	1.585E 07	1.339E 03	6.310E-08	7.466E-04
7.25	1.778E 07	1.408E 03	5.623E-08	7.102E-04
7.30	1.995E 07	1.480E 03	5.012E-08	6.755E-04
7.35	2.239E 07	1.556E 03	4.467E-08	6.426E-04
7.40	2.512E 07	1.636E 03	3.981E-08	6.113E-04
7.45	2.818E 07	1.720E 03	3.548E-08	5.814E-04
7.50	3.162E 07	1.808E 03	3.162E-08	5.531E-04
7.55	3.548E 07	1.901E 03	2.818E-08	5.261E-04
7.60	3.981E 07	1.998E 03	2.512E-08	5.005E-04
7.65	4.467E 07	2.101E 03	2.239E-08	4.760E-04
7.70	5.012E 07	2.208E 03	1.995E-08	4.528E-04
7.75	5.623E 07	2.322E 03	1.778E-08	4.307E-04
7.80	6.310E 07	2.441E 03	1.585E-08	4.097E-04
7.85	7.079E 07	2.566E 03	1.413E-08	3.898E-04
7.90	7.943E 07	2.697E 03	1.259E-08	3.707E-04
7.95	8.913E 07	2.836E 03	1.122E-08	3.527E-04

Table A-3 (continued)

x	10^x	e^x	10^{-x}	e^{-x}
8.00	1.000E 08	2.981E 03	1.000E-08	3.355E-04
8.05	1.122E 08	3.134E 03	8.913E-09	3.191E-04
8.10	1.259E 08	3.294E 03	7.943E-09	3.035E-04
8.15	1.413E 08	3.463E 03	7.079E-09	2.887E-04
8.20	1.585E 08	3.641E 03	6.310E-09	2.747E-04
8.25	1.778E 08	3.828E 03	5.623E-09	2.613E-04
8.30	1.995E 08	4.024E 03	5.012E-09	2.485E-04
8.35	2.239E 08	4.230E 03	4.467E-09	2.364E-04
8.40	2.512E 08	4.447E 03	3.981E-09	2.249E-04
8.45	2.818E 08	4.675E 03	3.548E-09	2.139E-04
8.50	3.162E 08	4.915E 03	3.162E-09	2.035E-04
8.55	3.548E 08	5.167E 03	2.818E-09	1.935E-04
8.60	3.981E 08	5.432E 03	2.512E-09	1.841E-04
8.65	4.467E 08	5.710E 03	2.239E-09	1.751E-04
8.70	5.012E 08	6.003E 03	1.995E-09	1.666E-04
8.75	5.623E 08	6.311E 03	1.778E-09	1.585E-04
8.80	6.310E 08	6.634E 03	1.585E-09	1.507E-04
8.85	7.079E 08	6.974E 03	1.413E-09	1.434E-04
8.90	7.943E 08	7.332E 03	1.259E-09	1.364E-04
8.95	8.913E 08	7.708E 03	1.122E-09	1.297E-04
9.00	1.000E 09	8.103E 03	1.000E-09	1.234E-04
9.05	1.122E 09	8.519E 03	8.913E-10	1.174E-04
9.10	1.259E 09	8.955E 03	7.943E-10	1.117E-04
9.15	1.413E 09	9.414E 03	7.079E-10	1.062E-04
9.20	1.585E 09	9.897E 03	6.310E-10	1.010E-04
9.25	1.778E 09	1.040E 04	5.623E-10	9.611E-05
9.30	1.995E 09	1.094E 04	5.012E-10	9.142E-05
9.35	2.239E 09	1.150E 04	4.467E-10	8.697E-05
9.40	2.512E 09	1.209E 04	3.981E-10	8.272E-05
9.45	2.818E 09	1.271E 04	3.548E-10	7.869E-05
9.50	3.162E 09	1.336E 04	3.162E-10	7.485E-05
9.55	3.548E 09	1.404E 04	2.818E-10	7.120E-05
9.60	3.981E 09	1.476E 04	2.512E-10	6.773E-05
9.65	4.467E 09	1.552E 04	2.239E-10	6.443E-05
9.70	5.012E 09	1.632E 04	1.995E-10	6.128E-05
9.75	5.623E 09	1.715E 04	1.778E-10	5.829E-05
9.80	6.310E 09	1.803E 04	1.585E-10	5.545E-05
9.85	7.079E 09	1.896E 04	1.413E-10	5.275E-05
9.90	7.943E 09	1.993E 04	1.259E-10	5.017E-05
9.95	8.913E 09	2.095E 04	1.122E-10	4.773E-05
10.00	1.000E 10	2.203E 04	1.000E-10	4.540E-05

Table A-4 Trigonometric functions, multiples of π.

x-radians	$\sin(x)$	$\cos(x)$	$\tan(x)$
0	0.0000	1.0000	0.0000
$\pi/12$	0.2588	0.9659	0.2679
$\pi/6$	0.5000	0.8660	0.5774
$\pi/4$	0.7071	0.7071	1.0000
$\pi/3$	0.8660	0.5000	1.7320
$5\pi/12$	0.9659	0.2588	3.7320
$\pi/2$	1.0000	0.0000	—
$7\pi/12$	0.9659	−0.2588	−3.7321
$2\pi/3$	0.8660	−0.5000	−1.7321
$3\pi/4$	0.7071	−0.7071	−1.0000
$5\pi/6$	0.5000	−0.8660	−0.5774
$11\pi/12$	0.2588	−0.9659	−0.2679
π	0.0000	−1.0000	0.0000
$13\pi/12$	−0.2588	−0.9659	0.2679
$7\pi/6$	−0.5000	−0.8660	0.5773
$5\pi/4$	−0.7071	−0.7071	1.0000
$4\pi/3$	−0.8660	−0.5000	1.7321
$17\pi/12$	−0.9659	−0.2588	3.7320
$3\pi/2$	−1.0000	0.0000	—
$19\pi/12$	−0.9659	0.2588	−3.7321
$5\pi/3$	−0.8660	0.5000	−1.7321
$7\pi/4$	−0.7071	0.7071	−1.0000
$11\pi/6$	−0.5000	0.8660	−0.5774
$23\pi/12$	−0.2588	0.9659	−0.2679
2π	0.0000	1.0000	0.0000

Table A-5 The Trigonometric Ratios

Degrees	Radians	Sin	Cos	Tan	Cot	Sec	Csc		
0°	.0000	.0000	1.0000	.0000	--	1.000	--	1.5708	90°
1°	.0175	.0175	.9998	.0175	57.29	1.000	57.30	1.5533	89°
2°	.0349	.0349	.9994	.0349	28.64	1.001	28.65	1.5359	88°
3°	.0524	.0523	.9986	.0524	19.08	1.001	19.11	1.5184	87°
4°	.0698	.0698	.9976	.0699	14.30	1.002	14.34	1.5010	86°
5°	.0873	.0872	.9962	.0875	11.43	1.004	11.47	1.4835	85°
6°	.1047	.1045	.9945	.1051	9.514	1.006	9.567	1.4661	84°
7°	.1222	.1219	.9925	.1228	8.144	1.008	8.206	1.4486	83°
8°	.1396	.1392	.9903	.1405	7.115	1.010	7.185	1.4312	82°
9°	.1571	.1564	.9877	.1584	6.314	1.012	6.392	1.4137	81°
10°	.1745	.1736	.9848	.1763	5.671	1.015	5.759	1.3963	80°
11°	.1920	.1908	.9816	.1944	5.145	1.019	5.241	1.3788	79°
12°	.2094	.2079	.9781	.2126	4.705	1.022	4.810	1.3614	78°
13°	.2269	.2250	.9744	.2309	4.331	1.026	4.445	1.3439	77°
14°	.2443	.2419	.9703	.2493	4.011	1.031	4.134	1.3265	76°
15°	.2618	.2588	.9659	.2679	3.732	1.035	3.864	1.3090	75°
16°	.2793	.2756	.9613	.2867	3.487	1.040	3.628	1.2915	74°
17°	.2967	.2924	.9563	.3057	3.271	1.046	3.420	1.2741	73°
18°	.3142	.3090	.9511	.3249	3.078	1.051	3.236	1.2566	72°
19°	.3316	.3256	.9455	.3443	2.904	1.058	3.072	1.2392	71°
20°	.3491	.3420	.9397	.3640	2.747	1.064	2.924	1.2217	70°
21°	.3665	.3584	.9336	.3839	2.605	1.071	2.790	1.2043	69°
22°	.3840	.3746	.9272	.4040	2.475	1.079	2.669	1.1868	68°
23°	.4014	.3907	.9205	.4245	2.356	1.086	2.559	1.1694	67°
24°	.4189	.4067	.9135	.4452	2.246	1.095	2.459	1.1519	66°
25°	.4363	.4226	.9063	.4663	2.145	1.103	2.366	1.1345	65°
26°	.4538	.4384	.8988	.4877	2.050	1.113	2.281	1.1170	64°
27°	.4712	.4540	.8910	.5095	1.963	1.122	2.203	1.0996	63°
28°	.4887	.4695	.8829	.5317	1.881	1.133	2.130	1.0821	62°
29°	.5061	.4848	.8746	.5543	1.804	1.143	2.063	1.0647	61°
30°	.5236	.5000	.8660	.5774	1.732	1.155	2.000	1.0472	60°
31°	.5411	.5150	.8572	.6009	1.664	1.167	1.942	1.0297	59°
32°	.5585	.5299	.8480	.6249	1.600	1.179	1.887	1.0123	58°
33°	.5760	.5446	.8387	.6494	1.540	1.192	1.836	.9948	57°
34°	.5934	.5592	.8290	.6745	1.483	1.206	1.788	.9774	56°
35°	.6109	.5736	.8192	.7002	1.428	1.221	1.743	.9599	55°
36°	.6283	.5878	.8090	.7265	1.376	1.236	1.701	.9425	54°
37°	.6458	.6018	.7986	.7536	1.327	1.252	1.662	.9250	53°
38°	.6632	.6157	.7880	.7813	1.280	1.269	1.624	.9076	52°
39°	.6807	.6293	.7771	.8098	1.235	1.287	1.589	.8901	51°
40°	.6981	.6428	.7660	.8391	1.192	1.305	1.556	.8727	50°
41°	.7156	.6561	.7547	.8693	1.150	1.325	1.524	.8552	49°
42°	.7330	.6691	.7431	.9004	1.111	1.346	1.494	.8378	48°
43°	.7505	.6820	.7314	.9325	1.072	1.367	1.466	.8203	47°
44°	.7679	.6947	.7193	.9657	1.036	1.390	1.440	.8029	46°
45°	.7854	.7071	.7071	1.0000	1.000	1.414	1.414	.7854	45°
		Cos	Sin	Cot	Tan	Csc	Sec	Radians	Degrees

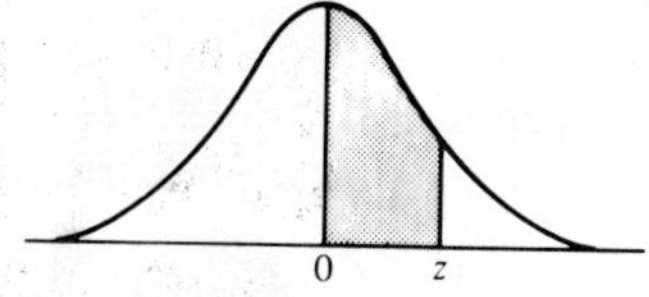

Table A-6 Normal Probability Curve

z	.00	.01	.02	.03	.04	.05	.06	.07	.08	.09
0.0	.0000	.0040	.0080	.0120	.0160	.0199	.0239	.0279	.0319	.0359
0.1	.0398	.0438	.0478	.0517	.0557	.0596	.0636	.0675	.0714	.0753
0.2	.0793	.0832	.0871	.0910	.0948	.0987	.1026	.1064	.1103	.1141
0.3	.1179	.1217	.1255	.1293	.1331	.1368	.1406	.1443	.1480	.1517
0.4	.1554	.1591	.1628	.1664	.1700	.1736	.1772	.1808	.1844	.1879
0.5	.1915	.1950	.1985	.2019	.2054	.2088	.2123	.2157	.2190	.2224
0.6	.2257	.2291	.2324	.2357	.2389	.2422	.2454	.2486	.2517	.2549
0.7	.2580	.2611	.2642	.2673	.2704	.2734	.2764	.2794	.2823	.2852
0.8	.2881	.2910	.2939	.2967	.2995	.3023	.3051	.3078	.3106	.3133
0.9	.3159	.3186	.3212	.3238	.3264	.3289	.3315	.3340	.3365	.3389
1.0	.3413	.3438	.3461	.3485	.3508	.3531	.3554	.3577	.3599	.3621
1.1	.3643	.3665	.3686	.3708	.3729	.3749	.3770	.3790	.3810	.3830
1.2	.3849	.3869	.3888	.3907	.3925	.3944	.3962	.3980	.3997	.4015
1.3	.4032	.4049	.4066	.4082	.4099	.4115	.4131	.4147	.4162	.4177
1.4	.4192	.4207	.4222	.4236	.4251	.4265	.4279	.4292	.4306	.4319
1.5	.4332	.4345	.4357	.4370	.4382	.4394	.4406	.4418	.4429	.4441
1.6	.4452	.4463	.4474	.4484	.4495	.4505	.4515	.4525	.4535	.4545
1.7	.4554	.4564	.4573	.4582	.4591	.4599	.4608	.4616	.4625	.4633
1.8	.4641	.4649	.4656	.4664	.4671	.4678	.4686	.4693	.4699	.4706
1.9	.4713	.4719	.4726	.4732	.4738	.4744	.4750	.4756	.4761	.4767
2.0	.4772	.4778	.4783	.4788	.4793	.4798	.4803	.4808	.4812	.4817
2.1	.4821	.4826	.4830	.4834	.4838	.4842	.4846	.4850	.4854	.4857
2.2	.4861	.4864	.4868	.4871	.4875	.4878	.4881	.4884	.4887	.4890
2.3	.4893	.4896	.4898	.4901	.4904	.4906	.4909	.4911	.4913	.4916
2.4	.4918	.4920	.4922	.4925	.4927	.4929	.4931	.4932	.4934	.4936
2.5	.4938	.4940	.4941	.4943	.4945	.4946	.4948	.4949	.4951	.4952
2.6	.4953	.4955	.4956	.4957	.4959	.4960	.4961	.4962	.4963	.4964
2.7	.4965	.4966	.4967	.4968	.4969	.4970	.4971	.4972	.4973	.4974
2.8	.4974	.4975	.4976	.4977	.4977	.4978	.4979	.4979	.4980	.4981
2.9	.4981	.4982	.4982	.4983	.4984	.4984	.4985	.4985	.4986	.4986
3.0	.4987	.4987	.4987	.4988	.4988	.4989	.4989	.4989	.4990	.4990

SELECTED ANSWERS

EXERCISE SET 1

1.1 a) $\{y \mid y \text{ is an ancestor of } x\}$ c) $\{x \mid 0°\text{C} \leq x \leq 100°\text{C}\}$
e) $\{x \mid x = 5n, \text{ for } n \in I\}\{0, \pm 5, \pm 10, \pm 15, \ldots\}$

1.2 a) $[0, 8]$ c) $\{x \mid 2 < x < 4\} = (2, 4)$ e) $(3, +\infty)$

1.3 a) $\{1, 2\}$ c) $\varnothing$ or $\{\ \}$ e) $\{1, 2\}$

1.5 a) Domain $= \{x \in I \mid 1 \leq x \leq 10\}$, Range $= \{x \in I \mid 1 \leq x \leq 15\}$
c) Domain $= I$, Range $= \{0, 1\}$
e) Domain $= \{0, 1, 2, 3\}$, Range $= \{0, 1, 2\}$

1.6 Example 1.7 III: Domain $= \{A, B, C\}$, Range $= \{a, c\}$
Example 1.7 VI: Domain $= \{A, B, C\}$, Range $= \{a, b, c\}$

1.7 Example 1.8 a) Domain = Range = R c) Domain $= N$, Range $= \{0, 1\}$

1.9 a) and b) are graphs of functions; c) and d) are not.

1.11 b), d), f), g)

1.12 a) $[3, 4]$ c) ϕ e) $(2, 4)$

1.13 a) Domain $= \{0, \frac{1}{2}, 1, 1\frac{1}{2}, 2\}$, Range $= \{0, \frac{1}{4}, \frac{1}{2}, \frac{3}{4}, 1\}$

EXERCISE SET 2

2.1 a) $\text{Dom}(f) = \{1, 2, 3, 5\}$, $\text{Dom}(g) = \{3, 5, 6\}$, $\text{Dom}(f + g) = \{3, 5\}$
c) $\text{Dom}(f) = \text{Dom}(g) = \text{Dom}(f + g) = [0, 3]$

2.2 a) $f + g = \{(3, 0), (5, 6)\}, f \cdot g = \{(3, -1), (5, 0)\}$

2.4 a) $-f = \{(1, -2), (2, -3), (3, 1), (5, -6)\}$, $1/g = \{(3, 1), (6, 1)\}$

2.5 a) $x^2 + 3x - 1$ c) $3x^{\frac{3}{2}} + x^{\frac{1}{2}}$ d) $3x^3 + x^2 - 6x - 2$
e) $2 - x^2$ g) $1/\sqrt{x}$

2.6 b) $\bar{u} = (A^{0.5}E)/(0.291(e_s - e_a))$

2.7 a) $E(3) = \frac{1}{2}$ $g(3) = \frac{2}{3}$ $d(3) = \frac{1}{6}$
b) $d(n) = (2/n) - (1/2)$
c) $A(2) = 2, A(3) = \frac{3}{2}, A(4) = 1, A(5) = \frac{1}{2}, A(6) = 0$

2.8 b) $b(n) = 2 + 2n$ $d(n) = n$ $\Delta(n) = 2 + n$
$b(1) = 4$ $d(1) = 1$ $\Delta(1) = 3$
$b(2) = 6$ $d(2) = 2$ $\Delta(2) = 4$
$b(3) = 8$ $d(3) = 3$ $\Delta(3) = 5$
c) Population will be 202 at generation 12.

2.9 If L is the side length of the cube then $0 < L < 6$.

2.13 a) $f(2) = 8$ c) $g(2) = 4$
e) $f(2 + x) = 3x + 8$ g) $g(x + 2) = x^2 + 6x + 4$
i) $f(x + h) = 3x + 3h + 2$ k) $f(2x) = 6x + 2$

2.15 a) $\{x|x \neq -\frac{1}{2}\}$ c) $\{x|x \neq 2 \quad \text{and} \quad x \neq -1\}$
e) $\{x|x > 0 \quad \text{and} \quad x \neq 2\}$

EXERCISE SET 3

3.1 For any x_1 and $x_2 \in (0, +\infty) x_1 < x_2$ implies $x_1{}^2 < x_2{}^2$ or $f_2(x_1) < f_2(x_2)$.

3.2 Decreasing on $(-5, 0)$ and increasing on $(0, 5)$. Hence neither on $(-5, 5)$.

3.3 $f(-x) = (-x)^2 = x^2 = f(x)$ for all x.

3.7 a) degree 1 c) degree 1 e) degree 2

3.8 a) Slope $= 5$, y-int. $= -5$, equation: $y = 5x - 5$
c) Slope $= 0$, y-int. $= 1$, equation: $y = 1$

3.9 a) $y = 6x - 15$ c) $y = (-\frac{1}{6})x + \frac{7}{2}$

3.12 $\overline{N} = 10^{12}$

3.13 $m_4 < m_3 < 0 < m_2 < m_1$ and $m_4 = -m_1$

3.16 a) $y = (x - 3)^2 + 1$ c) $y = (x - 2)^2 - 1$ e) $y = -\frac{1}{2}(x + \frac{1}{4})^2 + \frac{1}{32}$
g) $y = (x - \frac{1}{2})^2 - \frac{1}{4}$

3.17 a) $y = x^3 - 1$ c) $y = x^3 - 2x + 3$ e) $y = -(x + \frac{1}{3})^3 + \frac{4}{3}x + \frac{1}{27}$

3.20 a) $s(n) = N - A(n)$
d) after day $n = 7$, $A(n) = 28$, $S(n) = 72$, and $\Delta(n) = 7$ Note: $\Delta(-1) = 2$

3.21 a) $R([x]) = 0.5(0.3 - [x])(0.7 - [x])$ c) $R_{\max} = R(0) = R(1) = 0.105$

EXERCISE SET 4

4.1 a) $(f \circ g) = \{(2, 4), (4, 4), (-1, 2)\}$, $(g \circ f) = \{(1, 3), (2, 3), (3, 3), (4, 1)\}$
c) $(f \circ g)(x) = (x - 3)^2 + 2$, $(g \circ f)(x) = x^2 - 1$
e) $(f \circ g)(x) = (x + 1)^3 - 1$, $(g \circ f)(x) = x^3$

4.2 a) $(f \circ g)(x) = (2x - 1)^3$, $(g \circ f)(x) = 2x^3 - 1$
c) $(f \circ g)^{-1}(x) = \left(\dfrac{x^{\frac{1}{3}} + 1}{2}\right)$, $(g \circ f)^{-1}(x) = \left(\dfrac{x + 1}{2}\right)^{\frac{1}{3}}$

4.3 a) $B_g = mI + b$; m, b constants

$$U_g = \begin{cases} 0 & \text{if} \quad B_g < M \\ K(B_g - M) & \text{if} \quad B_g \geq M \end{cases}$$

4.4 $F = (10^{-2}F_8{}^3)^2 = 10^{-4}F_8{}^6$

4.5 a) $W(t) = -\frac{1}{25}t^2 + 2t$

4.6 a) $D = \{2, 4, -1\}$ c) $D = (-\infty, +\infty)$ e) $D = (-\infty, +\infty) = R$

4.7 b), d), and e) have inverses, a) does not; c) and f) are not graphs of functions

4.8 a) $y = \frac{1}{2}x$ c) $y = (x/2)^{\frac{1}{2}}$ e) $y = (x/2)^{\frac{1}{3}}$
g) $y = (x + 2)/3$ i) $y = 1/x$

4.9 d) and f) do not have inverses

4.10 a) $f(x) = x + 2, g(x) = x^2$ c) $f(x) = x^5, g(x) = x - 3$
e) $f(x) = x + 1/x, g(x) = x^2$

4.11 a) 2 c) -1 e) $\frac{1}{9}$ g) not defined

4.12 a) 2 c) $3^{\frac{1}{3}}$ e) $(0.6)^{\frac{1}{2}}$ g) $(-1)^2$ i) $(-32)^{\frac{2}{5}}$

4.13 $64h^3 - 400h^2 + 832h - 576$

4.14 a) $F(O_2) = -0.04(O_2 - 20)^2 + 16$ $O_2(x) = -0.1(x - 10)^2 + 10$

4.16 a) $A = (0.1)(64)^{\frac{3}{4}}(81)^{\frac{1}{4}} \simeq (0.1)(22.64)(3) \simeq 6.8$
c) $H = (10A)^{\frac{4}{3}}w^{-1/3}$

EXERCISE SET 5

5.1 a) 2^5 c) $\sqrt{10^x} > e^x$ for $x > 0$; $e^x > \sqrt{10^x}$ for $x < 0$. e) 3^5
g) 3 i) log 5

5.2 a) 2 days is 144 generation; $N(48) = 3000 \times 2^{144}$ c) $3000(1.6)^{144}$

5.4 $n(10) = 400(\frac{1}{2})^{\frac{2}{3}}, n(20) = 400(\frac{1}{2})^{\frac{4}{3}}$

5.5 a) $k = \ln(\frac{5}{2})/3$ c) $N_0 = (25)(2.5)^{\frac{2}{3}}$

5.6 a) $t_{\frac{1}{4}} = -\ln(\frac{3}{4})/k$

5.7 a) $t = -50\ln(\frac{25}{37}) \simeq 19.6$ min

5.8 a) $A(t) = A_0 e^{-\ln(2)t/8}; \dfrac{(A(t) - A(t+1))}{A(t)} = 1 - e^{-\ln(2)/8} \simeq 8.3\%$ c) 0.415 units

5.10 a) $-\ln(3.5)/2$ c) $\ln(7)/3$ e) $1/\log(5)$ g) $\frac{9}{5}$ i) $3(10^{-7})$

5.11 $\varepsilon \simeq 0.25$

5.12 $\varepsilon = (1 - \log(1.6))/0.36 \simeq 2.21$; using a 0.18 *molar* solution.

5.13 a) $k = (\log(10.5))/3 \simeq 0.34040$

5.15 $pH = 2$

5.16 b) $K_a = 10^{-7}(1.88)$

5.19 a) 1.0986 c) 1 e) -3.3222 g) 0.33 i) 2.2146

5.20 a) 649 c) 0.000439

5.21 a) 650 c) 649

5.22 b) $L = 10 \log 3 + 30 = 34.771DB.$

5.23 b) $I(7.2)/I(4.6) = 10^{2.6} \simeq 398$

EXERCISE SET 6

6.2

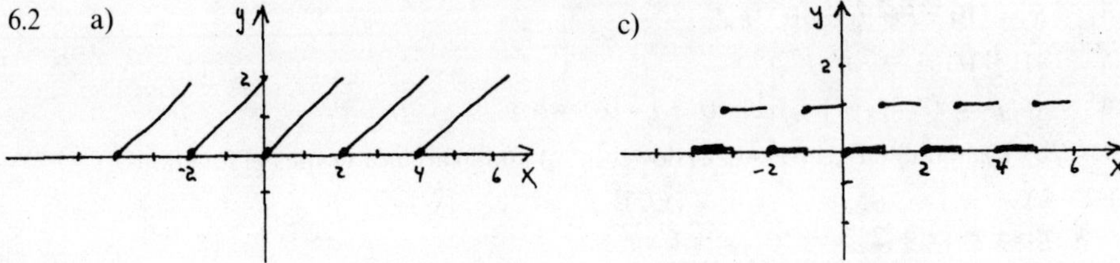

e)

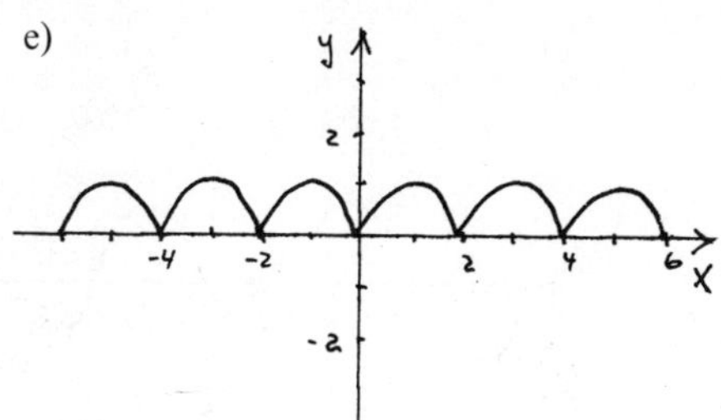

6.3 a)

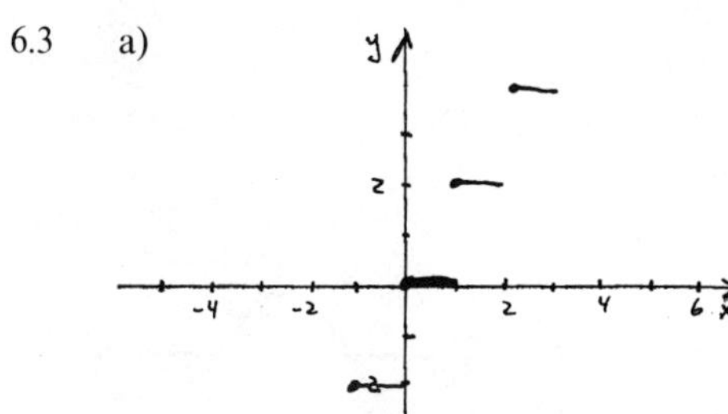

c)

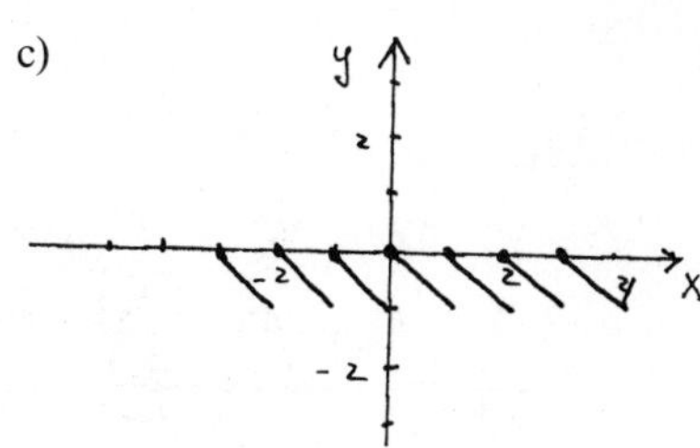

e)

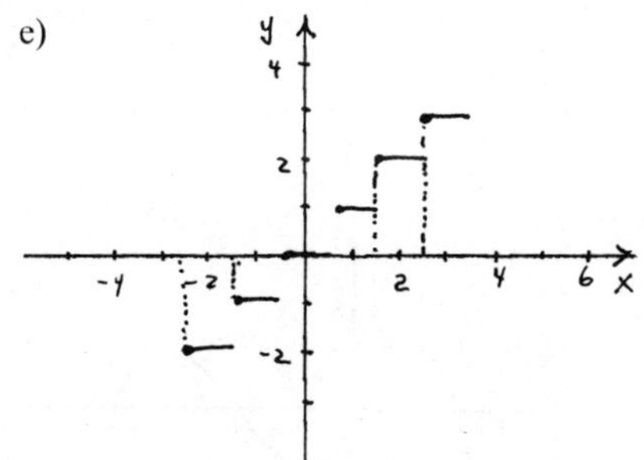
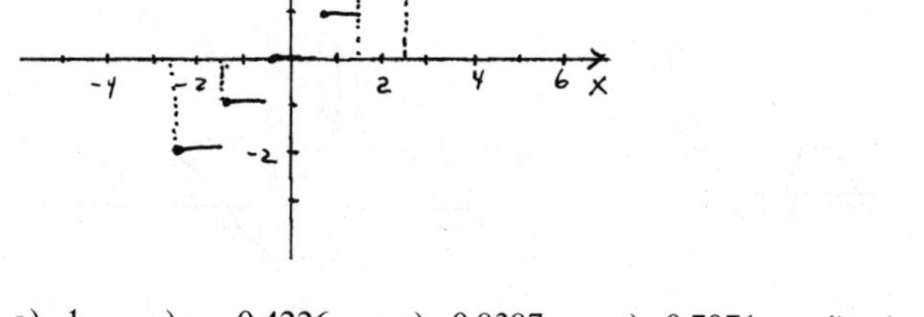

6.4 a) 1 c) -0.4226 e) 0.8387 g) 0.7071 i) 1 k) 1.1547

6.5 a)

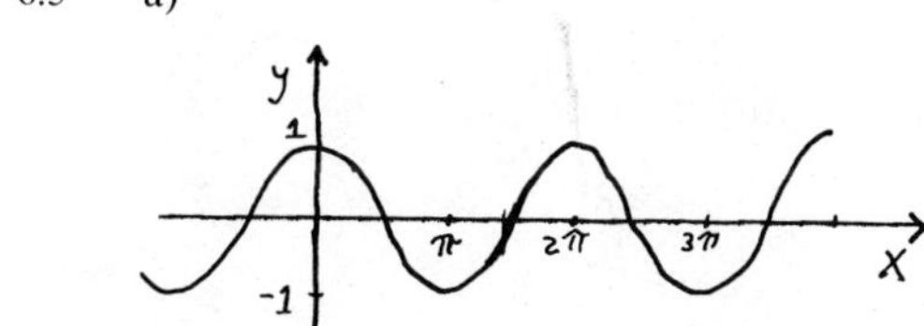

c)

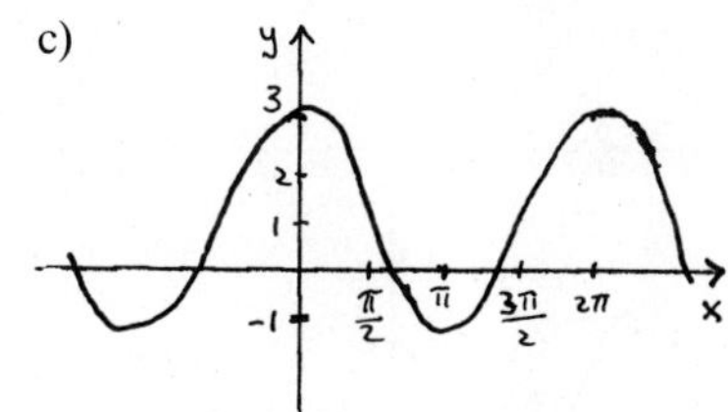

e)

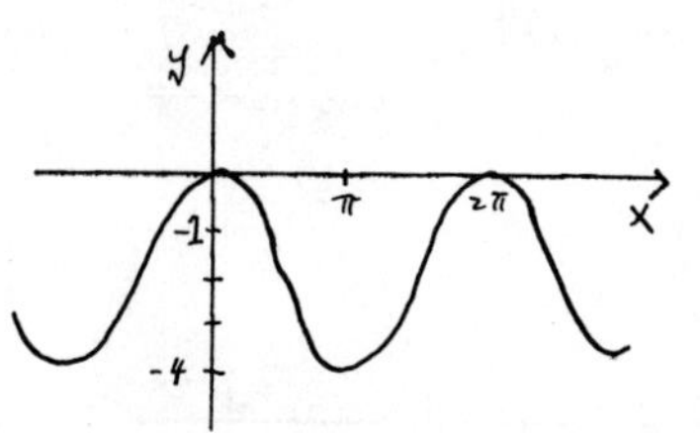

g)

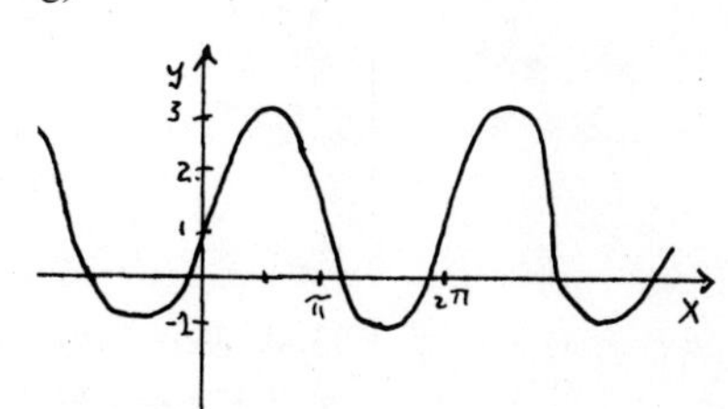

6.6 a)

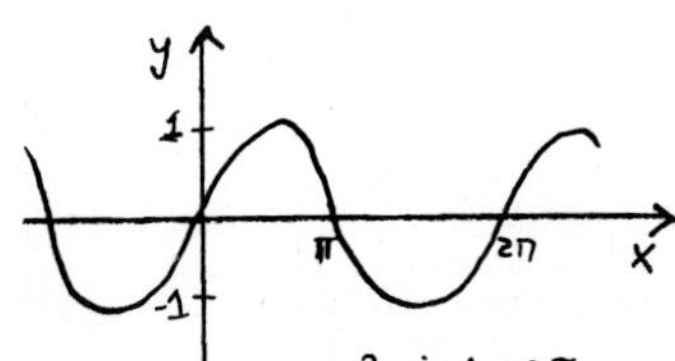

c)

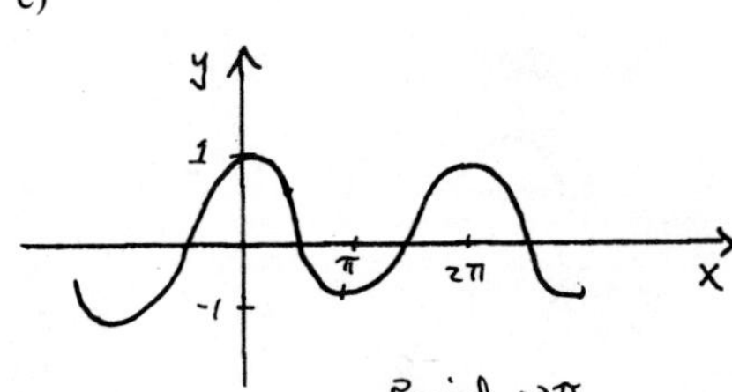

e)

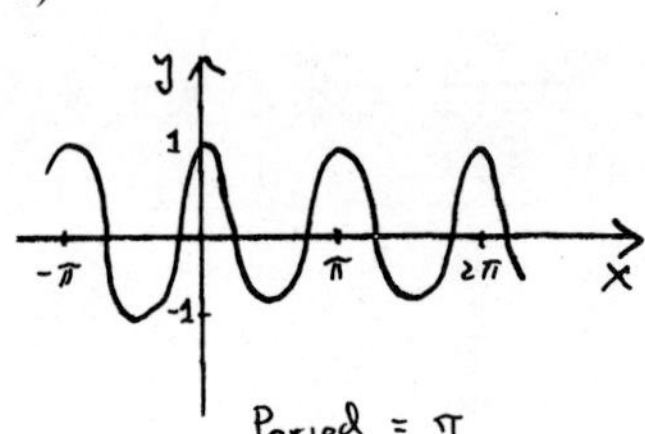

g)

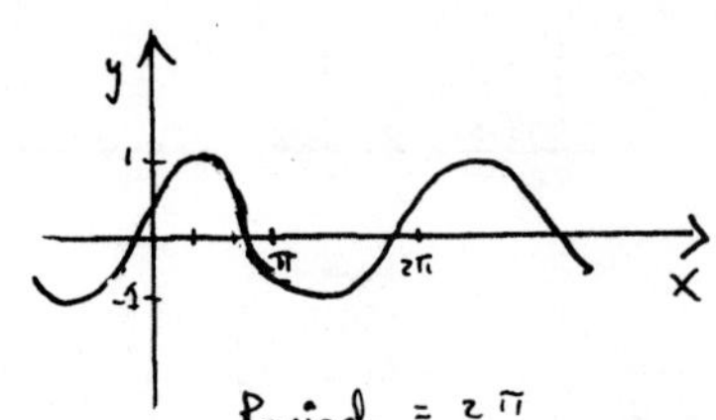

i)

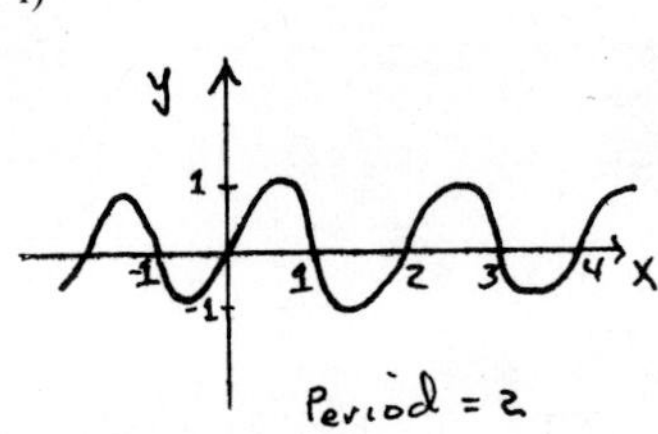

k)

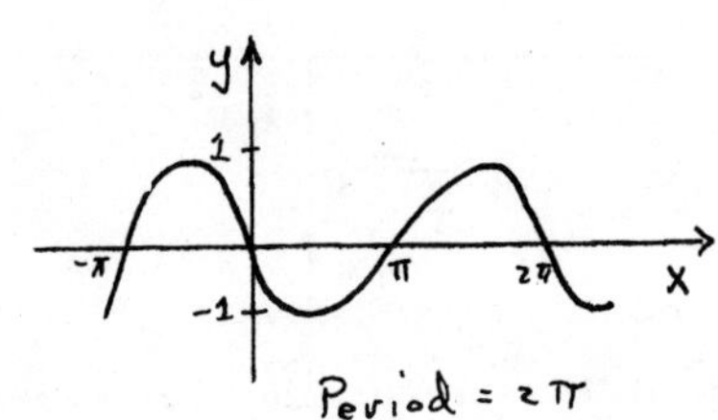

m)

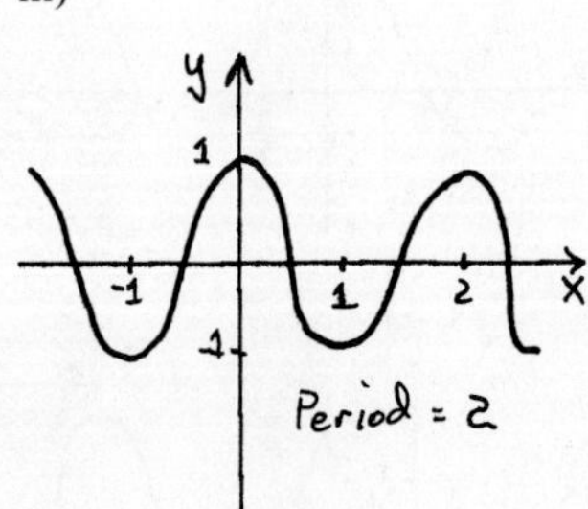

o)

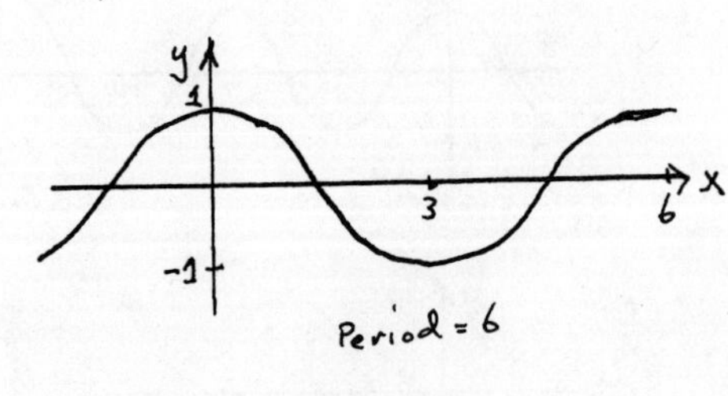

6.7 a) $\cos(x^2 + 1)$; 0.5446 c) $\sin(x - \pi)$; 0
e) $\cos^2(x - \pi) + 1$; 1 g) $\cos^2(x) + 1$; 2 i) $\sin(\cos(x))$; 0

6.8 a)

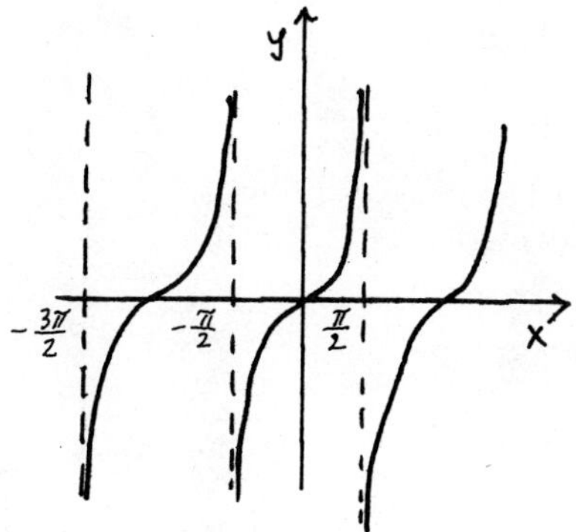

c)

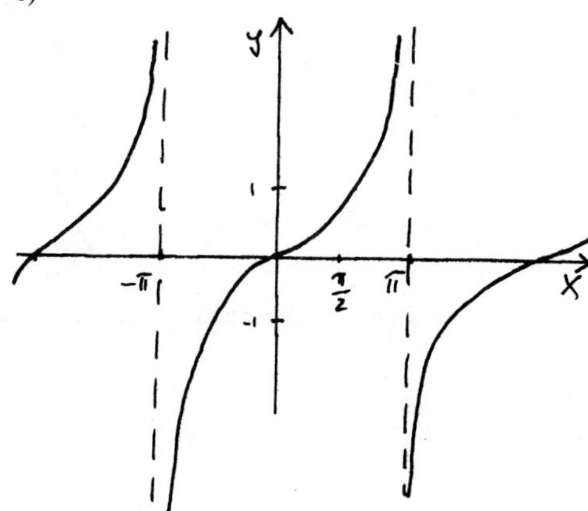

e)

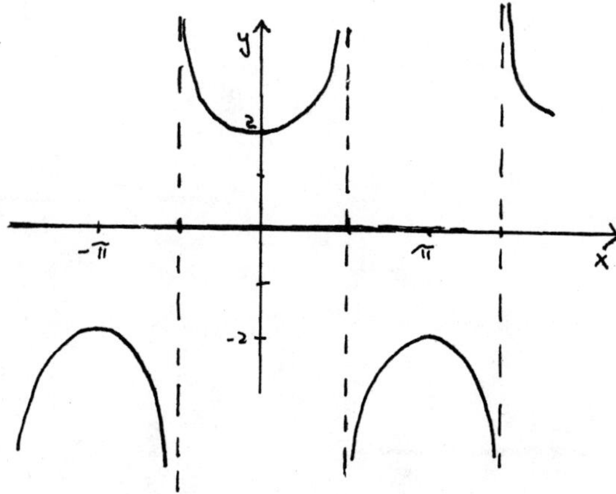

6.9 b) $v(t) = \begin{cases} a\sin((\pi/3)t), & \text{if} \quad 0 \le t < 3 \\ 0, & \text{if} \quad 3 \le t < 5 \end{cases}$

6.10 a) $\dfrac{\sqrt{4 - x^2}}{2}$ c) $\dfrac{x}{\sqrt{16 - x^2}}$

6.15 a) Fundamental tone: $y = 2\sin(t)$, second overtone: $y = 2\sin(3t)$
First overtone: $y = 2\sin(2t)$, third overtone: $y = 2\sin(4t)$

EXERCISE SET 7

7.1 a) c)

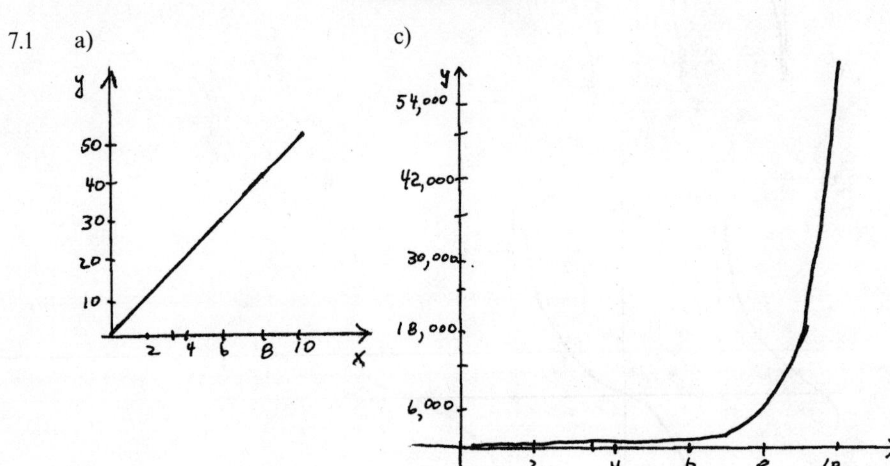

e) g)

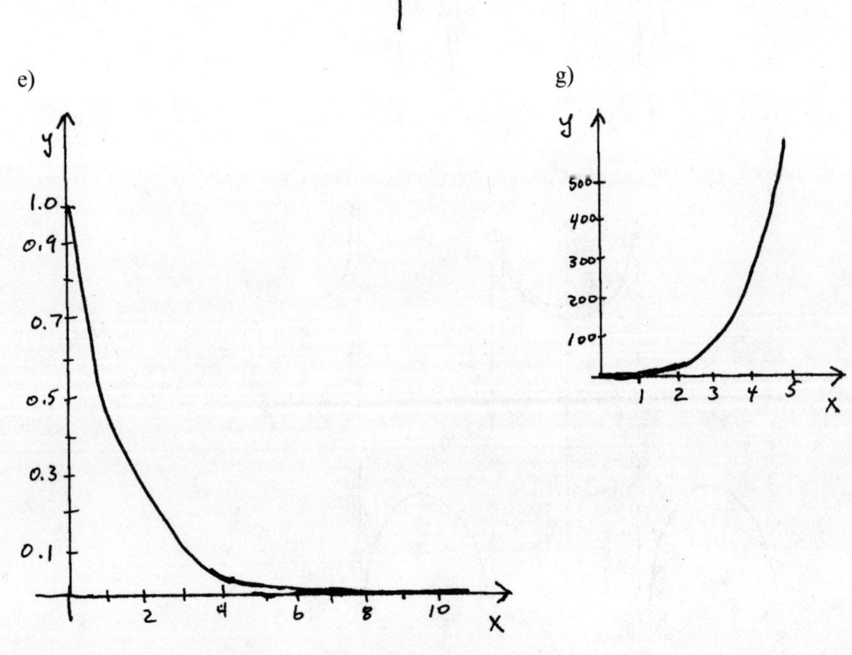

i)

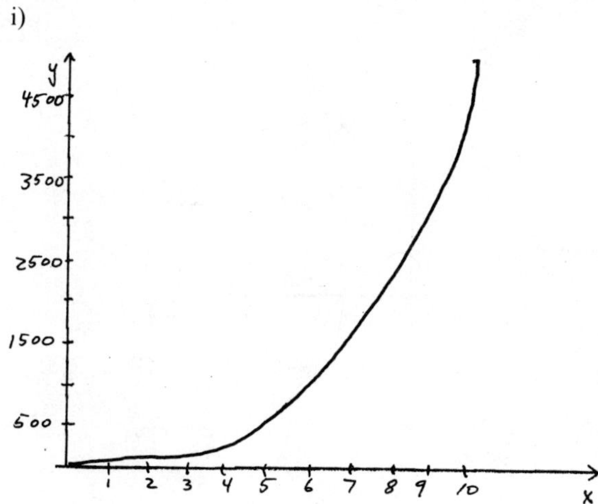

7.2 a) Using points (0.5, 20) and (1, 40): $a = 40$, $b = 0$
c) Using the point (2, 33): $a = \log_2(33) = \log(33)/\log(2) \simeq 5.044$

7.4 d) i) $u = x/6$, $v = y/5$ iii) $u = x/0.2$, $v = y/2.6$

7.7 a) (2, 3) c) (15,3) e) (40, 6) g) (1.5, 8)
i) (70, 2.5) k) (3, 1)

7.9 a) (0, 2) c) (2, 5) e) (2.5, 1.5) g) (3.75, 4) i) (6,10)
k) (4.5, 1)

7.11 a) $\log(y) = \log(2)x$; semi-log
c) $\log(y) = 2\log(e)x + \log(3)$; semi-log
e) $\log(y) = 5\log(x) + \log(4)$; log–log
g) $\log(x) + \log(y) = \log(4)$; log–log
i) $\log(y) = \log(e)x + \log(2) - 2\log(e)$; semi-log
k) $\log(y) = \frac{1}{2}\log(u)$, $u = x + 2$; log–log

7.12 a)

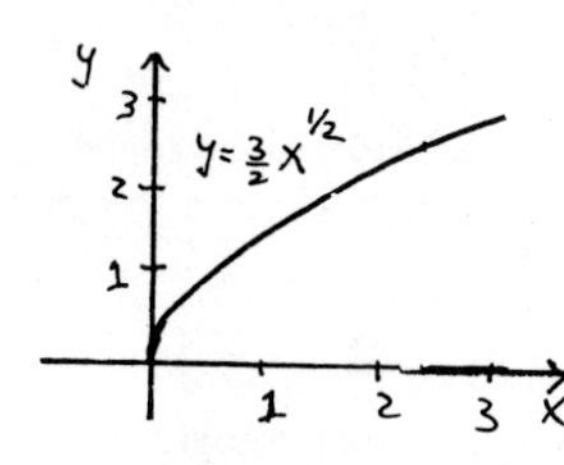

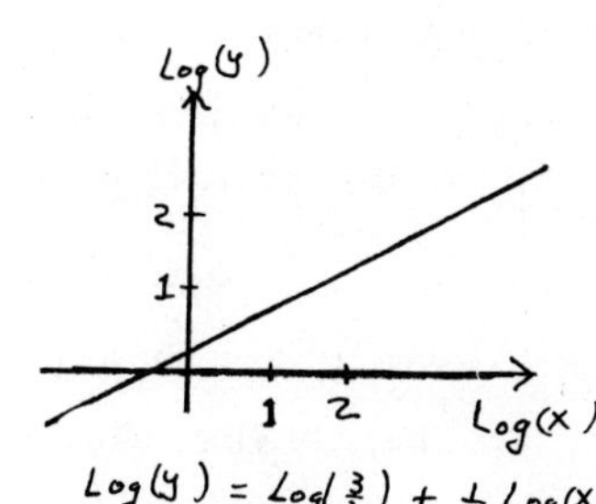

c)

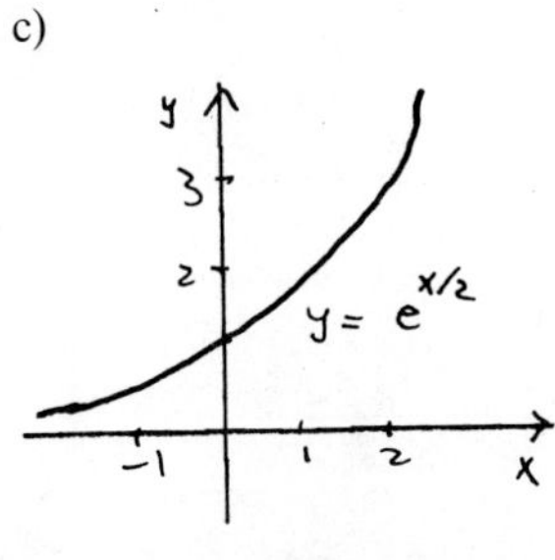

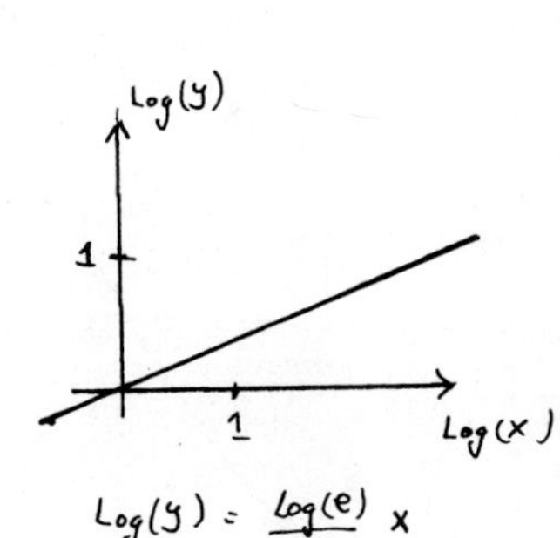

e)

$y = \frac{4}{x^2}$

Log(y) = log(4) − 2 log(x)

or

(Log(y) = Log(4) − 2 Log(−x))

Log(4)

Log(x) (or Log(−x))

g)

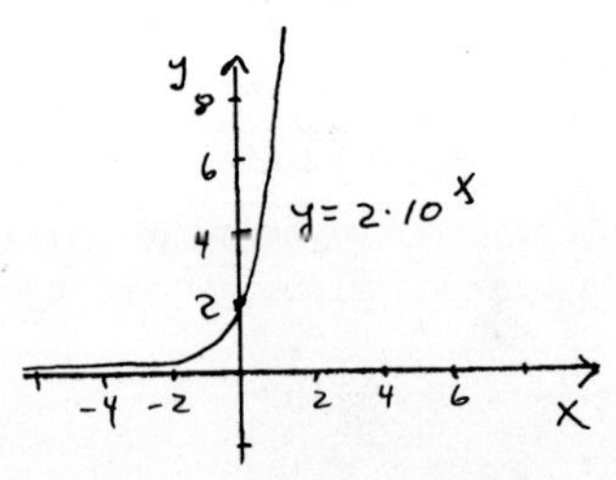

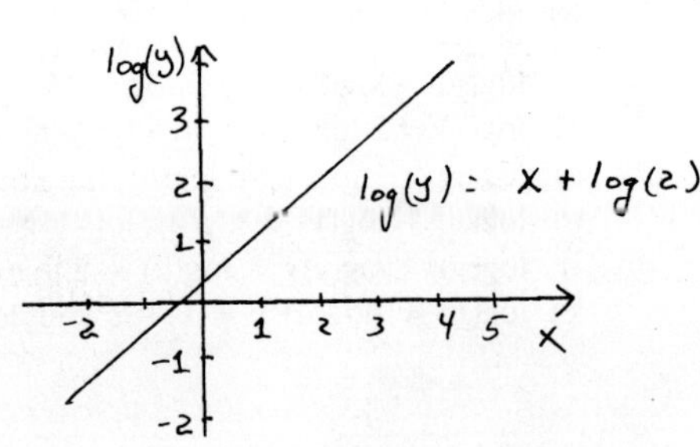

7.13 a) Points (log(3), log(7)), (log(1), log(1.5)):

$$y = \tfrac{3}{2}(x)^{\log(7/1.5)/\log(3)} = \tfrac{3}{2}x^{1.402}$$

c) Points (log(10), log(6)), (log(1), log(2)):

$$y = 2x^{\log(3)} = 2x^{0.4771}$$

e) Points (log(100), log(2)), (log(200), log(5)):

$$\log(y) = [\log(\tfrac{5}{2})/\log(2)][\log(x) - \log(100)] + \log(2)$$
$$y = 2[x/100]^{1.3219}$$

7.14 a) Points (0, log(2)), (2, log(5)):

$$\log(y) = \tfrac{1}{2}\log(\tfrac{5}{2})x + \log(2),$$
$$y = 2e^{(0.4581)x};\ 0.4581 \simeq 0.5\ln(2.5)$$

c) Points (0.5, log(1.5)), (0, log(30)):

$$y = 30e^{-2\cdot\ln(20)x} = 30e^{-(5.9915)x}$$

e) Points (3, log(2)), (4.5, log(40)).

$$y = (\tfrac{1}{200})e^{[\ln(20)/1.5]x} = 0.005e^{1.9972x}$$

7.15 a) $k = 10^{y/c}/x$ c) $y = x^{a\ln(10)}$ e) $y = 10^{3x+7}$

7.17 Using points (0.96, 1.7), (−0.84, 0.5),

$$\log(R) = \tfrac{2}{3}\log(S) + 1.06,\ R = 11.5S^{\frac{2}{3}}$$

7.19

a)

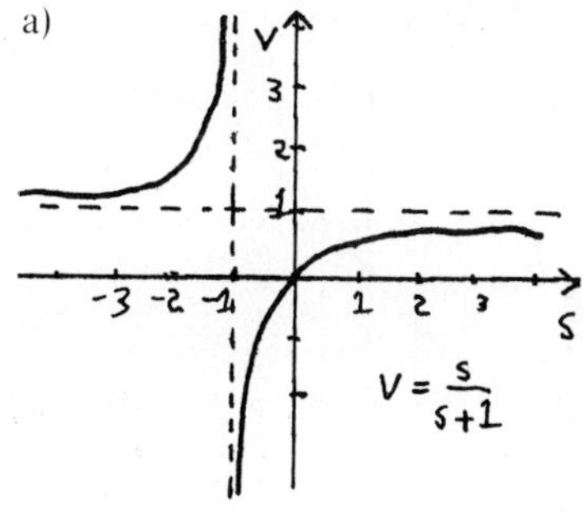

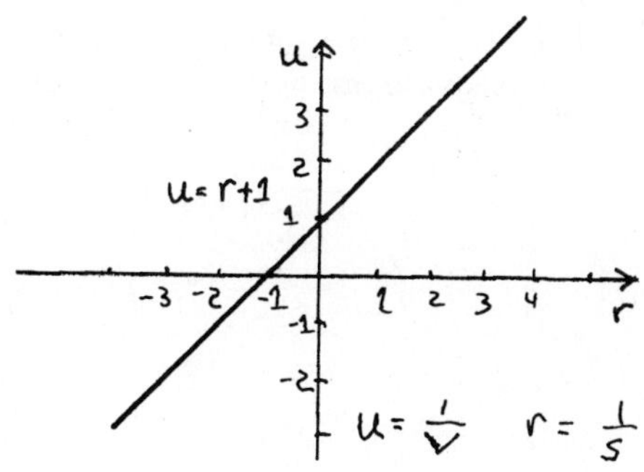

EXERCISE SET 8

8.1 The graphs should pass through the points indicated in the following table:

t	a) y	c) y	e) y	g) y
−1.0	0.18	0.74	7.39	0.90
−0.5	0.30	1.21	2.72	0.95
0	0.5	2.0	1.0	1.0
0.5	0.82	3.30	0.37	1.05
1.0	1.36	5.44	0.14	1.11

8.4 a) $y = e^{-2(t-\ln(\sqrt{3}))}$ c) $y = e^{5(t+\ln(3)/5)}$
e) $y = e^{t-(2-\ln(3))}$ f) $y = e^{2(t-\ln(3/\sqrt{5})}$

8.5 a) $y = 10/(1 + 2e^{-t})$
$B = 10,\ c = 2,\ \lambda = \frac{1}{10}$
y-intercept $= B/(1 + c) = 10/3$
$t_{max} = \ln(c)/\lambda B = \ln(2) = 0.69$
occurs at $y = B/2 = 5$

8.6 a) $y = 10e^{-e^{-t}}$
$c = 10,\ k = 1,\ \lambda = 1$
y-intercept $y_0 = 10e^{-1} \simeq 3.68$
max increase at $f(t_{max}) = c/e = 10/e = 10e^{-1}$
hence $t_{max} = 0$

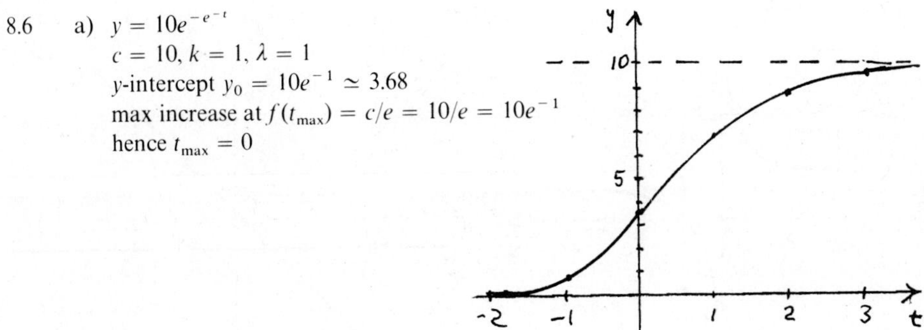

8.8 $t_{max} = 1/\lambda \ln(1/k)$: a) 0 b) -0.69 d) 0.69

8.10 a) $t_{max} = 1,\ t_i = 2$ c) $t_{max} = 2,\ t_i = 4$ e) $t_{max} = 2,\ t_i = 4$

8.11 a) $y = (\frac{10}{6} \cdot e)te^{-t/6} = \frac{10}{6}te^{1-t/6}$ c) $y = 400te^{1-5t}$

8.12 a) $y = 50(e^{-t} - e^{-2t})$
$t_{max} = \ln(2) \simeq 0.69$
$y_{max} = 12\frac{1}{2}$

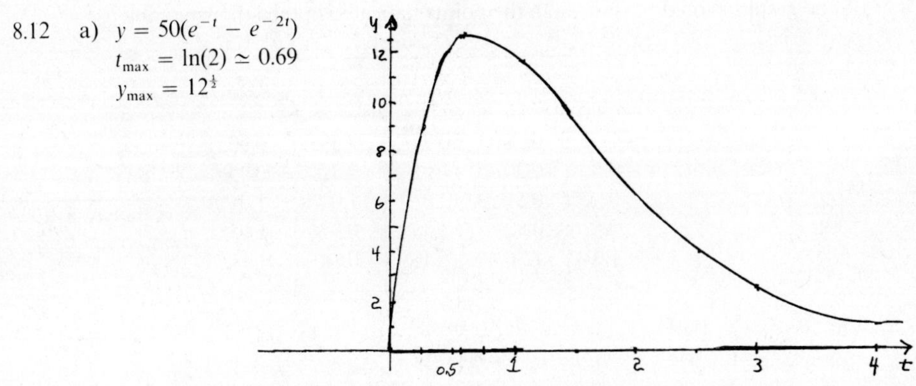

8.13 The graph should pass through the points.
a) (0, 0), (0.25, 8.61), (0.5, 11.93), (ln(2), 12.5) max pt, (1, 11.62), (2, 5.85), (3, 2.36), (5, 0.33)
e) (0, 0), (1, 20.55)($\frac{4}{3}$ln(4), 23.62) max pt, (2, 23.56), (3, 21.13), (5, 13.99), (8, 6.75), (12, 2.48), (16, 0.92)

8.14 a) $t_{max} = 0.69$ $t_i = 1.39$ c) $t_{max} = 1.39$ $t_i = 2.77$
e) $t_{max} = 1.85$ $t_i = 3.70$

8.15 a) The graph decreases initially and then increases and approaches the t-axis asymptotically.

8.16 a) $c = (m/N(0)) - 1$ c) $\lambda = \ln(4)/5000 \simeq 0.00028$

8.17 $y = B/(1 + e^{-\lambda Bt})$

8.18 a) $B \simeq 13.62$ $\lambda \simeq 0.08$

8.21 a) $N(t) = 13/(1 + 2\,9e^{-0.22t})$

EXERCISE SET 9

9.1 a) 1 c) $x_5 = 3.1$ e) $t_5 = 21\pi/2$ g) $b_5 = 55$

9.2 a) 1 c) $x_{25} = 3.5$ e) $t_{25} = 101\pi/2$ g) $b_{25} = 5525$

9.3 a) {1, 3, 7, 15, 31, 63} c) {log(1), log(2), . . . , log(6)}
e) $\{\sqrt{2}, \sqrt{5}, 2\sqrt{2}, \sqrt{11}, \sqrt{14}, \sqrt{17}\}$ g) {3, 1, 3, 1, 3, 1}

9.4 a) Finite, $N = 20$, $a_{19} = 19$, $a_{20} = 20$
c) Infinite, $S_1 = 1$, $S_{10} = 55$
e) Finite, $N = 3$, $a_2 = (3, 4)$, $a_3 = (5, 6)$

9.5 a) Arithmetic; $a_n = 1 + (n - 1)$ c) Neither
e) Arithmetic; $a_n = 100 + (n - 1)(-5)$ g) Geometric; $g_n = 1.75(0.25)^{n-1}$

9.6 a) {5, 8, 11, 14, 17, 20, . . .} c) $\{\frac{3}{2}, \frac{3}{4}, \frac{3}{8}, \frac{3}{16}, \frac{3}{32}, \frac{3}{64}, \ldots\}$
e) {0, 2, 0, 2, 0, 2, . . .} g) {−1, 1, −1, 1, −1, 1, . . .}

9.7 a) Yes; $A_n = -n$ c) $f(x) = A + d(x - 1)$

9.8 a) Yes; $g_n = (-1)^n$
c) $y = f(x)$, $y = g(x)$, where $f(x) = gr^{x-1}$ and $g(x) = \begin{cases} g_x, & \text{if} \quad x \in N \\ 0, & \text{if} \quad x \notin N \end{cases}$

9.9 a) {0.8415, 0.9093, 0.1411, −0.7568, −0.9589, −0.2794}
c) e) and g) {−1, 1, −1, 1, −1, 1}

9.10 a) 3 c) 6 e) 2

9.11 a) {2, 1, 4, 4} c) {−1, 5, 2, 6} e) {4, 1, 7, 6} g) {−1, 1, 2, 2}
i) {5, 17, 8, 16}

9.12 a) {95, 90, 85, 80, 75} c) {3, 1, 4, 5, 9}
e) {1.9, 3.6, 6.8, 12.8, 24} i) {98.02, 91.04, 89.08, 85.16, 84.32}

9.13 a) $+\infty$ b) 0 c) $+\infty$ d) 0 e) 0 f) does not exist
g) 0 h) 0 i) does not exist j) 1

9.14 a) $\lim_{n\to\infty} A_n = \infty$

9.15 a) 1 c) 0 e) -1

9.16 a) two roots c) infinite e) 18 roots

9.17 $\{1, 1, \frac{3}{4}, \frac{5}{8}, \frac{1}{2}, \frac{13}{32}, \frac{21}{64}, \frac{17}{64}, \frac{55}{256}, \frac{89}{512}\}$

9.18 $\{1, \frac{3}{4}, \frac{5}{6}, \frac{4}{5}, \frac{13}{16}, \frac{21}{26}, \frac{17}{21}, \frac{55}{68}, \frac{89}{110}, \frac{72}{89}\}$

9.19 $\{1, 0, -\frac{1}{3}, 0, \frac{1}{5}, 0, -\frac{1}{7}, 0, \frac{1}{9}, 0\}$ converges to zero

9.20 a) length is 13 and $a_7 = \{T\}$

EXERCISE SET 10

10.1 a) 27 c) 715 e) 35

10.2 a) $\frac{1}{2} + \frac{1}{3} + \cdots + \frac{1}{21}$ c) $x_1 {*} \Delta(1) + x_2 {*} \Delta(2) + \cdots + x_N {*} \Delta(N)$
e) $2 + 2 + 2 + 2 + 2 + 2 + 2 + 2$ g) $t_1 + t_2 + t_3 + t_4$

10.3 a) $\sum_{i=1}^{5} i^2$ c) $\sum_{i=1}^{5} 3 + 2i$ e) $\sum_{i=1}^{4} (ai + 2 + i)$ g) $\sum_{i=5}^{8} (i - (i - 1))$

10.4 a) 210 c) 9455 e) 1365 g) 37355 i) 10415

10.5 a) 3069 c) $2(1 - (-\frac{1}{2})^{10})$ e) $a^0 + a^1 + a^2 + \cdots + a^{10} = (1 - a^{11})/(1 - a)$

10.6 a) 381 c) -4.860032 e) 271

10.7 a) $s(10) = 10$ c) $s(10) = 2046$, $s(100) = 2(2^{100} - 1)$
e) $s(n) = (n + 3)/2$

10.8 I, II, III, and IV are not convergent. V is convergent when $d = 0$.
VI is convergent when $-1 < r < 1$.

10.9 a) $S_{10} = 385$ c) $S(10) \simeq 2.93$, diverges
e) $S_{10} \simeq -0.3330078$, converges to $-0.333\ldots$.

10.11 $2147483647 \simeq 2.1 \times 10^9$

10.12 $\frac{635}{32}$ grams (approximately 19.84 g.)

10.14 $P(N) = 102 - 2N + (N(N + 1)(2N + 1)/6)$

10.15 For special case $n = 1$, closed form is $a_N - a_0$.
In general case, closed form is $a_N - a_{n-1}$.

10.16 a) 10.8125 mg

EXERCISE SET 11

11.1 a) 1, linear c) 2, nonlinear e) 2, linear g) 0, not a difference equation

11.2 a) 65 c) $(756030)^2 - 756030 \simeq 5.7 \times 10^{11}$

11.3 a) 19 c) 1 e) 54

11.4 a) $x_n = x_0 3^n + 1 - 3^n$ c) $x_n = x_0(0.5)^n + 2(1 - (0.5)^n)$
e) $x_n = x_0 + 5n$ g) $x_n = 6g + (x_0 - 6)(\frac{1}{2})^n$.

11.5 a) $1 + 4(3^{10})$ c) $2 + 3(\frac{1}{2})^{10}$ e) 55 g) $6 - (\frac{1}{2})^{10}$

11.6 a) $X_{n+1} = X_n + c$ c) $X_{n+1} = X_n + \Delta_0 r^n$
e) $X_{n+1} = X_n + f(X_{n-t})$ g) $X_{n+1} = (p + 1)X_n$

11.7 a) $\Delta(X_n) = X_{n-2} + 1 - X_n$ c) $\Delta(X_n) = X_{n-1} + X_n(2h - 1)$
e) $\Delta(X_n) = 3(n-3)^2 X_{n-2} + X_{n-3} - X_n$
g) $\Delta(X_n) = X_n{}^2 - X_n - 2(X_{n-2})^2$

11.8 a) $X_n = X_0 + nc$ c) $X_n = X_0 + \Delta_0((1 - r^n)/(1 - r))$

11.9 a) $X_0 = 1000(\frac{2}{3})^{10}$ c) the first $n \geq \log(2)/\log(k) + 1$

11.10 a) $y_n = 5(3^n) + 5$ c) $X_n = 70(2^n) + 30$ e) $s_n = [(2 + h)/(2 - h)]^n$

11.11 a) $y_{n+1} = y_n(1 + ah) + bh$

11.13 Let X_n be the number of infectives on day n. The number of new infectives is $\Delta X_n = 0.2X_n$. Thus $X_n = X_0(1.2)^n$.

11.14 a) $x_0 = 5$, $\bar{X} = 5$ if $X_0 = 5$, $+\infty$ if $x_0 > 5$

11.15 a) $-\infty$ c) -20 e) 1 g) $+\infty$

11.16 c) 2 g) 6

11.17 $X_n = 8500(0.5)^n + 1500$

11.18 50

11.19 a) 2.2360 c) 3 e) oscillates

11.20 a) 2.0 c) $10^{\frac{1}{3}} \simeq 2.154$ e) $(-4)^{\frac{1}{3}} \simeq -1.587$

11.21 a) 1 c) $x_1 = 1.9, x_2 = 3.439, \ldots, x_5 = 9.88209815.$
e) $x_5 = 17.70622918$

11.22 $X_n = A + (x_0 - A)(1 - k)^n$

11.23 a) $X_n = c_1(-3 + 2\sqrt{3})^n + c_2(-3 - 2\sqrt{3})^n$

c) $X_n = c_1\left[\frac{5 + \sqrt{21}}{2}\right]^n + c_2\left[\frac{5 - \sqrt{21}}{2}\right]^n$

e) $X_n = c_1(-\frac{1}{4})^n + c_2$

g) $X_n = c_1\left[\frac{1 + \sqrt{5}}{2}\right]^n + c_2\left[\frac{1 - \sqrt{5}}{2}\right]^n$

11.24 a) $X_n = c_1 2^n + c_2(-3)^n - 1$
c) $Z_n = c_1 + c_2 n + n^2/2$

e) $Y_n = c_1\left[\frac{1 + \sqrt{17}}{2}\right]^n + c_2\left[\frac{1 - \sqrt{17}}{2}\right]^n - \frac{3}{4}$

11.25 a) $X_n = (\frac{11}{5})2^n + (\frac{4}{5})(-3)^n - 1$
c) $Z_n = 1 - (\frac{3}{2})^n + n^2/2$

e) $Y_n = \left(\frac{11 + \sqrt{17}}{8}\right)\left[\frac{1 + \sqrt{17}}{2}\right]^n + \left(\frac{11 - \sqrt{17}}{8}\right)\left[\frac{1 - \sqrt{17}}{2}\right]^n - \frac{3}{4}$

11.26 a) $\Delta X_n = 3\Delta X_{n-1} + 5$ or $X_{n+1} - 4X_n + 3X_{n-1} = 5$

11.27 a) Separate roots, $\lambda_1 = -1$ and $\lambda_2 = 3$: $X_{n+2} - 2X_{n+1} - 3X_n = 0$
c) Complex roots: $\sqrt{\beta} = 2$, $\cos(\pi/2) = -\alpha/2\sqrt{\beta}$ thus $X_{n+2} + 4X_n = 0$
e) Double root, $\lambda_1 = \lambda_2 = \frac{1}{2}$: $X_{n+2} - X_{n+1} + (\frac{1}{4})X_n = (\frac{1}{2})$

11.28 a) $c_1 = -(5 + 2\sqrt{5})/10$, $c_2 = -(5 - 3\sqrt{5})/10$

11.29 a) $f_{n+2} - f_{n+1} - f_n = 0; f_n = c_1\left(\frac{1 + \sqrt{5}}{2}\right)^n + c_2\left(\frac{1 - \sqrt{5}}{2}\right)^n$

EXERCISE SET 12

12.1 a) 5 c) -1 e) 2 g) -1 i) 0, if $x \to 0^+$; -1 if $x \to 0^-$

12.2 a) $\delta = 0.033$ c) $\delta = 0.02$ e) $\delta = 0.1$

12.3 a) $\lim_{x\to 1^+} f(x) = \lim_{x \to 1^-} f(x) = 0$ c) $\lim_{x\to 0^+} f(x) = \lim_{x\to 0^-} f(x) = 0$

e) $\lim_{x\to -1^+} f(x) = 1$, $\lim_{x\to -1^-} f(x) = 0$,
$\lim_{x\to 0^+} f(x) = 0$, $\lim_{x\to 0^-} f(x) = 0$,
$\lim_{x\to 1^+} f(x) = 2$, $\lim_{x\to 1^-} f(x) = 1$

12.4 a) continuous at $x_0 = 1$ c) not continuous at $x_0 = 0$
e) continuous at $x_0 = 0$, but not at $x_0 = -1$ or $x_0 = 1$

12.5 a), b), e), and f)

12.6 a)

Continuous everywhere.

c)

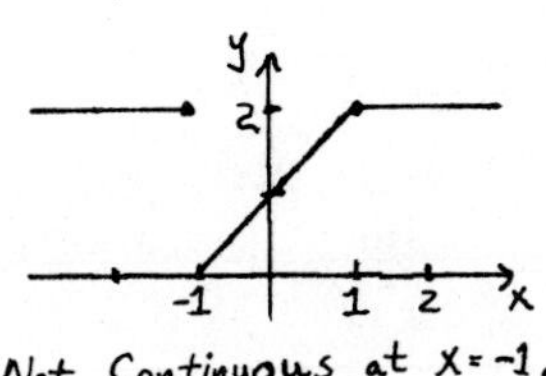

e)

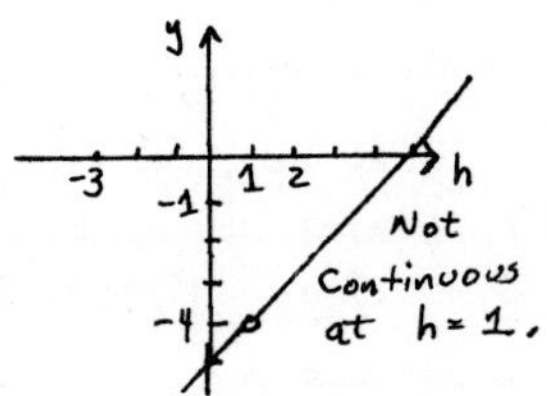

g)

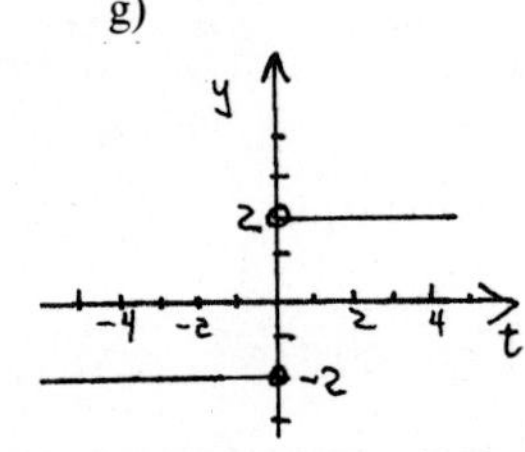

i)

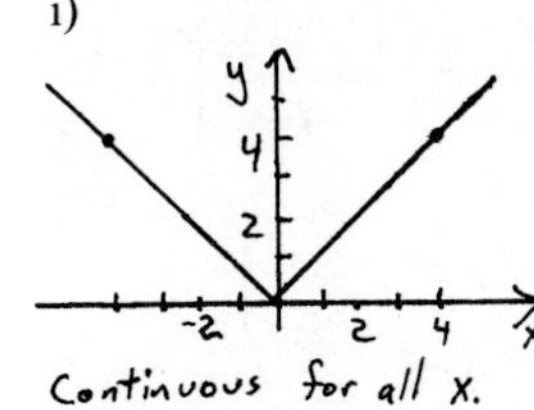

k)

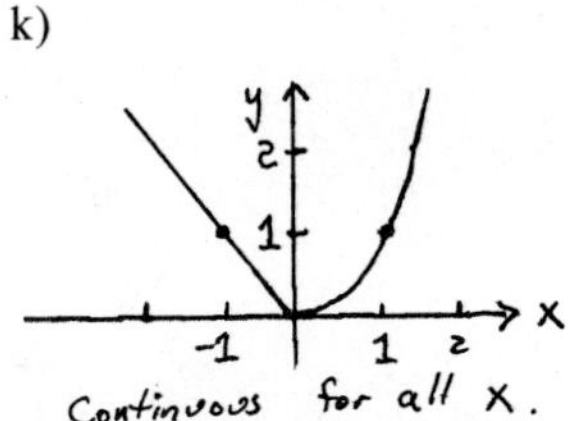

12.7 a) 0 c) 0 e) $+\infty$ g) 0 i) $\sin(1) = 0.84147$

12.8 a) and c) are not continuous functions

12.9 Examples of functions satisfying the conditions are

a) $f(x) = \begin{cases} 3 & \text{if} \quad x > 2 \\ 1 & \text{if} \quad x \leq 2 \end{cases}$ c) $f(x) = (x^2 + 2x - 3)/(x - 1)$

12.10 a) $\lim_{t\to 4^-} A(t) = 2/e^2$

c) Residual at the end of the nth dosage is $\sum_{i=0}^{n} 2e^{-2i}$

12.14 a) 14 c) 0 e) $-\infty$

12.15 a) redefine $f(3) = 6$ c) In the case $f(\pi/2) = 1$;
to make f continuous, at $\pi/2$ redefine $f(\pi/2) = 0$.
e) cannot redefine continuously since $\lim_{x\to 0} f(x)$ does not exist.

EXERCISE SET 13

13.1 a) 2 c) $5(e^{0.2} - 1) \simeq 1.1070$ e) -800

13.2 a) 2 c) $\dfrac{-1}{(t + \Delta t)t}$ e) $\dfrac{\sqrt{z + \Delta z} - \sqrt{z}}{\Delta z}$ g) $e^x(e^{\Delta x} - 1)/\Delta x$

13.3 a) 4 c) 0 g) $(e^3 - e)/2$

13.4 $\frac{3}{2}$ km.

13.5 1.3005 m b) 0.16683 m/wk; 0.02383 m/day

13.6 The graph passes through the points in the following table:

t	10	20	30	40	50	60	70	80	90	100	110	120
$V(t)$	2	17	37	55	73	89	103	111	113	113	103	68

13.7 a) $y = x + 3$ c) $y = 10^{-6}x + 3$

13.8 For $\Delta t = 1$

a) $y = 3t - 2$ b) $y = -\frac{1}{2}t + \frac{3}{2}$ c) $y = 0$
d) $y = \log_3(2)(t - 1)$ e) $y = 2t$ f) $y = 2$
g) $y = e^2(t - 1) + e(2 - t)$ h) $y = e^4(t - 1) + e^2(2 - t)$

13.9 a) $y = (2/\pi)x$ c) $y = 2x + 1$ e) $y = (8/\pi)(\sqrt{2} - 1)x + 2$

13.11 a) $\Delta Q(t)/\Delta t = 2.8\pi\Delta c(x)/\Delta x$

c) If $c(x) = mx + b$, then $\Delta c(x)/\Delta x = m$ and $\Delta Q(t)/\Delta t = DAm$. Hence, with $t_0 = 0$ and $\Delta t = t$, $(Q(t) - Q(0)/t) = DAm$ gives $Q(t) = Q(0) + DAm \cdot t$, a linear function in t.

13.13 a) $H(1) = (11, 0)$, $F(1) = (\sqrt{2}, 10 - \sqrt{2})$

c) $H(1) = (11, 0)F(1) = (22/\sqrt{221}, 10 - 20/\sqrt{221})$

13.14 a)

t	1	2	3	4	5	6	7	8	...	16	17
$c(t)mg$	50	54	57	59	60	60	59		...	5	-6

13.15 $t_{\frac{2}{3}} = 5\ln(4)$; average rate of change is $\frac{1}{5}\ln(4) \simeq 0.14$

13.16 10 jumps, longest horizontal distance $\sqrt{2.5} \simeq 0.14$

13.17 $h = 1$, $x(1) = 2$; $h = \frac{1}{2}$, $x(1) = \frac{9}{4}$;
$h = \frac{1}{100}$, $x(1) = (1.01)^{100} \simeq 2.7048$; note $e \simeq 2.7183$

13.18 $h = 1$, $x(1) = 40$; $h = \frac{1}{2}$, $x(1) = 62.5$, $h = \frac{1}{100}$,
$x(1) = 10(1.03)^{100} \simeq 192.19$; note $10e^3 \simeq 200.86$

EXERCISE SET 14

14.1 a) $y = 3x - 1$ c) $y = t - 2$ e) $y = 2t + 1$

14.2 a) $f'(2) = 4, f'(-2) = -4$ c) $M'(2) = 12$ e) $g'(0) = 2, g'(2) = -2$

14.4 $f'(0) = 1, f'(2) = 5$

14.5 $f'(1) = \frac{1}{2}, f'(3) = 1/2\sqrt{3}$

14.6 a) $M'(4) = 6$ c) $t = 5$

14.7 Tangent at $(-2, -1)$ is $y = 9x + 17$ and at $(1, -1)$ is $y = -1$.

14.8 a) $y' = 6x - 2$ c) $y' = 9x^2 - 70x^9$ e) $y' = \frac{3}{2}x^{\frac{1}{2}}$
g) $y' = \frac{1}{4}x^{-\frac{3}{4}} + 1$ i) $y' = 3x^{-\frac{1}{4}} - 2x$ k) $y' = {}^{-}\frac{2}{3}x^{-\frac{5}{3}} + \frac{2}{3}x^{-\frac{1}{3}}$

14.9 a) $9x^2 - 4x + 3$ c) $\dfrac{2}{x^6}(5 - 6x)$ e) $\dfrac{-2x}{(x^2 + 1)^2}$ g) $5x^4$
i) $\dfrac{(3 + 4x - 3x^2)}{(x^2 + 1)^2}$ k) $x^4(7x^2 + 18x - 5)$

14.10 a) $x^2 + 2x - 3$ c) $e^x + 1 - 2x$ e) $x^2 - 2 - 2e^x$

14.11 a) $-\sin(x) + \sec^2(x)$ c) $\cos^2(\theta) - \sin^2(\theta)$
e) $(3 + \sin(x))/(\cos^2(x))$ g) $2t\sin(t) + t^2\cos(t)$.

14.12 a) $3e^{3x}$ c) $-3e^{-3x}$ e) $4e^{4x+1}$ g) $(x + 1)e^x$ i) $e^x(\ln(x) + 1/x)$

14.13 a) $3/x$ c) $2/x$ e) $1/x$ g) $1 + \ln(x)$ i) $2x\ln(x) + x$

14.14 a) $9x^2 + 4x - 3$ c) $4t^3$ e) $2(t + 1)$

14.15 a) $v(0) = 1, v(\pi/2) = 0, v(\pi) = -1, a(0) = 0, a(\pi/2) = -1, a(\pi) = 0$

14.16 -2.4

14.17 $(0.15)\rho\pi$

14.18 $dv/ds = KV/(s + K)^2$

14.20 c) $dv/dl = P/3\eta l^2(r^2 - R^2)$

14.21 a) $L'(\theta) = k(\cos(\theta) - \sin(\theta)\theta)$ c) θ approximately 0.85 radians

14.22 a) $V = \pi r^2 l$ c) $V = \sqrt{3}\pi r^3, dV/dr = 3\sqrt{3}\pi r^2,$

14.23 a) $l'(\theta) = 10 - \sec(\theta) - \theta\sec(\theta)\tan(\theta)$

14.25 a) $D_x{}^+f(x_0) = D_x{}^-f(x_0) = 2$ c) $D_x{}^+f(x_0) = 1, D_x{}^-f(x_0) = -1$
e) $D_x{}^+f(x_0) = -2, D_x{}^-f(x_0) = 2$

14.27 $A'(l) = -54.04l^{-1.4}$

14.28

	$h = 0.1$	$h = 0.01$	$h = 0.001$	$f'(x_0)$
a) $\Delta f/\Delta x$	4.1	4.01	4.001	4
c) $\Delta f/\Delta x$	2.5893	2.3293	2.3052	2.3026
e) $\Delta f/\Delta x$	0.9531	0.9950	0.9995	1

EXERCISE SET 15

15.1 a) $4u + 3$ c) $2(2v + 1)^4[7v^2 + v - 10]$ e) $\frac{1}{2}u^{-\frac{1}{2}} + 2u$
g) $7(u^2 + 5u - 1)^6(2u + 5)$ i) e^u

15.2 a) $f \circ g(x) = (2x)^2 + 2$; $g \circ f(x) = 2(x^2 + 2)$
c) $f \circ g(x) = (x + 2)^2 + |x + 2|$; $g \circ f(x) = (x + \sqrt{x} + 2)^2$
e) $f \circ g(x) = 2\ln(x)$; $g \circ f(x) = (\ln(x + 1))^2 - 1$
g) $f \circ g(x) = (3x^{\frac{1}{2}} - 3x^{-\frac{1}{2}} - 2)^3$; $g \circ f(x) = (3x - 2)^{3/2} - (3x - 2)^{-3/2}$

15.3 a) $f(x) = x^2 + x^{\frac{1}{2}}; g(x) = 2x + 1$ c) $f(x) = xe^x; g(x) = x^2 + 1$
e) $f(x) = (x - 2)/x; g(x) = x^2 + 1$ g) $f(x) = x(1 + e^x); g(x) = e^x$

15.4 a) $6(3x + 1)$ c) $15x^4(x^5 + 1)^2$ e) $\frac{1}{2}(x^2 + 3x - 1)^{-\frac{1}{2}}(2x + 3)$
g) $6(x + 3)^5$ i) $8(2x + 1)^3 + 6x(x^2 - 1)^2$
k) $(3x^2 + 20x + 118)/(5x + 1)^4$
m) $[(10x)(x^5 - 1)(x^2 - 1)^4][2x^5 - x^3 - 1]$
o) $(-3x + 1)/[(x + 1)^3(x - 1)^2]$

15.5 a) $-2\cos(x)\sin(x)$ c) $2x\cos(1 + x^2)$ e) $-2\sec^2(1 - 2x)$
g) $\cos(t)\cos(\sin(t))$

15.6 a) $2e^{2x}$ c) $4xe^{2x^2}$ e) $(2 - 3x^2)e^{2x - x^3}$
g) $2(x - 1)e^{(x-1)^2}$ i) $\cos(x)e^{\sin(x)}$ k) $(1 + 2x^2)e^{x^2}$
m) $(\cos(x) - \sin(x))e^{-x}$ o) $e^x\cos(e^x)$ q) $e^{x + e^x}$
s) $6e^{3x + 2e^{3x}}$

15.7 a) $y = -(1/\sqrt{3})x + (2/\sqrt{3})$ c) $y = x - 1$ e) $y = 2x + 1$

15.8 a) $8x$ c) $2x + 5$ e) $2/x$ g) $\frac{9}{2}(3x - 2)^{\frac{1}{2}} + \frac{9}{2}(3x - 2)^{-\frac{5}{2}}$

15.9 a) $8x + 4 + (2x + 1)^{-\frac{1}{2}}$ c) $2x(x^2 + 2)e^{x^2 + 1}$ e) $-4x/(x^2 + 1)^2$
g) $[1 + (1 + x)e^x]e^x$

15.10 $D_t\hat{H}(28) = \frac{256}{243}M > 0$; therefore the population is increasing.

15.11 a) $d\hat{R}/dv = 6nv^{n-1}/(v^n + 2)^2$

15.12 c) $d\hat{g}(r)/dr = 2(40r - r^2 - 10)(40 - 2r) - 40 + 2r$

15.13 a) 15,000 ft gives 75 percent O_2 saturation, which corresponds to the app. 42 mm pressure.

15.15 a) 6 c) $\frac{3}{8}x^{-\frac{5}{2}}$ e) $-3x(1 - x^2)^{-\frac{5}{2}}$
g) $-\cos(x)$ i) $6\sec^4(x) - 4\sec^2(x)$ k) $(12x + 8x^3)e^{x^2}$

15.16 a) $v(t) = \sin(t) + t\cos(t)$, $a(t) = 2\cos(t) - t\sin(t)$

15.17 $dF/dx = dF/dO_2 \cdot dO_2/dx = -2A[O_2 - k](-2B)(x - c)$

15.19 $\mathring{X}(t) = -12 \cdot 10^{-4}$ cm/min

15.20 $dh/dt = 1.34 \times 10^{-3}\ m/hr$

15.21 300 w^2d, where d is constant density.

15.24 a) $d = \sqrt{h^2 + 30h + 144}$

15.25 a) $f_B(x) = 50 - 2.5x$ c) -7

15.26 a) $-a/\ln(10)W$.

15.27 b) i) $1/\sqrt{1 + x^2}$ iii) $\sqrt{1 + (x^2/4)}$ v) $\sqrt{1 - x^2}/x$
d) i) $3/\sqrt{1 - 9x^2}$ iii) $-1(1 - 2x + 2x^2)$ v) $4/[(4x - 1)\sqrt{(4x - 1)^2 - 1}]$

EXERCISE SET 16

16.1 I = increasing; D = decreasing:
a) I c) I e) D g) D i) D k) I m) I o) D

16.2 a) $(\pi/2 + 2N\pi, 3\pi/2 + 2N\pi)$ c) $(-\infty, [3 - \sqrt{33}]/12) \cup ([3 + \sqrt{33}]/12, +\infty)$

16.3 a) min at $x = 0$ c) max at $x = -1/\sqrt{3}$ e) no min or max
g) max at $x = \ln(2)/2$
i) max at $x = 2N\pi$, min at $x = (2N - 1)\pi$
k) max at $x = 0$, min at $x = \pm 2$ m) max at $x = 2$
o) min at $x = 0$

16.4 a) $M = 8$ at $x = 5$, $m = -3$ at $x = 0$
c) $M = 27$ at $x = 5$, $m = -8$ at $x = 0$, inflection at $x = 2$
e) $M = 9$ at $x = -1$, $m = 0$ at $x = 0$, relative max at $x = 1$
g) $M = 1$ at $x = 0$, $m \simeq -0.5874$ at $x = 2$, relative min at $x = -1$
i) $M = 1$ at $x = 0$, $m = 3^{-1} \simeq 0.3679$ at $x = \pm 1$
k) $M = \ln(5) \simeq 1.6094$ at $x = 0$, $m = 0$ at $x = \pm 2$

16.5 a) $M = 24$ at $x = 4$, $m = -56$ at $x = -4$, Loc max 4.0606 at $x \simeq -1.1196$, loc min -8.2088 at $x \simeq 1.7863$
c) $M = \pi/2$ at $t = \pm\pi/2$, $m = 0$ at $t = 0$
e) $M \simeq 0.5307$ at $t = 1/\ln(2)$, $m \simeq -2.3087$ at $t = -\sqrt{2}$
g) $M = 0.5$ at $t = \pi/4$, $m = -0.5$ at $t = -\pi/4$

16.7 Maximum area is $\frac{25}{6}$ sq km with dimensions $\frac{5}{2}$ by $\frac{5}{3}$ km.

16.9 a) $F(w) = w^3 - 7w^2$ b) $\frac{28}{3}$ by $\frac{14}{3}$ by $\frac{7}{3}$

16.11 $n = 69.5$ for maximum harvest; both $n = 69$ and $n = 70$ result in a harvest of 19,320 lb.

16.13 a) Set $dP/dJ = 0$ to obtain $J = (B^{\frac{1}{2}}V^2/A^{\frac{1}{2}}L) - 1$
Note V^2 is an increasing function, whereas $1/L$ is a decreasing function.
c) As the size L increases, J decreases. Note that for the same step size S, this requires a longer floating distance F for an elephant than for a mouse.

16.14 a) $x_1 = 2$, min c) $x_1 = 2$, max e) $x_1 = 2$, max
g) $x_1 = \ln(\frac{4}{3})$, min i) $x_1 = \sqrt{\frac{1}{2}}$, max $x_2 = -\sqrt{\frac{1}{2}}$, min

16.15

	x	y
	0	0.1
max	3	$\frac{57}{30} \simeq 1.90$
inf	4	$\frac{55}{30} \simeq 1.83$
min	5	$\frac{53}{30} \simeq 1.77$
	6	$\frac{57}{30} \simeq 1.90$

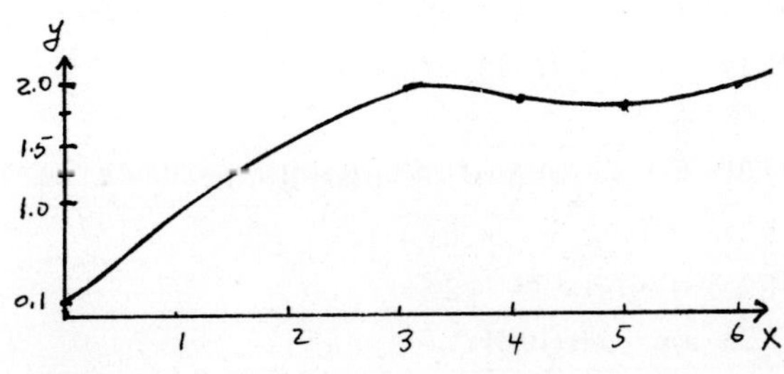

16.17

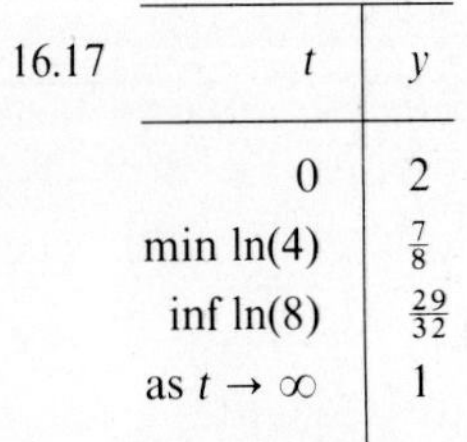

	t	y
	0	2
min	ln(4)	$\frac{7}{8}$
inf	ln(8)	$\frac{29}{32}$
as $t \to \infty$		1

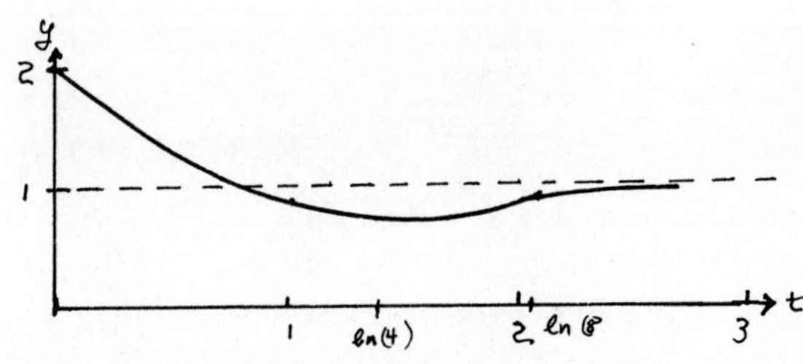

16.19 Not defined at $t = 0$
No inflection points
Max at $-\sqrt{2}$
Min at $\sqrt{2}$
Assymptotic to $y = t$ as $t \to \infty$ and to y axis as $t \to 0$.

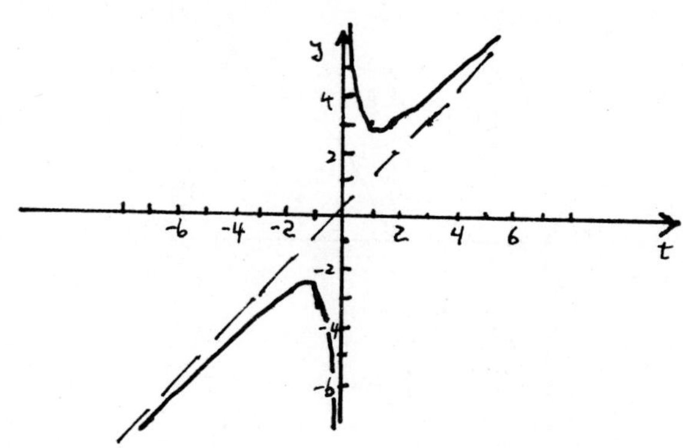

16.21 Max at $t = 0$
Inflection points $t = \pm\sqrt{\frac{1}{2}}$
Asymptotic to x axis as $t \to \infty$.

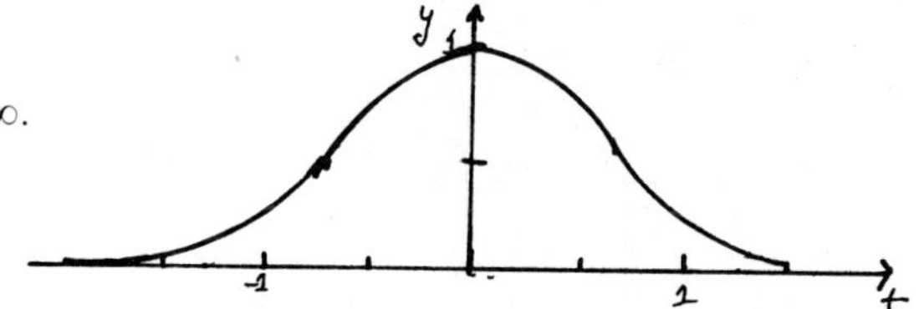

16.22 Inflection point $t = \ln(0.01) \simeq -4.6$
Asymptotic $y = 10$ as $t \to +\infty$, and $y = 0$ as $t \to -\infty$
y-intercept $\simeq 9.9$
$y(-10) \simeq 2 \cdot 10^{-95}$

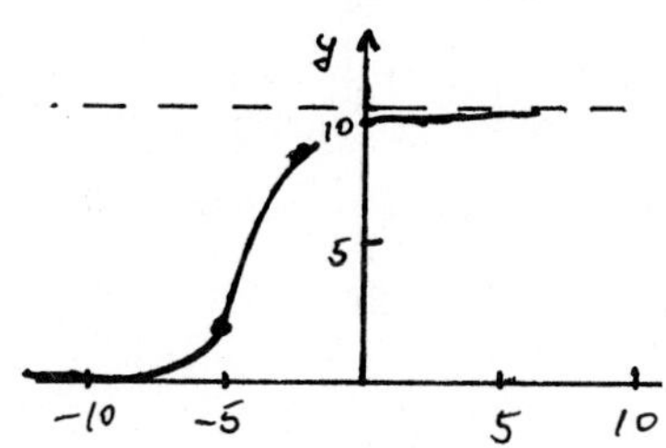

16.30 a)

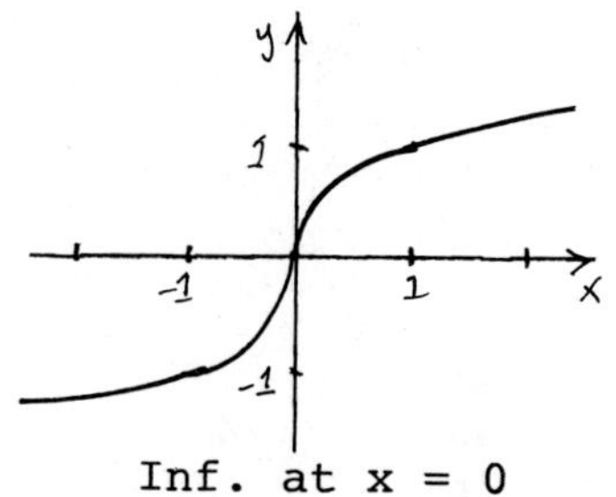

c)

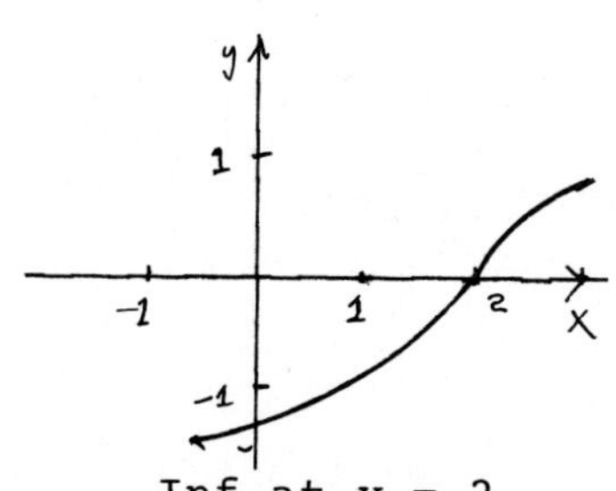

e)

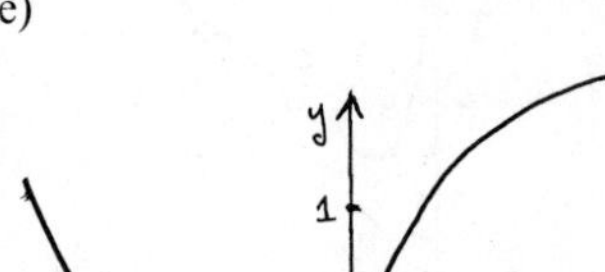
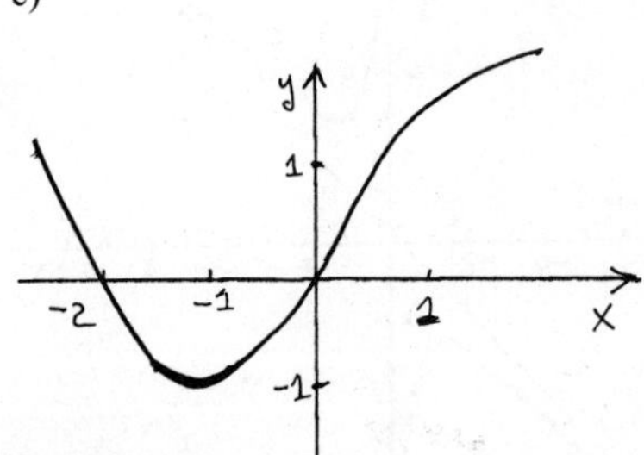

min. at x = 1
Inf. at x = -2 and x = 0

g)

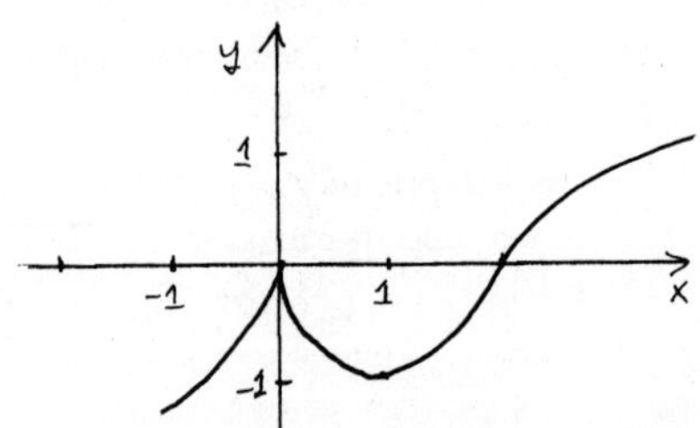

max. at x = 0, min. at x = 20/19
Inf. at x = 2

16.31 a)

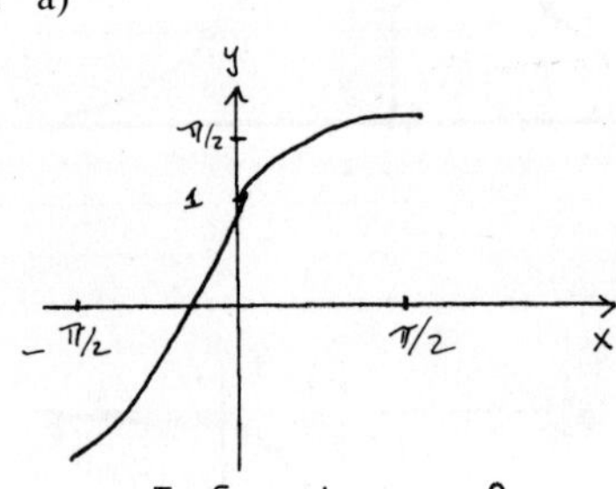

Inf. at x = 0

c)

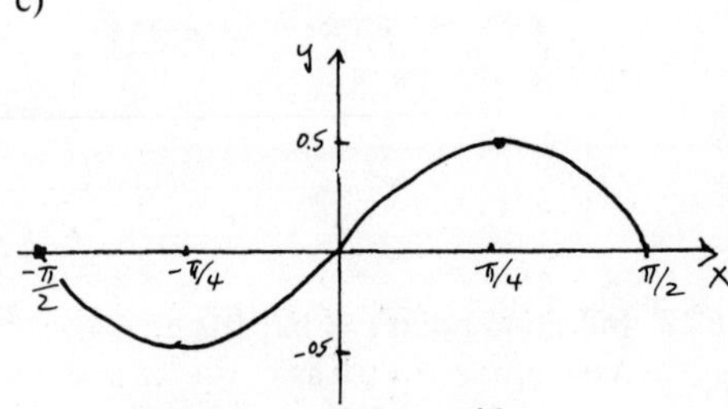

max. at x = π/4,
min. at x = π/4
Inf. at x = 0

e)

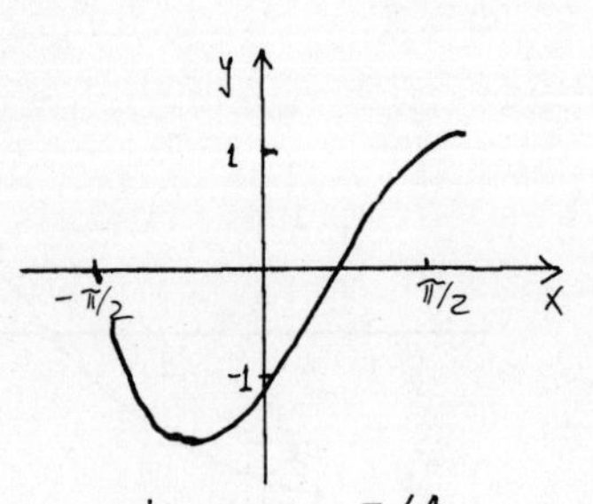

min. x = -π/4
Inf. x = π/4

g)

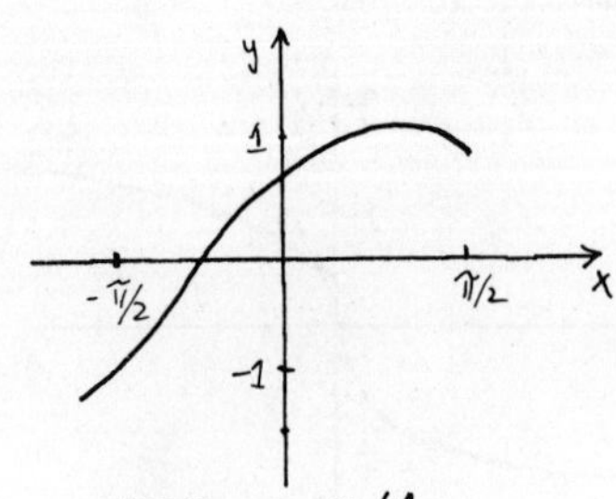

max. = π/4
Inf. x = -π/4

EXERCISE SET 17

17.1

	$dx = 0.1$			$dx = -0.1$			$dx = 0.5$		
	df	Δf	E	df	Δf	E	df	Δf	E
a)	1.20	1.2610	0.0610	−1.20	−1.410	0.059	6.00	7.625	1.625
c)	0.02	0.0198	0.0002	−0.02	−0.020	0.0002	0.10	0.095	0.005
e)	2.217	2.3710	0.1540	−2.217	−2.075	0.1420	11.08	15.680	4.600
g)	0.000	−0.003	0.000	−0.003	−0.003	−0.0030	0.00	−0.076	0.076
i)	0.000	−0.14	0.014	0.000	−0.019	0.0140	0.00	−0.313	0.313

17.2 a) $\frac{97}{14} \simeq 6.93$ c) 2.125 e) 0.6 g) 4.125 i) 0.1 k) 1.9375

17.4 $-0.3m^3$

17.6 a) $x_1 = -1, \quad x_2 = -0.2$ c) $x_1 = 5.4, \quad x_2 = 4.31$
e) $x_1 = 7.32, \quad x_2 = 5.55$

17.7 with $x_0 = 1$, $x_4 = 1.18920711$ valid to 7 places

17.9 with $x_0 = 1$, $x_4 = 1.763222834$ valid to 9 places

17.11 a) $x_{n+1} = 0.5(x_n + (\alpha/x_n))$ c) $x_{n+1} = 0.2(4x_n + (\alpha/x_n{}^4))$

17.13 a) $2udu - dv$ c) $(2uvdu - u^2dv)/v^2$ e) $(1 + 2xy^3)dx + 3x^2y^2dy$
g) $e^x[dy + ydx]$ i) $-\sin(xy)(xdy + ydx)$ k) $2\sin(x)\cos(x)dx$

17.14 a) $-\frac{3}{2}$ c) $(1 - y)/(x - 1)$ e) $(e^{-xy} - y)/x$ g) y/x i) $(x - 1)/xe^y$
k) $-\sin(x)/\cos(y)$ m) $\cos(x)\sin(x)/\cos(y)\sin(y)$

17.15 a) $y = \frac{14}{13}x + \frac{12}{13}$ c) $y = -1.5x - \sqrt{2}$ e) $y = x/3$ g) $y = 4x/\pi$

17.16 a) $y' = 0.75$ e) y' is not finite

17.17 a) $dQ \div Q$ = relative error = 12 percent b) $dQ \div Q = -3$ percent

17.19 If $l_1 = l_1 + dl$ and $w_1 = w + dw$ then $A_1 = l_1 \cdot w_1$
$= l \cdot w + l \cdot dw + w \cdot dl + dl \cdot dw = A + dA_w + dA_l + dldw.$

EXERCISE SET 18

18.1 a) Does c) No, $f(7) \neq 0$ e) Does

18.2 a) $\xi = 1.5$ c) $\xi = \frac{8}{3}$ e) None since $f'(1)$ does not exist g) $\xi = 0$

18.3 a) $1 - 3x + 2x^2$ c) $-2 + 5x^2$ e) $e[1 + x + (x^2/2) + (x^3/6)]$
g) $1 + x^2$ i) $1 - x + (x^2/2) - (x^3/6)$

18.4

For $x = 1$	a)	c)	e)	g)	i)
$f(x)$	0	4	7.389	2.718	0.368
$P_3(x)$	0	3	7.249	2.333	0.333
$E = \lvert f(x) - P_3(x)\rvert$	0	1	0.140	0.385	0.035

For $x = 0.01$	a)	c)	e)	g)	i)
$f(x)$	0.99	−1.999	2.746	1.000	0.990
$P_3(x)$	0.99	−1.999	2.763	1.000	0.993
$E = \lvert f(x) - P_3(x)\rvert$	0	10^{-8}	0.017	3.3×10^{-7}	0.003

18.5 a) $(x-1)+2(x-1)^2$ c) $4+14(x-1)+11(x-1)^2+4(x-1)^3$
e) $e^2[1+(x-1)+(\frac{1}{2})(x-1)^2+(\frac{1}{6})(x-1)^3]$
g) $e[1+2(x-1)+3(x-1)^2+3(x-1)^3]$
i) $e^{-1}[1+(x-1)+(\frac{1}{2})(x-1)^2+(\frac{1}{6})(x-1)^3]$

18.7 $P_4(x)=4(x-1)+6(x-1)^2+4(x-1)^3+(x-1)^4$
$f(2)=15$, $P_1(2)=4$, $P_2(2)=10$, $P_3(2)=14$, $P_4(2)=15$

18.9 $P_5(x)=x-x^2+(x^3/2)-(x^4/6)+(x^5/24)$
$P_1(0.5)=0.5$, $P_2(0.5)=0.25$, $P_3(0.5)=0.3125$
$P_4(0.5)=0.3021$, $P_5(0.5)=0.3034$, while $f(0.5)=0.3033$

18.11 a) $\mathring{N}=F(N_0)+F'(N_0)(N-N_0)$
b) $F(0)=0$ and $\mathring{N}=F'(0)N$ has the exponential solution $N(t)=ce^{F'(0)t}$
$F(N_E)=0$ and $\mathring{N}=F'(0)(N-N_E)$ has the solution $N(t)=N_e+ce^{F'(0)t}$

EXERCISE SET 19

19.1 a) $C=-5$ c) $C=21$ e) $C=7$

19.2 a) $x^3/3+c$ c) $x^6/2+2x+c$ e) $5x^4/4-x^6/3+c$
g) $-x^{-6}-x^7+c$ i) $(\frac{2}{3})x^{\frac{3}{2}}+(\frac{4}{7})x^{\frac{7}{4}}+c$

19.3 a) $(\frac{1}{3})x^3-x^2-2x+8$ c) $(\frac{5}{6})x^6+2x+1$ e) $2x^{\frac{1}{2}}-x+3$
g) $\tan(x)$ i) $3\ln(x)+2$

19.4 a) $(\frac{1}{5})(x+3)^5+c$ c) $(x^2+x+2)^3+c$ e) $(\frac{1}{36})(2x^3-1)^6+c$
g) $\frac{1}{11}(x^2+7x-2)^{11}+c$ i) $(\frac{1}{126})(3x^2+6)^{21}+c$

19.5 a) $u=x^2-3$, $v=x$, $r=\frac{1}{2}$, $D_x^{-1}(y)=(x^2-3)^{\frac{3}{2}}/3$
c) $u=x^3+4.5x^2-6x$, $v=x^2+3x-2$, $r=5$
$D_x^{-1}(y)=(x^3+4.5x^2-6x)^6/18$

19.6 a) $e^{2x}+c$ c) $0.5e^{2x+1}+c$ e) $0.5e^{x^2}+c$ g) $(\frac{1}{6})e^{3x^2+1}+c$
i) $(\frac{1}{6})e^{2x^3+3x^2+6x-2}+c$

19.7 a) $2\ln(x)+c$ c) $\ln(x^2)+c$ e) $0.5\ln(x^2-4)+c$
g) $0.5\ln(x^2-4x+7)+c$ i) $\ln(x+\sin(x))+c$

19.8 a) $(\frac{5}{2})x^2+c$ c) $(\frac{1}{6})(2x+1)^3+c$ e) $(\frac{4}{3})t^{\frac{3}{2}}+c$
g) $(\frac{1}{5})(x^2-3)^5+c$ i) $(\frac{1}{5})x^5+x^4+(\frac{2}{3})x^3-2x^2+x+c$

19.9 a) $(\frac{1}{3})\sin(3x+\pi)+c$ c) $\tan(x)-\sec(x)+c$
e) $-(3/\pi)\cos(\pi x)+c$ g) $(\frac{1}{4})\sin^4(x)+c$
h) $(\frac{1}{2})\tan^2(x)+c$ j) $-\cos(\sin(x))+c$

19.10 a) $2e^x+c$ c) $(\frac{1}{4})e^{2x^2}+c$ e) $(\frac{1}{3})e^{x^3+6x-2}+c$
g) $-e^{\cos(\theta)}+c$

19.11 a) $\ln(x^2+1)+c$ c) $(\frac{1}{4})\ln(2x^2-4x+1)+c$ e) $\ln(5x^4+5x^3-2x)+c$

19.13 a) $N(t)=ce^{3t}$
b) $N(10)=3.2\times10^4\times e^{30}$ c) $N(10)=5\times10^4\times e^{24}$

19.15 $f(t)=-3t^{\frac{1}{2}}$, $F(t)=-2t^{\frac{3}{2}}+c$, $F(0)=0$ gives $c=10$; so $F(t)=10-2t^{\frac{3}{2}}$ mg
$F(t_1)=0$, where $t_1=5^{\frac{2}{3}}\simeq 2.9$ hrs

19.17 $P(t)=\int r(t)dt=cos^2(t/4)+c$. $P(0)=96$ gives $c=95$. P(2) $\simeq$ 95.17 mg Hg. $P(t)=96$, where $t=4n\pi$, $n=1,2,\ldots$

19.19 $V(t) = \int e^{-kt}dt$; $V(t) = c - (1/k)e^{-kt}$; c denotes the total volume of water that will pass over the dam.

EXERCISE SET 20

20.1 a) $\Delta t_1 = \frac{1}{8}, I_1 = [0, \frac{1}{8}]; \Delta t_2 = \frac{3}{8}, I_2 = [\frac{1}{8}, \frac{1}{2}];$
$\Delta t_3 = \frac{1}{4}, I_3 = [\frac{1}{2}, \frac{3}{4}]; \Delta t_4 = \frac{1}{4}, I_4 = [\frac{3}{4}, 1].$
c) $\Delta t_n = \frac{3}{4}$ for $n = 1, 2, 3, 4$; $I_1 = [2, \frac{11}{4}], I_2 = [\frac{11}{4}, \frac{7}{2}], I_3 = [\frac{7}{2}, \frac{17}{4}], I_4 = [\frac{17}{4}, 5].$
e) $\Delta t_1 = \frac{1}{2}, I_1 = [0, \frac{1}{2}]; \Delta t_2 = \frac{1}{4}, I_2 = [\frac{1}{2}, \frac{3}{4}];$
$\Delta t_3 = \frac{3}{4}, I_3 = [\frac{3}{4}, \frac{3}{2}]; \Delta t_4 = 1, I_4 = [\frac{3}{2}, \frac{5}{2}]; \Delta t_5 = \frac{1}{2}, I_5 = [\frac{5}{2}, 3]$

20.2 For the partition a). Right endpoints: $\frac{1}{8}, \frac{1}{2}, \frac{3}{4}, 1$. Left endpoints: $0, \frac{1}{8}, \frac{1}{2}, \frac{3}{4}$. Midpoints: $\frac{1}{16}, \frac{5}{16}, \frac{5}{8}, \frac{7}{8}$.

20.3 4.4

20.5 $104g$.

20.7 a) 11.2 b) 6.2 c) 9.0

20.9 Let N be arbitrary, $t_n{}^*$ be the right endpoint. Then $\Delta t_n = 3/N$, $t_0 = 0$, and $t_n = t_n{}^* = n \cdot 3/N$ for $n = 1, 2, \ldots, N$.

$$P(3) \simeq \sum_{n=1}^{N} 2t_n{}^*\Delta t_n = \sum_{n=1}^{N} (2 \cdot 3 \cdot n/N) \cdot (3/N)$$

$$\text{(By formula (10.5))} = (18/N^2) \sum_{n=1}^{N} n = (18/N^2)(N(N-1)/2) = 9[1 - (1/N)].$$

20.12 a) $T = 1$ b) $t = 1$ c) $Q(0) + \frac{3}{8}$
d) $N = 3, Q(T) = Q(0) + \frac{13}{27}$; $N = 4, Q(T) = Q(0) + \frac{17}{32}$
e) $Q(0) + 0.615$

EXERCISE SET 21

21.1 a) 6 c) 12 e) $\frac{1}{3}$ g) $0.5h^2 + h^5$
i) $[x^4 - 8x - 585]/4$ k) 9280 m) $\frac{8}{3}$ o) $[(3T^2 + 2)^{\frac{3}{2}} - 2\sqrt{2})]/9$

21.2 a) 2 c) 0 e) 1

21.3 a) FTC applies
c) Not continuous at $\theta = \pi/2$

21.4 a) First one: 2191.39 c) First one: 0.5

21.5 a) $\int_0^6 (t^2 + 3t - 6)dt$ c) $\int_3^6 (t + 7)dt$ e) $2\int_3^4 e^x dx$

21.6 a) $e - 1$ c) $e^{-1} - 1$ e) 0 g) 1 i) $\ln(11)$ k) $-3\ln(2)$.

21.9 c) Excess heat over $[0, 3]$ is 62.4 and over $[2, 4]$ is 62.93.

21.11 $23k$ cal.

21.14 a) $2x^2 + x^3$ c) $4s^2 - 2$ e) $2x^3 + 8x$
g) $2[(t + 1)^3 - (t + 1)^5]$ i) $-2x[2x^2 - 2 - (x^2 - 1)^4]$

21.15 a) $\sec^2(\theta)$ c) $2(1 + \theta^2)^{-\frac{1}{2}}$ e) $-\sin(\theta) - \sin(\pi - \theta)$

21.17 7.6

21.18 b) $V(t) = t^3/3 - 2t^2 + 6t + 10$

21.19 a) $\int_0^{x(2-x)} (t-3)^2 dt$ b) $(-x^2 + 2x - 3)^2(2 - 2x)$

EXERCISE SET 22

22.1 Left endpoints 3; right endpoints 5

22.2 a) $e^2 - 3$ c) 4.5 e) 6

22.4 c) Using a uniform partition with $N = 5$ and left endpoints yields $A(2) < 0.745$. Using the rectangle of height 0.5 yields $A(2) > 0.5$.
d) Using the rectangle over [1, 2] of height 1 and over [2, 3] of height 0.5 yields $A(3) < 1.5$.
e) Using a uniform partition with $N = 8$ and right endpoints yields $A(3) > 1.0198$.

22.6 a) $(2, 5), (-2, 5); \frac{32}{3}$ c) $(\pm\sqrt{2}, 3); 16\sqrt{2}/3$
e) $(-3\pi/4, 1/\sqrt{2}), (\pi/4, 1/\sqrt{2}); 2\sqrt{2}$ g) $(0, 0), (1, 1); \frac{1}{3}$

22.7 a) $\int_0^{\frac{4}{3}} 2x - (x/2)dx + \int_{\frac{4}{3}}^{\frac{8}{3}} (4 - x) - (x/2)dx = \frac{8}{3}$
c) $\int_{1-\pi/2}^{0} (2x + 1) - (x^2 - \pi^2/4)dx + \int_0^{\pi/2} \cos(x) - (x^2 - \pi^2/4)dx$
$= [-\frac{5}{3} + \pi + (\pi/2)^2 - (\frac{4}{3})(\pi/2)^3] + [1 + (\frac{2}{3})(\pi/2)^3] \simeq 2.358$
e) $\int_0^a 2x\, dx + \int_a^4 2\sqrt{4 - x}\, dx \simeq 7.52; a = (\sqrt{17} - 1)/2$

22.8 a) 14.5 c) 15
e) $\int_0^2 3t + 5dt \simeq (2/N) \sum_{n=0}^{N-1} (3 \cdot 2n/N + 5)$, which simplifies using formulas (10.4) and (10.5) to $16 - (3/N)$.

22.9 a) $\frac{21}{12}$ c) $\frac{53}{27}$
e) $\int_0^2 t^2 dt \simeq (2/N) \sum_{n=0}^{N-1} (2n/N)^2$, which simplifies using formula (10.6) to $(\frac{8}{3}) - (4/N) + (4/3N^2)$.

22.11 $(\pi/6) \sum_{n=0}^{5} \sin(n\pi/6) \simeq 1.954$

22.13 $\log(39916800) \simeq 7.60$

22.15 $(\frac{1}{2})[1 + (\frac{4}{9}) + (\frac{1}{4}) + (\frac{4}{25})] \simeq 0.927$

22.18 $N = 4, 32\ell^3$; $N = 8, 31.5\ell^3$

22.22 a) 3.988981 b) 3.725597 c) 3.734152

22.24 a) 2.857332 b) 3.279882 c) 3.307415

22.25 a) 5.8 b) 5.575 c) 5.216667

22.27 a) 1.7083 c) $P_3(t) = 1 - t^3$; 0.75

22.28 a) $x^2/2 - x^4/24$ c) $x - (\pi x)^3/3! + (\pi x)^4/4!$
e) $x - (x - 1)^2/2 + (x - 1)^3/3 - (x - 1)^2/4 + (x - 1)^2/5$
g) $x + x^2 + 2x^3/3 + x^4/3 + 2x^5/15$

EXERCISE SET 23

23.2 a) $3e^{3x}$ c) $2xe^{x^2+2}$ e) $10(2x-1)^4e^{(2x-1)^5}$
g) $-(1/x^2)e^{1/x}$ i) $(2x+2x^3)e^{x^2}$ k) $2xe^{x^2}-(2x+3)e^{x^2+3x}$

23.3 a) $2x/(x^2+1)$ c) $(4-2t)/(4t-t^2)$ e) $\ln(t)+1$
g) $[(4/x)\ln(x+1)-(4/(x+1)\ln(x)][\ln(x+1)]^{-2}$

23.4 a) $\sec^2(\theta)e^{\tan(\theta)}$ c) $-\tan(\theta)$ e) $[1+\theta\cos(\theta)]e^{\sin(\theta)}$
g) $(\ln(\theta)-1/\theta)e^{\theta}(\ln(\theta))^{-2}$

23.5 a) $\ln(10)\cdot 10^x$ c) $\ln(2)\cdot 2^t$ e) $\ln(10)\cdot(2x-2)\cdot 10^{x^2-2x+3}$
g) $5t^4$ i) $[2t+4t^2\ln(3)]3^{4t}$ k) $\ln(3)$

23.6 a) $(x\ln(10))^{-1}$ c) $3(x\ln(10))^{-1}$ e) $(t\ln(2))^{-1}$
g) $(1+x)/x\ln(10)$ i) $\log(ex)$

23.7 $y'(x)=x^x[\ln(x)+1]$

23.9 In all cases, A_2 is growing faster. A_1 has growth rate of 100 when $l=\ln(100)+2$ *and* A_2 when $l=\frac{1}{20}$.

23.11 $w'(t)=[A-w(t)]k(1+e^{-t/\tau})$.

23.13 a) $d\lambda/dw_2=(2.3/t)\{w_2{}^{-1}-(w_1+w_2)^{-1}+w_1[c_t/c_0-w_1(w_1+w_2)]^{-1}\}$
b) $dk_1/dw_2=[w_2/(w_2+w_1)]d\lambda/dw_2-\lambda w_1(w_2+w_1)^{-2}$

23.17 a) $3^{-t}(\ln(3))^2$ c) $\dfrac{Bc\lambda^2(e^{\lambda t}-c^2e^{-\lambda t})}{(e^{\lambda t}+2c+c^2e^{-\lambda t})^2}$
e) $(\ln(3))^2(Ak_1{}^2 3^{-k_1 t}-Bk_2{}^2e^{-k_2 t})$
g) $k\ln(2)[-2+k\cdot\ln(2)]2^{-kt}$ i) $-2/\ln(10)x^2$
k) $-2/\ln(10)(1+t)^2$ m) $(1+\ln(2)e^t)\ln(2)e^t2^{e^t}$

23.18 a) e^t+c c) $0.5e^{x^2-2}+c$ e) $0.5e^{(x+3)^2}+c$ g) $(\frac{1}{6})e^{3x^2-2}+c$

23.19 a) $3\ln(x)+c$ c) $(\frac{1}{3})\ln(x^3-1)+c$ e) $\ln(1+e^x)+c$
g) $-\ln(-x+x^{-2})+c$ i) $\ln(1-x^{-1})+c$

23.20 a) $2^x/\ln(2)$ c) $3^{x+1}/\ln(3)+c$ e) $2^{e^x}/\ln(2)+c$
g) $-1/2^x\ln(2)+c$

23.21 a) e^5-1 c) $(e^4-e)/2$ e) $\ln(6)$ g) $0.5\ln(2.5)$ i) $8/\ln(3)$ k) 0

23.23 a) $e^{\tan(x)}+c$ c) $-e^{\cot(\theta)}+c$ e) $2^{\sin(\theta)}/\ln(2)+c$
g) 0 i) $\ln(\sin(\theta))+c$ k) $\ln(1+\tan(\theta))+c$

23.29 a) $y'(x)=y(x)[3/(3x+1)+1/x+-1/(-x+6)]$
c) $y'(x)=y(x)[2x/(x^2+1)+2x/(x^2-1)+2/(2x+3)]$
e) $y'(x)=y(x)[1/(x+1)+1/(x+2)+1/(x+3)+1/(x+4)]$
g) $y'(x)=y(x)[2x/(2x^2-1)+(2x+3)/(x^2-3x-7)+5/4x]$
i) $y'(x)=y(x)[2x/(x^2+1)+8x/(2x^2+1)+18x/(3x^2+1)+32x/(4x^2+1)]$

23.31 $\hat{p}=0.7$

EXERCISE SET 24

(*Note*: All the indefinite integrals are determined up to an arbitrary constant c, which is not indicated.)

24.1 a) $x\ln(x)-x$ c) $\frac{1}{2}e^{x^2}(x^2-1)$ e) $-(1/R^2)-e^{-kx}(kx+1)$
g) $\frac{1}{2}(5+x)\ln(5+x)-\frac{1}{2}x$ i) $-\frac{1}{9}e^{-6x}(6x+1)$
k) $x(\ln(x))^3-3x(\ln(x))^2+6x\ln(x)-6x$ m) $\frac{3}{4}(x^{\frac{4}{3}}\ln(x)-\frac{3}{4}x^{\frac{4}{3}})$

24.2 a) $\cos(x) + x\sin(x)$ c) $\frac{1}{3}x\tan(3x) + \frac{1}{9}\ln(\cos(3x))$
e) $\frac{1}{2}(x^2\sin(x^2) + \cos(x^2))$ g) $\frac{1}{2}e^x(\sin(x) - \cos(x))$
i) $\frac{1}{5}e^x(\sin(2x) - 2\cos(2x))$

24.3 a) $-\frac{1}{6}\cos(3x)\sin(3x) + \frac{1}{2}x$ c) $\frac{1}{3}\cos^3(x) - \cos(x)$ e) $\tan(x) - x$

24.4 a) $\frac{3}{2}\ln(x-1) + \frac{1}{2}\ln(x+1)$ c) $\frac{5}{3}\ln(x-1) - \frac{2}{3}\ln(x+2)$
e) $\frac{5}{12}\ln(x-1) - \frac{14}{3}\ln(x+2) + \frac{29}{4}\ln(x+3)$
g) $-\frac{1}{2}\ln(x-1) + [(9 - 2\sqrt{3})/(6 - 2\sqrt{3})]\ln(x - \sqrt{3})$
$+ [(7\sqrt{3} - 3)/4\sqrt{3}]\ln(x + \sqrt{3})$

24.5 a) $x^2/2 + 2\ln x$ c) $\frac{1}{3}x^3 + x^2 + 3x + 7\ln(x-2)$
e) $x + \ln(x - 1/x + 1)$ g) $\frac{1}{2}x^2 - 3x + 7\ln(x+2)$

24.6 a) $-2/(x-2) + \ln(x-2)$ c) $-2/(x-1) + \ln(x-1)$
e) $\frac{1}{2}x^2 + 2x - 1/(x-1) + 3\ln(x-1)$
g) $-(1/8(x-1)) + (\ln(x-1)/16) - (5/8(x+3)) - (\ln(x+3)/16)$

24.7 a) $\ln(x^2+1) + \ln(x)$ c) $\ln(x^2 - x + 1) + \ln(x+1)$
e) $\frac{3}{2}\ln(x^2 + x + 1) - (5/\sqrt{3})\arctan((2x+1)/\sqrt{3}) - 2\ln x$

24.8 a) $2(x+3)^{\frac{1}{2}}$ c) $2(x+3)^{\frac{1}{2}} + (3/\sqrt{3})\ln((\sqrt{x+3} - \sqrt{3})/(\sqrt{x+3} + \sqrt{3}))$
e) $-\frac{1}{20}(1-2x)^{\frac{5}{2}} + \frac{1}{12}(1-2x)^{\frac{3}{2}} - \frac{1}{4}(1-2x)^{\frac{1}{2}}$
g) $-2(2+x)^{\frac{1}{2}}\cos((2+x)^{\frac{1}{2}}) + 2\sin((2+x)^{\frac{1}{2}})$ i) $\frac{3}{5}(x+3)^{\frac{5}{3}} - \frac{9}{2}(x+3)^{\frac{2}{3}}$

24.10 a) $\frac{1}{2}\arcsin(x) + \frac{1}{2}\sqrt{1-x^2}$ c) $\arctan(x)$ e) $\arcsin(x/2)$
g) $\frac{1}{2}\ln((1 - \sqrt{1-x^2})/(1 + \sqrt{1-x^2}))$

24.12 a) $\arcsin((2x+1)/3)$
c) $\sqrt{x^2+2x} - \frac{1}{2}\ln[1 + x\sqrt{x^2+2x}]$

EXERCISE SET 25

25.1 a) 1 c) $\pm\infty$

25.2 a) $-\infty$ c) 0 e) $\mp\infty$

25.3 a) 0 c) 0

25.4 a) 0 c) ∞ e) 0

25.5 a) 0 c) 0 e) 0

25.6 a) 0 c) 1 e) 1 g) e^2

25.7 a) $\ln(\frac{3}{5})$ c) $-\infty$

25.9 L'Hospital's rule applies for Exercise (e) only, the limit is $+\infty$. The other actual limits are a) $-\infty$, c) 0, g) ∞. To evaluate g), see Exercise 25.11.

25.11 a) ∞ c) $\frac{1}{2}$ *Hint*: Cross multiply, then use L'Hospital's rule twice. e) $+\infty$

25.12 a) Does not exist c) $\frac{5}{2}e^{-4}$ e) Does not exist g) $9 - 3\log_3 e$ i) $e^{-\frac{1}{5}}$
k) 0 m) 0 o) Does not exist

25.13 a) Does not exist c) $\frac{1}{2}$
e) Does not exist because $\lim_{t\to 0}\cos(1/t)$ is not defined.
g) Does not exist because $\lim_{t\to\infty} e^{\sin(t)}$ is not defined.
i) Does not exist k) $-e^{\sin(1)}$ m) -1 o) 0

25.15 a) A/k c) $A/3$ e) 595

EXERCISE SET 26

26.1 a) $y = t^2 - e^t + c$ c) $y = \frac{1}{2}\sin(2t) + c$ e) $y = ce^{\frac{1}{3}e^{3t}}$
g) $y = 1 - ce^{-\frac{1}{2}t^2}$ i) $y = c(t + 1)$

26.2 a) $y = ce^{3t}$ c) $y = ce^{3t^2/2}$ e) $y = ct^5$
g) $y = ce^{-\cos(t)}$ i) $y = \sqrt{2(\sin(t) - t\cos(t)) + c}$

26.3 a) $y = e^{7(t-1)}$ c) $y = (\frac{1}{3})e^{3t-15}$ e) $y = \sqrt{t^2 + 3}$
g) $y = 0.5e^{5(t-2)}$

26.4 a) $y = (11e^{3t} - 2)/3$ c) $y = (19e^{3t} - 3t - 1)/9$
e) $y = (e^{-t} + \cos(t) + \sin(t))/2$ g) $y = t^2 - t$
i) $y = (1/t)(1 - e^t) + e^t$

26.5 a) $\mathring{T} = -k(T - \hat{T})$ b) $T(t) = ce^{-kt} + \hat{T}$
c) approximately 0.35 hrs, or 21 min.

26.7 a) $X(t)$ is the amount of salt (Kg) at time t. $\mathring{X} = -25t/1000$ or $\mathring{X} = -0.025t$.
b) $X(15) = 10e^{-\frac{3}{8}} \simeq 6.9$ kg, $X(30) \simeq 4.7$ kg, $X(60) \simeq 2.2$ kg
c) 64.4 sec

26.9 a) $\mathring{I} = kIS$ b) $\mathring{S} = -k(N - S)S$
c) $S = N/(1 + ce^{Nkt})$ d) $\mathring{S}(t) = -k(N - S(t - \sigma))S(t)$.

26.11 a) $x = L/k$.
b) on the interval $[0, 1)$, $\mathring{x} = -kx$, $x(0) = x_0 + l$. On the next interval, $[1, 2)$, $\mathring{x} = -kx$ with $x(1)$ equal to the limiting amount of litter left over from the first year, plus the new addition of l units; $x(1) = \lim_{t \to 1^-} x(t) + l$.

26.13 The solution has the form $X(t) = e^{K(t)}/(1 + e^{K(t)})$, where

$$K(t) = \int_0^t k(t)dt = \{[K_{\max} - ca^2 - (cb^2/2]t + (240cab/\pi)[\cos(\pi t/120) - 1] + (60cb^2/\pi)\cos(\pi t/120)\sin(\pi t/120)\}.$$

26.15 a) $N = (2n + 1)\pi$, n = integer are stable equilibrium
$N = 2n\pi$ is unstable equilibrium.
b) $N = 1$ is semistable.
c) $N = n\pi$ is semistable.
d) $N = -1$ is unstable; $N = 1$ is stable.
e) $N = 0$ is stable and if $N_1 = 2\sin(N_1)$, $N = \pm N_1$ is unstable.

26.17 a) $y = c_1e^{-4t} + c_2e^t$ c) $y = c_1e^t + c_2e^{-2t}$
e) $y = c_1e^{2t} + c_2e^{-2t}$ g) $y = [c_1\cos(\sqrt{7}t/4) + c_2\sin(\sqrt{7}t/4)]e^{t/4}$
i) $y = c_1e^{(-3+\sqrt{21})t/2} + c_2e^{(-3-\sqrt{21})t/2}$

26.18 a) $y = \frac{3}{4}e^{3t} + \frac{1}{4}e^{-t}$ c) $y = \cos(\sqrt{5}t) + (2/\sqrt{5})\sin(\sqrt{5}t)$ e) $y = 3e^t - 2e^{t/2}$

26.19 a) $y = \frac{2}{3}\sin(3t) + \cos(3t)$ c) $y = [1 - (\frac{1}{2})(1 + e^2)t]e^{-t}$
e) $y = c\sin(\lambda t)$, c arbitrary
g) $y = \cos(\lambda t)$

26.20 a) $y'' + 3y' + 2y = 0$ c) $y'' - y' - 2y = 0$
e) $y'' + 9y = 0$ g) $y'' + (k_1 + k_2)y' + k_1k_2y = 0$

26.22 $x_1(t) = (ce^{-(k_1+k_2)t} + 5k_2)/(k_1 + k_2)$. Steady state $\bar{x} = 5k_2/(k_1 + k_2)$ is approached as $t \to \infty$

26.23 and 26.24 The basic solution method is as follows.
Given (1) $x_1' = ax_1 + bx_2$ and (2) $x_2' = cx_1 + dx_2$. Differentiate (1): $x_1'' = ax_1' + bx_2'$. Substitute from (2) $x_1'' = ax_1' + bcx_1 + bdx_2$. Solve (1) for x_2: $x_2 = (1/b)x_1' - (a/b)x_1$. Substitute for x_2 in last equation: $x_1'' = ax_1' + bcx_1 + dx_1' - adx_1$ or $x_1'' - (a + d)x_1' + (ad - bc)x_1 = 0$. Solve this second order equation for x_1 as in Theorem 26.1.

26.24 a) $x_1(t) = c_1 + c_2e^{-3t}$; $x_2(t) = 2c_1 - c_2e^{-3t}$
c) $x_1(t) = c_1 + c_2e^{-4t}$; $x_2(t) = 3c_1 - c_2e^{-4t}$
e) $x_1(t) = [c_1 \cos(\sqrt{3}x) + c_2 \sin(\sqrt{3}x)]e^{-3x}$
$x_2(t) = [\sqrt{3}c_2 \cos(\sqrt{3}x) - 3c_1 \sin(\sqrt{3}x)]e^{-3x}$
g) $x_1 = c_1e^{(1+\sqrt{3})t} + c_2e^{(1-\sqrt{3})t}$
$x_2 = \sqrt{3}c_1e^{(1+\sqrt{3})t} - \sqrt{3}c_2e^{(1-\sqrt{3})t}$

26.25 a) 1.6 c) 2.0625 e) 1.399

26.26 a) 2.563 c) 2.9605 e) 1.348

26.27 and 26.28 a) $N = 5$, 1.405; $N = 10$, 1.428
c) $N = 5$, 1.010; $N = 10$, 1.010

EXERCISE SET 27

27.1 a) 0.6 b) 0.4 c) 0.4

27.3 a) 0.078 b) 0.174
c) approximately 23 cases in 1000 will involve bleeding; approximately 5.

27.5 a) S = {bass, pike, trout}; P(bass) $\simeq$ 0.084, P(pike) $\simeq$ 0.091, and P(trout) $\simeq$ 0.824 *Note*: the sum is only 0.999 due to roundoff error.)
b) S = {heads, tails}; P(heads) = 0.41, P(tails) = 0.59
c) S = {positive, negative, doubtful}; P(positive) $\simeq$ 0.11, P(negative) $\simeq$ 0.70, P(doubtful) $\simeq$ 0.19

27.7 a), b), d), and e) are probability functions.

27.9 a) $0 < P(x) < 1$ for each x.
$P(1) + P(2) + P(3) + P(4) + P(5) = \frac{1}{15} + \frac{2}{15} + \frac{3}{15} + \frac{4}{15} + \frac{5}{15} = 1$
b) i) 0.2 ii) 0.6 iii) 0.4 c) 0.4

27.11 a) $S = \{2, 3, \ldots, 11, 12\}$ b) There are 36 different ways the pair can land, 6 of which add to 7; thus $P(7) = \frac{6}{36} \simeq 0.167$.
c) $P(2) = P(12) = \frac{1}{36}$; $P(3) = P(11) = \frac{2}{36}$; $P(4) = P(10) = \frac{3}{36}$; $P(5) = P(9) = \frac{4}{36}$; $P(6) = P(8) = \frac{5}{36}$, $P(7) = \frac{6}{36}$
d) $A = \{2, 3, 4\}$; $P(A) = \frac{1}{6}$
$B = \{8, 9, 10, 11, 12\}$; $P(B) = \frac{5}{12}$
$C = \{7, 11\}$; $P(C) = \frac{2}{9}$
e) i) $\frac{7}{12}$, ii) $\frac{7}{18}$, iii) $\frac{7}{16}$, iv) 0 v) $\frac{13}{36}$ vi) $\frac{1}{6}$, vii) $\frac{7}{12}$, viii) $\frac{5}{12}$

27.13 0.6

27.15 a) $S = \{HHH, HHT, HTH, HTT, THT, THH, TTH, TTT\}$
b) 0.5 c) 0.25 d) 0.125

27.17 a) 0.25 b) 0.15 c) 0.64

27.19 a) 0.5 b) 0.5 c) 0.5

27.21 a) $P(G) = 0.0000025$ b) $P(V|G) = 0.5$ $P(V \cap G) = 0.00000125$
c) $P(V) = 0.06666$, hence $P(V) \neq P(V|G)$.
$P(G|V) = P(V|G)/P(V) = 0.00001875$. Thus $P(G|V) \simeq 7.5P(G)$.

27.23 a) 0.08 b) 0.32 c) 0.52

27.25 a) 0.5 b) 0.666 c) 0.5 d) 0.5

27.27 0.333

27.29 0.34

27.31 a) $\bar{x} = 3.5$, $s = 1.870$ b) $\bar{x} = 3.33$, $s = 2.338$
c) $\bar{x} = 3.417$, $s = 1.975$ d) $\bar{x} = 2.75$, $s = 1.420$
e) $\bar{x} = 0$, $s = 2.739$ f) $\bar{x} = 0.6$, $s = 1.342$

27.33 a) $\mu = 0.6$, $\sigma = 1.78$ b) $\mu = 1.0$, $\sigma = 1.58$
c) $\mu = 2.25$, $\sigma = 0.968$ d) $\mu = 1.4$, $\sigma = 1.624$

27.37 a) 1.25 b) $\bar{x} = 1.1$, $s = 1.29$ $\bar{x} = 25.2$

27.39 The mean of Data Set 2 is highest, $\bar{x} = 25.2$, while Data Set 3 has the least variance $V = 14.666\ldots$

EXERCISE SET 28

28.1 a) No c) No e) Yes g) Yes i) No

28.2 a) No c) Yes e) No g) Yes

28.3 a) $\frac{1}{9}$ c) $\frac{4}{9}$ e) 1

28.4 a) 0.25 c) 0.75 e) 0.1875

28.5 a) 0.2955 c) 0.7045 e) 0.7071

28.6 a) $x/2$; 0.5 c) $x^3/4$; 0.3125 e) $e^x/(e^2 - 1)$; 0.4434
g) $2xe^{x^2}/(e^4 - 1)$; 0.1531 i) $(\pi/4)\sin(\pi x/2)$; $1/\sqrt{2} \simeq 0.7071$
k) $(2 - x)3x/4$; 0.6875

28.7 a) 1.0 c) 0.5 e) 0.0625

28.8 a) $f(x) = 1$ c) $f(x) = 2x$
e) $f(x) = 0$ if $0 \leq x \leq 0.5$ and $f(x) = 2$ if $0.5 < x \leq 1$. g) $f(x) = 4x - 1$

28.9 a) $F(x) = 2x - x^2$ c) $F(x) = (x^3 + 5x)/6$

e) $F(x) = \begin{cases} 2x^2 & \text{if} \quad 0 \leq x \leq 0.5 \\ 4x - 2x^2 & \text{if} \quad 0.5 \leq x \leq 1 \end{cases}$

g) $F(x) = \begin{cases} 3.5(x - x^2) & \text{if} \quad 0 \leq x \leq 0.5 \\ \frac{7}{8} - (0.5 - x)^3 & \text{if} \quad 0.5 \leq x \leq 1 \end{cases}$

28.11 a) $\displaystyle\int_0^M f(t)dt = \ln(1 + t^2)^{\frac{1}{2}} \Big|_0^M = 1$

b) 0.8540 c) $\ln(\sqrt{\frac{5}{2}}) = 0.4581$

28.14 a) $\mu = 4.5$ $\sigma = 1.06$ c) $\mu = 0.75$ $\sigma = 0.19$
e) $\mu = 5$ $\sigma = \sqrt{5} \simeq 2.236$ g) $\mu = 2.857$ $\sigma = 2.319$

28.15 a) $P \leq \frac{1}{9} \simeq 0.11$ c) $P \geq \frac{63}{64} \simeq 0.98$

28.16 b) $F(t) = 1 - e^{-0.5t}$; $S(t) = e^{-0.5t}$ c) $\lambda = \frac{1}{2}$ e) $S(2) \cdot F(1) = 0.145$

28.19 a) $F(t) = 1 - e^{-\lambda t^2}$; $S(t) = e^{-\lambda t^2}$ c) $\lambda(t) = 2\lambda t$

28.20 a) 0.4772 c) 0.4960 e) 0.4960 g) 0.9544
i) 0.1359 k) 0.8413 m) 0.0013 o) 0.5762 q) 0.9974

28.21 a) 0.153 c) 1.96 e) 1.96 g) 1.6545

28.22 a) 0.3413 c) 0.4452 e) 0.1841 g) 0.9032 i) 0.9544

28.23 with $\sigma = 2$ a) 0.1915 c) 0.2881 e) 0.3264 g) 0.7422 i) 0.6826

28.25 $P(X \geq 600) = 0.9772$; approximately 75 percent of the salmon will be longer than 633 mm.

28.27 a) 1.30 b) 0.2687 c) (11.52, 13.28)

28.29 0.2203

28.31 a) [7476, 7524] b) 6256 to 6303

MODELS USED IN EXAMPLES AND EXERCISES

INDEX